Rhizosphere Biology

Series Editor

Anil Kumar Sharma, Biological Sciences, CBSH, G.B. Pant University of Agriculture & Technology, Pantnagar, Uttarakhand, India

The Series **Rhizosphere Biology**, emphasizes on the different aspects of Rhizosphere. Major increase in agricultural productivity, to meet growing food demands of human population is imperative, to survive in the future. Along with methods of crop improvement, an understanding of the rhizosphere biology, and the ways to manipulate it, could be an innovative strategy to deal with this demand of increasing productivity. This Series would provide comprehensive information for researchers, and encompass all aspects in field of rhizosphere biology. It would comprise of topics ranging from the classical studies to the most advanced application being done in the field. Rhizoshpere is a dynamic environment, and a series of processes take place to create a congenial environment for plant to grow and survive. There are factors which might hamper the growth of plants, resulting in productivity loss, but, the mechanisms are not very clear. Understanding the rhizosphere is needed, in order to create opportunities for researchers to come up with robust strategies to exploit the rhizosphere for sustainable agriculture.

There are titles already available in the market in the broad area of rhizosphere biology, but there is a major lack of information as to the functions and future applications of this field. These titles have not given all the up-to-date information required by the today's researchers and therefore, this Series aims to fill out those gaps.

Piyush Mathur • Rupam Kapoor • Swarnendu Roy
Editors

Microbial Symbionts and Plant Health: Trends and Applications for Changing Climate

Springer

Editors
Piyush Mathur
Department of Botany
University of North Bengal
Siliguri, West Bengal, India

Rupam Kapoor
Department of Botany
University of Delhi
New Delhi, Delhi, India

Swarnendu Roy
Department of Botany
University of North Bengal
Siliguri, West Bengal, India

ISSN 2523-8442 ISSN 2523-8450 (electronic)
Rhizosphere Biology
ISBN 978-981-99-0029-9 ISBN 978-981-99-0030-5 (eBook)
https://doi.org/10.1007/978-981-99-0030-5

This Springer imprint is published by the registered company Springer Nature Singapore Pte Ltd.
The registered company address is: 152 Beach Road, #21-01/04 Gateway East, Singapore 189721, Singapore

Preface

Global climate change has led to variability in the number of abiotic and biotic factors that notably affect plant health. Microbial symbionts confer pronounced benefits to the host plants that enormously improve plant health. This book comprehensively describes the functional role played by different classes of microbial symbionts such as endophytes, arbuscular mycorrhizal fungi (AMF), and plant growth-promoting rhizobacteria (PGPRs) for the augmentation of plant growth and yield under the current changing climatic conditions. The book has been framed in a manner to cover a wide aspect of microbial symbionts starting from their ecological niche, their diversity and association with different hosts, and also the insights into the techniques for their identification. The major part of the book has stressed on the application of these microbial symbionts in the regulation of plant health in the presence of various abiotic and biotic stresses. Furthermore, the book also covers the bioprospection of these microbial symbionts for the development of biofertilizers. The book has delivered the basic concepts related to plant and microbial sciences as well as provides a plethora of knowledge regarding the recent advances in the field of microbial and agricultural biotechnology, nanobiotechnology, climate-resilient agriculture, etc. for enhancing plant productivity and crop protection. Lastly, the book is a cumulative outcome of the efforts of all the researchers working in diverse areas whose contribution has immensely helped in making this book a success.

Siliguri, India — Piyush Mathur
Delhi, India — Rupam Kapoor
Siliguri, India — Swarnendu Roy

Contents

Editors and Contributors

About the Editors

Piyush Mathur works as an Assistant Professor in the Department of Botany, University of North Bengal, West Bengal, India. He has been teaching for the past 6 years and has taught both at the undergraduate and postgraduate level. He has done his Ph.D. in Botany with a specialization in Mycology and Plant Pathology. Presently, he is supervising four Ph.D. research scholars who are working on microbial diversity, bioprospecting, and plant disease management. He has published quite a number of articles and papers in well-known international journals of repute. He has also contributed chapters to books published internationally and nationally. He has recently edited a book entitled *Plant Stress: Challenges and Management in the New Decade* published by Springer under ASTI series and is working on other books too with Springer. Dr. Mathur is an active member of the Society of Rapeseed Mustard Research (SRMR), Bharatpur, and Delhi University Botanical Society (DUBS) Delhi, India.

Rupam Kapoor is presently working as a Senior Professor in Department of Botany, University of Delhi. She is an active researcher working in the field of plant–fungus interactions for many years. She has been listed in the global ranking of the top 2% scientist for her significant contribution for two consecutive years 2020–21. She has supervised 12 Ph.D. thesis, 10 M. Phil. Theses, and mentored a number of M.Sc. dissertations. She has publications in well-known peer-reviewed National and International journals of high repute. A number of articles have also been published in the form of book chapters in International and National books. She is presently serving as an Associate Editor of "3Biotech" an International journal from Springer. She is also working as a review editor on the editorial board of a journal. She is a member of various learned societies like the International Mycorrhiza Society, International Symbiosis Society as well as a life member of Mycological Society of India, Indian Society of Mycology and Plant Pathology, Indian Botanical Society, and many more.

Swarnendu Roy is presently working as an Assistant Professor at the University of North Bengal, Siliguri, India. He has more than 12 years of teaching experience. His doctoral thesis involved understanding the mechanism of salinity stress tolerance in grasses mainly *Cynodon dactylon* and *Imperata cylindrica* by comparative evaluation with the model grass rice. He has published more than 35 articles in reputed journals with high impact factors, such as *Scientific Reports* and many others. He has recently edited a book entitled *Plant Stress: Challenges and Management in the New Decade* published by Springer and has also published several book chapters in edited volumes to date. Presently, he is actively engaged in research topics encompassing stress management in plants by engineered nanoparticles and the synthesis of starch-based biofilms and edible coatings. He is presently supervising four research scholars. He received International Travel Support (ITS) from DST, Govt. of India in 2019 to present his research paper at International

Conference on Integrative Plant Physiology at Sitges, Spain. He is a life member of prestigious societies like "The Indian Science Congress Association" and "Association of Food Scientists and Technologists (India)."

Contributors

Rajsekhar Adhikary Plant and Microbial Physiology and Biochemistry Laboratory, Department of Botany, University of Gour Banga, Malda, West Bengal, India

Sanghmitra Aditya Division of Plant Pathology, Indian Agricultural Research Institute, New Delhi, India

Dinesh Kumar Agrawala Botanical Survey of India, Salt Lake, Kolkata, India

Akshitha Ramachandra Amin St Aloysius College (Autonomous), Mangaluru, Karnataka, India

T. Archana Department of Plant Pathology, Tamil Nadu Agricultural University, Coimbatore, Tamil Nadu, India

Mayura Asra St Aloysius College (Autonomous), Mangaluru, Karnataka, India

Meenakshi Attri Centurion University of Technology and Management, Paralakhemundi, Odisha, India
Sher-e-Kashmir University of Technology and Management, Jammu, Jammu and Kashmir, India

Kavya Bakka Department of Microbiology, School of Life Sciences, Central University of Tamil Nadu, Thiruvarur, India

Purabi Banerjee Department of Agronomy, Bidhan Chandra Krishi Viswavidyalaya, Nadia, West Bengal, India
Division of Crop Sciences, ICAR-Central Research Institute for Dryland Agriculture, Hyderabad, India

Binit Baraik Department of Botany, Dr. Shyama Prasad Mukherjee University, Ranchi, Jharkhand, India
Department of Botany, J. N College Ranchi, Ranchi, Jharkhand, India

Anamika Barman Division of Agronomy, ICAR-IARI, New Delhi, India

Geetanjali Baruah Department of Biotechnology, The Assam Kaziranga University, Jorhat, Assam, India

Indrani Baruah Plant Breeding and Genetics Department, Assam Agricultural University, Jorhat, Assam, India
DBT-NECAB, Assam Agricultural University, Jorhat, Assam, India

Anurag Bera Department of Agronomy, Institute of Agricultural Sciences, Banaras Hindu University, Varanasi, Uttar Pradesh, India

N. Bhagyalakshmi St Aloysius College (Autonomous), Mangaluru, Karnataka, India

Parijat Bhattacharya Department of Agricultural Chemistry and Soil Science, Bidhan Chandra Krishi Viswavidyalaya, Mohanpur, Nadia, West Bengal, India
School of Agriculture, Swami Vivekananda University, Parganas, West Bengal, India

Bhagyashree Bhatt Shoolini University, Solan, Himachal Pradesh, India

V. Bhuvaneswari Assam Agricultural University, Jorhat, India

Jishusree Bhuyan Department of Biotechnology, The Assam Kaziranga University, Jorhat, Assam, India

Bedika Boruah Department of Biotechnology, The Assam Kaziranga University, Jorhat, Assam, India

Arka Pratim Chakraborty Department of Botany, Raiganj University, Raiganj, West Bengal, India

Oindrila Chakraborty Botanical Survey of India, Howrah, Kolkata, West Bengal, India

Dinakar Challabathula Department of Life Sciences, School of Life Sciences, Central University of Tamil Nadu, Thiruvarur, India

Chandrani Choudhuri Department of Botany, North Bengal St. Xaviers' College, Jalpaiguri, West Bengal, India

Divya Chouhan Microbiology Laboratory, Department of Botany, University of North Bengal, Rajarammohunpur, Darjeeling, West Bengal, India

K. Darshan Forest Protection Division, ICFRE-Tropical Forest Research Institute, Jabalpur, Madhya Pradesh, India

Keshab Das Department of Botany, Balurghat Mahila Mahavidyalaya, Balurghat, Dakshin Dinajpur, West Bengal, India

Liza Devi Department of Biotechnology, The Assam Kaziranga University, Koraikhowa, Jorhat, Assam, India

Rajkumari Soniya Devi Department of Biotechnology, The Assam Kaziranga University, Koraikhowa, Jorhat, Assam, India

D. Durgadevi Department of Plant Pathology, Tamil Nadu Agricultural University, Coimbatore, Tamil Nadu, India

Rajeswari Emani Acharya N.G. Ranga Agricultural University, Lam, Guntur, Andhra Pradesh, India

Dinkar Jagannath Gaikwad Centurion University of Technology and Management, Paralakhemundi, Odisha, India

Mateti Gayithri Assam Agricultural University, Jorhat, India

Bhanu Pratap Ghasil Centurion University of Technology and Management, Paralakhemundi, Odisha, India
Sri Karan Narendra Agriculture University, Jobner, Rajasthan, India

Saurav Anand Gurung Department of Botany, School of Life Sciences, Sikkim University, Tadong, Sikkim, India

Saurav Anand Gurung Department of Botany, School of Life Sciences, Sikkim University, Tadong, Sikkim, India

Shyamalina Haldar Department of Biochemistry, Asutosh College, Kolkata, West Bengal, India

S. Harish Department of Oilseeds, Tamil Nadu Agricultural University, Coimbatore, Tamil Nadu, India

Manisha Hazarika Department of Biotechnology, The Assam Kaziranga University, Koraikhowa, Jorhat, Assam, India

Gurudatt M. Hegde University of Agricultural Sciences, Dharwad, Karnataka, India

Sukanya Hembrom Department of Microbiology, Dr. Shyama Prasad Mukherjee University, Ranchi, India

Akbar Hossain Division of Agronomy, Bangladesh Wheat and Maize Research Institute, Dinajpur, Bangladesh

G. Karthikeyan Department of Plant Pathology, Tamil Nadu Agricultural University, Coimbatore, Tamil Nadu, India

R. Karthik Assam Agricultural University, Jorhat, Assam, India

R. Kavitha Department of Plant Pathology, Tamil Nadu Agricultural University, Coimbatore, Tamil Nadu, India

B. S. Kavya Department of Plant Pathology, University of Agricultural Sciences, Bangalore, Karnataka, India

Charishma Krishnappa Division of Plant Pathology, ICAR-Indian Agricultural Research Institute, New Delhi, India

H. M. Akshay Kumar Division of Plant Pathology, ICAR-Indian Agricultural Research Institute, New Delhi, India

Shalini Lal Department of Microbiology, Dr. Shyama Prasad Mukherjee University, Ranchi, Jharkhand, India
Department of Botany, Dr. Shyama Prasad Mukherjee University, Ranchi, India

Sagar Maitra Department of Agronomy, Centurion University of Technology and Management, Paralakhemundi, Odissa, India

Jayjit Majumdar Department of Ecological Studies, University of Kalyani, Kalyani, Nadia, West Bengal, India

Vivekananda Mandal Plant and Microbial Physiology and Biochemistry Laboratory, Department of Botany, University of Gour Banga, Malda, West Bengal, India

Piyush Mathur Microbiology Laboratory, Department of Botany, University of North Bengal, Rajarammohunpur, Darjeeling, West Bengal, India

Prashanta Kumar Mitra Plant and Microbial Physiology and Biochemistry Laboratory, Department of Botany, University of Gour Banga, Malda, West Bengal, India
Department of Botany, Kalyani University, Nadia, West Bengal, India

Yalavarthi Nagaraju ICAR-National Bureau of Agriculturally Important Microorganisms, Kushmaur, Mau, Uttar Pradesh, India

P. Shree Naveena Kerala Agricultural University, Thiruvananthapuram, India

Dinesh Panwar Chaudhary Charan Singh University, Meerut, Uttar Pradesh, India

Bappa Paramanick Department of Soil Science, Dakshin Dinajpur KVK, Uttar Banga Krishi Viswavidyalaya, Majhian, West Bengal, India

Smita Paul Department of Biotechnology, The Assam Kaziranga University, Koraikhowa, Jorhat, Assam, India

Biswajit Pramanick Department of Agronomy and Horticulture, University of Nebraska-Lincoln, Scottsbluff, NE, USA
Department of Agronomy, Dr. Rajendra Prasad Central Agricultural University, Pusa, Bihar, India

M. Vani Praveena Acharya N.G. Ranga Agricultural University, Guntur, Andhra Pradesh, India

Preeti Chaudhary Charan Singh University, Meerut, Uttar Pradesh, India

T. Raguchander Department of Plant Pathology, Tamil Nadu Agricultural University, Coimbatore, Tamil Nadu, India

Arun Kumar Rai Department of Botany, School of Life Sciences, Sikkim University, Tadong, Sikkim, India

L. Rajendran Department of Plant Pathology, HRS, Tamil Nadu Agricultural University, Ooty, India

B. Rajeshwaran Kerala Agricultural University, Thiruvananthapuram, India

G. Rajeshwar Rao Forest Protection Division, ICFRE-Tropical Forest Research Institute, Jabalpur, Madhya Pradesh, India

Geddi Purna Dattha Reddy Banana Research Station, Pulivendula, Dr. YSR Horticultural University, Venkataramannagudem, Andhra Pradesh, India

Priya Reddy Faculty of Biological Science, Friedrich Schiller University, Jena, Germany

Yaratha Nishith Reddy Division of Plant Pathology, Indian Agricultural Research Institute, New Delhi, India

Swarnendu Roy Plant Biochemistry Laboratory, Department of Botany, University of North Bengal, Siliguri, India

Lalichetti Sagar Centurion University of Technology and Management, Paralakhemundi, Odisha, India

Priyanka Saha Division of Agronomy, ICAR-IARI, Pusa, New Delhi, India

Ranjan Kumar Sahoo Centurion University of Technology and Management, Paralakhemundi, Odisha, India

Upasana Sahoo Centurion University of Technology and Management, Paralakhemundi, Odisha, India

Bandana Saikia Assam Agricultural University, Jorhat, India

Tinamoni Saikia Department of Biotechnology, The Assam Kaziranga University, Koraikhowa, Jorhat, Assam, India

Poonam Saini Rakesh P.G College, Pilani, Rajasthan, India

Masina Sairam Centurion University of Technology and Management, Paralakhemundi, Odisha, India

Ajit Kumar Savani ICM Division, ICAR-NAARM, Hyderabad, India

V. Sendhilvel Department of Plant Pathology, ICAR-KVK, Vellore, India

Sanghamitra Sengupta Department of Biochemistry, University of Calcutta, Kolkata, West Bengal, India

Kruti Shah P.D. Patel Institute of Applied Sciences, CHARUSAT, Anand, Gujarat, India

Tanmoy Shankar Centurion University of Technology and Management, Paralakhemundi, Odisha, India

Aaina Sharma Punjab Agriculture University, Ludhiana, Punjab, India

Supriya Sharma Assam Agricultural University, Jorhat, India

Sana Sheikh St Aloysius College (Autonomous), Mangaluru, Karnataka, India

Anupama Shrivastav Center of Research for Development (CR4D), Parul Institute of Applied Sciences (PIAS), Parul University (DSIR-SIRO Recognized), Vadodara, Gujarat, India

Ajay Singh Sindhu Division of Nematology, Indian Agricultural Research Institute, New Delhi, India

Geetanjali Singh Department of Botany, Dr. Shyama Prasad Mukherjee University, Ranchi, Jharkhand, India

Sultan Singh Sri Karan Narendra Agriculture University, Jobner, Rajasthan, India

Kiran Sunar Department of Botany, Balurghat Mahila Mahavidyalaya, Balurghat, Dakshin Dinajpur, West Bengal, India

Sweta Sushmita Tigga Department of Microbiology, Dr. Shyama Prasad Mukherjee University, Ranchi, Jharkhand, India

Vijay Jagdish Upadhye Center of Research for Development (CR4D), Parul Institute of Applied Sciences (PIAS), Parul University (DSIR-SIRO Recognized), Vadodara, Gujarat, India

Jitendra Kumar Vats Jai Prakash University, Chapra, Bihar, India

Varsha Venugopalan Department of Microbiology, School of Life Sciences, Central University of Tamil Nadu, Thiruvarur, India

Vineet Vishal Department of Botany, Dr. Shyama Prasad Mukherjee University, Ranchi, Jharkhand, India
Department of Botany, Bangabasi Evening College, Kolkata, West Bengal, India

Chapter 1
Global Climate Perturbations: Sustainable Microbial Mitigation Strategies

Ajit Kumar Savani, Yalavarthi Nagaraju, Rajeswari Emani, Geddi Purna Dattha Reddy, and M. Vani Praveena

Abstract Climate change insists on the change in temperatures, precipitation, and crop production and its dependent factors. Crop failure, irregular precipitation, and droughts are the standard drivers of crop failure, a consequence of climate change. The global climate scenario is drastically changing, and the results are astronomically vivid, unpredictable, and primarily affecting the susceptible populations in the ecosystem. The future of ecosystem balance highly depends on the environment, and climate change forges the new species by threatening other species. Change in the climate is not only driving speciation, and it is laying the path to extinction by eliminating ecological diversity. The current conservation methods are unarguably futile as they are not future-proof. Microbes are the key players in climate change through photosynthetic and respiratory by-products. However, microbes play a key role in driving climate change to neutrality through mitigation strategies.

Keywords Climate change · Crop failure · Ecosystem · Ecological diversity · Microorganisms

---

A. K. Savani
ICM Division, ICAR-NAARM, Rajendranagar, Hyderabad, India

Y. Nagaraju (✉)
ICAR-National Bureau of Agriculturally Important Microorganisms, Kushmaur, Mau, Uttar Pradesh, India
e-mail: yalavarthi.nagaraju@icar.gov.in

R. Emani · M. Vani Praveena
Acharya N.G. Ranga Agricultural University, Lam, Guntur, Andhra Pradesh, India

G. P. D. Reddy
Banana Research Station, Pulivendula, Dr. YSR Horticultural University, Venkataramannagudem, Andhra Pradesh, India

P. Mathur et al. (eds.), *Microbial Symbionts and Plant Health: Trends and Applications for Changing Climate*, Rhizosphere Biology,
https://doi.org/10.1007/978-981-99-0030-5_1

Abbreviations

CFC	Chlorofluor carbons
eCO_2	Elevated carbon dioxide
ET	Ethylene
FAO	Food and Agriculture Organization
GHI	Global Hunger Index
HGT	Horizontal gene transfer
JA	Jasmonic acid
MAT	Mean annual temperature
OMZ	Oxygen minimum zone
RCP	Representative Concentration Pathway
SII	Sea Ice Index

1.1 Introduction

Humans interact, understand, evolve, and exploit natural resources, increasing soil, air, and water pollution around the globe, and the remedial measures are unsatisfactory (Connor 2015). Increased temperatures; abundance of gases like carbon dioxide (CO_2), methane (CH_4), nitrous oxide (N_2O), and chlorofluorocarbons (CFCs); ill distribution of rains; forest blazes; changes in seasons; natural disasters such as floods, drought, and cyclones; and habitat loss escalated during the twenty-first century due to human intervention. Burning of fossil fuels, industrial activity, and intensive agriculture practices increased global greenhouse gas concentrations; it was estimated that the present CO_2 concentrations are 14% advanced (Stocker et al. 2013). Marine environments are also important in the turnover of these vital elements. Marine nitrogen fixation is 30% higher than terrestrial, whereas denitrification in the oceans more than doubles the terrestrial.

Climate change affects microbial processes such as methanogenesis, nutrient cycling, and decomposition of organic matter. Decreased soil microbial diversity limits the capacity of soil to support plant growth due to the decrease in the overall functionality of soil (Jing et al. 2015). Fungal-based food webs adapt better to drought conditions than bacterial-based food webs (de Vries et al. 2018). Increased temperatures and fertilisation strengthen the growth of cyanobacterial blooms (Lehman et al. 2017). Some human pathogens reported increased antibiotic resistance due to climate change. An increase in ocean surface temperatures correlates with cholera infections in Bangladesh (Pascual et al. 2000). Higher temperatures favour the horizontal gene transfer (HGT) of mobile genetic elements of resistance (MacFadden et al. 2018). Population growth and distribution also have a positive effect due to the segregation of the population into susceptible and resistant individuals.

Methane is a simple hydrocarbon that is 30 times more potent than carbon dioxide as a greenhouse gas. The atmospheric levels of CH_4 increased several folds during 2014–2017, but the reasons are uncertain until now. However, the causes can be determined from prior information, including human activities, anaerobic wastewater deposits, oxygen minimum zones (OMZs) in oceans, ruminant animals, and rice fields (Gålfalk et al. 2016). Rice is a staple crop and feeds almost half of the population in the world, necessitating its cultivation through the seasons and years. Anthropogenic activities like rice cultivation and meat production from the ruminants leave large amounts of carbon footprints. Rice fields contribute to ~20% and ruminant animals up to 19–48% of the CH_4 emissions (Ripple et al. 2014). It is currently estimated that while about 90% of the methane released into the atmosphere by these activities plus other natural processes has a biological origin, its biosynthesis is conducted only by microorganisms. Methane is degraded by bacterial and archaeal methanotrophs via aerobic and anaerobic processes, while oxidation takes place in the atmosphere by hydroxyl radicals. Resilience to changing conditions is the characteristic feature of all microorganisms; modulation of physiology and genomes help in adoption. Their short generation time leads to rapid evolution and immediate response to the conditions. However, artificial fertilisers significantly increase environmental availability, disrupt the biogeochemical cycle, and jeopardise ecological sustainability (Greaver et al. 2016).

Metagenomics, shotgun metagenomics, and high-throughput sequencing help scientists better interpret hidden microorganisms' ecological roles. While several adaptations and mitigation techniques have been used over the last few decades, less focus has been placed on the microbial adaptability responsible for climate change. The two main issues are the lack of knowledge and industrial agriculture's use of pesticides and fertilisers, damaging the agroecosystem's soil organic matter. Recently, scientists have begun to understand how important microbes are to preserving soil health. To promote sustainable agriculture and the environment, this chapter will examine how microorganisms might help mitigate the effects of climate change.

1.2 Global Climate Change and Consequences

Only by guaranteeing that everyone has access to high-quality, nutrient-dense foods that form the basis of a balanced diet will we be able to achieve food security and, in turn, provide nutritional security for everyone on the planet. Based on the Global Hunger Index (GHI 2020), in 2019, 8.9% of the world's population, or nearly 690 million people, suffered from malnutrition. By 2030, more than 840 million people will be undernourished if the current trend persists. These developments increase the likelihood of food insecurity, which also impacts people's health in several ways (FAO 2020). In order to do this, we must develop comprehensive answers to the current crises and advance in ways that promote the transition of the present food system into one that is more inclusive, sustainable, and resilient. To support human existence, a variety of food grain crops are grown. Global agriculture systems are largely harmed by climate change. Typhoons, heavier rainfall, high heat stress, and protracted droughts are just a few examples of extreme weather occurrences brought on by rising sea levels. Major cereal crops are affected by climate change in a predictable manner; for example, the grain production of rice (*Oryza sativa* L.) harvests decreased by 10% for every 1 °C increase in the growing season minimum temperature during the dry season, although the influence of crop output on maximum temperature was negligible. Additionally, the earth's living soil ecosystem and soil microbiome are also being altered by climate change (Meena et al. 2022).

1.3 Current Global Climate Scenario and Status

According to the fifth Assessment Report (AR5)-IPCC, current estimates of climate change indicate an increase in global mean annual temperatures of 1 °C by 2025 and 3 °C by 2100. Variability in rainfall patterns and intensity is expected to be high (IPCC 2021). Greenhouse gases (CO_2 and O_3) would increase global precipitation by 2 ± 0.5 °C per 1 °C warming. The global average surface mole fraction for carbon dioxide (CO_2), in 2019, was 410.5 ± 0.2 ppm (WMO 2019). Future climate-changing events are illustrated in the form of the Representative Concentration Pathway (RCP), a greenhouse gas concentration (not emissions) trajectory adopted by the IPCC. In order to project the changes in average surface temperatures, the Representative Concentration Pathway (RCP) is considered a standard parameter by the IPCC (Fig. 1.1). Changes in these elements result in (i) warmer and more

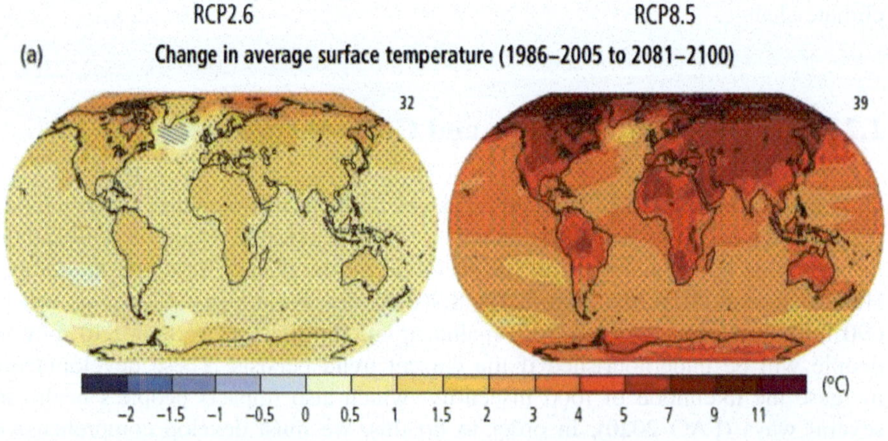

Fig. 1.1 Current global climate scenario showing future average surface temperature (Source: IPCC 2021)

frequent hot days and nights and (ii) erratic rainfall distribution patterns leading to drought or high precipitation that can influence plant diseases and insect pests.

1.4 Spatial and Temporal Changes in the Soil Microflora as Affected by Global Climate Change

Rapid increases in global temperature have the potential to disrupt many ecological processes. Warming due to climate change has caused an increase of 0.85 °C in global surface temperatures over the past century (Voosen 2021). Most living organisms, including microorganisms, have an optimal living temperature at which they thrive (Ratkowsky et al. 1983: Savage et al. 2004). As global temperatures increase, many species interactions will be disrupted (Ogilvie et al. 1997). Recently, soil regeneration and saving soil campaigns are gaining importance in the scenario of reduced crop yields due to loss of soil health and fertility (Zhang et al. 2014; https://consciousplanet.org). The functional diversity of the soil microbiota is affected by geographical locations (Dubey et al. 2019) and cropping patterns (Wieland et al. 2001). Changes in the soil microbiome should not be solely attributed to the single term climate change since it is a complex phenomenon and can be exuberated with elevated CO_2, drought, permafrost thaw, increased temperature, and higher temperature precipitation levels which further lead to floods.

1.4.1 Elevated CO_2

Elevated CO_2 level indirectly affects the soil bacterial population driven by tree species-specific root exudation. Vegetation properties also affect this process. Elevated CO_2 level leads to microbial shift as the population of soil copiotrophic bacteria increases over specialised oligotrophic acidobacteria.

Very less is known about the belowground soil microorganisms' response towards elevated levels of CO_2 across different soil depth profiles in different forest soil ecosystems. Ten years of elevated CO_2 exposure study revealed a remarkable shift in the microbial community compositions, comprising phylogenetic and functional genes of forest soil microbial communities across different soil depths.

1.4.2 Drought

An increase in drought decreases microbial functionality, which affects ecosystem sustainability. Less water content in soil pores results in soil dryness which further forms disconnected resource islands; subsequently, as the soil becomes drier, there is less water, resulting in less soil organic carbon decomposition (Schimel 2018). Soil microbial communities that have earlier been exposed to drought seem to show resilience towards subsequent drought events due to the accumulation of drought-resistant genes, which is an adaptive mechanism for the environment changes by the microbiota (Preece et al. 2019).

1.4.3 Permafrost Thaw and Soil Microbiome

Permafrost thaw leads to rapid shifts in the function of the boreal ecosystem; it changes soil hydrology and affects soil carbon turnover. The composition of the fungal community varies significantly between the thawed and intact permafrost sites, but soil nutrient content did not. With increasing permafrost thaw, the relative abundance of the mycorrhizal fungal group decreases while the relative abundance of putative fungal pathogens increases, thus mediating plant community response to permafrost thaws.

1.4.4 Effect of Temperature on Soil Microbiome

The temperature may greatly influence the soil microbiome profile by altering microbial communities' enzymatic activity and respiration with decreasing mean annual temperature (MAT). The temperature sensitivity of microbial communities may vary widely with time. The microbial communities of the warmest and driest sites have the most prominent temperature optimum, while the communities of the coolest and wettest sites lack a temperature optimum (Alster et al. 2016).

1.5 Effect of Climate Change on Plant-Microbe Interaction

Climate change greatly impacts plants, microbes, and plant-microbe interaction locally to globally. Disturbed climatic conditions like elevated CO_2, temperature, and drought affect plants' and microbes' physiology and ecological role. These interactions manipulate plant growth and development, contributing to shaping the ecosystems (Harris et al. 2020). Plant-microbe interactions respond to the climate in two ways: (1) pathogens or mutualist microbes amend the primary production of plants and affect the carbon sink in soil (Sulman et al. 2019), and (2) the decomposition of soil carbon is influenced due to affected respiration and the release of carbon into the atmosphere in the form of CO_2 (Classen et al. 2015).

1.5.1 Influence on Inter-Kingdom Interactions or Trophic-Level Interactions

The effects of climate change may be seen in rising temperatures and the melting of polar icebergs, but there is less consensus over how changes in microbial metabolism affect higher taxa growth and performance (Bang et al. 2018). Due to climate change, temperature changes were more pronounced in northern regions than in areas close to the equator, and high-altitude boreal forests are particularly vulnerable to the impacts of rising temperatures. Additionally, as a result, it alters the ecology and structure of forests by fostering species invasions, fire dangers, and biotic stress caused by pest and disease assaults (King et al. 2018). Complementing it is the uneven distribution and level of precipitation, which promotes the self-transition from deserts to grasslands and forests, reversing the process when the environment is unfavourable. The current Kyoto Protocol insists on the decrease of greenhouse gas emissions. The forest cover is increasing in developing countries and decreasing in low-income countries, and it has been estimated that Brazil and Indonesia have been losing the majority of forest cover since the 1990s (Li et al. 2020). It is challenging to draw firm conclusions from the available data since the multitrophic interactions concerning microbial activities are poorly understood. However, this part tries to comprehend the multitrophic interactions, how the climate affects those interactions, and the results. A brief discussion of bacteria's function in helping higher taxa adapt to climate change was also included, and climate change enhances microbial metabolism and biogeochemical cycles (Meena et al. 2022).

Under typical circumstances, higher photosynthetic activity and carbon sequestration by photosynthesis demand higher water, which is reversed under drought conditions. In arid conditions, elevated carbon dioxide concentrations aid forest trees in effectively using limited water (Preece et al. 2019). The forest's net carbon balance is influenced by nutrient availability, and the sheer quantity of nutrients allows the forests to release more carbon (Fernández-Martínez et al. 2014). Nonetheless, autotrophic respiration by plants and heterotrophic respiration by microorganisms release 60 picograms C each back into the atmosphere (Ballantyne et al. 2017). Higher temperatures increase the rapid organic matter decomposition (Bradford et al. 2016). Earthworms help decompose organic matter, but increased temperatures and decreased rainfall negatively affect their feeding. Although earthworms are beneficial, the anaerobic gut harbours the denitrifying microorganisms that produce the N_2O, a greenhouse gas (Lubbers et al. 2013). The carbon budget is essential in non-forested, arid, and semi-arid regions due to the poor coverage of plants. The most oversized carbon sinks on the earth lie in the form of permafrost, followed by peat soils, which are larger than the respirational losses. Plant composition changes alter carbon availability, melt permafrost, and decrease moss productivity, acting as a source of carbon rather than a sink (Bragazza et al. 2013). The increased microbial decomposition in the thawed permafrost accelerates global climate change through CO_2 and CH_4 production (Knoblauch et al. 2018). Nonetheless, it can be noted that the microbial growth efficiency is lower at higher temperatures, so organisms release more amounts of carbon into the environment rather than immobilisation or fixation (Hagerty et al. 2014). Corroborating evidence suggests that the genes responsible for aerobic and anaerobic decomposition and nutrient cycling increased over 1.5 years in the permafrost melted area (Xue et al. 2016).

Additionally, it has a fertilisation impact, although it relies on the availability of other nutrients like nitrogen and phosphorous (Singh et al. 2010). However, despite the varied distribution, the total amount of biomass on earth stays the same. Forests react pessimistically to climate change, which impacts species terribly. The effect of dieback on neem (*Azadirachta indica*) trees in India has brought the issue to light. Changing climatic conditions favour successive pathogens with evolutionary advantages and better temperature tolerance. With the alterations in the temporal circumstances of the specific environment, the viruses' sporadic character transforms into an epidemic (Danovaro et al. 2011). The organisms' metabolic activity depends highly on the surrounding temperature and the available nutrient concentrations. Global climate perturbations result in increased photosynthesis which modulates the rhizodeposits and their dependent microflora. Plant metabolism changes alter rhizodeposits' composition, and substrate-induced microbial population dynamics are inevitable in the rhizosphere. The competition of the available nitrogen in the soil is increased under the elevated CO_2 concentration (Stevnbak et al. 2012). Increased temperatures result in moisture loss, which ultimately reduces the availability of nutrients in the soil resulting in poor growth and less belligerent towards biotic and abiotic stress.

There is increased herbivory by *Spodoptera litura* and *Achaea janata* on peanut and ramie to meet the nitrogen requirement as a repercussion of climate change. The plant pathogens and insect infestation were assumed necessary due to the compromised defence molecules reported in response to increased temperatures and carbon dioxide in the tomato and *Arabidopsis* plants. It was understood that enhanced photosynthesis increases the plant biomass through carbohydrate accumulation; nonetheless, the plant's nitrogen content remains the same; hence insects are instigated to increase herbivory to meet the nitrogen demand of the body (Peng et al. 2004). Grasshoppers reduce the plant biomass and nitrogen requirement, thereby increasing the soil microbial activity (Stevnbak et al. 2012). Conversely, in oceans, increased temperatures increase protein synthesis in eukaryotic phytoplanktons (Toseland et al. 2013).

Additionally, the nitrogen losses in the soil are aggravated by denitrification, nitrification, and fossil fuel burning. The by-products of dissimilatory processes release gases like NO (nitric oxide), which have a suppressive action on the ethylene (ET) signaling, influencing the JA signaling pathway in plants. It was reported that elevated carbon dioxide (eCO_2) directly impacts plants' jasmonic acid (JA) pathway. Silencing of the JA biosynthetic genes lipoxygenase 3 (LOX 3) or the JA signaling receptor coronatine insensitive 1 (COI1) will reduce the resistance and increase the feed intake of *Popillia japonica* (Japanese beetle), *Diabrotica virgifera* (corn rootworm), and *Manduca sexta* (tobacco hornworm). A species-specific inhibition is reported against *Spodoptera litura* in tobacco and tomato by altering the JA signaling pathway. Insect and plant defence interactions evolve throughout the evolution, and the winning factor differs from the surrounding environmental influences.

Increased temperatures enhance denitrification and elevate the emissions of N_2O, resulting in the loss of nitrogen and contributing to brisk global climate change. N_2O significantly affects global climate change due to its long lifetime in the atmosphere (116 years), increased global warming up to 310 times more than CO_2, and climate temperature change. The formation of soil organic carbon (SOC) and its mineralisation is crucial for climate control; however, changes in the climate, like drought, have a cynical influence on the rhizodeposits and soil microbial biomass. The loss of soil organic carbon is observed when the plants expand to the tundra regions, promoting microbial growth through rhizodeposits and microbial decomposition of native soil carbon (Hartley et al. 2012). Increased global temperatures may blaze the evergreen forests, which results in the permanent loss of habitat for wild animals, death of trees, and acceleration of nutrient cycling. Dispersal of pathogenic spores is expected to escalate during the dry and warmer environmental regimes, and further, it involves changes in phenology (life cycle events) and increases the risk of pathogen infestation. It also favours the long-term survivability of pathogens which fetches them a competitive advantage over the keystone species. In the Antarctic Peninsula and associated islands, the fungal diversity increased upon the increased temperature (Kleinteich et al. 2012). It also endangers the older forests and affects their regeneration after severe drought. The carbon sinks of forests are critically injured if the forests are under continuous fire.

The natural soil flora complies with the plants' need for nitrogen due to the metabolism of extracellular proteins and amino acids and allows the successful establishment (Hill et al. 2011). Furthermore, this also favours the growth of invasive species or activates the seed germination of some recalcitrant seeds with a greater capacity for dispersal and wider adaptability to environments that will allow them to annihilate other local species, sequentially reducing ecosystem diversity. The range expansion of Antarctic hair grass (*Deschampsia antarctica*) was reported in the Antarctic region due to the increased temperatures and outcompeting with the other species like moss *Sanionia uncinata* through nitrogen acquisition (Hill et al. 2011).

An increase in the temperatures and carbon dioxide shortens the life cycle of insects, increases the reproduction cycles and fecundity, and allows them to expand to new avenues. They also have pernicious effects on photosynthesis rates and the growth of plants. In soil, the decomposition of organic matter increases as the material's thermostability is compromised with the increased temperatures, and bacteria outcompete the other microorganisms. Lack of nitrogen in the soil impedes beneficial microflora and restricts plant growth. Hence, nutrient availability is limited through the immobilisation process, which hinders the primary productivity of forests.

A decreased water reserve in the soil is a rising concern, primarily affected by the ill-distribution of rains and increased temperatures. Poor water reserves impact nutrient cycling, and limited nitrogen prevents carbon uptake by most organisms.

A decrease in the rhizodeposits restricts the rhizo-decomposition process and enhances the microbial immobilisation of nutrients. From an ecological perspective, the competition for ecosystem services in the forest is increased due to human interventions. The interactions between pests, pathogens, and their hosts, as well as their ensuing relationships with mutualists, natural enemies, and wildlife animals, are all influenced by the perturbations in the host physiology brought on by climate change.

1.6 Microbiome Dynamics

The primary force behind shaping the biomes are climate and nutrients, and the changing climate significantly affects the functions, metabolism, mortality, and successions. Over the last few decades, the growing carbon emission into the atmosphere has been affecting the earth's climate, causing climate warming and, thus, impacting the environment and ecology. One significant effect of climate change is the thawing of permafrost, which raises concerns about the rise in global carbon output. Although increased temperatures may be the reason for permafrost's thawing, increased temperatures enhance photosynthesis and biological yield.

Oceans are teeming with microbial diversity, and the primary productivity of oceans is directly dependent on microbes as their population size and microbial biomass are outnumbered. In the deep depths of oceans, the mineralisation of organic matter is fueled by benthic organisms (Behrenfeld et al. 2017). The oceans majorly contribute to the global carbon dioxide production, and the turnover of the carbon dioxide produced by the phytoplankton is much higher than the terrestrial plants. Phytoplanktons are sensitive to changes in carbon dioxide concentrations and are hence considered indicator microorganisms for global climate change. However, the productivity of phytoplankton increases with the increase in carbon dioxide concentrations under ideal nutrient conditions (Toseland et al. 2013).

Since industrialisation, the oceans have become 0.1 pH units more acidic; by the end of this century, further acidification is predicted (Bunse et al. 2016). Given the unprecedented rate of pH change, it is imperative to find out how marine life reacts to it as soon as possible. The biodiversity and species distribution of coral reefs and salt marshes are declining sharply due to the increased temperatures, while mangroves are expanding due to the associated microbial photosynthetic mats playing a crucial role in nutrient availability (Lee and Joye 2006). Half of the total CO_2 fixed and half of the oxygen produced on the earth is contributed by marine photosynthetic microorganisms (Behrenfeld 2014). Due to the intricacy of primary phytoplankton production, which is surrounded by the bottom-up and top-down regulation of oceans, it is difficult to make meaningful conclusions when assessing the effects of global climate change on primary productivity (Hutchins and Boyd 2016). Marine viruses influence primary productivity and biogeochemical cycling of microorganisms through predator-prey interactions (Danovaro et al. 2011).

A field experiment revealed that *Vicicitus globosus*, a deadly microalga, benefits from rising CO_2 levels, disrupting the movement of organic matter across trophic levels. *Trichodesmium*, a marine cyanobacterial genus, responds to prolonged (4.5 years) exposure to increased CO_2 levels by undergoing irreversible genetic alterations that promote nitrogen fixation and growth. Elevated CO_2 levels promote growth, cell size, and carbon-to-nitrogen ratios in the photosynthetic green alga *Ostreococcus tauri*. With changes in ecotypes and niche occupancy, higher CO_2 levels also impact the population structure of *O. tauri*, which affects more enormous food webs and biogeochemical cycles. With rising ocean surface temperatures, a decline in the number of coral species has been observed. Increased coral infections, alterations in microbiomes, weakened immune systems by the acidification of oceans, and changed nutrient cycling are all strongly connected with a decline in coral populations (Bourne et al. 2016).

A perfect balance in the microbial community structure is crucial to controlling the planet's significant gases and imparting resilience to global climate change. Increased temperatures enhance transpirational losses, reduce water density, and impede water circulation and stratification of water bodies. Rivers, estuarine, precipitation, and winds contribute to the changes in the microbiomes of soil and oceans. In oceans, eutrophication, acidification, and warming reduce the coral reefs and increase the microalgae and benthic cyanobacterial mats. The worldwide sea ice (Sea Ice Index) is decreasing, which may result in increased light penetration and maybe more primary production. However, there are competing projections regarding the consequences of changing mixing patterns, changes in nutrient availability, and productivity trends in polar zones (Behrenfeld et al. 2017). An increase in the oxygen minimum zones (OMZs) in oceans increased the production of N_2 and N_2O. Fertilisation causes eutrophication, resulting in abundant cyanobacterial blooms in the lakes. However, an intriguing finding emerged from studies conducted in Lake Zurich, where non-nitrogen-fixing cyanobacteria (*Planktothrix rubescens*) predominate owing to lower eukaryotic phytoplankton and non-fertilisation of phosphorus (Posch et al. 2012).

Continuous cropping exhausts the soil's nutrients and makes it infertile under the conditions of improper management of crop residues. Conversely, it ensures continuous CO_2 harvesting but is challenged by single-crop cultivation, soil salinity, abiotic, biotic stresses, and crop holidays that threaten the global CO_2 cycle. The continuous combustion of fossil fuels exacerbates the disturbances in the CO_2 cycle. It also reduces the concentration of nitrogen and cycle in soil and negatively affects the diversity of dependent microflora. Meanwhile, increased global temperatures fasten the decomposition of organic matter and the emission of CO_2 (Schindlbacher et al. 2011). Species that are less adept to the conditions and can't regulate their internal metabolism are more prone to competitive exclusion in climate change scenarios (Flynn et al. 2012).

Through the repercussions, the loss of microbial biodiversity was overshadowed in favour of the extinction of species brought on by human activity. The soil nitrogen turnover is solely taken care of by the microorganisms; changes in the climate have negative consequences on the rhizo-deposits or carbon turnover with an ultimate effect on microbial diversity. Moreover, there is less or no discussion on the microbial dynamics under climate change and their role in climate change due to their intangible nature. Although they contribute to climate change resilience, much emphasis was never given to microorganisms as they are microscopic or submicroscopic. The global diversity of microorganisms must be maintained to balance the global ecosystem's functioning. They assume the importance of supporting the whole higher flora and fauna either directly or indirectly through microbial diversity. The diversity of the microorganisms results in the availability of diversified products for the higher taxa, which sustains the ecosystem's functioning in the long run. Limited resources and climate change are the major factors contributing to the loss of microbial diversity. Intentional bio-inputs (biofertilisers) may require additional attention as they may lead to unpredicted outcomes that severely affect the ecology and its function due to the highly competitive resource behaviour and secretion of growth-inhibiting compounds. Climate change increases the risk of diseases through the modulation of host physiology and parasite acclimatisation. Increased dry spells enhance the chances for the quick dispersal of pathogenic spores, thereby increasing the occurrence of pathogenesis (Raffel et al. 2013).

Some essential elements of many dryland ecosystems are biological soil crusts (biocrusts), made up of mosses, lichens, and cyanobacteria. Though there are few long-term studies to follow possible changes in these delicate soil-surface communities, it is believed that mosses and lichens are declining due to climate change and other anthropogenic disturbances.

1.7 Metabolic Modulation in the Microbiome

The ongoing changes in climate patterns have the potential to threaten the already vulnerable global food security in various ways, including exacerbating major plant diseases and creating weather conditions for devastating new diseases to emerge in critical food-producing regions. One can envision harnessing a community-defined microbiome's potential in suppressing disease and increasing environmental tolerance in the future to enhance crop resilience and productivity. Climate change may directly affect several aspects of the biology of host plants, including their phenology (including senescence), sugar and starch contents, nitrogen and phenolic contents, root and shoot biomass, number and size of leaves, amount and composition of wax on leaves, changes in stomatal densities, and conductance and root exudation. Any change in any of these areas may influence infection and colonisation by pathogens. For example, after infection, plant water potentials and nutritional content may affect the host tissues' colonisation rate, the production of new inoculum, and the expression of symptoms in the host (Colhoun 1973). The increase in fixed carbon may also affect microbial communities in the soil and the functioning of the microbial ecosystem. The concentration of CO_2, temperature levels, and nitrogen deposition are important factors affecting soil microbial communities (Garrett et al. 2006). Climate change will likely alter soil microflora's distribution and population levels, affecting both intensive and low-input agricultural production systems (Mazzola 2010). Climate changes will affect pathogens, hosts, and their interactions, resulting in changes in epidemics and their effects. In most of the pathosystems that have been examined, increases in disease severity have been predicted. Studies suggest that microbial communities may generally experience decreased levels of available N. However, plant species composition and soil type are expected to affect the type of responses observed (Hungate et al. 1996).

It is well known that inoculum production and dispersal are critical for disease epidemics in crop fields. Major causes for concern due to plant diseases in the context of climate change are attributed as follows.

1.7.1 Increased Temperatures

1.7.1.1 Incidence of Plant Diseases

In Asia, a greater incidence of rice bacterial and sheath blight could be observed as temperatures increase. In Europe, increasing temperatures will affect sugar beet production due to increased incidences of leaf spots and rhizomania. In Brazil, higher temperatures may increase sugarcane infection by *Colletotrichum falcatum*.

1.7.1.2 Pathogen Overwintering

In the United States, warm temperatures would allow Asian soybean rust to overwinter in higher latitudes and also favour the development of grey leaf spots in maize. In China, warm winters have already contributed to severe epidemics of wheat head blight in recent years. Further, increased precipitation levels lead to the potential threat to potato cultivation in the European Union.

Asexual zoospore production in the oomycete *Phytophthora infestans* requires low temperatures. Temperature regulation of zoospore production involves the expression of a protein phosphatase (PinifC3), which controls the phosphorylation status of RNA polymerase and other transcriptional regulators in this oomycete. PinifC3 transcription is induced at 10 °C, as the gene has a cold-induced promoter element (Tani and Judelson 2006). Thus, global climate changes appear to have pervasive effects on many steps of pathogen infection, microbial modulation ranging from pathogen sporulation, pathogen growth and virulence gene expression in plants, and the overwintering of inocula, which is critical for initiating infections in subsequent plant-growing seasons.

1.8 Microbial Strategies to Mitigate the Global Climate Change

When climate change is inevitable, the mitigation plans are unlimited beyond boundaries. One of the chief mandates of the mitigation is to manage crop production to ensure food security. Understanding plants' and microbes' structure, function, and metabolism are essential to structuring mitigation strategies that are future-ready and foolproof. Existing phenotype and metabolic plasticity to the changing environment decide the success of the organisms. Microorganisms are resilient to changes in climate; however, the change in the plant metabolism influences their diversity and succession in the rhizosphere. Microbial resilience and phenotype plasticity are due to short generation time, horizontal gene transfer (HGT) between closely related species, and high mutation rate (Bang et al. 2018). Climate change's dramatic causes and consequences are mitigated using different plant growth-promoting microorganisms (Geurts et al. 2012).

Despite extreme salt and drought stress, nitrogen fixation in the soil and leaves stimulates plant growth. Microbes can augment or regulate the plant's metabolic processes through their participation. Plants' steady development is facilitated by the biosynthesis of auxins and gibberellins, which induce the JA pathway and disrupt the ET pathway through the action of ACC deaminase produced by PGPR and PGPF (De Zelicourt et al. 2013). Nitric oxide, a diffusible gas produced by *Azospirillum lipoferum* induces root growth by activating the IAA biosynthetic pathway (Creus et al. 2005). The bacterium *Paenibacillus polymyxa* activated the drought-responsive gene, i.e. ERD15, in *Arabidopsis thaliana* (Timmusk and Wagner 1999).

A proper understanding of the parasitic relationship between *Wolbachia* and mosquitos has reduced Zika, chikungunya, and dengue viruses and reduced the number of mortalities (King et al. 2018). Options for reducing emissions in agriculture are made possible by advancements in our understanding of the ecophysiology of the bacteria that convert N_2O to innocuous N_2. N_2O emissions from soybean have been reduced by using bacterial strains with more significant N_2O reductase activity, and both naturally occurring and genetically engineered strains with higher N_2O reductase activity provide potential solutions for N_2O emission reduction. Reduced methane emission from cattle may be achieved by breeding programmes focusing on host genetic variables that alter microbial community responses and rumen microbiota manipulation.

In the second scenario, the goal would be to create cattle lines that support microbial communities that release less methane without compromising the health and productivity of the animals. By substituting fungi for meat, we may reduce our nutritional carbon footprint. One agricultural strategy for broadly and indirectly reducing the microbial consequences of climate change is biochar. When biomass is thermochemically converted under oxygen restriction, biochar is created, which aids in the stability and build-up of organic matter in soils with high iron content. Biochar promotes organic matter retention by lowering microbial mineralisation and the impact of root exudates on releasing organic material from minerals. This encourages grass development and lowers carbon release (Weng et al. 2017).

A paradigmatic shift in biomass from higher animals to microbes is acquiring prominence due to increased carbon emissions and deforestation, which encourage the microorganisms to fix more carbon in the oceans, augmented by eutrophication. Microbial-based food or single-cell protein consumption may balance the biomass accumulation from oceans to land. The phytoplanktons' production of dimethyl sulphate and its derivatives can anticipate how the phytoplanktons will contribute to the synthesis of aerosols, how the changing ocean conditions will impact them, and how these changes will affect the climate (Sanchez et al. 2018). Increasing public awareness of the crucial contributions that microbes make to global warming would help mobilise support for such initiatives, i.e. by fostering microbiological literacy in society.

1.9 Conclusions

To tackle climate change, one has to have a good sense of direction and drive. Policymakers must look back to the technologies available and their long-term sustainability. Most of the existing technologies need modifications or replacement with eco-friendly approaches. Eco-friendly approaches with future proof and fool-proof are the need of the hour. Microbial technologies are promising in the long term and environmentally benign without further harmful consequences. The impacts of microorganisms on climate change are little understood, which requires the most attention. Disturbances in the climate force the rapid speciation and annihilation of species, for the world, is unprepared. The future generations must have a safe and congenial life promised through the current practices.

References

Alster CJ, Koyama A, Johnson NG, Wallenstein MD, von Fischer JC (2016) Temperature sensitivity of soil microbial communities: an application of macromolecular rate theory to microbial respiration. J Geophys Res Biogeosci 121:1420–1433

Ballantyne A, Smith W, Anderegg W, Kauppi P, Sarmiento J, Tans P, Running S (2017) Accelerating net terrestrial carbon uptake during the warming hiatus due to reduced respiration. Nat Clim Chang 7(2):148–152

Bang C, Dagan T, Deines P, Dubilier N, Duschl WJ, Fraune S, Bosch TC (2018) Metaorganisms in extreme environments: do microbes play a role in organismal adaptation? Zoology 127:1–19

Behrenfeld MJ (2014) Climate-mediated dance of the plankton. Nat Clim Chang 4:880–887

Behrenfeld MJ, Hu Y, O'Malley RT, Boss ES, Hostetler CA, Siegel DA, Scarino AJ (2017) Annual boom–bust cycles of polar phytoplankton biomass revealed by space-based lidar. Nat Geosci 10:118–122

Bourne DG, Morrow KM, Webster NS (2016) Insights into the coral microbiome: underpinning the health and resilience of reef ecosystems. Annu Rev Microbiol 70:317–340

Bradford MA, Wieder WR, Bonan GB, Fierer N, Raymond PA, Crowther TW (2016) Managing uncertainty in soil carbon feedbacks to climate change. Nat Clim Chang 6:751–758

Bragazza L, Parisod J, Buttler A, Bardgett RD (2013) Biogeochemical plant–soil microbe feedback in response to climate warming in peatlands. Nat Clim Chang 3:273–277

Bunse C, Lundin D, Karlsson CM, Akram N, Vila-Costa M, Palovaara J, Pinhassi J (2016) Response of marine bacterioplankton pH homeostasis gene expression to elevated CO_2. Nat Clim Chang 6:483–487

Classen AT, Sundqvist MK, Henning JA, Newman GS, Moore JA, Cregger MA, Patterson CM (2015) Direct and indirect effects of climate change on soil microbial and soil microbial-plant interactions: what lies ahead? Ecosphere 6:1–21

Colhoun JC (1973) Effects of environmental factors on plant diseases. Annu Rev Phytopathol. (United States), 11

Connor R (2015) The United Nations world water development report 2015: water for a sustainable world, vol 1. UNESCO publishing

Creus CM, Graziano M, Casanovas EM, Pereyra MA, Simontacchi M, Puntarulo S, Barassi CA, Lamattina L (2005) Nitric oxide is involved in the Azospirillum brasilense-induced lateral root formation in tomato. Planta 221(2):297–303

Danovaro R, Corinaldesi C, Dell'Anno A, Fuhrman JA, Middelburg JJ, Noble RT, Suttle CA (2011) Marine viruses and global climate change. FEMS Microbiol Rev 35:993–1034

De Zelicourt A, Al-Yousif M, Hirt H (2013) Rhizosphere microbes as essential partners for plant stress tolerance. Mol Plant 6(2):242–245

Dubey A, Malla MA, Khan F, Chowdhary K, Yadav S, Kumar A, Khan ML (2019) Soil microbiome: a key player for conservation of soil health under changing climate. Biodivers Conserv 28:2405–2429

FAO (2020) Seguridad Alimentaria bajo la Pandemia de COVID-19. Seguridad Alimentaria Bajo La Pandemia de COVID-19. https://doi.org/10.4060/ca8873es

Fernández-Martínez M, Vicca S, Janssens IA, Sardans J, Luyssaert S, Campioli M, Peñuelas J (2014) Nutrient availability as the key regulator of global forest carbon balance. Nat Clim Chang 4:471–476

Flynn KJ, Blackford JC, Baird ME, Raven JA, Clark DR, Beardall J, Wheeler GL (2012) Changes in pH at the exterior surface of plankton with ocean acidification. Nat Clim Change 2:510–513

Gålfalk M, Olofsson G, Crill P, Bastviken D (2016) Making methane visible. Nature. Climate Change 6:426–430

Garrett KA, Dendy SP, Frank EE, Rouse MN, Travers SE (2006) Climate change effects on plant disease: genomes to ecosystems

Geurts R, Lillo A, Bisseling T (2012) Exploiting an ancient signalling machinery to enjoy a nitrogen fixing symbiosis. Curr Opin Plant Biol 15:438–443

GHI (2020) Global hunger index (GHI): Mozambique [WWW Document]. The Global Hunger Index. www.globalhungerindex.org

Greaver TL, Clark CM, Compton JE, Vallano D, Talhelm AF, Weaver CP, Haeuber RA (2016) Key ecological responses to nitrogen are altered by climate change. Nat Clim Change 6:836–843

Hagerty SB, Van Groenigen KJ, Allison SD, Hungate BA, Schwartz E, Koch GW, Dijkstra P (2014) Accelerated microbial turnover but constant growth efficiency with warming in soil. Nat Clim Change 4:903–906

Harris JM, Balint-Kurti P, Bede JC, Day B, Gold S, Goss EM, Alvarez ME (2020) What are the top 10 unanswered questions in molecular plant-microbe interactions? Mol Plant Microbe Interact 33:1354–1365

Hartley IP, Garnett MH, Sommerkorn M, Hopkins DW, Fletcher BJ, Sloan VL, Wookey PA (2012) A potential loss of carbon associated with greater plant growth in the European Arctic. Nat Clim Change 2:875–879

Hill PW, Farrar J, Roberts P, Farrell M, Grant H, Newsham KK, Jones DL (2011) Vascular plant success in a warming Antarctic may be due to efficient nitrogen acquisition. Nat Clim Chang 1: 50–53

Hungate BA, Canadell J, Chapin FS (1996) Plant species mediate changes in soil microbial N in response to elevated CO_2. Ecology 77:2505–2515

Hutchins DA, Boyd PW (2016) Marine phytoplankton and the changing ocean iron cycle. Nat Clim Chang 6:1072–1079

IPCC (2021) Climate Change2021: thephysical science basis. In: contribution of working group I to the sixth assessment report of the intergovernmental panel on climate change. Cambridge University Press, UK

Jing X, Sanders NJ, Shi Y, Chu H, Classen AT, Zhao K, He JS (2015) The links between ecosystem multifunctionality and above-and belowground biodiversity are mediated by climate. Nat Commun 6:1–8

King JG, Souto-Maior C, Sartori LM, Maciel-de-Freitas R, Gomes MGM (2018) Variation in Wolbachia effects on Aedes mosquitoes as a determinant of invasiveness and vectorial capacity. Nat Commun 9:1–8

Kleinteich J, Wood SA, Küpper FC, Camacho A, Quesada A, Frickey T, Dietrich DR (2012) Temperature-related changes in polar cyanobacterial mat diversity and toxin production. Nat Clim Change 2:356–360

Knoblauch C, Beer C, Liebner S, Grigoriev MN, Pfeiffer EM (2018) Methane production as key to the greenhouse gas budget of thawing permafrost. Nat Clim Change 8:309–312

Lee R, Joye S (2006) Seasonal patterns of nitrogen fixation and denitrification in oceanic mangrove habitats. Mar Ecol Prog Ser 307:127–141

Lehman PW, Kurobe T, Lesmeister S, Baxa D, Tung A, Teh SJ (2017) Impacts of the 2014 severe drought on the Microcystis bloom in San Francisco estuary. Harmful Algae 63:94–108

Li J, Bååth E, Pei J, Fang C, Nie M (2020) Temperature adaptation of soil microbial respiration in alpine, boreal and tropical soils: an application of the square root (Ratkowsky) model. Glob Chang Biol 27:1281–1292

Lubbers IM, Van Groenigen KJ, Fonte SJ, Six J, Brussaard L, Van Groenigen JW (2013) Greenhouse-gas emissions from soils increased by earthworms. Nat Clim Change 3:187–194

MacFadden DR, McGough SF, Fisman D, Santillana M, Brownstein JS (2018) Antibiotic resistance increases with local temperature. Nat Clim Chang 8:510–514

Mazzola M (2010) Management of resident soil microbial community structure and function to suppress: soilborne disease development. In: Reynolds MP (ed) Climate change and crop production. CABI, Wallingford, UK, pp 200–218

Meena M, Yadav G, Sonigra P, Nagda A, Mehta T, Swapnil P, Kumar S (2022) Multifarious responses of forest soil microbial community toward climate change. Microb Ecol 1-26

Ogilvie BG, Rutter M, Nedwell DB (1997) Selection by temperature of nitrate-reducing bacteria from estuarine sediments: species composition and competition for nitrate. FEMS Microbiol Ecol 23:11–22

Pascual M, Rodó X, Ellner SP, Colwell R, Bouma MJ (2000) Cholera dynamics and El Niño-southern oscillation. Science 289:1766–1769

Peng S, Huang J, Sheehy JE, Laza RC, Visperas RM, Zhong X, Cassman KG (2004) Rice yields decline with higher night temperature from global warming. Proc Natl Acad Sci 101:9971–9975

Posch T, Köster O, Salcher MM, Pernthaler J (2012) Harmful filamentous cyanobacteria favoured by reduced water turnover with lake warming. Nat Clim Chang 2:809–813

Preece C, Verbruggen E, Liu L, Weedon JT, Peñuelas J (2019) Effects of past and current drought on the composition and diversity of soil microbial communities. Soil Biol Biochem 131:28–39

Raffel TR, Romansic JM, Halstead NT, McMahon TA, Venesky MD, Rohr JR (2013) Disease and thermal acclimation in a more variable and unpredictable climate. Nat Clim Change 3:146–151

Ratkowsky DA, Lowry RK, McMeekin TA, Stokes AN, Chandler R (1983) Model for bacterial culture growth rate throughout the entire biokinetic temperature range. J Bacteriol 154:1222–1226

Ripple WJ, Smith P, Haberl H, Montzka SA, McAlpine C, Boucher DH (2014) Ruminants, climate change and climate policy. Nat Clim Chang 4:2–5

Sanchez KJ, Chen CL, Russell LM, Betha R, Liu J, Price DJ, Behrenfeld MJ (2018) Substantial seasonal contribution of observed biogenic sulfate particles to cloud condensation nuclei. Sci Rep 8:1–14

Savage VM, Gillooly JF, Brown JH, West GB, Charnov EL (2004) Effects of body size and temperature on population growth. Am Nat 163:429–441

Schimel JP (2018) Life in dry soils: effects of drought on soil microbial communities and processes. Annu Rev Ecol Evol Syst 49:409–432

Schindlbacher A, Rodler A, Kuffner M, Kitzler B, Sessitsch A, Zechmeister-Boltenstern S (2011) Experimental warming effects on the microbial community of a temperate mountain forest soil. Soil Biol Biochem 43(7):1417–1425

Singh BK, Bardgett RD, Smith P, Reay DS (2010) Microorganisms and climate change: terrestrial feedbacks and mitigation options. Nat Rev Microbiol 8:779–790

Stevnbak K, Scherber C, Gladbach DJ, Beier C, Mikkelsen TN, Christensen S (2012) Interactions between above-and belowground organisms modified in climate change experiments. Nat Clim Chang 2:805–808

Stocker BD, Roth R, Joos F, Spahni R, Steinacher M, Zaehle S, Prentice IC (2013) Multiple greenhouse-gas feedbacks from the land biosphere under future climate change scenarios. Nat Clim Chang 3:666–672

Sulman BN, Shevliakova E, Brzostek ER, Kivlin SN, Malyshev S, Menge DN, Zhang X (2019) Diverse mycorrhizal associations enhance terrestrial C storage in a global model. Glob Biogeochem Cycles 33:501–523

Tani S, Judelson H (2006) Activation of zoosporogenesis-specific genes in Phytophthora infestans involves a 7-nucleotide promoter motif and cold-induced membrane rigidity. Eukaryot Cell 5: 745–752

Timmusk S, Wagner EGH (1999) The plant-growth-promoting rhizobacterium Paenibacillus polymyxa induces changes in Arabidopsis thaliana gene expression: a possible connection between biotic and abiotic stress responses. Mol Plant-Microbe Interact 12(11):951–959

Toseland ADSJ, Daines SJ, Clark JR, Kirkham A, Strauss J, Uhlig C, Mock T (2013) The impact of temperature on marine phytoplankton resource allocation and metabolism. Nat Clim Chang 3: 979–984

Voosen P (2021) Global temperatures in 2020 tied record highs

de Vries FT, Griffiths RI, Bailey M, Craig H, Girlanda M, Gweon HS, Bardgett RD (2018) Soil bacterial networks are less stable under drought than fungal networks. Nat Commun 9:1–12

Weng SF, Reps J, Kai J, Garibaldi JM, Qureshi N (2017) Can machine-learning improve cardiovascular risk prediction using routine clinical data? PloS one 12(4):e0174944

Wieland G, Neumann R, Backhaus H (2001) Variation of microbial communities in soil, rhizosphere, and rhizoplane in response to crop species, soil type, and crop development. Appl Environ Microbiol 67:5849–5854

WMO (2020) confirms 2019 as second hottest year on record. Press release number: 01/15/2020. https://public.wmo.int/en/media/press-release/wmo-confirms-2019-second-hottest-year-record

Xue K, Yuan M, Shi Z, Qin Y, Deng YE, Cheng L, Zhou J (2016) Tundra soil carbon is vulnerable to rapid microbial decomposition under climate warming. Nat Clim Change 6:595–600

Zhang, Y., Wu, Z., Liu, M., He, J., Shi, K., Wang, M. and Yu, Z., 2014. Thermal structure and response to long-term climatic changes in Lake Qiandaohu, a deep subtropical reservoir in China. Limnology and Oceanography, 59(4), pp.1193-1202

Chapter 2
Soil Microflora and Their Interaction with Plants Under Changing Climatic Scenarios

Biswajit Pramanick, Anurag Bera, Priyanka Saha, Anamika Barman, Bappa Paramanick, Sagar Maitra, and Akbar Hossain

Abstract Microbes are considered one of the most important living organisms in agriculture, and changing climatic scenarios influence microbial diversity highly impacting overall agriculture. Soil microbial diversity is highly altered with changes in soil moisture, types of vegetation, soil reaction, soil organic matter, etc. Climate change results in increased concentrations of carbon dioxide in the environment that largely influences the soil microbial diversities and rhizosphere-dwelling microorganisms. All these ultimately impact the soil health vis-à-vis productivity of the crops grown in the soil. Upscaling of plant mutualism and elevation in the CO_2 level usually helped in gaining the total microorganisms' biomass and the richness of

B. Pramanick (✉)
Department of Agronomy and Horticulture, University of Nebraska-Lincoln, Scottsbluff, NE, USA

Department of Agronomy, Dr. Rajendra Prasad Central Agricultural University, Pusa, Bihar, India
e-mail: biswajit@rpcau.ac.in

A. Bera
Department of Agronomy, Institute of Agricultural Sciences, BHU, Varanasi, UP, India

P. Saha · A. Barman
Division of Agronomy, ICAR-IARI, Pusa, New Delhi, India

B. Paramanick
Department of Soil Science, Dakshin Dinajpur KVK, Uttar Banga Krishi Viswavidyalaya, Majhian, West Bengal, India

S. Maitra
Department of Agronomy, Centurion University of Technology and Management, Paralakhemundi, Odissa, India
e-mail: sagar.maitra@cutm.ac.in

A. Hossain
Department of Agronomy, Bangladesh Wheat and Maize Research Institute, Dinajpur, Bangladesh

P. Mathur et al. (eds.), *Microbial Symbionts and Plant Health: Trends and Applications for Changing Climate*, Rhizosphere Biology,
https://doi.org/10.1007/978-981-99-0030-5_2

mycorrhizal fungi. Besides these, an increase in temperature also influences microbes. Environmental, as well as soil heating due to climate change, also result in an abundance of nematodes in the soil and a decrease in the beneficial microbes. Micro-microbe interaction is another important concern. Soil microflora and their dynamics in the soil environment must be taken into serious consideration as this plays a huge role in agricultural production systems. In this chapter, efforts are made to highlight the distribution of soil microflora and their interactions under changing climatic scenarios; the influence of changing climatic scenarios on plant-microbial along with soil-microbial interactions; and micro-microbe interactions. A proper understanding of these might help in planning sustainable agricultural production systems under changing climatic scenarios.

Keywords Climate change · Microbial distribution · Soil microflora · Soil-microbe interaction

Abbreviations

ABA	Abscisic acid
AMF	Arbuscular *mycorrhizal* fungi
ANN	Artificial neural network
BNF	Biological nitrogen fixation
FACE	Free air carbon dioxide enrichment
GAM	Generalized additive models
GLM	Generalized linear model
SDMs	Species distribution models
VAM	Vesicular arbuscular *mycorrhizae*

2.1 Introduction

Under the changing climatic scenarios, many changes in the life cycle, as well as the crop ecologies, have been observed (Pramanick et al. 2014; Pramanick et al. 2018; Laik et al. 2021). Continuous increments in the surface temperature both on the soil surface and inside the soil put the soil's biological activity under great concern. Biological activities in the soil have been controlled by the soil microbes and rapid increases in the temperature make shifts in the soil microbial communities. The soil microbial activities in the tropical and sub-tropical soils are more comparing the temperate soils (Laik et al. 2021) making rapid changes in the labile soil organic matter pools. Under such conditions, an increment in the activities of soil microbes makes a further decrease in soil organic matter, which is a major concern for sustainable crop production in the long run (Singh et al. 2017; Dey et al. 2021a, b).

Plants and their growth and development are very much correlated with diverse soil microbial populations. Soil bacteria, fungi, actinobacteria, and nematodes are important soil microorganisms (Trivedi et al. 2020). Soil microbiomes and their interaction with plant roots create dynamic systems of plant-microbe interactions. Such interactions are very much crucial and make the plant continue its life cycle even under environmental stresses (Trivedi et al. 2020, 2021; Dey et al. 2021a). Thus, a better understanding of the soil microflora and their interaction with plants is very much needed to make agriculture sustainable. The importance of soil microbes on plant health, productivity, and growth is a well-known fact (Berg et al. 2016; Kumar et al. 2021). Microbiome regulates plant systems by influencing the metabolic activities of plants including the signaling system in the plants. Research on understanding the mechanism of plant-microbe interactions on mitigating the pathogenic impact of one microbe on plants by another beneficial one is gaining importance particularly under changing climatic scenarios. Many previous pieces of research explored the possible interaction between plants and microbiomes on signaling processes for plant-defense mechanisms, symbiosis effects, mitigating plant stresses, etc. (Jones et al. 2016). Hence, the chapter on "soil microflora and their interaction with plants under changing climatic scenarios" is very much important in the present context. This chapter includes the details of soil microflora and their distribution in the soil system, the impact of climate change on plant-microbial interaction, how climate change alters plant and microbial distribution, and micro-microbe interaction.

2.2 Soil Microflora and Their Distribution

2.2.1 Groups of Soil Microflora, Their Characteristics, and Distribution

Soil contains millions of living organisms, making it both a living and dynamic system. These organisms not only aid in soil development but also serve as the key driving factors of nutrient cycling and regulate the dynamics of the organic materials found in soil, increasing plant nutrient acquisition, conferring stress tolerance, resisting pathogens, and improving plant health (Hesammi et al. 2014; Jacoby et al. 2017). Soil-dwelling microorganisms are broadly classified into two groups: soil flora, i.e., belonging to the plant kingdom, and soil fauna, i.e., the animal forms. These groups are further divided into macro and micro groups based on their size (Lartey 2006). Microorganisms are those which are big enough to be seen by the naked eye and microorganisms are so small that they can be seen only after magnification using a microscope. In this chapter, we will majorly focus on soil microflora. Soil microflora is a group of soil microorganisms that primarily comprise a portion of the soil organic matter and a colloidal portion of the soil humus. This is a key contributor to determining soil biological properties. The organic matter quality in any soil is dependent upon the nature of the soil microflora present in it and its ability in biochemical transformations. Four groups of organisms primarily constitute the soil microflora population, viz., bacteria, fungi, actinomycetes, and algae (Allison 1973; Baldock and Skjemstad 2000). Each of these groups develops numerous species of their genera and thereby maintains the population dynamics. More embellished information about these organisms needed to be studied to know their distribution in soil. In Table 2.1, comparative studies on microflora distribution have been represented subtly.

2.2.2 *Factors Affecting the Soil Microflora Distribution*

Just like higher plants, soil microflora relies on the soil for their sustenance, growth, and activity. Soil factors that have a significant influence on microflora distribution are as followed.

2.2.2.1 Soil Moisture

Soil moisture is one of the most important factors determining microbial population and their activity. It serves as a source of nutrients for the microbes by providing hydrogen and oxygen and as a solvent and carrier of other dietary ingredients for microbes. Moisture levels between 20 and 60 percent are ideal for microbial activity and population growth. Under a waterlogged situation or when the pores are filled up in any soil, the anaerobic microflora becomes active and similarly vice versa, in aerated soil, aerobic microflora gets active. Thus, the proportion of air and water present in any soil can determine the types of microflora which would be dominant and which would be suppressed (Borowik and Wyszkowska 2016).

2.2.2.2 Soil Reaction or Soil pH

Soil pH has a significant impact on the composition of soil microorganisms, both quantitatively and qualitatively. Most soil bacteria, blue-green algae, diatoms, and protozoa prefer a neutral or slightly alkaline reaction between pH 4.5 and 8.0, while fungi thrive well in the acidic reaction between pH 4.5 and 6.5 and actinomycetes prefer slightly alkaline soil reactions. The type of microflora found in soil is also influenced by the pH level prevailing in soil reaction. For example, nitrifying bacteria (*Nitrosomonas* and *Nitrobacter*) and diazotrophs such as Azotobacter are absent or inactive at lower pH, whereas diazotrophs such as *Beijerinckia*, *Derxia*, and sulfur-oxidizing bacteria such as *Thiobacillus thiooxidans* are active in acidic soil (Rousk et al. 2009).

Table 2.1 Comparative study on characteristics and distribution of various soil microflora

	Bacteria	Fungi	Actinomycetes	Algae
Basic information	Very minute, unicellular, achlorophyllous, and the most numerous organisms present in the soil	Filamentous organisms with much larger cell widths comparing other soil microorganisms	These organisms are intermediate between fungi and bacteria. They share the characteristics of both bacteria and fungi	Unicellular, chlorophyll-containing organisms, less abundant
Distribution	Bacteria constitute the most abundant group of microorganisms in the soil. A wide range of factors influences bacterial density, including soil type, crop cover type, and seasonal climate. Generally, soils with low organic matter and sandy texture have a very low population of bacteria (Mhete et al. 2020)	The number of fungi in arable soil is lesser than in bacteria. Though they can grow in a wide range of soil pH, their population is highest under acidic conditions, because of severe competition with bacteria at neutral pH (Yuvaraj and Ramasamy 2020)	Actinomycetes are next to bacteria in numbers and fairly distributed in soils. They are more common in dry soils and in undisturbed pastures and grasslands. As they can withstand drought, they occur more frequently in soils undergoing dry spells (Ghorbani-Nasrabadi et al. 2013)	The most important species of algae present in soil, *Cyanophyta* or blue-green algae (BGA), loves a water-logged condition for their growth. In rice fields, the submerged condition provides an ideal condition for their growth (Pabbi 2015; Kumar et al. 2017)
Favorable pH level	Neutral to slightly alkaline condition close to 7.0	Acidic condition (pH 4.5–5.5). (Aguilera et al. 2017)	Slightly alkaline to alkaline condition (pH 6.5–8.5). (Ganesan and Rajarajan 2011)	Strongly acidic soil (pH 4.6–5.7) is preferred by green algae, while the dispersion of blue-green algae was more apparent in neutral to alkaline soil (pH 7.2 to 7.8) (Thomas et al. 2017)
Temperature	Most of the bacteria grow well in the mesophilic temperature range (20 to 45 °C).	The optimum temperature for fungal growth is in the range of 25–30 °C	The optimum temperature for its population is 25–30 °C	The optimum temperature range was found to be 20–30 °C among different algal species
Population/g of soil at 0–15 cm soil depth	10^8–10^9	10^5–10^6	10^7–10^8	10^4–10^5
Biomass (g/m^2) at 0–15 cm soil depth	40–500	100–1500	40–500	1–50

2.2.2.3 Soil Organic Matter

Organic matter binds soil particles into aggregates and influences the texture and structure of the soil, which alters the microbial habitat (Mohammadi et al. 2011). Large soil aggregates have higher nutrient concentrations than the rest of the soil. In aggregates, soil microorganisms can hide from predators and protect themselves from drying up (Foster 1988).

2.2.2.4 Types of Vegetation

To understand the effects of vegetation type on the characteristics of soil microbial communities, in 2007, Han et al. in North China assessed plate counts in five different plant communities and thereby reported that a high number of plant species, legumes, and natural vegetation types likely support soil microbial populations that are more effective. Legumes support the development and growth of the microorganisms present in the soil. They contribute in fostering the production of greater soil biomass by providing additional N. Soil microbes use the increased N to break down carbon-rich residues of crops like wheat or corn (Yuvaraj and Ramasamy 2020).

2.2.2.5 Spatial and Seasonal Variation

The quality and quantity of microorganisms present in any soil are greatly influenced by its location and climate. Changes in the type and numbers of different microflora are observed in soils of tropical, subtropical, and temperate regions. Shail and Dubey (1997) have found seasonal fluctuations and species diversity in fungi in banj-oak and chir-pine forest soils of the Kumaon Himalayan region which can be attributed to the topography of that region. A maximum number of fungal taxa and an average number of bacteria and fungi (per gram soil) were recorded in the rainy season and a minimum in the summer season from both soils.

2.3 Impact of Climate Change on Plant Microbial Interaction

Natural and human-caused increases in atmospheric CO_2 levels are predicted to elevate global surface temperatures by 1.8 to 3.6 degrees Celsius by the year 2100 (IPCC CWT 2007). Plant physiology and root exudation will likely change as a result of changing climatic conditions caused by global warming. Raising temperature is also projected to cause soil water content to decrease in some locations, resulting in increased drought in some parts of the world. As a result, significant climate change is currently occurring. These climate-altering variables are known to have an impact on terrestrial ecosystems, plants, and other microorganisms. Ecosystem processes such as mineral nutrient cycling, breakdown of biological material, and animal and plant diseases are governed by both bacteria as well as fungi, but the events of fungi and bacteria are affected by both the non-biotic and biotic factors. Global warming has an impact on both these factors of our ecosystem (Singh et al. 2021). Climate change has an impact on both the above- and below-ground terrestrial ecosystems, both directly and indirectly. The population of plant species above ground will be influenced by increased CO_2, erratic rainfall-pattern, rise in temperature, etc. due to climate change (Tylianakis et al. 2008).

2.3.1 Elevated CO_2 Impacts on Soil Microbes

The impacts of increased CO_2 concentration in the environment, on plant physiological parameters such as growth, reproduction, photosynthesis, and yield, are stimulatory. However, rising CO_2 levels would increase the frequency of risky condition weather like moisture stress and heat waves that become detrimental to crop production (Gray and Brady 2016). Similarly, rising CO_2 levels would exacerbate the effects of interactions between plants and pathogens (Eastburn et al. 2011), and the population of microorganisms in the rhizosphere (Gschwendtner et al. 2016), disrupting plant growth, development, and yield. Due to increased plant mutualism, elevated CO_2 levels usually increased the microorganisms' biomass and the richness of mycorrhizal fungi (Blankinship et al. 2011; Tresede 2004). As a result, heating has a positive impact on some microbes, such as more population of nematodes (Blankinship et al. 2011) as well as other microbe communities (Allison and Martiny 2008). In long-term field research, the impact of high CO_2 can be precisely predicted, and it has imparted highly in cycling the soil organic C, N-cycle, and plant output. Experiments on free air carbon dioxide enrichment (FACE), for example, are very crucial field studies for such investigations (Ruhil et al. 2015). Rhizobial interactions with legumes were triggered by an increase in CO_2 levels (Schortemeyer et al. 1996; Marilley et al. 1999; Montealegre et al. 2000). In addition to their ability to fix nitrogen in legumes, these rhizobia are well-known for their beneficial impacts on plant growth (Sessitsch et al. 2002). C4 plants potentially allocate more carbohydrates to mycorrhizal fungi present in the root rhizosphere to gain more benefits from them, and this may drive the selection of mycorrhizal fungi colonizers, whereas C3 plants employ additional carbon for biomass production. This example indicates that altered environmental conditions caused by elevated CO_2 levels can affect plant-microorganism interactions in general, as well as plant competition and community structure, in part due to altered plant–microorganism interactions. Some data indicate that elevated carbon dioxide improves the colonization behavior of these fungi (Tylianakis et al. 2008).

2.3.2 Influence of Soil Moisture Variation on Soil Microbes

Soil moisture is an important element of organic matter breakdown and soil respiration (Aanderud and Lennon 2011). Soil moisture regulates a variety of processes in the soil, including the movement of solute, water, and gaseous elements, as well as existence vis-à-vis the movement of microorganisms (Rodrigo et al. 1997). Moisture stress (drought stress) could be another side effect of global warming. Temperature changes are frequently combined with changes in soil moisture which could explain some unexpected results from experiments looking into how climate change affects microbial communities. *Azospirillum brasilense* inoculated maize seedlings resulted in higher values of relative water content than uninoculated treatments under drought stress. Similarly, during low water conditions, bacterial inoculation has been reported to improve total aerial biomass, root growth, foliar area, and proline accumulation in leaves and roots (Casanovas et al. 2002). Drought stress causes a rise in leaves' abscisic acid (ABA), lowering the cytokinin levels, amplifying ABA concentration, and leaving stomata vulnerable (Cowan et al. 1999; Figueiredo et al. 2008). The inoculation impacts of osmolyte-generating bacteria on dried root and shoot biomass of paddy and several tillers were more substantial than the uninoculated controls under extreme stress conditions (Yuwono et al. 2005; Pramanick et al. 2020). The faster-growing bacterial community may respond to water status alteration promptly, while more time might be taken by the slower-growing fungal community (Bell et al. 2008; Cregger et al. 2012, 2014). Drought also exacerbates the difference in soil microorganism groups that are sensitive to temperature (Briones et al. 2014).

2.3.3 Influence of Temperature Variation

The increased temperature had a beneficial effect on AMF colonization and hyphal length usually. However, no effects or detrimental effects of increased temperature on AMF have been found in other circumstances. The amount to which plant growth increases or decreases for the same plant genotype depends on the temperature fluctuations (Heinemeyer and Fitter 2004). However, warming by 5 degrees centigrade in a temperate forest, for example, altered the relative abundances of soil bacteria and increased the bacterial-to-fungal ratio of the community (DeAngelis et al. 2015). Some rhizosphere bacteria and endophytes have been demonstrated to help plants cope with heat stress, and these strains could promote crop growth in a variety of climates, soils, and temperatures (Bashan and Holguin 1998). Some strains of plant growth-promoting rhizobacteria can grow at a higher temperature than at a lower temperature, making them particularly interesting for use in agriculture that is subjected to higher temperatures. Current climate change is causing an increase in extreme heat events (Stocker et al. 2013), and inadequate soil management and vegetation management can lead to desertification and the establishment of arid or semiarid soils. Soils with little or no plant cover are subjected to intense solar radiation, which raises temperatures in the upper soil layers. Overall, these factors result in soil temperatures that are well above the optimum for commonly studied mesophilic soil bacteria; values above 40 °C are common, with measurements reaching 75 °C, and some researchers have reported temperatures as high as 90 °C in deserts (McCalley and Sparks 2009). The microbial decomposition of soil organic matter is highly sensitive to changes in surrounding environmental conditions (temperature rise), which has the potential to alter enzyme kinetics and associated nutrient availability in the soil system through changes in resource allocation strategy and soil biota community composition (Stone et al. 2012; Steinweg et al. 2012). The effective direction and net magnitude of C flux among the source-sink components of the global carbon cycle, as well as the status of soil C pools, available nutrient status, and soil C stock, may be determined by the modified dynamics of soil microbial activity in a warmer environment (Majumder et al. 2008; Wall et al. 2015).

2.4 Climate Change Alters Plant and Microbial Distribution

2.4.1 Climate Change Vis-a-Vis Plant Distribution

The impact of humans on the world's climate is undeniable, and it has only grown in recent decades. Climate change has influenced many ecosystems and plant distribution all over the world in recent years (Parmesan 2006). Climatic variability, i.e., increasing temperatures, CO_2 elevations, and the uncertainty of rainfall patterns, can have a substantial impact on the potential distribution of species range shifts, with some species' suitable habitats anticipated to disappear completely (Anderegg et al. 2015). The scientific consensus now is that due to dramatic changes in climatic conditions, species are changing and decreasing considerably faster than in the past (Chen et al. 2011; Dobrowski et al. 2013). Changing climatic conditions can alter the structure, function, and distribution of forest plants by changing the species, pathogen-associated, insect outbreak, drought, rainfall pattern, hurricane, and storm (Dale et al. 2001). As per the Millennium Ecosystem Assessment (MEA 2005), near to 50% of the earth's biodiversity has already been vulnerable to extinction.

Several studies involving the climate change impact on plants as well as soil microbes have been conducted using various models. Most studies were conducted in Europe and America on climate change impacts on hill ecologies (Theurillat and Guisan 2001; Feeley and Silman 2010). Various models characterize existing plant distributions concerning the current climate and then estimate geographical alterations in species' climatic envelopes in response to climate change scenarios (Thomas et al. 2004; Bakkenes et al. 2002). For studying the possible consequences of climate change on plant distribution and biodiversity, a variety of methodological

Table 2.2 Different types of modeling methods for predicting plant distribution and biodiversity due to climate change

Models	Study conducted by
1. BIOCLIM	Kadmon et al. (2003); Beaumont et al. (2005)
2. The Florida Model	Crumpacker et al. (2001)
3. GLM, GAM, CTA, ANN	Thuiller et al. (2005a, b)
4. Species distribution models (SDMs)	Manish et al. (2016); Sheppard et al. (2014)
5. MigClim	Engler and Guisan (2009)
6. SPECIES model	Pearson et al. (2002)

models are available (Table 2.2), i.e., correlative bioclimatic envelope models (Thuiller et al. 2005b), dynamic ecosystem and biogeochemistry models (Peng 2000), spatially explicit mechanistic models for single species range shifts (Hill et al. 2001), species distribution models (SDMs) (Manish et al. 2016), etc.

Changes in species assemblages, regional and temporal mismatches in nutrient dynamics, the transformation of species phenologies, shifts in species ranges and niches, and evolutionary changes, such as extinctions and selective adaption, are the most visible ecological effects of climate change (Parmesan and Yohe 2003; Post et al. 2009; Bellard et al. 2012; Telwala et al. 2013). Despite the global effects of changing climatic scenarios on varied ecologies, hill ecology is considered the most vulnerable. Montane ecosystems around the world typically have distinct biological communities and high levels of endemism due to their complicated topography and a biogeographic history that includes repeated altitudinal movements of vegetation zones in response to climatic change (Diaz et al. 2003). In recent decades, the flora over the hill areas of the Himalayas has shown a shift of their habitat in the altitude of the hills (Panigrahy et al. 2010; Negi 2012). The full extinction of boxwood trees, bolly beech trees, Japanese bay tree, kumkum trees, etc. from the valley of Kashmir after the Pleistocene glacial demonstrates the most significant influence of climatic change on flora in the Himalayas (Puri 1945a, b). According to studies, C4 plants were common in the arid ecology of the Himalayan region, and C3 plants became more common as humidity increased in the mid-Holocene. Himalayan pine, a significant component of the western Himalaya's upper tree line, prefers soil that is well drained and not excessively wet (Raizada and Sahni 1960). Different studies indicated the pine vegetation shifting in the Himalayas due to shifts in micro-climatic and biotic factors (Yadava et al. 2017). Many terrestrial plants' populations are shifting in latitude and elevation as a result of changing climate. Sun et al. (2021) used a meta-analysis to conclude that species distributions had recently faced a decadal shift of ~11 m and ~ 17 m concerning higher altitude and latitude, respectively. These rates are two to three times faster than those previously reported. They estimated the impact of climate change on plant species richness across drylands in China from the past to the present and into the future by using random forest models. They reported plant diversity has risen during the last interglacial, and this trend is expected to continue, climate change sensitivity differs between plant life forms, and herbaceous plants adapt to climate change faster than woody species. During the period 1849–2010, Telwala et al. (2013) found that indigenous plant species in the Sikkim Himalayan region moved northward at a rate of 27.53 meters per decade. Changes in the richness and distribution of plant communities have expressed themselves in the Himalayan ecosystem as a result of fast warming. Future vegetation communities in the Himalayas are anticipated to have more woody plants; however, regions that are currently habitable for herbaceous flora are likely to become habitable in the future. Climate change is anticipated to have a particularly negative impact on higher elevations in the north of the Sikkim Himalaya, with an 18% loss of indigenous species (Manish et al. 2016). Significant and irreversible repercussions on the natural environment are almost certain if current climatic trends and intensity do not change. Therefore, policymakers and planners must pay attention to climate change.

2.4.2 Climate Change Vis-à-Vis Microbial Distribution

The microbial community has a vital impact on the aspect of climate variations since it is a key component of the carbon and nitrogen cycles, as well as the production and removal of greenhouse gases like carbon dioxide and methane, which contribute to climate change. Climate change alters the abundance and distribution of soil microbial communities due to their difference in physiology, biology, growth rate, temperature sensitivity, etc. (Whitaker et al. 2014). As soil biota are poor dispersers, they may react to climate change at a slower rate than plants (Van der Putten 2012). It was reported that warming a temperate forest by 5 °C shifted the abundance and distribution of soil bacteria and increased the bacterial-to-fungal ratio (DeAngelis et al. 2015). Soil microbial communities may change their distribution into the soil layers owing to variations in the climate. The microbial population performs a lot of functions, including nitrification, methanogenesis, denitrification, nitrogen fixation, and phosphorus mobilization, among others. Thus, changes in the makeup of the microbial population are likely to result in changes in ecosystem function. Soil fungal communities may transition from one dominant member to another despite slight variations in soil moisture availability (30% loss in water holding capacity), whereas bacterial communities remain steady. The faster-growing bacterial communities may respond quickly to moisture pulses, while the slower-growing fungal communities may take longer (Cregger et al. 2014). During non-extreme wet-dry cycles, these patterns suggest that fungal adaptability is stronger than bacterial plasticity (Kaisermann et al. 2015). The rise in temperature controls microbial activities, their turnover, and processing. As a result, the microbial population shifts in favor of species that are better suited to higher temperatures and have faster development rates (Castro et al. 2010). Warming is anticipated to alter the abundance of both fungi and bacteria in this way. Climate change may alter the distribution, structure, and function of marine microbial communities, affecting the biological carbon pump's effect (Evans et al. 2011). In arid top soils of the western USA, there is evidence of changing climatic scenarios on two major blue-green algae in the upper soil layer (0–15 cm), *Microcoleus vaginatus* and *Microcoleus steenstrupii.* Global temperature rise resulted in an abundance of *Microcoleus steenstrupii* over *Microcoleus vaginatus* (Dutta and Dutta 2016). These bacteria are essential for sustaining the topsoil's microbial population, which is important for limiting soil erosion (DiGregorio 2015). The negative consequences of these unintended alterations could result in significant ecological losses, as some of them directly impair the growth and survival of beneficial microbes. Microbes must either inherit or develop resistance to physical changes and acclimate to preserve biodiversity and conservation in an ecosystem (Srivastava et al. 2022). It is mostly unknown how quickly isolated microbial communities can adapt to climate change. As a result, fundamental issues like when microbial dispersal limitation starts to matter for ecosystem function and how rapidly microorganisms adapt to changing climate still need to be answered.

2.5 Micro-Microbe Interaction

In conventional agriculture, the natural role and dynamics of microorganisms have been overlooked due to over-dependency on high-input synthetic agrochemicals (Aktar et al. 2009). However, with time, we started to realize the significance of maintaining sustainability in our ecosystem, which prompted us to shed light on soil microflora and their dynamics in the soil environment (Suman et al. 2022; Trivedi et al. 2017). Soil is the largest terrestrial ecosystem where different species of microorganisms interact with one another in complex ways, making it the most diverse ecosystem on earth. Soil microbes act as the pivotal element in keeping the soil fertile and keeping plant pathogens in check and thus bear sustainability in the crop ecosystem (Jacoby et al. 2017). The interrelationships and interactions among various microorganisms are in a dynamic state and maintain a biological equilibrium in the community. Numerous investigations have confirmed microbial efficacy in multiple plant growth-promoting activities and soil sustainability, but our understanding of the mechanisms that underpin their interactions remains limited now (Backer et al. 2018). To harness the potential use of soil microflora, the interactions between different species of organisms in soil are needed to be understood more vividly (Johansson et al. 2004). Interactions between any two species can take several forms, these can be well understood from Fig. 2.1. Among various types of soil microbial interactions, the important ones are discussed in this section.

2.5.1 Symbiotic Interaction

Among various types of interactions performed by soil microbes, specific symbiotic interactions between soil bacteria and legumes are given capital importance in agriculture. In this interaction, *Rhizobium*, one type of soil bacteria, can fix

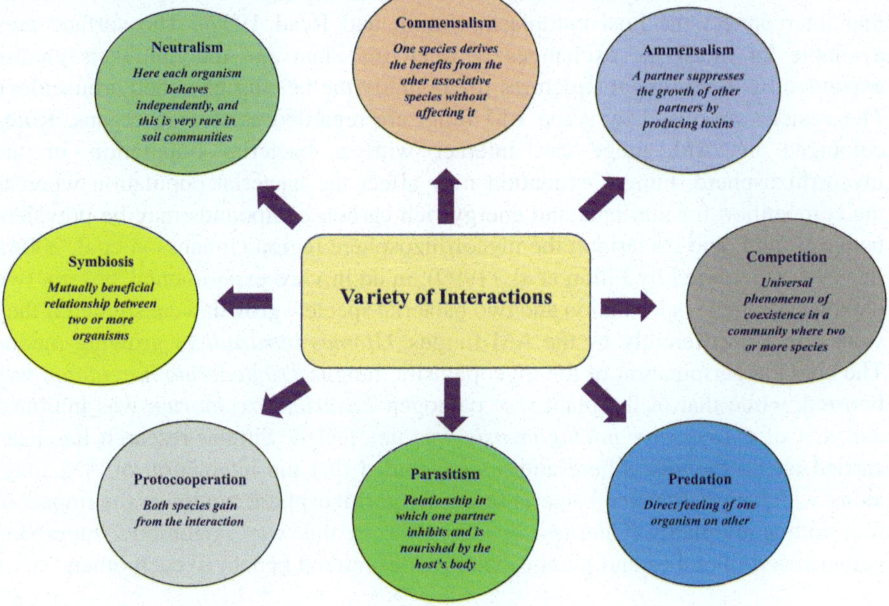

Fig. 2.1 Variety of interactions observed between different species in an ecosystem

atmospheric nitrogen in the root nodule of leguminous crops. This biological process is termed biological nitrogen fixation (BNF) (Zahran 1999; Brewin 2010). This biological nitrogen fixation (BNF) was the primary mechanism through which the biosphere acquired reactive N before the Industrial Revolution (Vitousek et al. 2013). Contribution by these nitrogen fixers is reported to be about 140 million tons every year, whereas N-input fertilizers are about 100 million tons N per year. The significance of these organisms in the nitrogen economy of soils thus becomes obvious. Some very different kinds of bacteria and archaea are all involved in biological N fixing. The application of genetic and molecular approaches in a variety of contexts has greatly expanded our understanding of the suite of species capable of BNF (Reed et al. 2011). In this interaction, some plants of the Fabaceae family form a symbiotic association with soil bacteria of the genus *Rhizobium*, which can fix atmospheric nitrogen. Specific rhizobia for different legumes infect the root moving to the root cortex, which results in the formation of a tiny growth called a root nodule. Atmospheric nitrogen is captured in this nodule by the *Rhizobium* and converted into a usable form. On the other hand, the plant roots provide the habitat for the rhizobacteria. In this way, legume-*Rhizobium* symbiosis is established through microbial interaction with the host plant (Beringer et al. 1979). Another symbiotic interaction that has much significance occurs between arbuscular mycorrhizal fungi (AMF) and higher plants or with other microorganisms, particularly bacteria. During the development of AM symbiosis, the fungus penetrates the root cortical cell membranes and develops haustoria-like structures (arbuscules or coils) that interconnect the host cytoplasm (Smith and Read 1997). The surface area available for metabolic exchanges between the plant and the fungus is greatly expanded by these fungal structures (particularly the heavily branched arbuscules). The vesicles produced by some AM fungi are regarded as storage organs. Roots colonized by AM fungi can interact with a bacterial population in the mycorrhizosphere. Fungal exudation may affect the bacterial population whereas the competition for nutrients and energy-rich carbon compounds may be prevalent between fungi and bacteria in the mycorrhizosphere region (Johansson et al. 2004). In 1999, as reported by Filion et al. (1999), in an in vitro experimental system, two pathogenic fungi's sporulation and two bacterial species' growth were studied if they were affected differently by the AM fungus, *Glomus intraradices* growing media. The conidial germination of the mycoparasitic fungus *Trichoderma harzianum* was boosted, while that of the plant root pathogen *Fusarium oxysporum* was inhibited and that of *Clavibacter michiganensis* was unaffected. Similar research has been carried out in the past where authors concluded that the interaction of AM fungi along with their associated bacteria in the mycorrhizosphere can lower the impact of root rotting of plants (Citernesi et al. 1996). In this way, symbiotic interaction between soil microbes and plant roots provides mutual benefit to each other.

2.5.2 Protocooperation Interaction

This is one type of positive synergistic interaction where both partners get a mutual benefit. But unlike symbiosis, protocooperation is not obligatory for survival (Selim and Zayed 2017). These types of interactions are bound to terrestrial ecosystems and are very useful in agriculture. The interaction between the *vesicular arbuscular mycorrhizae* (VAM)-legume plant and *Rhizobium* is one such example. In terms of nitrogen fixation and phosphorus availability, the efficiency and utility of mutualism among the three are significantly higher, leading to increased crop yields and improved soil fertility (Lindström and Mousavi 2020). The role of plant growth-promoting *rhizobacteria* (PGPR) is another good example of synergism, where rhizobacteria aid the plant by limiting the proliferation of pathogens in the rhizosphere through several methods and by providing some growth-promoting substances (Vejan et al. 2016). Association between sulfate-reducing bacteria, viz., *Desulfovibrio vulgaris*, and phototrophic bacteria, viz., *Chromatium vinosum*, is an example of a protocooperation type of interaction between the carbon cycle and the sulfur cycle. The rate of sulfate reduction depends on the rate of cellulose fermentation. In the presence of phototrophic bacteria, the end products of cellulose degradation and sulfate reduction disappeared from the medium and the protein content increased highly (Bharati et al. 1982).

2.5.3 Commensalism Interaction

In the case of commensalism, one organism is benefitted and another is neither benefitted nor harmed. This type of interaction is found in organic matter decomposition. The waste product of one microbe species might serve as a substrate for another microorganism species in a commensalism relationship. For example, true cellulose decomposers having C1 enzyme act on undegraded cellulose polysaccharides to break them down into smaller fragments. Later, the partially decomposed cellulose can be metabolized by many soil microorganisms (Jayasekara and Ratnayake 2019). Another evidence of commensalism can be observed during the nitrification process, where in the first step, *Nitrosomonas* sp. oxidizes ammonium to nitrite and then nitrite is oxidized into nitrate by *Nitrobacter* sp. Here *Nitrobacter* sp. takes benefits from their association by using nitrate to obtain energy for growth (Prescott et al. 2005).

2.5.4 Amensalism Interaction

It is one type of antagonistic interaction between two species. Here, one microbial population produces substances that are inhibitory to other microbial populations (Mougi 2016). For example, soil sulfur-oxidizing bacteria, i.e., *Thiobacillus thiooxidans*, produce sulfuric acid by oxidation of sulfur and thereby lowering the pH of soil solution which inhibits the growth of most other bacteria (Kumar et al. 2020). Besides that, sometimes, the growth of *Nitrobacter* and some fungi may be affected adversely by a large amount of NH_3 released during the decomposition of leguminous green manures (Pieters 1927; Shi 2013).

2.5.5 Competition, Parasitism, and Predation

In competition, two microbial populations are negatively correlated, and both populations suffer as a result in terms of survival and growth. Normally, severe competition between soil microbes for easily metabolizable carbon compounds is a rule rather than an exception, as it is readily utilized by almost all soil microbes. Therefore, the organisms having the inherent capability to grow fast are better competitors. Parasitism is probably widespread in soil communities. Bacteria, fungi, viruses, and other parasites can attack earthworms and other microorganisms. Predation is the direct feeding of any organism on another. For example, myxobacteria and slime molds feed on soil bacteria (Marsh 2011).

Soil is a habitat of many microorganisms that supports life forms and anchors the root system of plants. In changing climate conditions, the activity and interaction of soil microorganisms varied significantly. Few researchers on this topic reported that *arbuscular* and *ectomycorrhizal* fungi were shown to be more abundant in environments with higher CO_2 concentrations, but plant-beneficial fungi and bacteria were affected more by this change. Positive and neutral impacts of higher temperatures on plant-associated microbes were more common, while negative consequences were just as widespread and varied. Therefore, soil microorganisms associated with plants are an important factor to determine the response of plants toward climate change. To develop agricultural practices that are more adaptable to climate change, in-depth research on the relationships between these organisms is urgently required in the coming days.

2.6 Conclusion

Changing climatic scenarios highly impact the microbial diversity that governs many crucial metabolic activities in plant vis-à-vis soil. Soil moisture, types of vegetation, soil reaction, soil organic matter, etc. influence the soil microbial diversity. Changes in these parameters impact the soil microbes and their interaction with plant soil very much. Elevated CO_2 under changing climatic scenarios also impacts highly soil microbial diversities, plant-pathogen interactions, and rhizosphere-dwelling microorganisms, disrupting plant growth, development, and yield. Due to increased plant mutualism, rising CO_2 usually increased the total biomass of microorganisms and the richness of mycorrhizal fungi. Besides these, an increase in temperature also influences microbes. As a result, heating has a positive impact on some microbes, such as nematode abundance and the composition of other microbes. Micro-microbe interaction is another important concern. In conventional agriculture, the natural role and dynamics of microorganisms have been overlooked due to over-dependency on high-input synthetic agrochemicals. However, with time, we started to realize the significance of maintaining sustainability in our ecosystem, which prompted us to shed light on soil microflora and their dynamics in the soil environment. So, understanding all these would help to plan sustainable agriculture under changing climatic scenarios.

References

Aanderud ZT, Lennon JT (2011) Validation of heavy-water stable isotope probing for the characterization of rapidly responding soil bacteria. Appl Environ Microbiol 77(13):4589–4596

Aguilera P, Marín C, Oehl F, Godoy R, Borie F, Cornejo P (2017) Selection of aluminum tolerant cereal genotypes strongly influences the arbuscular mycorrhizal fungal communities in an acidic Andosol. Agric Ecosyst Environ 246:86–93

Aktar W, Sengupta D, Chowdhury A (2009) Impact of pesticides use in agriculture: their benefits and hazards. Interdiscip Toxicol 2(1):1–12

Allison FE (1973) Soil organic matter and its role in crop production. Elsevier, Amsterdam, p 637

Allison SD, Martiny JB (2008) Resistance, resilience, and redundancy in microbial communities. Proc Nat Acad Sci 105(supplement_1):11512–11519

Anderegg WRL, Hicke JA, Fisher RA, Allen CD, Aukema J, Bentz B et al (2015) Tree mortality from drought, insects, and their interactions in a changing climate. New Phytol 208:674–683

Backer R, Rokem JS, Ilangumaran G, Lamont J, Praslickova D, Ricci E, Subramanian S, Smith DL (2018) Plant growth-promoting rhizobacteria: context, mechanisms of action, and roadmap to commercialization of biostimulants for sustainable agriculture. Front Plant Sci 1473

Bakkenes M, Alkemade J, Ihle F, Leemans R, Latour J (2002) Assessing the effects of forecasted climate change on the diversity and distribution of European higher plants for 2050. Glob Chang Biol 8:390–407

Baldock JA, Skjemstad JO (2000) Role of the soil matrix and minerals in protecting natural organic materials against biological attack. Organic Geochem 31(7–8):697–710

Bashan Y, Holguin G (1998) Proposal for the division of plant growth-promoting rhizobacteria into two classifications: biocontrol-PGPB (plant growth-promoting bacteria) and PGPB. Soil Biol Biochem 30:1225–1228

Beaumont LJ, Hughes L, Poulsen M (2005) Predicting species distributions: use of climatic parameters in BIOCLIM and its impact on predictions of species' current and future distributions. Ecol Model 186:250–269

Bell C, McIntyre N, Cox S, Tissue D, Zak J (2008) Soil microbial responses to temporal variations of moisture and temperature in a Chihuahuan Desert grassland. Microbial Ecol 56(1):153–167

Bellard C, Bertelsmeier C, Leadley P, Thuiller W, Courchamp F (2012) Impacts of climate change on the future of biodiversity. Ecol Lett 15:365–377

Berg G, Rybakova D, Grube M, Köberl M (2016) The plant microbiome explored: implications for experimental botany. J Exp Bot 67:995–1002

Beringer JE, Brewin N, Johnston AW, Schulman HM, Hopwood DA (1979) The *Rhizobium*—legume symbiosis. Proc Royal Soc London Series B Biol Sci 204(1155):219–233

Bharati PL, Baulaigue R, Matheron R (1982) Degradation of cellulose by mixed cultures of fermentative bacteria and anaerobic sulfur bacteria. Zentralblatt für Bakteriologie Mikrobiologie und Hygiene: I. Abt Originale C: Allgemeine, angewandte und ökologische Mikrobiologie 3(4): 466–474

Blankinship JC, Niklaus PA, Hungate BA (2011) A meta-analysis of responses of soil biota to global change. Oecologia 165(3):553–565

Borowik A, Wyszkowska J (2016) Soil moisture as a factor affecting the microbiological and biochemical activity of soil. Plant Soil Environ 62:250–255

Brewin NJ (2010) Root nodules (Legume-*Rhizobium* symbiosis). Encyclopedia of life sciences. John Wiley and Sons, Ltd, Chichester, p 2010

Briones MJI, McNamara NP, Poskitt J, Crow SE, Ostle NJ (2014) Interactive biotic and abiotic regulators of soil carbon cycling: evidence from controlled climate experiments on peatland and boreal soils. Glob Chang Biol 20(9):2971–2982

Casanovas EM, Barassi CA, Sueldo RJ (2002) Azospiriflum inoculation mitigates water stress effects in maize seedlings. Cereal Res Commun 30(3):343–350

Castro HF, Classen AT, Austin EE, Norby RJ, Schadt CW (2010) Soil microbial community responses to multiple experimental climate change drivers. Appl Environ Microbiol 76(4): 999–1007

Chen IC, Hill JK, Ohlemüller R, Roy DB, Thomas CD (2011) Rapid range shifts of species associated with high levels of climate warming. Science 333(6045):1024–1026

Citernesi AS, Fortuna P, Filippi C, Bagnoli G, Giovannetti M (1996) The occurrence of antagonistic bacteria in Glomus mosseae pot cultures. Agronomie 16(10):671–677

Cowan AK, Cairns AL, Bartels-Rahm B (1999) Regulation of abscisic acid metabolism: towards a metabolic basis for abscisic acid-cytokinin antagonism. J Exp Bot 50(334):595–603

Cregger MA, Sanders NJ, Dunn RR, Classen AT (2014) Microbial communities respond to experimental warming, but site matters. PeerJ 2:e358

Cregger MA, Schadt CW, McDowell NG, Pockman WT, Classen AT (2012) Response of the soil microbial community to changes in precipitation in a semiarid ecosystem. Appl Environ Microbiol 78(24):8587–8594

Crumpacker DW, Box EO, Hardin ED (2001) Implications of climatic warming for conservation of native trees and shrubs in Florida. Conserv Biol 15:1008–1020

Dale VH, Joyce LA, McNulty S, Neilson RP, Ayres MP, Flannigan MD et al (2001) Climate change and forest disturbances: climate change can affect forests by altering the frequency, intensity, duration, and timing of fire, drought, introduced species, insect and pathogen outbreaks, hurricanes, windstorms, ice storms, or landslides. BioSci 51(9):723–734

DeAngelis KM, Pold G, Topçuoğlu BD, van Diepen LT, Varney RM, Blanchard JL et al (2015) Long-term forest soil warming alters microbial communities in temperate forest soils. Front Microbiol 6:104

Dey P, Mahapatra BS, Juyal VK, Pramanick B, Negi MS, Paul J, Singh SP (2021a) Flax processing waste—a low-cost, potential biosorbent for treatment of heavy metal, dye and organic matter contaminated industrial wastewater. Ind Crop Prod 174:114195. https://doi.org/10.1016/j.indcrop.2021.114195

Dey P, Mahapatra BS, Pramanick B, Kumar A, Negi MS, Paul J, Shukla DK, Singh SP (2021b) Quality optimization of flax fibre through durational management of water retting technology under sub-tropical climate. Ind Crop Prod 162:113277. https://doi.org/10.1016/j.indcrop.2021.113277

Diaz H, Grosjean M, Graumlich L (2003) Climate variability and change in high elevation regions: past, present and future. Clim Chang 59:1–4

DiGregorio BE (2015) Climate change affecting microbes in North America soils. Am Soc Microbiol. https://www.microbemagazine.org/index.php?option=com_contentandview=articleandid=6497:climatechangeaffectingmicrobesinnorthamericasoils

Dobrowski SZ et al (2013) The climate velocity of the contiguous United States during the 20th century. Glob Change Biol 19:241–251

Dutta H, Dutta A (2016) The microbial aspect of climate change. Energy Ecol Environ 1(4): 209–232

Eastburn DM, McElrone AJ, Bilgin DD (2011) Influence of atmospheric and climatic change on plant–pathogen interactions. Plant Pathol 60(1):54–69

Engler R, Guisan A (2009) MigClim: predicting plant distribution and dispersal in a changing climate. Divers Distribution 15(4):590–601

Evans C, Thomson PG, Davidson AT, Bowie AR, Van Den Enden R, Witte H, Brussaard CP (2011) Potential climate change impacts on microbial distribution and carbon cycling in the Australian Southern Ocean. Deep Sea Res II Top Stud Oceanogr 58(21–22):2150–2161

Feeley KJ, Silman MR (2010) Land-use and climate change effects on population size and extinction risk of Andean plants. Glob Change Biol 16:3215–3222

Figueiredo MV, Burity HA, Martinez CR, Chanway CP (2008) Alleviation of drought stress in the common bean (Phaseolus vulgaris L.) by co-inoculation with Paenibacillus polymyxa and *Rhizobium* tropici. Appl Soil Ecol 40(1):182–188

Filion M, St-Arnaud M, Fortin JA (1999) Direct interaction between the arbuscular mycorrhizal fungus Glomus intraradices and different rhizosphere microorganisms. New Phytol 141:525–533

Foster RC (1988) Microenvironments of soil microorganisms. Biol Fert Soil 6(3):189–203

Ganesan T, Rajarajan MD (2011) Feather degrading actinomycetes: a biotechnological approach to environmental clean-up. Department of Plant Sci Tagore Arts College, Puducherry 605:008

Ghorbani-Nasrabadi R, Greiner R, Alikhani HA, Hamedi J, Yakhchali B (2013) Distribution of actinomycetes in different soil ecosystems and effect of media composition on extracellular phosphatase activity. J Soil Sci Plant Nutr 13(1):223–236

Gray SB, Brady SM (2016) Plant developmental responses to climate change. Dev Biol 419(1): 64–77

Gschwendtner S, Engel M, Lueders T, Buegger F, Schloter M (2016) Nitrogen fertilization affects bacteria utilizing plant-derived carbon in the rhizosphere of beech seedlings. Plant Soil 407(1): 203–215

Han XM, Wang RQ, Jian L, Wang MC, Juan ZHOU, Guo WH (2007) Effects of vegetation type on soil microbial community structure and catabolic diversity assessed by polyphasic methods in North China. J Environ Sci 19(10):1228–1234

Heinemeyer A, Fitter AH (2004) Impact of temperature on the arbuscular mycorrhizal (AM) symbiosis: growth responses of the host plant and its AM fungal partner. J Exp Bot 55: 525–534

Hesammi E et al (2014) The role of soil organisms on soil stability (a review). Int J Curr Life Sci 4: 10328–10334

Hill J, Collingham YC, Thomas CD, Blakeley DS, Fox R, Moss D, Huntley B (2001) Impacts of landscape structure on butterfly range expansion. Ecol Lett 4:313–321

IPCC CWT (2007) Climate change 2007: synthesis report. Contribution of working groups I, II, and III to the fourth assessment report of the intergovernmental panel on climate change, pp 104

Jacoby R, Peukert M, Succurro A, Koprivova A, Kopriva S (2017) The role of soil microorganisms in plant mineral nutrition—current knowledge and future directions. Front Plant Sci 8:1617

Jayasekara S, Ratnayake R (2019) Microbial cellulases: an overview and applications. Cellulose 22

Johansson JF, Paul LR, Finlay RD (2004) Microbial interactions in the mycorrhizosphere and their significance for sustainable agriculture. FEMS Microbiol Ecol 48(1):1–13

Jones JDG, Vance RE, Dangl JL (2016) Intracellular innate immune surveillance devices in plants and animals. Science 354:6395

Kadmon R, Farber O, Danin A (2003) A systematic analysis of factors affecting the performance of climatic envelope models. Ecol Appl 13:853–867

Kaisermann A, Maron PA, Beaumelle L, Lata JC (2015) Fungal communities are more sensitive indicators to non-extreme soil moisture variations than bacterial communities. Appl Soil Ecol 86:158–164

Kumar M, Mitra S, Mazumdar SP, Majumdar B, Saha AR, Singh SR, Pramanick B, Gaber A, Alsanie WF, Hossain A (2021) Improvement of soil health and system productivity through crop diversification and residue incorporation under jute-based different cropping systems. Agronomy 11:1622. https://doi.org/10.3390/agronomy11081622

Kumar M, Zeyad MT, Choudhary P, Paul S, Chakdar H, Rajawat MVS (2020) Thiobacillus. In Beneficial Microbes in Agro-Ecology, Academic Press, pp 545–557

Kumar N, Mukherjee I, Sarkar B, Paul RK (2017) Degradation of tricyclazole: effect of moisture, soil type, elevated carbon dioxide and Blue Green Algae (BGA). J Hazard Mater 321:517–527

Laik R, Kumara BH, Pramanick B, Singh SK, Nidhi AM, Gaber A, Hossain A (2021) Labile soil organic matter pools are influenced by 45 years of applied farmyard manure and mineral nitrogen in the wheat—pearl millet cropping system in the sub-tropical condition. Agronomy 11:2190. https://doi.org/10.3390/agronomy11112190

Lartey RT (2006) Dynamics of soil flora and fauna in biological control of soil inhabiting plant pathogens. Plant Pathol J 5(2):125–142

Lindström K, Mousavi SA (2020) Effectiveness of nitrogen fixation in rhizobia. Microbial Biotechnol 13(5):1314–1335. https://doi.org/10.1111/1751-7915.13517

Majumder B, Mandal B, Bandyopadhyay PK (2008) Soil organic carbon pools and productivity in relation to nutrient management in a 20-year-old rice–berseem agroecosystem. Biol Fert Soil 44(3):451–461

Manish K, Telwala Y, Nautiyal DC, Pandit MK (2016) Modelling the impacts of future climate change on plant communities in the Himalaya: a case study from Eastern Himalaya, India. Model Earth Syst Environ 2(2):1–12

Marilley L, Hartwig UA, Aragno M (1999) Influence of an elevated atmospheric CO2 content on soil and rhizosphere bacterial communities beneath Lolium perenne and Trifolium repens under field conditions. Microbial Ecol 38(1):39–49

Marsh G (2011) Slime moulds prosper on the microfarm. Nature. https://doi.org/10.1038/news.2011.27

McCalley CK, Sparks JP (2009) Abiotic gas formation drives nitrogen loss from a desert ecosystem. Science 326(5954):837–840

MEA (2005) Millennium ecosystem assessment. Synthesis. Island Press, Washington, D.C, Ecosystems and Human Well-Being

Mhete M, Eze PN, Rahube TO, Akinyemi FO (2020) Soil properties influence bacterial abundance and diversity under different land-use regimes in semi-arid environments. Sci African 7:e00246

Mohammadi K, Heidari G, Khalesro S, Sohrabi Y (2011) Soil management, microorganisms and organic matter interactions: a review. African J Biotechnol 10:19840–19849. https://doi.org/10.5897/AJBX11.006

Montealegre CM, Van Kessel C, Blumenthal JM, Hur HG, Hartwig UA, Sadowsky MJ (2000) Elevated atmospheric CO_2 alters microbial population structure in a pasture ecosystem. Glob Change Biol 6(5):475–482

Mougi A (2016) The roles of amensalistic and commensalistic interactions in large ecological network stability. Sci Rep 6:29929. https://doi.org/10.1038/srep29929

Negi PS (2012) Climate change, alpine treeline dynamics and associated terminology: focus on northwestern Indian Himalaya. Trop Ecol 53:371–374

Pabbi S (2015) Blue green algae: a potential biofertilizer for rice. In: The algae world, Springer, Dordrecht, pp 449–465

Panigrahy S, Anitha D, Kimothi MM, Singh SP (2010) Timberline change detection using topographic map and satellite imagery. Trop Ecol 51:87–91

Parmesan C (2006) Ecological and evolutionary responses to recent climate change. Annu Rev Ecol Evol Syst 37:637–669

Parmesan C, Yohe G (2003) A globally coherent fingerprint of climate change impacts across natural systems. Nature 421(6918):37–42

Pearson RG, Dawson TP, Berry PM, Harrison PA (2002) SPECIES: a spatial evaluation of climate impact on the envelope of species. Ecol Model 154(3):289–300

Peng C (2000) From static biogeographical model to dynamic global vegetation model: a global perspective on modelling vegetation dynamics. Ecol Model 135:33–54

Pieters AJ (1927) Green manuring. Willey, New York

Post E, Forchhammer MC, Bret-Harte MS, Callaghan TV, Christensen TR, Elberling B, Fox AD, Gilg O, Hik DS, Høye TT, Ims RA, Jeppesen E, Klein DR, Madsen J, McGuire AD, Rysgaard S, Schindler DE, Stirling I, Tamstorf MP, Tyler NJC, van der Wal R, Welker J, Wookey PA, Schmidt NM, Aastrup P (2009) Ecological dynamics across the Arctic associated with recent climate change. Science 325:1355–1358

Pramanick B, Brahmachari K, Ghosh A, Zodape ST (2014) Foliar nutrient management through Kappaphycus and Gracilaria saps in rice-potato-greengram crop sequence. J Sci Ind Res 73:613–617

Pramanick B, Brahmachari K, Ghosh D, Bera PS (2018) Influence of foliar application seaweed (*Kappaphycus* and *Gracilaria*) saps in rice (*Oryza sativa*)–potato (*Solanum tuberosum*)–blackgram (*Vigna mungo*) sequence. Indian J Agron 63(1):7–12

Pramanick B, Brahmachari K, Kar S, Mahapatra BS (2020) Can foliar application of seaweed sap improve the quality of rice grown under rice–potato–greengram crop sequence with better efficiency of the system? J Appl Phycol 32:3377–3386. https://doi.org/10.1007/s10811-020-02150-z

Prescott LM, Harley JP, Klein AD (2005) Microorganism interaction and microbial ecology In Microbiology. Microbiology. McGraw-Hill, New York, pp 577–613

Puri GS (1945a) Some fossil leaves of *Litsea lanuginosa* Nees from the Karewa beds at Liddarmarg, Pir Panjal, Kashmir. J Ind Bot Soc 24:95–100

Puri GS (1945b) The genus *Quercus* in the Karewa deposits of Kashmir, with remarks on the oak forests of the Kashmir Valley during the Pleistocene. Proc Ind Acad Sci Sect B 22:232–256

Raizada MB, Sahni KC (1960) Living Indian Gymnosperms. Part 1 (Cycades, Ginkgoales and Coniferales). Ind For Rec 5L1–150

Reed SC, Cleveland CC, Townsend AR (2011) Functional ecology of free-living nitrogen fixation: a contemporary perspective. Annual Rev Ecol Evol Syst 42:489–512

Rodrigo A, Recous S, Neel C, Mary B (1997) Modelling temperature and moisture effects on C–N transformations in soils: comparison of nine models. Ecol Model 102(2–3):325–339

Rousk J, Brookes PC, Bååth E (2009) Contrasting soil pH effects on fungal and bacterial growth suggest functional redundancy in carbon mineralization. Appl Environ Microbiol 75(6): 1589–1596

Ruhil K, Ahmad A, Iqbal M, Tripathy BC (2015) Photosynthesis and growth responses of mustard (*Brassica juncea* L. cv. Pusa Bold) plants to free air carbon dioxide enrichment (FACE). Protoplasma 252(4):935–946

Schortemeyer M, Hartwig UA, Hendrey GR, Sadowsky MJ (1996) Microbial community changes in the rhizospheres of white clover and perennial ryegrass exposed to free air carbon dioxide enrichment (FACE). Soil Biol Biochem 28(12):1717–1724

Selim SM, Zayed MS (2017) Microbial interactions and plant growth. In Plant-microbe interactions in agro-ecological perspectives. Springer, Singapore, pp 1–15

Sessitsch A, Howieson JG, Perret X, Antoun H, Martinez-Romero E (2002) Advances in *Rhizobium* research. Crit Rev Plant Sci 21(4):323–378

Shail S, Dubey RC (1997) Seasonal changes in microbial community in relation to edaphic factors in banj-oak and chir-pine forests of Kumaun Himalaya. Himalayan Microbial Diversity 381–391

Sheppard CS, Burns BR, Stanley MC (2014) Predicting plant invasions under climate change: are species distribution models validated by field trials? Glob Change Biol 20(9):2800–2814

Shi J (2013) Decomposition and nutrient release of different cover crops in organic farm systems. Dissertations and Theses in Natural Resources 75. https://digitalcommons.unl.edu/natresdiss/75

Singh R, Pramanick B, Singh AP, Neelam KS, Kumar A, Singh G (2017) Bio-efficacy of Fenoxaprop-P-Ethyl for grassy weed control in onion and its residual effect on succeeding maize crop. Indian J Weed Sci 49(1):63–66. https://doi.org/10.5958/0974-8164.2017.00015.6

Singh VK, Gautam P, Nanda G, Dhaliwal SS, Pramanick B, Meena SS, Alsanie WF, Gaber A, Sayed S, Hossain A (2021) Soil test based fertilizer application improves productivity, profitability and nutrient use efficiency of rice (*Oryza sativa* L.) under direct seeded condition. Agronomy 11:1756. https://doi.org/10.3390/agronomy11091756

Smith SE, Read DJ (1997) Mycorrhizal Symbiosis. Academic Press, San Diego, p 605

Srivastava S, Chakraborty A, Suthindhiran K (2022) Microbial adaptation to climate change and its impact on sustainable development. Relationship Between Microbes and the Environment for Sustainable Ecosystem Services 1:85–105

Steinweg JM, Dukes JS, Wallenstein MD (2012) Modeling the effects of temperature and moisture on soil enzyme activity: linking laboratory assays to continuous field data. Soil Biol Biochem 55:85–92

Stocker BD, Roth R, Joos F, Spahni R, Steinacher M, Zaehle S et al (2013) Multiple greenhouse-gas feedbacks from the land biosphere under future climate change scenarios. Nat Clim Chang 3(7): 666–672

Stone MM, Weiss MS, Goodale CL, Adams MB, Fernandez IJ, German DP, Allison SD (2012) Temperature sensitivity of soil enzyme kinetics under N-fertilization in two temperate forests. Glob Change Biol 18(3):1173–1184

Suman J, Rakshit A, Ogireddy SD, Singh S, Gupta C, Chandrakala J (2022) Microbiome as a key player in sustainable agriculture and human health. Front Soil Sci 12

Sun Y, Sun Y, Yao S, Akram MA, Hu W, Dong L et al (2021) Impact of climate change on plant species richness across drylands in China: from past to present and into the future. Ecol Indic 132:108288

Telwala Y, Brook BW, Manish K, Pandit MK (2013) Climate induced elevational range shifts and increase in plant species richness in a Himalayan biodiversity epicenter. PLoS One 8:e57103

Theurillat J-P, Guisan A (2001) Potential impact of climate change on vegetation in the European Alps: a review. Clim Chang 50:77–109

Thomas BT, Rajan R, Saranyamol ST, Bhagya MV (2017) Influence of pH on the diversity of soil algae. J Glob Biosci 6(6):5088–5094

Thomas CD, Cameron A, Green RE et al (2004) Extinction risk from climate change. Nature 427: 145–148

Thuiller W, Lavorel S, Araújo MB (2005a) Niche properties and geographical extent as predictors of species sensitivity to climate change. Glob Ecol Biogeo 14:347–357

Thuiller W, Lavorel S, Araújo MB, Sykes MT, Prentice IC (2005b) Climate change threats to plant diversity in Europe. Proc Natl Acad Sci 102:8245–8250

Tresede KK (2004) A meta-analysis of mycorrhizal responses to nitrogen, phosphorus, and atmospheric CO_2 in field studies. New Phytol 164(2):347–355

Trivedi P, Delgado-Baquerizo M, Trivedi C, Hamonts K, Anderson IC, Singh BK (2017) Keystone microbial taxa regulate the invasion of a fungal pathogen in agro-ecosystems. Soil Biol Biochem 111:10–14

Trivedi P, Leach JE, Tringe SG, Sa T, Singh BK (2020) Plant–microbiome interactions: from community assembly to plant health. Nat Rev Microbiol 18:607–621

Trivedi P, Mattupalli C, Eversole K, Leach JE (2021) Enabling sustainable agriculture through understanding and enhancement of microbiomes. New Phytol 230:2129–2147

Tylianakis JM, Didham RK, Bascompte J, Wardle DA (2008) Global change and species interactions in terrestrial ecosystems. Ecol Lett 11(12):1351–1363

Van der Putten WH (2012) Climate change, aboveground-belowground interactions, and species' range shifts. Annu Rev Ecol Evol Syst 43(365):2012

Vejan P, Abdullah R, Khadiran T, Ismail S, Nasrulhaq Boyce A (2016) Role of plant growth promoting rhizobacteria in agricultural sustainability—a review. Molecules 21(5):573

Vitousek PM, Menge DN, Reed SC, Cleveland CC (2013) Biological nitrogen fixation: rates, patterns and ecological controls in terrestrial ecosystems. Philos Trans R Soc Lond Ser B Biol Sci 368(1621):20130119. https://doi.org/10.1098/rstb.2013.0119

Wall DH, Nielsen UN, Six J (2015) Soil biodiversity and human health. Nature 528(7580):69–76

Whitaker J, Ostle N, Nottingham AT, Ccahuana A, Salinas N, Bardgett RD et al (2014) Microbial community composition explains soil respiration responses to changing carbon inputs along an Andes-to-Amazon elevation gradient. J Ecol 102(4):1058–1071

Yadava AK, Sharma YK, Dubey B, Singh J, Singh V, Bhutiyani MR, Yadav RR, Misra KG (2017) Altitudinal tree line dynamics of Himalayan pine in western Himalaya, India. Quat Int 444:44–52

Yuvaraj M, Ramasamy M (2020) Role of fungi in agriculture. In: Mirmajlessi SM, Radhakrishnan R (eds) Biostimulants in plant science. IntechOpen. https://doi.org/10.5772/intechopen.89718

Yuwono T, Handayani D, Soedarsono J (2005) The role of osmotolerant rhizobacteria in rice growth under different drought conditions. Aust J Agric Res 56(7):715–721

Zahran HH (1999) *Rhizobium*-legume symbiosis and nitrogen fixation under severe conditions and in an arid climate. Microbiol Mol Biol Rev: MMBR 63(4):968–989. https://doi.org/10.1128/MMBR.63.4.968-989.1999

Chapter 3
Beneficial Microbial Consortia and Their Role in Sustainable Agriculture Under Climate Change Conditions

Kiran Sunar, Keshab Das, Arun Kumar Rai, and Saurav Anand Gurung

Abstract Rhizosphere is a complex and dynamic structure composed of both biotic and abiotic components. Rhizospheric microorganisms comprise a major group of biotic component and the microorganisms that contribute to the improvement of soil and plant health are known as "beneficial microorganisms". When beneficial microorganisms are applied singly or in combinations as "consortia", they can enhance most of the processes in the rhizosphere. Beneficial microorganisms mainly include plant growth-promoting rhizobacteria (PGPR), plant growth-promoting fungi (PGPF), arbuscular mycorrhiza fungi (AMF) and biocontrol agents (BCA) and have been studied and used in various ecosystems and in different combinations as "consortia". Their cumulative effects include plant growth promotion and mobilization of phosphate, nitrogen and other micronutrients in the rhizosphere; enhance root nodulation physiology; enhance photosynthesis, rate of transpiration, carbon dioxide assimilation and gas exchange; ameliorate salinity and drought stress; and induce resistance against phytopathogens. Rhizospheric microorganisms have developed mechanisms to survive under stressful environments like low and high temperatures as psychrophiles and thermophiles, under saline conditions as halophiles and in acid and alkaline conditions as acidophiles and alkaliphiles. The positive attributes of these microorganisms will be a basic tool to combat changes triggered by climate change and global warming. Future research should be focused on developing formulations based on nanoparticles, which can be used to modify microbe-based inoculums to enhance colonization and their overall effect on host and rhizosphere. Microbial consortia containing diverse microorganisms with promising

K. Sunar (✉) · K. Das
Department of Botany, Balurghat Mahila Mahavidyalaya, Balurghat, Dakshin Dinajpur, West Bengal, India

A. K. Rai · S. A. Gurung
Department of Botany, School of Life Sciences, Sikkim University, Tadong, Sikkim, India
e-mail: akrai@cus.ac.in

P. Mathur et al. (eds.), *Microbial Symbionts and Plant Health: Trends and Applications for Changing Climate*, Rhizosphere Biology,
https://doi.org/10.1007/978-981-99-0030-5_3

functionalities can be one of the strategies to manage rhizosphere functioning under climate change conditions.

Keywords AMF · Beneficial microorganisms · Climate change · PGPR · PGPF · Microbial consortia

Abbreviations

ACC-1	Aminocyclopropane-1-carboxylate
AIMs	Agriculturally important microorganisms
AMF	Arbuscular mycorrhiza fungi
BCA	Biocontrol agent
CERK1	Chitin elicitor receptor kinase 1
EFR	Elongation factor-TU receptor
FITC	Fluorescein isothiocyanate
FLS2	Flagellin-sensing 2
HCN	Hydrogen cyanide
IAA	Indole-3-acetic acid
ISR	Induced systemic resistance
K	Potassium
LYK5	Motif containing receptor-like kinase 5
MAMP	Microbe-associated molecular patterns
MTI	MAMP-triggered immunity
N	Nitrogen
P	Phosphorus
PAHs	Polycyclic aromatic hydrocarbons
PAL	Phenylalanine ammonia lyase
PGPB	Plant growth-promoting bacteria
PGPF	Plant growth-promoting fungi
PGPM	Plant growth-promoting microorganisms
PGPR	Plant growth-promoting rhizobacteria
PO	Peroxidase
PPO	Polyphenol oxidase
PRR	Pattern recognition receptors
ROS	Reactive oxygen species
VOC	Volatile organic compounds

3.1 Introduction

Soil is a complex and heterogeneous microhabitat – firstly the microbial population in soil is very diverse and secondly soil is a structured, heterogeneous and discontinuous system, generally poor in nutrients and energy sources. The chemical, physical and biological characteristics of these microhabitats differ greatly but are influenced by each other. It is now evident from various reports that soil microbial population is essential for maintenance of rhizosphere function, root health, nutrient uptake and tolerance of environmental stress and pathogen attack (Zake et al. 2011). Soil microflora plays the most important role in the soil region of the higher plants. The variable microflora changes the soil fertility conditions to a specific plant and in turn is dependent on the exudates of the roots in the rhizosphere. Microorganisms in soil are very essential for the maintenance of soil function in both natural and agricultural soils because of their involvement in key processes such as soil structure formation, decomposition of organic matter, toxin removal and the cycling of carbon, nitrogen, phosphorus and sulphur. In addition, microorganisms play key roles in suppressing soil-borne plant diseases, in promoting plant growth and changes in vegetation (Singh 2013).

The rhizosphere is composed of an extremely complex microbial community which includes saprophytes, endophytes, pathogens and beneficial microorganisms. In natural systems, these microbial communities exist in relative harmony where all populations generally balance each other out in their quest for food and space (Bélanger and Avis 2002). However, in "artificial" systems, i.e. agriculture, there is a modification in the natural balance of microorganisms that can significantly alter the microbial community and can lead to loss of beneficial microbes, increase the population of plant pathogens that may have a devastating effect on plant health and productivity. In these cases, the introduction of beneficial microorganisms in the rhizosphere can shift the balance of the microbial communities towards a population structure beneficial for plant health and productivity. Beneficial rhizosphere microorganisms are termed as agriculturally important microorganisms (AIMs) and are classified into two broad groups:

(i) Microorganisms with direct effects on plant growth promotion (plant growth-promoting microorganisms (PGPM))
(ii) Biological control agents (BCA) that indirectly assist with plant productivity through the control of plant pathogens

3.2 Players in Rhizosphere Function: The Rhizosphere Microbiome

It is now widely accepted that there are two major groups of microorganisms that provide mutualistic benefits to their hosts: the aboveground and belowground microflora. The aboveground microflora represents microorganisms that colonize the plants and can be categorized into epiphytes, endophytes, phyllospheric and rhizospheric. Rhizospheric microorganisms are considered the most dynamic and significantly impact the nutritional status of plant and its growth as well as inhibit a number of phytopathogens through direct and indirect mechanisms (Chakraborty et al. 2014, 2020; Sunar et al. 2015, 2017).

There are different beneficial microorganisms that utilize several strategies such as fixing, solubilizing, mobilizing and recycling nutrients in the soil which directly influence the plant growth and productivity (Bhowmik and Das 2018). The rhizosphere, a narrow zone of soil surrounding the root system, is colonized by a wide range of microorganisms, among these, bacteria and fungi are dominant and abundant (De la Fuente Cantó et al. 2020). Rhizospheric bacteria are known to colonize plant roots and facilitate plant growth and are designated as plant growth-promoting rhizobacteria; this unique group of bacteria promote plant growth through a wide array of direct and indirect mechanism (Sunar et al. 2014a, b, 2015; Khoshru et al. 2020; Basu et al. 2021). On the other hand, bacteria and fungi in the rhizosphere also function as biofertilizers and can protect plants against phytopathogens through mechanisms like competition, antagonisms as well as induction of defence mechanisms in the host plants (Sunar et al. 2020; Chakraborty et al. 2020). Rhizospheric microorganisms are also potential biopesticides and inhibit harmful pests through direct influence. They also have the ability to degrade and detoxify harmful organic as well as inorganic contaminants that accumulate in the soil including pesticides (Tarekegn et al. 2020). One of the main features of rhizospheric microorganisms is the solubilization of various macro and micro nutrients, which includes nitrogen fixation, phosphate and potassium solubilization or other micronutrients. These microorganisms are also known to secrete organic compounds to suppress plant pathogens or growth-enhancing substances to support plant growth (Gupta et al. 2015; Fasusi et al. 2021). Among the rhizospheric microorganisms, arbuscular mycorrhizal fungi (AMF) is a symbiosis between the roots/rhizoids of most of the plant species, including some of the most important crops, and specialized soil fungi (Brundrett and Tedersoo 2018). The AMF provide a direct interconnection between roots and soil as well as between root systems of different plant individuals belonging to the same or different plant species (Rezácová et al. 2018). The AM fungi exert several direct (e.g. enhanced nutrient acquisition, pollutant immobilization/detoxification, plant carbon reallocation, induced pathogen tolerance, signal transfer) and indirect (e.g. photosynthesis stimulation, drought tolerance, soil physical and microbial conditioning) effects on the plants, with possible consequences to yield and agricultural product quality, multitrophic interaction networks and soil quality (Diagne et al. 2020; Liu et al. 2021).

3.3 The Microbial Consortia/Microbiome

Diverse plant community serves as suitable hosts for many of the microbial communities that we know today. As we all know that a single plant is not inhabited by a single microbe, the belowground ecosystem is a network of many microorganisms and chemical signalling molecules. The aboveground plant diversity supports belowground microbial biodiversity, through root exudation, chemical signalling and rhizo-deposition; however most of them are host specific and selective in mode of action (Eisenhauer et al. 2017; Morella et al. 2020). A diverse and dynamic microorganisms are found to survive in the rhizosphere, which includes bacteria, fungi, oomycetes, nematodes, algae, protozoa, viruses and archaea (Bonkowski et al. 2009). Among all of these, mycorrhizae, *Rhizobium*, plant growth-promoting rhizobacteria (PGPR) and biocontrol agents are the most common and studied microorganisms which exert positive and beneficial effect on their host plants (Woo and Pepe 2018). These microorganisms that are predominantly associated with plants are known as microbiome. The structure and composition of plant microbiome is influenced by complex interactions between hosts, microbes and prevailing environmental factors. Some of the associations in the rhizosphere may be harmful and beneficial. When these beneficial microorganisms are isolated, characterized and used as bio-stimulants, they are known as beneficial microbial consortia. Plant biostimulants or plant growth-promoting microorganisms (PGPM) are used to improve soil health, nutrition availability, crop production and overall crop quality and production. They are now widely adopted in sustainable agricultural management practices and are useful tools for reducing chemical inputs, increasing productivity and recovering the natural equilibrium in agro-ecosystems (du Jardin 2015). Plant growth-promoting microorganisms play very important roles in growth promotion and improve plant resistance against biotic and abiotic stresses (Olenska et al. 2020). Until recent, the most widely used beneficial microorganisms include mycorrhizal fungi and rhizobia plant growth-promoting fungi and few actinomycetes. These microorganisms are found to possess several important traits like phosphate solubilization, nitrogen fixation, ACC deaminase production, siderophore production, biofilm formation, plant hormone production, biotic and abiotic stress tolerance or resistance and induction of defence mechanisms in the host plant by enhanced bio-signalling. Until now, most of the reports were based on the studies conducted in simple, single microbe-based systems in laboratory conditions; later it was also observed that these single microorganisms fail to perform in field conditions. In this context, recent studies show that application of beneficial microorganisms in groups (consortia) exerts more synergistic effects on the host plants. These consortia are also reported to be equally efficient in field conditions and have proven to be far better bioinoculants than those containing single microorganism (Parnell et al. 2016; Sarkar et al. 2018).

Application of microorganisms as effective bioinoculants to imitate the biological networks in soils and stimulate the recovery of functional, beneficial microbial groups responsible for soil fertility is an emerging approach termed as “rhizosphere engineering". These microorganisms when introduced in the rhizosphere of natural environment helps in rebuilding the microbial population lost due to cultivation and agricultural practices (Sunar et al. 2015; Leff et al. 2016; Stringlis et al. 2018). Introduction of beneficial microorganisms in the rhizosphere influences many key processes like nitrogen fixation, phosphate solubilization, siderophore, phytohormone and exo-polysaccharide production related to plant growth promotion and amelioration of many biotic and abiotic stresses.

Even though knowledge is limited on the survival of the microbial inoculants, the ability of rhizosphere-competent bacteria and fungi to establish close associations with the native microbiota and soil fauna has been sufficiently demonstrated (Bonanomi et al. 2018; de Vries and Wallenstein 2017). There are very less reports on the survival of introduced microorganisms in the rhizosphere; nevertheless there are sufficient evidences to support that microorganisms like bacteria and fungi introduced singly or in combinations as consortia can positively establish synergistic relationships with the native microflora and can establish new microbial communities. It is also now evident that the effects of co-applications are much higher and effective than those microorganisms applied singly. Studies have also suggested that microbial consortia can perform better than individual strains. Bioinoculants based on microbial consortia include bacteria of different species or strains; others may include both beneficial bacteria and fungi. The application of different microorganisms characterized with diverse mechanisms of action provides a wide spectrum of benefits for the plant, including direct stimulation of its growth and health, as well as increases in production as well as synchronized action against plant pathogens and induction of resistance in the host plants (Sunar et al. 2020; Behera et al. 2020; Bradácová et al. 2020).

Most of the studies on the role of beneficial microorganisms are based on interactions of single microorganisms with their hosts (Abhilash et al. 2016; Khan et al. 2019). However, rhizosphere is known to harbour numerous microorganisms and contribute collectively towards plant growth and other "rhizosphere function". Beneficial rhizospheric microorganisms like plant growth-promoting bacteria (PGPB), plant growth-promoting fungi (PGPF) and arbuscular mycorrhizal fungi (AMF) can establish multiple beneficial interactions with their host plant and among each other in the rhizosphere. The beneficial interaction ultimately promotes plant growth and development, induces plant defence system against pathogens, promotes nutrient mobilization and uptake and mediates the process of amelioration of environmental stress. The synergistic effect on the host is a result of combined mechanisms exerted by different types of beneficial microorganisms. Important functionalities include nutrient mobilization, regulation of plant hormones related to growth, plant defence against phytopathogens and influence on the native microbial population of the rhizosphere (Fig. 3.1) (Santoyo et al. 2021).

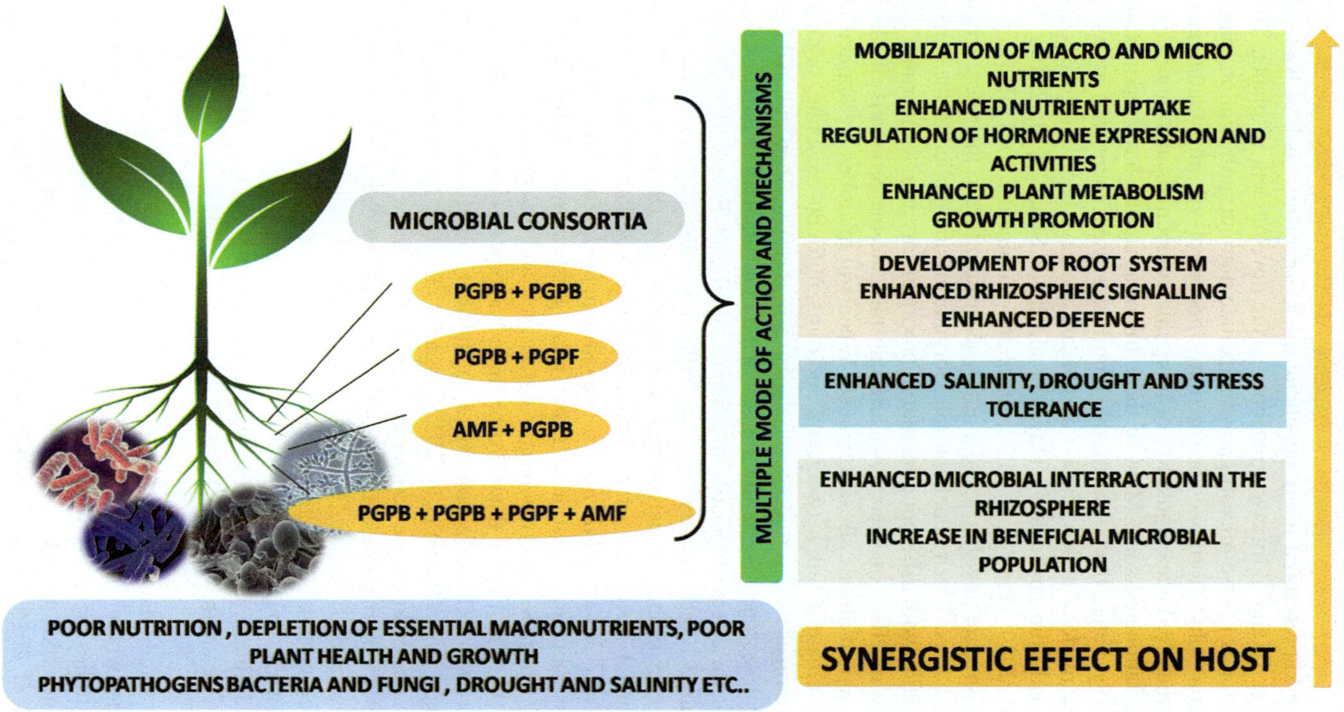

Fig. 3.1 Diagrammatic representation of the role of different types of rhizospheric microorganisms collectively known as "microbial consortia" towards plant growth promotion and suppression of phytopathogens in the rhizosphere. (*PGPB* plant growth-promoting bacteria, *PGPF* plant growth-promoting fungi, *AMF* arbuscular mycorrhizal fungi): The synergistic effect on the host is a result of combined mechanisms exerted by these different types of beneficial microorganisms. Important functionalities include nutrient mobilization, regulation of plant hormones related to growth, plant defence against phytopathogens and influence on the native microbial population of the rhizosphere (Santoyo et al. 2021 with modifications) *(Figure republished with modifications. Available under a Creative Commons Attribution 4.0 International License)*

3.4 Microbial Consortia and Their Diverse Roles

Till date, we have known about many beneficial microorganisms, such as bacteria and fungi that directly and indirectly contribute to the development of plants. The beneficial interactions of these microorganisms include nutrient mobilization, plant growth stimulation, phytohormone production, biocontrol of phytopathogens, improving soil structure, bioaccumulation of inorganic compounds and bioremediation of metal-contaminated soils. Most of these microorganisms have reported to perform remarkably better than few of the commonly used chemicals as well. According to their mode of action, these microorganisms have been categorized as "bio-fertilizers" and "bio-pesticides" (Hayat et al. 2010; Sunar et al. 2014a, b; Chakraborty et al. 2020; Khan et al. 2020; Ortiz and Sansinenea 2021). However these microorganisms are now known to exert synergistic effect to their host when applied in combinations. When beneficial microorganisms are applied to the rhizosphere in combination with different potential microorganisms as a "consortium", it shows wide range of effect on their hosts and can mitigate the problems of plant nutrition and biotic and abiotic stress tolerance. When soil microbes are applied to the rhizosphere, it creates a link between soil and roots, nutrients recycling, organic matter decomposition as well as tolerance towards various stress (Bhowmik and Das 2018). Numerous researches have shown that when a PGPR is used in combination with other PGPRs, the effects on host plant are quite significant. Few of the examples presented in Table 3.1 shows that *Thalassobacillus devorans* and *Oceanobacillus kapialis* increased germination percentage and rate in rice (Shah et al. 2017); *Bacillus subtilis* and *Arthrobacter* sp. increased dry biomass and total soluble sugars (Upadhyay et al. 2012); *Halomonas*, *Oceanobacillus* and *Zhihengliuella* sp. increased the root and shoot length and plant fresh weight (Orhan 2016); *Bacillus*, *Microbacterium*, *Methylophaga*, *Agromyces* and *Paenibacillus* (Bhise and Dandge 2019). When more than two or three bacteria are used together in a combination, it can significantly contribute to nutrient mobilization, ameliorate salinity and drought stress (Khan et al. 2019; Sagar et al. 2020; Sapre et al. 2022). PGPRs are also reported to induce resistance in crop plants more significantly when they are applied in combinations. In a study conducted by Sunar et al. (2017), *Bacillus altitudinis* and *B. pumilus* were applied in combination to the tea rhizosphere and showed systemic induction of defence enzyme (chitinase) starting from the roots to the leaves. The expression of the defence enzymes and other associated metabolites like chitinase, glucanase and phenylalanine ammonia lyase (PAL) were much higher in treatments where both PGPRs were applied in combination (Fig. 3.2).

On the other hand, the positive effect on host plants is also shown to be significantly higher when rhizobial and nonrhizobial (PGPR) are used in combination. Different studies have shown that such microbial consortia can enhance nodulation, enhance root nodulation physiology and mobilize nutrients in the soils (Egamberdieva et al. 2016; Korir et al. 2017; Tavares et al. 2018; Nascimento et al. 2020). Further combination of *Rhizobium* and PGPR can help to overcome different

Table 3.1 Combined inoculation of plant growth-promoting rhizobacteria (PGPR) in combinations as consortia and their effect on various crop plants

PGPR + PGPR	Crop/plant	Role/response	Reference
Brevibacterium epidermidis, *Bacillus aryabhattai*	Canola	Increased seed germination	Saravanakumar and Samiyappan (2007)
Bacillus sp., *Paenibacillus* sp.	*Oryza sativa*	Promotion of root and shoot growth	Beneduzi et al. (2008)
Pseudomonas aeruginosa, *Pseudomonas stutzeri*	Tomato	Enhanced root and shoot length	Tank and Saraf (2010)
Bacillus subtilis, *Arthrobacter* sp.	Wheat	Increased dry biomass, total soluble sugars	Upadhyay et al. (2012)
Bacillus cereus, *Pseudomonas* sp.	Rice	Increased N (26%), P (16%), K (31%)	Jha and Subramanian (2013)
Pseudomonas putida, *Pseudomonas aeruginosa*, *S. proteamaculans*	Wheat	Increased plant height, root length and grain yield	Nadeem et al. (2013)
Pseudomonas putida, *Enterobacter cloacae*, *Serratia ficaria*, *Pseudomonas fluorescens*	Wheat	Improved growth and yield	Singh and Jha (2015)
Pseudomonas syringae, *Pseudomonas fluorescens*, *Rhizobium phaseoli*	Mung bean	Improved seedling growth and nodulation	Goswami et al. (2015)
Thalassobacillus, *Bacillus*, *Halomonas*, *Oceanobacillus*, *Zhihengliuella* sp.	Wheat	Increased the root and shoot length and plant fresh weight	Orhan (2016)
Rhodotorula graminis, *Burkholderia vietnamiensis*, *Rhizobium tropici*, *Acinetobacter calcoaceticus*, *Rahnella* sp., *Burkholderia* sp., *Enterobacter asburiae*, *Sphingomonas yanoikuyae*, *Pseudomonas* sp. and *Curtobacterium* sp.	Poplar (cotton wood) under drought stress	Plant growth promotion (increased root dry weight, shoot dry weight, total dry weight, total nitrogen), reduced damage by reactive oxygen species (ROS)	Khan et al. (2016)
Bacillus cereus, *Bacillus* sp. and *Bacillus subtilis*	Wheat under salinity stress	Increased photosynthetic rate, the content of carotenoids and crude protein, higher grain yield	Shahzad et al. (2017)
Thalassobacillus devorans, *Oceanobacillus kapialis*	Rice	Increased germination percentage and rate	Shah et al. (2017)
B. altitudinis and *B. pumilus*	Tea	Increased nutrient uptake, promoted growth and induction of resistance against *Sclerotium* blight disease	Sunar et al. (2017)

(continued)

Table 3.1 (continued)

PGPR + PGPR	Crop/plant	Role/response	Reference
Sinorhizobium meliloti, *Paenibacillus yonginensis*	Lucerne and ginseng	Increased chlorophyll and carotenoid	Sukweenadhi et al. (2018)
Azotobacter chroococcum, *Lactobacillus* sp.	Lettuce	Increased root length at 50 and 100 mM NaCl	Hussein and Joo (2018)
Bacillus, *Pseudomonas*, *Enterobacter*, *Azotobacter*, *Rhizobium*	Strawberry	Increased plant height, biomass and yield	Rao et al. (2018)
Xanthomonas sp., *Stenotrophomonas* sp., *Microbacterium* sp.	*Arabidopsis thaliana* under biotic stress	Reduced the number of pathogen spores. Plant growth promotion (shoot fresh weight)	Berendsen et al. (2018)
Pseudomonas putida and *Bacillus subtilis*	Mung bean	ISR activation (peroxidase (PO), polyphenol oxidase (PPO), phenylalanine ammonia-lyase (PAL), β-1,3 glucanase and chitinase) and growth promotion	Sharma et al. (2018)
Ochrobactrum pseudogrignonense, *Pseudomonas* sp., *Bacillus subtilis*	Black gram and pea under drought stress	Plant growth promotion, elevated production of ROS-scavenging enzymes and cellular osmolytes and higher leaf chlorophyll content	Saikia et al. (2018)
Pseudomonas fluorescens, *P. fluorescens*, *P. migulae* and *Enterobacter hormaechei*	*Setaria italica* under drought stress	Enhanced seedling growth under drought stress via colonization of the root-adhering soil, increase soil moisture content	Niu et al. (2018)
Bacillus, *Microbacterium*, *Methylophaga*, *Agromyces*, *Paenibacillus*	Rice	Enhanced yield, photosynthesis and biomass	Bhise and Dandge (2019)
Alcaligenes, *Bacillus*, *Ochrobactrum*	Rice	Positive impact on germination percentage, shoot and root growth and chlorophyll content	Sagar et al. (2019)
Brevibacillus fluminis, *Brevibacillus agri*, *Bacillus paralicheniformis*	Brinjal, potato, tomato and chilli under salinity stress	Plant growth promotion shoot and root growth	Goswami et al. (2019)
Arthrobacter woluwensis, *Microbacterium oxydans*, *A. aurescens*, *Bacillus megaterium* and *B. aryabhattai*	*Glycine max* under salinity stress	Salt-tolerant gene GmST1 was highly expressed in AK1-treated plants	Khan et al. (2019)
Pseudomonas putida, *Pseudomonas fluorescens*	Rice	Promoted rice growth by colonizing rice roots	Sagar et al. (2020)
Acinetobacter bereziniae, *Enterobacter ludwigii*, *Alcaligenes faecalis*	*Pisum sativum* under salinity stress	Lower levels of electrolyte leakage and H_2O_2 contents under saline stress	Sapre et al. (2022)

types of biotic and abiotic stress through a series of integrated mechanisms like activation of different types of genes and their up- and downregulations (Table 3.2) (Korir et al. 2017; Htwe et al. 2018; Nascimento et al. 2020). The prospects of application of mycorrhiza as plant biostimulators has increased to a great extent and has shown promising results in both in vitro and in vivo experiments. AMF when applied in combinations with other AMF as consortium has shown enhanced efficacy, significant nutrient mobilization as well as enhanced biotic and abiotic tolerance (Zhang et al. 2018; Abdelhameed and Rabab 2019; Pellegrino et al. 2022). When AMF is applied in combination with rhizospheric bacteria (PGPR), then their efficacy is enhanced. Results have suggested that when potential AMFs are applied with PGPR, it can enhance drought tolerance capabilities of the host plant; this type of combinations can effectively modulate salt tolerance in the host plant (Ortiz et al. 2015; Hashem et al. 2016). *Glomus mosseae* and *Bacillus amyloliquefaciens* have shown to enhance photosynthesis, rate of transpiration, carbon dioxide assimilation and gas exchange and nutrient uptake (Pan et al. 2020). Combinations like *Glomus versiforme* and *Pseudomonas fluorescens* have also shown to play an important role in bioremediation where these microorganisms have shown to effectively remove polycyclic aromatic hydrocarbons (PAHs) from the soil (Tables 3.3 and 3.4) (Li et al. 2022). Apart from all these groups of beneficial microorganisms in the rhizosphere, there are some microorganisms that can mimic the role and mechanisms of plant growth-promoting rhizobacteria (PGPR) and can have positive effect on the host plant; this group of microorganisms are known as plant growth-promoting fungi (PGPF). Majority of PGPF include few species of the genera *Trichoderma*, *Penicillium*, *Aspergillus*, *Fusarium*, *Piriformospora*, *Phoma*, *Rhizoctonia*, *Talaromyces*, etc., these fungi have shown to have natural abilities to stimulate plant growth as well as suppress diseases and induce resistance in crop plants (Chakraborty et al. 2012, 2020). Many studies have shown that these fungi can stimulate plant growth when applied in the rhizosphere singly; however, these effects are much more enhanced when PGPF are applied as consortia or in combination with other PGPF. Data presented in Table 3.5 shows examples of few PGPF with multiple mechanisms utilized as bioinoculants for plant growth promotion and disease suppression. PGPF consortium like *Aspergillus niger*, *Penicillium citrinum* and *Trichoderma harzianum* can mimic the roles of PSBs (phosphate solubilizing bacteria) and can mobilize phosphate from rock phosphate and Ca-phosphate by organic acid production (Li et al. 2016). Similarly, *Purpureocillium lilacinum*, *Purpureocillium lavendulum* and *Metarhizium marquandii* have shown to enhance plant growth through production of IAA production and P-solubilization (Baron et al. 2020). Different species of *Trichoderma* are soil-inhabiting fungi commonly occurring in different types of soil and may constitute up to as much as 3% of the total microbial population in the rhizosphere. Recent studies have shown that many of the *Trichoderma* spp. form endophytic associations and can interact with other rhizosphere microorganisms; these combined interactions are reported to have positive effect on plant growth and disease suppression (Chakraborty et al. 2020). Several *Trichoderma* species are known to have beneficial effect on host plants and provides promising alternatives for plant growth promotion and induction of immunity, motivating mechanisms of

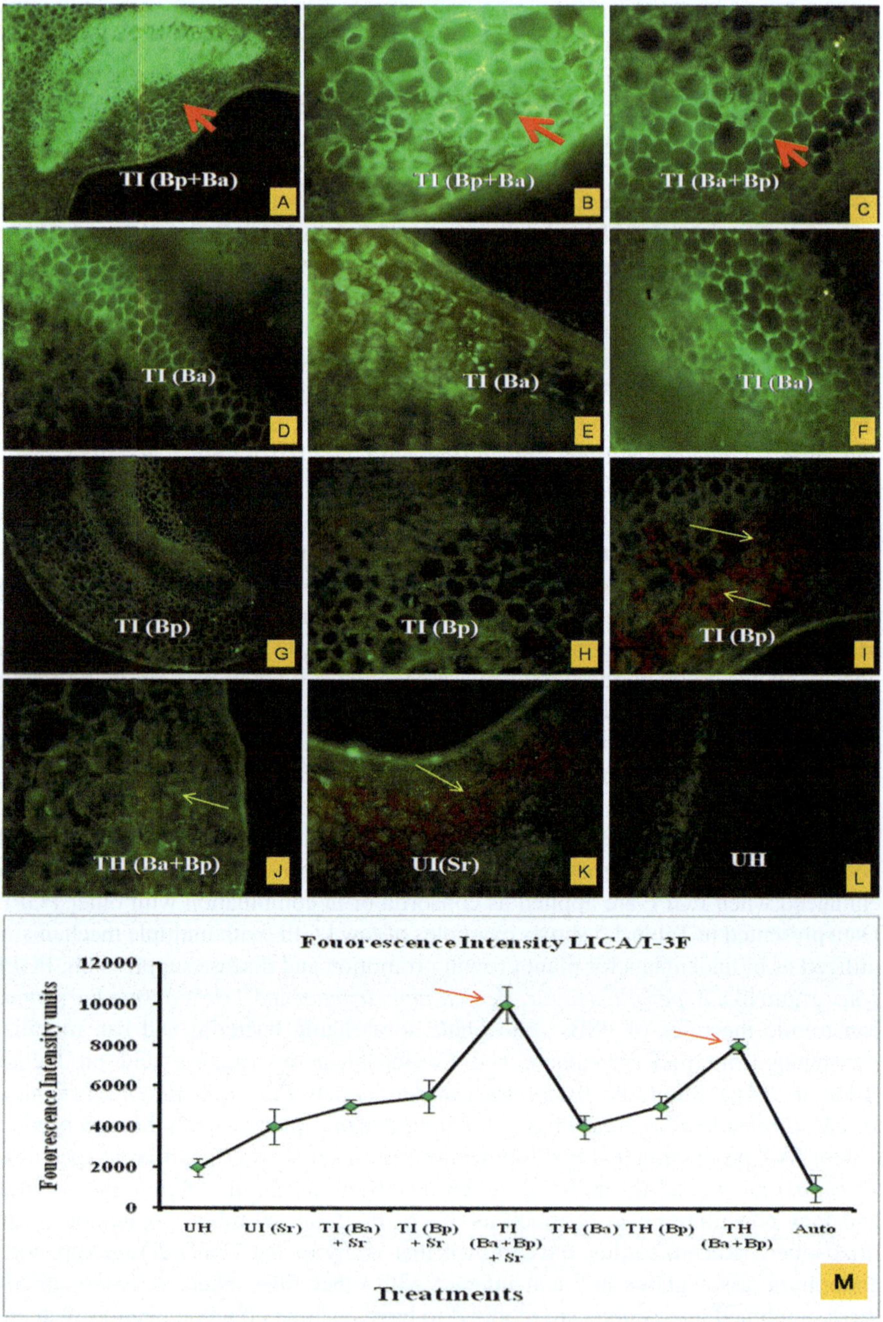

Fig. 3.2 Induction of defence enzyme (chitinase) in the leaf tissues of *Camellia sinensis* after treatment with PGPR singly and in combination. Expression of chitinase was analysed in the untreated and treated plants using polyclonal antibodies (PAb) raised against chitinase enzyme and labelled with FITC conjugate and observed under Leica Leitz Biomed Microscope (Germany) Filter I-3. Fluorescence intensity was significantly higher in plants treated with *B. altitudinis* and

plant defence, avoiding pathogen outbreaks and controlling plant disease. The most common species of *Trichoderma* are *T. harzianum*, *T. asperellum*, *T. asperelloides*, *T. viride* and *T. longibrachiatum* (Hewedy et al. 2020). Most of the *Trichoderma* spp. are now known to perform better in the rhizosphere when they are applied in combinations. Combinations of different species of *Trichoderma* like *Trichoderma harzianum*, *T. asperellum*, *T. virens*, *T. virens* and *T. virens* have shown to enhance disease resistance and improve plant health (Fiorentino et al. 2018). On the other hand, this PGPF has the potential of enhanced performance even if they are applied in combination with PGPR and AMF (Kumar et al. 2015; Dehariya et al. 2015). All these applications have proved that the efficacy of microbial consortium depends on the synergistic effects of two or more biocontrol agents and their mutualistic associations with the hosts.

3.5 Microbial Consortia and Rhizospheric Interactions

The rhizosphere is a dynamic structure and the microbial diversity within the rhizosphere is maintained by transfer of microorganisms from the environment to soil and vice versa, as well as from the plant parts to the soil and vice versa, and these population constitute the microbial reservoirs (Hardoim et al. 2015). The root microbiome is maintained through a systematic mechanism of horizontal transfer. Most of the dominant communities include *Acidobacteria*, *Bacteroidetes*, *Proteobacteria*, *Planctomycetes* and *Actinobacteria* (Fierer 2017). The transfer also takes place vertically, by seeds and other planting materials which is a reliable source of proliferating beneficial microbes that ultimately promote good health (Hardoim et al. 2012). Apart from these, microbial transfer of the rhizosphere itself is a highly active area for microbial movement, making it one of the most intricate environments. Other phenomenon that clearly defines microbial activities in the rhizosphere is the root exudations by the host plant. Root exudation is primarily the secretion of several complex chemical compounds by the roots into the rhizosphere which includes organic acids, sugars, amino acids, polyphenols, flavonoids, hormones and nutrients. These compounds act as chemical signalling for root colonization, rhizosphere habitation and nutrients for the microorganisms and are known as "rhizosphere effect" (Compant et al. 2019). Root exudates are important mediator of plant and microbiome within the rhizosphere and are a key mechanism by which the plants interact with soil microbes (Rajniak et al. 2018). Studies have

Fig. 3.2 (continued) B. pumilus in combination. TS of leaf treated with *B. pumilus* (**a**–**f**), TS of leaf treated with *B. altitudinis* (**g**–**k**); TS of leaf of untreated control plant (**l**); fluorescence intensity: Leica Leitz Biomed Microscope: Filter I-3 (M) (Treatments: *TI* treated inoculated, *TH* treated healthy, *UI* untreated inoculated, *UH* untreated healthy) (Sunar et al. 2017). *(A part of this data has been published by* Sunar et al. (2017)*; but the figure presented in this chapter is original data and unpublished)*

Table 3.2 Combined inoculation of rhizobia in combination with PGPR as consortia and their effect on various crop plants

Rhizobia and nonrhizobial bacteria	Crop/plant	Role/response	Reference
Rhizobium sp. and *Bacillus* sp.	*Phaseolus vulgaris*	Enhanced nodulation	Stajkovic et al. (2011)
R. phaseoli and *Pseudomonas syringae*, *Pseudomonas fluorescens*	Mung bean under salinity stress	Reduction of depressing effect of salinity on mung bean	Ahmad et al. (2012)
Bradyrhizobium japonicum and *Bacillus amyloliquefaciens*	*Glycine max*	Enhanced nodulation and growth	Masciarelli et al. (2014)
Ensifer adhaerens, *Bacillus anthracis*, *Paenibacillus taichungensis*, *Paenibacillus xylanilyticus*	*Vigna radiata*	Exploring plant growth promoting potential of nonrhizobial root nodules	Pandya et al. (2015)
B. japonicum and *Stenotrophomonas rhizophila*	Soya bean under salt stress	Enhanced root system physiology, nitrogen and phosphorus acquisition	Egamberdieva et al. (2016)
B. japonicum and *Pseudomonas putida*	Soya bean under salt stress	Alteration of root system architecture and induction of salt tolerance	Egamberdieva et al. (2017)
Rhizobium sp. and *Paenibacillus polymyxa* and *Bacillus megaterium*	*Phaseolus vulgaris* under low phosphorus soil	Synergistic effect on bean growth and improved functionality	Korir et al. (2017)
R. tropici and *Bacillus megaterium*	*Phaseolus vulgaris*	Enhanced growth and metabolism in a low phosphorus soil	Korir et al. (2017)
Azospirillum brasilense and *Rhizobium tropici*	Maize under salinity stress	Mitigation of salinity stress	Fukami et al. (2018)
B. elkanii and *Streptomyces griseoflavus*	*Glycine max*	Promoted plant growth, nodulation, nitrogen fixation, nutrient uptake and yield in field condition	Htwe et al. (2018)
R. tropici and *Serratia grimesii*	*Phaseolus vulgaris*	Growth promotion and early nodulation	Tavares et al. (2018)
R. tropici and *Pseudomonas fluorescens*	*Phaseolus vulgaris*	Growth and early nodulation	Nascimento et al. (2019)
Bradyrhizobium sp. and *Pantoea phytobeneficialis*	*Calopogonium mucunoides*	Plant growth and enhanced functionality	Nascimento et al. (2020)
Rhizobium spp., *Bacillus aryabhattai*, *Azotobacter vinelandii*	*Trifolium repens*	Nitrogen fixation and nutrient uptake in low phosphorus soil	Matse et al. (2020)

now confirmed that several plant microbe interaction results in the formation of short termed selective microbial niches that help in microbial growth and survival. However root exudates also act as limiting factors for microbial survival in the rhizosphere. Most of the beneficial microorganisms utilize the root exudates for

Table 3.3 Combined inoculation of arbuscular mycorrhizal fungi (AMF) in combination with other AMF as consortia and their effect on various crop plants

AMF + AMF	Crop/plant	Role/response	Reference
Glomus mosseae, *G. intraradices*	*Pepper*	Increased the growth and plants had higher P and chlorophyll content	Cekic et al. (2012)
Rhizophagus intraradices, *Glomus aggregatum*, *Glomus viscosum*, *Glomus etunicatum* and *Glomus claroideum*	Corn (*Zea mays* L.)	Improves crop growth, yield and grain quality	Berta et al. (2013)
Funneliformis mosseae, *Rhizophagus clarus* and *F. caledonium*	*Citrus aurantium*	Enhancement of shoot height, diameter and shoot and root dry matter	Ortas and Ustuner (2014)
Funneliformis mosseae, *Rhizophagus irregularis*	Chickpea (*Cicer arietinum* L.)	Increases plant biomass, production and grain quality	Pellegrino and Bedini (2014)
Glomus mosseae, *Glomus fasciculatum* and *Gigaspora decipiens*	*Triticum aestivum* under drought	Increased plant growth parameters and total chlorophyll pigments	Pal and Pandey (2016)
Rhizophagus irregularis, *Funneliformis mosseae*, *Funneliformis geosporum*	*Triticum aestivum* under temperature stress	Increased grain number, nutrient allocation and nutrient composition in root	Cabral et al. (2016)
Rhizophagus intraradices, *Funneliformis mosseae* and *F. geosporum*	*Zea mays* under high temperature stress	Increased leaf length, chlorophyll and photosynthetic rate, stomatal conductance and transpiration rate	Mathur et al. (2016)
Glomus etunicatum, *Glomus intraradices* and *Glomus mosseae*	*Cucumis sativus* under salinity stress	Increased biomass, photosynthetic pigment synthesis and enhanced antioxidant enzymes	Hashem et al. (2018)
Funneliformis mosseae, *Paraglomus occultum*	*Poncirus trifoliata* under drought	Increased hyphal length, hyphal water absorption rate and leaf water potential	Zhang et al. (2018)
Glomus monosporum, *G. clarum*, *Gigaspora nigra* and *Acaulospora laevis*	*Trigonella foenum-graecum* under heavy metal stress	Increased antioxidant enzymes activities	Abdelhameed and Rabab (2019)
Rhizophagus intraradices, *Funneliformis mosseae*, *Rhizophagus intraradices*, *Rhizophagus fasciculatus* and *Gigaspora* sp.	*Pisum sativum* under salt stress	Improved nutrient uptake, biochemical response, nodulation and growth	Parihar et al. (2020)
Funneliformis mosseae, *Rhizophagus irregularis*, *Funneliformis coronatum*, *Funneliformis geosporum*, *Rhizophagus clarus*, *Septoglomus viscosum*, *Claroideoglomus etunicatum*, *Diversispora spurca*, *Acaulospora rugosa*	Alfalfa (*Medicago sativa*)	Enhanced forage yield, N content, P content and total fatty acid content in the forage	Pellegrino et al. (2022)

Table 3.4 Combined inoculation of arbuscular mycorrhizal fungi (AMF) in combination with PGPR as consortia and their effect on various crop plants

AMF + PGPR	Crop/plant	Role/response	Reference
Rhizophagus intraradices and *Bacillus thuringiensis*, *B. megaterium*, *Pseudomonas putida*	*Trifolium repens* under drought stress	Drought tolerance under natural soil conditions	Ortiz et al. (2015)
Claroideoglomus etunicatum, *Rhizophagus intraradices*, *Funneliformis mosseae* and *B. subtilis*	*Acacia gerrardii* under salinity stress	Modulation of salt stress	Hashem et al. (2016)
Glomus mosseae and *Bacillus amyloliquefaciens*	*Elaeagnus angustifolia* under salinity stress	Enhanced photosynthesis, transpiration, carbon dioxide assimilation and gas exchange and nutrient uptake	Pan et al. (2020)
Funneliformis geosporum, *Claroideoglomus* sp. and *Pseudomonas tolaasii*, *Bacillus pumilus*	Bell pepper	Enhanced plant growth	Angulo-Castro et al. (2021)
Funneliformis mosseae, *Rhizophagus irregularis* and *Pseudomonas fluorescens*, *P. putida*	Myrtle (oil plant)	Improved leaf physiology, reduced electrolyte leakage, malondialdehyde and proline concentrations and mitigated oxidative pigment losses under drought condition	Azizi et al. (2021)
Glomus versiforme and *Pseudomonas fluorescens*	Bioremediation	Removal of removing polycyclic aromatic hydrocarbons (PAHs) from the soil	Li et al. 2022
Funneliformis mosseae and *Bacillus megaterium*	*Elymus nutans*	Improved plant growth regulation, nutrient acquisition and soil properties	Yu et al. (2022)

successful establishment and colonization and are regarded as successful rhizosphere colonizers; the microorganisms on the other hand secrete chemicals that will facilitate the formation of new niches for the rest of the rhizosphere microorganisms – a process known as "cross-feeding approach" (Jacoby and Kopriva 2019). For example, benzoxazinoids secreted by maize roots are reported to change the structure of root microbiome and influence the group of *Actinobacteria* and *Proteobacteria* within the rhizosphere (Hu et al. 2018). In a similar study, Zhalnina et al. (2018) reported that the amalgamation of root exudate composition and substrate selectivity significantly modified the assemblage of bacterial population in rhizosphere in *Avena barbata* roots. It is now widely accepted that the spatiotemporal organization of the rhizosphere and changes in physicochemical conditions are the major factors that control the rhizospheric microbial consortia formation and survival (Vetterlein et al. 2020).

Table 3.5 Combined inoculation of plant growth promoting fungi (PGPF) as consortia and their effect on various crop plants

PGPF strain	Broad function	Specific activities	References
Aspergillus niger, *Penicillium citrinum*, *Trichoderma harzianum*	Phosphate solubilization	Solubilized P from rock phosphate and Ca-P by organic acid	Li et al. (2016)
T. harzianum, *Phoma multirostrata*, *T. harzianum* and *Penicillium chrysogenum*	Suppression of deleterious pathogens	Suppressed bacterial wilt disease caused by *Ralstonia solanacearum*	Jogaiah et al. (2013)
A. niger, *Mucor circinelloides*, *A. flavus* and *P. oxalicum*	Suppression of fungal disease	Recovered the damage to morphological traits, photosynthetic pigments' total carbohydrate and total soluble protein of infected plants	Attia et al. (2022)
Aspergillus fumigatus and *Rhizopus oryzae*	*Fusarium* wilt in tomato	Increased photosynthetic pigments, total soluble carbohydrate and total soluble protein, whereas content of free proline, total phenols and the activity of antioxidant enzymes activity increased under infection	Attia et al. (2022)
Acremonium sp. and *Sarocladium* sp.	Biocontrol potential	Biocontrol against fungal pathogen, induction of resistance	Vivas et al. (2018)
Purpureocillium lilacinum, *Purpureocillium lavendulum* and *Metarhizium marquandii*	Plant growth and development	Production of IAA and solubilize P from fluorapatite	Baron et al. (2020)
Trichoderma koningii and *Macrophomina phaseolina*	Growth promotion of plants	Growth promotion, mineral solubilization and siderophore production.	Suebrasri et al. (2020)
Trichoderma harzianum, *Bacillus subtilis* and *Pseudomonas putida*	Biocontrol of late blight of tomato	Disease suppression and induction of resistance in seedlings	Kumar et al. (2015)
Trichoderma harzianum, *T. virens* and *T. viride*, *Funneliformis mosseae*, *Glomus cerebriforme*, *Rhizophagus irregularis*, *Fusarium udum*	Fusarium wilt pigeon pea	Disease suppression and induction of resistance in seedlings	Dehariya et al. (2015)
Trichoderma harzianum, *T. asperellum*, *T. virens*, *T. virens and T. virens*	Damping-off and root rot tomato	Mycoparasitism, suppression of disease, induction of resistance in seedlings	Fiorentino et al. (2018)

3.6 Microbial Consortia-Interaction-Establishment and Responses

There are various factors that can influence positive microbial colonization and establishment in the rhizosphere; this includes host roots, soil structure and chemical and mineral compositions of the soil. Microorganisms such as biological control agents (BCA), plant growth-promoting rhizobacteria (PGPR) and plant growth-promoting fungi (PGPF) when introduced to the rhizosphere initially colonize the roots firstly in response to the root exudates and secondly are influenced by soil texture, water content, soil temperature and pH value. It has now been known that colonization by introduced microorganisms in the rhizosphere is mainly influenced by biotic and abiotic factors. Biotic factors include nature of root exudates produced by the host plants, physiological characteristics of introduced microorganisms and interactions between introduced microorganisms and native microbes. Abiotic factors include soil environmental conditions, soil texture, water content, temperature and pH (Hartman and Tringe 2019). On the other hand, arbuscular mycorrhizal fungi respond to the biotic and biotic factors differentially. Environmental variation and host phenology can influence the AMF colonization as well as the spore density of the introduced AMF can also influence colonization and diversity. It has also been observed that soil and root habitats as well as soil characteristics, especially pH, nitrogen and micronutrients (Zn and Cu), as well as soil moisture can influence AMF colonization and communities significantly (Xu et al. 2017; Melo et al. 2019).

Among the biotic factors, root exudates are the most prominent compounds in the rhizosphere that have been studied and explored for their effect on introduced microorganisms. Root exudates are the compounds that allow specific microorganisms in its rhizosphere, and hence introduced microorganisms as "beneficial consortia" should respond positively to these exudates. Root exudates also provide specific signals to the microorganisms for mucilage production and for biofilm production (Venturi and Keel 2016; Gouda et al. 2018; Hartman and Tringe 2019). Host plants produce and exude specific molecules within the rhizosphere, and these metabolites can affect the assembly and colonization of the introduced microbiome. This exudation is greatly influenced by the age of the plant, and therefore subsequently the microbes are influenced by it as well. Studies have shown that root exudation pattern change during the growth cycle of host plants, and it has been reported that sucrose levels are high during the initial growth of the host while specific defence-related enzymes, metabolites and amino acids are released at the later stage of the development. Exometabolomic studies have shown that selected metabolites like nicotinic acid, shikimic acid, salicylic acid, cinnamic acid, IAA, etc. are responsible for the colonization and survival of specific microorganisms around the roots during the different growth stages of the host plant (Sasse et al. 2017; Zhalnina et al. 2018). On the other hand, root exudate compounds vary with plant genotype and are primary determinants for microbial colonization. Therefore, microbial consortia containing multiple microorganisms can be an answer to this chemical barrier for host colonization (Hernández-Montiel et al. 2017; Arora et al. 2020).

Another important biotic factor that can influence the microbial activities in the rhizosphere is the structural defence mechanisms of the host plant. Firstly introduced microorganisms have to overcome the niche competition with native microorganisms in the rhizosphere, and secondly they have to respond to the different types of root exudates produced by the host plant. The other biotic factor that these microorganisms have to overcome is the armoury of structural and chemical defence mechanisms of the host. Most of the structural defence components of the roots include the cuticle, lignin, suberin and deposition of callose. These compounds fortify the lignin, they make endodermis a barrier between the xylem and the soil, and in some cases, cutin acts as a waxy polymer of the cuticle coating the epidermis, which has a barrier-like properties present in the primary and lateral roots. Other structural modifications of the roots like emergence of lateral roots or formation of root hairs are also involved in creating micro-niches which also have a great influence on colonization and rhizosphere activity (Senthil-Kumar and Mysore 2013; Geldner 2013; Berhin et al. 2019).

3.7 Microbial Consortia and Overcoming the Host Immune Barrier

Irrespective of the type of introduced microbial consortia and the beneficial effect of these microorganisms to their host plants, the host plants initially respond in a way they respond to a pathogenic microorganism. As a result, the host subsequently reprogram their defence strategies to allow or block their colonization. In order to effectively and timely perceive microbial signals, plants have established an innate multilayered detection system that leads to the activation of downstream defence responses (Dodds and Rathjen 2010; Yu et al. 2019). The first layer of host defence are complex molecules known as surface-localized pattern recognition receptors (PRRs) and are responsible for recognizing molecules originating from microorganisms or microbe-associated molecular patterns (MAMPs). Plants are continuously monitoring the presence of all types of microorganisms within its territory to establish an adapted response. MAMP detection finally leads to the establishment of a plant defence program called MAMP-triggered immunity (MTI) (Trdá et al. 2015; Couto and Zipfel 2016). In the case of bacteria, studies have shown that bacterial flagellin and the immunogenic epitope of flagellin peptide flg22 are perceived by receptor kinase flagellin-sensing 2 (FLS2); on the other hand, elongation factor-TU receptor (EFR) recognizes bacterial elongation factor Tu and its derived immunogenic peptide elf18 (Gomez-Gomez and Boller, 2000). In the case of fungal isolates, chitin elicitor receptor kinase 1 (CERK1) and lysin motif-containing receptor-like kinase 5 (LYK5) recognize hepta- or octamers of the fungal elicitor chitin (Cao et al. 2014). In general, the host plant recognizes pathogenic as well as beneficial microorganisms initially as harmful invader in order to limit microbial spread and proliferation. This primary recognition is an outcome of multiple chemical signalling of plant immune system (Pel and Pieterse 2013). An efficient host immune response helps plants to achieve self-protection, and as a result, a stable microbial community is maintained within the rhizosphere (Yin et al. 2021); however, excessive immune responses of the host plants lead to the inhibition other beneficial microorganisms (Ma et al. 2021). Studies have now shown that beneficial microorganisms have adopted a number of strategies to avoid recognition by PRRs that occur at different levels. The first level is to prevent MAMP production through changes in protein structures and downregulation of MAMP production. The second level is through blocking the release of MAMP by inhibiting host hydrolases and disintegrating host-derived hydrolases. And the third level is by preventing MAMP perception through degradation of MAMPs and sequestering the released MAMPs (Buscaill and Hoorn 2021). The other host factor that favours beneficial microorganisms is the lower immune response of roots in comparison to the shoots (Beck et al. 2014; Emonet et al. 2021). Studies have shown that beneficial microorganisms can suppress root immunity by producing chemicals like gluconic acid and can also interfere with host signalling components (Yu et al. 2019).

3.8 Microbial Consortia and Abiotic Rhizospheric Factors

Colonization of microorganisms within the rhizosphere is also greatly affected by abiotic stress that interferes with the host metabolism. Abiotic stresses are reported to modify root exudation patterns and as a result rhizosphere microbiome, plant-microbe interactions and, consequently, the impact of potential benefit of PGPM to host plant are impaired. Changes in root exudation pattern due to abiotic factors like change in pH, temperature, drought, salinity and flood will cause the same types of genotypes to interact differently, and such changes affect the microbial colonization in the rhizosphere (Oosten et al. 2017; Hartman and Tringe 2019). Earlier studies have also suggested that soil characteristics and the geographic factors are the most important factors in shaping the structure of the soil microbial communities. Studies conducted in the Canadian Arctic, Antarctic soils and Tibetan permafrost soils have shown that the soil moisture content greatly influences the microbial community structure and function in the rhizosphere and can also hamper movement of microorganisms within the rhizosphere (Singh et al. 2009; Chu et al. 2011; Zhang et al. 2013). Other than this, soil aggregates are also an important element for the selection and survival of certain microbial groups; for example, the division *Acidobacteria* is often found in soil macro-aggregates but not soil micro-aggregates. Moreover, communities can also vary according to the size of the pore dwellings which influences carbon mineralization. Studies have also shown that in forest soils, species richness was shown to be modified according to the soil horizon, which promotes an organic layer for bacteria and a mineral layer that is mostly inhabited by *Archaea* (Ruamps et al. 2013; Uroz et al. 2009). The other abiotic factor that greatly influences microbial community in the rhizosphere is the soil nutrition status. The effectiveness of bacterial communities in promoting plant growth has been investigated in different soils. The studies have shown that *Pseudomonas*, *Bacillus* and *Mycobacterium* that are more efficient in stimulating the uptake of N, P and K in plants were more likely to grow in nutrient-deficient soils compared to nutrient-rich soils; moreover, C and N content are markers for soil quality which has a direct impact on the performance of beneficial microorganisms like PGPR. Soil nutrients and their bioavailability have also shown to have both direct and indirect (through plants) effects on the diversity and abundance of the rhizosphere microbiome

(Berendsen et al. 2012). The effect of macronutrients on microbial colonization is well documented and studied, and nitrogen enrichment is a predominant factor in some soil types that can have substantial effects on both plant productivity and the composition and colonization of microbial communities within the rhizosphere. Carbon is also one of the main determinants of the structure and function of microbial communities in the soil and can have positive and negative effect on the root exudation pattern and colonization structure of many beneficial microorganisms. Another important soil nutrient is phosphorus, which is also a modulating factor of the rhizosphere microbiome, and higher P content in the rhizosphere has shown adverse effect in the root colonization patterns of beneficial microorganism like AMF in the rhizosphere (Ahmed et al. 2008; Beauregard et al. 2010). Studies have shown that when P is applied to plants, the plant regulates colonization by AMF such that colonization decreases with increasing available P (Salvioli di Fossalunga and Novero 2019). Apart from this, pH of the rhizospheric soil plays a very important role in the establishment of microbial colonies, and soil pH is regarded as one of the main elements defining the structure of microbiome communities. Soil microbes show a wide range of optimal pH tolerance, and therefore soil pH can affect microbial communities. Several studies have shown that there is a strong relationship between soil pH and microbial communities which demonstrate that pH was a main factor responsible for microbial association with the hosts and rhizosphere (Lauber et al. 2009; Zhalnina et al. 2014). It is reported that neutral pH favours bacterial colonization, whereas acidic pH favours fungal colonization in the soil. pH also interferes in plant metabolism, which leads to the alteration of root exudation and changes in the biological activities, inhibiting the microorganisms that inhabit the rhizosphere. Therefore, the changes in plant metabolism, composition of root exudates and rhizosphere environment caused by abiotic factors can negatively affect successful colonization of the host by microorganisms and can affect the functioning of microbial consortia (Salwan et al. 2019; Msimbira and Smith 2020). On the other hand, soil pH is directly related to the availability of nutrients for plants by controlling the chemical forms of the soil compounds. This has also been suggested to be an indirect limiting factor for formation and colonization of soil microorganisms (Zhalnina et al. 2014).

3.9 Microbial Consortia and Diverse Mechanisms for Tolerance Against Climate Change

Application of microorganisms in agriculture is now considered to be an eco-friendly approach for developing alternative measures for sustainable agriculture. However application of microorganisms to meet the nutrient demands, promote plant growth and suppress plant disease have to face the challenges associated with changing climatic conditions, which is now coined as “global climate change”. Climate change is defined as a significant and constant change in the global climate.

As a result in this change, there are possibilities of rise in temperature which in turn would cause droughts, erratic rainfalls and even change in climate patterns and biogeochemical cycling of a particular region or area. This eventually would have a drastic impact on the biological process including microorganisms (Santoyo et al. 2017, 2021). Moreover, climate change will have great impacts on plant biology, leading to significant changes in the rhizosphere microbiome. One of the prominent effects of climate change within the host system is the change in the root exudation pattern of the host (Badri et al. 2013). The adverse effect of climate change can be mitigated by exploiting and utilizing the diverse functionalities of rhizospheric microorganisms. Rhizospheric microorganisms are found to display phenotypic variability that can build resilience caused by climate changes (Bang et al. 2018). The microbial consortia that have been applied to the rhizosphere should be able to utilize and recognize root exudates for colonization, proliferation and competition with the native microbiota and adapt to environmental changes to mitigate abiotic stresses in plants (Mimmo et al. 2018; Msimbira and Smith 2020). It is now evident from various researches that rhizospheric microorganisms have developed mechanisms to survive under stressful environments like low and high temperatures as psychrophiles and thermophiles, under saline conditions as halophiles and in acid and alkaline conditions as acidophiles and alkaliphiles. Other mechanisms include modification of cell wall, alteration in metabolic responses and gene expression against environmental stress (Khoshru et al. 2020). Rhizospheric signalling among PGPR includes quorum sensing; this is a communication system which is highly specific and sensitive and can enable microorganisms within the rhizosphere to modify their behaviour pattern under stress conditions. Microorganisms are also reported to secrete volatile metabolites (VOC), such as alkyl sulphides, indole and terpenes. These VOCs are reported to promote microbe-microbe signalling and microbe-host plant signalling under stress conditions. Reports have also shown that microbes can accumulate amino acids and reduce their water potential to avoid dehydration and death under low soil moisture. Whenever there are changes in pH due to various biochemical changes within the rhizospheric soil, beneficial microorganisms are reported to use proton transfer systems in their cytoplasm to maintain osmotic balance, control metabolic activities, and maintain their cellular vitality (de Souza et al. 2015). Some microbes, such as *Azospirillum*, *Pseudomonas* and *Bacillus*, can even influence the micronutrient status of the soil through solubilization, chelation and oxidation reduction reactions and thereby alter soil pH through acidification of their surroundings (Abhilash et al. 2016; Oosten et al. 2017). One of the main adaptations of rhizospheric microorganisms is through the horizontal gene transfer where the genetic materials like plasmids, transposons and phases are exchanged between closely related species, and eventually these microorganisms are able to tolerate different types of effects of climate change through beneficial mutations (Aminov 2011).

Under climate change conditions, atmospheric carbon dioxide (CO_2) will be a driving factor that can change the biomass and diversity of microorganisms in the rhizosphere (Paterson et al. 1997). Studies have shown that rhizodeposition of carbon (C) increases under elevated atmospheric CO_2 (eCO2). Microorganisms that rely on carbon from root exudates will show enhanced colonization under elevated CO_2 (Eisenhauer et al. 2012). There are reports of impact of atmospheric CO_2 on plant-beneficial microorganisms in the rhizosphere; however, it remains difficult to explain clearly the mechanisms related to plant responses to altered colonization by these microbes. Most studies on the effects of CO_2 gradients on rhizosphere microbes have shown correlation with the other abiotic factors such as nutrient availability, soil moisture, temperature, soil pH as well as the type and nature of host plant species present. These studies show that enhanced CO_2 in the rhizosphere will have impact in the abiotic factors of the rhizosphere and thereby influence the overall colonization of microorganisms in the rhizosphere (Castro et al. 2010; Classen et al. 2015; Dam et al. 2017).

3.10 Conclusion and Future Perspectives

The rhizosphere microbiome consists largely of bacteria, fungi, actinomycetes and AMF that benefit plant growth and health. Multiple direct or indirect mechanisms like nutrient mobilization, production of siderophores, phytohormones, volatile compounds, ACC deaminase, HCN, IAA as well as suppression of phytopathogens in the rhizosphere and induction of resistance in the host have been reported which are directly related to plant growth and benefit agriculture production (Santoyo et al. 2012). It is also essential to understand the abiotic and biotic interactions to successfully exploit the rhizosphere microbiome to benefit agricultural production. Continuous environmental changes and impacts of anthropogenic effects on the environment and climate must also be taken into account for future agricultural practices. Few researchers have proposed to manipulate the rhizosphere by introducing genetically modified transgenic plants with altered exudation patterns. Alteration in root exudation will then influence microbial communities in the rhizosphere (Chaparro et al. 2014). However, this approach will require the release of a genetically modified organism and is currently restricted by law in several countries. Another alternative and much more sustainable approach will be the microbiome-based bioinoculants. Co-inoculation of microorganisms that will complement each other in functionalities within the rhizosphere will be an environment-friendly approach to improve rhizospheric soil health and crop production. The current research involving microbial consortia should be focused on targeted laboratory studies of candidate microorganisms that have potential to mitigate various stress. Laboratory probing of microbial responses should ultimately assess environmentally relevant conditions, adopt a "microbcentric" view of environmental stressors and be followed up by field tests. Field experiments and validations are therefore important for understanding community-level responses to real environmental conditions (Cavicchioli et al. 2019). Understanding of microbial interactions within the rhizosphere would certainly help us to design specific and effective measures to mitigate and control climate change and its effects. Microbial biotechnology can provide solutions for sustainable development (Timmis et al. 2017), including in the provision and regulation of ecosystem services for humans, animals and plants. Microbial technologies provide practical solutions for achieving many of the 17 United Nations Sustainable Development Goals and can be an efficient tool for addressing poverty, hunger, health, clean water, clean energy, economic growth, industry innovation, sustainable cities, responsible consumption, climate action, life below water and life on land.

Climate change is a prevalent phenomenon and can have direct influence on the demand for food supply to a growing population. The major consequence of climate change includes enhanced concentration of CO_2 in the atmosphere, elevated temperature and change in rainfall patterns and alteration of various climatic conditions. Reports have confirmed that climate change has a direct effect on soil microbe and their interactions (Verheijen et al. 2015). One of the best approaches to overcome the effects of climate change is to adopt alternatives provided by modern technology. In this connection, microbial mediated green nanotechnology has emerged as a reliable tool to mediate climate resilience. Microbes used in such process include *Bacillus*, *Fusarium*, *Pseudomonas*, *Aspergillus*, *Candida albicans*, *Caulerpa racemosa*, *Tobacco mosaic virus* and many more microorganisms. Nanoparticles are carefully selected, synthesized, characterized and then introduced to the rhizosphere along with rhizospheric microorganisms as carriers; these fabricated microorganisms have shown promising results for crop improvement and stress tolerance (Kashyap et al. 2018).

The changing climatic conditions also have adverse effect on rhizospheric and phyllospheric biology. Studies conducted on plant-microbe interactions have shown various aspects of symbiotic association of plant with microbes and their beneficial roles in plant health. However, engineering microbial communities through biotechnological approaches should be the future aspects to overcome adverse effect of climate change (Hakim et al. 2021).

Acknowledgement The authors sincerely acknowledge the Research Grant Assistance from the Department of Biotechnology, Ministry of Science and Technology, Government of India.

References

Abdelhameed RE, Rabab AM (2019) Alleviation of cadmium stress by arbuscular mycorrhizal symbiosis. Int J Phytoremediation 21:1–9

Abhilash PC, Dubey RK, Tripathi V, Gupta VK, Singh HB (2016) Plant growth-promoting microorganisms for environmental sustainability. Trends Biotechnol 34:847–850

Ahmad M, Zahir ZA, Asghar HN, Arshad M (2012) The combined application of rhizobial strains and plant growth promoting rhizobacteria improves growth and productivity of Mung bean (*Vigna radiata* L.) under salt-stressed conditions. Ann Microbiol 62:1321–1330

Ahmed Z, Lim BR, Cho J, Song KG, Kim KP, Ahn KH (2008) Biological nitrogen and phosphorus removal and changes in microbial community structure in a membrane bioreactor: effect of different carbon sources. Water Res 42:198–210

Aminov RI (2011) Horizontal gene exchange in environmental microbiota. Front Microbiol 2:158

Angulo-Castro A, Ferrera-Cerrato R, Alarcón A, Almaraz-Suárez J, Delgadillo-Martínez J, Jiménez-Fernández M, García-Barradas O (2021) Improved growth of bell pepper (*Capsicum annuum*) plants by inoculating arbuscular mycorrhizal fungi and beneficial rhizobacteria. Scientia fungorum 51:e1299

Arora NK, Fatima T, Mishra I, Verma S (2020) Microbe-based inoculants: role in next green revolution. In: Environmental Concerns and Sustainable Development, vol 2. Springer, Singapore, pp. 191–245

Attia MS, Abdelaziz AM, Al-Askar AA, Arishi AA, Abdelhakim AM, Hashem AH (2022) Plant growth-promoting fungi as biocontrol tool against *Fusarium* wilt disease of tomato plant. J Fungi 8:775

Azizi S, Kouchaksaraei MT, Hadian J, Nosrat Abad ARF, Sanavi M, Ammer C, Bader M (2021) Dual inoculations of arbuscular mycorrhizal fungi and plant growth promoting rhizobacteria boost drought resistance and essential oil yield of common Myrtle. For Ecol Manag 497:119478

Badri DV, Chaparro JM, Zhang R, Shen Q, Vivanco JM (2013) Application of natural blends of phytochemicals derived from the root exudates of *Arabidopsis* to the soil reveal that phenolic-related compounds predominantly modulate the soil microbiome. J Biol Chem 288:4502–4512

Bang C, Dagan T, Deines P, Dubilier N, Duschl JW, Fraune S, Hentshel U, Hirt H, Hulter N, Lachnit T, Picazo D, Pita L, Porgoreul ZC, Radecker N, Saad MM, Schmitz AR, Schulenburg H, Voolstra RC, Bosch GCT (2018) Metaorganisms in extreme environments: do microbes play a role in organismal adaptation? Zoology 127:1–19

Baron NC, Pollo SA, Rigobelo EC (2020) *Purpureocillium lilacinum* and *Metarhizium marquandii* as plant growth-promoting fungi. Peer J 8:e9005

Basu A, Prasad P, Das SN, Kalam S, Sayyed RZ, Reddy MS, Enshasy EH (2021) Plant growth promoting rhizobacteria (PGPR) as green bioinoculants: recent developments, constraints and prospects. Sustainability 13:1140

Beauregard MS, Hamel C, Arnaud SM (2010) Long-term phosphorus fertilization impacts soil fungal and bacterial diversity but not AM fungal community in *Alfalfa*. Microb Ecol 59:379–389

Beck M, Wyrsch I, Strutt J, Wimalasekera R, Webb A, Boller T, Robatzek S (2014) Expression patterns of FLAGELLIN SENSING 2 map to bacterial entry sites in plant shoots and roots. J Exp Bot 65:6487–6498

Behera B, Das TK, Raj R, Ghosh S, Raza MB, Sen S (2020) Microbial consortia for sustaining productivity of non-legume crops: prospects and challenges. Agric Res 10:1–14

Bélanger RR, Avis TJ (2002) Ecological processes and interactions occurring in leaf surface fungi. In: Phyllosphere Microbiology, APS Press, St. Paul, MN, pp 193–207

Beneduzi A, Peres D, Vargas LK, Bodanese-Zanettini MH, Passaglia LM (2008) Evaluation of genetic diversity and plant growth promoting activities of nitrogen-fixing Bacilli isolated from Rice fields in South Brazil. Appl Soil Ecol 39:311–320

Berendsen RL, Pieterse CM, Bakker PA (2012) The rhizosphere microbiome and plant health. Trends Plant Sci 8:478–486

Berendsen RL, Vismans GYK, Song Y, de Jonge R, Burgman WP, Burmølle M, Herschend J, Bakker PAHM, Pieterse CMJ (2018) Disease-induced assemblage of a plant-beneficial bacterial consortium. ISME J 12:1496–1507

Berhin A, De Bellis D, Franke RB, Buono RA, Nowack MK, Nawrath C (2019) The root cap cuticle: a cell wall structure for seedling establishment and lateral root formation. Cell 176: 1367–1378

Berta G, Copetta A, Gamalero E, Bona E, Cesaro P, Scarafoni A, D'Agostino G (2013) Maize development and grain quality are differentially affected by mycorrhizal fungi and a growth-promoting pseudomonad in the field. Mycorrhiza 24:161–170

Bhise KK, Dandge PB (2019) Alleviation of salinity stress in Rice plant by encapsulated salt tolerant plant growth promoting bacteria *Pantoea agglomerans* strain KL and its root colonization ability. Arch Agron Soil Sci 65:1955–1968

Bhowmik SN, Das A (2018) Biofertilizers: a sustainable approach for pulse production. In: Legumes for Soil Health and Sustainable Management, Springer, Singapore, pp 445–485

Bonanomi G, Lorito M, Vinale F, Woo SL (2018) Organic amendments, beneficial microbes, and soil microbiota: toward a unified framework for disease suppression. Annu Rev Phytopathol 56: 1–20

Bonkowski M, Villenave C, Griffiths B (2009) Rhizosphere fauna: the functional and structural diversity of intimate interactions of soil fauna with plant roots. Plant Soil 321:213–233

Bradácová K, Kandeler E, Berger N, Ludewig U, Neumann G (2020) Microbial consortia inoculants stimulate early growth of maize depending on nitrogen and phosphorus supply. Plant Soil Environ 66:000105–000112

Brundrett MC, Tedersoo L (2018) Evolutionary history of Mycorrhizal symbioses and global host plant diversity. New Phytol 220:1108–1115

Buscaill P, Hoorn RAL (2021) Defeated by the nines: nine extracellular strategies to avoid MAMP recognition in plants. Plant Cell 33:2116–2130

Cabral C, Sabine R, Ivanka T, Bernd W (2016) Arbuscular Mycorrhizal Fungi modify nutrient allocation and composition in wheat (*Triticum aestivum* L.) subjected to heat-stress. Plant Soil 408:385–399

Cao Y, Liang Y, Tanaka K, Nguyen CT, Jedrzejczak RP, Joachimiak A (2014) The kinase LYK5 is a major chitin receptor in *Arabidopsis* and forms a chitin-induced complex with related kinase CERK1. elife 3:e03766

Castro HF, Classen AT, Austin EE, Norby RJ, Schadt CW (2010) Soil microbial community responses to multiple experimental climate change drivers. Appl Environ Microbiol 76:999–1007

Cavicchioli R, Ripple WJ, Timmis KN, Azam F, Bakken RL, Baylis M, Behrenfeld JM, Boetius A, Boyd WP, Classen AT, Crowther WT, Danovaro R, Foreman MC, Huisman J, Hulchins AD, Janson KJ, Karl MD, Koslkella B, Welch MBD, HBJ M, Moran AM, Orphan JV, Read SD, Remais VJ, Rich IV, Singh KB, Stein YL, Stewart JF, Sullivan BM, Oppen VHJM, Weaver CS, Webb AE, Webster SN (2019) Scientists' warning to humanity: microorganisms and climate change. Nat Rev Microbiol 17:569–586

Cekic FO, Unyayar S, Ortas I (2012) Effects of Arbuscular Mycorrhizal inoculation on biochemical parameters in *Capsicum annuum* grown under long term salt stress. Turk J Bot 36:63–72

Chakraborty BN, Chakraborty U, Dey PL, Sunar K (2014) Evaluation of PGPR traits of native bacterial strains from Darjeeling and analysis of their diversity. J Bot Soc Bengal 68:159–176

Chakraborty BN, Chakraborty U, Sunar K (2020) Induced immunity developed by *Trichoderma* species in plants. In: *Trichoderma* rhizosphere biology, Springer, Singapore, pp. 125–147

Chakraborty BN, Chakraborty U, Sunar K, Dey PL (2012) Evaluation of plant growth promoting and antifungal activities of *Talaromyces flavus* (NAIMCC-F-01948) against *Sclerotium rolfsii*. Ind Phytopathol 65:258–263

Chaparro JM, Badri DV, Vivanco JM (2014) Rhizosphere microbiome assemblage is affected by plant development. ISME J 8:790–803

Chu H, Neufeld JD, Walker VK, Grogan P (2011) The influence of vegetation type on the dominant soil bacteria, archaea, and fungi in a low arctic tundra landscape. Soil Biol Biochem 75:1756–1765

Classen AT, Sundqvist MK, Henning JA, Newman GS, Moore JAM, Cregger MA, Moorhead CL, Paterson MC (2015) Direct and indirect effects of climate change on soil microbial and soil microbial-plant interactions: what lies ahead. Ecosphere 6:130

Compant S, Samad A, Faist H, Sessitsch A (2019) A review on the plant microbiome: ecology, functions and emerging trends in microbial application. J Adv Res 20:29–37

Couto D, Zipfel C (2016) Regulation of pattern recognition receptor signaling in plants. Nat Rev Immunol 16:537–552

Dam M, Bergmark L, Vestergård M (2017) Elevated CO_2 increases fungal based micro-food webs in soils of contrasting plant species. Plant Soil 415:549–561

De la Fuente Cantó C, Simonin M, King E, Moulin L, Bennett MJ, Castrillo G, Laplaze L (2020) An extended root phenotype: the rhizosphere, its formation and impacts on plant fitness. Plant J 103:951–964

Dehariya K, Sheikh IA, Vyas D, Shukla A (2015) *Trichoderma* and arbuscular mycorrhizal fungi based biocontrol of *Fusarium udum* Butler and their growth promotion effects on Pigeon Pea. J Agric Sci Technol 17:505–517

Diagne N, Ngom M, Djighaly PI, Fall D, Hocher V, Svistoonoff S (2020) Roles of arbuscular mycorrhizal fungi on plant growth and performance: importance in biotic and abiotic stressed regulation. Diversity 12:370

Dodds PN, Rathjen JP (2010) Plant immunity: towards an integrated view of plant-pathogen interactions. Nat Rev Genet 11:539–548

Egamberdieva D, Jabborova D, Berg G (2016) Synergistic interactions between *Bradyrhizobium japonicum* and the endophyte *Stenotrophomonas rhizophila* and their effects on growth and nodulation of Soybean under salt stress. Plant Soil 405:35–45

Egamberdieva D, Wirth S, Jabborova D, Räsänen LA, Liao H (2017) Coordination between *Bradyrhizobium* and *Pseudomonas* alleviates salt stress in Soybean through altering root system architecture. J Plant Interact 12:100–107

Eisenhauer N, Cesarz S, Koller R, Worm K, Reich PB (2012) Global change belowground: impacts of elevated CO_2, nitrogen, and summer drought on soil food webs and biodiversity. Glob Change Biol 18:435–447

Eisenhauer N, Lanoue A, Strecker T, Scheu S, Steinauer K, Thakur MP, Mommer L (2017) Root biomass and exudates link plant diversity with soil bacterial and fungal biomass. Sci Rep 7:1–8

Emonet A, Zhou F, Vacheron J, Heiman CM, Dénervaud Tendon V, Ma KW, Lefert SP, Keel C, Geldner N (2021) Spatially restricted immune responses are required for maintaining root meristematic activity upon detection of bacteria. Curr Biol 31:1012–1028

Fasusi OA, Cruz C, Babalola OO (2021) Agricultural sustainability: microbial biofertilizers in rhizosphere management. Agriculture 11:163–181

Fierer N (2017) Embracing the unknown: disentangling the complexities of the soil microbiome. Nat Rev Microbiol 15:579–590

Fiorentino N, Ventorino V, Woo SL, Pepe O, De Rosa A, Gioia L, Romano I, Lombardi N, Napolitano M, Colla G, Rouphael Y (2018) *Trichoderma*-based biostimulants modulate rhizosphere microbial populations and improve N uptake efficiency, yield, and nutritional quality of leafy vegetables. Front Plant Sci 9:743

Fukami J, de la Osa C, Ollero FJ, Megías M, Hungria M (2018) Co-inoculation of maize with *Azospirillum brasilense* and *Rhizobium tropici* as a strategy to mitigate salinity stress. Funct Plant Biol 45:328–339

Geldner N (2013) The endodermis. Annu Rev Plant Biol 64:531–558

Gomez-Gomez L, Boller T (2000) FLS2: an LRR receptor–like kinase involved in the perception of the bacterial elicitor flagellin in *Arabidopsis*. Mol Cell 5:1003–1011

Goswami D, Parmar S, Vaghela H, Dhandhukia P, Thakker JN (2015) Describing *Paenibacillus mucilaginosus* strain N3 as an efficient plant growth promoting rhizobacteria (PGPR). Cogent Food Agric 1:1000714

Goswami SK, Kashyap PL, Awasthi S (2019) Deciphering rhizosphere microbiome for the development of novel bacterial consortium and its evaluation for salt stress management in solanaceous crops in India. Indian Phytopathol 72:479–488

Gouda S, Kerry RG, Das G, Paramithiotis S, Shin H, Patra JK (2018) Revitalization of plant growth promoting rhizobacteria for sustainable development in agriculture. Microbiol Res 20:131–140

Gupta G, Parihar SS, Ahirwar NK, Snehi SK, Singh V (2015) Plant growth promoting rhizobacteria (PGPR): current and future prospects for development of sustainable agriculture. J Microb Biochem Technol 7:96–102

Hakim S, Naqqash T, Nawaz MS, Laraib I, Siddique MJ, Zia R, Mirza SM, Imran A (2021) Rhizosphere engineering with plant growth-promoting microorganisms for agriculture and ecological sustainability. Front Sustain Food Syst 5:617157

Hardoim PR, Hardoim CC, van Overbeek LS, van Elsas JD (2012) Dynamics of seed borne Rice endophytes on early plant growth stages. PLoS One 7:e30438

Hardoim PR, van Overbeek LS, Berg G, Pirttilä AM, Compant S, Campisano A (2015) The hidden world within plants: ecological and evolutionary considerations for defining functioning of microbial endophytes. Microbiol Mol Biol Rev 79:293–320

Hartman K, Tringe SG (2019) Interactions between plants and soil shaping the root microbiome under abiotic stress. Biochem J 476:2705–2724

Hashem A, Adb Allah EF, Alqarawi AA, Al-Huqail AA, Shah MA (2016) Induction of osmoregulation and modulation of salt stress in *Acacia gerrardii* Benth by arbuscular mycorrhizal fungi and *Bacillus subtilis* (BERA 71). Biomed Res Int 2016:6294098

Hashem A, Alqarawi AA, Radhakrishnan R, Al-Arjani AF, Aldehaish HA, Egamberdieva D, Abd-Allah FE (2018) Arbuscular mycorrhizal fungi regulate the oxidative system, hormones and ionic equilibrium to trigger salt stress tolerance in *Cucumis sativus L.* Saudi J Biol Sci 25: 1102–1114

Hayat R, Ali S, Amara U, Khalid R, Ahmed I (2010) Soil beneficial bacteria and their role in plant growth promotion: a review. Ann Microbiol 60:579–598

Hernández-Montiel LG, Chiquito-Contreras CJ, Murillo-Amador B, VidalHernández L, Quiñones-Aguilar EE, Chiquito-Contreras RG (2017) Efficiency of two inoculation methods of *Pseudomonas putida* on growth and yield of tomato plants. Soil Sci Plant Nutr 17:1003–1012

Hewedy OA, Abdel Lateif KS, Seleiman MF, Shami A, Albarakaty FM, M El-Meihy R (2020) Phylogenetic diversity of *Trichoderma* strains and their antagonistic potential against soil-borne pathogens under stress conditions. Biology (Basel) 9:189

Htwe AZ, Moh SM, Moe K, Yamakawa T (2018) Effects of co-inoculation of *Bradyrhizobium japonicum* SAY3-7 and *Streptomyces griseoflavus* P4 on plant growth, nodulation, nitrogen fixation, nutrient uptake, and yield of soybean in a field condition. Soil Sci Plant Nutr 64:222–229

Hu L, Robert CAM, Cadot S, Zhang X, Ye M, Li B, Manzo D, Chervet N, Steinger T, van der Heijden AGM, Schlaeppi K, Erb M (2018) Root exudate metabolites drive plant-soil feedbacks on growth and defense by shaping the rhizosphere microbiota. Nat Commun 9:2738

Hussein KA, Joo JH (2018) Plant growth-promoting rhizobacteria improved salinity tolerance of *Lactuca sativa* and *Raphanus sativus*. J Microbiol Biotechnol 28:938–945

Jacoby RP, Kopriva S (2019) Metabolic niches in the rhizosphere microbiome: new tools and approaches to analyse metabolic mechanisms of plant-microbe nutrient exchange. J Exp Bot 70: 1087–1094

du Jardin P (2015) Plant biostimulants: definition, concept, main categories and regulation. Sci Hortic 196:3–14

Jha Y, Subramanian R (2013) Paddy plants inoculated with PGPR show better growth physiology and nutrient content under saline condition. Chilean J Agric Res 73:213–219

Jogaiah S, Abdelrahman M, Tran LS, Ito SI (2013) Characterization of rhizosphere fungi that mediate resistance in tomato against bacterial wilt disease. J Exp Bot 64:3829–3842

Kashyap PL, Rai P, Kumar R, Sharma S, Jasrotia P, Srivastava AK, Kumar S (2018) Microbial nanotechnology for climate resilient agriculture, In: Microbes for Climate Resilient Agriculture 1st Edn, John Wiley & Sons Inc, pp 279–344

Khan A, Singh J, Upadhayay VK, Singh AV, Shah S (2019) Microbial biofortification: a green technology through plant growth promoting microorganisms. In: Sustainable green technologies for environmental management, Springer, Singapore, pp. 255–269

Khan N, Bano A, Curá JA (2020) Role of beneficial microorganisms and salicylic acid in improving rainfed agriculture and future food safety. Microorganisms 8:1018

Khan Z, Rho H, Firrincieli A, Hung SH, Luna V, Masciarelli O, Kim SH, Doty SL (2016) Growth enhancement and drought tolerance of hybrid poplar upon inoculation with endophyte consortia. Curr Plant Biol 6:38–47

Khoshru B, Mitra D, Khoshmanzar E, Myo EM, Uniyal N, Mahakur B, Mohapatra DKP, Panneerselvam P, Boutaj H, Alizadeh M, Cely TVM, Senapati A, Rani A (2020) Current scenario and future prospects of plant growth-promoting rhizobacteria: an economic valuable resource for the agriculture revival under stressful conditions. J Plant Nutr 43:3062–3092

Korir H, Mungai NW, Thuita M, Hamba Y, Masso C (2017) Co-inoculation effect of *Rhizobia* and plant growth promoting rhizobacteria on common bean growth in a low phosphorus soil. Front Plant Sci 8:141

Kumar SM, Chowdappa P, Krishna V (2015) Development of seed coating formulation using consortium of *Bacillus subtilis* OTPB1 and *Trichoderma harzianum* OTPB3 for plant growth promotion and induction of systemic resistance in field and horticultural crops. Indian Phytopathol 68:25–31

Lauber CL, Hamady M, Knight R, Fierer N (2009) Pyrosequencing-based assessment of soil pH as a predictor of soil bacterial community structure at the continental scale. Appl Environ Microbiol 75:5111–5120

Leff JW, Lynch RC, Kane NC, Fierer N (2016) Plant domestication and the assembly of bacterial and fungal communities associated with strains of the common Sunflower, *Helianthus annuus*. New Phytol 214:412–423

Li L, Li YQ, Jiang Z, Gao R, Nimaichand S, Duan YQ, Egamberdieva D, Chen W, Li WJ (2016) *Ochrobactrum endophyticum* sp. nov. isolated from roots of *Glycyrrhiza uralensis*. Arch Microbiol 198:171–179

Li W, Wb L, Xing L, Guo S (2022) Effect of arbuscular mycorrhizal fungi (AMF) and plant growth-promoting rhizobacteria (PGPR) on microorganism of phenanthrene and pyrene contaminated soils. J Phytoremed 13:1–12

Liu RC, Xiao ZY, Hashem A, AbdAllah EF, Wu QS (2021) Mycorrhizal fungal diversity and its relationship with soil properties. Camellia Oleifera Agric 11:470

Ma KW, Niu Y, Jia Y, Ordon J, Copeland C, Emonet A, Geldner N, Guan R, Stolze CS, Nakagami H, Oter GR, Lefert SP (2021) Coordination of microbe–host homeostasis by crosstalk with plant innate immunity. Nat Plants 7:814–825

Masciarelli O, Llanes A, Luna V (2014) A new PGPR co-inoculated with *Bradyrhizobium japonicum* enhances soybean nodulation. Microbiol Res 169:609–615

Mathur S, Sharma MP, Jajoo A (2016) Improved photosynthetic efficacy of maize *Zea mays* plants with arbuscular mycorrhizal fungi (AMF) under high temperature stress. J Photochem Photobiol B 180:149–154

Matse DT, Huang HC, Huang MY, Yen MY (2020) Nitrogen uptake and growth of white clover inoculated with indigenous and exotic *Rhizobium* strains. J Plant Nutr 13:2013–2027

Melo CD, Walker C, Krüger C, Borges PAV, Luna S, Mendonca D, Fonseca HMAC, Machado AC (2019) Environmental factors driving arbuscular mycorrhizal fungal communities associated with endemic woody plant *Picconia azorica* on native forest of Azores. Ann Microbiol 69: 1309–1327

Mimmo T, Pii Y, Valentinuzzi F, Astolfi S, Lehto N, Robinson B, Brunetto G, Terzano R, Cesco S (2018) Nutrient availability in the rhizosphere: a review. Acta Hortic 1217:13–28

Morella NM, Weng FCH, Joubert PM, Metcalf CJE, Lindow S, Koskella B (2020) Successive passaging of a plant-associated microbiome reveals robust habitat and host genotype-dependent selection. Proc Natl Acad Sci 117:1148–1159

Msimbira LA, Smith DL (2020) The roles of plant growth promoting microbes in enhancing plant tolerance to acidity and alkalinity stresses. Front Sustain Food Syst 4:1–14

Nadeem SM, Zahir ZA, Naveed M, Nawaz S (2013) Mitigation of salinity-induced negative impact on the growth and yield of wheat by plant growth-promoting rhizobacteria in naturally saline conditions. Ann Microbiol 63:225–232

Nascimento FX, Hernandez AG, Glick BR, Rossi MJ (2020) The extreme plant-growth-promoting properties of *Pantoea phytobeneficialis* MSR2 revealed by functional and genomic analysis. Environ Microbiol 22:1341–1355

Nascimento FX, Tavares MJ, Franck J, Ali S, Glick BR, Rossi MJ (2019) ACC deaminase plays a major role in *Pseudomonas fluorescens* YsS6 ability to promote the nodulation of alpha- and *Betaproteobacteria rhizobial* strains. Arch Microbiol 201:817–822

Niu X, Song L, Xiao Y, Ge W (2018) Drought-tolerant plant growth-promoting rhizobacteria associated with Foxtail Millet in a semi-arid agro ecosystem and their potential in alleviating drought stress. Front Microbiol 8:2580

Olenska E, Małek W, Wójcik M, Swiecicka I, Thijs S, Vangronsveld J (2020) Beneficial features of plant growth-promoting rhizobacteria for improving plant growth and health in challenging conditions: a methodical review. Sci Tot Environ 32:429–448

Oosten MJV, Pepe O, Pascale S, Silletti S, Maggio A (2017) The role of biostimulants and bioeffectors as alleviators of abiotic stress in crop plants. Chem Biol Technol Agric 4:5

Orhan F (2016) Alleviation of salt stress by halotolerant and halophilic plant growth-promoting bacteria in wheat (*Triticum aestivum*). Braz J Microbiol 47:621–627

Ortas I, Ustuner O (2014) The effects of single species, dual species and indigenous Mycorrhiza inoculation on *Citrus* growth and nutrient uptake. Eur J Soil Biol 63:64–69

Ortiz A, Sansinenea E (2021) Recent advancements for microorganisms and their natural compounds useful in agriculture. Appl Microbiol Biotechnol 105:891–897

Ortiz N, Armada E, Duque E, Roldán A, Azcón R (2015) Contribution of arbuscular mycorrhizal fungi and/or bacteria to enhancing plant drought tolerance under natural soil conditions: effectiveness of autochthonous or allochthonous strains. J Plant Physiol 174:87–96

Pal A, Pandey S (2016) Role of arbuscular mycorrhizal fungi on plant growth and reclamation of barren soil with wheat (*Triticum aestivum* L.) crop. Int J Soil Sci 12:25–31

Pan J, Huang C, Peng F, Zhang W, Luo J, Ma S, Xue X (2020) Effect of arbuscular mycorrhizal fungi (AMF) and plant growth-promoting bacteria (PGPR) inoculations on *Elaeagnus angustifolia* L. in saline soil. Appl Sci 10:945

Pandya M, Rajput M, Rajkumar S (2015) Exploring plant growth promoting potential of non rhizobial root nodules endophytes of *Vigna radiata*. Microbiol (Russ Fed) 84:80–89

Parihar M, Rakshit A, Rana K, Meena RP, Joshi DC (2020) A consortium of arbuscular Mycorrhizal fungi improves nutrient uptake, biochemical response, nodulation and growth of the pea (*Pisum sativum L.*) under salt stress. Rhizosphere 15:100235

Parnell JJ, Berka R, Young HA, Sturino JM, Kang Y, Barnhart D, DiLeo MV (2016) From the lab to the farm: an industrial perspective of plant beneficial microorganisms. Front Plant Sci 7:1110

Paterson E, Hall JM, Rattray EAS, Griffiths BS, Ritz K, Killham K (1997) Effect of elevated CO_2 on rhizosphere carbon flow and soil microbial processes. Glob Change Biol 3:363–377

Pel MJ, Pieterse CM (2013) Microbial recognition and evasion of host immunity. J Exp Bot 64: 1237–1248

Pellegrino E, Bedini S (2014) Enhancing ecosystem services in sustainable agriculture: biofertilization and biofortification of chickpea (*Cicer arietinum* L.) by arbuscular mycorrhizal fungi. Soil Biol Biochem 68:429–439

Pellegrino E, Nuti M, Ercoli L (2022) Multiple arbuscular mycorrhizal fungal consortia enhance yield and fatty acids of *Medicago sativa*: a two-year field study on agronomic traits and tracing of fungal persistence. Front Plant Sci 13:814401

Rajniak J, Giehl RFH, Chang E, Murgia I, Von Wiren N, Sattely ES (2018) Biosynthesis of redox-active metabolites in response to iron deficiency in plants. Nat Chem Biol 14:442–450

Rao KP, Lal AM, Abraham G, Ramteke PW (2018) Threshold capacity of strawberry cultivars to salinity and rhizosphere bacterial population for tolerance. Indian J Agric Biochem 31:65–70

Rezácová V, Slavíková R, Zemková L, Konvalinková T, Procházková V, Štovícek V, Hršelová H, Beskid O, Hujslová M, Gryndlerová H, Gryndler M, Püschel D, Jansa J (2018) Mycorrhizal symbiosis induces plant carb on reallocation differently in C3 and C4 *Panicum grasses*. Plant Soil 425:441–456

Ruamps LS, Nunan N, Pouteau V, Leloup J, Raynaud X, Roy V, Chenu C (2013) Regulation of soil organic C mineralization at the pore scale. FEMS Microb Ecol 86:26–35

Sagar A, Sayyed RZ, Ramteke PW, Sharma S, Marraiki N, Elgorban AM, Syed A (2020) ACC deaminase and antioxidant enzymes producing halophilic *Enterobacter sp.* ameliorates salt stress and promotes the growth of Rice and Millets under salt stress. Physiol Mol Biol Plants 26:1847–1854

Sagar A, Shukla PK, Sayyad RZ, Ramteke PW (2019) Stimulation of seed germination and growth parameters of Rice var. Sahbhagi by *Enterobacter cloacae* (KP226569) PR4 in presence of Ammonia Sulphate as substitute of ACC. In: Plant Growth Promoting Rhizobacteria (PGPR): Prospects in sustainable agriculture, Springer, Singapore, pp 117–124

Saikia J, Sarma RK, Dhandia R, Yadav A, Bharali R, Gupta VK, Saikia R (2018) Alleviation of drought stress in pulse crops with ACC deaminase producing rhizobacteria isolated from acidic soil of Northeast India. Sci Rep 8:1–16

Salvioli di Fossalunga A, Novero M (2019) To trade in the field: the molecular determinants of arbuscular mycorrhiza nutrient exchange. Chem Biol Technol Agric 6:12

Salwan R, Sharma A, Sharma V (2019) Microbes mediated plant stress tolerance in saline agricultural ecosystem. Plant Soil 442:1–22

Santoyo G, Guzmán-Guzmán P, Parra-Cota FI, Sdl S-V, Orozco-Mosqueda MC, Glick BR (2021) Plant growth stimulation by microbial consortia. Agronomy 11:219

Santoyo G, Hernández PC, Hernández SJ, Hernández LR (2017) The role of abiotic factors modulating the plant-microbe-soil interactions: toward sustainable agriculture—a review. Span J Agric Res 15:e03R01

Santoyo G, Orozco-Mosqueda MC, Govindappa M (2012) Mechanisms of biocontrol and plant growth-promoting activity in soil bacterial species of *Bacillus* and *Pseudomonas*—a review. Biocon Sci Technol 22:855–872

Sapre S, Gontia-Mishra I, Tiwari S (2022) Plant growth-promoting rhizobacteria ameliorates salinity stress in pea (*Pisum sativum*). J Plant Growth Regul 41:647–656

Saravanakumar D, Samiyappan R (2007) ACC deaminase from *Pseudomonas fluorescens* mediated saline resistance in groundnut (*Arachis hypogea*) plants. J Appl Microbiol 102:1283–1292

Sarkar A, Pramanik K, Mitra S, Soren T, Maiti TK (2018) Enhancement of growth and salt tolerance of Rice seedlings by ACC deaminase-producing *Burkholderia sp.* MTCC 12259. J Plant Physiol 231:434–442

Sasse J, Martinoia E, Northen T (2017) Feed your friends: do plant exudates shape the root microbiome? Trends Plant Sci 23:25–41

Senthil-Kumar M, Mysore KS (2013) Non-host resistance against bacterial pathogens: retrospectives and prospects. Annu Rev Phytopathol 51:407–427

Shah G, Afreen Jan M, Anees M, Rehman M, Daud SMK, Malook I, Jamil M (2017) Halophilic bacteria mediated phytoremediation of salt-affected soils cultivated with rice. J Geochem Explor 174:59–65

Shahzad R, Khan AL, Bilal S, Waqas M, Kang SM, Lee IJ (2017) Inoculation of abscisic acid-producing endophytic bacteria enhances salinity stress tolerance in *Oryza sativa*. Environ Exp Bot 136:68–77

Sharma CK, Vishnoi VK, Dubey RC, Maheshwari DK (2018) A twin rhizospheric bacterial consortium induces systemic resistance to a phytopathogen *Macrophomina phaseolina* in Mung bean. Rhizosphere 5:71–75

Singh B, Dawson LA, McDonald CA, Buckland SM (2009) Impact of biotic and abiotic interaction on soil microbial communities and functions: a field study. Appl Soil Ecol 41:239–248

Singh JS (2013) Plant growth promoting rhizobacteria potential microbes for sustainable agriculture. Resonance 3:275–281

Singh RP, Jha PN (2015) The plant-growth-promoting bacterium *Klebsiella* sp. SBP-8 confers induced systemic tolerance in Wheat (*Triticum aestivum*) under salt stress. J Plant Physiol 184: 57–67

de Souza R, Ambrosini A, Passaglia LMP (2015) Plant growth promoting bacteria as inoculants in agricultural soils. Genet Mol Biol 38:401–419

Stajkovic O, Delic D, Josic D, Kuzmanovic D, Rasulic N, Vukcevic KJ (2011) Improvement of common bean growth by co-inoculation with *Rhizobium* and plant growth-promoting bacteria. Rom Biotechnol Lett 16:5919–5926

Stringlis IA, Zhang H, Pieterse CMJ, Bolton MD, de Jonge R (2018) Microbial small molecules—weapons of plant subversion. Nat Prod Rep 35:410–433

Suebrasri T, Somteds A, Harada H, Kanokmedhakul S, Jogloy S, Ekprasert J, Lumyong S, Boonlue S (2020) Novel endophytic fungi with fungicidal metabolites suppress *Sclerotium* disease. Rhizosphere 16:100250

Sukweenadhi J, Balusamy SR, Kim Y-J, Lee CH, Kim Y-J, Koh SC, Yang DC (2018) A growth-promoting bacteria, *Paenibacillus yonginensis* DCY84T enhanced salt stress tolerance by activating defense-related systems in *Panax ginseng*. Front Plant Sci 9:813

Sunar K, Chakraborty U, Chakraborty BN (2014a) Exploitation of native microflora of Darjeeling hills for Sclerotial rot management and growth promotion in pulses. Indian Phytopathol 67:59–69

Sunar K, Chakraborty U, Chakraborty BN (2014b) Harnessing beneficial microorganisms from Darjeeling hills and development of strategies for their utilization in management of root diseases. J Mycol Plant Pathol 44:25–40

Sunar K, Chakraborty U, Chakraborty BN (2017) Influence of indigenous bacilli isolated from Darjeeling Hills on phosphate mobilization and induction of resistance against sclerotial blight disease of tea cultivars. Int J Basic Appl Biol 4:128–134

Sunar K, Chakraborty U, Chakraborty BN (2020) *Bacillus altitudinis* and *Bacillus pumilus* induce resistance in *Brassica juncea* L. against *Thanatephorus cucumeris* causing seedling blight. J Mycopathol Res 58:167–179

Sunar K, Dey PL, Chakraborty U, Chakraborty BN (2015) Biocontrol efficacy and plant growth promoting activity of *Bacillus altitudinis* isolated from Darjeeling hills, India. J Basic Microbiol 55:91–104

Tank N, Saraf M (2010) Salinity-resistant plant growth promoting rhizobacteria ameliorates sodium chloride stress on tomato plants. J Plant Interact 5:51–58

Tarekegn MM, Salilih FZ, Ishetu AI (2020) Microbes used as a tool for bioremediation of heavy metal from the environment. Cogent Food Agric 6:1783174

Tavares MJ, Nascimento FX, Glick BR, Rossi MJ (2018) The expression of an exogenous ACC deaminase by the endophyte *Serratia grimesii* BXF1 promotes the early nodulation and growth of common bean. Lett Appl Microbiol 66:252–259

Timmis K, Vos DMW, Ramos LJ, Vlaeminck EK, Prieto A, Danchin A, Verstraete W, Lorenzo de V, Lee YS, Brussow H, Timmis KJ, Singh KB (2017) The contribution of microbial biotechnology to sustainable development goals. Microb Biotechnol 10:984–987

Trdá L, Boutrot F, Claverie J, Brulé D, Dorey S, Poinssot B (2015) Perception of pathogenic or beneficial bacteria and their evasion of host immunity: pattern recognition receptors in the frontline. Front Plant Sci 6:219

Upadhyay SK, Maurya SK, Singh DP (2012) Salinity tolerance in free-living plant growth promoting rhizobacteria. Indian J Res 3:73–78

Uroz S, Calvaruso C, Turpault M-P, Frey-Klett P (2009) The microbial weathering of soil minerals: ecology, actors and mechanisms. Trends Microbiol 17:378–387

Venturi V, Keel C (2016) Signaling in the rhizosphere. Trends Plant Sci 21:187–198

Verheijen LM, Aerts R, Brovkin V, Cavender-Bares J, Cornelissen J, Kattge J, Bodegom PM (2015) Inclusion of ecologically based trait variation in plant functional types reduces the projected land carbon sink in an earth system model. Glob Change Biol 21:3074–3086

Vetterlein D, Carminati A, Kögel KI, Bienert GP, Smalla K, Oburger E, Schnepf A, Banitz T, Tarkka TM, Schluter S (2020) Rhizosphere spatiotemporal organization–a key to rhizosphere functions. Front Agron 2:8

Vivas JM, Silveira SF, Santos PH, Carvalho BM, Poltronieri STP, Jorge TS, Santos JS, Kurosawa RD, Moraes R (2018) Antagonism of fungi with biocontrol potential of papaya black spot caused by *Asperisporium caricae*. Aust J Crop Sci 12:827–833

de Vries FT, Wallenstein MD (2017) Below-ground connections underlying above-ground food production: a framework for optimising ecological connections in the rhizosphere. J Ecol 105: 913–920

Woo SL, Pepe O (2018) Microbial consortia: promising probiotics as plant biostimulants for sustainable agriculture. Front Plant Sci 9:1801

Xu X, Chen C, Zhang Z, Sun Z, Chen Y, Jiang J, Shen Z (2017) The influence of environmental factors on communities of arbuscular mycorrhizal fungi associated with *Chenopodium ambrosioides* revealed by MiSeq sequencing investigation. Sci Rep 7:45134

Yin C, Casa Vargas JM, Schlatter DC, Hagerty CH, Hulbert SH, Paulitz TC (2021) Rhizosphere community selection reveals bacteria associated with reduced root disease. Microbiome 9:86

Yu K, Pieterse CMJ, Bakker PAHM, Berendsen RL (2019) Beneficial microbes going underground of root immunity. Plant Cell Environ 42:2860–2870

Yu L, Zhang H, Zhang W, Liu K, Liu M, Shao (2022) Cooperation between arbuscular mycorrhizal fungi and plant growth-promoting bacteria and their effects on plant growth and soil quality. Peern J 10:e13080

Zake DR, Pregitzer KS, Burton AJ, Edwards HK (2011) Microbial responses to a changing environment: implications for the future functioning of terrestrial ecosystems. Fungal Ecol 4: 386–395

Zhalnina K, Dias R, de Quadros PD, Davis-Richardson A, Camargo OAF, Clark ML, Mcgrath PS, Hirsch RP, Triplett WE (2014) Soil pH determines microbial diversity and composition in the park grass experiment. Microb Ecol 69:395–406

Zhalnina K, Louie KB, Hao Z, Mansoori N, Da Rocha UN, Shi S, Cho H, Karaoz U, Loque D, Bowen PB, Firestone KM, Norhthen RT, Brodie LE (2018) Dynamic root exudate chemistry and microbial substrate preferences drive patterns in rhizosphere microbial community assembly. Nat Microbiol 3:470–480

Zhang F, Jia-Dong HE, Qiu-Dan NI, Qiang-Sheng WU, Zou YN (2018) Enhancement of drought tolerance in trifoliate Orange by mycorrhiza: changes in root sucrose and proline metabolisms. Not Bot Horti Agrobot Cluj-Napoca 46:270

Zhang XF, Zhao L, Xu SJ, Liu HY, Cheng GD (2013) Soil moisture effect on bacterial and fungal community in Beilu river (Tibetan plateau) permafrost soils with different vegetation types. J Appl Microbiol 114:1054–1065

Chapter 4
Unfolding the Role of Beneficial Microbes and Microbial Techniques on Improvement of Sustainable Agriculture Under Climatic Challenges

Indrani Baruah, Geetanjali Baruah, Smita Paul, Liza Devi, Bedika Boruah, Rajkumari Soniya Devi, Manisha Hazarika, Tinamoni Saikia, and Jishusree Bhuyan

Abstract Sustainable agriculture has the basic principle of utilizing the resources mindfully at present to fulfill current needs without hampering the requirements of future generations. It also provides equal importance to the economic, environmental, and societal aspects of agriculture. There is a growing concern regarding climatic challenges and the harm caused by them. The non-judicious usage of chemical fertilizers has also led to non-fertile soils and added issues, thereby causing a drastic loss in yield. Beneficial microbes and other microbial techniques hence present an excellent alternative to chemical fertilizers for improvement of the current scenario in maintaining sustainability in agriculture. Various microbial formulations can be utilized with advance biotechnological tools to increase the yield of crops which are discussed in this chapter. This chapter summarizes various categories or different categories of agriculturally important microbes and the application of potential microbial techniques to sustain crop productivity under adverse climatic conditions.

Keywords Biocontrol agents · Biofertilizer · Biopesticides · Mycorrhizae · PGPR · Sustainable

---

Indrani Baruah and Geetanjali Baruah: Equal contribution/Co-First authors

---

I. Baruah (✉)
Plant Breeding and Genetics Department, Assam Agricultural University, Jorhat, Assam, India

DBT-NECAB, Assam Agricultural University, Jorhat, Assam, India

G. Baruah · S. Paul · L. Devi · B. Boruah · R. S. Devi · M. Hazarika · T. Saikia · J. Bhuyan
Department of Biotechnology, The Assam Kaziranga University, Koraikhowa, Jorhat, Assam, India

P. Mathur et al. (eds.), *Microbial Symbionts and Plant Health: Trends and Applications for Changing Climate*, Rhizosphere Biology,
https://doi.org/10.1007/978-981-99-0030-5_4

Abbreviations

ABA	Abscisic acid
ACC-1	Aminocyclopropane-1-carboxylate
ACCD	ACC deaminase
AMF	Arbuscular mycorrhizal fungi
BNF	Biological N_2 fixation
CRISPR/Cas	Clustered regularly interspaced short palindromic repeats and CRISPR-associated protein
DNA	Deoxyribonucleic acid
ePGPR	Extracellular PGPR
FAO	Food and Agriculture Organization
FFA	Free fatty acid
GA	Gibberellic acid
IAA	Indole-3-acetic acid
IPCC	Intergovernmental Panel on Climate Change
iPGPR	Intracellular PGPR
ISR	Induction of systemic resistance
MAGs	Metagenome-assembled genome
MS	Mass spectrometry
NMR	Nuclear magnetic resonance
PGPMs	Plant growth-promoting microbes
PGPR	Plant growth-promoting rhizobacteria
PRRs	Protein recognition receptors
PSM	Phosphorus solubilizing microbes
RNAi	RNA interference
SMs	Secondary metabolites
VOCs	Volatile organic compounds

4.1 Introduction

For all living beings, throughout the world, the primary source of nutrition is agriculture. Since the mid-twentieth century, the agricultural sector has experienced phenomenal growth. However, the phase of agricultural growth faces a serious challenge in terms of sustainability. Sustainability is based on the principle that the needs of the present must be met, without compromising the ability of the future generation to meet their own needs. Under the changing agricultural scenario, the conditions for sustainable agriculture are becoming more and more favorable. With the growing population of the world and the loss of agricultural land due to climate change, the world is facing a huge problem. It is estimated by the World Bank that the production of food will have to increase by 70% by 2050 (van Dijk et al. 2021). That is where the importance of sustainable agriculture comes in. Some of the reasons behind the importance of sustainable agriculture are as follows:

(1) sustainable agriculture uses natural fertilizers and crop rotation method, which ultimately nourishes and restores the soil and make the soil toxic-free. (2) Sustainable agriculture attempts to minimize the use of energy at all levels of production. Some of the strategies to attain sustainable agriculture include crop rotation, organic farming, introducing plant growth-promoting microorganisms (PGPMs), and so on. The field of biotechnology mainly deals with the improvement of crops by genetic manipulation. Through genetic engineering, we can modify the crop plant characteristics like disease resistance, drought resistance, etc. Recombinant DNA technology can help to overcome the limitations of traditional genetic engineering methods.

The human population is increasing drastically, and this rapidly growing population leads to a remarkable demand for food and high production of agricultural products. To fulfill this demand, the Green Revolution comes into play by using fertilizers and pesticides on a large scale to lower the pathogen and weeds of crops (Pingali 2012). Gradual and immoderate use of chemical agents in the agriculture field leads to the constant devastation of the environment and animal and human health (Nicolopoulou-Stamati et al. 2016). India's economy is mainly based on agriculture. Biofertilizers are a highly potent alternative to chemical fertilizers because of their eco-friendly, non-toxic, easy-to-apply, and cost-effective property. Biofertilizers make nutrients that exist or are present in nature in large amounts and that are useful for plants. India is one of the biofertilizer-based countries, and that is why the Indian Government is trying to increase the application of biofertilizers along with modern agrochemicals.

Bioremediation is the utilization of living organisms or microbes to remove contaminants, pollutants, etc. from soil and water. One of the microorganisms that have bioremediating potentials is the PGPR (plant growth-promoting rhizobacteria). It can help in detoxifying pollutants like pesticides, and heavy metals, and it can control a range of phytopathogens such as biopesticides. However, to overcome all this havoc, an alternative solution is taken by using microbial biotechnology, which is eco-friendly, non-toxic, and sustainable agriculture (Glick 2014). Among the strategies used by microbial biotechnology for agriculture production are genetically modified organisms or plants, biopesticides, biofertilizers, or PGPM. The roots of plants are generally known as storehouses of microbes or rhizospheres (Gouda et al. 2017). Some plant growth-promoting microbes, such as actinomycetes, rhizobacteria, fungi, arbuscular mycorrhizal fungi, and endophytes, reside in a microbe accumulation known as the phytomicrobiome (Kundan et al. 2015; Knack et al. 2015; Sagar et al. 2021). Some biological processes, such as nitrogen fixation and phosphate solubilization, stress alleviation through the modulation of ACC deaminase expression, and production of phytohormones and siderophores provide nutrients for the growth of a plant using PGPM (Hakim et al. 2021). PGPM works as a biofertilizer. The use of heavy amounts of artificial fertilizer restores soil nitrogen and phosphorus, but on the other hand, it gradually introduces xenobiotics to the soil. PGPM contributes N2-fixing and phosphorus solubilizing bacteria which act as an important factor for the growth of plant nutrition in an agrosystem (Elhaissoufi et al. 2021; Singh 2013).

4.2 Plant Growth-Promoting Rhizobacteria

The term "PGPR" refers to a group of soil bacterial species that are rapidly expanding in plant rhizosphere and which develop in, on, or around plant tissues and stimulate plant development through a variety of ways, i.e. plant growth-promoting rhizobacteria (Backer et al. 2018). PGPR mainly promote plant growth by generating or changing the concentrations of growth regulator, i.e., cytokinins, gibberellic acid (GA), and auxin (Tsukanova et al. 2017). PGPR also promote symbiotic nitrogen fixation, decrease ethylene synthesis in roots, augment production of antibiotics, siderophores, fluorescent pigments, β-1-3-glucanases, chitinases, cyanide, phosphates, zinc, and other micro-macromolecules (Basu and Mandal 2017; Olanrewaju et al. 2017). According to the level of bacterial proximity to the root and affinity of its association, the range of PGPR varies (Grover et al. 2021). On the basis of its ranges, PGPR can be differentiated into extracellular (ePGPR) and intracellular (iPGPR) (Figueiredo et al. 2011). A few examples of ePGPR are *Agrobacterium*, *Arthrobacter*, *Azotobacter*, *Azospirillum*, *Bacillus*, *Caulobacter*, *Chromobacterium*, *Erwinia*, *Flavobacterium*, *Pseudomonas*, *Serratia*, etc. (Bhadrecha et al. 2020; Bhattacharyya and Jha 2012). *Allorhizobium*, *Azorhizobium*, *Bradyrhizobium*, *Mesorhizobium*, and *Rhizobium* are a few examples of the iPGPR.

4.2.1 Nitrogen Fixation

In nitrogen fixation, organic or atmospheric nitrogen (N_2) is converted into ammonia. It is then converted into nitrites (NO_2) and then into nitrate (NO_3) with the help of biological N_2 fixation (BNF) using a complex enzyme system known as nitrogenase (Signorelli et al. 2020). Biological N_2 fixation is a vital aspect of sustainable agriculture for plant growth and productivity, but the Haber-Bosch process, or the production of industrial nitrogen, acts as a torpedo of the human population (Smil 2002). Nitrogen-fixing microorganisms are generally differentiated into two categories, viz., symbiotic N_2-fixing bacteria (e.g., *Rhizobiaceae* family, which forms a symbiosis with leguminous plants) and non-symbiotic or free-living bacteria (Ahemad and Khan 2009). Non-symbiotic, nitrogen-fixing bacteria include *Anabaena*, *Nostoc*, *Azospirillum*, *Azotobacter*, *Gluconacetobacter diazotrophicus*, *Azocarus*, and others (Santi et al. 2013). Nitrogenase (nif) genes are responsible for nitrogen fixation, which is present in both symbiotic and free-living systems (Haskett et al. 2022). These genes activate iron, molybdenum, Fe protein, cofactor biosynthesis, electron donation, and regulatory genes required for the synthesis and function of the enzyme (Burén et al. 2020; Ahemad and Kibret 2013). Nitrogen fixation can also be performed by using genetically modified bacteria. For example, ammonium excreting *Azospirillum*, *Pseudomonas*, and *Azotobacter* (Haskett et al. 2022; Van Dommelen et al. 2009; Setten et al. 2013; Geddes et al. 2015; Ambrosio et al. 2017).

4.2.2 Phosphorus Solubilizing Bacteria

Phosphorus is the second most essential element required for the growth of plants (Tak et al. 2012). The concentration of phosphorus available for plant growth is too minimal (Zhu et al. 2011). The deficiency of phosphorus in the agricultural world is generally overcome by using phosphorus fertilizers. Regular use of phosphorus fertilizers creates a critical condition for environmental health, so to minimize the rate of hazards nowadays, a set of bacteria called phosphorus solubilizing microbes (PSM) are used. PSMs are a group of beneficial microbes that have the capability to transform organic or inorganic insoluble phosphorus into soluble form (Bhattacharyya and Jha 2012). It provides a desired amount of phosphorus to the plant and also works as a replaceable option in place of phosphorus fertilizers (Khan et al. 2006). Bacteria responsible for phosphorus solubilization and mineralization are *Azotobacter*, *Bacillus*, *Enterobacter*, *Kushneria*, *Erwinia*, *Paenibacillus*, *Ralstonia*, *Rhizobium*, *Rhodococcus*, *Serratia*, *Bradyrhizobium*, *Salmonella*, *Sinomonas*, and *Thiobacillus* (Jahan et al. 2013; Kumar et al. 2014; Zhu et al. 2011; Chakraborty et al. 2009; Fernández Bidondo et al. 2011; Tajini et al. 2012).

4.2.3 Plant Growth-Promoting Mycorrhizal Bacteria

A biofertilizer known as arbuscular mycorrhizal fungi (AMF) provides capacity for the growth of plants in harsh conditions like heat, salinity, drought, metals, extreme temperature, and low nutrients, and it provides a perfect assimilating between plant and fungus interaction (Gianinazzi et al. 2002; Allen 2011). AMF is also used to protect plants from fungal infections (Smith and Read 2008). So, AMFs are indispensable endosymbionts playing a crucial role in plant growth and the operation of the ecosystem. They are vital factors for sustainable crop improvement (Gianinazzi et al. 2010). The majority of AMF are members of the phylum Mucoromycota's sub-phylum Glomeromycotina. Some studies have found that (Berruti et al. 2016) under drought conditions, a specific fungal association, *Antirrhinum majus*, can increase the concentration of macronutrients such as N, P, Ca, and Mg. AMF also limits the accumulation of Mn, Mg, Na, and Fe in roots (Bati et al. 2015).

4.3 Effect of Climate Change on Agriculture

Agriculture and climate change are linked in many ways. Climate change is a major source of biotic and abiotic pressures that tend to adversely affect local agriculture. Regions and their agriculture are affected by climate change in many ways, for example, variations in annual precipitation; average temperature; heatwaves; changes in weeds, pests, or microbes; global changes in atmospheric CO_2 or ozone levels; and changes in sea level. The threat of a fluctuating global climate has received much attention from scientists as these changes adversely affect global crop production and threaten food security around the world. Agriculture is considered to be the most vulnerable activity to climate change. To date, food security and ecological resilience remain the top concerns around the world (Raza et al. 2019). Crop yield losses caused by abiotic stresses are expected to increase as climate change and other factors create harsher environmental conditions in areas traditionally used for cultivation. Breeding and genetically modified and engineered organisms have produced many cultivars with greater abiotic stress tolerance, but their practical utilization remains a long process of biological cycles and regulatory issues (Godoy et al. 2021). Climate variability has had a significant impact on plant physiology in several ways. Climate variability and environmental extremes increased the likelihood of multiple stressors on plants (Thornton et al. 2014). Boyer reported that climate changes have reduced crop yield by up to 70% since 1982 (Boyer 1982). According to a study by the Food and Agriculture Organization (FAO) in 2007, all cultivated areas in the world are affected by climatic changes and only 3.5% of areas are safe from environmental limitations (Van Velthuizen 2007). An unusual increase in maximum and minimum temperature with departure from mean exceeding by more than 5 °C was observed in most parts of India in March and April 2022 (Bal et al. 2022). Water scarcity and temperature extremes brought on by climate change have an impact on plant growth's reproductive stage. The initiation and inflorescence of flowers in grains are adversely affected by water stress (Raza et al. 2019). Inadequate water content in the plant due to drought stress causes stomatal closure resulting into a reduction of photosynthetic rate and alteration in leaf water potential and carbon fixation (Wijewardana et al. 2019). Abiotic stress conditions like heat and drought salinity cause lots of physiological changes depending on the plant variety. Such physiological changes include initiation of seed dormancy, reduction in early-stage germination growth, and impaired mitosis and cell elongation in pea (*Pisum sativum* L.), alfalfa (*Medicago sativa* L.), and rice (*Oryza sativa* L.) (Fahad et al. 2017). Drought stress influences wheat during all developmental stages, but grain formation and the reproductive stage are the most critical ones (Pradhan et al. 2012). Wheat yield decreased from 1% to 30% during the mild drought stress at post-anthesis while this reduction increased up to 92% in case of prolonged mild drought stress at flowering and grain formation (Araus et al. 2002; De Oliveira et al. 2013). Important grain legumes have had a significant reduction in output due to drought stress. Mashbean (*Vigna mungo* L.) yield has been reduced by drought stress from 31% to 57% during the flowering stage while a 26% reduction was reported by drought stress during the reproductive phase (Baroowa and Gogoi 2014). Maleki et al. (2013) reported that the soybean yield has been largely affected by drought stress and a 42% reduction was observed during the grain-filling stage of soybean. Schlenker and Roberts (2009) described that maize yield was increased at an optimum temperature of 29 °C but a further increase in temperature hampered the yield of maize (Schlenker and Roberts 2009). Every 1 ° C rise in temperature was found to negatively influence the maize yield (Lobell et al. 2011). Similarly, it was reported that yield in maize decreased by 8.3% with every

1 °C rise in temperature from the optimum growth temperature (Lobell and Field 2007). Brown et al. (2009) reported that wheat yield decreased by 10% with every 1 °C increase in temperature (Brown 2008). In another report, it was revealed that a 3–4% reduction in wheat yield takes place for every 1 °C increase in temperature (Ray et al. 2015). Easterling et al. (2007) described that a 2 °C increase in temperature caused a 7% reduction in yield while a further increase in temperature to 4 °C decreased the yield by up to 34% in wheat. Similarly, rice yield decreased by 2.6% for every 1 °C rise in temperature. Food security is impacted by climate change in a very complex way. It hampers the agricultural yield directly using disturbing the agro-ecological environment and indirectly by putting pressure on the growth and circulation of income and consequently increasing the necessity of agricultural products (Intergovernmental Panel on Climate Change (IPCC) 2007).

One of the most significant issues currently facing agriculture is abiotic stress. It causes serious losses in crop production worldwide and reduces planted acreage. Amid a growing population and climate change, this scenario becomes increasingly complex. Because the world population is forecast to increase from 7 to 9–10 billion people by 2050, an increase of between 60% and 110% in global food production will be required (Rockström et al. 2017). Growing stress-tolerant plants and comprehending their reactions to various stress circumstances are essential for sustainable agriculture and food safety for a growing global population. Plants respond differently to different climatic conditions in terms of gene expression, physiology, and metabolism. It was reported that plants can sense any variation in surrounding environmental signals, but despite many studies, only some reputed sensors have been recognized (Zhu 2016). The organs and tissues of plants are harmed by numerous pressures, and they react accordingly. For instance, individual cells have varied transcriptional responses to various stresses (Dinneny et al. 2008). The cellular signals that are created in response to salinity, drought, and chemical effluence include the production of stress-responsive proteins, high levels of related solutes, and higher antioxidant ratios. These stresses are regarded as primary stresses, and they generate secondary stresses like oxidative and osmotic stress (Carvalho and Amâncio 2018).

4.3.1 Drought

Drought refers to a situation in which the amount of available water through rainfall and/or irrigation is insufficient to meet the evapotranspiration needs of the crop (Kumaraswamy and Shetty 2016). Changes in water availability (volumes and seasonal distribution), as well as water demand from competing industries like agriculture, are what cause climate change. The impending climate change adversities are known to alter the abiotic stresses like variable temperature regimes and their associated impacts on water availability leading to drought, increased diseases and pest's incidence, and extreme weather events at the local to regional scale (Kumaraswamy and Shetty 2016). Moisture or drought stress accounts for about

30–70% loss of productivity of field crops during the crop growth period (Kumaraswamy and Shetty 2016). Drought stress can induce abscisic acid (ABA) accumulation in guard cells to trigger stomatal closure (Li and Cui 2014). Drought also results in abnormal metabolism that may reduce plant growth and/or cause the death of an entire plant. Drought has different effects at different stages of plant growth with the most sensitive growth stage being the flowering period.

4.3.2 Heat Stress

Heat stress is the rise in temperature beyond a threshold level for a period sufficient to cause permanent damage to plant growth and development (Hasanuzzaman et al. 2013). The Intergovernmental Panel on Climate Change (IPCC) projected a rise of the temperature by 3–4° by 2050 (IPCC 2002; Meeh et al. 2007). High-temperature regimes due to climate change affect the percentage of seed germination, photosynthetic efficiency, crop phenology, reproductive biology, flowering times, pollen viability, and pollinator populations (Descamps et al. 2021). Under heat stress at the reproductive growth stage, the increase in temperature prevents the swelling of pollen grains, which results in the poor release of pollen from the anther at dehiscence. Plant development, particularly the formation and operation of reproductive organs, is negatively impacted by heat stress. Additionally, the unpredictable spread of disease outbreaks throughout geographical locations in the world is due to changing temperature regimes. Heat stress contributed about 40% to the overall yield loss of wheat, 1.0–1.7% yield loss per day in maize for every rise in temperature above 30 °C (Lobell et al. 2011).

4.3.3 Cold Stress

Cold or chilling stress experienced by plants from 0 to 15 °C leads to major crop losses (Yadav 2010). Non-freezing low temperatures harm or destroy a variety of crops with tropical or subtropical origins, and they show varied signs including poor germination, stunted seedlings, chlorosis, or a slowdown in growth, reduced leaf expansion, withering, and necrosis. In general, plants adapt to low temperatures by altering their pattern of gene expression and protein synthesis (Sanghera et al. 2011). Usually, plans which are native to temperate climates are considered to be freezing-tolerant plants that undergo a process known as cold acclimation when exposed to temperatures just above 0 °C. Plants that have been exposed to cold are prepared for freeze stress in subzero temperatures as well as any additional stresses that may come with low temperatures (Collin et al. 2021).

4.3.4 Soil Properties

The top layer of the earth's crust, or soils, is created primarily by the weathering of rocks, the production of humus, and material movement. They differ in terms of their place of origin, appearance, traits, and level of production. The capacity of soil to provide nutrients necessary for the optimum growth of a particular crop is known as soil fertility. One of the most crucial elements in crop production is soil fertility (Wang 2014). Soil fertility can influence the physical, chemical, and biological attributes of crop production. Soil is a significant source of micronutrients (Fe, B, Cl, Mn, Zn, and others) and macronutrients (N, P, and Cu). It also provides nutrients (K, Ca, S, Mg, C, O, and H) necessary for plant growth. Soil nutrient deficiencies may lead to various diseases and toxicities in plants.

4.3.4.1 Soil Salinity and Acidity Stress

In about 30% of irrigated crops and 7% of dry land agriculture worldwide, salinity stress has an impact on crop productivity (Schroeder et al. 2013). One of the main issues influencing crop output worldwide is that 33% of irrigated land and 20% of cultivated land are both affected by salt (Machado and Serralheido 2017). Crop plants experience ionic toxicity and osmotic stress due to salt. Under normal circumstances, the greater osmotic pressure in plant cells allows water and important nutrients to be absorbed into the root cells from a soil solution. However, under salt stress conditions, the high concentration of salts in the soil solution prevents the absorption of water and essential minerals but will facilitate the entry of Na^+ and Cl^- ions into the cells, which will have direct toxic effects on cell membranes as well as on metabolic activities in the cytosol (Kumar 2013).

4.3.4.2 Over Usage of Chemical Fertilizers Causes Loss of Soil Fertility Resulting in Crop Yield Loss

In order to meet increasing public needs and to promote crop products, the use of high inputs of chemicals in the soil in the form of fertilizers, pesticides, fungicides, insecticides, nematicides, and weedicides, along with intensive irrigation practices, helped to achieve the target to a certain stage. However, despite the fertilizer being applied, crop yields decreased. Chemical fertilizers have a negative impact on the environment since they process substances that produce hazardous compounds or gases that pollute the air, such as NH4, CO2, CH4, etc. (Bisht and Chauhan 2020). And when industrial trash is dumped untreated in surrounding water bodies, it will pollute the water. It also involves the most detrimental effect of chemical waste accumulation in water bodies, namely, water eutrophication. Additionally, continuous usage of it damages the health and condition of the soil when applied to it, resulting in soil pollution (Bisht and Chauhan 2020). Overuse of chemical fertilizers can cause soil acidification and soil crust, which lowers the amount of organic matter, humus, and beneficial organisms in the soil, stunts plant development, changes the soil's pH, feeds pests, and even triggers the release of greenhouse gases. The soil's acidity lowers phosphate uptake by plants, increases the number of detrimental ions present in the soil, and prevents plants from growing (Chandini et al. 2019). Humus depletion reduces the soil's ability to hold onto nutrients. Large-scale nitrogen fertilizer application to fields over time destroys the equilibrium between the three macronutrients, N, P, and K, resulting in lower crop yields. Heavy metals including arsenic, cadmium, and uranium may accumulate toxically in the soil as a result of repeated chemical fertilizer applications. These dangerous heavy metals not only contaminate the soil but also build up in cereals, fruits, and vegetables. Fertilizers such as triple superphosphate have trace elements such as cadmium and arsenic that accumulate in plants and enter humans via food chains that can cause health problems (Sonmez Kaplan and Sonmez 2007). Application of fertilizers without the recommendation of soil testing can lead to implications such as soil degradation, nutrient imbalance, soil structure destruction, and bulk density increase (Savci 2012).

4.4 Plant Growth-Promoting Microorganisms (PGPMs)

The region of the soil containing microbes referred to as rhizosphere is in contact with the plant roots that affect the root activity (Gouda et al. 2017). Plant interaction with rhizosphere microbe's mainly rhizobacteria, actinomycetes, fungi, arbuscular mycorrhiza fungi, and endophytes helps in plant growth by the uptake of water and nutrients from the rhizospheric zone and soil (Gamini Seneviratne et al. 2010). This interaction with PGPMs can be symbiotic or non-symbiotic (Kundan et al. 2015). Hence, rhizosphere's microbes directly affect the soil quality and plant growth. This characteristic feature fascinates the people involved in enhancing the plant growth using microorganisms for sustainable development (Gamini Seneviratne et al. 2010). There are several types of PGPMs available in nature which are discussed in this chapter.

4.4.1 Plant Growth-Promoting Rhizobacteria (PGPR)

Plant growth-promoting rhizobacteria (PGPR) stimulates plant growth, nutrition, and production by a wide variety of mechanisms like phytohormones production, 1-aminocycloprppane-1-carboxylate (ACC) deaminase production, induction of systemic resistance (ISR), phosphate solubilization, siderophore production, biological nitrogen fixation (BNF), rhizosphere engineering, quorum sensing (QS) signal interference and inhibition of biofilm formation, exhibiting antifungal activity production of volatile organic compounds (VOCs), promoting beneficial plant-microbe symbioses, interference with pathogen toxin production, etc. (Bhattacharyya and Jha 2012; Kumar and Singh 2020). Some common examples of PGPR genera that exhibit plant growth-promoting activity for enhancing plant health and developing crop production are *Pseudomonas*, *Azotobacter*, *Bacillus*, *Rhizobium, Klebsiella*, *Azospirillum*, *Enterobacter*, *Caulobacter*, *Micrococcus*, *Chromobacterium*, *Alcaligenes*, *Burkholderia*, *Serratia*, and *Arthrobacter* (Swarnalakshmi et al. 2020; Sharma et al. 2022). Many rhizosphere bacteria can produce and alter the plant growth regulators concentration like cytokinins, gibberellic acid (GA), and indole-3acetic acid (IAA) to boost growth and stress response (Tsukanova et al. 2017; Seema et al. 2018). PGPR that produce auxins induce transcriptional changes in hormone, defense-related, and cell wall-related genes, induce longer roots, increase root biomass, and decrease stomata size and density (Spaepen and Vanderleyden 2011; Llorente et al. 2016). Many PGPR produce cytokinins and gibberellins that enhance the root exudate production and plant shoot growth, respectively (Jha and Saraf 2015; Ruzzi and Aroca 2015). Ethylene is generally considered as "a stress hormone" that accumulates in response to any kind of stress and induce drought salinity stress signaling pathway (Verma et al. 2016). After many studies of PGPR function, it has been seen that they provide greater growth stimulation under stressful conditions (Rubin et al. 2017). PGPR produces ACC deaminase that reduces ethylene production of plant (Glick 2014). About 2–5% rhizobacteria are involved in the PGP activity (Antoun and Kloepper 2001). If a bacterial strain can fulfill at least two of the three criteria: aggressive colonization, plant growth stimulation, or biocontrol then they are considered as PGPR (Vessey 2003). Among these microbial populations in the rhizospheric zone, actinobacteria or actinomycetes have a crucial job due to their potential action as PGPR and soil nutrient cycling (Yadav et al. 2018; Franco-Correa et al. 2010). Actinobacteria are Gram-positive bacteria with high GC content and produce metabolites like antibiotics, enzymes, and vitamins (Ventura et al. 2007; Saxena 2014; Terkina et al. 2006). About two-third of natural antibiotics are produced by actinomycetes out of which 75% is produced by the *Streptomyces* genus (Yekkour et al. 2012).

4.4.2 Plant Growth-Promoting Fungus (PGPF)

Plant growth-promoting fungi (PGPF) include non-pathogenic soil-borne filamentous fungi that are beneficial for plant growth and sustainable agriculture (Kumari et al. 2021). PGPF produces hormones that help in the interaction between plants and soil for organic matter decomposition through mineral solubilization (Hossain and Sultana 2020). This microbial interaction is enhanced by root colonization which is one of the important characteristic features of PGPF (Hossain and Sultana 2020). Some of the important PGPF genera used for stimulating plant growth are *Penicillium*, *Fusarium*, *Penicillium*, *Mycorrhiza*, *Phoma*, *Gliocladium*, *Metarhizium*, *Trichoderma*, *Clonostachys rosea*, *Pythium*, and *Aspergillus* (Thakur and Sagar 2019).

Among the PGPF, arbuscular mycorrhizal fungi (AMF) enhance plant growth in stress conditions and helps in the interaction between plant and soil (Jeffries and Barea 2000; Gianinazzi et al. 2002). There is a symbiotic association between mycorrhizal PGPMs and plant roots that can be either intracellular colonization as in AMF or extracellular association as in ectomycorrhizal fungi with the host plant. About 80% of terrestrial host species are in mutualistic symbiotic association (MSA) with AMF (Bharadwaj et al. 2008). AMF forms a link between plant roots and soil mineral nutrients, and this mycorrhizal colonization increases the uptake of micro-macro nutrients and water by the roots (Igiehon and Babalola 2018). Mycorrhizal plants are stress tolerant, have better access to phosphorus, and are less vulnerable to root-feeding nematodes (Augé 2001; Neumann and George 2004; Khaosaad et al. 2007). AMF interacts with soil rhizobacteria that affects the development and symbiotic establishment, and this interaction can be positive, negative, and neutral (Söderberg et al. 2002; Igiehon and Babalola 2018). PGPR strain of *Pseudomonas putida* improves AMF root colonization (Yu et al. 2022). Mycorrhiza-associated bacteria (MAB) are in obligatory symbiotic association with AMF in nature (Mitra et al. 2019). It has been reported that AMF in rhizosphere replaces the facultative anaerobic bacteria with fluorescent pseudomonads to shift the diversity in specific groups of bacteria (Etesami et al. 2021). This change in bacterial population is due to soil nutrient content modification and architecture changes carried by AMF and released by the extra radical mycelium of AMF as well as mycorrhizal roots (Azcón-Aguilar and Barea 2015).

4.4.3 Plant Growth-Promoting Endophytes (PGPE)

In the term endophyte, endo implies "within" and phyte implies "plant"; therefore it refers to all the organisms living within the plant (Gouda et al. 2016). Generally, it describes the type of interaction the fungi or bacteria have with their host which does not elicit the symptom of the disease. Endophytic fungi protect the host from disease-causing pathogens (Fontana et al. 2021). The apoplectic space (xylem vessel) and parenchymal tissue in the host plant are occupied by the endophytes (Bortolami et al. 2019).

Endophytes rather than epiphytes are present entirely within the host plant substrate. Endophytic PGPM colonization stimulates plant growth by nitrogen fixation, provides tolerance against biotic and abiotic stress, helps in the production of phytohormones such as cytokines and IAA, antagonistic activity of plant pathogens by the formation of antifungal bacterial agents, nutrient competition, siderophore production and induction of the acquired host resistance, or improvement of the bio-availability of minerals and nutrients (Glick 2015; Suman et al. 2001; Nath et al. 2013; Suman et al. 2016). Therefore, they can be used as bio-regulators to induce resistance against diseases, as biological control agents against certain pathogens, and also in the biological control of undesirable weeds (Morelli et al. 2020). These endophytes prove to be a major source of new bioactive compounds used in the pharmaceutical, agricultural, as well as food industries (Ortega et al. 2020). Some of the important PGPE genera that promote plant growth activity are *Streptomyces*, *Enterobacter*, *Acinetobacter*, *Pseudomonas*, *Ochrobactrum*, *Bacillus*, *Piriformospora*, *Sinorhizobium*, *Ascomycota*, *Basidiomycota*, and *Zygomycota* (Jafari et al. 2018; Park et al. 2017).

4.5 Formulation of Plant Growth-Promoting Microorganisms (PGPMs)

Public health and safety concerns have led to the wide application of formulated plant beneficial microorganisms than chemical agro-products in the last 10–15 years. The methods used for formulation depending on the PGPMs (viruses, bacteria, yeast or fungi, and nematodes) varies such that bacteria and yeast are usually produced in liquid fermentation while fungi are produced in a solid-state fermentation (Montesinos 2003). Formulation of PGPM inoculants is a multistep process starting from the isolation of a pure culture, screening of its PGP or antagonistic traits employing an array of in vitro and in vivo bioassays followed by demonstration under greenhouse and field conditions ((Montesinos 2003). In microbial inoculant production and application, two issues are being faced:

(a) Sustainability based on the overall production cost during the microbial formulation
(b) The creation of commercial goods with a high potential for soil-plant colonization under controlled conditions but ineffective in the field for nutrient mobilization and/or plant pathogen defense

To solve these problems, microbial formulations produced by immobilization methods are gaining attention as they are advantageous over solid and liquid formulations (Vassilev et al. 2020; Nakkeeran et al. 2005). The formulation includes an active ingredient (microbe or microbial product) in a suitable carrier material (sterile or non-sterile) with additives, which help in the stabilization and protection of the microbial cells during storage, transport, and at the target site (Gopalakrishnan et al. 2016). The microbial formulation can be of different types – dry products (such as granules and dust), liquid products (such as emulsions, oil, and water), and microencapsulation (Nakkeeran et al. 2005).

4.5.1 Ingredients Used in the Formulation

Additives in formulation maintain the bioactivity of PGPMs which is important for multiplying PGPMs that improve the physical, chemical, or nutritional properties of the formulated biomass. Such ingredients are stickers/binders (corn flour, gum arabic, and CMC); surfactants (Tween 80); dispersants (microcrystalline cellulose);

thickeners (xanthan gum); desiccants (silica gel) stabilizers (lactose and sodium benzoate); and UV protectants (Schisler et al. 2004).

4.5.2 Types of Formulation

The most widely used formulations for PGPMs are liquid-based, talc-based, sawdust-based, fly ash-based, encapsulation-based, and peat-based.

4.5.2.1 Liquid-Based Formulation

In liquid-based formulation, PGPMs are formulated in a liquid buffer with or without added protectants such as sugars. The liquid formulation is easier as the preparation is simple and cheaper, and no cells are killed during the formulation; it has the disadvantage of shorter shelf life, especially when stored at high temperatures (Melin et al. 2011). The addition of sucrose or glycerol improves the survival of rhizobia, phosphate-solubilizing bacteria, and *Bacillus amyloliquefaciens* (Taurian et al. 2010). It has been reported that liquid formulation has enhanced agricultural productivity under field conditions, for example, inoculation with *Azospirillum brasilense* enhanced vegetative growth and also harvested grains in wheat (Diaz-Zorita and Fernandez-Canigia 2009).

4.5.2.2 Talc-Based Formulation

Talc or steatite or soapstone or magnesium silicate-based formulation of bacterial inoculants is reported effective against plant diseases (Nakkeeran et al. 2005; Meena et al. 2002). Talc-based formulations of *Streptomyces griseus* have an enhanced shelf life of up to 105 days. According to Nandakumar et al. (2001), *Bacillus subtilis* and *P. fluorescens* in talc-based formulations can prevent rice sheath blight and early blight of tomatoes caused by *Rhizoctonia solani*.

4.5.2.3 Sawdust-Based Formulation

Sawdust has high organic matter and good water-holding capacity; hence it is used extensively (Arora et al. 2008). For both mono-inoculants and co-inoculant, the sawdust-based formulation is proved to enhance the seedling biomass and fertility of rhizosphere soil (Arora et al. 2008). Recently, it was found that sawdust could be used as a carrier material for five cellulolytic bacteria including *Chaetomium globosum*, *C. crispatum*, *C. olivaceum*, *C. nigricolor*, and for *P. fluorescens* and *B. cereus* (Ambardar and Sood 2010). Sawdust-based formulations for *Bacillus amyloliquefaciens*, *Serratia marcescens*, and *Bacillus pumilus* reported a shelf life of up to 9 months of storage (Chakraborty et al. 2013).

4.5.2.4 Fly Ash-Based Formulation

Fly ash is used as a carrier material as it contains good mineral contents for the bio-formulation development of PGP microbes. It promotes crop growth and improves soil structure (Biswal and Panda 2018). For example, fly ash-based formulation for *Bacillus* spp., *Azotobacter* spp., and *Pseudomonas* spp. has encouraging results (Kumar 2014). The advantage of fly ash-based formulation is that it increases soil pH and aids in nutrient availability (Paul et al. 2019). Fly ash alone and in combination with other materials was demonstrated in the bio-formulation of *Rhizobium* and *Trichoderma viride* and *T. harzianum* (Kumar and Gupta 2008).

4.5.2.5 Encapsulation-Based Formulation

Encapsulation of PGPMs in polymers of polysaccharide (alginate or gluten) helps the cells to remain viable for longer duration (Balla et al. 2022). Encapsulation of PGPMs protect them from harsh environment under field conditions, reduce natural microbial inhabitant competition in soils, and facilitate colonization on host plant roots (Souza et al. 2015). Encapsulation-based formulation for *Bacillus megaterium* and *T. viride* enhances seedling emergence (Soundara Rajan et al. 2020). Encapsulation of *B. subtilis*, in alginate beads enriched with humic acid, helps in establishment in the rhizosphere (Young et al. 2006). Encapsulation-based formulation is advantageous over other free cell formulations as it protects from biotic stress, abiotic stress, inhibitory effect of toxic compounds, enhanced survival, and improved physiological activity and supply of encapsulated nutritional additives (Chaudhary et al. 2020).

4.5.2.6 Peat-Based Formulation

Peat is the surface organic layer that consists of carbonized vegetable tissue formed in wet conditions by the slow decay of aquatic and semiaquatic plants such as sedges, rushes, reeds, and mosses (Nakkeeran et al. 2005). Peat-based formulation is most commonly used in rhizobia inoculation industry. In peat-based formulations, during the storage period, bacteria remain metabolically active until nutrients, moisture, and the optimum temperatures are maintained (Lobo et al. 2019). Peat-based formulations are coated on seeds or pelleted for sowing in furrows for rhizobia (Zhou et al. 2017). Peat-based formulation enhances the stability and effectiveness of the biocontrol agents (Ardakani et al. 2010). Under greenhouse conditions, *P. fluorescens* in peat formulation enhances soybean plant to grow (Habazar et al. 2014). A disadvantage of peat formulation is that it is unavailable in various countries.

4.6 Survival of PGPMs in Formulation

Under field condition, survival and establishment of PGPMs in the rhizospheric zone in competition with native microbial flora is absolutely essential in order to avail the maximum benefits. When inoculated under field condition, PGPMs find difficult to survive in the unpredictable environmental condition and among the predators and competitors. The key factor that declines the inoculated microbial density is sunlight; it reduces the bioactivity of aerial PGPMs like bacteria, virus, and fungi to field crops (Slininger et al. 2003; Trabelsi and Mhamdi 2013). For microbial formulation to be commercially competent, it must provide a shelf life of a minimum of 6 months to 1 year. For example, *Bacillus*, *Pseudomonas*, and *Ochrobactrum* formulations have a shelf life of 1 year or more in several bio-formulations (Trivedi et al. 2005). Sawdust, talc powder, and rice husk were used as bio-formulations for *B. amyloliquefaciens*, *Serratia marcescens*, and *B. pumilus*, which have a shelf life of up to 9 months of storage (Chakraborty et al. 2013).

4.7 Interaction of Beneficial Microbes with Crops

The current scenario of the agricultural sector is vastly dependent on chemical-based fertilizers and pesticides, which affect the nutritional quality, fitness, and productivity of the crops. Furthermore, when these chemicals are released continuously, toxic compounds such as metals accumulate in the soil and are transferred to the plants over a long period, which eventually transfers into living organisms as food sources. Therefore, it becomes necessary to bring out alternatives to chemical pesticides/fertilizers to improve agricultural production. The rhizosphere (plant-root interface) is an important part of the plant root system which is enriched by the population of special microorganisms (McNear Jr 2013). They have the properties to increase plant growth development, prevent disease, remove toxic compounds, and absorb nutrients to plants. Plants are densely inhabited by microbes both under the ground and above the ground that serve their mutual benefit. The microbes that are colonized in the plants can be classified as endophytes, which are located inside the plant tissue, and phyllosphere—which are present on the leaf and rhizospheric soils that surround the root of a plant (Mendes et al. 2013). Microbial interactions with different host plants are shown in Table 4.1.

4.7.1 Endophytic Microbiomes

Endophytic microbiomes are microorganisms beneficial to plant development that enter the plant's internal tissue (i.e., roots, stem flower, fruit, or seed) either vertically or horizontally (White et al. 2019). A large number of endophytic microbiomes have

Table 4.1 List of microbial techniques used in a variety of crops applied and results reported

Sl no.	Microbial techniques used	Crop where applied	Result reported	References
1	*Rhizobium*	Cow pea, gram, pea, soybean	Maintain soil fertility, increase the yield	Mahmud et al. (2021)
2	*Azotobacter*	Mustard sun-flower, grapes, sugarcane, papaya, coconut	Promotion of growth sub-stances such as indole acetic acid, gibberellic acid, etc., destroyed plant pathogen	Mahmud et al. (2021)
3	*Azospirillum*	Maize, rice, wheat, oil seeds	Enhance the intake of both minerals and water, increase roots developments, and improves the vegetative growth and yield of crops	Mahmud et al. (2021)
4	*Azolla*	Rice	Use as green manure due to their large biomass	Mahmud et al. (2021)
5	*Blue-green algae*	Rice, banana	Promote the production of growth substance	Mahmud et al. (2021)
6	*Phosphate solu-bilizing micro-organism (PSM)*	Wheat, potato, rice, millet	Promote the production of phytohormone	Mahmud et al. (2021)
7	PGPR	Used in all crops	Promotes plant growth through increased phytohor-mones and water and min-erals uptake	Mahmud et al. (2021)
8	*Pseudomonas fluorescens*	Soybean	Increased plant growth	Khan et al. (2009)
9	*Trichoderma asperellum*	Blackberry	Growth promotion, increase fruit weight and yield	Yarzábal and Chica (2021)
10	*Rhizobacteria*	Oca, ullucu, mashua	Rhizobacteria diversity	Yarzábal and Chica (2021)
11	*Consortia of microorganisms*	Strawberry	Increased root growth and leaves per plant	Yarzábal and Chica (2021)
12	*Native rhizo-bium* spp.	Common bean	Increased nodulation, bio-mass increased	Yarzábal and Chica (2021)
13	*Ochrobactrum intermedium*	Sunflower	Increased plant growth and decreased Cr uptake	Khan et al. (2009)
14.	*Pseudomonas* spp.	Soyabean, mungbean, wheat	Promotes growth of plants	Khan et al. (2009)

been identified belonging to different species of genera including *Azoarcus*, *Achromobacter*, *Burkholderia*, *Nocardioides*, *Herbaspirillum*, *Pantoea*, *Klebsiella*, *Gluconacetobacter*, *Enterobacter*, *Microbispora*, *Micromonospora*, *Planomonospora*, *Pseudomonas*, and *Streptomyces* (Kandel et al. 2017).

4.7.1.1 Applications

In natural communities, plants maintain symbiotic associations with endophytic microbes, which support growth and protect plants from biotic and abiotic stresses (Zhang et al. 2022). However, symbiotic microorganisms may be lost as a result of domestication and extensive cultivation. In a crop experiment with *Nicotiana attenuate*, an annual wild tobacco plant, it was discovered that ongoing cultivation and seed cleaning led to the loss of symbiotic microbes and an increase in diseases caused by fungi including *Fusarium* and *Alternaria* (White et al. 2019). Investment of those endophytic microbes from wild populations of the tobacco was applied to seedlings in cultivation which resulted in resistance to the disease.

4.7.1.2 Mechanism

One of the many ways endophytes improve plant health is by reducing the growth and capability of pathogens. There are several mechanisms involved, including direct antagonism with pathogens for space and nutrients through the production of antimicrobial metabolites and the upregulation of host defense genes (Irizarry and White 2017). Bacterial endophytes of the genus *Pseudomonas*, including *P. aeruginosa* and *P. fluorescens*, produce a variety of antifungal compounds including phenazine-1-carboxylic acid, 2,4-diacetylphloroglucinol, pyrrolnitrin, and volatiles like hydrogen cyanide compounds that significantly inhibit the growth of fungal pathogens (Tamošiūnė et al. 2018). Genus *Bacillus* species are important disease control agents because they organize a variety of biologically active molecules that are potential inhibitors of phytopathogens (Miljaković et al. 2020). Different types of lipopeptides they produce induce leakage into the fungal hyphal membrane which greatly reduces their virulence as plant pathogens (Fira et al. 2018). This can lead to a "quorum-quenching effect" where pathogenic fungi persist rather than cause disease (White et al. 2019). Several antifungal compounds produced by endophytes target the membranes of fungi, reducing nutrient leakage, which results in reduced virulence of the fungi. Endophytic symbionts can also improve plant resistance and protect plants against a broad spectrum of pathogens pathways and ethylene or PR proteins (White et al. 2019).

4.7.2 Phyllospheric Microbiome

The phyllosphere (stem-colosphere, leaf-phylloplane, flower-anthosphere, and fruit-carposphere) is the common site for interaction between beneficial microbes and plants (Yadav 2021). Phyllospheric microbiomes are most adapted to the plant surface and can tolerate more abiotic stress of UV radiation and high temperature (30–50 °C) as compared to endophytic and rhizospheric microbiomes (Yadav 2020).

An enormous amount of research has been done on the phyllospheric microbiome and its applications for PGP and crop protection through various plant growth-promoting mechanisms. Numerous beneficial microbes such as *Achromobacter*, *Acinetobacter*, *Agrobacterium*, *Arthrobacter*, *Bacillus*, *Beijerinckia*, *Burkholderia*, *Flexibacterium*, *Methylobacterium*, *Micrococcus*, *Micromonospora*, *Nocardioides*, *Pantoea*, *Penicillium*, *Planomonospora*, *Pseudomonas*, *Streptomyces*, and *Xanthomonas* are reported to be an inhabitant of phyllosphere part (Sivakumar et al. 2020).

4.7.2.1 Mechanism

Microbial diversity and abundance in the phyllosphere are determined by leaf physiology (Sivakumar et al. 2020). It establishes the microhabitat where microorganisms adapt their physiology to survive in the habitat. Epiphytic microbes were formed in a colonial form, which provided the microorganisms with protection from harsh microorganisms. Typically, bacteria develop sizable colonial associations on the leaf surface, particularly on trees of epidermal cells along the veins (Hoagland et al. 2007). Sugar and water are particularly abundant as nutrient content in epidermal grooves. Normally, the entire leaf surface is covered in a waxy cuticle that prevents the permeability and wetting of the leaf surface and regulates the colonization of microbes on the phyllosphere. This section, however, has a less waxy cuticle (Carlos et al. 2016). Water droplets on the leaf surface stretch the waxy cuticle and improve the permeability by which compounds are diffused from the apoplast to the phyllosphere surface (Sivakumar et al. 2020). The compounds and water are filtrates on the phyllosphere, making nutrients available to microorganisms. The flow of water from the stomata (transpiration) to the guard cells and their surface cuticle favours aggregation of microrganisms on leaf surface (Schönherr 2006). Therefore, the high permeation of the cuticle layer allows microbes to live in dense populations. In addition, surface bacterial produce certain compounds such as bio-surfactants (syringafactin produced by *Pseudomonas syringae*) which can modify the leaf cuticle surfaces and establish its association (Van der Wal and Leveau 2011). This may facilitate water availability and replace sugar availability which may improve living conditions for the growth of epiphytic bacteria. Epiphytes such as *Pseudomonas* sp., *Stenotrophomonas* sp., and *Achromobacter* increase the water permeability of lipophilic cuticles present in *Hedera* and *Prunus*, which increases the availability in the phyllosphere that will improve epiphytic fitness at the leaf surface (Schreiber 2005).

4.7.3 Rhizospheric Microbiome

The most important relationship between plants and microbes is the interaction of soil microbes with the root ecosystem of the plant as in the rhizospheric microbiome.

Microbes that are present in the soil ecosystem are attracted to the rhizospheric region due to the release of roots. Rhizospheric microbes such as *Methylobacterium*, *Pseudomonas*, *Serratia*, *Rhizobium*, *Paenibacillus*, *Erwinia*, *Enterobacter*, *Flavobacterium*, *Bacillus*, *Azospirillum*, *Burkholderia*, *Arthrobacter*, *Alcaligenes*, and *Acinetobacter* allow the growth of a large number of plant species (Sivakumar et al. 2020).

4.7.3.1 Mechanism

Plant species present in an ecosystem exhibit many distinctive traits, such as growth patterns, behavior under stress, and its mitigation. These characteristics of plant species allow occupying different niches in space and time. This causes a high diversity of plant species, which can exist in correlation to a given habitat (Kraft et al. 2015). Sometimes, interaction of microbes with plants induce defense responses and signalingcascades in plant system. The defense of the plant immune system is based on the recognition of response pattern-triggered immunity (Denancé et al. 2013). The first line of defense action is thought to be PTI, involving protein recognition receptors (PRRs) that are present on the surface of the cells. The conserved pattern known as the pathogen which is associated molecular pattern serves as the binding sites for the PRR initiating the microbial growth. However, some of the pathogens can cause downregulations of PTI by secreting effector proteins (Gao et al. 2013). This can lead to the activation of the second line of defensive actions, i.e., ETI, where nucleotides are present.

Advances in the associations of plants and microbes in the rhizosphere have increased the demands for the development and commercialization of microbe-based inoculants/formulations (Yadav and Chandra 2014). Microbial inoculants are agricultural amendments that can be applied to soil or plant to increase crop productivity. These inoculants contain one or more microbes. This can be facilitated in some ways, including the introduction of new microbial species into the rhizosphere, manipulating environmental parameters such as moisture, pH, temperature, etc., and soil microbial growth (Finkel et al. 2017; Pineda et al. 2017). During the inoculation of bacterial formulations into the rhizosphere, sophisticated and complex interactions between plant-microbe and microbe-microbe occur, which would have been controlled by the establishment of chemical communication in the rhizosphere. The process of root exudation actively engages itself in indicated cascades in the rhizosphere due to inoculation. These associations confer resistance to plant pathogens. A competitive pressure always exist among the introduced microbial inoculants for nutrients selectivity, chemotaxis, and root colonization to make its place in the rhizosphere along with native microbial communities. (Sikora and Dababat 2007). The plant-microbe interaction is shown in Fig. 4.1.

There are two types of microbial interactions, viz., phyllosphere and rhizosphere. Phyllosphere is those interactions that occur above the ground, and rhizosphere are those interactions that occur below the ground. In phyllosphere, if interactions occur on leaf surface, it is called phylloplane, and if in fruits and flowers, it is called

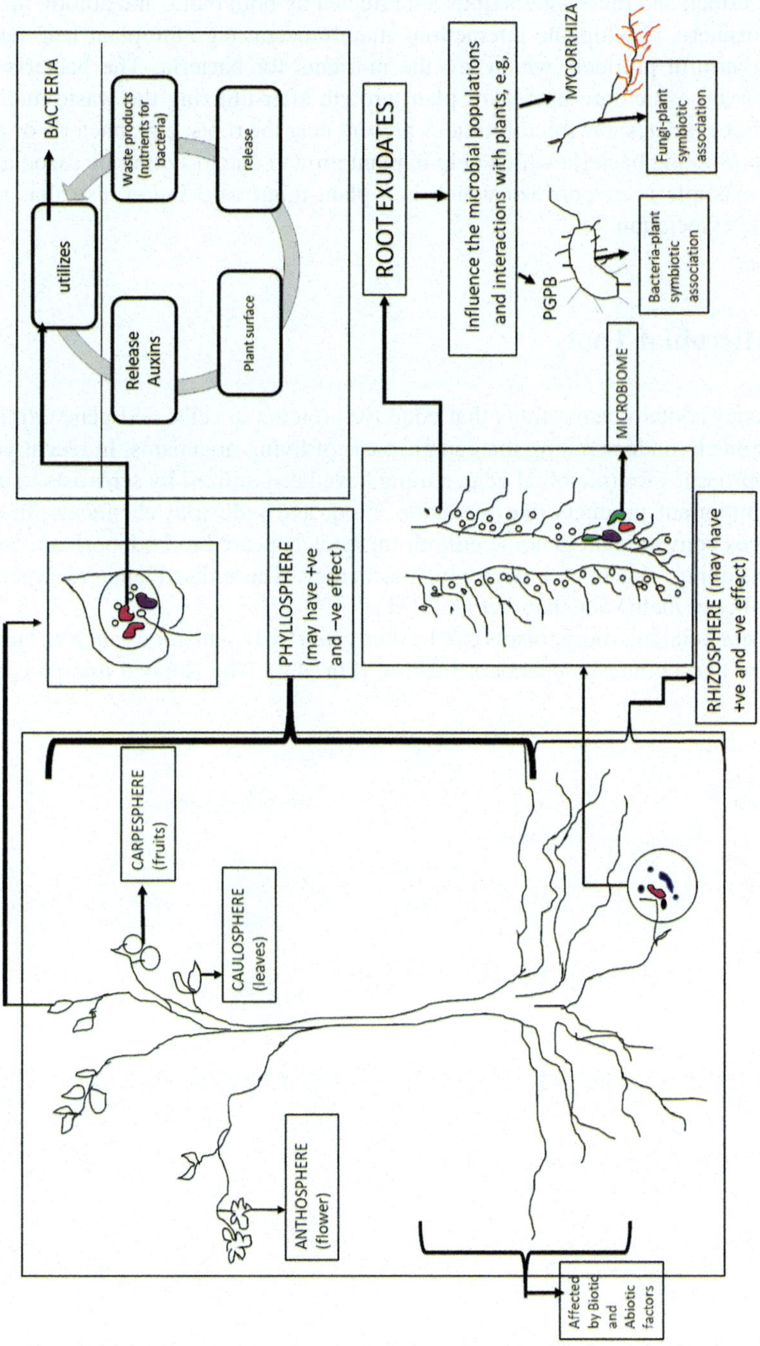

Fig. 4.1 Schematic diagram of plant-microbe interaction

carposphere and anthosphere. Plant-microbe interactions may have a positive or a negative effect, and these interactions are affected by both biotic and abiotic factors. In phyllosphere, phylloplane interactions mainly occur, e.g., the plant leaf surface releases harmful products which are the nutrients for bacteria. The bacteria will release auxins which are useful for plant growth after utilizing the waste material. Due to root exudates, the microbiome is present near the roots, e.g., bacteria or plant growth-promoting bacteria which help in plant growth due to symbiotic association; another example is mycorrhiza which is a plant-fungi association also that has a symbiotic association.

4.8 Microbial Tools

Many genes contain instructions that code for proteins in cells, and gene editing is the keystone to understanding the genetic basis of living organisms. In recent years, many approaches for microbial gene editing have been utilized by scientists to make several important products like medicine, food, and industrial chemicals, in such approaches conventional genome editing methods appeared to be laborious, costly, time-consuming, and have a degree of inaccuracy. These disadvantages urged the need for better methods (Zhang et al. 2021).

The beneficial microorganisms can be cultivated by two methods such as culture-dependent and culture-independent method (Fig. 4.2). The isolated microbes from

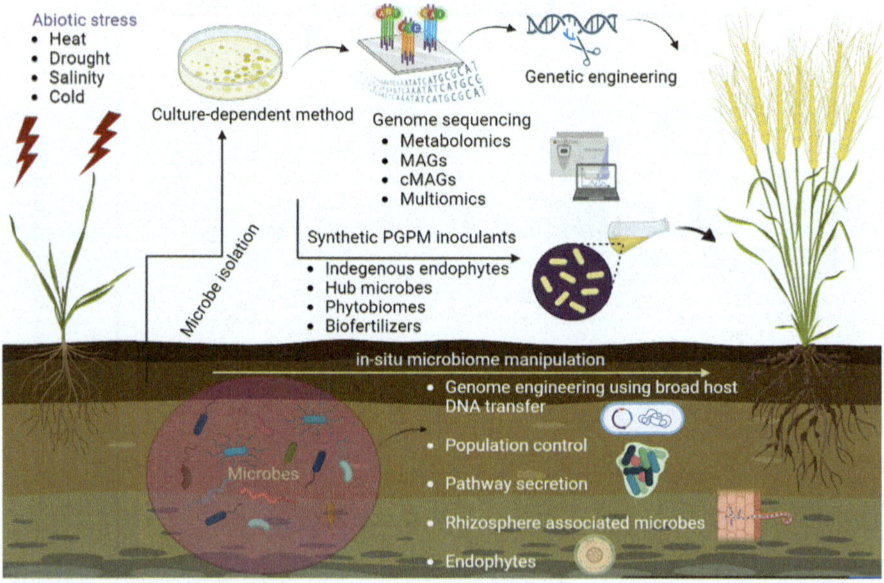

Fig. 4.2 Application of different microbial tools in agricultural crop development and challenges in the development of sustainable crops using microbial tools

culture-dependent method can then be further analyzed for whole genome sequencing and library constructions through next-generation sequencing tools. In bacteria and archaea, the CRISPR system is an adaptive immune system that protects from invaders (Grissa et al. 2007). Among the six major types of CRISPR cas system, cas9 (Type II), cas12a (Type V), and their mutant variants showed potential application in genome editing, gene regulation, DNA detection, DNA imaging, etc. (Tang and Fu 2018). The advantage of the CRISPR-Cas system is that genome editing at multi-loci is concise in one transformation without the involvement of marker genes for selection, thus saving time and labor in metabolic engineering work. For example, GTR-CRISPR is employed for lipid metabolism in *S. cerevisiae* for the production of free fatty acid (FFA) (Zhang et al. 2019). Secondary metabolites such as alkaloids, pigments, antimicrobials, etc. are used in food and biochemical industries. Microbial secondary metabolite (SM) isolation is tedious; hence the development of omics-based techniques like genomics and metagenomics enables the complete identification and structural information of novel SMs coded in microbe's DNA using highly sophisticated instrumentation (Tong et al. 2019). Metabolomics tools such as mass spectrometry (MS) and nuclear magnetic resonance (NMR) have been used to identify both targeted and untargeted secondary metabolites (Dettmer et al. 2007). However, these days many metabolomics databases such as MELTIN, MassBank, and Biological Magnetic Resonance Data Bank are reused for the identification of metabolites (Brown et al. 2009). With the discovery of high-throughput sequencing techniques such as illumine, metagenomics assembly of composite genomes called metagenome-assembled genomes (MAGs) allows to investigate of complex microbial communities and their interaction with the environment they live in (Sharon and Banfield 2013). In such cases, to resolve lineage of closely related organisms that complicate metagenome assembly and prevent complete metagenome-assembled genome, the MAG Phase is developed that separate lineages by discerning variant haplotypes across hundreds of kilobases of genome sequences (Bowers et al. 2017). Thus by using genome editing and omics approaches, microbial organisms can be exploited to develop multiple stress-tolerant crops.

Microbial too is a great approach to sustaining the agricultural productivity. To obtain high-yield agricultural output, however, there are some obstacles associated with the use of microbial tools that must be solved. The accuracy determination on quantification of the microbial inoculum, the reliability of the microbial formulation, and the difficulty to prove the efficiency of the homemade microbial formulation as compared to commercial formulation are the major challenges in the usage of microbial tools in agricultural development (Koskey et al. 2021). The microbiome inoculants containing numerous exotic microbial strains are highly in demand to develop engineered bio-stimulator or biofertilizers for industrial purposes (Lobo et al. 2019). However, the limitation of knowledge on the toxic effect of microbial strains could cause the emergence of phytopathogens leading to huge crop loss (Schwartz et al. 2006). Therefore, proper quality control standards and storage measurements should be taken care of to avoid any hazardous effect on crop production as well as on farmer's livelihoods (Raimi et al. 2019). Moreover, the usage of PGPM inoculants in smallholder contexts is mostly unaccountable despite their significance in scaling up agriculture, and more studies should concentrate on quantifying their use, adoption, and overall impact on soil, crops, and farmers' livelihood (Oruru and Njeru 2016). The industrial development of bioinoculants that could be applied to a wide range of crops cultivated in varied geographic and climatic regions faces difficulty (Santos et al. 2019). The selection of specific microbial consortia for a wide range of crops for a specific location is quite challenging, and therefore, a lot more research studies require resolving this problem. Farmers sometimes utilize "blanket" methods to address their field challenges, so it is important to promote the use of specific approaches that offer particular solutions in particular places (Koskey et al. 2021). Thus, it is very much essential to formulate some innovative microbial inoculants which are viable for long-term usage and which could be used in various geographic soil ranges under climatic changes.

4.9 Future Perspectives and Conclusion

The advancement in biotechnological research has made benchmark progress in the development of agricultural product development from the past few years. China, for instance, registered more than 800 inoculant-related patents in 2019, while India had more than 100 patents (Santos et al. 2019). Therefore, it is anticipated that additional bioinoculants will be developed in the upcoming years through advanced genetic engineering and microbial research. Although the use of biofertilizers is rapidly increasing, it still has several loopholes which need to overcome to sustain crop productivity. Along with microbial strain specificity, several other factors including environmental factors and soil composition as well influence the effectiveness of bio-formulation on crop improvement or soil quality improvement. In order to preserve prolonged shelf life and microbial viability without the need for a specialized storage facility, a good microbial inoculum should be packaged in a carrier material that offers an ideal microenvironment (pH, water, and carbon content) for bacteria (Soumare et al. 2020).

Understandably, creating a universal microbial inoculum is not feasible to apply for all plant species. However, formulation of the spectrum of bioinoculants is possible in today's date for different climatic conditions, varietal crops, and soil conditions through genetic engineering or several in situ microbial approaches like tillage. The soil beneficial bacterium can be isolated and treated through culture-dependent and culture-independent methods. The microbes having plant disease resistance can be utilized through next-generation whole genome sequencing followed by omics techniques or genetic engineering approaches. Several advanced genome editing tools like CRISPR/Cas and RNAi have made it easier to develop sustainable disease or pest-resistant crop development. Prior information on microbial interaction with the host plant and other indigenous microbes is a much considered practice in the case of developing synthetic microbial inoculum. Thus, by assembling beneficial microbial communities with advanced genomic tools, sustainable agricultural products can be developed which would be disease/pest resistant as well as can endure adverse climatic conditions.

References

Ahemad M, Khan MS (2009) Effect of insecticide-tolerant and plant growth promoting Mesorhizobium on the performance of chickpea grown in insecticide stressed alluvial soils. J Crop Sci Biotechnol 12:213–222

Ahemad M, Kibret M (2013) Mechanisms and applications of plant growth promoting rhizobacteria: current perspective, Journal of King Saud University Science 1018–3647

Allen MF (2011) Linking water and nutrients through the vadose zone: a fungal interface between the soil and plant systems. J Arid Land 3:155–163

Ambardar VK, Sood AK (2010) Suitability of different growth substrate for mass multiplication of bacterial antagonists. Indian Phytopathol 63:380–383

Ambrosio R, Ortiz-Marquez JC, Curatti L (2017) Metabolic engineering of a diazotrophic bacterium improves ammonium release and biofertilization of plants and microalgae. Metab Eng 40: 59–68

Antoun H, Kloepper JW (2001) Plant-growth-promoting rhizobacteria and medicinal plants, New York, 1477–1480

Araus J, Slafer G, Reynolds M, Royo C (2002) Plant breeding and drought in C3 cereals: what should we breed for? Ann Bot 89:925–940

Ardakani SS, Heydari A, Khorasani N, Arjmandi R (2010) Development of new bio-formulations of Pseudomonas fluorescens and evaluation of these products against damping-off of cotton seedlings. J Plant Pathol 92:83–88

Arora NK, Khare E, Naraian R, Maheswari DK (2008) Saw-dust as a superior carrier for production of multipurpose bio-inoculant using plant growth promoting rhizobial and Pseudomonad strains and their impact on productivity of Trifolium repense. Curr Sci 95:90–94

Augé RM (2001) Water relations, drought and vesicular arbuscular mycorrhizal symbiosis. Mycorrhiza 11(1):3–42

Azcón-Aguilar C, Barea JM (2015) Nutrient cycling in the mycorrhizosphere. J Soil Sci Plant Nutr 15(2):372–396

Backer R, Rokem JS, Ilangumaran G, Lamont J, Praslickova D, Ricci E, Subramanian S, Smith DL (2018) Plant growth-promoting rhizobacteria: context, mechanisms of action, and roadmap to commercialization of biostimulants for sustainable agriculture. Front Plant Sci 9:1473

Bal SK, Prasad JVNS, Singh VK (2022) Heat wave 2022 Causes, impacts and way forward for Indian Agriculture. Technical Bulletin No ICAR/CRIDA/ TB/01/2022 ICAR-Central Research Institute for Dryland Agriculture Hyderabad Telangana India p50

Balla A, Silini A, Cherif-Silini H, Ali CB, Alenezi FN, Lassaad B (2022) Recent advances in encapsulation techniques of plant growth-promoting microorganisms and their prospects in the sustainable agriculture. Appl Sci 12:9020

Baroowa B, Gogoi N (2014) Biochemical changes in black gram and green gram genotypes after imposition of drought stress. J Food Legum 27:350–353

Basu BS, Mandal NC (2017) Field application of two plant growth promoting rhizobacteria with potent antifungal properties. Rhizosphere 3(1):170–175. ISSN 2452-2198

Bati CB, Santilli E, Lombardo L (2015) Effect of arbuscular mycorrhizal fungi on growth and on micronutrient and macronutrient uptake and allocation in olive plantlets growing under high total Mn levels. Mycorrhiza 25(2):97–108

Berruti A, Lumini E, Balestrini R, Bianciotto V (2016) Arbuscular mycorrhizal fungi as natural biofertilizers: let's benefit from past successes. Front Microbiol 6:1559

Bhadrecha P, Bala M, Khasa YP, Arshi A, Singh J, Kumar M (2020) *Hippophae rhamnoides* L. rhizobacteria exhibit diversified cellulase and pectinase activities. Physiol Mol Biol Plants 26(5):1075–1085

Bharadwaj DP, Lundquist PO, Alström S (2008) Arbuscular mycorrhizal fungal spore-associated bacteria affect mycorrhizal colonization, plant growth and potato pathogens. Soil Biol Biochem 40(10):2494–2501

Bhattacharyya PN, Jha DK (2012) Plant growth-promoting rhizobacteria (PGPR): emergence in agriculture. World J Microbiol Biotechnol 28:1327–1350

Bisht N, Chauhan PS (2020) Excessive and disproportionate use of chemicals cause soil contamination and nutritional stress. In: Larramendy ML, Soloneski S (eds) Soil contamination threats and sustainable solutions. IntechOpen, London. https://doi.org/10.5772/intechopen.94593

Biswal T, Panda RB (2018) Impact of fly ash on soil properties and productivity. JAEB 11:275–283. https://doi.org/10.30954/0974-1712.04.2018.8

Bortolami G, Gambetta GA, Delzon S, Lamarque LJ, Pouzoulet J, Badel E, Burlett R, Charrier G, Cochard H, Dayer S, Jansen S, King A, Lecomte P, Lens F, Torres-Ruiz JM, Delmas CEL (2019) Exploring the hydraulic failure hypothesis of esca leaf symptom formation. Plant Physiol 181(3):1163–1174

Bowers RM, Kyrpides NC, Stepanauskas R et al (2017) Minimum information about a single amplified genome (MISAG) and a metagenome-assembled genome (MIMAG) of bacteria and archaea. Nat Biotechnol 35:725–731

Boyer JS (1982) Plant productivity and environment. Science 218:443–448. https://doi.org/10.1126/science.218.4571.443

Brown LR (2008) Plan B 3.0: mobilizing to save civilization (substantially revised). WW Norton & Company, New York, NY

Brown M, Dunn WB, Dobson P, Patel Y, Winder CL, Francis-McIntyre S et al (2009) Mass spectrometry tools and metabolite-specific databases for molecular identification in metabolomics. Analyst 134(7):1322–1332

Burén S, Jiménez-Vicente E, Echavarri-Erasun C, Rubio LM (2020) Biosynthesis of nitrogenase cofactors. Chem Rev 120(12):4921–4968. https://doi.org/10.1021/acs.chemrev.9b00489

Carlos A, Ruiz-Pérez SR, Zambranoa MM (2016) Microbial and functional diversity within the phyllosphere of Espeletia species in an Andean high-mountain ecosystem. Appl Environ Microbiol 82:6

Carvalho LC, Amâncio S (2018) Cutting the gordian knot of abiotic stress in grapevine: from the test tube to climate change adaptation. Physiol Plant. https://doi.org/10.1111/ppl.12857

Chakraborty U, Chakraborty BN, Basnet M, Chakraborty AP (2009) Evaluation of *Ochrobactrum anthropi* TRS-2 and its talc based formulation for enhancement of growth of tea plants and management of brown root rot disease. J Appl Microbiol 107:625–634. https://doi.org/10.1111/j.1365-2672.2009.04242.x

Chakraborty U, Chakraborty BN, Chakraborty AP, Sunar K, Dey PL (2013) Plant growth-promoting rhizobacteria mediated improvement of health status of tea plants. Indian J Biotechnol 12:20–31

Chandini KR, Kumar R, Prakash O (2019) The impact of chemical fertilizers on our environment and ecosystem. Research trends in environmental Sciences, 2nd edition, pp 69–86

Chaudhary T, Dixit M, Gera R, Shukla AK, Prakash A, Gupta G, Shukla P (2020) Techniques for improving formulations of bioinoculants. 3 Biotech 10(5):199. https://doi.org/10.1007/s13205-020-02182-9

Collin LJ, diCenzo GC, Walker VK (2021) Cold acclimation and prospects for cold-resilient crops. Plant Stress 2:100028., ISSN 2667-064X. https://doi.org/10.1016/j.stress.2021.100028

De Oliveira ED, Bramley H, Siddique KH, Henty S, Berger J, Palta JA (2013) Can elevated CO2 combined with high temperature ameliorate the effect of terminal drought in wheat? Funct Plant Biol 40:160–171. https://doi.org/10.1071/FP12206

Denancé N, Sánchez-Vallet A, Goffner D, Molina A (2013) Disease resistance or growth: the role of plant hormones in balancing immune responses and fitness costs. Front Plant Sci 4:155. https://doi.org/10.3389/fpls.2013.00155

Descamps C, Quinet M, Jacquemart AL (2021) Climate change-induced stress reduce quantity and Alter composition of nectar and pollen from a bee-pollinated species (*Borago officinalis*, Boraginaceae). Front Plant Sci 11(12):755843. https://doi.org/10.3389/fpls.2021.755843

Dettmer K, Aronov PA, Hammock BD (2007) Mass spectrometry based metabolomics. Mass Spectrom Rev 26(1):51–78

Diaz-Zorita M, Fernandez-Canigia M (2009) Field performance of a liquid formulation of *Azospirillum brasilense* on dryland wheat productivity. Eur J Soil Biol 45:3–11

van Dijk M, Morley T, Rau ML, Saghai Y (2021) A meta-analysis of projected global food demand and population at risk of hunger for the period 2010–2050. Nat Food 2:494–501

Dinneny JR, Long TA, Wang JY, Jung JW, Mace D, Pointer S, Barron C, Brady SM, Schiefelbein J, Benfey PN (2008) Cell identity mediates the response of Arabidopsis roots to abiotic stress. Science 320:942–945

Easterling WE, Aggarwal PK, Batima P, Brander KM, Erda L, Howden SM, Kirilenko A, Morton J, Soussana JF, Schmidhuber J (2007) Food, fibre and forest products. Clim Chang 273:313

Elhaissoufi W, Ghoulam C, Barakat A, Zeroual Y, Bargaz A (2021) Phosphate bacterial solubilization: a key rhizosphere driving force enabling higher P use efficiency and crop productivity. J Adv Res 23(38):13–28

Etesami H, Jeong BR, Glick BR (2021) Contribution of arbuscular Mycorrhizal fungi, phosphate-solubilizing bacteria, and silicon to P uptake by plant. Front Plant Sci 12:699618

Fahad S, Bajwa AA, Nazir U, Anjum SA, Farooq A, Zohaib A et al (2017) Crop production under drought and heat stress: plant responses and management options. Front Plant Sci 8:1147. https://doi.org/10.3389/fpls.2017.01147

Fernández Bidondo L, Silvani V, Colombo R, Pérgola M, Bompadre J, Godeas A (2011) Pre-symbiotic and symbiotic interactions between *Glomus intraradices* and two *Paenibacillus* species isolated from AM propagules. *In vitro* and *in vivo* assays with soybean (AG043RG) as plant host. Soil Biol Biochem 43:1866–1872. https://doi.org/10.1016/j.soilbio.2011.05.004

Figueiredo MVB, Seldin L, Araujo FF, Mariano RLR (2011) Plant growth promoting rhizobacteria: fundamentals and applications. In: Maheshwari DK (ed) Plant growth and health promoting bacteria. Springer-Verlag, Berlin Heidelberg, pp 21–42

Finkel OM, Castrillo G, Herrera Paredes S, Salas González I, Dangl JL (2017) Understanding and exploiting plant beneficial microbes. Curr Opin Plant Biol 38:155–163. https://doi.org/10.1016/j.pbi.2017.04.018

Fira DJ, Dimkić I, Berić T, Lozo J, Stanković S (2018) Biological control of plant pathogens by *Bacillus* species. *J Biotechnol* 285:44–55

Fontana DC, de Paula S, Torres AG, de Souza VHM, Pascholati SF, Schmidt D, Dourado Neto D (2021) Endophytic fungi: biological control and induced resistance to phytopathogens and abiotic stresses. Pathogens 10(5):570

Franco-Correa M, Quintana A, Duque C et al (2010) Evaluation of actinomycete strains for key traits related with plant growth promotion and mycorrhiza helping activities. Appl Soil Ecol 45(3):209–217

Gamini Seneviratne MLMAW, Weerasekara KACN, Seneviratne JS et al (2010) Importance of biofilm formation in plant growth promoting rhizobacterial action. In: Maheshwari DK (ed) Plant growth and health promoting bacteria, microbiology monographs. SpringerVerlag, Berlin Heidelberg, p 18

Gao X, Chen X, Lin W, Chen S, Lu D, Niu Y et al (2013) Bifurcation of Arabidopsis NLR immune signaling via Ca2C-dependent protein kinases. PLoS Pathog 9:e1003127. https://doi.org/10.1371/journal.ppat.1003127

Geddes BA, Ryu MH, Mus F et al (2015) Use of plant colonizing bacteria as chassis for transfer of N2-fixation to cereals. Curr Opin Biotechnol 32:216–222

Gianinazzi S, Golotte A, Binet MN, Van Tuinen D, Redecker D, Wipf D (2010) Agroecology: the key role of arbuscular mycorrhizas in ecosystem services. Mycorrhiza 20:519–530

Gianinazzi S, Schuepp H, Barea JM, Haselwandter K, Mycorrhizal (2002) Technology in agriculture: from genes to bioproducts, Birkha user, Basel, 296

Glick BR (2014) Bacteria with ACC deaminase can promote plant growth and help to feed the world. Microbiol Res 169:30–39

Glick BR (2015) Introduction to plant growth-promoting bacteria. In: Beneficial plant-bacterial interactions, Springer Int Publishing, Switzerland, 1–28

Godoy F, Olivos-Hernández K, Stange C, Handford M (2021) Abiotic stress in crop species: improving tolerance by applying plant metabolites. Plants (Basel) 10(2):186

Gopalakrishnan S, Sathya A, Vijayabharathi R, Srinivas V (2016) Formulations of plant growth-promoting microbes for field applications. In: Singh D, Singh H, Prabha R (eds) Microbial inoculants in sustainable agricultural productivity. Springer, New Delhi

Gouda S, Das G, Sen SK, Shin HS, Patra JK (2016) Endophytes: a treasure house of bioactive compounds of medicinal importance. Front Microbiol 29(7):1538

Gouda S, Kerry RG, Das G, Paramithiotis S, Shin HS, Patra JK (2017) Revitalization of plant growth promoting rhizobacteria for sustainable development in agriculture. Microbiol Res 206: 131–140

Grissa I, Vergnaud G, Pourcel C (2007) CRISPR finder: a web tool to identify clustered regularly interspaced short palindromic repeats. Nucleic Acids Res 35:W52–W57

Grover M, Shrey B, Sharma A, Sharma P, Singh J, Nain L (2021) PGPR mediated alterations in root traits: way toward sustainable crop production frontiers in sustainable food systems 4 ISSN 2571–581

Habazar T, Yanti Y, Ritonga C (2014) Formulation of indigenous rhizobacterial isolates from healthy soybean's root, which ability to promote growth and yield of soybean. Int J Adv Sci Eng Inform Tech 4:5

Hakim S, Naqqash T, Nawaz M, Laraib I, Siddique M, Zia R, Mirza M, Imran A (2021) Rhizosphere engineering with plant growth-promoting microorganisms for agriculture and ecological sustainability. Front Sustain Food Syst 5:617157

Hasanuzzaman M, Nahar K, Alam MM, Roychowdhury R, Fujita M (2013) Physiological, biochemical and molecular mechanisms of heat stress tolerance in plants. Int J Mol Sci 14:9643–9684

Haskett TL, Karunakaran R, Bueno Batista M, Dixon R, Poole PS (2022) Control of nitrogen fixation and ammonia excretion in Azorhizobium caulinodans. PLoS Genet 18(6):e1010276

Hoagland RE, Douglas Boyette C, Weaver MA, Abbas HK (2007) Bioherbicides: research and risks. Toxin Rev 26:313–342

Hossain MM, Sultana F (2020) Application and mechanisms of plant growth promoting fungi (PGPF) for Phytostimulation in S. In: Das K (ed) Organic Agriculture. IntechOpen, London. https://doi.org/10.5772/intechopen.92338

Igiehon NO, Babalola OO (2018) Below-ground-above-ground plant-microbial interactions: focusing on soybean, rhizobacteria and Mycorrhizal fungi. Open Microbiol J 31(12):261–279

Intergovernmental Panel on Climate Change (IPCC) (2007) Climate change: impacts, adaptation, and vulnerability. Cambridge University Press; Cambridge, UK: Contribution of Working Group II to the Fourth Assessment Report of the Intergovernmental Panel on Climate Change

IPCC (2002) Climate change and biodiversity. In: IPCC Technical Paper V. pp. 1–86

Irizarry I, White JF (2017) Application of bacteria from non-cultivated plants to promote growth, alter root architecture and alleviate salt stress of cotton. J Appl Microbiol 122:1110–1120

Jafari M, Yari M, Ghabooli M, Sepehri M, Ghasemi E, Jonker A (2018) Inoculation and co-inoculation of alfalfa seedlings with root growth promoting microorganisms (Piriformospora indica, Glomus intraradices and Sinorhizobium meliloti) affect molecular structures, nutrient profiles and availability of hay for ruminants. Anim Nutr 4:90–991

Jahan M, Mahallati MN, Amiri MB, Ehyayi HR (2013) Radiation absorption and use efficiency of sesame as affected by biofertilizers inoculation in a low input cropping system. Ind Crop Prod 43:606–611

Jeffries P, Barea JM (2000) Arbuscular mycorrhiza key component of sustainable plant-soil ecosystems. In: Hock B (ed) The Mycota, vol IX. Fungal Associations, Springer Verlag, pp 95–113

Jha CK, Saraf M (2015) Plant growth promoting rhizobacteria (PGPR): a review. E3. J Agric Res Dev 5:108–119

Kandel SL, Joubert PM, Doty SL (2017) Bacterial endophyte colonization and distribution within plants. Microorganisms 5(4):77. https://doi.org/10.3390/microorganisms5040077

Khan MS, Zaidi A, Wani PA (2006) Role of phosphate solubilizing microorganisms in sustainable agriculture—a review. Agron Sustain Dev 27:29–43

Khan MS et al (eds) (2009) Microbial strategies for crop improvement. Springer-Verlag Berlin Heidelberg

Khaosaad T, Garcia-Garrido JM, Steinkellner S, Vierheilig H (2007) Take all diseases is systemically reduced in roots of mycorrhizal barley plants. Soil Biol Biochem 39:727–734

Knack J, Wilcox L, Delaux PM et al (2015) Microbiomes of streptophyte algae and bryophytes suggest that a functional suite of microbiota fostered plant colonization of land. Int J Plant Sci 176:405–420

Koskey G, Mburu SW, Richard A, Mugendi NE, Maingi JM (2021) Potential use of beneficial microorganisms for soil amelioration, phytopathogen biocontrol, and sustainable crop production in smallholder agroecosystems. Front Sustain Food Syst 5

Kraft NJB, Godoy O, Levine JM (2015) Plant functional traits and the multidimensional nature of species coexistence. Proc Natl Acad Sci U S A 112:797–802. https://doi.org/10.1073/pnas.1413650112

Kumar A, Singh J (2020) Biofilms forming microbes: diversity and potential application. Plant Microbiomes Sustain Agric 25:173

Kumar M (2013) Crop plants and abiotic stress abiotic and abiotic stresses. Biomol Res Ther 3(1):1000e125

Kumar S, Bauddh K, Barman SC, Singh RP (2014) Amendments of microbial bio fertilizers and organic substances reduces requirement of urea and DAP with enhanced nutrient availability and productivity of wheat (*Triticum aestivum* L.). Ecol Eng 71:432–437

Kumar V (2014) Characterization, bio-formulation development and shelf-life studies of locally isolated bio- fertilizer strains. Octa J Environ Res 2:32–37

Kumar V, Gupta P (2008) Efficacy of flyash based Rhizobium on growth and incidence of powdery mildew in pea. Ann Plant Protect Sci 16:248–249

Kumaraswamy S, Shetty PK (2016) Critical abiotic factors affecting implementation of technological innovations in rice and wheat production: a review. Agric Rev 37(4):268–278

Kumari P, Singh A, Kharwar RN (2021) Chapter 18—Phytostimulation and ISR responses of fungi. In: Sharma VK, Shah MP, Parmar S, Kumar A (eds) Fungi bio-prospects in sustainable agriculture, environment and nano-technology. Academic Press, pp 459–473. ISBN 9780128213940

Kundan R, Pant G, Jadon N, Agrawal PK (2015) Plant growth promoting rhizobacteria: mechanism and current prospective. J Fertil Pestic 6(2):9

Li W, Cui X (2014) A special issue on plant stress biology: from model species to crops. Mol Plant 7:755–757

Llorente BE, Alasia MA, Larraburu EE (2016) Biofertilization with Azospirillum brasilense improves in vitro culture of Handroanthus ochraceus, a forestry, ornamental and medicinal plant. New Biotechnol 33:32–40. https://doi.org/10.1016/j.nbt.2015.07.006

Lobell DB, Banziger M, Magorokosho C, Vivek B (2011) Nonlinear heat effects on African maize as evidenced by historical yield trials. Nat Climate Change 1:42–45

Lobell DB, Field CB (2007) Global scale climate–crop yield relationships and the impacts of recent warming. Environ Res Lett 2:014002

Lobo CB, Tomás MSJ, Viruel E, Ferrero MA, Lucca ME (2019) Development of low-cost formulations of plant growth-promoting bacteria to be used as inoculants in beneficial agricultural technologies. Microbiol Res 219:12–25. ISSN 0944-5013

Machado RMA, Serralheido RP (2017) Soil salinity: effect on vegetable crop growth. Management practices to prevent and mitigate soil salinization. Horticulture 3(30):1–13

Mahmud AA, Upadhyay SK, Srivastava AK, Bhojiya AA (2021) Biofertilizers: a Nexus between soil fertility and crop productivity under abiotic stress. Curr Res Environ Sustain 3:100063

Maleki A, Naderi A, Naseri R, Fathi A, Bahamin S, Maleki R (2013) Physiological performance of soybean cultivars under drought stress. Bull Environ Pharmacol Life Sci 2:38–44

McNear DH Jr (2013) The rhizosphere—roots, soil and everything in between. Nat Educ Knowledge 4(3):1

Meeh GA, Stocker TF, Collins WD (2007) Climate change: the physical science basis. In: Fourth Assessment Report of the Intergovernmental Panel on Climate Change. Cambridge, UK: University Press

Meena B, Radhajeyalakshmi R, Marimuthu T, Vidhyasekaran P, Velazhahan R (2002) Biological control of groundnut late leaf spot and rust by seed and foliar application of a powder formulation of Pseudomonas fluorescens. Biocontrol Sci Tech 12:195–204

Melin P, Schnurer J, Hakansson S (2011) Formulation and stabilization of the biocontrol yeast Pichia anomala. Antonie Van Leeuwenhoek 99:107–112

Mendes R, Garbeva P, Raaijmakers JM (2013) The rhizosphere microbiome: significance of plant beneficial, plant pathogenic, and human pathogenic microorganisms. FEMS Microbiol Rev 37(5):634–663

Miljaković D, Marinković J, Balešević-Tubić S (2020) The significance of *Bacillus* spp. in disease suppression and growth promotion of Field and vegetable crops. Microorganisms 8(7):1037

Mitra D, Uniyal N, Panneerselvam P, Senapati A, Ganeshamurthy A, Kumar V, Jain D (2019) Role of mycorrhiza and its associated bacteria on plant growth promotion and nutrient management in sustainable agriculture. Int J Life Sci Appl Sci 1(1):1–11

Montesinos E (2003) Development, registration and commercialization of microbial pesticides for plant protection. Int Microbiol 6:245–252

Morelli M, Bahar O, Papadopoulou KK, Hopkins DL, Obradović A (2020) Editorial: role of endophytes in plant health and defense against pathogens. Front Plant Sci 11:1312

Nakkeeran S, Fernando WGD, Siddiqui ZA (2005) Plant growth-promoting rhizobacteria formulations and its scope in commercialization for the management of pests and diseases. In: Siddiqui ZA (ed) PGPR: biocontrol and biofertilization. Springer, Dordrecht, pp 257–296

Nandakumar R, Babu S, Viswanathan R, Sheela J, Raguchander T, Samiyappan R (2001) A new bio-formulation containing plant growth-promoting rhizobacterial mixture for the management of sheath blight and enhanced grain yield in rice. Biol Control 46:493–510

Nath R, Sharma G, Barooah M (2013) Screening of endophytic bacterial isolates of tea (Camellia sinensis L.) roots for their multiple plant growth promoting activities. Int J Agric Environ Biotechnol 6(2):211–215

Neumann E, George E (2004) Colonisation with the arbuscular mycorrhizal fungus Glomus mosseae (Nicol. & Gerd.) enhanced phosphorus uptake from dry soil in Sorghum bicolor (L.). Plant Soil 261:245–255

Nicolopoulou-Stamati P, Maipas S, Kotampasi C, Stamatis P, Hens L (2016) Chemical pesticides and human health: the urgent need for a new concept in agriculture. Front Public Health 4:148

Olanrewaju OS, Glick BR, Babalola OO (2017) Mechanisms of action of plant growth promoting bacteria. World J Microbiol Biotechnol 33(11):197

Ortega HE, Torres-Mendoza D, Cubilla-Rios L (2020) Patents on endophytic fungi for agriculture and bio- and phytoremediation applications. Microorganisms 8(8):1237

Oruru MB, Njeru EM (2016) Upscaling arbuscular mycorrhizal symbiosis and related agroecosystems services in smallholder farming systems. Biomed Res Int:1–12

Park YH, Kim Y, Mishra RC, Bae H (2017) Fungal endophytes inhabiting mountain-cultivated ginseng (Panax ginseng Meyer): diversity and biocontrol activity against ginseng pathogens. Sci Rep 7(1):16221

Paul S, Acharya GC, Pradhan A (2019) Fly ash utilization and its potential benefits in agriculture: a review. 47:87–95

Pineda A, Kaplan I, Bezemer TM (2017) Steering soil microbiomes to suppress aboveground insect pests. Trends Plant Sci 22:770–778

Pingali PL (2012) Green revolution: impacts, limits, and the path ahead. Proc Natl Acad Sci U S A 109:12302–12308

Pradhan GP, Prasad PV, Fritz AK, Kirkham MB, Gill BS (2012) Effects of drought and high temperature stress on synthetic hexaploid wheat. Funct Plant Biol 39:190–198

Raimi A, Ezeokoli O, Adeleke R (2019) High-throughput sequence analysis of bacterial communities in commercial biofertiliser products marketed in South Africa: an independent snapshot quality assessment. 3 Biotech 9:108

Ray DK, Gerber JS, MacDonald GK, West PC (2015) Climate variation explains a third of global crop yield variability. Nat Commun 6:5989

Raza A, Razzaq A, Mehmood SS, Zou X, Zhang X, Lv Y, Xu J (2019) Impact of climate change on crops adaptation and strategies to tackle its outcome: a review. Plants (Basel) 8(2):34

Rockström J, Williams J, Daily G et al (2017) Sustainable intensification of agriculture for human prosperity and global sustainability. Ambio 46:4–17

Rubin RL, Van Groenigen KJ, Hungate BA (2017) Plant growth promoting rhizobacteria are more effective under drought: a meta-analysis. Plant Soil 416:309–323. https://doi.org/10.1007/s11104-017-3199-8

Ruzzi M, Aroca R (2015) Plant growth-promoting rhizobacteria act as biostimulants in horticulture. Sci Hortic 196:124–134. https://doi.org/10.1016/j.scienta.2015.08.042

Sagar A, Rathore P, Ramteke PW, Ramakrishna W, Reddy MS, Pecoraro L (2021) Plant growth promoting rhizobacteria, arbuscular Mycorrhizal fungi and their synergistic interactions to counteract the negative effects of saline soil on agriculture: key macromolecules and mechanisms. Microorganisms 1 9(7):1491

Sanghera GS, Wani SH, Hussain W, Singh NB (2011) Engineering cold stress tolerance in crop plants. Curr Genomics 12:30–43

Santi C, Bogusz D, Franche C (2013) Biological nitrogen fixation in non-legume plants. Ann Bot 111(5):743–767

Santos MS, Nogueira MA, Hungria M (2019) Microbial inoculants: reviewing the past, discussing the present and previewing an outstanding future for the use of beneficial bacteria in agriculture. AMB Express 9:1–13. https://doi.org/10.1186/s13568-019-0932-0

Savci S (2012) Investigation of effect of chemical fertilizers on environment. Apcbee Procedia 1: 287–292

Saxena S (2014) Microbial metabolites for development of ecofriendly agrochemicals. Allelopath J 33:1–24

Schisler DA, Slininger PJ, Behle RW, Jackson MA (2004) Formulation of bacillus spp. for biological control of plant diseases. Phytopathology 94:1267–1271

Schlenker W, Roberts MJ (2009) Nonlinear temperature effects indicate severe damages to US crop yields under climate change. Proc Natl Acad Sci U S A 106:15594–15598

Schönherr J (2006) Characterization of aqueous pores in plant cuticles and permeation of ionic solutes. J Exp Bot 57:2471–2491

Schreiber L (2005) Polar paths of diffusion across plant cuticles: new evidence for an old hypothesis. Ann Bot 95:1069–1073. https://doi.org/10.1093/aob/mci122

Schroeder JI, Delhaize E, Frommer WB et al (2013) Using membrane transporters to improve crops for sustainable food production. Nature 497:60–66

Schwartz MW, Hoeksema JD, Gehring CA, Johnson NC, Klironomos JN, Abbott LK et al (2006) The promise and the potential consequences of the global transport of mycorrhizal fungal inoculum. Ecol Lett 9:501–515

Seema K, Mehta K, Singh N (2018) Studies on the effect of plant growth promoting rhizobacteria (PGPR) on growth, physiological parameters, yield and fruit quality of strawberry cv. Chandler. J Pharmacogn Phytochems 7(2):383–387

Setten L, Soto G, Mozzicafreddo M, Fox AR et al (2013) Engineering Pseudomonas protegens Pf-5 for nitrogen fifixation and its application to improve plant growth under nitrogen-deficient conditions. PLoS One 8:e63666

Sharma S, Singh S, Dhanjal DS, Kumar A, Jan S, Ramamurthy PC, Singh J (2022) 15—role of rhizobacteria from plant growth promoter to bioremediator. In: Kumar V, Shah MP, Shahi SK (eds) Phytoremediation technology for the removal of heavy metals and other contaminants from soil and water. Elsevier, pp 309–328. ISBN 9780323857635

Sharon I, Banfield JF (2013) Genomes from metagenomics. Science 342:1057–1058. https://doi.org/10.1126/science.1247023

Signorelli S, Sainz M, Tabares-da Rosa S, Monza J (2020) The role of nitric oxide in nitrogen fixation by legumes. Front Plant Sci 3(11):521

Sikora R, Dababat AEF (2007) Influence of the mutualistic endophyte Fusarium oxysporum 162 on Meloidogyne incognita attraction and invasion. Nematology 9:771–776. https://doi.org/10.1163/156854107782331225

Singh JS (2013) Plant growth promoting rhizobacteria. Reson 18:275–281

Sivakumar N, Sathishkumar R, Selvakumar G, Shyamkumar R, Arjunekumar K (2020) Phyllospheric microbiomes: diversity, ecological significance, and biotechnological applications. Plant Microbiomes Sustain Agric 7(25):113–172

Slininger PJ, Behle RW, Jackson MA, Schisler DA (2003) Discovery and development of biocontrol agents to control crop pests. Neotropical Entomol 32:183–195

Smil V (2002) Nitrogen and food production: proteins for human diets. Ambio 31:126–131

Smith S, Read D (2008) Mycorrhiza symbiosis, 3rd edn. Academic Press, San Diego, CA

Söderberg KH, Olsson PA, Bååth E (2002) Structure and activity of the bacterial community in the rhizosphere of different plant species and the effect of arbuscular mycorrhizal colonization. FEMS Microbiol Ecol 40:223–231

Sonmez Kaplan M, Sonmez S (2007) An investigation of seasonal changes in nitrate contents of soils and irrigation waters in greenhouses located in Antalya-Demre region. Asian J Chem 19(7):5639

Soumare A, Boubekri K, Lyamlouli K, Hafidi M, Ouhdouch Y, Kouisni L (2020) From isolation of phosphate solubilizing microbes to their formulation and use as biofertilizers: status and needs. Front Bioeng Biotechnol 7:425

Soundara Rajan S, Daivasikamani S, Manoharan A (2020) Bio encapsulation of bio control agents for effective management of collar rot (Rhizoctonia solani) in coffee (Coffea arabica). Int J Curr Microbiol App Sci 9(6):1158–1166

Souza RD, Ambrosini A, Passaglia LM (2015) Plant growth-promoting bacteria as inoculants in agricultural soils. Genet Mol Biol 38(4):401–419

Spaepen S, Vanderleyden J (2011) Auxin and plant-microbe interactions. Cold Spring Harb Perspect Biol 3:a001438

Suman A, Shasany AK, Singh M, Shahi HN, Gaur A, Khanuja SPS (2001) Molecular assessment of diversity among endophytic diazotrophs isolated from subtropical Indian sugarcane. World J Microbiol Biotechnol 17(1):39–45

Suman A, Yadav AN, Verma P (2016) Endophytic microbes in crops: diversity and beneficial impact for sustainable agriculture, In Microbial inoculants in sustainable agricultural productivity, Springer, New Delhi, 117–143

Swarnalakshmi K, Yadav V, Tyagi D, Dhar DW, Kannepalli A, Kumar S (2020) Significance of plant growth promoting rhizobacteria in grain legumes: growth promotion and crop production. Plants (Basel) 9(11):1596

Tajini F, Trabelsi M, Drevon JJ (2012) Combined inoculation with Glomus intraradices and Rhizobium tropici CIAT899 increases phosphorus use efficiency for symbiotic nitrogen fifixation in common bean (Phaseolus vulgaris L.). Saudi J Biol Sci 19:157–163

Tak HI, Ahmad F, Babalola OO, Inam A (2012) Growth, photosynthesis and yield of chickpea as influenced by urban wastewater and different levels of phosphorus. Int J Plant Res 2:6–13

Tamošiūnė I, Stanienė G, Haimi P, Stanys V, Rugienius R, Baniulis D (2018) Endophytic Bacillus and Pseudomonas spp. modulate apple shoot growth, cellular redox balance, and protein expression under in vitro conditions. Front Plant Sci 28(9):889. https://doi.org/10.3389/fpls.2018.00889

Tang Y, Fu Y (2018) Class 2 CRISPR/Cas: an expanding biotechnology toolbox for and beyond genome editing. Cell Biosci 8:59. https://doi.org/10.1186/s13578-018-0255-x

Taurian T, Anzuay MS, Angelini JG, Tonelli ML, Ludueña L, Pena D, Ibáñez F, Fabra A (2010) Phosphate-solubilizing peanut associated bacteria: screening for plant growth-promoting activities. Plant Soil 329:421–431

Terkina IA, Parfenova VV, Ahn TS (2006) Antagonistic activity of actinomycetes of Lake Baikal. Appl Biochem Micro 42(2):173–176

Thakur J, Sagar A (2019) Diversity of rhizosphere fungi and VAM fungal spores associated with Murraya koenigii (L.) Spreng. From Hamirpur District of Himachal Pradesh. Int J Agric Innovat Res 8:4. ISSN (Online) 2319-1473

Thornton PK, Ericksen PJ, Herrero M, Challinor AJ (2014) Climate variability and vulnerability to climate change: a review. Glob Chang Biol 20:3313–3328. https://doi.org/10.1111/gcb.12581

Tong Y, Weber T, Lee SY (2019) CRISPR/Cas-based genome engineering in natural product discovery. Nat Prod Rep 36:1262–1280

Trabelsi D, Mhamdi R (2013) Microbial inoculants and their impact on soil microbial communities: a review. Biomed Res Int 2013:863240. https://doi.org/10.1155/2013/863240

Trivedi P, Pandey A, Palni LMS (2005) Carrier-based preparations of PGP bacterial inoculants suitable for use in cooler regions. World J Microbiol Biotechnol 21:941–945

Tsukanova KA, Chebotar VK, Meyer JJM, Bibikova TN (2017) Effect of plant growth-promoting rhizobacteria on plant hormone homeostasis. South African J Bot 113:91–102. ISSN 0254-6299

Van der Wal A, Leveau JHJ (2011) Modelling sugar diffusion across plant leaf cuticles: the effect of free water on substrate availability to phyllosphere bacteria. Environ Microbiol 13:792–797

Van Dommelen A, Croonenborghs A, Spaepen S, Vanderleyden J (2009) Wheat growth promotion through inoculation with an ammonium-excreting mutant of *Azospirillum brasilense*. Biol Fertil Soils 45:549–553

Van Velthuizen H (2007) Mapping biophysical factors that influence agricultural production and rural vulnerability. Food & Agriculture Organization, Rome, Italy

Vassilev N, Vassileva M, Martos V, Garcia Del Moral LF, Kowalska J, Tylkowski B, Malusá E (2020) Formulation of microbial inoculants by encapsulation in natural polysaccharides: focus on beneficial properties of carrier additives and derivatives. Front Plant Sci 11:270

Ventura M, Canchaya C, Tauch A et al (2007) Genomics of actinobacteria: tracing the evolutionary history of an ancient phylum. Microbiol Mol Biol Rev 71:495–548

Verma V, Ravindran P, Kumar PP (2016) Plant hormone-mediated regulation of stress responses. BMC Plant Biol 16:86

Vessey JK (2003) Plant growth promoting rhizobacteria as biofertilizers. Plant Soil 255:571–586

Wang CH (2014) Farming methods effects on the soil fertility and crop production under a rice vegetables cropping sequences. J Plant Nutr 37:1498–1513

White JF, Kingsley KL, Zhang Q (2019) Review: endophytic microbes and their potential applications in crop management. Pest Manag Sci 75(10):2558–2565

Wijewardana C, Alsajri FA, Trenton Irby J, Jason Krutz L, Golden B, Brien Henry W, Gao W, Reddy KR (2019) Physiological assessment of water deficit in soybean using midday leaf water potential and spectral features. J Plant Interact 14(1):533–543

Yadav AK, Chandra K (2014) Mass production and quality control of microbial inoculants. Proc Indian Natl Sci Acad 80:483–489

Yadav AN (2020) Plant microbiomes for sustainable agriculture: current research and future challenges. In: Yadav A, Singh J, Rastegari A, Yadav N (eds) Plant microbiomes for sustainable agriculture. Sustainable development and biodiversity, vol 25

Yadav AN (2021) Beneficial plant microbe interaction for agriculture sustainability. J Appl Biol Biotechnol 9(1):i–iv

Yadav AN, Verma P, Kumar S, Kumar V, Kumar M, Kumari Sugitha TC et al (2018) Actinobacteria from Rhizosphere, in: Elsevier, editor. In New and future developments in microbial biotechnology and bioengineering p 13–41

Yadav SK (2010) Old stress tolerance mechanisms in plants. A review. Agron Sustain Dev 30:515–527

Yarzábal LA, Chica EJ (2021) Microbial-based technologies for improving smallholder agriculture in the Ecuadorian Andes: current situation, challenges, and prospects. Front Sustain Food Syst 5:617444

Yekkour A, Sabaou N, Zitouni A et al (2012) Characterization and antagonistic properties of Streptomyces strains isolated from Saharan soils and evaluation of their ability to control seedling blight of barley caused by Fusarium culmorum. Lett Appl Microbiol 55(6):427–435

Young CC, Rekha PD, Lai WA, Arun AB (2006) Encapsulation of plant growth-promoting bacteria in alginate beads enriched with humic acid. Biotechnol Bioeng 95:76–83

Yu L, Zhang H, Zhang W, Liu K, Liu M, Shao X (2022) Cooperation between arbuscular mycorrhizal fungi and plant growth-promoting bacteria and their effects on plant growth and soil quality. PeerJ 10:e13080

Zhang D et al (2021) CRISPR/Cas: a powerful tool for gene function study and crop improvement. J Adv Res 29:207–221

Zhang J, Ahmed W, Dai Z, Zhou X, He Z, Wei L, Ji G (2022) Microbial consortia: an engineering tool to suppress clubroot of Chinese cabbage by changing the rhizosphere bacterial community composition. Biology (Basel) 11(6):918

Zhang Y, Wang J, Wang Z, Zhang Y, Shi S, Nielsen J et al (2019) A gRNA-tRNA array for CRISPR-Cas9 based rapid multiplexed genome editing in Saccharomyces cerevisiae. Nat Commun 10:1053

Zhou J, Deng B, Zhang Y, Cobb AB, Zhang Z (2017) Molybdate in rhizobial seed-coat formulations improves the production and nodulation of alfalfa. PLoS One 12:e0170179

Zhu F, Qu L, Hong X, Sun X (2011) Isolation and characterization of a phosphate solubilizing halophilic bacterium Kushneria sp. YCWA18 from Daqiao Saltern on the coast of yellow sea of China. Evid Based Complement Alternat Med 2011:615032

Zhu JK (2016) Abiotic stress signaling and responses in plants. Cell 167:313–324

Chapter 5
Microbes and Their Role in Alleviation of Abiotic and Biotic Stress Tolerance in Crop Plants

Sana Sheikh, Akshitha Ramachandra Amin, Mayura Asra, and N. Bhagyalakshmi

Abstract Climatic changes influence a number of biotic and abiotic factors in the environment. This influence severely affects the growth, development, and productivity of crop plants. Plants encounter stress that enhances the susceptibility of plants to occurrences of various diseases by a number of pests and pathogens. As plants have been exposed to various stress, they have evolved well-developed mechanisms to respond to the stress signal and enable response. In recent years, advances in physiology, molecular biology, and genetics have greatly improved our understanding of crop response to these stresses and the basis of varietal differences in tolerance. This chapter will depict the role of microbes in alleviating stress tolerance and their impacts on agricultural productivity.

Keywords Crop · Drought · Microbes · PGPR · Salinity

Abbreviations

ABA	Abscisic acid
EPS	Exopolysaccharides
HSFs	Heat shock transcription factors
PGPB	Plant growth-promoting bacteria
PGPR	Plant growth-promoting rhizobacteria
PR	Pathogenesis-related
ROS	Reactive oxygen species

S. Sheikh (✉) · A. R. Amin · M. Asra · N. Bhagyalakshmi
Department of Botany, St Aloysius College (Autonomous), Mangaluru, Karnataka, India

P. Mathur et al. (eds.), *Microbial Symbionts and Plant Health: Trends and Applications for Changing Climate*, Rhizosphere Biology,
https://doi.org/10.1007/978-981-99-0030-5_5

5.1 Introduction

Stress influences productivity of crop plants. Stress triggers a wide range of plant responses like altered gene expression, cellular metabolism, changes in growth rates, crop yields, etc., adversely affecting plant growth and development. Climatic change is a key for the induction of stress. Plants encounter stress that enhances the susceptibility of plants to occurrences of various diseases by a number of pests and pathogens. Stress also affects the defense response of plants (Pandey et al. 2017). In their natural environment, plants are often exposed to a range of biotic stresses such as bacteria, fungi, nematodes, *Phytoplasma*, *Mycoplasma*, etc. The influence of biotic factors may be positive or negative depending on the type of interaction. The interaction may be a mutualism type where the plant will be benefited because of the association; may be a parasitic type, where the associated organism will harm the plant; or may be a competitive interaction between the plants because of the release of the toxic chemicals (Ali et al. 2020).

5.2 Types of Stress

Plant stress in general is classified into two types: biotic stress and abiotic stress (Fig. 5.1).

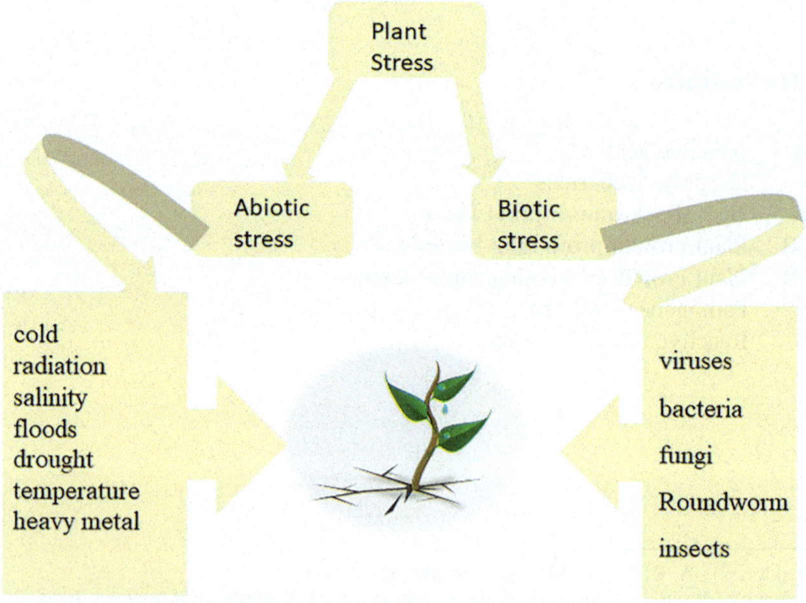

Fig. 5.1 Types of stress affecting plant

5.2.1 *Biotic Stress and Crop Plants*

Biotic or biological stress is triggered by living organisms such as viruses, bacteria, fungi, roundworms (nematodes), insects, arachnids, etc. Biological stressors often act as pests to the host plant, directly starving the host plant for nutrients and possibly causing death of plants. Biological stress can be intensified by pre- and post-harvest losses of crop plants (Kumar and Nautiyal 2022). Although they lack an immune system, plants combat biological stresses by developing complex defensive strategies. The defense mechanisms that work against these stresses are genetically controlled.

5.2.2 *Abiotic Stress and Crop Plants*

The abiotic stress causes the loss of major crop plants worldwide and includes cold, radiation, salinity, floods, drought, extremes in temperature, heavy metals, etc. (Gull et al. 2019). Plants are encountered by a number of abiotic stresses which impact on the crop productivity worldwide. These abiotic stresses are interconnected with each other and may occur in the form of osmotic stress, malfunction of ion distribution, and plant cell homeostasis. The growth rate and productivity are affected by a response caused by a group of genes by changing their expression patterns (Prelich 2012). So, the identification of responsive genes against abiotic stresses is necessary in order to understand the abiotic stress response mechanisms in crop plants.

5.2.2.1 Cold

Cold stress as an abiotic stress has been found to be the main abiotic stress reducing agricultural crop yields by affecting crop quality and postharvest shelf life. Plants immobile in nature are always busy modifying their mechanisms to guard against such stresses. In temperate conditions, plants face cold and freezing conditions, which are very harmful to plants such as stress. To adapt, plants acquire cold and frost tolerance against these deadly cold pressures through a process called acclimatization (Gull et al. 2019). Abiotic stress from cold affects the cellular functions of plants in all their aspects. There are several signal transduction pathways through which these cold strains are transduced such as components of ROS, protein kinase, protein phosphate, ABA and Ca2+, etc., and out of these, ABA turns out to be the best.

5.2.2.2 Salt/Salinity

Soil salinity poses a global threat to global agriculture by reducing crop yields and ultimately crop yields in salt-affected areas. Salt stress reduces plant growth and yield in a number of ways. The two main effects on crops are due to salinity: osmotic stress and ion toxicity (Heba et al. 2022). The osmotic pressure of salinization in the soil solution exceeds the osmotic pressure in the plant cells due to the presence of more salts, thus limiting their ability to take in water and minerals such as K^+, K^+, and Ca^{2+} of the tree. The primary effects of salinity stress induce secondary effects such as anabolic production, decreased cell expansion and membrane function, and decreased cellular metabolism.

5.2.2.3 Drought

The worldwide climate has changed as temperatures and CO_2 levels in the atmosphere continue to rise (Yang et al. 2019). The uneven distribution of rainfall due to climate change causes major stressors such as drought. Soil water available to plants steadily increased due to severe drought conditions and caused early plant death. Once a drought has hit a crop, stopping growth is the first experience the plant will experience. Plants reduce their shoot growth under drought conditions and reduce their metabolic demands. The protective compounds are then synthesized by plants during drought by mobilizing metabolites required for their osmotic regulation (Seleiman et al. 2021).

5.2.2.4 Heat or Temperature

Rising temperatures worldwide have become a major concern, affecting not only the growth of plants but also their yield, especially for agricultural crops. When plants are subjected to heat stress, seed germination rate, photosynthetic efficiency, and yield decrease. The exposition to thermal stress can cause enormous damage to the cell membrane and the protein conformation, leading to ROS production, triggering oxidative stress. In addition, heat stress decreased protein synthesis, transcription, and translation of heat shock proteins (HSPs); production of phytohormones and antioxidants; and changes in the organization of cell structures leading to alterations in hormonal homeostasis (Li et al. 2021).

5.2.2.5 Toxin

Agriculture's increasing reliance on chemical fertilizers and wastewater irrigation systems and rapid industrialization has added toxic metals to farmland, causing adverse effects on soil environmental systems.

5.3 Role of Microbes in Stress Tolerance in Crop Plants

Our agro-ecosystems are constantly affected by abiotic and biotic stress that directly alters crop yield and soil health and fertility. Various stress factors negatively affect the growth and productivity of crop plants. The main effect of these stresses is the loss of microbial diversity in the soil, fertility, and competition for nutrients (Chodak et al. 2015).

The only possible alternatives are plant-associated microbial communities, such as mycorrhizal fungi and plant growth-promoting bacteria (PGPB), which help plants grow and develop under various forms of abiotic and biotic stress. Effective application of microorganisms such as rhizobacteria (PGPR) and mycorrhizal fungi is very useful to strengthen and improve agriculture and environmental stability. Various genera of *Pseudomonas, Enterobacter, Bacillus, Variovorax, Klebsiella, Burkholderia, Azospirillum, Serratia*, and *Azotobacter* are termed as PGPR which promote plant growth and development under both normal and stress conditions. The PGPM and mycorrhizae maintain plant fitness and health under abiotic and biotic stress environments (Vimal et al. 2017). The plant associated with beneficial microbes enhances the efficiency of their growth and development under abiotic and biotic stress conditions.

5.4 Soil Microorganisms and their Role in Abiotic Stress Management

The soil is heaving with millions of living organisms which make it living as well as dynamic system. Under a microscope, it reveals a complex arrangement of soil particles and pore spaces filled with air and water. It is in these pore-spaces that plant roots and millions of organisms develop, ranging from sub-microscopic to macroscopic in size. These organisms not only help in the development of soil but are also the primary driving agent of nutrient cycling, regulating the dynamics of soil organic matter, enhancing the amount of nutrient acquisition by vegetation, conferring stress tolerance, resisting pathogens and improving plant health (Selvakumar et al. 2012).

5.5 Microbes as Stress-Alleviating Agents under Various Stress Situation

5.5.1 Drought Stress

Alleviation of drought stress by rhizobacterial species has received much attention in the past. Under transient water stress, the ACC deaminase producing PGPR *Achromobacter piechaudii* ARV8 significantly increased the fresh and dry weights of crop tomato and pepper seedlings and reduced the ethylene production under transient drought stress. Indigenous bacterial strains *Pseudomonas putida*, *Pseudomonas* sp., and *Bacillus megaterium* isolated from water-stressed soil could stimulate plant growth under dry conditions (Marulanda et al. 2009). Plant growth-promoting bacteria (PGPB) can help mitigate drought stress in various ways.

Phytohormone production, 1-aminocyclopropane-1-carboxylate deaminase exopolysaccharide activation, solute production, chlorophyll synthesis, and increased mineral solubilization can help reduce stress (Shaffique et al. 2022).

5.5.2 *High/Low Temperature Stress*

In the context of climate change, microbial-mediated alleviation of high/low temperature stress are receiving recent attention. *Pseudomonas* AKM-P6 has the ability to alleviate heat stress in sorghum seedlings.

5.5.3 *Soil/Salinity*

Soil salinity has a profound effect on seed germination, which is the most vital aspect of successful crop production. Under such situations, it is essential to stimulate seed germination and seedling growth. The most appropriate solution in such situation is to use salt-tolerant bacterial inoculants that produce auxins and gibberellins and promote plant growth under salinity conditions. PGPB and endophytes also play a significant role in inducing plant signaling under salinity stress conditions (Kumar et al. 2020). For instance, PGPR such as *Arthrobacter protophormiae* (SA3) and *Dietzianatronolimnaea* (STR1) enhanced salinity stress tolerance in wheat plants by modulating the expression of a regulatory component CTR1 (Constitutive Triple Response1) of the ethylene signaling pathway and DREB2 TF (Barnawal et al. 2017).

5.5.4 *Heavy Metals*

In metal-contaminated soil, the poor performance of plant growth and root development are major limiting factors for phytoaccumulation of metals. To overcome these problems, improvement of the microbial activity in rhizosphere in addition to organic amendments is necessary. Numerous plant-associated microbes, namely, bacteria and fungi, are known to exhibit plant growth-promoting traits under heavy metal stress. These microbes impart favorable effects on plants via several direct and indirect mechanisms such as biofilm formation, siderophores, exopolysaccharide, and phytohormones production (Tiwari et al. 2016).

5.5.5 *Nutrient Deficiency-Associated Stresses*

Many rhizospheric bacteria enhance the uptake of plant nutrients from the soil by facilitating improved root growth under low-input crop production situations. In cucumber plants under Cu deficiency or dual deficiency of Cu and Fe, plant stress symptoms were alleviated by the plant-beneficial bacteria *A. brasilense*, accompanied by better root development and nutrient uptake (Marastoni et al. 2019).

5.6 Regulatory Mechanism in Plants in Response to Stress

As plants have been exposed to various stress, they have evolved well-developed mechanisms to respond to the stress signal and enable response. There are a number of factors that enable the plant to adapt and respond to such adverse conditions (Verma et al. 2016). Plants are challenged by various pathogens and insects. To survive these difficulties, plants have developed elaborate sensing mechanisms that are mediated by signaling cascades and gene transcription networks that respond to environmental cues.

5.6.1 *Plant Hormones and Transcription Factors*

Phytohormones play a crucial role in mediating defense responses. Plants produce a number of hormones like auxin, gibberellin, cytokinin, abscisic acid, ethylene, strigolactone, salicylic acid, jasmonates, and brassinosteroids that help them in growth and development. Among the above examples, salicylic acid, jasmonic acid, and ethylene play crucial roles in response to various pathogens and pets. While salicyclic acid is involved in the activation of defense response against biotrophic and hemi-biotrophic pathogens, jasmonic acid and ethylene are responsible for defense against necrotrophic pathogens and herbivorous insects. With the attack of pathogenic organism, it has been observed that the salicylic concentration increases, which further influences the activation of pathogenesis-related (PR) genes. It has been observed that the concentration of jasmonic acid increases when the plant is attacked by pathogens like caterpillars, spider mites, beetles, thrips, and mirid bugs (Verma et al. 2016). With the pathogenic attack or treatment with jasmonic acid in tobacco plants, nicotine concentration increases in the vacuole. The nicotine concentration protects the plant from subsequent attack and enhances the memory for any further attack. Expression of transcription factors such as WRKY transcription factors also increases in response to the attack.

5.6.2 Transcription Factors

Transcription factors are proteins that control the transcription rate in cell. A number of studies have revealed that transcription factor plays a significant role in defense mechanism. With more than 58 families, around 6 transcription factor families are closely related to defense signaling mechanism and response. The six families of transcription factor are AP2/ERF (APETALA2/ethylene responsive factor), bHLH (basic helix-loop-helix), MYB (myeloblastosis related), NAC (no apical meristem (NAM), *Arabidopsis* transcription activation factor (ATAF1/2), and cup-shaped cotyledon (CUC2)), WRKY, and bZIP (basic leucine zipper). While AP2/ERF family is involved in abiotic stress response, the bHLH and MYB are involved in jasmonate-responsive gene expression. Along with DNA binding, NAC transcript factor is involved in dimer formation and nuclear localization and bZIP f and WRKY functions by protein dimerization and DNA binding (Danny Ng et al. 2018). The response is mainly mediated by a transcription factor JASMONATE INSENSITIVE 1/MYC2 (JIN1/MYC2). With the increase in concentration of ethylene, transcription factor ETHYLENE INSENSITIVE3 (EIN3) was suggested to induce ERF1 gene expression, which further activates defense response against the attack. Other than the above three, there are several other plant hormones that are involved in triggering the response (Verma et al. 2016).

Small RNA and defense response in plants – Micro RNA (miRNA) – miRNA are small, few nucleotides long, single-stranded RNA that take part in defense response. Through up- and downregulation of miRNA, plants regulate the gene expression and counter the attack of pathogen. The miRNA take part in translational repression or degradation of targeted mRNA. MYB, NAC, TCP bHLH, NYF, AP2, and HD-zip are the target transcription factors of miRNA and have a varied range of defense genes such as MAPK genes, peroxidase, BB-LRRs, and ABC transporters (Ali et al. 2020). Small interfering RNA (siRNA) is an endogenous or exogenous RNA that i about 20–24 nucleotide long. siRNA functions by methylation, transcriptional gene silencing, or post-transcriptional gene silencing. siRNAs are known to be the fundamental antiviral pathways in insects. RNAi or RNA interference is a double-stranded RNA, 20–24 nucleotide long, which recognizes the target mRNA and proceeds with the homology-dependent degradation. They also silence the target mRNA by translational repression and chromatin remodeling (Ali et al. 2020).

5.6.3 Heat Shock Proteins

Plants produce a set of genes, which are called as heat shock proteins. The proteins are constitutively expressed in cells in the absence of environmental stress stimuli but often undergo degradation by molecular chaperons present in the cell. As the stress signal is perceived, the genes are expressed and their concentration increases. The transcription of HSPs is primarily regulated by heat shock transcription factors

(HSFs). A number of heat shock transcription factors are activated upon infection. HSFs bind to HSEs; 5′ -AGAAnnTTCT-3′ sequence that are a highly conserved motifs of the promoter regions of HSP genes known as heat stress elements. The binding results in increased tolerance to biotic and abiotic stresses (Nejat and Mantri 2017).

5.6.4 *Receptor Proteins*

Plants produce surface-localized recognition receptors that are called as plant pattern recognition receptors (PRRs). These receptors are synthesized in endoplasmic reticulum and then are transported to the plasma membrane. PRRs belong to the type of receptor-like kinases (RLK) and receptor-like proteins (RLPs). Through their extracellular binding domain, they perceive a wide range of signal and trigger a downstream intracellular signaling cascade (Nejat and Mantri 2017).

5.6.5 *Epigenetic Changes*

The changes in the genome that are heritable are called as epigenetic changes (Nejat and Mantri 2017). They are inherited through cell division and are transmitted to the next generation and therefore lead to stress memory in plants. They play vital role in adaptation of plants to stress. The genome undergoes epigenetic modification such as DNA methylation, remodeling of chromatin, non-coding RNAs, etc. (Nejat and Mantri 2017). DNA methylation pattern has been observed which is termed as differentially methylated regions (DMRs) that influence stress responsive gene expression pattern in plants. Stress-induced differentially methylated regions (DMRs) regulate the expression of stress-responsive genes. Factors such as AhHDA1, an RPD3/HDA1-like superfamily histone deacetylase (HDAC), regulate the expression of stress-responsive genes (Nejat and Mantri 2017).

5.7 Microbial Application in Agricultural Sustainability

5.7.1 *Microbes and Drought Stress Tolerance*

Drought-tolerant microorganisms have the ability to improve plant growth and develop in conditions of water scarcity. Microbes have evolved, adapted, and/or developed a tolerance mechanism to survive under low water potential. They can form a thick wall or dormant stage, can accumulate osmolytes, and produce exopolysaccharides (EPS). These plant-associated microbes have various mechanisms to cope with the negative effects of drought on crops as well as on soil.

Regardless of water content, they provide nutrients and better environment conditions for the plant to continue to grow. Beneficial microorganisms colonize around the biosphere, promoting plant growth and development through different direct and indirect mechanisms. Potential mechanisms include the production of phytohormones such as indole-3-acetic acid (IAA), cytokinin and abscisic acid (ABA), exopolysaccharides, and ACC deaminase which induce systemic tolerance. Phytohormones produced by plants play an important role in growth and development (Farooq et al. 2009). In addition, PGPR has the ability to synthesize plant hormones that stimulate plant growth and split under pressure. IAA is a highly active auxin that regulates vascular, root, and lateral root tissue differentiation, cell division, and shoot growth during water stress (Goswami et al. 2015). ABA is an important growth regulator during water stress. When seeds or plants are transplanted with PGPR, ABA concentration increases and regulates plant physiology to be tolerant to drought stress. ABA improves water stress by regulating transcription of drought-related genes and root hydraulic conductivity (Jiang et al. 2013). For example, *Azospirillum brasilense* improves *Arabidopsis thaliana*'s response to drought mainly by improving ABA levels. 1-Aminocyclopropane-1-Carboxylate (ACC) is an immediate precursor of ethylene during stress. Bacterial ACC deaminase hydrolyzes ACC to ammonia and alpha-ketobutyrate (Bal et al. 2013). Water stress tolerance and PGPR increase biomass and water potential and reduce water loss in maize under stressful conditions. These preparations reduce antioxidant activity and also improve the production of proline, free amino acids, and sugars in plants (Vardharajula et al. 2011). Furthermore, the combination of endogenous PGPR and rhizome improves stress tolerance. Exopolysaccharide produced by microorganisms improves the drought tolerance of some crops. For example, three strains of drought-tolerant bacteria *Proteus penneri* (Pp1), *Pseudomonas aeruginosa* (Pa2), and *Alcaligenes faecalis* (AF3) which were inoculated with maize plants have been shown to have the ability to increase relative water, protein, and sugar content due to the proline content (Naseem and Bano 2014). Surviving in such dry conditions, bacteria provide a variety of physiological, biochemical, and molecular mechanisms for defense unfavorable conditions. They synthesize EPS, compatible solutes, and spore formation (Chithrashree et al. 2011). Bacteria that produce EPS make plants water resistant under water pressure. Compatible solutes such as glycine, proline, and betaine accumulate during water stress and help bacteria maintain membrane permeability, enzymes to maintain their integrity, and proteins in their functional form. The combination of mycorrhizal inoculation with specific bacteria improves plant growth and nutrient absorption compared to water content to reduce the impact of drought. The relationships of drought-tolerant microbial communities to plants can maintain proper growth and survival in arid conditions.

5.7.2 *Microbes and Salinity Stress Tolerance*

Saline soil is a difficult task for farmers. Accumulation of toxic Na and Cl ions and nutritional imbalance in soil has a serious impact on plant growth and microbial activities. It has been widely reported that inoculation of PGP and endophyte microorganisms minimizes the negative effects of salt on plants. PGP microorganisms can promote plant growth in drought and salinity conditions through different direct and indirect mechanisms. Beyond that, the biofilm formed by PGPB under salinity pressure is effective in mitigating harmful effects (Kasim et al. 2016). Lettuce seeds inoculated with *Azospirillum* showed better germination and better vegetative growth than the control in salinity status (Barassi et al. 2006). In another study, implanting plants with the growth-promoting bacteria *Pseudomonas stutzeri* in salt-tolerant and salt-sensitive pepper reduces the negative impact on the soil salinity (Bacilio et al. 2016). While some species of microorganisms enhance salt stress biofilm-forming activity on cereals (Kasim et al. 2016), co-culture of AM fungi with salt-tolerant bacteria significantly improves the salt tolerance of some crops because, for example, the co-culture of *R. intraradices* and *Massilia* sp. RK4 renews root colonization of arbuscular mycorrhizal fungi (AMF) and accumulates nutrients under the action of salt in corn plants. These fungal and microbial associations have a significant impact on salinity tolerance in maize (Krishnamoorthy et al. 2016). AM fungi have also been considered an important factor in improving soil salinity. It was reported this AM inoculation improves plant growth under saline stress conditions. They inhibit the absorption of Na or Cl in citrus fruits in saline solution state (Navarro et al. 2014). Therefore, the implantation of PGPR and the bacteria can serve as potential tools for mitigating salinity stress in cultures sensitive to salt.

5.7.3 Microbes and Heavy Metal Stress Tolerance

Continuous industrialization, intensive agricultural practices, and human activities lead to soil pollution by heavy metals. These heavy metals have serious impacts on plants and human health. Heavy metals are metallic elements with a density greater than 4 g/cm^3, non-biodegradable, and also toxic at low concentrations (Duruibe et al. 2007). Phytoremediation, a growing technique, uses plants and their related microorganisms to clean heavy metal toxins from the soil. During the expansion, it adopted a viable and sustainable approach to the expulsion of crushed metal (Chirakkara et al. 2016). Microorganisms are more sensitive than any other living organisms and can be a good indicator of heavy metal stress (Broos et al. 2004). The use of microbial diversity to help tackle heavy metals over the past year was achieved due to an inexpensive, environmentally friendly and aesthetic approach and equally applicable in different situations. A wide range of microorganisms and plants tolerate heavy metal-related microbes such as rhizobacteria, mycorrhizae, and firmicutes which have the ability to promote the growth and development of plants in the metallurgical process stress state. These microbes are associated with mechanisms such as flow, metal impermeability, evaporation, EPS isolation, metal complexation, and enzymatic detoxification. These plant-related microorganisms promote plant growth and growth by reducing the concentration of ethylene, producing plant growth regulators such as IAA, ACC deaminase and disease inhibitors (Glick 2010). In addition, nitrogen fixation, nutrient mobilization, lateral cell and phosphate solubility, improves both plant growth and heavy metal removal (Verma et al. 2013). Bioaccumulation by microorganisms is a powerful method to remove heavy metals from contaminated soil. Here, it has been reported that proteobacteria, firmicutes, and actinobacteria are able to remove higher concentrations of Mn, Pb, and As from metal-contaminated soil (Zhang et al. 2015). Fatnassi et al. (2015) in their study revealed that above a concentration of 1 mM copper (Cu) is harmful to plant growth due to *Viciafaba*, but when inoculated with rhizobium and PGPR, reduced effect. AM fungi ameliorate the deleterious effects of cadmium stress by reducing malonaldehyde and hydrogen peroxide (Hashem et al. 2016). Removal of Cd, Pb, and Zn metals from contaminated soil, (Jing et al. 2014) demonstrated that *Enterobacter* p. And *Klebsiella* sp. efficiently tolerate metals by producing plant growth agents. Phytoremediation is more advanced than traditional methods, and efficiency can be improved through the use of PGPM to eliminate metals from contaminated soil (Glick 2010). This method is the most effective, innovative, and healthy for removing heavy metals. Some microorganisms have the ability to completely degrade heavy metal. For example, PGPM like *Pseudomonas* sp. MBR shows biotransformation of Fe (III)-, Zn-, and Cd-citrate complexes and their removal (Qian et al. 2012).

5.7.4 Microbes and Temperature Stress Tolerance

Climate change has increased the frequency and intensity of temperature stress. Temperature stress and cold stress are serious abiotic stress conditions that affect crop yields and food security worldwide. The main effect of temperature stress is the change in plasma membrane, water content (transpiration), altered photosynthetic activity, enzyme activity, cell division, and plant growth. The greatest impacts of climate change are found in tropical and subtropical regions, including India (Alam et al. 2017). Temperature can affect various components of cells and cell membranes. For example, heat can increase fluidity while cold makes it harder. Heat stress is the result of many physiological and organic resorts if not managed properly. Heat stress is one of the severe abiotic stresses and causes many changes in phytohormones' focus and response. Plants have a complex regulatory mechanism to direct plant tolerance. Countless species of plants have adapted to low and high temperatures. These changing environmental conditions cause many physiological changes in the plant. Plants use a variety of mechanisms to overcome heat stress including production and accumulation of enzymes and osmolytes. Heat shock proteins (HSP20, HSP 60, HSP70, HSP 90, HSP100) and reactive oxygen species (ROS) enzymes (ascorbate peroxidase and catalase) are the major functional proteins (Kotak et al. 2007). But most crops cannot withstand the extremes of both heat stress and cold shock. Thereby, there is an urgent need for such a mechanism to withstand extreme temperatures, an approach focused on the use of microorganisms to reduce the harmful effects of hot and cold stress. Temperature plays an important role in regulating the physiology and metabolism of microorganisms in extreme temperatures. Microbial enzymes to help microorganisms adapt to low and high temperature. These microbes are effective mechanisms that protect their proteins, membranes, and nucleic acids for survival in such conditions. Gene expression of heat- and cold-tolerant proteins and enzymes is enhanced under these conditions. Molecular chaperones are among the most effective antipyretics. Based on growth, microorganisms are divided into two groups, the macrobiotic group and the heterotrophic group. The development of psychrophilic lies maximal at or below 15 while the growth of psychrotrophic bacteria is greater than 15 °C. Heat stress-induced expression of the gene is responsible for the persistence of virus. The DnaK gene in *Alicyclobacillus acidoterrestris* is increased in expression during heat stress to encode HSP, which protects bacteria from heat. Expression of HSP is a strategy for dealing with high temperatures. Under heat stress conditions, they are maintained by obtaining efficiently absorbing nutrients and water and improving photosynthesis. Trehalose synthesis is induced during heat stress and protects microorganisms from heat and cold shock damage and oxidative stress. The accumulation of trehalose in bacteria and fungi increases many times during heat stress. During heat stress and cold shock, trehalose accumulates in microbial cells that protect against heat damage (Li et al. 2009). It plays a major role in stabilizing proteins in cells. In fungi, trehalose reduces thermal stress-induced denaturation and agglomeration, so the protein remains in its original form. Trehalose can also provide some protection to proteins against heat-induced protein denaturation. Trehalose has been reported to be the most active against frozen and dried conditions. Microbes suitable for low temperature represent the properties of plant growth at low temperatures. Yadav et al. (2014) reported that *Pseudomonas cedrina*, *Brevundimonas terrae*, and *Arthrobacter nicotianae* are adapted to low temperatures. PGPR isolated from Root nodules in peas at low temperatures have the ability to effectively biofertilizer at low temperatures (Meena et al. 2015). Further, Javani et al. (2015) reported that psychrophilic bacteria isolated from Antarctica showed antibacterial activity. On the other hand, the cultivation of phosphate-solubilizing bacteria in the agricultural field acts as a multifunctional bio-fertilizer. It acts as a biogeochemical phosphorus cycle in agriculture. The function of phosphate-degrading bacteria is the transformation of insoluble phosphorus to the soluble form by acidification (Chang and Yang 2009).

5.8 Microbes and Biotic Stress

In nature, soil and plant roots are the habitat for the colonization of plant pathogens and beneficial microorganisms in the soil. Secretion from the roots of plants and other chemicals produced by plants attract microbial diversity. Plant pathogens, e.g., bacteria, fungi, viruses, and pests, have destroyed crop yields (Ramegowda and Senthil 2015). The common effects of these biological factors include hormonal imbalance regulation, nutritional imbalance, and physiological disorders. Further, the increasing cost of pesticides and their harm to the soil can be felt. Many plant species have the ability to alter gene expression and deal with these stresses through adaptation and adaptation while others cannot. However, non-pathogenic bacteria have shown the ability to prevent many diseases caused by these pathogens. To use beneficial microbes as a biological control, PGPM has been considered as an alternative and sustainable approach to replace pesticides and chemical fertilizer. Related bacteria and fungi naturally colonize the root hair and promote plant growth and development. PGPM has been considered an environmentally friendly and cost-effective way to control diseases. They provide protection against pathogens through the activation of cellular components, including cell proliferation, cell wall consolidation, and accumulation of secondary metabolites. Defense-related hormones including JA, ethylene, and salicylic acid (SA) play a key role in signaling and protection mechanisms (Bari and Jones 2009). Co-implantation of PGPR with mycorrhiza also ameliorated the deleterious effects of biotic stress. They protect plants from pathogens by improving growth attributes and reducing disease susceptibility (Dohroo and Sharma 2012). Interactions between plants and microorganisms in natural habitats are important for proper growth and development. They play an important role in nutrient mobilization and defense against pathogens (Shoebitz et al. 2009). Biological control of soil-borne diseases as an alternative contributed significantly to crop yield under abiotic stress conditions. Bacterial interactions with plant release induce and trigger physiological and biochemical changes in plants. These changes lead to disease resistance of the plant for several months.

Microorganism infections, for example, bacteria, fungi, and viruses, can cause plants to develop resistance to a future induced systemic resistance (Heil 2001). Systemic resistance caused by phytopathogens immunizes plants against broad-spectrum pathogens, which causes systemic resistance accompanied by PGPM due to heterotrophic compound production, ecological competition, and nutrients. Biochemical substances such as extra cells and antibiotics are effective against pathogens and inhibit their growth (Choudhary and Johri 2009). The defense mechanism induced by PGPM was first reported in response to the pathogen *Fusarium* sp. causing wilt of carnation (*Dianthus caryophillus*) and cucumber (*Cucumis sativus*) in response to the pathogen *Colletotrichum orbiculare*, causing leaf disease (Compant et al. 2005). In his study, the induced systemic resistance in *Panax ginseng* against *Phytophthora* Lee et al. (2015) reported that the link root *B. amyloliquefaciens* strain HK34 induced effective resistance against *P. cactorum*. In addition, strains of *Pseudomonas* and *Bacillus* manage plant disease in many cultures through the induced systemic resistance. *Paenibacillus*, P16, shows effective biocontrol agent (BCA) in cabbage against black rot (*Xanthomonas campestris*) and potentially of induced systemic resistance (Ghazalibiglar et al. 2016). PGPB species such as *Bacillus* strains cause systemic resistance in rice to leaf blight caused by *Xanthomonas oryzae pv. oryzae* (Chithrashree et al. 2011).

5.9 Conclusion

Different types of biotic and abiotic stress affect plant growth, yield, and viability properties. Plants undergo stressful conditions that can alter their physiological and biological properties by the expression of cold, heat, drought, salty, and alkaline-tolerant protein. These constraints are a major constraint on crop productivity, food quality, and global food security. Hormonal imbalance, nutrient mobilization, ionic toxicity, and disease susceptibility continue to affect plant growth and development due to different stresses in the present scenario. The only alternative to plant stress is to grow. Microbiological tools and techniques of plant-microbe-soil interaction. The application of stress-tolerant microbial consortium of PGPM strains and mycorrhizal fungi may be used for enhancing plant growth under abiotic and biotic stress conditions. These microbes could promote plant growth by regulating plant hormones, improve nutrition and siderophore production, and enhance the antioxidant system. Using microorganisms has the potential to solve future problems in food security and maintaining soil health. Therefore, under current assessment, microbiology can play an important role as an ecological engineer to solve problems of environmental stress. Therefore, with a current scenario, future research is needed to identify stress-tolerant PGPMs. Of course, the diversity of microorganism strains needs to be tested to form effective microbial populations to overcome the negative impact of environmental change.

References

Alam MA, Seetharam K, Zaidi PH, Dinesh A, Vinayan MT, Nath UK (2017) Dissecting heat stress tolerance in tropical maize (Zea mays L.). Field Crop Res 204:110–119

Ali M, Javaid A, Naqvi HS, Batcho A, Kayani KW, Aamir L, Imtiaz SA, Joy NO (2020) Biotic stress triggered small RNA and RNAi defense response in plants. Mol Biol Rep 47:5511–5522. https://doi.org/10.1007/s11033-020-05583-4

Bacilio M, Moreno M, Bashan Y (2016) Mitigation of negative effects of progressive soil salinity gradients by application of humic acids and inoculation with pseudomonas stutzeri in a salt-tolerant and a salt-susceptible pepper. Appl Soil Ecol 107:394–404

Bal HB, Nayak L, Das S, Adhya TK (2013) Isolation of ACC deaminase producing PGPR from rice rhizosphere and evaluating their plant growth promoting activity under salt stress. Plant Soil 366(1):93–105

Barassi CA, Ayrault G, Creus CM, Sueldo RJ, Sobrero MT (2006) Seed inoculation with Azospirillum mitigates NaCl effects on lettuce. Sci Hortic 109(1):8–14

Bari R, Jones JD (2009) Role of plant hormones in plant defence responses. Plant Mol Biol 69(4): 473–488

Barnawal D, Bharti N, Pandey SS, Pandey A, Chanotiya CS, Kalra A (2017) Plant growth promoting rhizobacteria enhance wheat salt and drought stress tolerance by altering endogenous phytohormone levels and TaCTR1/TaDREB2 expression. Physiol Plant 161:502–514. https://doi.org/10.1111/ppl.12614

Broos K, Uyttebroek M, Mertens J, Smolders E (2004) A survey of symbiotic nitrogen fixation by white clover grown on metal contaminated soils. Soil Biol Biochem 36(4):633–640

Chang CH, Yang SS (2009) Thermo-tolerant phosphate-solubilizing microbes for multi-functional biofertilizer preparation. Bioresour Technol 100(4):1648–1658

Chirakkara RA, Cameselle C, Reddy KR (2016) Assessing the applicability of phytoremediation of soils with mixed organic and heavy metal contaminants. Rev Environ Sci Biotechnol 15(2): 299–326

Chithrashree AC, Udayashankar S, Chandra N, Reddy MS, Srinivas C (2011) Plant growth-promoting rhizobacteria mediate induced systemic resistance in rice against bacterial leaf blight caused by *Xanthomonas oryzae pv. Oryzae*. Biol Control 59(2):114–122. https://doi.org/10.1016/j.biocontrol.2011.06.010

Chodak M, Gołębiewski M, Morawska-Płoskonk J, Kuduk K, Niklińska M (2015) Soil chemical properties affect the reaction of forest soil bacteria to drought and rewetting stress. Ann Microbiol 65(3):1627–1637

Choudhary DK, Johri BN (2009) Interactions of Bacillus spp. and plants–with special reference to induced systemic resistance (ISR). Microbiol Res 164(5):493–513

Compant S, Duffy B, Nowak J, Clément C, Barka EA (2005) Use of plant growth-promoting bacteria for biocontrol of plant diseases: principles, mechanisms of action, and future prospects. Appl Environ Microbiol 71(9):4951–4959

Danny Ng W-K, Abeysinghe KJ, Kamali M (2018) Regulating the regulators: the control of transcription factors in plant defense signaling. Int J Mol Sci 19:3737

Dohroo A, Sharma DR (2012) Role of plant growth promoting rhizobacteria, arbuscular mycorrhizal fungi and their helper bacteria on growth parameters and root rot of apple. WJST 2(12): 35–38

Duruibe JO, Ogwuegbu MOC, Egwurugwu JN (2007) Heavy metal pollution and human biotoxic effects. Int J Phys Sci 2(5):112–118

Farooq M, Wahid A, Kobayashi N S, Fujita DB, Basra S MA (2009) Plant drought stress: effects, mechanisms and management. In *Sustainable agriculture* (pp. 153–188). Springer, Dordrecht

Fatnassi IC, Chiboub M, Saadani O, Jebara M, Jebara SH (2015) Impact of dual inoculation with Rhizobium and PGPR on growth and antioxidant status of Vicia faba L. under copper stress. C R Biol 338(4):241–254

Ghazalibiglar H et al (2016) Is induced systemic resistance the mechanism for control of black rot in Brassica oleracea by a Paenibacillus sp.? Biol Control 92:195–201

Glick BR (2010) Using soil bacteria to facilitate phytoremediation. Biotechnol Adv 28(3):367–374

Goswami D, Thakker JN, Dhandhukia PC (2015) Simultaneous detection and quantification of indole-3-acetic acid (IAA) and indole-3-butyric acid (IBA) produced by rhizobacteria from l-tryptophan (Trp) using HPTLC. J Microbiol Methods 110:7–14

Gull A, Lone AA, Wani NUI (2019) Biotic and Abiotic Stresses in Plants. In (Ed.), Abiotic and Biotic Stress in Plants. Intech Open. https://doi.org/10.5772/intechopen.85832

Hashem A, Abd-Allah EF, Alqarawi AA, Al Huqail AA, Egamberdieva D, Wirth S (2016) Alleviation of cadmium stress in Solanum lycopersicum L. by arbuscular mycorrhizal fungi via induction of acquired systemic tolerance. Saudi J Biol Sci 23(2):272–281

Heba IM, Tony KS, Roshan S, Rami B, Youssef NS, Abdul B (2022) Plant-mediated copper nanoparticles for Agri-ecosystem applications. In: Abd-Elsalam KA, Periakaruppan R, Rajeshkumar S (eds) Nanobiotechnology for plant protection, agri-waste and microbes for production of sustainable nanomaterials, pp 79–120

Heil M (2001) Induced systemic resistance (ISR) against pathogens–a promising field for ecological research. Perspect Plant Ecol Evol Syst 4(2):65–79

Javani S, Marín I, Amils R, Abad JP (2015) Four psychrophilic bacteria from Antarctica extracellularly biosynthesize at low temperature highly stable silver nanoparticles with outstanding antimicrobial activity. Colloids Surf A Physicochem Eng Asp 483:60–69

Jiang S, Zhang D, Wang L, Pan J, Liu Y, Kong X, Li D (2013) A maize calcium-dependent protein kinase gene, ZmCPK4, positively regulated abscisic acid signaling and enhanced drought stress tolerance in transgenic *Arabidopsis*. Plant Physiol Biochem 71:112–120

Jing YX, Yan JL, He HD, Yang DJ, Xiao L, Zhong T, Li SB (2014) Characterization of bacteria in the rhizosphere soils of *Polygonum pubescens* and their potential in promoting growth and Cd, Pb, Zn uptake by *Brassica napus*. Int J Phytoremediation 16(4):321–333

Kasim WA, Gaafar RM, Abou-Ali RM, Omar MN, Hewait HM (2016) Effect of biofilm forming plant growth promoting rhizobacteria on salinity tolerance in barley. Ann Agric Sci 61(2): 217–227

Kotak S, Larkindale J, Lee U, von Koskull-Döring P, Vierling E, Scharf KD (2007) Complexity of the heat stress response in plants. Curr Opin Plant Biol 10(3):310–316

Krishnamoorthy R, Kim K, Subramanian P, Senthilkumar M, Anandham R, Sa T (2016) Arbuscular mycorrhizal fungi and associated bacteria isolated from salt-affected soil enhances the tolerance of maize to salinity in coastal reclamation soil. Agric Ecosyst Environ 231:233–239

Kumar A, Singh S, Gaurav K, Srivastava S, Verma JP (2020) Plant growth-promoting bacteria: biological tools for the mitigation of salinity stress in plants. Front Microbiol, 07 Sec. Microbiotechnology. https://doi.org/10.3389/fmicb.2020.01216

Kumar V, Nautiyal CS (2022) Plant abiotic and biotic stress alleviation: from an endophytic microbial perspective. Curr Microbiol 79:311. https://doi.org/10.1007/s00284-022-03012-2

Lee BD, Swarnalee D, HojinRyu S-JY, Dong-Sang S, Kyungseok P (2015) Induction of systemic resistance in *Panax ginseng* against *Phytophthora cactorum* by native *Bacillus amyloliquefaciens* HK34. J Ginseng Res 39(3):213–220. https://doi.org/10.1016/j.jgr.2014.12.002

Li L, Ye Y, Pan L, ZhuY ZS, Lin Y (2009) The induction of trehalose and glycerol in Saccharomyces cerevisiae in response to various stresses. Biochem Biophys Res Commun 387(4): 778–783

Li NE, Cha JY, Lin Z, Lu M, Huang LJ, Kim WY (2021) Plant hormone-mediated regulation of heat tolerance in response to global climate change. Front Plant Sci 11:2318

Marastoni L, Pii Y, Maver M, Valentinuzzi F, Cesco S, Mimmo T (2019) Role of Azospirillumbrasilense in triggering different Fe chelate reductase enzymes in cucumber plants subjected to both nutrient deficiency and toxicity. Plant Physiol Biochem 136:118–126. https://doi.org/10.1016/j.plaphy.2019.01.013

Marulanda A, Barea JM, Azcón R (2009) Stimulation of plant growth and drought tolerance by native microorganisms (AM fungi and bacteria) from dry environments: mechanisms related to bacterial effectiveness. J Plant Growth Regul 28:115–124. https://doi.org/10.1007/s00344-009-9079-6

Meena RK, Singh RK, Singh NP, Meena SK, Meena VS (2015) Isolation of low temperature surviving plant growth–promoting rhizobacteria (PGPR) from pea (*Pisum sativum* L.) and documentation of their plant growth promoting traits. Biocatal Agric Biotechnol 4(4):806–811

Naseem H, Bano A (2014) Role of plant growth-promoting rhizobacteria and their exopolysaccharide in drought tolerance of maize. J Plant Interact 9(1):689–701

Navarro JM, Pérez-Tornero O, Morte A (2014) Alleviation of salt stress in citrus seedlings inoculated with arbuscularmycorrhizal fungi depends on the rootstock salt tolerance. J Plant Physiol 171(1):76–85

Nejat N, Mantri N (2017) Plant immune system: crosstalk between responses to biotic and abiotic stresses the missing link in understanding plant defence. Plant Immune Syst Mol Biol:1–16

Pandey P, Irulappan V, Bagavathiannan MV, Senthil-Kumar M (2017) Impact of combined abiotic and biotic stresses on plant growth and avenues for crop improvement by exploiting physio-morphological traits. Front Plant Sci 8:537

Prelich G (2012) Gene overexpression: uses, mechanisms, and interpretation. Genetics 190(3): 841–854. https://doi.org/10.1534/genetics.111.136911. PMID: 22419077; PMCID: PMC3296252

Qian J, Li D, Zhan G, Zhang L, Su W, Gao P (2012) Simultaneous biodegradation of Ni–citrate complexes and removal of nickel from solutions by *Pseudomonas alcaliphila*. Bioresour Technol 116:66–73

Ramegowda V, Senthil KM (2015) The interactive effects of simultaneous biotic and abiotic stresses on plants: mechanistic understanding from drought and pathogen combination. J Plant Physiol 176:47–54

Seleiman MF, Al-Suhaibani N, Ali N, Akmal M, Alotaibi M, Refay Y, Dindaroglu T, Abdul-Wajid HH, Battaglia ML (2021) Drought stress impacts on plants and different approaches to alleviate its adverse effects. Plants (Basel) 10(2):259. https://doi.org/10.3390/plants10020259. PMID: 33525688; PMCID: PMC7911879

Selvakumar G, Panneerselvam P, Ganeshamurthy A (2012) Bacterial mediated alleviation of abiotic stress in crops. https://doi.org/10.1007/978-3-642-23465-1_10

Shaffique S, Khan MA, Imran M, Kang SM, Park YS, Wani SH, Lee IJ (2022) Research progress in the field of microbial mitigation of drought stress in plants. Front Plant Sci:13

Shoebitz M, Ribaudo CM, Pardo MA, Cantore ML, Ciampi L, Curá JA (2009) Plant growth promoting properties of a strain of *Enterobacter ludwigii* isolated from *Lolium perenne* rhizosphere. Soil Biol Biochem 41(9):1768–1774

Tiwari S, Lata C, Chauhan PS, Nautiyal CS (2016) *Pseudomonas putida* attunes morphophysiological, biochemical and molecular responses in *Cicer arietinum* L. during drought stress and recovery. Plant Physiol Biochem 99:108–117. https://doi.org/10.1016/j.plaphy.2015.11.001

Vardharajula S, Zulfikar AS, Grover M, Reddy G, Bandi V (2011) Drought-tolerant plant growth promoting Bacillus spp.: effect on growth, osmolytes, and antioxidant status of maize under drought stress. J Plant Interact 6(1):1–14

Verma JP, Yadav J, Tiwari KN, Kumar A (2013) Effect of indigenous Mesorhizobium spp. and plant growth promoting rhizobacteria on yields and nutrients uptake of chickpea (Cicer arietinum L.) under sustainable agriculture. Ecol Eng 51:282–286

Verma V, Ravindran P, Kumar P, Prakash (2016) Plant hormone-mediated regulation of stress responses. BMC Plant Biol 16:86

Vimal SR, Singh JS, Arora NK, Singh S (2017) Soil-plant-microbe interactions in stressed agriculture management: a review. Pedosphere 27(2):177–192

Yadav J, Verma JP, Jaiswal DK, Kumar A (2014) Evaluation of PGPR and different concentration of phosphorus level on plant growth, yield and nutrient content of rice (*Oryza sativa*). Ecol Eng 62:123–128

Yang H, Huntingford C, Wiltshire A, Sitch S, Mercado L (2019) Compensatory climate effects link trends in global runoff to rising atmospheric CO_2 concentration. Environ Res Lett 14:124075. https://doi.org/10.1088/1748-9326/ab5c6f

Zhang J, Wang LH, Yang JC, Liu H, Dai JL (2015) Health risk to residents and stimulation to inherent bacteria of various heavy metals in soil. Sci Total Environ 508:29–36

Chapter 6
Plant-Microbe Interaction and Their Role in Mitigation of Heat Stress

Divya Chouhan, Piyush Mathur, and Chandrani Choudhuri

Abstract Out of all the abiotic stress, heat stress is considered one of the detrimental factors that limit crop yield worldwide. Temperature modulates the normal physiology, morphology, genetic behavior, and a series of biochemical events in plants. Researchers are endlessly putting their efforts into solving this global problem through biological means. The use of beneficial microorganisms for the elevation of heat stress in plants is an emerging horizon of today's research. Interaction of plants with soil microbes such as plant growth-promoting rhizobacteria (PGPR), arbuscular mycorrhizal fungi (AMF), and bacterial or fungal endophytes helps in the mitigation of heat or chilling stress and promotes plant growth. These microorganisms in interaction with plants stimulate the production of essential phytohormones, secondary metabolites, organic acids, and amino acids that helps the plant to overcome heat stress. This chapter mainly focuses on the amelioration of temperature stress through the use of beneficial soil microbes.

Keywords Arbuscular mycorrhizal fungi · Endophytes · Heat shock protein · Plant growth-promoting rhizobacteria · Rhizosphere

Abbreviations

ABA	Abscisic acid
AMF	Arbuscular mycorrhizal fungi
APX	Ascorbate peroxidase
GR	Glutathione reductase

D. Chouhan · P. Mathur
Microbiology Laboratory, Department of Botany, University of North Bengal, Darjeeling, West Bengal, India
e-mail: rs_divyac@nbu.ac.in; piyushmathur316@nbu.ac.in

C. Choudhuri (✉)
Department of Botany, North Bengal St. Xaviers' College, Jalpaiguri, West Bengal, India

P. Mathur et al. (eds.), *Microbial Symbionts and Plant Health: Trends and Applications for Changing Climate*, Rhizosphere Biology,
https://doi.org/10.1007/978-981-99-0030-5_6

HSF3	Heat shock factor 3
HsfB1	Heat inducible transcription factor
HSP	Heat shock protein
IAA	Indole-acetic-acid
MBF1c	Multi-protein bridging factor
PGPB	Plant growth-promoting bacteria
PGPR	Plant growth-promoting rhizobacteria
PIP	Plasma membrane intrinsic protein
ROS	Reactive oxygen species
SA	Salicylic acid
SAMS	S-adenosylmethionine synthetase
SOD	Superoxide dismutase

6.1 Introduction

Soil shelters a broad range of microorganisms like fungi, bacteria, viruses, protists, pests, and nematodes that may be beneficial, neutral, or pathogenic. Soil microbiomes are the key indicators of plant health and play a pivotal role in the management of growth and productivity (Berg et al. 2016). Moreover, it plays an important role in the regulation of plant health by synchronizing their antagonistic and symbiotic relationships with plants. In particular, metabolic cooperation, microbial signaling-induced responses, and microbial dysbiosis influence the overall plant health. Researchers have established the basic principle of plant-microbe interaction, including the signaling responses that influence the plant-microbiome symbiotic relationship (Jones et al. 2016), the microbial genetic system transporting signaling molecules that stimulate host cell performances (Hwang et al. 2017), and certain binary and community level that disputes between plant-microbiome interaction (Hacquard et al. 2017). Microbes differ in their metabolism and sensitivity to altered climatic conditions. This may affect the assemblage of plant microbiome and their symbiotic interactions. Microbes decipher dynamic interactions with plants and can improve their competence during changing environmental stress (Trivedi et al. 2020) Microbes are also known to leach extracellular polymeric substances into the soil to boost water retention capacity. Soil microbes harbor carbon as a source of cellular biomass which are fetched into stable metabolites. Sequestration of carbon in the soil occurs as dead microbial biomass or necromass (Deltedesco et al. 2020). Moreover, the plant growth-promoting rhizobacteria which are symbiotic nitrogen fixers enhance nutrient uptake with the help of mycorrhizal fungi and regulate growth-promoting hormones like indole-acetic-acid (IAA) (Pati and Padhi, 2022). Through differential nutrient cycling in the soil, roots directly control the microbial diversity and its load. The most beneficial contribution of climate change on microbiome diversity is its genetical activation for the growth and productivity of plants (Bradford et al. 2008).

A number of natural and anthropogenic activities have resulted in the rise of the average temperature and levels of carbon dioxide which are responsible for the change in the global climatic condition. Changes in the overall climatic conditions may result in undistributed rainfall, overindulgent release of greenhouse gases that brought about rise in air temperature, and increased dryness with low moisture level resulting in drought-like conditions (Sharma et al. 2022). These climatic variations affect food security and crop yield differently in regions of higher and lower latitudes. Studies revealed that variations in the climatic condition resulted in excessive heat stress leading to extensive crop failure (Huntington et al. 2017). These gross alternations in the climatic condition enormously influence the growth pattern of various crop species. It mainly impacts on the acquisition of nutrients and minerals in the plants. Due to the excessive heat stress condition, plants showed imbalanced morphological and physiological changes that aid to phenotypic and productivity alterations. Increment in the oxidative stress due to raised temperature, drought, and other climatic variations may hinder the production of proteins, carbohydrates, and secondary metabolites in plants (Sharma et al. 2022).

Heat stress also influences the course of irrigation, soil fertility, pests and disease incident that ultimately affects the increasing global food demand and trade opportunities (Raza et al. 2019). It is often studied that increasing global warming decreases the moisture content of the soil and obstruct the microbial dispersion, survival and colonization in the soil space (Carson et al. 2010). A sudden rise in the ambient temperature leads to excessive heating of the soil that can modify the rhizospheric microbiome structure developed post-interaction with plants (Okubo et al. 2014). Due to the increased temperature, heat stress has become one of the detrimental factors causing increased evaporation of soil moisture and reducing the establishment of extra mycorrhizal mycelium (Wu and Xia 2006). Researchers have also confirmed that extensive heat dryness of the soil causes loss of photosynthates in the plants formed during photosynthesis (Figueiredo et al. 2008). These comprehensive climatic variations cause a negative impact on the plant growth and developmental pattern, as well as on the microbiome present in the soil. Excessive heat followed by drought and elevated CO_2 levels impacts on soil microbial dynamics, soil biomass development, and microbial community construction (Singh et al. 2019).

Beneficial soil microbiome such as plant growth-promoting rhizobacteria (PGPR), fungi, algae, actinomycetes, cyanobacteria, etc., protect plants from heat stress by modulating various physiochemical and biochemical processes of plants. Therefore, this chapter aims in elaborating the effects of increasing temperature on the plant-microbe interaction mediating ecological corollary. The existence of plant-0beneficial microbial population in the soil and change of their functioning with climatic resilience are described in a designed framework for the better understanding of the collaboration of plant and microbes.

6.2 Plant and Soil Microbiome Interaction

Plants and microbes are the two highly interacting biological components of the terrestrial ecosystem. Their mode of interaction may be mutualistic, parasitic, or competitive that further facilitates certain ecological services like mineralization, decomposition, and nutrient cycling (Brundrett 2009). Plants and soil generally shield a huge range of microbiomes that may present either in the endosphere or episphere of plant tissue. The microbiome structure differs notably from rhizosphere to endosphere and phyllosphere in the ecosystem (Trivedi et al. 2020). Plants and diversified microbiomes circumscribe to form holobiont, which is defined as the congregation of various species with the surrounding host, thus forming an ecological unit. Plant-microbiome interaction serves differential ecological functions according to their location above or below the ground level. Their site of habitation potentially controls the resistance to climatic resilience and various stress responses of plants (Sharma et al. 2022).

The aerial segment of the plant termed as phyllosphere contains various microorganisms like fungi, bacteria, actinomycetes, viruses, and others. The composition of the microbiome may also vary according to the plant species. The phyllospheric community is very dynamic, and its function depends upon various factors like temperature, moisture, precipitation, light, etc. The bacterial community present in the phyllosphere is generally hyperdiverse and non-pathogenic. Microorganisms also render anchorage support to the roots of plants. Roots in association with the soil contain manifold rhizospheric as well as endophytic microfauna. The varied microbiota of the rhizosphere may disperse through the soil, water, air, insects, and animals. Secretion of roots or root exudates such as a range of organic acids, amino acids, and certain metabolites have uttered importance in the formation of the microenvironment surrounding the root. Beneficial microorganisms penetrate the root through the root tip region and facilitate processes like symbiotic nitrogen fixation or may help the plant to overcome various stress responses like drought, salinity, etc. (Menge et al. 2019). Irrespective of pathogenic ones, the microbiome that establishes a beneficial and mutualistic relationship with the plants includes genera of *Agrobacterium*, *Rhizobium*, *Pseudomonas*, and *Cupriavidus* (Jin et al. 2018; Xu et al. 2018).

Scientists have confirmed that PGPR plays a pivotal role to eliminate the toxic effect of contaminants from the environment, i.e., phytoremediation. A study conducted on *Helianthus tuberosus* affirmed that PGPR as an endophyte facilitates elevated resistance capacity against raised concentrations of cadmium and zinc in the soil (Montalban et al. 2017). Also, the microfauna of the soil surrounding the *Populus* genus contains members of Acidobacteria and Actinobacteria which helps in the mitigation of certain abiotic stress responses like excessive heat, water limitation, and light limitation (Timm et al. 2018). Besides that, a few species of orchids harbor members of Ascomycota and Basidiomycota as beneficial endophytes and are generally called as Orchidaceous mycorrhizal fungi. Hence, the microbiome association may also differ geographically (Rudgers et al. 2020).

These organisms convert organic matters into simpler and usable forms to the plants. Certain filamentous fungi and bacteria emit secondary metabolites for regulating plant growth and some phenological responses (Sharma et al. 2022). Rhizobacteria helps in nutrient acquisition by regulating the production of phytohormones – auxin and cytokinins (Kudoyarova et al. 2014; Egamberdieva et al. 2017). According to a study, it is revealed that species of *Trichoderma* produces a range of secondary metabolites such as α-pyrone, trichocarenes, harzianic acid, and harzianolide, which are known to induce plant growth. Research has been also conducted on turmeric rhizome colonized by several endophytic and rhizospheric bacteria stimulating the production of curcumin for modulating growth and management of diseases (Kumar et al. 2017). Studies showed that cereal, pulses, and oilseeds endure both mycorrhizal and endophytic assemblages of enhanced nutritional importance (Bhattacharyya et al. 2016). The association of a range of mycorrhizal bacterium including species of *Bacillus*, *Rhizobium*, *Mycobacterium*, and *Arthrobacter* are found colonizing in the roots of soybean promoting their growth (Egamberdieva et al. 2016). Besides that, orchid rhizosphere contains species of *Mycobacterium*, while when rhizosphere of wheat is harbored with species of *Azospirillum*, *Azotobacter*, *Rahnella*, and *Mycoplana*, they confirm resistance against drought (Tsavkelova et al. 2007). Another endophytic fungus named *Sinorhizobium meliloti* found in the root nodules of *Medicago sativa* provides resistance against drought conditions (Naya et al. 2007). *Bacillus* sp. and *Leifsonia* sp. found in maize rhizosphere promote cadmium tolerance and induce root-shoot growth (Ahmad et al. 2016).

There are multiple algal species which serve as biocontrol agent, such as *Anabaena*, *Calothrix*, and *Chlorella fusca*, which provide protection against phytopathogens (Prasanna et al. 2008, 2013; Chaudhary et al. 2012; Lee et al. 2016; Kim et al. 2018). Algae present in the soil microbiome protects plant against nematode infection (Khan et al. 2005). A range of cyanobacteria acts as biofertilizer enhancing plant growth and yield, often fixes atmospheric nitrogen in rice fields, or used as green manure. Their presence in the soil increases its porosity and aggregation properties (Vaishampayan et al. 2001).

There are many algal species reported which provide resistance against abiotic stress. *Scytonema hofmanni* facilitates salt tolerance when exposed to rice fields (Rodríguez et al. 2006). In *Erinus alpinus*, senescence is delayed by the application of *Chlorella fusca* and strain ABC001 and HS2 (Lee et al. 2020). It has been reported that *Desmodesmus pleiomorphus*, *Chlorella vulgaris*, *Ulva reticulata*, *Spirogyra insignis*, and *Cladophora fascicularis* are efficient in removing metal toxicity of the soil laden with heavy metals like cadmium, zinc, and copper (Hassan et al. 2017). Henceforth, both rhizobacteria and rhizofungi have essential role in regulating defense mechanism of plants against biotic and abiotic stress through the stimulation of defense hormones, bioactive metabolites, and synthesis of defensive proteins and enzymes (Rashid and Chung 2017).

6.3 Effect of Elevated Temperature on Plant-Microbe Interactions

Temperature stands as an important fundamental factor that regulates plant growth and development by stimulating its metabolism and physiology. This factor also affects the conglomeration of plant-microbe association in the phyllospheric and rhizospheric region. The density of the microbial population is primarily affected by the increase or decrease of this factor (Kashyap et al. 2017; Angilletta 2009). Increased emission of greenhouse gases like CO_2, methane, water vapor, etc., and certain human interferences with the environment are mainly responsible for the elevation of temperature. A study suggested that the mean temperature will raise by 1.8–3.6 °C by the year 2100 that brought about conditions of excessive heat, water deficiency, and drought (Compant et al. 2019). Both phenotypic and genotypic characters of the plant are influenced by the raise in temperature. This raise in the temperature exponentially increases rate of microbial respiration (Classen et al., 2015). Researchers have confirmed that the utilization of organic matter by the soil microbiome is a temperature-dependent process (Frey et al., 2013). Temperature also plays an important role in the establishment of pathogen within the host to form the disease, while the beneficial associations are also temperature-dependent (Vela'squez et al. 2018). Excessive warming of the soil hampers nutrient uptake and mobilization and exchange of carbohydrate in between plant and rhizosphere (Newsham et al. 1995). Arbuscular mycorrhizal fungi like *Glomus intraradices* and *G. mossae* exhibit increased rate of colonization with the increasing temperature (Sharma et al. 2022). Upliftment of the temperature than the normal increases the virulence of the soft rot causing bacteria – *Pectobacterium atrosepticum* – and stimulates the degradation of cell wall enzymes (Hasegawa et al. 2005). Elevated temperature accomplished with excessive heat waves take part in mineralization of microbial necromass (Donhauser et al., 2020). Due to rising temperature the relative abundance of certain methanotrophs such as, *Methylosinus* and *Methylocystis* increases in the soil microbiome (Okubo et al. 2014). Elevation in the mean temperature effects the PGPR population in the soil. Under such conditions, PGPR population contributes to increase the organic carbon containing root exudates (Kachhap et al. 2015). High temperature often impacts on the expression of proteins involved in carbon utilization (Mosier et al. 2015). High temperature influences pre-host-rhizobia interaction stages through molecular signal exchange between nodules and rhizosphere (Hungria and Vargas 2000). During excessive heat stress, trehalose synthase complex is formed by the microbes which are responsible for the breakdown of this carbohydrate reserve food—trehalose. This trehalose cycling helps in the activation of anabolic and catabolic enzymes, thereby helping in overcoming heat stress condition, observed for the first time in yeast, *Saccharomyces cerevisiae* (Hottiger et al. 1987).

6.4 Microbes as a Stress Ameliorating Agent under Temperature Stress

Microbes are established biological control agents that can combat heat stress. Inoculation with microbes helps in ameliorating temperature stress. Microbes are known to release exoploysaccharides into the soil which contain 97% water and maintain the soil structure under heat stress (Shaffique et al. 2022). Microbes contribute to the biomass production and higher the net photosynthetic production rate (Zhu et al. 2017). Different microorganisms confirm differently to thermotolerance (Fig. 6.1). The three major categories of thermotolerant microbes are discussed below.

6.4.1 PGPR

Plant root colonizing bacteria along the rhizospheric axis, promoting to plant growth are defined as plant growth promoting rhizobacteria. They regulate plant growth directly by the solubilization of phosphate, nitrogen fixation, iron sequestration and overall maintaining the nutritional status of the plant. They help in the synthesis of various heat stress regulators, viz., hormones like IAA, gibberellin, cytokinin, etc.; they help plants indirectly by boosting immunity to combat against biotic or abiotic stress. PGPR helps in the secretion of compatible solutes, such as proline, glycine betaine, sugars, and organic acids (Shaffique et al. 2022; Basu et al. 2021; El-Sawah et al. 2021).

Researchers have proved that PGPR are helpful in the mitigation of heat stress. In an experiment, *Triticum aestivum* was inoculated with *Bacillus amyloliquefaciens* UCMB5113 and given exposure of short-term heat stress. Results revealed that inoculated plants showed reduced ascorbate peroxidase (APX1), glutathione reductase (GR), S-adenosylmethionine synthetase (SAMS1), and heat-shock protein (HSP17) expression. PGPR-inoculated plants showed significantly more survival rates than the non-inoculated ones (Issa et al. 2018). Another report suggests that PGPR helps to ameliorate heat stress generated in *Triticum aestivum* by the application of *Pseudomonas aeruginosa* strain 2CpS1 (Meena et al. 2015). Rhizobia stimulates the activation of heat shock proteins (HSPs) that facilitates thermotolerance (Alexandre and Oliveira 2011). Soybean plants when inoculated with *Bacillus cereus* stimulate thermoregulation under heat stress and enhance the levels of chlorophyll a and b and other accessory pigments like carotenoids, proteins, and stress-related enzymes (Shaffique et al. 2022; Bisht et al. 2020). In recent years, a range of PGPRs such as *Pseudomonas putida*, *P. fluorescens*, *B. cereus*, and *Serratia liquifaciens* applied on *Triticum* spp., *Glycine max*, *Solanum lycopersicum* L., and *Cajanus cajan* raised the production of essential phytohormones, antioxidant enzymes, and ACC deaminase for the mitigation of heat stress in plants (Shaffique et al. 2022) (Table 6.1). There are reports for the application of *Burkholderia*

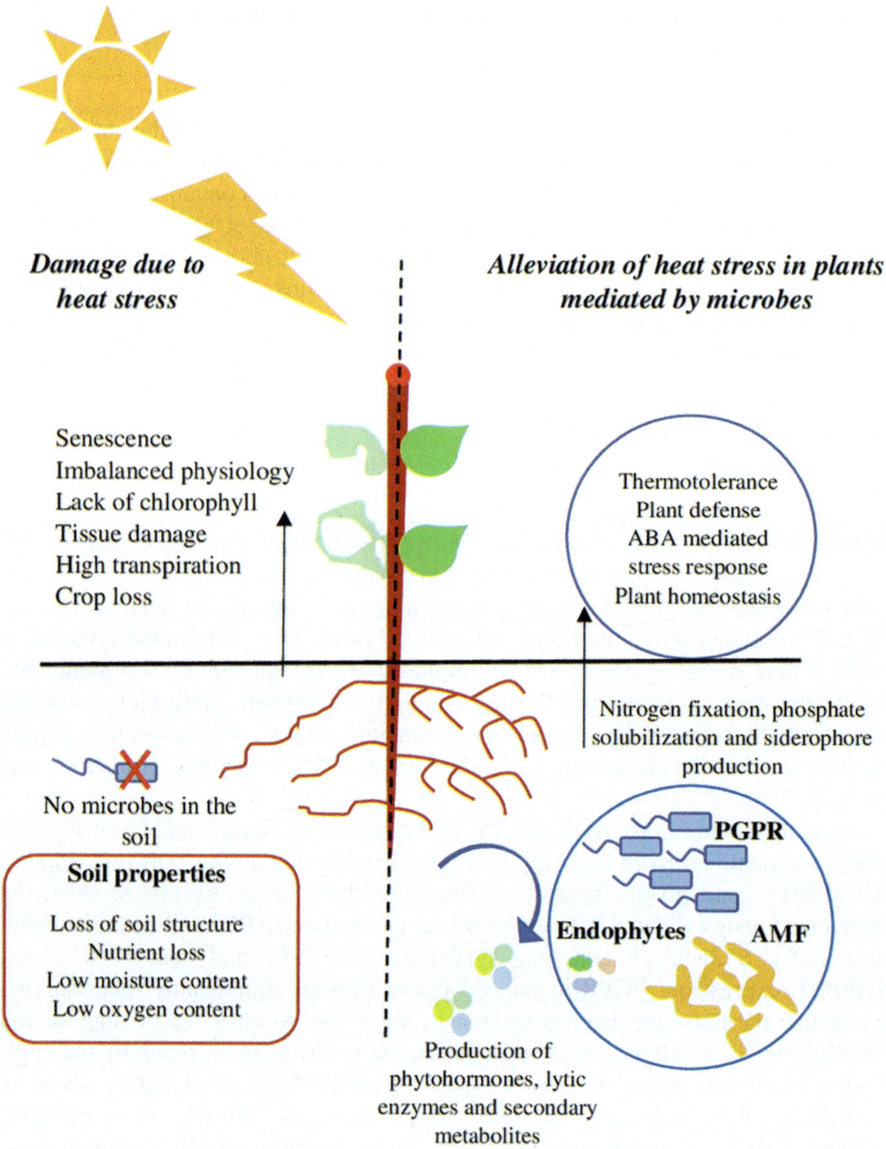

Fig. 6.1 Role of different soil microbes in alleviation of heat stress in plants

phytofirmans PsJN and *Curvularia proturberata* Cp4666D on *T. aestivum*, *S. lycopersicum* L., and *Dichanthelium lanuginosum* for the enhancement of thermotolerance in plants. They suggested that microbes increased the production of IAA, cytokines, pathogenesis-related protein, chlorophyll levels, and generation of free radicals (Rana et al. 2021). It is said that soybean is highly sensitive to temperature stress. In an experiment conducted at Republic of South Korea, heat

Table 6.1 Application of different beneficial microorganisms for the mitigation of heat stress

Host plant	Symbiont microbes	Model	Stress exposure	Outcome	Related HSF	Reference
Triticum spp. and *Arabidopsis thaliana*	*Enterobacter* SA187	Field experiment	22 °C/ 37 °C/ 44 °C	Increased biomass, plant height, grain yield, and seed weight	APX2, HSP18.2, transcription factors HSFA1A, B, D and E	Shekhawat et al. (2020)
Arabidopsis thaliana	*Nostoc muscorum*	Lab experiment	25, 35, 45, 50, 55, 65, 75, or 85 °C	Increased proline content	AtProT1, AtProT2 and AtProT3	Chua et al. (2020)
Glycine max	*Bacillus tequilensis* SSB07	Lab experiment	28 °C and 24 °C	Increased gibberellins, IAA and ABA, jasmonic acid, and salicylic acid contents	GA1, GA3, GA5, GA8, GA19, GA24, and GA53	Kang et al. (2019)
Glycine max	*B. cereus* SA1	Lab experiment	28 ± 1 °C	Accumulation of glucose, fructose, glucitol, oleic acid, gulonic acid, raffinose, inositol	ITPK1, PFK1	Bisht et al. (2020)
Solanum lycopersicum	*Septoglomus deserticola* and *Septoglomus constrictu*	In vitro	26/20 °C	Improved stomal conductance, water content, and leaf water content	SlLOXD, SlPIP2.7	Duc et al. (2018)
Lycopersicon esculentum	*Paraburkholderia phytofirmans* PsJN	Greenhouse experiment	32 °C	Increased growth, biomass chlorophyll content	CHI3, TIV1, Frk2, Hxk1 and Hxk2RbcS, RbcL	Issa et al. (2018)
Cucumis sativus	*Thermomyces* sp.	Field experiment	45 °C	Metabolites and stress-related enzymes are increased	HSPs	Ali et al. (2017)
Glycine max	*Bacillus aryabhattai* SRB02	Lab experiment	38 °C /30 °C	Increased IAA, Jasmonic acid, and growth parameters	GA12, GA4, and GA7	Park et al. (2017)
Triticum aestivum (HUW-234)	*Pseudomonas aeruginosa* 2CpS1	Net house experiment	30 °C–47 °C	Increased plant height, root length, and chlorophyll content	ACC deaminases	Meena et al. (2015)

(continued)

Table 6.1 (continued)

Host plant	Symbiont microbes	Model	Stress exposure	Outcome	Related HSF	Reference
Triticum aestivum	*Bacillus amyloliquefaciens* UCMB5113	Field experiment	28 °C/ 45 °C	Survival rate increased	HSP17, APX, GR	Abd El-Daim et al. (2014)
Triticum aestivum	*Pseudomonas putida* AKMP7	Pot experiment	40 °C	Increased root and shoot length, biomass, CAT, SOD and APX	IAA, ACC deaminase and siderophores	Ali et al. (2011)
Glycine max	*Aeromonas hydrophilla, Serratia liquefaciens, S. proteamaculans*	In vitro	Range of high to low temperature	Exopolysaccharide production	Phytohormones	Maitra et al. (2021)
Sorghum	*Pseudomonas* AKM-P6	Lab experiment	47–50 ° C day/ 30–33 °C night	Increase in cellular metabolites	HSP101, HSP70	Ali et al. (2009)

stress was given in a day-night cycle of 16 h at 38 °C and 8 h at 30 °C for 7 days. For ameliorating heat stress, the bacterial strain *B. tequilensis* SSB07 was inoculated to stimulate the production of various phytohormones and growth of the seedling. Increased shoot length and biomass of the plants were observed (Kang et al. 2019). Sorghum seeds were inoculated with *Pseudomonas* sp. AKM-P6 strain at the concentration of 108 cells/gm before sowing. Temperature stress were given at a cycle of 47–50 °C at the day time and 30–33 °C at the night for a period of 10 days. It was observed that the PGPR-inoculated seedlings showed enhanced growth and biochemical parameters even at the 15th, day but the non-inoculated seedlings died at the fifth day due to heat stress. It was inspected that due to bacterial inoculation, high molecular weight proteins were synthesized in the leaves. These proteins modulate the physiology and biochemical parameters of the sorghum seedlings that help the plant to survive under heat stress (Ali et al. 2009). Once, the seeds of *T. aestivum* were soaked with two strains of PGPR (1×10^7 cfu/mL) at 28 °C for 2 h and observed for 12 days in growth chamber under 45 °C. The treatment enhanced the level of APX1, SAMS1, HSP17.8, heat shock factor 3 (HSF3), heat inducible transcription factor (HsfB1), and multi protein bridging factor (MBF1c). Bacterial inoculation leads to increased level of transcription factors related to heat stress (Abd El-Daim et al. 2014).

6.4.2 *Arbuscular Mycorrhizal Fungi (AMF)*

For several years, it has been studied that arbuscular mycorrhiza forms a symbiotic relation with plants and benefits them with water access and nutrients, extending their growth. AMF contributes to plant biomass production and photosynthetic factors (Fig. 6.1). It also contributes to the rigidity of the plasma membrane, when injured by heat stress. AMF colonization in the roots of the host plant reduces the generated oxidative stress through different enzymatic reactions. Compatible osmolytes were also found to be increased in the host plant that increases thermotolerance. This symbiosis often enhances carbohydrate metabolism and stimulates the production of secondary metabolites that relaxes the host plant during high or low-temperature stress (Zhu et al. 2017).

There are several reports that arbuscular mycorrhiza (AM) plants have greater root and shoot dry weight than the plants without AM (Table 6.1). AMF increases the leaf area, leaf number, and crown diameter of *Fragaria ananassa* under temperature stress (Matsubara et al. 2004). Scientists have observed that AM plant *Dichanthelium lanuginosum* showed increased root length and root diameter when exposed to temperature stress (Bunn et al., 2009). *Rhizophagus irregularis* colonized on the roots of *Medicago truncatula* increases the seed number and plant biomass under low-temperature stress of 1.53 °C (Hu et al. 2015). *Glomus intraradices* and *G. ambisporum* inoculated on *Gossypium hirsutum* increase plant growth and availability of phosphorus, zinc, copper, and manganese content (Smith and Roncadori 1986). *G. etunicatum* when applied on the roots of *Fraxinus pennsylvanica* during low-temperature stress of 12, 16, and 20 °C short increased relative growth rate of the leaf area, plant weight, plant-to-leaf weight ratio, and mean leaf size (Andersen et al. 1987).

Water plays the key role in the management of growth and physiology of the plant. With the change in ambient temperature, the uptake of the water by the plants is an important variable. AMF colonizing on the maize plant showed relatively higher water holding capacity and water conservation in the soil during heat stress (Zhu et al. 2010a, b; Liu et al. 2014a, b). During chilling stress, AMF colonized on bean plants showed greater leaf water potential. AMF enhances the hydraulic conductivity of the root and increases water extraction by their external hyphal structure. The maize-AMF symbiosis also contributes to the stomatal opening in the leaf along with stomatal conductance and increased transpiration rate during high- or low-temperature stress (Zhu et al. 2017). A group of researchers suggested that AMF facilitates the regulation of aquaporins in their host plant. It also regulates the expression of genes related to plant aquaporin for the passive flow of water through the membrane. During low-temperature stress, the plasma membrane intrinsic protein (PIP) is synthesized, and PIP1, PIP2, and PIP3 genes are expressed (Liu et al. 2014a, b).

AMF averts protein misfolding under heat-stress conditions (Estrada et al. 2013). AMF facilitates the accumulation of trehalose in the rhizosphere and provides thermotolerance (Lenoir et al. 2016). Plants are known to synthesize various organic compounds for the mitigation of temperature stress, suitably osmolytes and amino acids. Many plants accumulate proline to scavenge hydroxyl radicals and acidity in the cell. Proline acts as a sink of energy for regulating redox potential. Stress generated due to low temperature in maize showed accumulation of proline in the leaves (Zhu et al. 2017). It was suggested that change in the proline level of leaves reflects the level of injury in the plants due to temperature stress (Zhu et al. 2010a, b). The colonization of AMF in cucumber plant denotes the reduction of H_2O_2 in the leaves produced during temperature stress (Fig. 6.1). Lower H_2O_2 level acts as signaling molecule for defense response during heat stress (Chen et al. 2013). It is known that NADPH oxidase produces $O_2^{\cdot-}$ and converts it to H_2O_2 by the action of superoxide dismutase (SOD) in plants. AMF reduces the NADPH oxidase activity and inhibits H_2O_2 accumulation in plants during temperature stress (Liu et al. 2014a, b).

Adverse temperature stress also affects the synthesis of chlorophyll and other accessory pigments in the plants. Various authors reported that AMF increases the chlorophyll biosynthesis and protects the light harvesting complex in plants. AMF ameliorates the damage caused in mesophyll chloroplast due to excessive heat and increases the photosynthetic efficiency. AMF symbiosis retards the toxic effect of temperature stress from PSII reaction center and improves electron transport rate and quantum yield in plants (Zhu et al. 2017). A group of researchers reported that the decrease in the ratio of maximum quantum efficiency of PSII (Fv/Fm) and that of potential photochemical efficiency of PSII (Fv/Fo) are potential indicators of damage caused due to temperature stress. This ratio was found comparatively higher in maize plants inoculated with AMF (Zhu et al. 2010a, b, 2017). Thus, the presence of

AMF supports thermotolerance in plants and leads to a sustainable agricultural system.

6.4.3 Endophytes

Microorganisms that live within the plant cells forming biofilms and producing exudates are generally called endophytes. Endophytes act as biostimulants for producing various compounds in plants. The majority of the endophytes are inaccessible due to their localization within the plant tissue. Similar to AMF and PGPR, endophytes also maintain symbiotic relationship with plants. They are known to produce various kinds of bioactive compounds that stimulate the growth and development of the plant. They often produce biochemicals that cannot be synthesized within the plant body (Shaffique et al. 2022). Researchers have explored the mode of action of endophytes and their growth-promoting traits in plants in consideration to heat stress (Fig. 6.1).

In a field experiment, an endophytic strain of *Enterobacter* SA187 was inoculated on *Arabidopsis thaliana* and wheat plants. Both inoculated and non-inoculated plants were given temperature exposure of 44 °C to induce heat stress. The inoculated plants showed increment in the plant biomass, plant height by 14%, yield of grain by 40%, and increase in the seed weight by 12%, whereas, the non-inoculated plants were not thermotolerant resulting in the death of the plants due to high heat conditions. *Enterobacter* SA187 were used in various agronomic traits for thermopriming to induce thermotolerance (Shaffique et al. 2022; Shekhawat et al. 2020). *Bacillus cereus* SA1, a bacterial endophyte applied on soybean plant, showed pronounced thermotolerance activity by increasing the salicylic acid (SA), SOD, and glutathione content (Khan et al. 2020). Moreover, *B. paramycoides* isolated from an invasive weed *Parthenium hysterophorus* exhibited membrane integrity to wheat seeds injured during heat stress. It also increases the basic proline level and different antioxidant enzymes and enhances germination parameters to sustain against heat stress (Dubey et al. 2022).

In another study, *Septoglomus deserticola* and *S. constrictu* inoculated in tomato seedlings during heat and drought stress showed improved cellular parameters, leaf water content, and stomatal conductance. The inoculation enhanced the physiological attributes of the plants under combined stress conditions (Duc et al. 2018).

A group of researchers has isolated an endophytic fungus, identified as *Thermomyces lanuginosus*, from a desert plant *Cullen plicata* Delile. The inoculation of this hot desert fungi in plants showed efficient growth-promoting activity would and resist the plant against heat stress. This thermophilic fungus facilitates the ecophysiological mechanism of heat tolerance in plant (Ali et al. 2019). The non-clavicipitaceous fungi including the members of Ascomycota and Basidiomycota, namely, *Phoma* sp., *Arthrobotrys* sp., and *Mycophycia ascophylli*, inhibiting in the root shoot and rhizome of the plants facilitate increased biomass production and provide tolerance against heat and desiccation stress. These endophytes produce phytohormones and various secondary metabolites like alkaloids, terpenoids, steroids, flavonoids, benzopyranones, phenolic acids, and hydrocarbons in response to various abiotic stress (Verma et al. 2022). Another endophytic fungus *Aspergillus foetidus* AdR-13, isolated from *Adiantum capillus-veneris* L., was examined against *Glycine max* and *Helianthus annuus* and should increase plant biochemical profile such as elevated chlorophyll level and increased proline and abscisic acid (ABA) content along with the increment in total sugar lipid and protein content. The activity of reactive oxygen species (ROS) degrading antioxidant enzymes such as glutathione reductase, superoxide dismutase, ascorbic acid oxidase, catalase, and peroxidase was found to be increased in endophyte-inoculated plants maintained under heat stress. The application with AdR-13 significantly decreases lipid peroxidation and generation of ROS. AdR-13 inoculation also increases the biomass of the crop species (Hamayun et al. 2021). Furthermore, *Aspergillus japonicus* EuR-26 isolated from wild *Euphorbia indica* L. mitigates heat stress in soybean and sunflower plant. Similar to the previous study, the inoculation increases the concentration of SA, phenolics, flavonoid, and IAA to release the plant from thermal stress in semi-arid and arid regions (Hamayun et al. 2018) (Table 6.1).

6.5 Genetic Perspectives of Plant-Microbe Interaction

In recent years, multidimensional studies on the plant-microbe interaction at molecular level are conducted for better understanding of the signaling pathways involved in the establishment of their beneficial association and alleviation of various biotic and abiotic stress. Various researches have been conducted in this genomic era that aimed to resolve the complexity at cellular level and modulation of the interaction of components between the host and the rhizosphere (Collakova et al. 2012). Recently, DNA genome sequencing have been used to study the expression of genes encoded by plant growth-promoting bacteria (PGPB), followed by the use of proteomics, metabolomics, and transcriptomics. Current methods include genome editing of PGPB and use them as inoculants into the soil for plant growth promotion and stress tolerance (Gamalero et al. 2022). Scientists have used whole genome sequencing technique to study and characterize the different phenotypic traits responsible for combating heat stress. For example, *Brevibacterium frigoritolerans* ZB201705 can be efficiently used as an inoculant in the crop field to combat certain abiotic stress (Zhang et al. 2019).

For several years, researchers are trying to understand the cascade of events for thermotolerance in plants. An experiment on a molecular level in *Arabidopsis* revealed that different phytohormones like ABA, ethylene, and SA along with secondary metabolites act as signaling molecule and pursue stimulation activity to activate different factors for thermotolerance. An in-depth study revealed that MBF1c and HOT2 genes encode for chitinase protein facilitating hemostatic heat tolerance (Shaffique et al. 2022). Microarrays study revealed that heat stress leads to the development of HSPs and APX. There is a multiple range of HSP factors, namely, HSP20b, HSP60, HSP70, and HSP90. Among these factors, HSP70 and HSP60 are thermodynamically more potential (Shaffique et al. 2022). HSP 70 sequences in *Arabidopsis* contain 1.2% cysteine and are involved in nitrosylation and sulfenylation. When HSP70 undergoes sulfenylation, it activates heat shock transcription and induces thermotolerance. HSP70 regulates the transport of SOD into mitochondria and is directly involved in the metabolism of reactive oxygen (Berka et al. 2022). Heat stress disorganized chromatin organization in plant cells. HPS20b is involved in homooligomerization. A histone demethylase protein JUMONJI (JMJ) is responsible for chromatin organization in plants (Shaffique et al. 2022). Demethylases are enzymes that eliminate methyl groups and provide structural support to the chromosomes. The plant possesses heat stress memory due to the expression of histone H3 lysine 27 trimethylation (H2K27me3). This protein controls heat shock genes and provides heat stress tolerance to plants through a memory mechanism (Yamaguchi et al. 2021). HSP factors range from a molecular mass of 10–100 kDa. A detailed study of the heat shock genes revealed that heat shock elements are present in the TATA box with five proximal flanking regions. There are reports that senescence-related genes (*sag*), HSP stay green (*sgr*), and dehydrins (*dhn*) are also involved in the relaxation of heat stress in plants (Li et al. 2021). These thermotolerant genes are majorly activated by different organic acids and phytohormones produced by the application of PGPR, endophytes, and AMF into the rhizosphere.

6.6 Conclusion and Future Aspects

With the progress in climatic abnormalities, challenges for developing sustainable agricultural systems have emerged. As discussed earlier, excessive heat stress leads to agricultural disturbances. The use of biological organisms increases the efficiency of sustainable agriculture multi-folds. PGPR inoculants facilitate diversified beneficial interactions with plants giving promising solutions for sustainable and eco-friendly agricultural practices. The application of beneficial bacterial populations in the rhizospheric soil imparts plant growth-promoting effects under both laboratory and greenhouse conditions. In recent times, efforts have been made to the establishment of genetically engineered PGPRs for the remediation of soil contaminants and increasing crop yield and production. PGPRs are exploited to be used as biofertilizers, mitigating negative effects on the plants generated due to various biotic and abiotic stress and satisfying the food demands. Besides that, conventional agricultural practices like crop rotation, cover cropping, intercropping, organic composting, etc., are often used as alternative methods to develop sustainable agriculture (Singh and Singh 2017).

Microbial engineering is nowadays an expertise tool that helps to develop feasible agriculture to meet the global demand for food. Beneficial microbes that are engineered can be used in soil amendments and applied in the rhizosphere of a crop to promote growth. Beneficial microbes are used as a microbial consortium and microbial breeding for developing climate-resilient agricultural crop systems (Arif et al. 2020). Plants are also rendered with the production of essential phytohormones, bioactive secondary metabolites, and various other secretions through roots into the rhizosphere that facilitates nutrient availability and uptake (Hakim et al., 2021). Thus, microbial engineering is a way to fight against climatic calamities faced by crops. But studies dealing with the phyllospheric and rhizospheric modifications need to be explored (Wu et al. 2017). The rhizospheric engineering and approaches to synthetic biology open up new dimensions for exploring plant-microbe beneficial interactions and quick adaptation of plants with the changing climatic variabilities.

The climatic adversities and consequences to food security are a matter of global concern. Adverse climate also affects the compositions of soil, availability of nutrients to the plants, and functioning of the ecological niches. With the application of beneficial microbiome into the soil, the extent of overcoming stress responses needs to be investigated. Host genetic factor is another major concern that needs to be explored which could modulate the plant-microbe interaction and generation of resistance factors to different stress responses. Molecular approaches need to be studied to render host-specific interactions. Emerging technologies like meta-omics, transcriptomics, and bioinformatics tools are novel approaches that could interlink the beneficial association favoring the ecosystem and microbial community.

References

Abd El-Daim IA, Bejai S, Meijer J (2014) Improved heat stress tolerance of wheat seedlings by bacterial seed treatment. Plant Soil 379:337–350

Ahmad I, Akhtar MJ, Asghar HN, Ghafoor U, Shahid M (2016) Differential effects of plant growth-promoting rhizobacteria on maize growth and cadmium uptake. J Plant Growth Regul 35:303–315

Alexandre A, Oliveira S (2011) Most heat-tolerant rhizobia show high induction of major chaperone genes upon stress. FEMS Microbiol Ecol 75:2836

Ali AH, Abdelrahman M, Radwan U, El-Zayat S, El-Sayed MA (2017) Effect of *Thermomyces* fungal endophyte isolated from extreme hot desert adapted plant on heat stress tolerance of cucumber. Appl Soil Ecol 124:155–162

Ali AH, Radwan U, El-Zayat S, Sayed MAE (2019) The role of the endophytic fungus, *Thermomyces lanuginosus*, on mitigation of heat stress to its host desert plant *Cullen plicata*. Biologia futura 70:1–7

Ali SZ, Sandhya V, Grover M, Kishore N, Rao LV, Venkateswarlu B (2009) *Pseudomonas* sp. strain AKM-P6 enhances tolerance of sorghum seedlings to elevated temperatures. Biol Fertil Soils 46:45–55

Ali SZ, Sandhya V, Grover M, Linga VR, Bandi V (2011) Effect of inoculation with a thermotolerant plant growth promoting *pseudomonas* putida strain AKMP7 on growth of wheat (*Triticum* spp.) under heat stress. J Plant Interact 6:239–246

Andersen CP, Sucoff EI, Dixon RK (1987) The influence of low soil temperature on the growth of vesicular-arbuscular mycorrhizal *Fraxinus pennsylvanica*. Can J For Res 17:951–956

Angilletta MJJ (2009) Thermal adapatation. A theoretical and empirical synthesis. Oxford University Press, Oxford

Arif Y, Singh P, Siddiqui H, Bajguz A, Hayat S (2020) Salinity induced physiological and biochemical changes in plants: an omic approach towards salt stress tolerance. Plant Physiol Biochem 156:64–77

Basu A, Prasad P, Das SN, Kalam S, Sayyed R, Reddy M, El Enshasy H (2021) Plant growth promoting rhizobacteria (PGPR) as green bioinoculants: recent developments, constraints, and prospects. Sustainability 13:1140

Berg G, Rybakova D, Grube M, Köberl M (2016) The plant microbiome explored: implications for experimental botany. J Exp Bot 67:995–1002

Berka M, Kopecka R, Berkova V, Brzobohaty B, Cerny M (2022) Regulation of heat shock proteins 70 and their role in plant immunity. J Exp Bot 73(7):1894–1909

Bhattacharyya PN, Goswami MP, Bhattacharyya LH (2016) Perspective of beneficial microbes in agriculture under changing climatic scenario: a review. J Phytology 8:26–41

Bisht N, Mishra SK, Chauhan PS (2020) *Bacillus amyloliquefaciens* inoculation alters physiology of rice (*Oryza sativa* L. var. IR-36) through modulating carbohydrate metabolism to mitigate stress induced by nutrient starvation. Int J Biol Macromol 143:937–951

Bradford MA, Davies CA, Frey SD, Maddox TR, Melillo JM, Mohan JE, Reynolds JF, Treseder KK, Wallenstein MD (2008) Thermal adaptation of soil microbial respiration to elevated temperature. Ecol Lett 11:1316–1327

Brundrett MC (2009) Mycorrhizal associations and other means of nutrition of vascular plants, understanding the global diversity of host plants by resolving conflicting information and developing reliable means of diagnosis. Plant Soil 320:37–77

Bunn R, Lekberg Y, Zabinski C (2009) Arbuscular mycorrhizal fungi ameliorate temperature stress in thermophilic plants. Ecology 90(5):1378–1388

Carson JK, Gonzalez-Quinones V, Murphy DV, Hinz C, Shaw JA, Gleeson DB (2010) Low pore connectivity increases bacterial diversity in soil. Appl Environ Microbiol 76:3936–3942

Chaudhary V, Prasanna R, Nain L, Dubey S, Gupta V, Singh R et al (2012) Bioefficacy of novel cyanobacteria-amended formulations in suppressing damping off disease in tomato seedlings. World J Microbiol Biotechnol 28:3301–3310

Chen S, Jin W, Liu A, Zhang S, Liu D et al (2013) Arbuscular mycorrhizal fungi (AMF) increase growth and secondary metabolism in cucumber subjected to low temperature stress. Sci Hortic 160:222–229

Chua A, Sherwood OL, Fitzhenry L, Ng CKY, McCabe PF, Daly CT (2020) Cyanobacteria-derived proline increases stress tolerance in *Arabidopsis thaliana* root hairs by suppressing programmed cell death. Front Plant Sci 11:490075

Classen AT, Sundqvist MK, Henning JA, Newman GS, Moore JAM et al (2015) Direct and indirect effects of climate change on soil microbial and soil microbial-plant interactions: what lies ahead? Ecosphere 6(8):130

Collakova E, Yen JY, Senger RS (2012) Are we ready for genome-scale modeling in plants. Plant Sci 19:53–70

Compant S, Samad A, Faist H, Sessitsch A (2019) A review on the plant microbiome: ecology, functions, and emerging trends in microbial application. J Adv Res 19:29–37

Deltedesco E, Keiblinger KM, Piepho HP, Antonielli L, Potsch EM, Zechmeister-Boltenstern S, Gorfer M (2020) Soil microbial community structure and function mainly respond to indirect effects in a multifactorial climate manipulation experiment. Soil Biol Biochem 142:107704

Donhauser J, Qi W, Pinto BB, Frey B (2020) High temperature enhances the microbial genetic potential to recycle C and N from necromass in high mountain soils. Glob Chang Biol 27:1365–1386

Dubey A, Kumar K, Srinivasan T, Kondreddy A, Raja K, Kumar R (2022) An invasive weed-associated bacteria confers enhanced heat stress tolerance in wheat. Heliyon 8(7):e09893

Duc N, Zsolt C, Katalin P (2018) Arbuscular mycorrhizal fungi mitigate negative effects of combined drought and heat stress on tomato plants. Plant Physiol Biochem 132:297–307

Egamberdieva D, Wirth S, Behrendt U, Abd-Allah EF, Berg G (2016) Biochar treatment resulted in a combined effect on soybean growth promotion and a shift in plant growth promoting rhizobacteria. Front Microbiol 7:209

Egamberdieva D, Wirth SJ, Alqarawi AA, Abd Allah EF, Hashem A (2017) Phytohormones and beneficial microbes: essential components for plants to balance stress and fitness. Front Microbiol 8:2104

El-Sawah AM, El-Keblawy A, Ali DFI, Ibrahim HM, El-Sheikh MA et al (2021) Arbuscular mycorrhizal fungi and plant growth-promoting rhizobacteria enhance soil key enzymes, plant growth, seed yield, and qualitative attributes of guar. Agriculture 11:194

Estrada B, Aroca R, Maathuis FJM, Barea JM, Ruiz-lozano JM (2013) Arbuscular mycorrhizal fungi native from a Mediterranean saline area enhance maize tolerance to salinity through improved ion homeostasis. Plant Cell Environ 36:1771–1782

Figueiredo MVB, Burity HA, Martinez CR, Chanway CP (2008) Alleviation of drought stress in common bean (*Phaseolus vulgaris* L.) by co-inoculation with *Paenibacillus polymyxa* and *Rhizobium* tropici. Appl Soil Ecol 40:182188

Frey SD, Lee J, Melillo JM, Six J (2013) The temperature response of soil microbial efficiency and its feedback to climate. Nat Clim Chang 3:395–398

Gamalero E, Bernard E, Glick R (2022) Current techniques to study beneficial plant-microbe interactions. Microorganisms 10:1380

Hacquard S, Spaepen S, Garrido-Oter R, Schulze-Lefert P (2017) Interplay between innate immunity and the plant microbiota. Annu Rev Phytopathol 55:565–589

Hakim S, Naqqash T, Nawaz MS, Laraib I, Siddique MJ et al (2021) Rhizosphere engineering with plant growth-promoting microorganisms for agriculture and ecological sustainability. Front Sustain Food Syst 5:16

Hamayun I, Hussain A, Iqbal A, Khan SA, Lee IJ (2018) Endophytic fungus *Aspergillus japonicus* mediates host plant growth under Normal and heat stress conditions. Bio Med Int:7696831

Hamayun I, Hussain M, Iqbal A, Khan SA, Ahmad SA et al (2021) *Aspergillus foetidus* regulated the biochemical characteristics of soybean and sunflower under heat stress condition: role in sustainability. Sustainability 13:7159

Hasegawa H, Chatterjee A, Cui Y, Chatterjee AK (2005) Elevated temperature enhances virulence of *Erwinia carotovora* subsp. *carotovora* strain EC153 to plants and stimulates production of the quorum sensing signal, Nacyl homoserine lactone, and extracellular proteins. Appl Environ Microbiol 71:4655–4663

Hassan TU, Bano A, Naz I (2017) Alleviation of heavy metals toxicity by the application of plant growth promoting rhizobacteria and effects on wheat grown in saline sodic field. Int J Phytoremediation 19:522–529

Hottiger T, Boller T, Wiemken A (1987) Rapid changes of heat and desiccation tolerance correlated with changes of trehalose content in Saccharomyces cerevisiae cells subjected to temperature shifts. FEBS Lett 147:113–115

Hu Y, Wu S, Sun Y, Li T (2015) Arbuscular mycorrhizal symbiosis can mitigate the negative effects of night warming on physiological traits of *Medicago truncatula* L. Mycorrhiza 25(2): 131–142

Hungria M, Vargas MAT (2000) Environmental factors affecting N_2 fixation in grain legumes in the tropics, with an emphasis on Brazil. Field Crops Res 65:151–164

Huntington JL, Hegewisch KC, Daudert B, Morton CG, Abatzoglou JT, McEvoy DJ et al (2017) Climate engine: cloud computing and visualization of climate and remote sensing data for advanced natural resource monitoring and process understanding. Bull Am Meteorol Soc 98: 2397–2409

Hwang HH, Yu M, Lai EM (2017) *Agrobacterium*-mediated plant transformation: biology and applications. In Arabidopsis Book 15:0186

Issa A, Esmaeel Q, Sanchez L, Courteaux B, Guise JF, Gibon Y, Ballias P, Clément C, Jacquard C, Vaillant-Gaveau N (2018) Impacts of *Paraburkholderia phytofirmans* strain PsJN on tomato (*Lycopersicon esculentum* L.) under high temperature. Front. Plant Sci 9:1397

Jin X, Zhang Y, Zhang P, Trivedi P, Riera N, Wang Y et al (2018) The structure and function of the global citrus rhizosphere microbiome. Nat Commun 9:4894

Jones JDG, Vance RE, Dangl JL (2016) Intracellular innate immune surveillance devices in plants and animals. Science 354:6395

Kachhap S, Chaudhary A, Singh SD (2015) Response of plant growth promoting rhizobacteria (PGPR) in relation to elevated temperature conditions in groundnut (*Arachis hypogaea* L.). Ecoscan 9:771–778

Kang SM, Khan AL, Waqas M, Asaf S, Lee KE et al (2019) Integrated phytohormone production by the plant growth-promoting rhizobacterium *Bacillus tequilensis* SSB07 induced thermotolerance in soybean. J Plant Interact 14:416–423

Kashyap AS, Pandey VK, Manzar N, Kannojia P, Singh UB, Sharma PK (2017) Role of plant growth-promoting rhizobacteria for improving crop productivity in sustainable agriculture. In: Singh D, Singh H, Prabha R (eds) Plant-microbe interactions in AgroEcological perspectives. Springer, Singapore

Khan MA, Asaf S, Khan AL, Jan R, Kang SM et al (2020) Thermotolerance effect of plant growth-promoting *Bacillus cereus* SA1 on soybean during heat stress. BMC Microbiol 20:175

Khan Z, Park S, Shin S, Bae S, Yeon I, Seo Y (2005) Management of Meloidogyne incognita on tomato by root-dip treatment in culture filtrate of the blue-green alga, *Microcoleus vaginatus*. Bioresour Technol 96:1338–1341

Kim MJ, Shim CK, Kim YK, Ko BG, Park JH, Hwang SG et al (2018) Effect of biostimulator *Chlorella fusca* on improving growth and qualities of chinese chives and spinach in organic farm. Plant Pathol J 34:567

Kudoyarova GR, Melentiev AI, Martynenko EV, Timergalina LN, Arkhipova TN, Shendel GV et al (2014) Cytokinin producing bacteria stimulate amino acid deposition by wheat roots. Plant Physiol Biochem 83:285–291

Kumar A, Singh AK, Kaushik MS, Mishra SK, Raj P, Singh PK et al (2017) Interaction of turmeric (*Curcuma longa* L) with beneficial microbes: a review. 3 Biotech 7:357

Lee SM, Lee B, Shim CK, Chang YK, Ryu CM (2020) Plant anti-aging: delayed flower and leaf senescence in Erinus alpinus treated with cell-free chlorella cultivation medium. Plant Signal Behav 15:1763005

Lee YJ, Ko YJ, Jeun YC (2016) Illustration of disease suppression of anthracnose on cucumber leaves by treatment with *Chlorella fusca*. Res Plant Dis 22:257–263

Lenoir I, Fontaine J, Sahraoui ALH (2016) Arbuscular mycorrhizal fungal responses to abiotic stresses, a review. Phytochemistry 123:415

Li Z, Tang M, Hassan MJ, Zhang Y, Han L, Peng Y (2021) Adaptability to high temperature and stay-green genotypes associated with variations in antioxidant, chlorophyll metabolism, and γ-aminobutyric acid accumulation in creeping Bentgrass species. Front Plant Sci 12:750728

Liu A, Chen S, Chang R, Liu D, Chen H et al (2014a) Arbuscular mycorrhizae improve low temperature tolerance in cucumber via alterations in H_2O_2 accumulation and ATPase activity. J Plant Res 127:775–785

Liu ZL, Ma LN, He XY, Tian C (2014b) Water strategy of mycorrhizal rice at low temperature through the regulation of PIP aquaporins with the involvement of trehalose. Appl Soil Ecol 84: 185–191

Maitra S, Pramanick B, Dey P, Bhadra P, Shankar T, Anand K (2021) Thermotolerant soil microbes and their role in mitigation of heat stress in plants. In: Yadav AN (ed) Soil microbiomes for sustainable agriculture. Sustainable development and biodiversity. Springer, Cham, pp 203–242

Matsubara Y, Hirano I, Sassa D, Koshikawa K (2004) Alleviation of high temperature stress in strawberry plants infected with arbuscular mycorrhizal fungi. Environ Control Biol 42(2): 105–111

Meena H, Ahmed MA, Pravin P (2015) Amelioration of heat stress in wheat, *Triticum aestivum* by PGPR (*Pseudomonas aeruginosa* strain 2CpS1). Biosci Biotechnol Res Commun 8(2):171174

Menge DNL, Chisholm RA, Davies SJ, Abu Salim K, Allen D et al (2019) Patterns of nitrogen-fixing tree abundance in forests across Asia and America. J Ecol 107:2598–2610

Montalban B, Thijs S, Lobo MC, Weyens N, Ameloot M et al (2017) Cultivar and metal-specific effects of endophytic bacteria in *Helianthus tuberosus* exposed to Cd and Zn. Int J Mol Sci 18: 2026

Mosier AC, Li Z, Thomas BC, Hettich RL, Pan C, Banfield JF (2015) Elevated temperature alters proteomic responses of individual organisms within a biofilm community. ISME J 9:180–194

Naya L, Ladrera R, Ramos J, Gonzalez EM, Arrese-Igor C, Minchin FR et al (2007) The response of carbon metabolism and antioxidant defences of alfalfa nodules to drought stress and to the subsequent recovery of plants. Plant Physiol 144:1104–1114

Newsham KK, Fitter AH, Watkinson AR (1995) Multifunctionality and biodiversity in arbuscular mycorrhizas. Trends Ecol Evol 10:407–411

Okubo T, Tokida T, Ikeda S, Bao Z, Tago K, Hayatsu M et al (2014) Effects of elevated carbon dioxide, elevated temperature, and rice growth stage on the community structure of rice root associated bacteria. Microbes Environ 29:184–190

Park YG, Mun BG, Kang SM, Hussain A, Shahzad R et al (2017) *Bacillus aryabhattai* SRB02 tolerates oxidative and nitrosative stress and promotes the growth of soybean by modulating the production of phytohormones. PLoS One 12:e0173203

Pati B, Padhi S (2022) Indole-3acetic acid production by plant growth promoting rhizobacteria of saline soil, Odisha and its role in root elongation in rice plant. Sch Acad J Biosci 10(5):100–106

Prasanna R, Chaudhary V, Gupta V, Babu S, Kumar A, Singh R et al (2013) Cyanobacteria mediated plant growth promotion and bioprotection against *Fusarium* wilt in tomato. Eur J Plant Pathol 136:337–353

Prasanna R, Nain L, Tripathi R, Gupta V, Chaudhary V, Middha S et al (2008) Evaluation of fungicidal activity of extracellular filtrates of cyanobacteria–possible role of hydrolytic enzymes. J Basic Microbiol 48:186–194

Rana S, Chauhan R, Walia A, Sharma G, Datt N (2021) Beneficial microbes in agriculture under abiotic stress conditions: an overview. Pharma Innov 10:360–368

Rashid MH, Chung YR (2017) Induction of systemic resistance against insect herbivores in plants by beneficial soil microbes. Front Plant Sci 8:1816

Raza A, Razzaq A, Mehmood SS, Zou X, Zhang X, Lv Y et al (2019) Impact of climate change on crops adaptation and strategies to tackle its outcome: a review. Plan Theory 8:34

Rodríguez A, Stella A, Storni M, Zulpa G, Zaccaro M (2006) Effects of cyanobacterial extracellular products and gibberellic acid on salinity tolerance in *Oryza sativa* L. Saline Syst 2:7

Rudgers JA, Afkhami ME, Bell-Dereske L, Chung YA, Crawford KM, Kivlin SN et al (2020) Climate disruption of plant-microbe. Annu Rev Ecol Evol Syst 51:561–586

Shaffique S, Khan MA, Wani SH, Pande A, Imran M et al (2022) A review on the role of endophytes and plant growth promoting rhizobacteria in mitigating heat stress in plants. Microorganisms 10:1286

Sharma B, Singh BN, Dwivedi P, Rajawat MVS (2022) Interference of climate change on plant-microbe interaction: present and future prospects. Front Agron 3:725804

Shekhawat K, Sheikh A, Mariappan K, Jalal R, Hirt H (2020) *Enterobacter* sp. SA187 mediates plant thermotolerance by chromatin modification of heat stress genes. Biorxiv:1–44

Singh R, Singh GS (2017) Traditional agriculture: a climate-smart approach for sustainable food production. Energ Ecol Environ 2:296–316

Singh VK, Kumar A, Singh AK (2019) Impact of climate change on plant microbe interactions under agroecosystems. In Climate change and agricultural ecosystems. Elsevier, 153–179

Smith GS, Roncadori RW (1986) Responses of three vesicular-arbuscular mycorrhizal fungi at four soil temperatures and their effects on cotton growth. New Phytol 104:89–95

Timm CM, Carter KR, Carrell AA, Se-Ran J, Jawdy SS, Velez JM et al (2018) Abiotic stresses shift belowground *Populus*-associated bacteria toward a core stress microbiome. mSystems 3: e00070–e00017

Trivedi P, Leach JE, Tringe SG, Sa T, Singh BK (2020) Plant–microbiome interactions: from community assembly to plant health. Nat Rev Microbiol 18:607–621

Tsavkelova EA, Cherdyntseva TA, Klimova SY, Shestakov AI, Botina SG, Netrusov AI (2007) Orchid-associated bacteria produce indole-3-acetic acid, promote seed germination, and increase their microbial yield in response to exogenous auxin. Arch Microbiol 188:655–664

Vaishampayan A, Sinha RP, Hader DP, Dey T, Gupta AK, Bhan U et al (2001) Cyanobacterial biofertilizers in rice agriculture. Bot Rev 67:453–516

Vela'squez AC, Castroverde CDM, He SY (2018) Plant pathogen warfare under changing climate conditions. Curr Biol 28:619–634

Verma A, Nowsheen S, Singh JH, Eetela S, Parray JA et al (2022) Fungal endophytes to combat biotic and abiotic stresses for climate-smart and sustainable agriculture. Front Plant Sci 13: 953836

Wu QS, Xia RX (2006) Arbuscular mycorrhizal fungi influence growth, osmotic adjustment and photosynthesis of citrus under well-watered and water stress conditions. J Plant Physiol 163(4): 417–425

Wu Y, Lin H, Lin Y, Shi J, Xue S, Hung Y et al (2017) Effects of biocontrol bacteria *Bacillus amyloliquefaciens* LY-1 culture broth on quality attributes and storability of harvested litchi fruit. Postharvest Biol Technol 132:81–87

Xu L, Naylor D, Dong Z, Simmons T, Pierroz G, Hixson KK et al (2018) Drought delays development of the sorghum root microbiome and enriches for monoderm bacteria. Proc Natl Acad Sci 115:E4284–E4293

Yamaguchi N, Matsubara S, Yoshimizu K, Seki M, Hamada K et al (2021) H3K27me3 demethylases alter *HSP22* and *HSP17.6C* expression in response to recurring heat in *Arabidopsis*. Nat Commun 12:3480

Zhang C, Li X, Yin L, Liu C, Zou H, Wu Z, Zhang Z (2019) Analysis of the complete genome sequence of *Brevibacterium frigoritolerans* ZB201705 isolated from drought- and salt-stressed rhizosphere soil of maize. Ann Microbiol 69:1489–1496

Zhu X, Song F, Liu F (2017) Arbuscular mycorrhizal fungi and tolerance of temperature stress in plants. In Arbuscular mycorrhizas and stress tolerance of plants, Springer Nature Singapore Pt Ltd. 163–194

Zhu XC, Song FB, Xu HW (2010a) Arbuscular mycorrhizae improves low temperature stress in maize via alterations in host water status and photosynthesis. Plant Soil 331:129–137

Zhu XC, Song FB, Xu HW (2010b) Influence of arbuscular mycorrhizae on lipid peroxidation and antioxidant enzyme activity of maize plants under temperature stress. Mycorrhiza 20:325–332

Chapter 7
Role of Soil Microbes against Abiotic Stresses Induced Oxidative Stresses in Plants

Lalichetti Sagar, Sultan Singh, Aaina Sharma, Sagar Maitra, Meenakshi Attri, Ranjan Kumar Sahoo, Bahnu Pratap Ghasil, Tanmoy Shankar, Dinkar Jagannath Gaikwad, Masina Sairam, Upasana Sahoo, Akbar Hossain, and Swarnendu Roy

Abstract The global increase in the frequency of climatic extremes exposed the plants to various abiotic stresses. Assuring food as well as nutritional security of the increasing climate change-induced abiotic stress turned out to be a major challenge for the sustainability of crop production in the current era. To cope with this situation, researchers have already recommended many climate-smart strategies. However, these alternatives are sometimes technically complex, costly, and nonrenewable. Therefore, there is an urgent requirement to move into an

---

L. Sagar · S. Maitra (✉) · R. K. Sahoo · T. Shankar · D. J. Gaikwad · M. Sairam · U. Sahoo
Centurion University of Technology and Management, Paralakhemundi, Odisha, India
e-mail: lalichetti.sagar@cutm.ac.in; sagar.maitra@cutm.ac.in; ranjan.sahoo@cutm.ac.in; tanmoy@cutm.ac.in; gaikwad@cutm.ac.in; sairam.masina@cutm.ac.in; upasana.sahoo@cutm.ac.in

S. Singh
Sri Karan Narendra Agriculture University, Jobner, Rajasthan, India

A. Sharma
Punjab Agriculture University, Ludhiana, Punjab, India

M. Attri
Sher-e-Kashmir University of Technology and Management, Jammu, Jammu and Kashmir, India

B. P. Ghasil
Centurion University of Technology and Management, Paralakhemundi, Odisha, India

Sri Karan Narendra Agriculture University, Jobner, Rajasthan, India

A. Hossain
Bangladesh Maize and Wheat Research Institute, Dinajpur, Bangladesh

S. Roy
Plant Biochemistry Laboratory, Department of Botany, University of North Bengal, Siliguri, India
e-mail: swarnendubotany@nbu.ac.in

P. Mathur et al. (eds.), *Microbial Symbionts and Plant Health: Trends and Applications for Changing Climate*, Rhizosphere Biology,
https://doi.org/10.1007/978-981-99-0030-5_7

ecologically friendly, cost-effective, and sustainably appropriate alternative in the current situation of climate change. It was well-established recently that microbes can be considered in agriculture as environmentally friendly and cost-effective biotechnological resources. Microorganisms are unraveling their unique properties in recent times for the mitigation of the negative effects of abiotic stresses on crops. In this chapter, the prospects of the most prominent beneficial microorganisms (viz., plant growth-promoting rhizobacteria, mycorrhizal fungi, actinomycetes, and cyanobacteria) in terms of their innate stress-tolerant mechanisms such as biosynthesis of phytohormones, ion homeostasis, induction and accumulation of osmoprotectants, nutrient uptake enhancement, and antioxidant mechanisms have been discussed elaborately to understand the potential of soil microorganisms in imparting plant stress tolerance.

Keywords Abiotic stresses · Beneficial microorganisms · Rhizosphere · Stress tolerance

Abbreviations

ACC-1	Aminocyclopropane-1-carboxylic acid
AMF	Arbuscular mycorrhizal fungi
IAA	Indole acetic acid
IPCC	Intergovernmental panel on climate change
PEG	Polyethylene glycol
PGPR	Plant growth-promoting rhizobacteria
PS	Photosystem

7.1 Introduction

Globally, a sharp rise in population accompanied by large-scale industrial development and urbanization conferred significant climate change (Chapman et al. 2017). Among various sectors, agriculture is the most vulnerable sector to climate change (Malhi et al. 2021; Bhadra et al. 2022). In this scenario, crops are frequently exposed to various stresses owing to stagnation in crop productivity (Majeed et al. 2018). Crop performance is mainly influenced by both time and magnitude of exposure of the plant to these external factors beyond their threshold levels (Raza et al. 2019). In general, every plant has an optimum range within which they perform ideally in terms of growth processes (Dambreville et al. 2015). However, when these external factors cross the optimum limit they impart stress, which in turn widens the gap between genetic yield potential and actual yield realized by the crop at a specified management level (Coyne et al. 2020).

In the current context of climatic aberrations, the farm sector is mainly endangered by a decline in the quantum of annual rainfall, an increase in rainfall intensity, increased periodicity of dry spells, temperature fluctuations, and salinity (Rahman et al. 2017). The IPCC estimated that the global mean temperature was reported to increase by 0.3 °C per decade, leading to the warming of the planet (Gulzar et al. 2018). Further, these adverse conditions are attributed to reducing plant growth and crop yields (Onwuka and Mang 2018). Although the individual impact of different abiotic stresses, viz., drought, heat, cold, salinity, heavy metal, and cold stresses on plants, was widely studied, but in nature, they usually occur in combination (Swamy et al. 2019). The combined influence of various abiotic stresses on plant metabolism varies significantly in their impact (Kumar 2020). The adverse effects of climatic variation in crops are more prominent in the tropics than in temperate regions (Raneesh 2017).

To fulfill the requirement of the increasing population and ensure global food security for all under this adverse climate, it is wise to adopt an ideal strategy for the mitigation of abiotic stresses (Hossain et al. 2021). However, most of the strategies have some potential in alleviating the impact of abiotic burdens caused to plants that are technically complex, costly, and nonrenewable (Sabki 2021). Therefore, there is a necessity to adopt ecologically friendly and cost-effective alternatives targeting agricultural sustainability. The use of beneficial soil microbes in the mitigation of the negative influences of abiotic burdens on plants is an alternative strategy that has a wider recognition (Rajput et al. 2021). This strategy is mainly based on the beneficial role of microbes in the growth regulation of plants and in making nutrients available to them (Jacoby et al. 2017). The most predominant groups of microorganisms intricate in the alleviation, as well as mitigation of abiotic stresses, including plant growth-promoting rhizobacteria (PGPR), actinomycetes, cyanobacteria, and mycorrhizal fungi (Shaffique et al. 2022) (Fig. 7.1).

These beneficial soil microbes conventionally ease the adverse possessions of environmental stresses attributed to the release of phytohormones, production of metabolites, regulating ion balance within the plants, signal transduction, etc. (Sharma et al. 2019). Based on these facts, the chapter focuses on the role of all the rhizosphere microorganisms playing a role in lessening and mitigating the adverse impacts of abiotic stresses on plants attributed to their multifaceted mechanisms.

7.2 Adverse Effects of Major Abiotic Stress on Plants

The most prominent abiotic factors, namely, drought, temperature, salinity, and heavy metal stresses impacting the morphology, physiology, and metabolism in the plant system, have been discussed below.

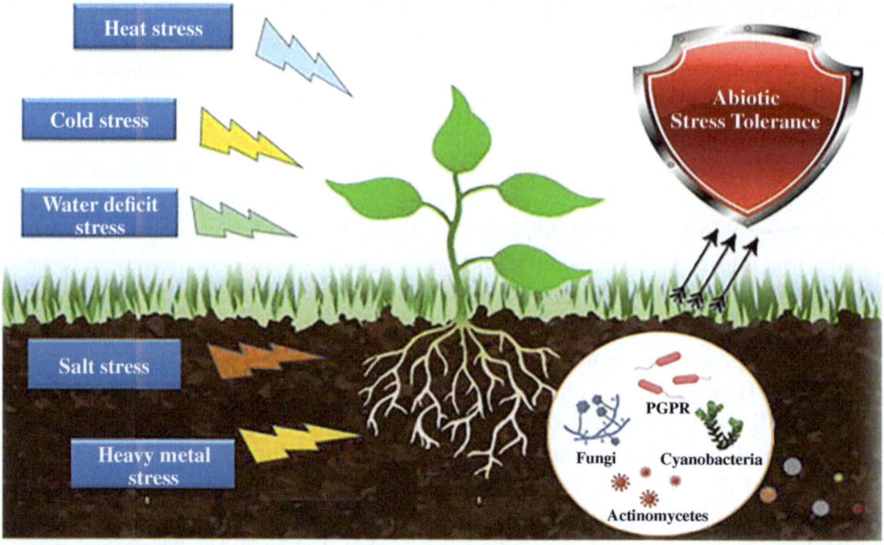

Fig. 7.1 Soil microbes-mediated abiotic stress alleviation

Fig. 7.2 Physiological and metabolic impacts of drought on plants

7.2.1 Drought

Water is an essential resource for the flourishing of the crop, and it is evident that water shortage for a prolonged period at any stage of the crop growth negatively influences the crop yield due to the changes manifested at its morpho-physiological and molecular levels, respectively (Begna 2020) (Fig. 7.2).

Crops seeded under water-deficit soils recorded low germination, poor vigor, and inferior crop stand (Abdoli and Saeidi 2012). In a study on wheat, the rate of germination, seedling vigor, and the length of the coleoptiles was significantly reduced in all the wheat genotypes subjected to drought stress induced by polyethylene glycol (PEG) (Kizilgeci et al. 2017). Conversely, root growth increases substantially in tolerant genotypes (Chen et al. 2021). Turgor pressure is essential to accomplish cell division and cell elongation attributed to crop growth (Kutschera and Niklas 2013). Under severe drought, due to insufficient water supply, substantial loss in cell turgor was observed, leading to diminished crop growth (Ndjiondjop et al. 2018).

In general, the yield is influenced by many physiological processes that are highly sensitive to soil moisture deficit. However, the quantum of drought impact depends on the extent, time of occurrence, and harshness (Bolat et al. 2014). In rice, exposure to drought during flowering was severely harmful as it reduced spikelets number (18%) and filled grains (19%) in YLY6 attributed to a 23.2% reduction in overall grain yield (Yang et al. 2019). Similarly, in maize, drought stress at the time of pollination desiccate the pollen grains and impair pollination drastically. Drought at the reproductive stage of the crop is dangerous compared with the vegetative stage (Bheemanahalli et al. 2022). Many studies indicated that reduction in vegetative growth due to water deficiency can be recovered if an adequate amount of water is supplied at the most sensitive period before flowering (Cui et al. 2015).

The plants under drought realized a substantial decline in relative water content and leaf water potential, triggering high canopy temperature (Pirzad et al. 2011). Further, this impairs plant transpiration rate, resulting in limited nutrient absorption by the root and their translocation (Alaoui et al. 2022). Stomata are the most sensitive organ that closes immediately to minimize the plant water demand as a defensive mechanism (Bertolino et al. 2019). However, shut down of stomata minimizes the intake of carbon dioxide, drastically affecting photosynthesis and leading to a substantial reduction in the growth and development of plants. In addition, acute water shortage is highly deleterious to the functioning of photosynthetic enzymes mainly due to the thickening of cytoplasm (Harrison et al. 2020). Malfunctioning of enzymes further impairs the electron transport chain, resulting in the gathering of reactive oxygen species (ROS). Consequently, the stability of membranes is adversely affected, leading to irreparable damage to photosynthetic organelles (Sharma et al. 2012).

7.2.2 High Temperature

Because of global warming, the high-temperature stress on the sustainability of crop production has widely been a concern across the world. In general, every variety will have three cardinal points, viz., maximum, optimum, and minimum temperature. The minimum or lower threshold temperature synonymously termed the base temperature is the reference point below which the growth of the plant curtails. On the other hand, the maximum or upper threshold temperature is the reference point above which the growth of the plant stops. The temperature rise in a crop-growing region above its upper threshold of a given variety for a considerable period is considered heat stress (Luo 2011; Maitra et al. 2021a). The magnitude of heat stress is increased with an enhancement in atmospheric temperature as well as the duration of exposure (Lundgren et al. 2013).

Under high-temperature conditions, plants show several marked changes in their morphology, anatomy, physiology, and biochemical behavior as a response to withstand adverse impacts of stress (Purnama et al. 2018; Maitra et al. 2021b). Morphologically, high-temperature stress confers leaf yellowing, scorching, senescence, and abscission (Nahar et al. 2015). Moreover, the plant growth rate and duration are mainly determined by the accumulation of heat units. To attain any phenological stage, a quantified accumulation of heat units specific to a variety is essential. Heat stress hastens this accumulation, bringing down the crop duration (Parthasarathi et al. 2013).

The crop sensitivity to high-temperature stress varies with the stage of the crop. In comparison, the reproductive stage of the crop is more highly sensitive to heat stress than the vegetative stage, triggering significant loss in grain yield (Hossain et al. 2012). During reproduction, even a short span of exposure to high temperatures above the upper threshold desiccates pollen, impairs pollination, and floral abortion and improper grain filling, resulting in a decline in crop productivity (Chaturvedi et al. 2021).

Heat stress influences the physiological processes, viz., photosynthesis, respiration, etc. The decline in the rate of photosynthesis due to heat stress is more prominent in C_3 plants than in C_4 plants (Yamori et al. 2014). Generally, when plants are bare to heat stress due to an increase in membrane fluidity the PS II complex dissociates impairing the electron transport system. In addition, high temperature also inactivates vital photosynthetic enzymes, and in extreme cases, dissociation of the oxygen-evolving complex was also reported (Song et al. 2014). All these imbalance the electron transport system and the gathering of ROS (Markulj Kulundzic et al. 2022). Similarly, the rate of respiration is enhanced with an increase in temperature, and when the rate of respiration exceeds the photosynthesis the carbohydrate reserves were reported to be lost, resulting in a rapid decline in the size and quality of the edible economic products (Gao et al. 2017).

7.2.3 *Low Temperature*

The cold effect is highly detrimental to crop growth and productivity. In general, the adverse impact of low-temperature stress on plant growth was mainly due to cellular dehydration and suboptimal temperatures. The low-temperature stress is mainly classified into two based on the magnitude of exposed temperature, viz., chilling and freezing stress. Chilling stress occurs when the plants are exposed to temperatures less than 10 °C while freezing stress occurs when the plants have exposed the temperatures below 0 °C.

Plants exposed to chilling stress show various morphological changes that include increased crop duration, reduced seed germination, reduced leaf expansion, thickening of leaves, chlorosis, and in severe cases might lead to leaf necrosis. Moreover, plants exposed to chilling stress for a considerable period show an adverse impact on various metabolic processes, viz., transpiration, photosynthesis, assimilate translocation, nutrient uptake, and cell division. Owing to a high percentage of saturated fatty acids than unsaturated fatty acids in the cell membrane of low-temperature sensitive plants confers reduced membrane fluidity and electrolyte leakage. In comparison, the reproductive stage of the crop is highly sensitive to chilling stress than other stages of growth. In rice, the spikelet fertility was reported to reduce significantly when the crop was exposed to chilling temperatures during anthesis, and it was mainly attributed to low pollen germination (Zeng et al. 2017).

Freezing stress is the most adverse form of chilling stress. It is more prominent in temperate regions than in tropical areas. When the sensitive plants are exposed to below 0 °C, they show twisting and curling of leaves, poor leaf expansion, delayed seed germination, and in severe cases cause plant death. The freezing stress is more severe than chilling stress since freeze induces cellular crystallization of cytoplasm. Cellular crystallization is further classified into two types: intercellular and intracellular. Initial exposure to freezing stress is reported to form ice crystals between the cells that can be recovered fully by warming. However, long-term exposure to freezing stress results in intracellular crystallization, which restricts the movement of water in the cytoplasm, causing excessive cellular dehydration and damage.

7.2.4 Salt

Globally, salinity stress is a predominant warning factor that affects the growth and productivity of crops (Yadav et al. 2019; Billah et al. 2021). Based on the nature and composition of the salt conferring stress, it has been classified into two forms, viz., salinity stress (predominance of soluble salts) and sodic stress (predominance of sodium carbonate or bicarbonate) (Gunasekaran et al. 2022). The induction of salt stress occurs both naturally and by human intervention. In general, inadequate rainfall, poor drainage, the composition of parent rock, and poor management of agricultural lands are the key factors responsible for its predominance across the world (Aslam et al. 2017).

The adverse effect of salinity on plants occurs due to the induction of drought or ion toxicity (Javed et al. 2019). In fact, water always moves from higher to lower potential, and under salt stress, a marked reduction in soil water potential with a subsequent increase in salt concentration was witnessed in many investigations (Forni et al. 2017; Jabro et al. 2020). Germination is the foremost and essential process in the crop life cycle that determines crop establishment, which was hampered seriously due to alteration in water potential (Kaur et al. 2020). Ehtaiwwesh and Emsahel (2020) observed a drastic reduction in germination percentage (37%) and germination rate (45%) in *Pisum sativum* seeds exposed to 150 mm NaCl.

Plant roots absorb water and uptake nutrients from the soil (Alaoui et al. 2022). During salt stress, due to an increase in salt concentration the water potential of the soil solution substantially decreased compared with the root and eventually led to slower plant growth (Nawaz et al. 2010). The turgid potential of the cell is the prime requisite for cell division and enlargement (Deng et al. 2011). Therefore, an inadequate supply of water due to alteration in the water potential gradient significantly limits the growth and development of the plant. Initially, due to an insufficient supply of water, the stomata close, leading to a substantial decrease in carbon assimilation. However, when salinity stress persists for a longer period disorganization of the lipid–protein ratio was reported in many studies disintegrating thylakoids (Gupta and Huang 2014).

In extreme severity, salts may enter the plant system through water, which is usually trapped inside a vacuole through a process called vacuolar compartmentalization (Tan et al. 2019). Further, if the salt load exceeds the cellular ability to compartmentalize, it is reported to cause ion toxicity within the plant. At this juncture, the cellular potassium concentration was found to be crucial in maintaining the membrane potential (Abbasi et al. 2016). Several studies indicated that the uptake of potassium ions was markedly restricted by the increased sodium concentration during salt stress attributed to the antagonistic effect of sodium and potassium (Eker et al. 2013).

7.2.5 Heavy Metals

In general, trace amounts of heavy metals, viz. cadmium (Cd), copper (Cu), lead (Pb), nickel (Ni), chromium (Cr), cobalt (Co), arsenic (As), and zinc (Zn), play a key role in crop growth and development (Arif et al. 2016). Heavy metal stress could be primarily induced due to the overuse of fertilizers (viz., phosphatic), improper application of sewage and sludge, irrigation using poor quality water, etc. (Shahbazi et al. 2017). Heavy metal contamination is a rising issue currently due to its significant influence in deteriorating the productivity of arable agricultural lands (Hu et al. 2014). The plants induced with heavy metal stress primarily trigger the production of ROS, which could be due to the induced disruption of electron transport chains adversely affecting the antioxidant activity in the plant (Shahid et al. 2014). In addition, this also induces membrane instability due to its significant role in lipid peroxidation and thereby resulting in leakage of protoplasm and cell death, respectively (Iuchi et al. 2021).

The crops cultivated on cadmium-rich soils hurt the photosynthetic mechanism, and water and nutrient uptake, thus reflecting crop growth inhibition (Nazar et al. 2012). Usually, cadmium is accumulated in agricultural soils through fertilizers or by using industrial wastewater for irrigation before treatment (Khan et al. 2019). The high mobility of cadmium in the plant system is attributed to its easy uptake and translocation, resulting in the yellowing of leaves, lipid peroxidation, impaired pollen tube formation and germination, inactivation of several enzymes involved in various metabolic activities, etc. (Scholz et al. 2020).

Copper plays a key role in cell signaling (Kardos et al. 2018). Plastocyanin is a copper-based mobile electron carrier that is involved in electron transfer during noncyclic electron transfer of photosynthesis, which in turn plays a key role in supplying electrons from PS-II to PS-I, respectively (Roach and Krieger-Liszkay 2014). Under high concentrations of accumulated copper, it gets oxidized to Cu^{2+}, resulting in the production of ROS, thereby causing oxidative stress (Duanghathaipornsuk et al. 2021). Copper-induced oxidative stress impairs DNA structure, stability of lipid membrane, etc. (Aboul-Ela et al. 2011).

Zinc is another important heavy metal that performs a vital role in plant metabolism (Ul Hassan et al. 2017) as it is a cofactor for many enzymes, and the deficiency of Zn causes a substantial yield reduction in crops (Alegre et al. 2020). Zinc toxicity has a severe impact on photosynthesis because of its ability to replace magnesium ions from the center of the chlorophyll molecule, thereby impairing chlorophyll structure; on the other hand, zinc regulates the movement of stomata and, upon an increase in zinc, the concentration reduces its conductivity, thereby accelerating the intake of carbon dioxide (Mostoni et al. 2019). A high concentration of cobalt also inhibits several metabolic activities and the growth of a plant. During prophase, the excess concentration of cobalt was reported to result in chromosome deserialization and impairs the structural integrity of the plastids (Potapova and Gorbsky 2017). Similarly, lead (Pb) plays a vital role in the biosynthesis of chlorophyll and inhibits the activity of catalase enzyme attributed to increased ROS production (Chen et al. 2018). Additionally, lead also inhibits key enzymes involved in the Calvin cycle attributed to poor photosynthesis (Yang et al. 2017).

7.3 Beneficial Microorganisms Save Plants from Abiotic Stress-Induced Oxidative Stress

The group of microorganisms predominant in easing the negative effects of abiotic stress, viz., plant growth-promoting rhizobacteria (PGPR), mycorrhizal fungi, cyanobacteria, and actinomycetes, was discussed briefly in this section.

7.3.1 Plant Growth-Promoting Bacteria

The plant growth-promoting rhizobacteria is a group of rhizosphere-colonizing bacteria that facilitate better crop growth, leading to improving the desirable productivity of crops (Elshahat et al. 2016). The most predominant genera in this group are *Bacillus, Enterobacter, Pseudomonas, Azotobacter,* etc. (Singh et al. 2017). These bacteria play a vital role in inducing plant growth through nutrient mobilization, phytohormone secretion, regulation of nutrient and hormonal balance, etc. (Kudoyarova et al. 2015; Maitra et al. 2021c; Hossain et al. 2022). The PGPR are

Table 7.1 Plant growth-promoting bacteria-induced abiotic stress tolerance in plants

Scientific name	Strain	Stress	Reference
Azospirillum spp.	Az19 and Az39	Drought	Garcia et al. (2017)
Pseudomonas spp.	BV-P13	Drought	Sandhya et al. (2010)
Acinetobacter calcoaceticus	SAVSo04	Drought	Leontidou et al. (2020)
Pseudomonas putida	GAP-P45	Drought	Sandhya et al. (2010)
Bacillus spp.	KB122, KB129, KB133 and KB14	Drought	Grover et al. (2014)
Burkholderia phytofirmans	PsJN	Drought	Naveed et al. (2014)
Bacillus cereus	PGPR (58)[195]	Drought	Debasis et al. (2019)
Agrobacterium tumefacien	ISSDS-425	PGPR isolates and abiotic stress	Gururani et al. (2013)
Bacillus thuringiensis	BS-45	Drought	Getahun et al. (2020)
Pseudomonas fluorescens	PGPR	Drought	Kavino et al. (2010)
Pseudomonas fluorescens	PGPR	Water stress	Agami et al. (2016)
Pseudomonas putida	MTCC5279	Drought	Tiwari et al. (2016)
Azospirillum brasilense	PGPR	Drought	Cohen et al. (2015)
Pseudomonas chlororaphis	PGPR	Salinity	Egamberdieva (2012)

mainly classified into two types, viz., symbiotic rhizobacteria and free-living rhizobacteria (Odoh 2017). Symbiotic rhizobacteria are those bacteria that establish a mutualistic relationship with crop plants and influence crop growth directly through stress mitigation, biofertilization, etc., while free-living rhizobacteria colonize in soils influencing nutrient availability, disease resistance, etc. (Ojuederie et al. 2019).

Due to successful research progress in this direction, the function of PGPRs is widely gaining attention in plant science; some of the shreds of evidence are provided in Table 7.1. Further, their inherent metabolic and genotypic ability to withstand unfavorable environments made them the most powerful entities to attribute tolerance to abiotic stresses (Dar et al. 2021; Maitra et al. 2021d). However, the mechanism of imparting stress tolerance by PGPR includes both direct and indirect mechanisms. Direct mechanisms involve rapid stimulation of plant growth through increased phytohormone synthesis, improved nutrient and water uptake, etc., whereas indirect mechanisms mainly involve antagonism, production of hydrolytic enzymes, release of compatible solutes, etc. (Fahad et al. 2015). Moreover, PGPR was also reported to produce sigma factors that are involved in altering gene expression and attributing to the mitigation of the negative impacts of environmental factors (Ali et al. 2022).

In addition, PGPR was reported to minimize the crop nitrogen demand by possessing the ability to fix atmospheric nitrogen, altering the root morphology and increasing its surface area (Gallart et al. 2021). Similarly, phosphorus-solubilizing and potassium-solubilizing bacteria in the rhizospheric region of the plant increase the bioavailability of phosphorus and potassium, respectively (Panhwar et al. 2011). Several species of rhizobacteria were reported to induce siderophore production. Siderophores are low-molecular-weight organic molecules that help in improving the bioavailability of iron to plants while depriving it to plant pathogens (Wang et al. 2022).

7.3.2 *Mycorrhizal Fungi*

The beneficial soil mycorrhizal fungi (MF) increase the plants' nutrient absorption under abiotic burdens such as drought, salinity, heavy metals, and adverse temperature conditions, and provide tolerance against the negative effects (Sun et al. 2018). MF fungi assist the affected plants in the upregulation of lenience mechanisms and prevent the downregulation of key metabolic pathways. MF being natural root symbionts can facilitate the absorption of nutrients to affected plants and, thus, improve growth when the plants are under stress. MF used as biofertilizers can support the adaptability of plants to perform under extreme environmental conditions. MF play the role of growth regulator in several plants and can improve plant growth. They also facilitate soil moisture absorption, and thus, increase the ability of tolerance in plants against drought and salinity. In drought stress, mycorrhizal fungi can improve root and shoot biomass production. Worldwide, soil salinity causes various environmental problems to global food security. Several research studies reported that MF can improve plant growth and their yielding under salinity stress. Ait-El-Mokhtar et al. (2019) observed the positive impacts of MF symbiosis on photosynthesis, stomatal conductance, and leaf water relations under salinity stress. MF can increase the production of betaine and confirm its role in plant osmoregulation under salinity stress.

Mycorrhizal fungi helped the plants to uptake nutrients from low-fertile soils and reduced metal toxicity in plants. Mycorrhizal fungi-inoculated pistachio plants exhibited high levels of phosphorus, potassium, zinc, and manganese under drought conditions. It has been widely accepted that symbiotic MF take a considerable amount of nitrogen from dead and decomposed organic bodies and facilitate the plants to grow healthier and stay alive. MF cuts down the uptake of heavy metals by plants from the contaminated growing media.

Unlike other abiotic stresses, heat stress hampers seed germination, seedling vigor, and growth processes and, thus, decreases the assimilated output, leaves to

Table 7.2 Mycorrhizal fungi-induced abiotic stress tolerance in plants

Scientific name	Crops	Stresses	Reference
Funneliformis mosseae	Orange	Drought	Zhang et al. (2018)
Rhizophagus intraradices	Maize	Drought	Zhao et al. (2015)
Rhizophagus irregularis	Wheat	Heat	Cabral et al. (2016)
Rhizophagus irregularis	Tomato	Heat	Calvo-Polanco et al. (2016)
Acaulospora laevis	Trigonella	Heavy metal	Abdelhameed and Metwally (2019)
Rhizophagus irregularis	Pigeon pea	Heavy metal	Garg and Singh (2017)
Rhizophagus irregularis	Tomato	Salinity	Khalloufi et al. (2017)
Claroideoglomus etunicatum	Aeluropus	Salinity	Hajiboland et al. (2015)
Glomus lamellosum	Cinnamon	Drought	Liao et al. (2021)
Streptomyces pactum	Wheat	Drought	Li et al. (2020)

burn, wilt, and causes damage to reproductive parts. It also reduces the crop yield, damages the fruits, and leads to cell death. An earlier study conducted by Maya and Matsubara (2013) stated an association of MF that boosts the plant immunity against fungal pathogens and changes the growth under high temperatures (Maya and Matsubara 2013). In extremely cold environments, mycorrhizal fungi store moisture in the host plant and trigger the production of secondary metabolites. Further, MF enhanced protein content in plants (Begum et al. 2019). The positive influences of MF symbiosis can protect plants against several stresses by triggering various physiological and metabolic processes such as enhanced photosynthesis, biomass production and nutrient uptake, gathering of osmoprotectants, and change in the rhizosphere ecosystem by upregulation of enzyme activity. Table 7.2 shows the beneficial impact of MF on different crops under various abiotic stresses.

7.3.3 Cyanobacteria

Cyanobacteria are the photosynthetic, oxygenic, and prokaryotic organisms found abundantly in terrestrial, freshwater, and marine waterbodies, ice shelves, bare rocks, hot springs, and Arctic and Antarctic lakes. They can fix nitrogen biologically, solubilize phosphorus, and release mineral nutrients to plants that ultimately enhance soil quality and crop yields. Besides, cyanobacteria show defense mechanisms against abiotic stresses (Gupta et al. 2022). Based on the temperature, cyanobacteria can be divided into three groups, namely, (1) psychrotrophics, which can tolerate cold temperatures (below 15 °C), (2) mesophilics, having a tolerance to temperatures up to 50 °C, and (3) thermophilic, with a tolerance level above 80 °C temperature. Cyanobacteria can adjust membrane fluidity in low-temperature conditions by altering fatty acids. They perform two mechanisms to ensure the protection of plants under low-temperature stress. They desaturate the membrane fatty acids and induce some enzymes that increase the efficiency of transcription and translation. Cyanobacteria protect against cold plants by changing their transcription and translation machines and membranes. Cyanobacteria have multiprotein complexes and are highly responsible for photosynthetic electron transport. Another important pigment in cyanobacteria is phycobilin, which is normally associated with proteins to form phycobiliproteins. Some other types of cyanobacteria can produce valuable plant compounds such as chlorophyll, β-carotene, astaxanthin, xanthophylls, and phycobiliprotein, and some also produce polysaccharides, lipids, proteins, vitamins, sterols, and enzymes. Cyanobacteria can produce extracellular phosphatases to fight phosphorus deficiency and also tolerate the high concentration of Cd or Zn in soil. Cyanobacteria altered their metabolites and respond toward abiotic stress factors. Kollmen and Strieth (2022) reported that cyanobacteria provide some signals to plants, by which the plants can easily recognize them as their beneficial partner and allow them to make necessary changes in the host. Cyanobacterial signals induce mitotic division in the host plants, for example, in Gunnera (angiosperm) cells. Among the multiple environmental stresses, cyanobacteria are responsible for nutrient depletion, which limits their growth. The experimental results supported the existence of vitamins in cyanobacteria. An increase in the root growth of rice seedlings by the extract of C. *muscicola* was compared to pure cyanocobalamin (vitamin B12) and folic acid.

7.3.4 Actinomycetes

Actinomycetes are Gram-positive bacteria that have the potential to alleviate the abiotic stresses in plants. Many actinomycetes have plant growth-promoting (PGP) properties such as indole acetic acid (IAA) production, phosphate solubilization, siderophore production, biocontrol of phytopathogens, and 1-aminocyclopropane-1-carboxylic acid (ACC) deaminase activity. These soil bacteria can survive under abiotic burdens such as drought, heat, saltiness, etc. (Table 7.3). Recently several studies stated the role of actinomycetes in alleviating salinity and drought stress in crop plants (Grover et al. 2016). Actinomycetes are eco-friendly, improving soil

Table 7.3 Actinomycetes-induced abiotic stress tolerance in plants

Scientific name	Strains	Stresses	Crops	References
Streptomyces rochei	SM-3	Salinity	Chickpea	Srivastava and Kumar (2015)
Streptomyces spp.	PGPA-39	Salinity	Tomato	Palaniyandi et al. (2014)
Citricoccus zhacaiensis B-4	MTCC-12119	Drought	Onion	Selvakumar et al. (2015)
Streptomyces coelicolor	DE-07	Drought	Wheat	Yandigeri et al. (2012)
Streptomyces spp.	KLBMP-5084	Salinity	Tomato	Gong et al. (2020)
Streptomyces spp.	IT-25	Drought	Tomato	Abbasi et al. (2020)
Streptomyces spp.	C-2012	Salinity	Wheat	Akbari et al. (2020)
Streptomyces rimosus	C-2012	Salinity	Mentha	Esmaeil Zade et al. (2019)

fertility and supporting a yield increase of crops. These are widely spread in soil and are involved in dead organic matter decomposition, nitrogen fixation, and phosphate solubilization. Actinomycetes can produce antibiotics, biocontrol agents, and plant growth-promoting substances (Behie et al. 2017). Actinomycetes conquer the pathogens such as *Pythium ultimum* and *Erwinia carotovora*, causing damping-off and post-harvest rot diseases, respectively, in plants (AbdElgawad et al. 2020). Actinobacteria were evidenced to perform as facilitators in growth enhancement in plants. Actinomycetes support the plant in increasing chlorophyll content and biomass production (Chukwuneme et al. 2020).

7.4 Mechanisms of Stress Alleviation by Microbes

Microorganisms dwelling in the plant rhizosphere possess a special ability to withstand extreme environmental conditions and play a vital role in the alleviation of various forms of plant stresses. The role of soil microbes in the alleviation of stress is mainly attributed to their direct or indirect involvement in various mechanisms, viz., production of hormones, production of protective metabolites, ion homeostasis, nutrient uptake enhancement, and antioxidant mechanisms.

7.4.1 Hormones

Microorganisms help in altering the adverse impact of abiotic burdens on the growth processes of plants by employing various biomolecular interactions between plants and microorganisms. The microbial density at the plant root zone is extremely high compared with the rest of the soil. Exudates released by plant roots are most vital for microbial concentration in the rhizosphere. Recently, several roots-associated microorganisms were reported to produce plant growth-promoting hormones such as salicylic acid, cytokinins, gibberellins, auxins, and abscisic acid that influence the change in root morphology attributed in plants to tolerate abiotic stresses (Bagautdinova et al. 2022; Kim et al. 2022). These hormones help the plants to cope with stress through stimulation of rapid root growth, which enhances the water and nutrient uptake, proper stomatal regulation, upregulation of the antioxidant system, accumulation of compatible osmolytes, increase in membrane stability, etc. Keeping this in view, the production of growth-promoting hormones was found to be a predominant mechanism that commonly helps the rhizosphere-inhabiting microbes to cope with abiotic stress.

Among various soil-inhabiting microbes, plant growth-promoting rhizobacteria (PGPR) and arbuscular mycorrhizal fungi (AMF) have been widely investigated considering their positive role in the alleviation of stress. These soil microorganisms usually form colonies in the plant root zone and help enhance plant growth. In general, AMF establishes a symbiotic relationship with the roots of vascular plants and attributes to alter root morphology to improve water and nutrient acquisition from the soil. Microbes enhance lateral root formation by altering IAA transport and signaling. Arbuscular mycorrhizal fungi were reported to confer high-temperature stress tolerance by stimulating phytohormone signaling and preventing premature plant senescence through an increased accumulation of secondary metabolites, viz., proline and anthocyanin (Kapoor and Hasanuzzaman 2020). Fukami et al. (2017) reported that inoculation with *Azospirillum*, a plant growth promoter, consistently led to changes in root morphology, which has been linked to the growth hormones production, with auxin being the most important ascribed to tolerate drought and high-temperature stress. Similarly, under salinity stress, many salt-tolerant microorganisms were found to tolerate the negative impact of salinity in plants by stimulating the production of phytohormones as signaling molecules have been widely documented.

7.4.2 Protective Metabolites

Accumulation of protective metabolites is one of the most potential stress defense strategies that help sustain the normal photosynthetic rate even under exposure to unfavorable conditions (Isah 2019). Stomata are highly sensitive to water availability and usually close at reduced turgor and vice versa. Stomatal opening is a key adaptive mechanism to tolerate stress since the duration of stomatal opening determines crop productivity. However, alteration in osmotic potential critically limits the water uptake in sensitive plants. At this juncture, resistant plants reported an accumulation of osmotically active compounds that contribute to maintaining the cell turgor and maintaining osmotic balance. These osmotically active compounds are also called compatible solutes that include various compounds such as sugars, amino acids, organic acids, polyols, and quaternary ammonium compounds. Glycine betaine is the most abundant osmoprotective quaternary ammonium compound that plays a vital role in osmotic adjustment.

Proline, an amino acid, is most likely to accumulate in higher plants when subjected to stress. Proline, being osmotically active, mitigates cell membrane disruption due to salt stress, thereby preserving the stability of cell membranes (Jamil et al. 2012; Mushtaq et al. 2020). Under abiotic stress, the synthesis of proline was stimulated by beneficial bacteria, viz., Burkholderia Arthrobacter and Bacillus (Gimenez et al. 2018; Kayum et al. 2016). Similarly, Ali et al. (2022) found increased production of free proline in *Arabidopsis thaliana* induced with the application of *Bacillus subtilis*, which is attributed to enhanced tolerance to salt stress in transgenic plants. In another study, the formation of *Azospirillum* colonies in the rhizosphere of wheat crops imparted the enhanced accumulation of proline (Zarea et al. 2012). Further, increased proline content is attributed to inducing osmotolerance in wheat (Upadhyay et al. 2012). In soybean, salinity tolerance was linked to the gathering of proline stimulated by *Bacillus* strain SJ-5 (Kumari et al. 2015). Yasin et al. (2018) registered overexpression of stress-related genes attributed to proline accumulation induced by *Bacillus fortis* strain SSB21 that assured better growth of chilli plants under salinity stress.

7.4.3 Ion Homeostasis

High salt concentrations in the rhizosphere decrease soil water potential, resulting in reduced water absorption. The osmotic shock disrupts membrane permeability, reduction in relative water content, a slower rate of transpiration, low water usage, and water retention (Dotaniya and Meena 2015). At this stage, plants could not maintain cell turgor, leading to stomatal closure and reduced carbon assimilation. Salinity stress results increased influx of sodium ions in the cells, causing ionic imbalance. Plant tolerance to salinity stress is correlated with the decreased loading of sodium ions into the xylem and increased exclusion from shoots (Yun et al. 2018, Munns and Raddatz et al. 2020). The expression and function of plant vacuolar transporters involved in Na^+ detoxification can be controlled by microbes. The internal regulation of ionic balance is essential for a variety of plant metabolic processes and may be regulated by controlling the influx–efflux of sodium and potassium ions through microbial inoculation. Salt stress adaptations in plants may benefit from the decreased outflow of K^+ ions from roots and increased concentration in aboveground plant parts.

Due to their capacity for high salt concentration, halotolerant bacteria may quickly multiply in soils with varying salinity levels. Mukhtar et al. (2020) revealed that Azospirillum inoculation in wheat may limit the inflow of Na^+ from the roots. Similarly, Dal Cortivo et al. (2018) reported that inoculation of *Bacillus subtilis* GB03 in Arabidopsis enhanced salt tolerance by controlling the potassium transporter HKT1. In addition, Kasotia et al. (2015) revealed that inoculation with *Pseudomonas koreensis* in soybean under salt stress decreased Na^+ levels and raised K^+ levels in aboveground plant parts. Inoculation of *Piriformospora indica* increased salt tolerance in maize through enhanced functioning of stomata with a greater rate of K^+ delivery into the shoots, and higher K^+ loading on the shoots by restricting K^+ efflux from the roots (Yun et al. 2018). Chen et al. (2017) found that spermidine production by *Bacillus amyloliquefaciens* was reported to enhance salt tolerance in maize and Arabidopsis that in turn upregulated vacuolar transporters responsible for sodium accumulation into the vacuoles and exclusion of sodium ions from the cell to reduce ion toxicity. Chatterjee et al. (2018) demonstrated that *Brevibacterium linens* RS16 controlled salt buildup by modifying vacuolar H^+ ATPase activity, which induced salt tolerance in rice.

Recently, the arbuscular mycorrhizal fungus has been reported to enhance salt tolerance in certain plants. This trait of AM fungus was confirmed by its role in altering the ratio of K+/Na + in plant cells, vacuolar accumulation of ions, phytohormone production, and improving soil and rhizospheric conditions (Sofy et al. 2021). Additionally, AM acts as an osmoregulatory by boosting sugar and electrolyte concentrations, which reduces the harmful consequences of salinity (Bahadur et al. 2019). Hashem et al. (2018) recorded an increased tolerance against salinity as a result of AMF inoculation in cucumber by restricting sodium absorption and ion homeostasis.

7.4.4 Nutrient Uptake Enhancement

Soil fertilization is essential for agricultural output; however, it may also lead to nitrate and phosphate buildup that eventually contaminates surface and groundwater resources. Increased fish mortality is a result of the eutrophication of surface water, which is associated with phosphate run-off (Orandi et al. 2021). These environmental outcomes related to improper fertilizer application can be ascribed to low nutrient uptake efficiency by crops. For instance, phosphorous is strongly reactive with iron, aluminum, and calcium in soils, which can cause up to 90% of the phosphorous to precipitate out and be mainly inaccessible to plants (Ashekuzzaman et al. 2021). PGPB can help in preserving appropriate plant nutrition and minimize the harmful impacts of fertilizers on the environment. The common mycorrhizal network has a significant impact on the supply of nitrogen and phosphate to plants by fungi-mediated transport, which supports plant development in adverse environmental circumstances.

Some PGPRs have been linked to enhanced phosphate absorption and solubilization, which in turn promotes plant development (Wu et al. 2019). One of the distinctive qualities of mycorrhizal fungus that enhances plant nutrient absorption is phosphorous mobilization. Shokri and Maadi (2009) reported that mycorrhization might counteract the drop in phosphorus content in the plants caused by salt in *Trifolium alexandrinum*. Additionally, it has been observed that PGPR also influences plant nitrate absorption (Kowalska et al. 2015). Current attempts to extend these findings under controlled environment conditions and open-field settings have utilized combinations of PGPR strains with symbiotic nitrogen-fixing rhizobia or with mycorrhizal fungi (Diagne et al. 2020). Rhizobia are vulnerable to drought, leading to a noteworthy reduction in N_2 fixation due to low soil moisture. *Rhizobium tropici* and two strains of *P. polymyxa,* when co-inoculated into beans (*Phaseolus vulgaris* L.) under drought stress, lead to enhanced plant height, shoot dry weight, and nodule number (Shetta 2015).

In addition to an increase in general plant growth, some PGPRs encourage root development and change in the architecture of roots by producing phytohormones, leading to an increase in the number of root hairs and their surface area (Li et al. 2022). It is believed that one way that PGPR contributes to enhanced nutrient absorption is by promotion of root growth as the root is the primary site for nutrient uptake. Such root stimulation can help plants defend themselves against diseases. In addition, Goswami et al. (2014) observed that groundnut plants treated with *B. licheniformis* and grown at 50 mm NaCl produced longer roots. *Bradyrhizobium japonicum* and salt-tolerant *P. putida* inoculation in soybean improved salt tolerance by altering the root architecture, which in turn made it easier to acquire nitrogen and phosphorus and form nodules (Egamberdieva et al. 2017). Wang et al. (2018) found that bacterial inoculation in chilli grown under saline environments might reduce the stress by improving root length by 146%.

A number of researchers are currently examining the possibility of PGPR to sustain adequate output with tailored doses of fertilizer application and noted encouraging results in this direction. Basu et al. (2021), in a study conducted on wheat, found that the grain yield from 75% recommended dose of fertilizers in addition to a PGPR strain was comparable with a full dose of fertilizers without PGPR. In another study, Hernandez and Mrkovacki et al. (2016) noted a significantly higher dry weight in tomatoes cultivated with two PGPR strains and 75% RDF than with 100% RDF and without PGPR under controlled environments.

Another popular theory holds that plant growth-promoting rhizobacteria can help in enhancing nutrient accumulation in fertilized soils when employed as part of integrated nutrient management systems. A recent study (Altaf et al. 2019) indicated a significant increase in levels of nitrogen, phosphorus, and potassium was associated with maize crops treated with PGPR with or without arbuscular mycorrhizae, leading to substantial improvements in grain yield. Therefore, PGPR dramatically reduced the amount of nutrient buildup in the soil under the studied nutrient management system. There is a need to clarify the role that PGPR plays in nutrient management plans that reduce fertilizer application rates and agricultural runoff of nutrients.

7.4.5 Antioxidant Mechanisms

An increase in ROS production is one of the most common responses of plants exposed to environmental stress that can lead to membrane lipid peroxidation, proteins, and nucleic acid disintegration (Francis et al. 2020). A study indicated the role of PGPR strains in improving stress tolerance by stimulating the production of ROS scavengers (Hasanuzzaman et al. 2021). Moreover, microorganisms-induced stress tolerance involves the elevation of ROS scavenging enzymes, viz., superoxide dismutase, peroxidase, glutathione reductase, etc., within the plant system, eventually preventing oxidative stress damage. In another study, Misra and Chauhan (2020) found that maize plants treated with *Bacillus* spp. could mitigate abiotic stress by regulating ethylene production by metabolizing ACC into alpha-ketobutyrate and ammonia. In another research, AM-augmented antioxidation was recorded to induce salt stress tolerance in plants such as clover, lettuce, maize, tomato, and cucumber (Bahadur et al. 2019). During heavy metal stress, membrane fluidity and metabolism were sustained normally due to an increase in osmolyte accumulation and activation of antioxidant enzymes conferred due to the application of arbuscular mycorrhizae (Riaz et al. 2021). In their studies on wheat, Chakraborty et al. (2013) found that superoxide dismutase (SOD) and catalase (CAT) decreased in susceptible varieties GY and MW of wheat under drought stress and with the application of bacterial treatments, viz., *Bacillus safensis* or *Ochrobactrum pseudogrignonense,* attributed to enhance the production of these enzymes in both susceptible varieties conferring alleviation of drought stress.

The innate ability of soil rhizospheric microorganisms in withstanding drought-induced oxidative stress conferred tolerance to stress as a mechanism of drought relief. Under extreme drought conditions, inoculation of lettuce with *Pseudomonas mendocina* and *Glomus intraradices* stimulated the activity of the catalase enzyme, indicating their potential in minimizing drought-induced oxidative damage (Kamal et al. 2015). Gururani et al. (2013) reported enhanced tolerance of potatoes to 200 mm NaCl salt when inoculated with *Bacillus* spp. through the stimulation of antioxidant enzymes, viz., catalase, ascorbate peroxidase, and peroxidase. Islam et al. (2016) revealed improved development of mung bean inoculated with *Bacillus cereus* under salt stress, which was accompanied by enhanced peroxidase, superoxide dismutase, and catalase activities. Under salt stress, *Stenotrophomonas*-inoculated sorghum showed an increase in antioxidant enzyme activity (Singh and Jha 2016). Rice infected with *Curtobacterium albidum* recorded increased antioxidant enzyme activities that were associated with a reduction in salt stress (Vimal et al. 2019).

7.5 Conclusion

This chapter revealed the hidden potential of rhizospheric microorganisms in improving the plant's tolerance to abiotic burdens. The utilization of soil microbes in combating adverse impacts of abiotic factors in plants is an environmentally friendly and sustainable strategy. Although the importance of soil microbes is well known for ages, recent progress in research unraveled certain traits like the production of indole acetic acid and siderophores that is attributed to promote normal plant growth even under unfavorable environments. Similarly, the role of beneficial soil microorganisms in stimulating abiotic stress persistence is mainly associated with ACC deaminase production, regulation of stress-responsive proteins, and stimulation of osmoprotectants and ROS. These findings exploited the potential of soil microbes suitable for stress mitigation. However, many mechanisms associated with soil microbes-induced stress tolerance are underexploited. A proper understanding of these mechanisms could help achieve enhanced crop productivity under climatic change-induced abiotic stress.

References

Abbasi H, Jamil M, Haq A, Ali S, Ahmad R, Malik Z, Parveen Z (2016) Salt stress manifestation on plants, mechanism of salt tolerance and potassium role in alleviating it: a review. Zemdirbyste-Agriculture 103(2):229–238

Abbasi S, Sadeghi A, Safaie N (2020) Streptomyces alleviate drought stress in tomato plants and modulate the expression of transcription factors ERF-1 and WRKY-70 genes. Sci Hortic 265: 109206

AbdElgawad H, Abuelsoud W, Madany MMY, Selim S, Zinta G, Mousa ASM, Hozzein WN (2020) Actinomycetes enrich soil rhizosphere and improve seed quality as well as productivity of legumes by boosting nitrogen availability and metabolism. Biomol Ther 10(12):1675

Abdelhameed RE, Metwally RA (2019) Alleviation of cadmium stress by arbuscular mycorrhizal symbiosis. Int J Phytoremediation 21(7):663–671

Abdoli M, Saeidi M (2012) Effects of water deficiency stress during seed growth on yield and its components, germination and seedling growth parameters of some wheat cultivars. Int J Agric Crop Sci 4(15):1110–1118

Aboul-Ela HM, Saad AA, El-Sikaily AM, Zaghloul TI (2011) Oxidative stress and DNA damage in relation to transition metals overload in Abu-Qir Bay. Egypt J Genet Eng Biotechnol 9(1):51–58

Agami RA, Medani RA, Abd El-Mola IA, Taha RS (2016) Exogenous application with plant growth promoting rhizobacteria (PGPR) or proline induces stress tolerance in basil plants (*Ocimum basilicum* L.) exposed to water stress. Int J Environ Agric Res 2(5):78

Ait-El-Mokhtar M, Laouane RB, Anli M, Boutasknit A, Wahbi S, Meddich A (2019) Use of mycorrhizal fungi in improving tolerance of the date palm (*Phoenix dactylifera* L.) seedlings to salt stress. Sci Horic 253:429–438

Akbari A, Gharanjik S, Koobaz P, Sadeghi A (2020) Plant growth promoting streptomyces strains are selectively interacting with the wheat cultivars especially in saline conditions. Heliyon 6(2): e03445

Alaoui I, Serbouti S, Ahmed H, Mansouri I, El Kamari F, Taroq A, Farah A (2022) The mechanisms of absorption and nutrients transport in plants: a review. Trop J Nat Prod Res 6(1):8–14

Alegre ML, Steelheart C, Baldet P, Rothan C, Just D, Okabe Y, Bartoli CG (2020) Deficiency of GDP-L-galactose phosphorylase, an enzyme required for ascorbic acid synthesis, reduces tomato fruit yield. Planta 251(2):1–10

Ali B, Wang X, Saleem MH, Hafeez A, Afridi MS, Khan S, Ali S (2022) PGPR-mediated salt tolerance in maize by modulating plant physiology, antioxidant defense, compatible solutes accumulation and bio-surfactant producing genes. Plan Theory 11(3):345

Altaf MA, Shahid R, Qadir A, Naz S, Ren MX, Ejaz S, Shakoor A (2019) Potential role of plant growth promoting rhizobacteria (PGPR) to reduce chemical fertilizer in horticultural crops. Int J Res Agric For 6:21–30

Arif N, Yadav V, Singh S, Singh S, Ahmad P, Mishra RK, Chauhan DK (2016) Influence of high and low levels of plant-beneficial heavy metal ions on plant growth and development. Front Environ Sci 4:69

Ashekuzzaman SM, Fenton O, Meers E, Forrestal PJ (2021) Differing phosphorus crop availability of aluminium and calcium precipitated dairy processing sludge potential recycled alternatives to mineral phosphorus fertiliser. Agronomy 11(3):427

Aslam M, Ahmad K, Akhtar MA, Maqbool MA (2017) Salinity stress in crop plants: effects of stress, tolerance mechanisms and breeding strategies for improvement. J Agric Basic Sci 2(1): 70–85

Bagautdinova ZZ, Omelyanchuk N, Tyapkin AV, Kovrizhnykh VV, Lavrekha VV, Zemlyanskaya EV (2022) Salicylic acid in root growth and development. Int J Mol Sci 23(4):2228

Bahadur A, Batool A, Nasir F, Jiang S, Mingsen Q, Zhang Q, Pan J, Liu Y, Feng H (2019) Mechanistic insights into arbuscular mycorrhizal fungi-mediated drought stress tolerance in plants. Int J Mol Sci 20:4199

Basu A, Prasad P, Das SN, Kalam S, Sayyed RZ, Reddy MS, El Enshasy H (2021) Plant growth promoting rhizobacteria (PGPR) as green bioinoculants: recent developments, constraints, and prospects. Sustainability 13(3):1140

Begna T (2020) Effects of drought stress on crop production and productivity. Intl J Res Stud Agric Sci 6:34–43

Begum N, Qin C, Ahanger MA, Raza S, Khan MI, Ashraf M, Zhang L (2019) Role of arbuscular mycorrhizal fungi in plant growth regulation: implications in abiotic stress tolerance. Front Plant Sci 10:1068

Behie SW, Bonet B, Zacharia VM, McClung DJ, Traxler MF (2017) Molecules to ecosystems: actinomycete natural products in situ. Front Microbiol 7:2149

Bertolino LT, Caine RS, Gray JE (2019) Impact of stomatal density and morphology on water-use efficiency in a changing world. Front Plant Sci 10:225

Bhadra P, Maitra S, Shankar T, Hossain A, Praharaj S, Aftab T (2022) Climate change impact on plants: plant responses and adaptations. In: Aftab T, Roychoudhury A (eds) Plant perspectives to global climate changes. Elsevier Inc., Academic Press, pp 1–24. https://doi.org/10.1016/B978-0-323-85665-2.00004-2

Bheemanahalli R, Ramamoorthy P, Poudel S, Samiappan S, Wijewardane N, Reddy KR (2022) Effects of drought and heat stresses during reproductive stage on pollen germination, yield, and leaf reflectance properties in maize (*Zea mays* L.). Plant Direct 6(8):e434

Billah M, Aktar S, Brestic M, Zivcak M, Khaldun ABM, Uddin MS, Bagum SA, Yang X, Skalicky M, Mehari TG, Maitra S (2021) Progressive genomic approaches to explore drought-and salt-induced oxidative stress responses in plants under changing climate. Plan Theory 10(9):1910

Bolat I, Dikilitas M, Ercisli S, Ikinci A, Tonkaz T (2014) The effect of water stress on some morphological, physiological, and biochemical characteristics and bud success on apple and quince rootstocks. Sci World J 8:769732

Cabral C, Sabine R, Bernd WIT (2016) Arbuscular mycorrhizal fungi modify nutrient allocation and composition in wheat (*Triticum aestivum* L.) subjected to heat-stress. Plant Soil 408(1–2):385–399

Calvo-Polanco M, Sanchez-Romera B, Aroca R, Asins MJ, Declerck S, Dodd IC, Ruiz-Lozano JM (2016) Exploring the use of recombinant inbred lines in combination with beneficial microbial inoculants (AM fungus and PGPR) to improve drought stress tolerance in tomato. Environ Exp Bot 131:47–57

Chakraborty U, Chakraborty BN, Chakraborty AP, Dey PL (2013) Water stress amelioration and plant growth promotion in wheat plants by osmotic stress tolerant bacteria. World J Microbiol Biotechnol 29(5):789–803

Chapman S, Watson JE, Salazar A, Thatcher M, McAlpine CA (2017) The impact of urbanization and climate change on urban temperatures: a systematic review. Landsc Ecol 32(10):1921–1935

Chatterjee P, Samaddar S, Niinemets U, Sa TM (2018) *Brevibacterium linens* RS16 confers salt tolerance to *Oryza sativa* genotypes by regulating antioxidant defense and H^+ ATPase activity. Microbiol Res 215:89–101

Chaturvedi P, Wiese AJ, Ghatak A, Zaveska Drabkova L, Weckwerth W, Honys D (2021) Heat stress response mechanisms in pollen development. New Phytol 231(2):571–585

Chen GE, Canniffe DP, Barnett SF, Hollingshead S, Brindley AA, Vasilev C, Hunter CN (2018) Complete enzyme set for chlorophyll biosynthesis in Escherichia coli. Sci Adv 4(1):eaaq1407

Chen L, Liu Y, Wu G, Zhang N, Shen Q, Zhang R (2017) Beneficial rhizobacterium *Bacillus amyloliquefaciens* SQR9 induces plant salt tolerance through spermidine production. Mol Plant-Microbe Interact 30:423–432

Chen X, Zhu Y, Ding Y, Pan R, Shen W, Yu X, Xiong F (2021) The relationship between characteristics of root morphology and grain filling in wheat under drought stress. PeerJ 9:e12015

Chukwuneme CF, Babalola OO, Kutu FR, Ojuederie OB (2020) Characterization of actinomycetes isolates for plant growth promoting traits and their effects on drought tolerance in maize. J Plant Interact 15:93–105

Cohen AC, Bottini R, Pontin M, Berli FJ, Moreno D, Boccanlandro H, Piccoli PN (2015) Azospirillum brasilense ameliorates the response of Arabidopsis thaliana to drought mainly via enhancement of ABA levels. Physiol Plantarum 153(1):79–90

Coyne CJ, Kumar S, von Wettberg EJ, Marques E, Berger JD, Redden RJ, Smykal P (2020) Potential and limits of exploitation of crop wild relatives for pea, lentil, and chickpea improvement. Leg Sci 2(2):e36

Cui Y, Tian Z, Zhang X, Muhammad A, Han H, Jiang D, Dai T (2015) Effect of water deficit during vegetative growth periods on post-anthesis photosynthetic capacity and grain yield in winter wheat (*Triticum aestivum* L.). Acta Physiol Plant 37(10):1–10

Dal Cortivo C, Barion G, Ferrari M, Visioli G, Dramis L, Panozzo A, Vamerali T (2018) Effects of field inoculation with VAM and bacteria consortia on root growth and nutrients uptake in common wheat. Sustainability 10(9):3286

Dambreville A, Lauri PE, Normand F, Guedon Y (2015) Analysing growth and development of plants jointly using developmental growth stages. Ann Bot 115(1):93–105

Dar MH, Bano DA, Waza SA, Zaidi NW, Majid A, Shikari AB, Singh US (2021) Abiotic stress tolerance-progress and pathways of sustainable rice production. Sustainability 13(4):2078

Debasis M, Snezana A, Panneerselvam P, Manisha SA, Tanja V, Ganeshamurthy AN, Divya J (2019) Plant growth promoting microorganisms (PGPMs) helping in sustainable agriculture: current perspective. Int J Agric Sci Vet Med Bd 7(2):50–74

Deng Y, Sun M, Shaevitz JW (2011) Direct measurement of cell wall stress stiffening and turgor pressure in live bacterial cells. Phys Rev Lett 107(15):158101

Diagne N, Ndour M, Djighaly PI, Ngom D, Ngom MCN, Ndong G, Cherif-Silini H (2020) Effect of plant growth promoting rhizobacteria (PGPR) and arbuscular mycorrhizal fungi (AMF) on salt stress tolerance of Casuarina obesa (Miq.). Front Sustain Food Syst 4:601004

Dotaniya ML, Meena VD (2015) Rhizosphere effect on nutrient availability in soil and its uptake by plants: a review. Proc Natl Acad Sci, India Section B: Biol Sci 85(1):1–12

Duanghathaipornsuk S, Farrell EJ, Alba-Rubio AC, Zelenay P, Kim DS (2021) Detection technologies for reactive oxygen species: fluorescence and electrochemical methods and their applications. Biosensors 11(2):30

Egamberdieva D (2012) Pseudomonas chlororaphis: a salt-tolerant bacterial inoculants for plant growth stimulation under saline soil conditions. Acta Physiol Plantarum 34:751–756

Egamberdieva D, Wirth S, Jabborova D, Rasanen LA, Liao H (2017) Coordination between Bradyrhizobium and Pseudomonas alleviates salt stress in soybean through altering root system architecture. J Plant Interact 12(1):100–107

Ehtaiwwesh AF, Emsahel MJ (2020) Impact of salinity stress on germination and growth of pea (*Pisum sativum* L) plants. Al-Mukhtar J Sci 35(2):146–159

Eker S, Heybet EH, Barut H, Erdem H (2013) Effects of zinc on growth and sodium, potassium and calcium concentrations of shoot in bread wheat under salt stress. Fresenius Environ Bull 22: 1622–1627

Elshahat MR, Ahmed AA, Enas AH, Fekria MS (2016) Plant growth promoting rhizobacteria and their potential for biocontrol of phytopathogens. Afr J Microbiol Res 10(15):486–504

Esmaeil Zade NS, Sadeghi A, Moradi P (2019) Streptomyces strains alleviate water stress and increase peppermint (*Mentha piperita*) yield and essential oils. Plant Soil 434(1):441–452

Fahad S, Hussain S, Bano A, Saud S, Hassan S, Shan D, Huang J (2015) Potential role of phytohormones and plant growth-promoting rhizobacteria in abiotic stresses: consequences for changing environment. Environ Sci Pollut Res 22(7):4907–4921

Forni C, Duca D, Glick BR (2017) Mechanisms of plant response to salt and drought stress and their alteration by rhizobacteria. Plant Soil 410(1):335–356

Francis F, Jacquemyn H, Delvigne F, Lievens B (2020) From diverse origins to specific targets: role of microorganisms in indirect pest biological control. Insects 11(8):533

Fukami J, Ollero FJ, Megias M, Hungria M (2017) Phytohormones and induction of plant-stress tolerance and defense genes by seed and foliar inoculation with Azospirillum brasilense cells and metabolites promote maize growth. AMB Express 7(1):1–13

Gallart M, Paungfoo-Lonhienne C, Gonzalez A, Trueman SJ (2021) Nitrogen source influences the effect of plant growth-promoting rhizobacteria (PGPR) on Macadamia integrifolia. Agronomy 11(6):1064

Gao X, Endo H, Nagaki M, Agatsuma Y (2017) Interactive effects of nutrient availability and temperature on growth and survival of different size classes of saccharina japonica (Laminariales, Phaeophyceae). Phycologia 56(3):253–260

Garcia JE, Maroniche G, Creus C, Suarez-Rodriguez R, Ramirez JA, Groppa MD (2017) In vitro PGPB properties and osmotic tolerance of different Azospirillum native strains and their effects on growth of maize under drought stress. Microbiol Res 202:21–29

Garg N, Singh S (2017) Arbuscular mycorrhiza Rhizophagus irregularis, and silicon modulate growth, proline biosynthesis and yield in *Cajanus cajan,* L. Millsp. (pigeon pea) genotypes under cadmium and zinc stress. J Plant Growth Regul 37:1–18

Getahun A, Muleta D, Assefa F, Kiros S (2020) Plant growth-promoting rhizobacteria isolated from degraded habitat enhance drought tolerance of acacia (*Acacia abyssinica* Hochst. ex Benth.) seedlings. Int J Microbiol 2020

Gimenez E, Salinas M, Manzano-Agugliaro F (2018) Worldwide research on plant defense against biotic stresses as improvement for sustainable agriculture. Sustainability 10(2):391

Gong Y, Chen LJ, Pan SY, Li XW, Xu MJ, Zhang CM, Qin S (2020) Antifungal potential evaluation and alleviation of salt stress in tomato seedlings by a halotolerant plant growth-promoting actinomycete *streptomyces* spp. KLBMP-5084. Rhizosphere 16:100262

Goswami D, Pithwa S, Dhandhukia P, Thakker JN (2014) Delineating *Kocuria turfanensis* 2M4 as a credible PGPR: a novel IAA-producing bacteria isolated from saline desert. J Plant Interact 9(1):566–576

Grover M, Bodhankar S, Maheswari M, Srinivasarao C (2016) Actinomycetes as mitigators of climate change and abiotic stress. In Plant Growth Promoting Actinobacteria (pp. 203–212). Springer, Singapore

Grover M, Madhubala R, Ali SZ, Yadav SK, Venkateswarlu B (2014) Influence of *Bacillus* spp. strains on seedling growth and physiological parameters of sorghum under moisture stress conditions. J Basic Microbiol 54(9):951–961

Gulzar A, Mehmood M, Ganie S, Showqi I (2018) A brief review on global warming and climate change: consequences and mitigation strategies. Int J Adv Res Sci Eng 7:2146–2156

Gunasekaran Y, Thiyageshwari S, Ariyan M, Choudhury AR, Park JH, Selvi D, Chithra L, Anandham R (2022) Alleviation of sodic stress in rice by exploring the exopolysaccharide-producing sodic-tolerant bacteria. Agriculture 12:1451

Gupta A, Mishra R, Rai S, Bano A, Pathak N, Fujita M, Hasanuzzaman M (2022) Mechanistic insights of plant growth promoting bacteria mediated drought and salt stress tolerance in plants for sustainable agriculture. Int J Mol Sci 23(7):3741

Gupta B, Huang B (2014) Mechanism of salinity tolerance in plants: physiological, biochemical, and molecular characterization. Int J Genom 1:701596

Gururani MA, Upadhyaya CP, Baskar V, Venkatesh J, Nookaraju A, Park SW (2013) Plant growth-promoting rhizobacteria enhance abiotic stress tolerance in *Solanum tuberosum* through inducing changes in the expression of ROS-scavenging enzymes and improved photosynthetic performance. J Plant Growth Regul 32(2):245–258

Hajiboland R, Dashtebani F, Aliasgharzad N (2015) Physiological responses of halophytic C4 grass, Aeluropus littoralis to salinity and arbuscular mycorrhizal fungi colonization. Photosynthetica 53(4):572–584

Harrison EL, Arce Cubas L, Gray JE, Hepworth C (2020) The influence of stomatal morphology and distribution on photosynthetic gas exchange. Plant J 101(4):768–779

Hasanuzzaman M, Parvin K, Bardhan K, Nahar K, Anee TI, Masud AAC, Fotopoulos V (2021) Biostimulants for the regulation of reactive oxygen species metabolism in plants under abiotic stress. Cell 10(10):2537

Hashem A, Alqarawi AA, Radhakrishnan R, Al-Arjani AF, Aldehaish HA, Egamberdieva D, Abd-Allah EF (2018) Arbuscular mycorrhizal fungi regulate the oxidative system, hormones and ionic equilibrium to trigger salt stress tolerance in *Cucumis sativus* L. Saudi J Biol Sci 25: 1102–1114

Hossain A, da Silva JAT, Lozovskaya MV, Zvolinsky VP (2012) The effect of high temperature stress on the phenology, growth and yield of five wheat (*Triticum aestivum* L.) varieties. Asian Australas J Plant Sci Biotechnol 6(1):14–23

Hossain A, Pamanick B, Venugopalan VK, Ibrahimova U, Rahman MA, Siyal AL, Maitra S, Chatterjee S, Aftab T (2022) Emerging roles of plant growth regulators for plants adaptation to abiotic stress–induced oxidative stress. In: *Emerging Plant Growth Regulators in Agriculture*, Academic Press, pp. 1–72

Hossain A, Skalicky M, Brestic M, Maitra S, Ashraful Alam M, Syed MA, Islam T (2021) Consequences and mitigation strategies of abiotic stresses in wheat (*Triticum aestivum* L.) under the changing climate. Agronomy 11(2):241

Hu H, Jin Q, Kavan P (2014) A study of heavy metal pollution in China: current status, pollution-control policies and countermeasures. Sustainability 6(9):5820–5838

Isah T (2019) Stress and defense responses in plant secondary metabolites production. Biol Res 52: 1

Islam F, Yasmeen T, Arif MS, Ali S, Ali B, Hameed S, Zhou W (2016) Plant growth promoting bacteria confer salt tolerance in Vigna radiata by up-regulating antioxidant defense and biological soil fertility. Plant Growth Regul 80:23–36. https://doi.org/10.1007/s10725-015-0142-y

Iuchi K, Takai T, Hisatomi H (2021) Cell death via lipid peroxidation and protein aggregation diseases. Biology 10(5):399

Jabro JD, Stevens WB, Iversen WM, Allen BL, Sainju UM (2020) Irrigation scheduling based on wireless sensors output and soil-water characteristic curve in two soils. Sensors 20(5):1336

Jacoby R, Peukert M, Succurro A, Koprivova A, Kopriva S (2017) The role of soil microorganisms in plant mineral nutrition-current knowledge and future directions. Front Plant Sci 8:1617

Jamil M, Ashraf M, Rehman S, Ahmad M, Rha ES (2012) Salinity induced changes in cell membrane stability, protein and RNA contents. Afr J Biotechnol 11(24):6476–6483

Javed Q, Azeem A, Sun J, Ullah I, Jabran K, Anandkumar A, Du D (2019) Impacts of salt stress on the physiology of plants and opportunity to firewater the stressed plants with diluted water: a review. Appl Ecol Environ Res 17(5):12583–12604

Kamal R, Gusain YS, Sharma IP, Sharma S, Sharma AK (2015) Impact of arbuscular mycorrhizal fungus, *Glomus intraradices*, *Streptomyces* and *Pseudomonas* spp. strain on finger millet (*Eleusine coracana* L.) cv Korchara under water deficit condition. Afr J Biotechnol 14(48): 3219–3227

Kapoor RT, Hasanuzzaman M (2020) Exogenous kinetin and putrescine synergistically mitigate salt stress in Luffa acutangula by modulating physiology and antioxidant defense. Physiol Mol Biol Plants 26(11):2125–2137

Kardos J, Heja L, Simon A, Jablonkai I, Kovacs R, Jemnitz K (2018) Copper signalling: causes and consequences. Cell Commun Signal 16(1):1–22

Kasotia A, Varma A, Choudhary DK (2015) Pseudomonas-mediated mitigation of salt stress and growth promotion in *Glycine max*. Agric Res 4:31–41

Kaur G, Singh G, Motavalli PP, Nelson KA, Orlowski JM, Golden BR (2020) Impacts and management strategies for crop production in waterlogged or flooded soils: a review. Agron J 112(3):1475–1501

Kavino M, Harish S, Kumar N, Saravanakumar D, Samiyappan R (2010) Effect of chitinolytic PGPR on growth, yield and physiological attributes of banana (*Musa* spp.) under field conditions. Appl Soil Ecol 45(2):71–77

Kayum MA, Kim HT, Nath UK, Park JI, Kho KH, Cho YG, Nou IS (2016) Research on biotic and abiotic stress related genes exploration and prediction in *Brassica rapa* and *B. oleracea*: a review. Plant Breed Biotechnol 4(2):135–144

Khalloufi M, Martinez-Anduja C, Lachaal M, Karray-Bouraoui N, PerezAlfocea F, Albacete A (2017) The interaction between foliar GA3 application and arbuscular mycorrhizal fungi inoculation improves growth in salinized tomato *Solanum lycopersicum* L. plants by modifying the hormonal balance. J Plant Physiol 214:134–144

Khan ZI, Nisar A, Ugulu I, Ahmad K, Wajid K, Bashir H, Dogan Y (2019) Determination of cadmium concentrations of vegetables grown in soil irrigated with wastewater: evaluation of health risk to the public. Egypt J Bot 59(3):753–762

Kim G, Ryu H, Sung J (2022) Hormonal crosstalk and root suberization for drought stress tolerance in plants. Biomol Ther 12(6):811

Kizilgeci F, Tazebay N, Namli M, Albayrak O, Yildirim M (2017) The drought effect on seed germination and seedling growth in bread wheat (*Triticum aestivum* L.). Int J Agric Environ Food Sci 1(1):33–37

Kollmen J, Strieth D (2022) The beneficial effects of cyanobacterial co-culture on plant growth. Life 12(2):223

Kowalska I, Konieczny A, Gastol M (2015) Effect of mycorrhiza and the phosphorus content in a nutrient solution on the yield and nutritional status of lettuce grown on various substrates. J Elem 20:3

Kudoyarova GR, Arkhipova TN, Melent'ev AI (2015) Role of bacterial phytohormones in plant growth regulation and their development. In Bacterial metabolites in sustainable agroecosystem (pp. 69–86). Springer, Cham

Kumar S (2020) Abiotic stresses and their effects on plant growth, yield and nutritional quality of agricultural produce. Int J Food Sci Agric 4:367–378

Kumari A, Das P, Parida AK, Agarwal PK (2015) Proteomics, metabolomics, and ionomics perspectives of salinity tolerance in halophytes. Front Plant Sci 6:537

Kutschera U, Niklas KJ (2013) Cell division and turgor-driven stem elongation in juvenile plants: a synthesis. Plant Sci 207:45–56

Leontidou K, Genitsaris S, Papadopoulou A, Kamou N, Bosmali I, Matsi T, Mellidou I (2020) Plant growth promoting rhizobacteria isolated from halophytes and drought-tolerant plants: genomic characterisation and exploration of phyto-beneficial traits. Sci Rep 10(1):1–15

Li H, Guo Q, Jing Y, Liu Z, Zheng Z, Sun Y, Lai H (2020) Application of Streptomyces pactum Act12 enhances drought resistance in wheat. J Plant Growth Regul 39(1):122–132

Li M, Zhu Y, Li S, Zhang W, Yin C, Lin Y (2022) Regulation of phytohormones on the growth and development of plant root hair. Front Plant Sci 797

Liao X, Chen J, Guan R, Liu J, Sun Q (2021) Two arbuscular mycorrhizal fungi alleviates drought stress and improves plant growth in *Cinnamomum migao* seedlings. Mycobiology 49(4): 396–405

Lundgren K, Kuklane K, Gao C, Holmer I (2013) Effects of heat stress on working populations when facing climate change. Ind Health 51(1):3–15

Luo Q (2011) Temperature thresholds and crop production: a review. Clim Chang 109(3):583–598

Maitra S, Bhadra P, Yadav AN, Palai JB, Jena J, Shankar T (2021b) The omics strategies for abiotic stress responses and microbe-mediated mitigation in plants. In: Yadav AN (ed) Soil microbiomes for sustainable agriculture, sustainable development and biodiversity, vol 27, pp 315–377. https://doi.org/10.1007/978-3-030-73507-4_12

Maitra S, Brestic M, Bhadra P, Shankar T, Praharaj S, Palai JB, Shah MMR, Barek V, Ondrisik P, Skalicky M, Hossain A (2021c) Bioinoculants—natural biological resources for sustainable plant production. Microorganisms 10(1):51

Maitra S, Hossain A, Banerjee P, Bhadra P (2021a) The role of phytohormones in heat stress tolerance in plants. In: Plant growth regulators for climate-smart agriculture (Eds. Fahad, S., Sönmez, O., Saud, S., Wang, D., Wu, C., Adnan, M., and Turan, V., CRC Press, https://doi.org/10.1201/9781003109013, 145–164

Maitra S, Pramanick B, Dey P, Bhadra P, Shankar T, Anand K (2021d) Thermotolerant soil microbes and their role in mitigation of heat stress in plants. In: Yadav AN (ed) Soil microbiomes for sustainable agriculture. Sustainable development and biodiversity, vol 27. Springer, Cham, pp 203–242. https://doi.org/10.1007/978-3-030-73507-4_8. (29 June 2021)

Majeed A, Muhammad Z, Ahmad H (2018) Plant growth promoting bacteria: role in soil improvement, abiotic and biotic stress management of crops. Plant Cell Rep 37(12):1599–1609

Malhi GS, Kaur M, Kaushik P (2021) Impact of climate change on agriculture and its mitigation strategies: a review. Sustainability 13(3):1318

Markulj Kulundzic A, Viljevac Vuletic M, Matosa Kocar M, Antunovic Dunic J, Varga I, Zdunic Z, Lepedus H (2022) Effect of elevated temperature and excess light on photosynthetic efficiency, pigments, and proteins in the field-grown sunflower during afternoon. Horticulturae 8(5):392

Maya MA, Matsubara Y (2013) Influence of arbuscular mycorrhiza on the growth and antioxidative activity in Cyclamen under heat stress. Mycorrhiza 23(5):381–390

Misra S, Chauhan PS (2020) ACC deaminase-producing rhizosphere competent *Bacillus* spp. mitigate salt stress and promote *Zea mays* growth by modulating ethylene metabolism. 3 Biotech 10(3):1–14

Mostoni S, Milana P, Di Credico B, D'Arienzo M, Scotti R (2019) Zinc-based curing activators: new trends for reducing zinc content in rubber vulcanization process. Catalysts 9(8):664

Mrkovacki N, Dalovic I, Josic D, Bjelic D, Brdar-Jokanovic M (2016) The effect of PGPR strains on microbial abundance in maize rhizosphere in field conditions. Ratarstvo i povrtarstvo/Field Veg Crops Res 53(1):15–19

Mukhtar S, Hirsch AM, Khan N, Malik KA, Humm EA, Pellegrini M, Mehnaz S (2020) Impact of soil salinity on the cowpea nodule-microbiome and the isolation of halotolerant PGPR strains to promote plant growth under salinity stress. Phytobiomes J 4(4):364–374

Mushtaq Z, Faizan S, Gulzar B (2020) Salt stress, its impacts on plants and the strategies plants are employing against it: a review. J Appl Biol Biotechnol 8(3):8–1

Nahar K, Hasanuzzaman M, Ahamed KU, Hakeem KR, Ozturk M, Fujita M (2015) Plant responses and tolerance to high temperature stress: role of exogenous phytoprotectants. In Crop production and global environmental issues (pp. 385–435). Springer, Cham

Naveed M, Hussain MB, Zahir ZA, Mitter B, Sessitsch A (2014) Drought stress amelioration in wheat through inoculation with Burkholderia phytofirmans strain PsJN. Plant Growth Regul 73(2):121–131

Nawaz K, Hussain K, Majeed A, Khan F, Afghan S, Ali K (2010) Fatality of salt stress to plants: morphological, physiological and biochemical aspects. Afr J Biotechnol 9(34):5475–5480

Nazar R, Iqbal N, Masood A, Khan MIR, Syeed S, Khan NA (2012) Cadmium toxicity in plants and role of mineral nutrients in its alleviation. Am J Plant Sci 3(10):1476–1489

Ndjiondjop MN, Wambugu PW, Sangare JR, Gnikoua K (2018) The effects of drought on rice cultivation in sub-Saharan Africa and its mitigation: a review. Afr J Agric Res 13(25): 1257–1271

Odoh CK (2017) Plant growth promoting rhizobacteria (PGPR): a bioprotectant bioinoculant for sustainable agrobiology. A review. Int J Adv Res Biol Sci 4(5):123–142

Ojuederie OB, Olanrewaju OS, Babalola OO (2019) Plant growth promoting rhizobacterial mitigation of drought stress in crop plants: implications for sustainable agriculture. Agronomy 9(11):712

Onwuka B, Mang B (2018) Effects of soil temperature on some soil properties and plant growth. Adv Plants Agric Res 8(1):34–37

Orandi J, Mwonga S, Ojiem J, Lauren J (2021) Effect of phosphorus fortified compost on growth and yield of maize (*Zea mays* L.) and Lablab (*Lablab purpureus* L.) intercropped maize in acidic soils of Western Kenya. Afr J Agric Res 17(2):329–336

Palaniyandi SA, Damodharan K, Yang SH, Suh JW (2014) Streptomyces sp. strain PGPA39 alleviates salt stress and promotes growth of 'Micro Tom'tomato plants. J Appl Microbiol 117(3):766–773

Panhwar QA, Radziah O, Rahman AZ, Sariah M, Razi IM, Naher UA (2011) Contribution of phosphate-solubilizing bacteria in phosphorus bioavailability and growth enhancement of aerobic rice. Span J Agric Res 9(3):810–820

Parthasarathi T, Velu G, Jeyakumar P (2013) Impact of crop heat units on growth and developmental physiology of future crop production: a review. J Crop Sci Technol 2(1):2319–3395

Pirzad A, Shakiba MR, Zehtab-Salmasi S, Mohammadi SA, Darvishzadeh R, Samadi A (2011) Effect of water stress on leaf relative water content, chlorophyll, proline and soluble carbohydrates in *Matricaria chamomilla* L. J Med Plants Res 5(12):2483–2488

Potapova T, Gorbsky GJ (2017) The consequences of chromosome segregation errors in mitosis and meiosis. Biol 6(1):12

Purnama PR, Purnama ER, Manuhara YSW, Hariyanto S, Purnobasuki H (2018) Effect of high temperature stress on changes in morphology, anatomy and chlorophyll content in tropical seagrass Thalassia hemprichii. AACL Bioflux 11(6):1825–1833

Raddatz N, Morales de los Rios L, Lindahl M, Quintero FJ, Pardo JM (2020) Coordinated transport of nitrate, potassium, and sodium. Front Plant Sci 11:247

Rahman MA, Kang S, Nagabhatla N, Macnee R (2017) Impacts of temperature and rainfall variation on rice productivity in major ecosystems of Bangladesh. Agric Food Secur 6(1):1–11

Rajput VD, Minkina T, Kumari A, Singh VK, Verma KK, Mandzhieva S, Keswani C (2021) Coping with the challenges of abiotic stress in plants: new dimensions in the field application of nanoparticles. Plan Theory 10(6):1221

Raneesh KY (2017) A review on the impact of climate change on crop production in humid tropics. SF J Glob Warm 1:1–3

Raza A, Razzaq A, Mehmood SS, Zou X, Zhang X, Lv Y, Xu J (2019) Impact of climate change on crops adaptation and strategies to tackle its outcome: a review. Plan Theory 8(2):34

Riaz M, Kamran M, Fang Y, Wang Q, Cao H, Yang G, Deng L, Wang Y, Zhou Y, Anastopoulos I, Wang X (2021) Arbuscular mycorrhizal fungi-induced mitigation of heavy metal phytotoxicity in metal contaminated soils: a critical review. J Hazard Mater 402:123919

Roach T, Krieger-Liszkay A (2014) Regulation of photosynthetic electron transport and photoinhibition. Curr Protein Pept Sci 15(4):351–362

Sabki MH (2021) A review on abiotic stress tolerance and plant growth metabolite framework by plant growth-promoting bacteria for sustainable agriculture. Chem Eng Trans 83:367–372

Sandhya VSKZ, Ali SZ, Grover M, Reddy G, Venkateswarlu B (2010) Effect of plant growth promoting Pseudomonas spp. on compatible solutes, antioxidant status and plant growth of maize under drought stress. Plant Growth Regul 62(1):21–30

Scholz P, Anstatt J, Krawczyk HE, Ischebeck T (2020) Signalling pinpointed to the tip: the complex regulatory network that allows pollen tube growth. Plan Theory 9(9):1098

Selvakumar G, Bhatt RM, Upreti KK, Bindu GH, Shweta K (2015) Citricoccus zhacaiensis B-4 (MTCC-12119) a novel osmotolerant plant growth promoting actinobacterium enhances onion (*Allium cepa* L.) seed germination under osmotic stress conditions. World J Microbiol Biotechnol 31(5):833–839

Shaffique S, Khan MA, Imran M, Kang SM, Park YS, Wani SH, Lee IJ (2022) Research progress in the field of microbial mitigation of drought stress in plants. Front Plant Sci 13:870626

Shahbazi F, Ghasemi S, Sodaiezadeh H, Ayaseh K, Zamani-Ahmadmahmoodi R (2017) The effect of sewage sludge on heavy metal concentrations in wheat plant (*Triticum aestivum* L.). Environ Sci Pollut Res 24(18):15634–15644

Shahid M, Pourrut B, Dumat C, Nadeem M, Aslam M, Pinelli E (2014) Heavy-metal-induced reactive oxygen species: phytotoxicity and physicochemical changes in plants. Rev Environ Contamin Toxicol 232:1–44

Sharma A, Shahzad B, Kumar V, Kohli SK, Sidhu GPS, Bali AS, Zheng B (2019) Phytohormones regulate accumulation of osmolytes under abiotic stress. Biomol Ther 9(7):285

Sharma P, Jha AB, Dubey RS, Pessarakli M (2012) Reactive oxygen species, oxidative damage, and antioxidative defense mechanism in plants under stressful conditions. J Bot 1:26

Shetta ND (2015) Influence of drought stress on growth and nodulation of Acacia origena (Hunde) inoculated with indigenous rhizobium isolated from Saudi Arabia. Am-Eurasian J Agric Environ Sci 15(5):699–706

Shokri S, Maadi B (2009) Effects of arbuscular mycorrhizal fungus on the mineral nutrition and yield of *trifolium alexandrium* plants under salinity stress. J Agron 8:79–83

Singh R, Singh J, Deval R, Upadhyay S, Kumar D (2017) The potentiality of selected strain of PGPR: azotobacter, for sustainable agriculture in India. G-J of Environ Sci Technol 4(6): 49–53

Singh RP, Jha PN (2016) The multifarious PGPR *Serratia marcescens* CDP-13 augments induced systemic resistance and enhanced salinity tolerance of wheat (*Triticum aestivum* L.). PLoS One 11:e0155026

Sofy MR, Aboseidah AA, Heneidak SA, Ahmed HR (2021) ACC deaminase containing endophytic bacteria ameliorate salt stress in *Pisum sativum* through reduced oxidative damage and induction of antioxidative defense systems. Environ Sci Pollut Res 28:40971–40991

Song Y, Chen Q, Ci D, Shao X, Zhang D (2014) Effects of high temperature on photosynthesis and related gene expression in poplar. BMC Plant Biol 14(1):1–20

Srivastava P, Kumar R (2015) Soil salinity: a serious 745 environmental issue and plant growth-promoting bac-746 teria as one of the tools for its alleviation. Saudi J Biol Sci 22(2):123–131

Sun Z, Song J, Xin X, Xie X, Zhao B (2018) Arbuscular mycorrhizal fungal proteins 14-3-3- are involved in arbuscule formation and responses to abiotic stresses during AM symbiosis. Front Microbiol 5:9–19

Swamy MK, Nalina N, Nalina D, Akhtar MS, Purushotham B (2019) Heavy metal stress and tolerance in plants mediated by rhizospheric microbes. In Salt stress, microbes, and plant interactions: causes and solution (pp. 181–198). Springer, Singapore

Tan X, Li K, Wang Z, Zhu K, Tan X, Cao J (2019) A review of plant vacuoles: formation, located proteins and functions. Plan Theory 8(9):327

Tiwari S, Lata C, Chauhan PS, Nautiyal CS (2016) Pseudomonas putidaattunes morphophysiological, biochemical and molecular responsesin *Cicer arietinum* L. during drought stress and recovery. Plant Physiol Biochem 99:108–117

Ul Hassan Z, Ali S, Rizwan M, Hussain A, Akbar Z, Rasool N, Abbas F (2017) Role of zinc in alleviating heavy metal stress. Essential plant nutrients: uptake, use efficiency, and management; springer: Cham, Switzerland 351-366

Upadhyay SK, Singh JS, Saxena AK, Singh DP (2012) Impact of PGPR inoculation on growth and antioxidant status of wheat under saline conditions. Plant Biol 14:605–611

Vimal SR, Patel VK, Singh JS (2019) Plant growth promoting *Curtobacterium albidum* strain SRV4: an agriculturally important microbe to alleviate salinity stress in paddy plants. Ecol Indic 105:553–562

Wang W, Wu Z, He Y, Huang Y, Li X, Ye BC (2018) Plant growth promotion and alleviation of salinity stress in Capsicum annuum L. by *Bacillus* isolated from saline soil in Xinjiang. Ecotoxicol Environ Saf 164:520–529

Wang Y, Kang Y, Zhong M, Zhang L, Chai X, Jiang X, Yang X (2022) Effects of iron deficiency stress on plant growth and quality in flowering Chinese cabbage and its adaptive response. Agronomy 12(4):875

Wu F, Li J, Chen Y, Zhang L, Zhang Y, Wang S, Liang J (2019) Effects of phosphate solubilizing bacteria on the growth, photosynthesis, and nutrient uptake of Camellia oleifera Abel. Forests 10(4):348

Yadav SP, Bharadwaj R, Nayak H, Mahto R, Singh RK, Prasad SK (2019) Impact of salt stress on growth, productivity and physicochemical properties of plants: a review. Int J Chem Stud 7(2): 1793–1798

Yamori W, Hikosaka K, Way DA (2014) Temperature response of photosynthesis in C3, C4, and CAM plants: temperature acclimation and temperature adaptation. Photosynth Res 119(1): 101–117

Yandigeri MS, Meena KK, Singh D, Malviya N, Singh DP, Solanki MK, Arora DK (2012) Drought-tolerant endophytic actinobacteria promote growth of wheat (*Triticum aestivum*) under water stress conditions. Plant Growth Regul 68(3):411–420

Yang B, Liu J, Ma X, Guo B, Liu B, Wu T, Chen F (2017) Genetic engineering of the Calvin cycle toward enhanced photosynthetic CO2 fixation in microalgae. Biotechnol Biofuels 10(1):1–13

Yang X, Wang B, Chen L, Li P, Cao C (2019) The different influences of drought stress at the flowering stage on rice physiological traits, grain yield, and quality. Sci Rep 9(1):1–12

Yasin NA, Akram W, Khan WU, Ahmad SR, Ahmad A, Ali A (2018) Halotolerant plant-growth promoting rhizobacteria modulate gene expression and osmolyte production to improve salinity tolerance and growth in *Capsicum annum* L. Environ Sci Pollut Res 25(23):23236–23250

Yun P, Xu L, Wang SS, Shabala L, Shabala S, Zhang WY (2018) Piriformospora indica improves salinity stress tolerance in *Zea mays* L. plants by regulating Na+ and K+ loading in root and allocating K+ in shoot. Plant Growth Regul 86(2):323–331

Zarea MJ, Hajinia S, Karimi N, Goltapeh EM, Rejali F, Varma A (2012) Effect of *Piriformospora indica* and *Azospirillum* strains from saline or non-saline soil on mitigation of the effects of NaCl. Soil Biol Biochem 45:139–146

Zeng Y, Zhang Y, Xiang J, Uphoff NT, Pan X, Zhu D (2017) Effects of low temperature stress on spikelet-related parameters during anthesis in indica-japonica hybrid rice. Front Plant Sci 8: 1350

Zhang F, Jia-Dong HE, Qiu-Dan NI, Qiang-Sheng WU, Ying-Ning ZOU (2018) Enhancement of drought tolerance in trifoliate orange by mycorrhiza: changes in root sucrose and proline metabolisms. Not Bot Horti Agrobot Cluj-Napoca 46(1):270–276

Zhao R, Guo W, Bi N, Guo J, Wang L, Zhao J, Zhang J (2015) Arbuscular mycorrhizal fungi affect the growth, nutrient uptake and water status of maize (*Zea mays* L.) grown in two types of coal mine spoils under drought stress. Appl Soil Ecol 88:41–49

Chapter 8
An Overview of the Multifaceted Role of Plant Growth-Promoting Microorganisms and Endophytes in Sustainable Agriculture: Developments and Prospects

Shyamalina Haldar and Sanghamitra Sengupta

Abstract Plant growth-promoting microorganisms (PGPM) and endophytes are naturally plant-inhabiting microbial species that contribute as a resource for innumerable metabolites, agriculturally important promoters, and stress-adaptive molecules in their host plants and confer benefits to their hosts with respect to growth enhancement, a surge of metabolic competences, and conferring stress resistance. However, the interaction between the symbionts is a complex phenomenon. It is imperative to explore the host–microbial crosstalk at the root–surface and internal plant tissues to identify the cellular and molecular aspects of the alliance. This chapter focuses on the critical evaluation of the colonization and multifaceted interactions of PGPM and endophytes in the host, their impacts on the health management of plants, crop productivity, multistress tolerance, and production of agriculturally and biotechnologically important metabolites. Given that the literature is replete with various information regarding the beneficial roles of PGPMs and endophytes in plant health, the possibilities of commercialization of PGPMs in sustainable agriculture are critically apprised in this synthesis.

Keywords Endophytes · Plant growth-promoting microorganisms · Plant–microbe interaction · Plant protection · Soil health management

S. Haldar
Department of Biochemistry, Asutosh College, Kolkata, India

S. Sengupta (✉)
Department of Biochemistry, University of Calcutta, Kolkata, India

P. Mathur et al. (eds.), *Microbial Symbionts and Plant Health: Trends and Applications for Changing Climate*, Rhizosphere Biology,
https://doi.org/10.1007/978-981-99-0030-5_8

Abbreviations

ACC-1	Aminocyclopropane-1-carboxylase
AHL	Acyl-homoserine lactones
DAPG	2,4-Diacetylphloroglucinol
DDE	2,2-bis(p-Chlorophenyl)-1,1-dichloro-ethylene
DIMBOA	2,4-Dihydoxy-7-methoxy-1,4-benzoxazine-3-one
DMDS	Dimethyl disulfide
DMHDA	Dimethyl hexadecylamine
HCN	Hydrogen cyanide
HPLC	High-performance liquid chromatography
JA	Jasmonic acid
LPS	Lipopolysaccharides
MAMPS	Microbe-associated molecular patterns
OTUs	Operational taxonomic units
PAH	Polycyclic aromatic hydrocarbons
PGP	Plant growth-promoting
PGPM	Plant growth-promoting microorganisms
PGPR	Plant growth-promoting rhizobacteria
RNS	Reactive nitrogen species
ROS	Reactive oxygen species
TTSS	SPI1-type III secretion system in *Salmonella* pathogenicity island 1
VOC	Volatile organic compounds

8.1 Introduction

The investigation of structural dynamicity and functional attributes of the host-associated microbiome has been one of the pertinent foci of biological research in recent past decades. The main theme of these studies revolves around the chain of communication between the microbes and their respective hosts and the impacts of this crosstalk upon the nutritional status, timed growth and development, disease resistance, behavioral patterns, and ultimate maintenance of the overall physiological healthy living system throughout the plant and animal kingdoms. The mechanisms of these interactions along the various living species are intriguing but complex and challenging to appraise due to the involvement of diverse abiotic and biotic interactive factors (Berendsen et al. 2012; Pieterse et al. 2014; Marin et al. 2021).

The plant–rhizosphere and endophytic interactions are the obvious instances of such complex living associations that comprise multifarious intradomain and interdomain interactions, affected by challenges from climate diversity, edaphic factors, physical structures developing innumerable micro-niches, chemical abundance, and spatiotemporal variation. The effects of the interplay between plant and the allied microbiome determine the ontogeny of the interactive participants and develop the fundamental basis for both plant and microbial ecology, adaptation, and evolutionary processes (Chaparro et al. 2013; Durán et al. 2021). The presence of a large number of carbon-rich compounds in the root exudates stimulates the colonization of the varied rhizospheric microbial taxa with a different composition from that of the surrounding soil. The abundance of rhizospheric bacteria is influenced by the order of the root branching as the branching order determines the mineral content and exposure to the surrounding soil. The small root environment enhances the abundance of *Bradyrhizobium*, *Burkholderia*, *Pseudomonas*, *Sphingomonas*, and *Streptomyces* sp. (Chaparro et al. 2013; Pervaiza et al. 2020). The microbial population utilizes 15% of the root exudates due to their metabolic flexibility to regulate the quantity and the nature of the exudates for their use (Li et al. 2016a, b).

The holistic approaches explain the plant–microbiome as a whole, generally referred to as holobiont, nullifying the interfering factors to diminish the environmental variation and investigate the modal operation under natural circumstances. These observations are based on culture-independent high-throughput genomics techniques (Marin et al. 2021). Contrastingly, the reductionist approach analyzes the specific plant–microbiome interactions through culture-based techniques where the co-partners are known as in plant–mycorrhizal associations, plant–pathogen protection, and the interactions between plants with symbiotic nitrogen-fixing or plant growth–promoting microorganisms [PGPM]. The ultimate goal of these studies is to use this PGPM and endophytes as the substitute stratagems for agrochemicals and pesticides to boost the yields of agriculture in an environmentally sustainable approach to meet up with the demands of foods for the ever-increasing world population (Saeed et al. 2021).

With this background, this chapter focuses on the critical evaluation of the colonization and multifaceted interactions of PGPM and endophytes in the host, their impacts on the health management of plants, crop productivity, multistress tolerance, and production of agriculturally and biotechnologically important metabolites. The chapter ends with the possibilities and challenges of the commercialization of PGPMs in sustainable agriculture.

8.2 PGPM Vs. Endophytes

The plant root is the prime spot of communication between plants and allied microorganisms and creates the core machinery of plant microbiomes that influence agriculture (Vandana et al. 2021). The term "rhizosphere," first introduced in 1904 by Lorenz Hiltner, demarcates the region of soil closely associated with plant roots. The term 'rhizobacteria' suggests the assembly of rhizospheric microbes around the plant root surfaces (McNear Jr 2013; Rosier et al. 2018). Plant growth-promoting (PGP) components are supplied by these beneficial microbes residing within or around the roots of plants and hence are known as "plant growth-promoting rhizobacteria" (PGPR) (Meena et al. 2017). Kloepper and Schroth defined "plant growth-promoting rhizobacteria (PGPR)" as rhizospheric bacteria colonizing the plant root surfaces and enhances plant growth (Kloepper and Schroth 1978). However, around 2–5% of rhizospheric bacteria belong to the group of PGPR. *Acinetobacter*, *Arthrobacter*, *Azospirillum*, *Azotobacter*, *Bacillus*, *Bradyrhizobium*, *Burkholderia*, *Caulobacter*, *Chromobacterium*, *Enterobacter*, *Erwinia*, *Flavobacterium*, *Frankia*, *Klebsiella*, *Micrococcus*, *Pseudomonas*, *Rhizobium*, and *Serratia* are the common genera of PGPR reported to date (Vandana et al. 2021). The PGPR attached to the surface of plant roots is termed extracellular PGPR while those localized inside the plant cells and develop nodules in symbiotic associations are known as intracellular PGPR or endophytes. The presence of nutrients derived from plant roots results in 5–10 times more fungi and 10–50 times more bacteria within the rhizosphere than in ordinary soil (Li et al. 2016a, b; Vandana et al. 2021).

Endophytic bacteria are plant-associated bacteria that reside in the internal tissues of the plant without harming it. They are located in intra- and intercellular areas or the vascular tissue and colonize the aerial parts or the roots. However, the nature of their mutualistic association depends on their location in the plant tissue, either intercellularly or intracellularly (Potshangbam et al. 2017) They are grouped as obligate or facultative and local or systemic (Castanheira et al. 2017; Rangjaroen et al. 2017; Eid et al. 2020). Additionally, it is fascinating that the establishment of endophytes within plant tissues influences the relative composition and abundance of native microbial strata and overpowers the harmful microorganisms within the plant tissues (Vandana et al. 2021). The endophytes characterized till now belong to *Actinobacteria*, *Firmicutes*, and *Proteobacteria*, including species such as *Achromobacter* sp., *Agrobacterium* sp., *Acinetobacter* sp., *Bacillus* sp., *Brevibacterium* sp., *Microbacterium* sp., *Pseudomonas* sp., and *Xanthomonas* sp.

The endophytes are characterized by numerous plant species and organs, including stems, leaves, and roots, and be recruited from the root rhizosphere of the plants. Therefore, bacteria inhabiting the rhizosphere are the source of the formation of endophytic bacterial communities, and in plants, roots have the highest occurrence of endophytic bacteria (Vandana et al. 2021). The chemical nature of the root cell walls is modulated by the PGPR, which facilitates the progression of the bacterial cells between root cortex cells to grow as endophytes. These modifications are permanent as they involve changes in gene expression within the plants. Rice roots inoculated with endophytic PGPR *Azospirillum irakense* were observed to increase the expression of polygalacturonase genes. The endophytes then penetrate through epidermal cracks in the roots of the plants, particularly through lateral roots or root-hair cells, colonize within the cytoplasm of the cells, and release metabolites that influence the gene expression and functioning in the plants directly (Thomas and Reddy 2013; Thomas and Sekhar 2014). Interestingly, they are capable of both inter- and intracellular colonization. The endophytic bacteria colonizing internally in the plant roots have been observed to deliver different functions, adaptations, specialization, and competence compared with other tissue endophytes or PGPR (Vandana et al. 2021). The inhabitation of root endophytes at the localized entry point or inside the plant tissues is ensured with the use of various hydrolytic enzymes like cellulose-degrading enzymes (as cellulase as in *Enterobacter asburiae* JM22 in cotton plants) and bioactive metabolites produced by the endophytes themselves (Vandana et al. 2021). The motility of the bacteria, the roots developing passive colonization, and molecular and cellular shifts during plant growth and development are the determinants of successful colonization of root endophytes (Costa and Melo 2012). Though the PGPR and the endophytes employ comparable mechanisms for plant growth promotion, the latter are not exposed to changes in the external environment such as soil pH, temperature, and water content directly. Therefore, the rhizospheric bacteria are more prone to competition to binding sites on the root surfaces due to the constant changes in the immediate environments and also due to the presence of numerous competitive species in the surrounding soil that are also constantly trying to develop associations with the root surfaces. In comparison to the PGPR, the persistent nature of the endophytes is an advantage for them to be used in biofertilization and/or bioremediation (Glick 2012; Santoyo et al. 2016).

8.3 Colonization and Rhizospheric Competence

The plant rhizosphere is the niche that creates selective pressure toward attracting a few specific microbial communities from the bulk soil. The 16S rRNA gene sequencing comparison of microbial communities in the rhizosphere of 19 plant species recently demonstrated that the selected operational taxonomic units (OTUs) varied from 18 to a little more than 100 in each analyzed plant species among more than 1000 found (Dawson et al. 2017). The nutrient-rich photosynthates composed of carbohydrates, amino acids, organic acids, sugars, sterols, phenolics, fatty acids, vitamins, nucleic acids, and plant growth regulators secreted by the plant roots termed rhizodeposits act as the main driver for this selectivity and foster the development of a highly competitive ecosystem for the microorganisms due to its diverse chemical profile (Sasse et al. 2018; Hassan et al. 2019).

8.3.1 Mechanism of and Factors Controlling PGPR Colonization

The microbial acquisition and colonization in the plant rhizosphere are explained by a two-step approach, where the first step includes cellular recognition, cell wall and cell membrane characterization, and chemotrophic movements on the basis of the type and quantity of the substrate present in the niche while in the second step the genotype of the plants chooses, selects, and recruits specific groups of microbes on the basis of positive and negative selection of the microbial phylotypes using their exudate profile (Matthews et al. 2019). This has been observed in plants like *Arabidopsis thaliana*, *Medicago truncata*, *Zea mays* (maize), and *Solanum lycopersicum* (tomato) where the rhizodeposits such as malic acid and

2,4-dihydroxy-7-methoxy-1,4-benzoxazine-3-one (DIMBOA) attract only the beneficial bacteria like *Bacillus* sp. and inhibit the recruitment of the pathogenic species and differentially impact the rhizo-residents and nonresident microbial communities (Santoyo et al. 2021). Therefore, the determining factor for the recruitment of plant-specific PGPR is the capability of acquiring and utilizing the plant-secreted nutrients by those bacterial species. The mechanisms of rhizosphere colonization by PGPR with specific examples are provided in Table 8.1. These mechanisms include the main three components of (1) iron acquisition by siderophore production as found in *Pseudomonas* sp. aiding to plant growth or *Burkholderia cenocepacia* strain XXVI, which produces iron-chelating siderophores with biocontrol activity against the fungal pathogen *Colletotrichum lindemuthianum* ATCC MYA 456 or *Pseudomonas* sp., (2) phosphate solubilization as in the case of mycorrhizal fungi colonizing 90% of the vascular plants sequester and solubilize phosphorus from the soil and deliver to the plants for growth, and (3) nitrogen fixation and metabolism as done by the "rhizobia" (*Bradyrhizobium* sp., *Mesorhizobium* sp., and *Rhizobium* sp.) with the help of the nod factors produced by the stimulation of flavonoids in the root exudates (de Los Santos-Villalobos et al. 2012; Sulochana et al. 2014).

The resistance to oxidative stress and detoxification of reactive oxygen species (ROS) is another characteristic important for colonization of PGPR. The *Gluconacetobacter diazotrophicus* PAL5 mutants of ROS-detoxifying enzymes, glutathione reductase, and superoxide dismutase were observed to fail in the colonization of *Oryza sativa* IR-42 (rice) roots (Alquéres et al. 2013). This can be explained by the fact that the abiotic stresses developing inside the rhizosphere microenvironment due to variations of soil pH, temperature, salinity, water content, and the presence of heavy metals or pollutants develop oxidative stress and damage the bacterial cells. The successful colonization by the PGPR is possible by mitigating these stressors using enzymatic and nonenzymatic pathways in which they (1) either quench these molecules and/or transform them to less toxic products by the enzymes; (2) produce nonenzymatic antioxidant compounds like vitamins (A, C, E), carotenoids, flavonoids, and trehalose, which can inhibit or delay the oxidation of oxidizable substrates of bacterial cells; (3) migrate to lacunae with low incoming solar radiation by producing pigments that can absorb the harmful radiation; and (4) captivate the DNA with chromatin and proteins to offer alternative sites for ROS attack. This defense mechanism toward ROS is facilitated in the presence of microbial consortia forming biofilms. Therefore, the formation of an extracellular matrix or biofilm composed of lipopolysaccharides (LPS), proteins, nucleic acids, lipids, and carbohydrates plays a crucial role in the stress alleviation, maintenance of strong attachments onto the plant surfaces with the high cellular levels, and also provides a protective layer upon the plant surfaces. The PGP activities have been found to increase in the microbes, which are efficient in producing biofilms (Paulucci et al. 2015; Meena et al. 2017; Prieto-Barajas et al. 2018; Banerjee et al. 2019; Rojas-Solís et al. 2020; Zboralski and Filion 2020; Santoyo et al. 2021).

The colonization of PGPR is also controlled by the production of (1) volatile organic compounds (VOCs) and (2) antimicrobial compounds. The low-molecular-weight VOCs have been reported to aid in the plant growth promotion, and

Table 8.1 Mechanisms of colonization of host plants by PGPR and endophytes

PGPR/endophytes	Mechanism of colonization	Reference
Arthrobacter agilis strain UMCV2	Iron uptake and hypermotility by producing VOC (DMHDA)	Orozco-Mosqueda et al. (2013), Hernández-Calderón et al. (2018)
Bacillus subtilis	Induce systemic resistance and inhibit *Ralstonia solanacearum* by producing VOC (2,3-butanediol and DMDS)	Yi et al. (2016)
Bacillus phytofirmans	Induction of systemic resistance by activating defense genes	Trdá et al. (2014)
Bradyrhizobium japonicum	Induction of systemic resistance by activating defense genes	Piromyou et al. (2015)
Burkholderia cenocepacia strain XXVI	(1) Siderophore production (2) Biocontrol activity against *Colletotrichum lindemuthianum* ATCC MYA 456 and/ or *Pseudomonas* sp.	de los Santos-Villalobos et al. (2012)
Gluconacetobacter diazotrophicus PAL5	(1) Resistance to oxidative stress (2) Modulation of ethylene production (3) Production of cell-degrading enzymes (4) Detoxification of ROS and RNS	Alquéres et al. (2013), Vandana et al. (2021)
Herbaspirillum rubrisulbalbicans HCC103	(1) Modulation of ethylene production (2) Production of cell-degrading enzymes (3) Detoxification of ROS and RNS	Vandana et al. (2021)
Klebsiella pneumoniae 342	Induction of systemic resistance by activating defense genes	Vandana et al. (2021)
Rhizobia (*Bradyrhizobium* sp., *Mesorhizobium* sp., and *Rhizobium* sp.)	Nitrogen fixation	Fujita et al. (2014)
Pseudomonas aeruginosa	(1) Biofilm production (2) Induction of systemic resistance by activating defense genes	Valentini and Filloux (2016), Lopez-Gomez et al. (2012)
Pseudomonas sp.	(1) Inhibit *Gaeumannomyces graminis* by producing DAPG (2) Antagonism by secretory system TS66 (3) Siderophore production (4) Inhibition of defense response in plants	Berendsen et al. (2012) Durán et al. (2021) Sulochana et al. (2014) Yadong et al. (2013)
Pseudomonas stutzeri E25	Antifungal VOC production	Rojas-Solís et al. (2018)
Salmonella enterica serovar typhimurium strain 14,028	Induction of systemic resistance by activating defense genes	Vandana et al. (2021)
Stenotrophomonas maltophilia CR71	Antifungal VOC production	Rojas-Solís et al. (2018)

development of systemic induced resistance, may behave as antimicrobial compounds, and act as signal transducers/inducers (Pieterse et al. 2014; Aviles-Garcia et al. 2016; Hernández-Calderón et al. 2018). The volatile dimethyl hexadecyl amine (DMHDA) synthesized by rhizobacterium *Arthrobacter agilis* strain UMCV2 in the iron-deficient environments leads to the enhanced yield of *Medicago truncatula* sp. and also stimulates the exudation of protons, resulting in rhizosphere acidification, promoting iron uptake under limiting conditions, and thereby increasing the iron content in the plants (Orozco-Mosqueda et al. 2013). Similarly, the volatile 2,3-butanediol produced by *Bacillus subtilis* was found to induce systemic responses in plants against the pathogen *Ralstonia solanacearum* and also triggered those root exudates that could modulate the soil microflora (Yi et al. 2016). VOC-like dimethyl disulfide (DMDS) produced by the same *Bacillus* sp. was found to inhibit the growth of fungal mycelia (Rojas-Solís et al. 2018).

Motility is a primary key to rhizosphere inhabitation. It is stimulated by the rhizodeposits. The presence of flagella and/or pili is the first critical step for successful attachment and adhesion onto the plant surface and finally colonizing in the particular niche. Therefore, the hypermotile phenotypes are successful in rhizosphere competence compared with others. VOC, along with the flagellar proteins, is involved in the regulation of motility as dimethylhexadecylamine (DMHDA) produced by *Arthrobacter agilis* affected the motility of other species also like *Bacillus* sp. (Hernández-Salmerón et al. 2016; Hernández-Salmerón et al. 2017; Bakker et al. 2020; Martínez-Cámara et al. 2020; Fernández-Llamosas et al. 2021).

The synthesis of antimicrobial compounds is targeted to inhibit the growth of the undesired microbes to prevent the nutritional competition within a specific niche (Santoyo et al. 2012). These antimicrobial compounds interfere with various metabolic pathways of the other microbes (e.g., production of organic acids) and that change the environmental properties (e.g., pH of the soil), and the microbes fail to grow within this niche and only specific microorganisms can be recruited in this way. In wheat, *Pseudomonas* sp. are recruited that produce the antifungal 2,4 diacetylphloroglucinol (DAPG) that inhibits the pathogen *Gaeumannomyces graminis* (Berendsen et al. 2012). Similarly, the secretory system TS66 is responsible for antagonistic actions against different microbes and aids in rhizosphere adaptation for Proteobacteria such as *Pseudomonas* sp. (Durán et al. 2021). The synthesis of lytic enzymes like cellulases, glucanases, and chitinases by the PGPR (e.g., *Bacillus thuringiensis*, *Pseudomonas fluorescence*) is another mode of inhibition toward the pathogenic fungi by destroying their cell walls and also helps the PGPR to colonize inside the plant tissues without harming the plants (Martinez-Absalon et al. 2012; Martínez-Absalón et al. 2014; Rojas-Solís et al. 2016; Bhagwat et al. 2019). Antibiotics (lipopeptides like surfactin, fengycin, and iturin A) are being produced by the PGPR (*Bacillus* sp.) against various pathogens (*Podosphaera fusca*) to inhibit the colonization of the pathogenic species and simultaneously developing their rhizosphere adaptation (Fira et al. 2018; Stokes et al. 2019).

The induction of the plant immune system as well as repression of the same are the techniques adopted by the PGPR during colonization. This is achieved by various mechanisms. The first one is through the response of the recognition of transmembrane pattern recognition receptors (PRRs) of the plants to the microbe-associated molecular patterns (MAMPs) secreted by the PGPR. Activation of defense genes during early phases of colonization in legumes is done by the *Bradyrhizobium japonicum*. The binding of the PRR to the antigenic epitopes (fgl22) and turning on of the plant's immune system as in the case of bacterial colonization in *Arabidopsis* sp. is another mechanism of colonization of PGPR. This binding is facilitated by the flagellin protein of the colonizing bacteria. This same phenomenon is observed in the case of *Pseudomonas aeruginosa* and *Lotus japonicum* and *Burkholderia phytofirmans* and grapevine symbioses. The third mechanism is the evasion of the plant immune system by hiding of MAMPS and thereby surpassing the immune attacks of the plants. Degradation of the flagellin protein by the alkaline phosphatase enzyme by *Pseudomonas* sp. during rhizosphere colonization results in the failure of the plant immune system to recognize the MAMPs and hence remains inactivated. The same *Pseudomonas* sp. can also suppress the immune genes (e.g., genes for type III secretion system) of *Arabidopsis* sp. and thereby inhibits the plant's immune response completely prior to the successful colonization (Lopez-Gomez et al. 2012; Yadong et al. 2013; Pel et al. 2014; Trdá et al. 2014; Berendsen et al. 2015; Couto and Zipfel 2016; Pfeilmeier et al. 2016; Santoyo et al. 2016; Valentini and Filloux 2016; Stringlis et al. 2019; Yu et al. 2019).

8.3.2 Mechanism of and Factors Controlling Endophytes Colonization

The colonization by the endophytes is engaged in intricate crosstalk of the microbe and the respective host plant initiated by the recognition of explicit root exudates (Vandana et al. 2021). Since the endophytic bacteria invade plant roots and infect the adjacent plant tissues inside the roots, these groups of microbes trigger the plant defense mechanisms. *Klebsiella pneumoniae* 342 activated the ethylene-signaling pathway in *Medicago truncatula,* whereas *Salmonella enterica* serovar typhimurium strain 14,028 colonization was found to be affected by both salicylic acid (SA)-dependent and -independent responses. Even the absence of flagella and type III secretion system in Salmonella pathogenicity island 1 (TTSS-SPI1) impacts the colonization of endophytes in *Medicago* sp. (Vandana et al. 2021). However, endophytic colonization is always preceded by the inhabitation of PGPR in the plant roots, which expands the functionality of the plant tissues by enhancing the uptake of nutrients. Endophytic colonization is associated with the evoking of the immune system of the plants. The genes involved in ethylene production (e.g., ethylene receptor genes) in plants are modulated by the endophytes during colonization of *Gluconacetobacter diazotrophicus* PAL and *Herbaspirillum rubrisulbalbicans* HCC103 in sugarcane plants. The secretion of cell-degrading enzymes like endoglucanase, endopolygalacturonase, endo-β D-cellobiosidase, and exo-β-1,4-glucanase is the primary need for the endophytic colonization. Studies using green fluorescence protein-tagged endophytic bacteria have shown that the colonization of endophytes takes place inside the xylem vessels and the endodermis barrier and they move from the roots to the seeds (Vandana et al. 2021). The endophytes after colonization modify the host tissues through upregulation of various genes as found for the diazotrophic endophytes, resulting in the upregulation of nif genes for nitrogen fixation in grapevines (202). These endophytes are capable of detoxification of ROS and RNS (reactive nitrogen species) and thus can overcome the defense responses in the plants during colonization (Alquéres et al. 2013).

8.4 Role of PGPR and Endophytes toward Plant Physiology

The PGPR and endophytes offer various roles toward plant growth and development through either direct or indirect mechanisms (Fig. 8.1). The direct mechanisms include the production of growth stimulators like phytohormones, water, and nutrient acquisition while the indirect mechanisms involve the development of tolerance toward biotic (pathogens) and abiotic stresses (drought, salinity).

8.4.1 Nutrient Assimilation

The microbial partners of the plants contribute to the nutrient accumulation in plants by nitrogen fixation, solubilization of phosphorus, iron sequestration by siderophores, production of VOCs, HCN (hydrogen cyanide), organic acids, and increasing the surface area occupied by the plant roots. Though the best example of this interaction is provided by legume–rhizobia symbioses helping in nitrogen (N) fixation, the free-living soil diazotrophs (*Azoarcus* sp., *Azotobacter* sp., *Azospirillum* sp., *Bacillus polymyxa*, *Burkholderia* sp., *Diazotrophicus* sp., *Gluconacetobacter* sp., *Herbaspirillum* sp.) and non-nitrogen-fixing bacteria also have been found to increase N-uptake in the associated plants (Bashan and de-Bashan 2015; Beattie 2015). These endophytes produce cellulase by which they can colonize within xylem vessels where the exchange of minerals takes place and with the help of nitrogenase enzyme they help in the fixation of atmospheric nitrogen (Vandana et al. 2021). In addition, *Serratia* sp., *Pseudomonas* sp., and *Bacillus* sp. have been found to produce ammonia, indirectly contributing to the growth of plants (Agbodjato et al. 2015).

Nitrogen, phosphorus, iron, sulfur, and zinc are the limiting minerals for plants due to their unavailability from the soil because of their low solubility. The PGPR helps the plants to uptake these minerals through solubilization, enzyme degradation (phytase, phosphatase, lyase), mineralization, assimilation by the dissolution of complex compounds, and sequestration by forming chelating compounds (siderophores) or organic acids. These PGP activities have been observed in *Bacillus*

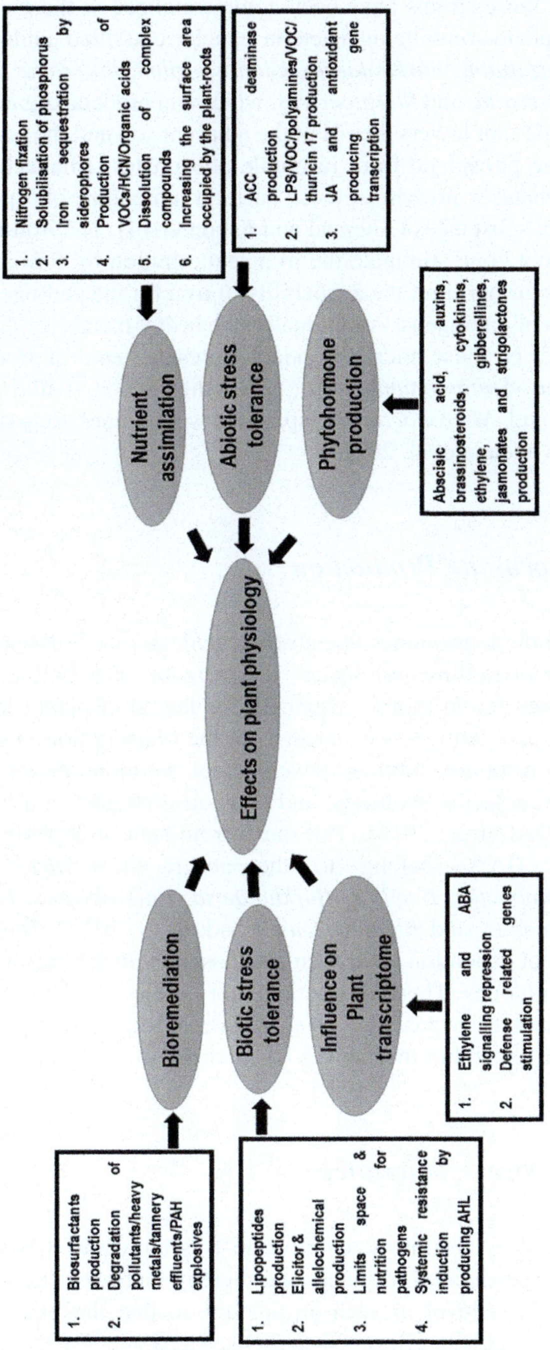

Fig. 8.1 Mechanisms of actions of PGPR and endophytes in plant growth and development

megaterium, *Bacillus polymyxa*, *Pseudomonas striata*, and *Acidithiobacillus ferrooxidans*, and these strains have been commercialized in the markets (Mehnaz 2016). The phosphate-solubilizing bacteria can be classified under *Azotobacter*, *Bacillus*, *Bradyrhizobium*, *Burkholderia*, *Cladosporium*, *Enterobacter*, *Pseudomonas*, *Rhizobium*, *Serratia*, and *Streptomyces*, which can produce organic acids during metabolism of sugar that lowers the pH of the adjacent soil and the organic acids can chelate and release phosphate from insoluble phosphates (Bhattacharyya and Jha 2012). The phosphatases present in these bacteria mineralize phosphates from the organic phosphatic substances (Ahemad and Kibret 2014). The iron-limiting conditions within the root zone stimulate the iron sequestration by forming siderophore complexes, which further limit the availability of iron for the pathogens and thereby restrict the growth of pathogens. Additionally, the heavy metals are also sequestered forming complexes by these microbial siderophores as found in *Methylobacterium mesophilicum* and *Sphingomonas* sp. absorbing nickel while *Pseudomonas fluorescens* G10 and *Microbacterium* sp. G16 were found to sequester lead in *Brassica napus* (Vandana et al. 2021).

8.4.2 Phytohormone Production

The production of phytohormones like abscisic acid, auxins, brassinosteroids, cytokinins, ethylene, gibberellins, jasmonates, and strigolactones by the PGPR and the endophytes has been found to have enhanced the length of plant roots and shoots, seed germination, alleviating stress, changes in the transcription rates of the genes involved in stress response, biomass development, hormone production, cell wall synthesis, defense molecule synthesis, and biocontrol (Brader et al. 2014; Santoyo et al. 2016; Shahzad et al. 2016). The most prominent endophytes that produce indole-acetic-acid (IAA) belong to the genera *Acetobacter*, *Alcaligenes*, *Azospirillum*, *Azotobacter*, *Bacillus*, *Burkholderia*, *Enterobacter*, *Herbaspirillum*, *Pantoea*, *Pseudomonas*, and *Rhodococcus* (Fouda et al. 2020). The alteration and enhancement of root exudation have been also observed in the presence of microbe-produced phytohormones. These phytostimulatory molecules help in the colonization of the microbes to bypass the defense mechanisms of the plants and bring physiological changes within themselves (as auxins).

8.4.3 Abiotic Stress Tolerance

The PGPR secreting 1-aminocyclopropane-1-carboxylase (ACC) deaminase decreases ethylene production in plants and thereby increases stress tolerance and maintains the ethylene level in such an amount so that the plant growth is not restricted under the ethylene effect (Glick 2014; Ruzzi and Aroca 2015; Heydarian et al. 2016; Vejan et al. 2016). Endophytic bacteria such as *Enterobacter*, *Klebsiella*, and *Pseudomonas* sp. have been characterized for ACC-deaminase activities (Jasim et al. 2013). The stress tolerance and plant biomass are also improved by the VOCs, polyamines, HCN, and signaling compounds like lipochitooligosaccharides and thuricin-17 secreted by PGPR and endophytes (Kumar et al. 2015; Zhou et al. 2016; Massalha et al. 2017; Zipfel and Oldroyd 2017). Additionally, the transcription of jasmonate (JA) synthesis and antioxidant producing genes was stimulated in droughts while root elongation was increased in the flooded conditions in the plants inoculated with *Pseudomonas* sp. (Etesami et al. 2014; Tiwari et al. 2016). *Bacillus thuringiensis* was also found to increase root biomass, nodule formation, and nitrogen and abscisic acid content in soybean plants under water-deficit conditions (Prudent et al. 2015). The *Variovorax paradoxus* 5C-2 strain inoculum in pea plants resulted in low effects of stress under high salinity, which was achieved by modulating ethylene production through ACC-deaminase activity, balancing ion homeostasis by decreasing sodium and increasing potassium flow in shoots and roots, respectively, increasing the rate of photosynthesis, and increasing the total biomass (Wang et al. 2016). Similar observations were noticed for maize seedlings inoculated with *Bacillus amyloliquefaciens* SQR9 and wheat plants inoculated with *Dietzia natronolimnaea* (Bharti et al. 2016; Chen et al. 2016). The *Serratia nematodiphila* inoculated pepper plants, grapevine inoculated with *Burkholderia phytofirmans* PsJN, and tomato plants inoculated with *Pseudomonas vancouverensis* OB155 and *P. frederiksbergensis* OS261 showed high tolerance under cold stress (Fernandez et al. 2012; Kang et al. 2015; Subramanian et al. 2015).

8.4.4 Biotic Stress Tolerance and Biocontrol

The tolerance in plants toward fungal pathogens is enhanced in the presence of PGPR and/or endophytes. The *Bacillus* sp. is the most potent candidate in this regard due to their abilities in the production of lipopeptides, which hydrolyze the fungal membranes, resulting in nutrient leakage, and thus the fungal virulence is reduced (Fouda et al. 2018; Lashin et al. 2021). *Bacillus amyloliquefaciens* (SN13) and *Bacillus* sp. colonizing rice and cotton plants increased the tolerance level of the plants toward *Rhizoctonia solani* and *Spodoptera exigua,* respectively, by increasing phytohormone production, elicitor, and allelochemical production, synthesis of scavengers,and scavenging the produced ROS (Srivastava et al. 2016; Zebelo et al. 2016). Similar responses were noticed in tomato, cucumber, and white barley plants inoculated with *Enterobacter asburiae* BQ9 and *Paenibacillus lentimorbus* B-30488 strains against yellow leaf curl virus and mosaic virus, respectively (Kumar et al. 2016; Li et al. 2016a, b).

The mechanisms by which the endophytes control the growth of plant pathogens are either direct or indirect processes. In direct mechanisms, the endophytes produce inhibitory molecules or outcompete the pathogens in competition for space and nutrients, thus limiting the growth of the pathogens. The endophytes outcompeted the *Ophiostoma novoulmi*, the causative agent for virulent Dutch elm disease, through their efficient utilizing capability of carbon compounds (Blumenstein et al. 2015). The growth of pathogens is inhibited and/or controlled by the release of VOC and antibiotics and production of a competitive environment by the endophytes within the plants that stimulate the propagation of the endophytes only. *Pseudomonas fluorescens* and *Pseudomonas aeruginosa* produce fungal antibiosis through the production of 2, 4-diacetylphloroglucinol (DAPG), pyoleutirin, pyrrolnitrin, hydrogen cyanide, and penazine-1-carboxylic acid (Lashin et al. 2021). The parasitic bacteria feed on phytopathogenic fungal cellular proteins, chitins, and glucan molecules through the production of LPS and lytic enzymes (Piromyou et al. 2015; Spadaro and Droby 2016).

In an indirect mechanism, the defense mechanisms of the plants are being stimulated. Acyl-homoserine lactones (AHL) were being produced by *Pseudomonas putida* IsoF and *Serratia liquefaciens* MG1 in tomato plants inducing systemic resistance against the pathogen *Alternaria alternate*. The root exudates are also found to produce AHL signal molecule mimicking chemicals that stimulate the PGPR and endophytes colonization and prohibits the association of the pathogenic microbes (Backer et al. 2018).

8.4.5 Impact on Plant Transcriptome

The endophytes and PGPR have been reported to overexpress or repress multiple genes in plants. The inoculation of *Pseudomonas putida* MTCC5279 and *Pseudomonas fluorescence* WCS417r resulted in the overexpression of 520 genes and differential expression of 97 genes in roots of *Arabidopsis* sp. Interestingly, gene expression was not affected in the leaves, proving that these microbes colonize the roots and affect the root structure and functioning only. These transcribed genes have functions related to ethylene and ABA signaling repression and induced systemic resistance signaling and induction (Van de Mortel et al. 2012; Vargas et al. 2012). Similarly, ethylene receptor genes were transcribed in rice roots inoculated with *Azospirillum brasilense* Sp245, *Azoarcus,* and *Herbaspirillum seropedicae* (Vandana et al. 2021). *Pseudomonas fluorescens* Q8r1-96 upregulated defense-related genes in wheat plants keeping the type III secretory and DAPG genes unaffected (Maketon et al. 2012).

8.4.6 PGPR and Endophytes-Mediated Phytoremediation

Endophyte-assisted bioremediation is commonly practiced in the removal of metals from contaminated soil and aquatic systems. Besides metals, heavy metals (uranium), organic compounds, explosives, hydrocarbons, herbicides, and tannery effluents are also being removed with the aid of endophytes (Ahsan et al. 2017; Feng et al. 2017; Ashraf et al. 2018). Zucchini plants grown in 2,2-bis(p-chlorophenyl)-

1,1-dichloro-ethylene (DDE)-contaminated soil showed enhanced growth in the presence of *Enterobacter aerogenes* UH1, *Sphingomonas tax* UH1, and *Methylobacterium radiotolerans* UH1 (Eevers et al. 2018). Increase in biomass was observed in sweet white clover plants grown in diesel-contaminated soil in the presence of *Pantoea* sp., *Stenotrophomonas* sp., *Pseudomonas* sp., and *Flavobacterium* sp. (Mitter et al. 2019). Endophytic *Bacillus safensis* could efficiently degrade C12–C32 n-alkanes of diesel oil and polycyclic aromatic hydrocarbons (PAH) under hypersaline conditions and also synthesized biosurfactants, which initiated higher growth and produced higher biomass in the plants inoculated with these strains (Wu et al. 2019).

8.4.7 Biotechnological and Industrial Applications of PGPR and Endophytes

The bioactive molecules or secondary metabolites produced by the PGPR and endophytes are of extensive importance in industries as they offer the source materials for medicinal drugs, cosmetics, foods, food supplements, detergents, and biopesticides. Extracellular cell-degrading enzymes, including cellulases, pectinases, amylases, chitanases, and proteases, are of tremendous industrial values in paper, jute, fuel, textile, beverage, pigment, cosmetics, paint, and waste-recycling industries. Endophytic bacteria from *Coffea arabica* L. produced enzymes capable of degrading caffeine and thus can be used for decaffeination of beverages. *Pseudomonas aeruginosa* isolated from the roots of *Phragmites australis* produced biosurfactants by degrading hydrocarbons.

The products from endophytes are used in the production of insecticides and antimicrobial agents with multiple targets in pathogens of plants, animals, and humans and thereby have prospects in human and veterinary medicines. Target-specific anticancer, antimalarial, immunosuppressive, antiviral, anti-dermatophytes, anti-inflammatory, antifungal, and antibacterial metabolic compounds have been identified from the PGPR and endophytes. It has been recently discovered that the chemical ingredients of the medicines deriving from medicinal plants are the by-products of the metabolism of the PGPR and endophytes or are derived from the interactomes of the plants and the PGPR and/or endophytes. Polyketides and small peptides with anti-tuberculosis properties have been isolated from the endophytes of medicinal plants while phenolic compounds with antioxidant and per-oxidant abilities have been identified from the profiles of secondary metabolites of *Acinetobacter baumannii* of *Capsicum annum* L. (Baker et al. 2012; Mitsuhashi 2014; Ntabo et al. 2018; Wu et al. 2018; Soliman et al. 2021).

The PGPR and endophytes synthesize nanoparticles with antimicrobial, antiallergic, anti-multidrug-resistant, larvicidal, anti-phytopathogenic, and in vitro cytotoxic properties (Eid et al. 2020). *Pseudomonas* sp. isolated from *Annona squamosa*, *Coffea arabica,* and *Allium sativum* were able to produce gold and silver nanoparticles with antimicrobial properties, whereas *Streptomyces* sp. isolated from *Convolvulus arvensis*, *Oxalis corniculata* L., *Ocimum sanctum*, *Achillea fragrantissima,* and *Mentha longifolia* L. were found to synthesize copper, magnesium, and silver nanoparticles with antimicrobial, larvicidal, and cytotoxic properties. *Bacillus* sp. from *Coriandrum sativum* synthesized silver nanoparticles with antibacterial properties (Baker and Satish 2015; El-Moslamy 2018; Ibrahim et al. 2019; Eid et al. 2020; Salem et al. 2020).

8.5 Strategies and Applications of PGPR and Endophytes

The "green revolution" of the twentieth century led to an immense increase in the production of food, fiber, timber, and fuel (Backer et al. 2018). However, the intensive use of chemical fertilizers and pesticides, on the one hand, and the changing climate scenario, on the other hand, have taken a great toll on human life and environment, leaving us to think in a new direction of green agricultural innovation. This new "bio-revolution" with greater sustainability and lower impacts on the environment can be achieved through utilization and manipulation of phytomicrobiome. This phyto-engineering though began in early twentieth century with rhizobial inoculation of the legumes, gained importance, and a new insight with the advent of technologies and discovery of other non-rhizobial PGPM and endophytes, including *Bacillus* sp., *Pseudomonas* sp., *Burkholderia* sp., *Paenibacillus* sp., *Lactobacillus* sp., and *Actinobacteria* sp. (Santoyo et al. 2012; Sivasakthi et al. 2014; Lamont et al. 2017; Shivlata and Satyanarayana 2017). The global warming, erosion of lands, rising sea levels, increase in salinity, and desertification are the present concerns for the less crop yields. Therefore, this phytomicrobiome will offer solution to this problem as they are responsible for the survival of the plants in extremely stressful environments of deserts and coastal areas (Coleman-Derr et al. 2016; Fonseca-García et al. 2016).

8.5.1 Strategies for Improving Rhizosphere Colonization

The rhizosphere colonization is highly impacted by various external factors, including edaphic factors (soil pH, temperature, moisture), geographical location, climate changes, soil physiology, presence of other microbial species, and anthropometric activity under natural field conditions. Therefore, effective inoculation strategies will help overcome these hindrances and will allow the effective colonization and growth of desirable microbial taxa. The compatibility, quality, and stability of the bioproducts used are the main determinants of not only the efficient microbial colonization but also the steady and long-term consistent performance of the microbial consortia in agriculture (Kristin and Miranda 2013; Lee et al. 2016).

The effective colonization of microbes is achieved using biofilms and biochar. The application of microbial inocula in biofilms in those soil environments where the microbes face great challenges and obstacles has been found to provide better results. This biofilm helps in the better survival of the target species, increased PGPR activities, and biocontrol of plant pathogens (Ricci 2015). The use of biofilms for the inoculation of seeds has been proven to have enhanced the seed germination, lengths of the roots and shoots, dry weights of the plants, and nitrogen accumulation in cotton, wheat, soybean, and maize plants (Mohd and Ahmad 2014).

Biochar is a carbon-rich mixture produced during pyrolysis where biomass is thermally decomposed at $<=700°$ C in the absence or limited supply of oxygen. It has large application in industry and environment as catalysts, porous materials, immobilizing agents for heavy metal remediation in soil, amendment of soil to improve soil health, remediation of toxic materials and organic contaminants, and nutrient and microbial carrier. Therefore, biochar forms a sustainable alternative for peat-based inoculants during microbial inoculation of seeds and provides early colonization of the rhizosphere by the beneficial microbes. The biochar develops ecological niche for the microbes protected from the predators by changing the parameters of the soil fertility like pH, temperature, water and organic matter content, bulk density, retention capacity of water and nutrients, and oxygen tension (Kameyama et al. 2012; Jaafar 2014; Ye et al. 2016; Backer et al. 2017; Backer et al. 2018).

8.5.2 Applications

Though the use of microbial consortia as inoculants for crops has produced inconsistent results, it is advantageous for the bioeconomy. This is achieved through two targets: (1) increasing the yield of the crops by lowering the use of fertilizers and (2) improving disease control by lessening the use of agrochemicals (Backer et al. 2018). The formulation of biofertilizers or biopesticides is based on (1) single and/or multispecies inoculum chosen from the PGPR and/or endophytes and/or (2) signal molecules deriving from the compounds involved in plant–plant, plant–microbe, and microbe–microbe interactions in any soil environment. These are termed bioeffectors: "When organic material and/or microorganisms applied to living plants or soil to enhance nutrient uptake, stimulate growth, improve stress tolerance and crop-quality traits, regardless of its nutrient contents" (Herrmann et al. 2022).

The mixed culture of *Trichoderma* sp. with *Bacillus amyloliquefaciens* or arbuscular mycorrhizal fungi or *Bradyrhizobium* sp. has been found to improve the yield of corn, tomato, soyabean, and also suppressed the plant pathogens present in the soil (Akladious and Abbas 2012; Molla et al. 2012). *Azospirillum* sp. singly and in consortium with *Azotobacter* sp. were found to enhance the yield of sesame and *Brassica* sp. (Nosheen et al. 2016; Shakeri et al. 2016). The yield was increased and quality became better for tomato after inoculation with consortia composed of *Bacillus subtilis*, *Bacillus cereus,* and *Serratia* sp. This additionally also decreased the root-knot nematodes in these plants (Niu et al. 2016). A similar inoculation of PGPR consortia in biofuel plants like switchgrass enhanced the quality and the yield. This can be of great use if PGPR-inoculated biofuel plants are grown in contaminated sites, then the PGPR will be able to perform remediation of the soil additionally (Ker et al. 2012; Ker et al. 2014; Evangelou and Deram 2014; Margaritopoulou et al. 2016; Shanta et al. 2016; Arunachalam et al. 2017; McCalmont et al. 2017).

The bioeffectors stimulate the acquisition of nutrients by the plants in nutrient-poor soil, thereby decreasing the requirement for inorganic fertilizers while in a nutrient-rich environment, the bioeffectors ease the use of chemical fertilizers and pesticides without harming the crop yields. A recent network meta-analysis study by Herrmann et al. with 1791 observations and 186 studies around the globe showed that bioeffectors positively increased the growth by 25%, yield by 30%, and nutrient contents (proteins by 55%, soluble solids by 75%, nitrogen content by 28%, and phosphorus by 40%) of crops like legumes, fruits, and herbs. This study also demonstrated that non-microbial products, including extracts and humic/amino acids, enhanced the plant biomass by 40–60% and phosphorus content by 54–110% (Herrmann et al. 2022). The derived substances consisting of amino acids, humic acids, organic acids, extracts from seaweeds, and botanicals present in the bioeffectors have diverse effects on plants by directly affecting plant growth through bioactive compounds and indirectly modulating the characteristics of the soil like water-holding capacity. These substances interact with the resident microflora of the soil and enhance the activities of the beneficial microbes. The bioeffectors also stimulate the proton pump of the plants aiding in the solubilization of the nutrients within the plants (Jindo et al. 2012). The tolerance of the plants toward abiotic and biotic stresses is enhanced in the presence of bioeffectors, which act in two ways: (1) adaptation of the microbes in the stressful environment and then increasing the tolerance level of the plants toward the abiotic stresses by improving the performance of the plants and (2) production of stress-adaptor substances (Van Oosten et al. 2017).

8.5.3 Applications of PGPR and Endophytes in Sustainable Agriculture under Climate Change

There is an upsurge toward the exploration of the effects of the climate change factors, including drought, global warming, and rising levels of CO_2 on PGPM activities (Compant et al. 2010). It has been found that elevated CO_2 positively impacts while the changes in temperature have variable effects on the plant–microbial interactions and the latter may vary between positive, negative, and neutral. The effects on the PGPM diversity, abundance, and activities can be attributed to the fact that the increased CO_2 results in differential allocation of carbon throughout the root zones, altered root-exudation patterns, change in the exudate composition, variation in the chemoattractants and plant signaling molecules, supply of nutrients, and ratio of carbon:nitrogen (Compant et al. 2010). Haase et al. have shown that few *Rhizobium leguminosarum* strains have been favored by the legumes under elevated CO_2 and the nodule production was increased by 17% by these isolates under those conditions (Haase et al. 2007). Besides, *Rhizobia*, *Actinobacteria*, *Deltaproteobacteria,* and *Pseudomonas* sp. have shown enhanced PGP activities in the presence of elevated CO_2. However, *Actinomycetes* and *Bacillus* sp. did not respond to elevated CO_2. This proves that different microbes respond differently under the variable CO_2 concentration (Compant et al. 2010). The plant genotype also influences this activity as it has been observed that the abundance of *Pseudomonas* sp. increased in rye plants but decreased in *Medicago* sp. and *Rhizobium* sp. increased in white clover under elevated CO_2 concentration (Compant et al. 2010).

PGPR and endophytes play an important role in alleviating the drastic effects of droughts and high temperatures on crops. However, these plant–microbe beneficial interactions are affected highly by the changes in the temperature too. Strains of *Mycobacterium* sp., *Pantoea agglomerans,* and *Pseudomonas fluorescens* enhanced shoot and root production in wheat plants at lower temperature (16 °C) while *Mycobacterium phlei* and *Mycoplana bullata* showed good PGP activities at both low and high temperatures (26^0 C). Therefore, the variation in global climate influences the plant–microbe crosstalk, ultimately influencing the plant diversity, the diversity and functioning of the soil microbiome, richness and the functions of the biocontrol microorganisms, and hence affecting the plant pathogen populations. Therefore, the change in the global climate induces adaptation mechanisms in plants and microorganisms and may lead to the selection of adapted plant cultivars in agriculture. However, adaptation processes or the use of cultivars that are not fully adapted to new environmental factors could be manipulated by calculative use of specific microbial consortia combinations, which would lead to better yield of crops under adverse climatic conditions. Since, under elevated CO_2 conditions, nutrient limitation (like limitation of nitrogen) calls for the requirement of enhanced inorganic fertilizers in agriculture, the use of PGPM to provide nutrients will not only result in betterment of agricultural crop production but will also alleviate the loss of crops under the adverse climatic changes.

8.5.4 Formulation and Commercialization of the Products

The bioformulation begins with the isolation of the species from the environmental ecosystem after exhaustive screening of plants from a wide range of eco-habitats, including deserts, wetlands, saline ecosystems, humid soil, soil from cold regions, and growing the species under laboratory conditions. In the laboratories, the PGPR activities of the isolated species are being tested for growth promotion in the model plants like *Arabidopsis* sp. or agricultural crop plants under both favorable and stressful plant growth conditions (Subramanian et al. 2016). To test for the biocontrol activities of the microbial species against plant pathogens, petri-plate assays with the measurement of the zones of inhibition created by the strains are evaluated (Ilangumaran and Smith 2017).

This is followed by the testing of the species for PGPR activities in natural field environment. In that case, it should be kept in mind that the highest potential of bioeffectors depends largely on the type of the crops. This is because of the symbiotic relationship of the plants with the microbes in a few cases and lack of the same in others (Herrmann et al. 2022). The activities of bioeffectors are also dependent on the soil characteristics as it is reported that the highest positive effects of the bioeffectors are found in soil with acidic pH (Herrmann et al. 2022). This might be because BEs act as a counterpart against nutrient fixation in soil by acidification, which is regarded as the most crucial impact of soil microbes (Raymond et al. 2020). Then, the possible combinations of the species are determined and, finally, the species are refined, tested for toxic effects on the environment, and delivered in the form of peat, granules, and liquid or wettable powder. The final step is to obtain the registration and regulatory approval for the product and make it available in the market (Backer et al. 2018).

For bioformulation of the signaling compounds, the microbial isolates are grown in liquid cultures (broths) and the supernatant that is believed to have the signal compound is collected through centrifugation and filtration of the cells, and the supernatant is tested for the ability to enhance plant growth promotion and biocontrol actions. To obtain the optimum synthesis of the signal compound at this stage, the triggering compounds that are involved in plant–microbe signaling under natural environments are used. The production of LCOs in *Rhizobia* is stimulated using isoflavonoids or jasmonates (Mabood et al. 2014; Smith et al. 2015). Then, the liquid is fractionated through high-performance liquid chromatography (HPLC), and each fraction obtained is again tested for plant growth promotion and biocontrol activities. The fraction showing the highest activities is isolated, purified, and identified through biophysical and biochemical analyses, including mass spectrometry.

For even distribution of the PGPR or the signaling compound in the fields, solid carriers like sterilized peats or alginate inoculated with seeds are used, which adhered to the sticking agents during sowing (Bashan et al. 2014; Bashan 2016). The liquid inoculants as in the case of signaling molecules are sprayed onto the seeds during sowing or are allowed for slow release through solid carriers. These are applied either regularly throughout the year (chronic) or once or for a limited number of times during sowing seasons or during droughts. The time of application of bioeffectors impacts the effectiveness of the substances. The best effect was found after the application of the bioeffectors during the sowing of seeds in a meta-analysis study conducted by Herrmann et al. with 186 groups around the world (Herrmann et al. 2022). Once the product is ready, it is registered following the regulatory laws of the specific country after the testing of efficacy and safety and is released in the market for commercial uses (Backer et al. 2018). Whether the effects of modes of application of bioeffectors determine their effectivity is debatable, it is well-established that the application techniques must be effective and easily adaptable. A meta-analysis study by Herrmann et al. showed that the application procedures have influenced the effectiveness of bioeffectors as foliar and soil application were most effective with respect to nutrient acquisition, improvement of the quality of the nutrition, and growth promotion (Herrmann et al. 2022).

8.5.5 Challenges

Although the technology of application of microbial inocula for agricultural improvement is an ancient one and has lots of prospects, some challenges and obstacles need to be addressed and overcome before leaping onto any great projects. Firstly, the choice of crops is still limited to cereals and legumes (Sessitsch and Mitter 2015). Secondly, the microbial species used presently belong to a few specific microbial taxa. In this context, the screening of new isolates with PGPR activities and purification of those cultures under laboratory conditions are needed. However, in many cases, though the strains produce promising results, reproducing similar results in field experiments has become cumbersome. Thirdly, the microbial species that do not produce interesting PGPR activities under laboratory conditions are discarded. But it should be kept in mind that these same species may have alternative pathways to produce PGP molecules or antibiotics that cannot be tested in the laboratory due to limited knowledge on this ground or due to tedious tasks that require enormous time and finance. Therefore, there is a chance that many PGPRs are being discarded based on their poor performance under the given laboratory conditions present and thus the agro-economy is facing a loss (Cardinale et al. 2015). Fourthly, inocula of microbes with high-shelf life and rhizosphere-colonizing capabilities are difficult to develop in sufficient quantities. Additionally, the appropriate carrier molecules are difficult to arrange, and thus these two result in ineffective colonization under field conditions where the other resident microflora cause additional hindrances. Fifthly, the soil and environmental factors influence the applied bioeffectors, thereby the outcomes become unreliable, thus limiting the practical applications of these techniques. Even the published research has shown varied observations regarding the efficacy of the bioeffectors on the different types of agricultural crops and lacks consistency, leading to inconclusive results (Herrmann et al. 2022). This problem is also difficult to address due to limited understanding of the modes of action of these bioeffectors (Schmidt and Gaudin 2018). Sixthly, the agriculture lands of high-value crops fumigated with pesticides undergo transformation in the soil health and microbial community structure, which leads to another challenge for the inocula to work efficiently. Seventhly, the microbiome-based plant breeding is not yet developed, which would have provided us with a heritable soil–microbial community. Therefore, altogether the development of soil-specific inocula, bypassing the environmental hazards, and proper hands-on training of the agricultural farmers and other staff are immediate necessities for proper synthesis, application, and commercialization of the bio-inocula (Bashan 2016; Parnell et al. 2016; Dangi et al. 2017; Trivedi et al. 2017; Itelima et al. 2018).

8.6 Conclusion

In the twenty-first century, the beneficial value of phytomicrobiome is well-known. But this is an extensively vast area and is still evolving. Therefore, to exploit this plant–microbiome association for sustainable agriculture, the knowledge and understanding available are not sufficient. Hence, exhaustive research is utmost necessary to explore all the avenues in this field, which then can be translated for economical betterment. In this regard, small steps can be taken at a time using a single microbial strain and/or consortia of a small group of microorganisms, and the studies can focus on two aspects separately—(1) promotion of plant growth under adverse situations and (2) control of plant diseases—as these two are the main concerns in agro-economy and are vital determinants for developing global food security. The focus of research should be on the harmonization of bioeffectors, modes of application, the mechanisms of the actions and formulation of nonmicrobial products, the study of characteristics of the various crops with emphasis on their physiological demands, problems hindering the proper yields, soil types, and the factors influencing them. However, the final concern should be the license from the common population concerning the commercialization of the bioproducts as the modern population is not yet fully convinced and accepting of the "bioproducts" with a clear perception. Therefore, a thorough systematic approach should be employed to develop the technology and apply it for the benefit of society.

References

Agbodjato NA, Noumavo PA, Baba-Moussa F, Salami HA, Sina H, Sèzan A, Bankolé H, Adjanohoun A, Baba-Moussa L (2015) Characterization of potential plant growth promoting rhizobacteria isolated from maize (*Zea mays* L.) in Central and Northern Benin (West Africa). Appl Environ Soil Sci 2015:1–9

Ahemad M, Kibret M (2014) Mechanisms and applications of plant growth-promoting rhizobacteria: current perspective. J King Saud Univ Sci 26:1–20

Ahsan MT, Najam-ul-haq M, Idrees M, Ullah I, Afzal M (2017) Bacterial endophytes enhance phytostabilization in soils contaminated with uranium and lead. Int J Phytoremediation 19:937–946

Akladious SA, Abbas SM (2012) Application of *Trichoderma harziunum* T22 as a biofertilizer supporting maize growth. Afr J Biotechnol 11:8672–8683

Alquéres S, Meneses C, Rouws L, Rothballer M, Baldani I, Schmid M, Hartmann A (2013) The bacterial superoxide dismutase and glutathione reductase are crucial for endophytic colonization of rice roots by *Gluconacetobacter diazotrophicus* PAL5. Mol Plant-Microbe Interact 26:937–945

Arunachalam S, Schwinghamer T, Dutilleul P, Smith DL (2017) Multiyear effects of biochar, lipo-chitooligosaccharide, thuricin 17, and experimental bio-fertilizer for switchgrass. Agron J 110: 77–84

Ashraf S, Afzal M, Naveed M, Shahid M, Ahmad Zahir Z (2018) Endophytic bacteria enhance remediation of tannery effluent in constructed wetlands vegetated with *Leptochloa fusca*. Int J Phytoremediation 20:121–128

Aviles-Garcia ME, Flores-Cortez I, Hernández-Soberano C, Santoyo G, Valencia-Cantero E (2016) La rizobacteria promotora del crecimiento vegetal *Arthrobacter agilis* UMCV2 coloniza endofíticamente a *Medicago truncatula*. Rev Argent Microbiol 48:4

Backer R, Rokem JS, Ilangumaran G, Lamont J, Praslickova D, Ricci E, Subramanian S, Smith DL (2018) Plant growth-promoting rhizobacteria: context, mechanisms of action, and roadmap to commercialization of biostimulants for sustainable agriculture. Front Plant Sci 9:1473

Backer RGM, Saeed W, Seguin P, Smith DL (2017) Root traits and nitrogen fertilizer recovery efficiency of corn grown in biochar-amended soil under greenhouse conditions. Plant Soil 415: 465–477

Baker S, Sahana S, Rakshith D, Kavitha H, Kavitha K, Satish S (2012) Biodecaffeination by endophytic Pseudomonas sp. isolated from *Coffee arabica* L. J Pharm Res 5:3654–3657

Baker S, Satish S (2015) Biosynthesis of gold nanoparticles by *Pseudomonas veronii* AS41G inhabiting *Annona squamosa* L. Spectrochim Acta Part A: Mol Biomol Spectrosc 150:691–695

Bakker PAHM, Berendsen RL, Van Pelt JA, Vismans G, Yu K, Li E, Van Bentum S, Poppeliers SWM, Sanchez Gil JJ, Zhang H et al (2020) The soil-borne identity and microbiome-assisted agriculture: looking back to the future. Mol Plant 13:1394–1401

Banerjee A, Sarkar S, Cuadros-Orellana S, Bandopadhyay R (2019) Exopolysaccharides and biofilms in mitigating salinity stress: the biotechnological potential of halophilic and soil-inhabiting PGPR microorganisms. In Microorganisms in saline environments: strategies and functions; Springer: Berlin/Heidelberg, Germany, pp. 133–153

Bashan N (2016) Inoculant formulations are essential for successful inoculation with plant growth-promoting bacteria and business opportunities. Indian Phytopathol 69:739–743

Bashan Y, de-Bashan LE (2015) Inoculant preparation and formulations for *Azospirillum sp*. In: Cassán FD, Okon Y, Creus CM (eds) Handbook for *Azospirillum*. Springer, Berlin, pp 469–485

Bashan Y, De-Bashan LE, Prabhu S, Hernandez JP (2014) Advances in plant growth-promoting bacterial inoculant technology: formulations and practical perspectives (1998–2013). Plant Soil 378:1–33

Beattie GA (2015) Microbiomes: curating communities from plants. Nature 528:340–341

Berendsen RL, Pieterse CMJ, Bakker PAHM (2012) The rhizosphere microbiome and plant health. Trends Plant Sci 17:478–486

Berendsen RL, van Verk MC, Stringlis IA, Zamioudis C, Tommassen J, Pieterse CMJ, Bakker PAHM (2015) Unearthing the genomes of plant-beneficial Pseudomonas model strains WCS358, WCS374 and WCS417. BMC Genomics 16:539

Bhagwat A, Collins CH, Dordick JS (2019) Selective antimicrobial activity of cell lytic enzymes in a bacterial consortium. Appl Microbiol Biotechnol 103:7041–7054

Bharti N, Pandey SS, Barnawal D, Patel VK, Kalra A (2016) Plant growth promoting rhizobacteria *Dietzia natronolimnaea* modulates the expression of stress responsive genes providing protection of wheat from salinity stress. Sci Rep 6:34768

Bhattacharyya PN, Jha DK (2012) Plant growth-promoting rhizobacteria (PGPR): emergence in agriculture. World J Microbiol Biotechnol 28:1327–1350

Blumenstein K, Albrectsen BR, Martín JA, Hultberg M, Sieber TN, Helander M, Witzell J (2015) Nutritional niche overlap potentiates the use of endophytes in biocontrol of a tree disease. BioControl 60:655–667

Brader G, Compant S, Mitter B, Trognitz F, Sessitsch A (2014) Metabolic potential of endophytic bacteria. Curr Opin Biotechnol 27:30–37

Cardinale M, Ratering S, Suarez C, Zapata Montoya AM, Geissler-Plaum R, Schnell S (2015) Paradox of plant growth promotion potential of rhizobacteria and their actual promotion effect on growth of barley (*Hordeum vulgare* L.) under salt stress. Microbiol Res 181:22–32

Castanheira NL, Dourado AC, Pais I, Semedo J, Scotti-Campos P, Borges N, Carvalho G, Crespo MTB, Fareleira P (2017) Colonization and beneficial effects on annual ryegrass by mixed inoculation with plant growth promoting bacteria. Microbiol Res 198:47–55

Chaparro JM, Badri DV, Bakker MG, Sugiyama A, Manter DK, Vivanco JM (2013) Root exudation of phytochemicals in Arabidopsis follows specific patterns that are developmentally programmed and correlate with soil microbial functions. PLoS One 8:e55731

Chen L, Liu YP, Wu GW, Njeri KV, Shen QR, Zhang N et al (2016) Induced maize salt tolerance by rhizosphere inoculation of bacillus amyloliquefaciens SQR9. Physiol Plant 158:34–44

Coleman-Derr D, Desgarennes D, Fonseca-Garcia C, Gross S, Clingenpeel S, Woyke T et al (2016) Plant compartment and biogeography affect microbiome composition in cultivated and native Agave species. New Phytol 209:798–811

Compant S, van der Heijden MG, Sessitsch A (2010) Climate change effects on beneficial plant-microorganism interactions. FEMS Microbiol Ecol 73(2):197–214

Costa FEC, Melo IS (2012) Endophytic and rhizospheric bacteria from Opuntia ficus-indica mill and their ability to promote plant growth in cowpea, *Vigna unguiculata* (L.) Walp Afr. J Microbiol Res 6:1345–1353

Couto D, Zipfel C (2016) Regulation of pattern recognition receptor signalling in plants. Nat Rev Immunol 16:537–552

Dangi S, Tirado-Corbalá R, Gerik J, Hanson B (2017) Effect of longterm continuous fumigation on soil microbial communities. Agronomy 7:37

Dawson W, Hör J, Egert M, van Kleunen M, Peste M (2017) A small number of low-abundance bacteria dominate plant species specific responses during rhizosphere colonization. Front Microbiol 8:975

Durán D, Bernal P, Arias DV, Romero EB, Sanz DG, Nieto MR, Rivilla R, Martín M (2021) *Pseudomonas fluorescens* F113 type VI secretion systems mediate bacterial killing and adaption to the rhizosphere microbiome. Sci Rep 5772

Eevers N, Hawthorne JR, White JC, Vangronsveld J, Weyens N (2018) Endophyte-enhanced phytoremediation of DDE-contaminated using *Cucurbita pepo*: a field trial. Int J Phytoremediation 20:301–310

Eid AM, Fouda A, Niedbała G, Hassan SED, Salem SS, Abdo AM, Hetta HF, Shaheen TI (2020) Endophytic *Streptomyces laurentii* mediated green synthesis of Ag-NPs with antibacterial and anticancer properties for developing functional textile fabric properties. Antibiotics 9:641

El-Moslamy SH (2018) Bioprocessing strategies for cost-effective large-scale biogenic synthesis of nano-MgO from endophytic *Streptomyces coelicolor* strain E72 as an anti-multidrug-resistant pathogens agent. Sci Rep 8:3820

Etesami H, Mirseyed Hosseini H, Alikhani HA (2014) Bacterial biosynthesis of 1-aminocyclopropane-1-caboxylate (ACC) deaminase, a useful trait to elongation and endophytic colonization of the roots of rice under constant flooded conditions. Physiol Mol Biol Plants 20:425–434

Evangelou MW, Deram A (2014) Phytomanagement: a realistic approach to soil remediating phytotechnologies with new challenges for plant science. Int J Plant Biol Res 2:1023

Feng NX, Yu J, Zhao HM, Cheng YT, Mo CH, Cai QY, Li YW, Li H, Wong MH (2017) Efficient phytoremediation of organic contaminants in soils using plant–endophyte partnerships. Sci Total Environ 583:352–368

Fernandez O, Theocharis A, Bordiec S, Feil R, Jacquens L, Clement C et al (2012) *Burkholderia phytofirmans* PsJN acclimates grapevine to cold by modulating carbohydrate metabolism. Mol Plant-Microbe Interact 25:496–504

Fernández-Llamosas H, Díaz E, Carmona M (2021) Motility, adhesion and c-di-gmp influence the endophytic colonization of rice by *Azoarcus sp. cib*. Microorganisms 9:554

Fira D, Dimkić I, Berić T, Lozo J, Stanković S (2018) Biological control of plant pathogens by *Bacillus* species. J Biotechnol 285:44–55

Fonseca-García C, Coleman-Derr D, Garrido E, Visel A, Tringe SG, Partida-Martínez LP (2016) The cacti microbiome: interplay between habitat-filtering and host-specificity. Front Microbiol 7:150

Fouda A, Hassan SED, Abdo AM, El-Gamal MS (2020) Antimicrobial, antioxidant and larvicidal activities of spherical silver nanoparticles synthesized by endophytic *Streptomyces spp*. Biol Trace Elem Res 195:707–724

Fouda A, Hassan SED, Eid AM, El-Din Ewais E (2018) The interaction between plants and bacterial endophytes under salinity stress. In: Jha S (ed) Endophytes and secondary metabolites. Springer International Publishing, Cham, Switzerland, pp 1–17

Fujita H, Aoki S, Kawaguchi M (2014) Evolutionary dynamics of nitrogen fixation in the legume-rhizobia symbiosis. PLoS One 9(4):e93670

Glick BR (2012) Plant growth-promoting bacteria: mechanisms and applications. Scientifica 2012: 1–15

Glick BR (2014) Bacteria with ACC deaminase can promote plant growth and help to feed the world. Microbiol Res 169:30–39

Haase S, Neumann G, Kania A, Kuzyakov Y, Römheld V, Kandeler E (2007) Elevation of atmospheric CO2 and N-nutritional status modify nodulation, nodule-carbon supply, and root exudation of Phaseolus vulgaris L. Soil Biol Biochem 39:2208–2221

Hassan MK, McInroy JA, Kloepper JW (2019) The interactions of rhizodeposits with plant growth-promoting rhizobacteria in the rhizosphere: a review. Agriculture 9:142

Hernández-Calderón E, Aviles-Garcia ME, Castulo-Rubio DY, Macías-Rodríguez L, Ramírez VM, Santoyo G, LópezBucio J, Valencia-Cantero E (2018) Volatile compounds from beneficial or pathogenic bacteria differentially regulate root exudation, transcription of iron transporters, and defense signaling pathways in *Sorghum bicolor*. Plant Mol Biol 96

Hernández-Salmerón JE, Hernández-León R, Orozco-Mosqueda MDC, Valencia-Cantero E, Moreno-Hagelsieb G, Santoyo G (2016) Draft genome sequence of the biocontrol and plant growth-promoting rhizobacterium Pseudomonas fluorescens strain UM270. Stand Genom Sci 2016:11

Hernández-Salmerón JE, Moreno-Hagelsieb G, Santoyo G (2017) Genome comparison of *Pseudomonas fluorescens* UM270 with related fluorescent strains unveils genes involved in rhizosphere competence and colonization. J Genom 5

Herrmann MN, Wang Y, Hartung J, Hartmann T, Zhang W, Nkebiwe PM, Chen X, Müller T, Yang H (2022) A global network meta-analysis of the promotion of crop growth, yield, and quality by bioeffectors. Front Plant Sci 13:816438

Heydarian Z, Yu M, Gruber M, Glick BR, Zhou R, Hegedus DD (2016) Inoculation of soil with plant growth promoting bacteria producing 1-aminocyclopropane-1-carboxylate deaminase or expression of the corresponding acds gene in transgenic plants increases salinity tolerance in camelina sativa. Front Microbiol 7:1966

Ibrahim E, Fouad H, Zhang M, Zhang Y, Qiu W, Yan C, Li B, Mo J, Chen J (2019) Biosynthesis of silver nanoparticles using endophytic bacteria and their role in inhibition of rice pathogenic bacteria and plant growth promotion. RSC Adv 9:29293–29299

Ilangumaran G, Smith DL (2017) Plant growth promoting rhizobacteria in amelioration of salinity stress: a systems biology perspective. Front Plant Sci 8:1768

Itelima JU, Bang WJ, Onyimba IA, Oj E (2018) A review: biofertilizer; a key player in enhancing soil fertility and crop productivity. J Microbiol Biotechnol Rep 2:22–28

Jaafar NM (2014) Biochar as a habitat for arbuscular mycorrhizal fungi in mycorrhizal fungi. In: Solaiman ZM, Abbott LK, Varma A (eds) Use in sustainable agriculture and land restoration. Springer, Berlin, pp 297–311

Jasim B, Jimtha CJ, Jyothis M, Radhakrishnan EK (2013) Plant growth promoting potential of endophytic bacteria isolated from *Piper nigrum*. Plant Growth Regul 71:1–11

Jindo K, Martim SA, Navarro EC, Pérez-Alfocea F, Hernandez T, Garcia C et al (2012) Root growth promotion by humic acids from composted and non-composted urban organic wastes. Plant Soil 353:209–220

Kameyama K, Miyamoto T, Shiono T, Shinogi Y (2012) Influence of sugarcane bagasse-derived biochar application on nitrate leaching in calcaric dark red soil. J Environ Qual 41:1131–1137

Kang SM, Khan AL, Waqas M, You YH, Hamayun M, Joo GJ et al (2015) Gibberellin-producing *Serratia nematodiphila* PEJ1011 ameliorates low temperature stress in *Capsicum annuum* L. Eur J Soil Biol 68:85–93

Ker K, Seguin P, Driscoll BT, Fyles JW, Smith DL (2012) Switchgrass establishment and seeding year production can be improved by inoculation with rhizosphere endophytes. Biomass Bioenergy 47:295–301

Ker K, Seguin P, Driscoll BT, Fyles JW, Smith DL (2014) Evidence for enhanced N availability during switchgrass establishment and seeding year production following inoculation with rhizosphere endophytes. Arch Agron Soil Sci 60:1553–1563

Kloepper JW, Schroth MN (1978) Plant growth-promoting rhizobacteria on radishes. In Proceedings of the 4th International Conf. on Plant Pathogenic Bacteria, Station de Pathologie Vegetable et Phytobacteriologie, INRA, Angers, France, 27 August–2 September 1978: pp. 879–882

Kristin A, Miranda H (2013) The root microbiota—a fingerprint in the soil? Plant Soil 370:671–686

Kumar A, Bahadur I, Maurya B, Raghuwanshi R, Meena V, Singh D et al (2015) Does a plant growth promoting rhizobacteria enhance agricultural sustainability. J Pure Appl Microbiol 9: 715–724

Kumar S, Chauhan PS, Agrawal L, Raj R, Srivastava A, Gupta S et al (2016) *Paenibacillus lentimorbus* inoculation enhances tobacco growth and extenuates the virulence of cucumber mosaic virus. PLoS One 11:e0149980

Lamont JR, Wilkins O, Bywater-Ekegard M, Smith DL (2017) From yogurt to yield: potential applications of lactic acid bacteria in plant production. Soil Biol Biochem 111:1–9

Lashin I, Fouda A, Gobouri AA, Azab E, Mohammedsaleh ZM, Makharita RR (2021) Antimicrobial and in vitro cytotoxic efficacy of biogenic silver nanoparticles (Ag-NPs) fabricated by callus extract of Solanum incanum L. Biomol Ther 11:341

Lee SK, Lur HS, Lo KJ, Cheng KC, Chuang CC, Tang SJ et al (2016) Evaluation of the effects of different liquid inoculant formulations on the survival and plant-growth-promoting efficiency of *Rhodopseudomonas palustris* strain PS3. Appl Microbiol Biotechnol 100:7977–7987

Li H, Ding X, Wang C, Ke H, Wu Z, Wang Y et al (2016b) Control of tomato yellow leaf curl virus disease by *Enterobacter asburiae* BQ9 as a result of priming plant resistance in tomatoes. Turk J Biol 40:150–159

Li Z, Zu C, Wang C, Yang J, Yu H, Wu H (2016a) Different responses of rhizosphere and non-rhizosphere soil microbial communities to consecutive Piper nigrum L. monoculture. Sci Rep 6:35825

Lopez-Gomez M, Sandal N, Stougaard J, Boller T (2012) Interplay of flg22-induced defence responses and nodulation in *Lotus japonicum.* J Exp Bot 63:393–401

de los Santos-Villalobos S, Barrera-Galicia GC, Miranda-Salcedo MA, Peña-Cabriales JJ (2012) *Burkholderia cepacia* XXVI siderophore with biocontrol capacity against *Colletotrichum gloeosporioides*. World J Microbiol Biotechnol 28:2615–2623

Mabood F, Zhou X, Smith DL (2014) Microbial signaling and plant growth promotion. Can J Plant Sci 94:1051–1063

Maketon C, Fortuna AM, Okubara PA (2012) Cultivar-dependent transcript accumulation in wheat roots colonized by *Pseudomonas fluorescens* Q8r1-96 wild type and mutant strains. Biol Control 60:216–224

Margaritopoulou T, Roka L, Alexopoulou E, Christou M, Rigas S, Haralampidis K et al (2016) Biotechnology towards energy crops. Mol Biotechnol 58:149–158

Marin O, González B, Poupin MJ (2021) From microbial dynamics to functionality in the rhizosphere: a systematic review of the opportunities with synthetic microbial communities. Front Plant Sci 12:650609

Martínez-Absalón S, Rojas-Solís D, Hernández-León R, Prieto-Barajas C, Orozco-Mosqueda MC, Peña-Cabriales JJ, Sakuda S, Valencia-Cantero E, Santoyo G (2014) Potential use and mode of action of the new strain bacillus thuringiensis UM96 for the biological control of the grey mould phytopathogen Botrytis cinerea. Biocontrol Sci Tech 24:1349–1362

Martinez-Absalon SC, Orozco-Mosqueda MC, Martinez-Pacheco MM, Farias-Rodriguez R, Govindappa M, Santoyo G (2012) Isolation and molecular characterization of a novel strain of bacillus with antifungal activity from the sorghum rhizosphere. Genet Mol Res 11:2665–2673

Martínez-Cámara R, Montejano-Ramírez V, Moreno-Hagelsieb G, Santoyo G, Valencia-Cantero E (2020) The volatile organic compound dimethylhexadecylamine affects bacterial growth and swarming motility of bacteria. Folia Microbiol 65

Massalha H, Korenblum E, Tholl D, Aharoni A (2017) Small molecules below-ground: the role of specialized metabolites in the rhizosphere. Plant J 90:788–807

Matthews A, Pierce S, Hipperson H, Raymond B (2019) Rhizobacterial community assembly patterns vary between crop species. Front Microbiol 10:581

McCalmont JP, Hastings A, Mcnamara NP, Richter GM, Robson P, Donnison IS et al (2017) Environmental costs and benefits of growing miscanthus for bioenergy in the UK. Glob Chang Biol Bioenergy 9:489–507

McNear DH Jr (2013) The rhizosphere-roots, soil and everything in between. Nat Educ 1:30

Meena KK, Sorty AM, Bitla UM, Choudhary K, Gupta P, Pareek A, Singh DP, Prabha R, Sahu PK, Gupta VK et al (2017) Abiotic stress responses and microbe-mediated mitigation in plants: the omics strategies. Front Plant Sci 8:172

Mehnaz S (2016) In: Arora NK, Mehnaz S, Balestrini R (eds) An overview of globally available bioformulations in bioformulations: for sustainable agriculture. Springer, Berlin, pp 267–281

Mitsuhashi S (2014) Current topics in the biotechnological production of essential amino acids, functional amino acids, and dipeptides. Curr Opin Biotechnol 26:38–44

Mitter EK, Kataoka R, de Freitas JR, Germida JJ (2019) Potential use of endophytic root bacteria and host plants to degrade hydrocarbons. Int J Phytoremediation 21:928–938

Mohd MA, Ahmad I. (2014) "A novel strain of Pseudomonas fluorescens WS1 Forms biofilm on root surface and enhances growth of wheat plant," in 2nd International Conference on Agricultural & Horticultural Sciences, Hyderabad

Molla AH, Haque MM, Haque MA, Ilias G (2012) Trichoderma enriched biofertilizer enhances production and nutritional quality of tomato (*Lycopersicon esculentum* mill.) and minimizes NPK fertilizer use. Agric Res 1:265–272

Niu DD, Zheng Y, Zheng L, Jiang CH, Zhou DM, Guo JH (2016) Application of PSX biocontrol preparation confers root-knot nematode management and increased fruit quality in tomato under field conditions. Biocontrol Sci Tech 26:174–180

Nosheen A, Bano A, Yasmin H, Keyani R, Habib R, Shah ST et al (2016) Protein quantity and quality of safflower seed improved by NP fertilizer and Rhizobacteria (*Azospirillum* and *Azotobacter sp.*). Front Plant Sci 7:104

Ntabo RM, Nyamache AK, Lwande W, Kabii J, Nonoh J (2018) Enzymatic activity of endophytic bacterial isolates from selected mangrove plants in Kenya. Open Microbiol J 12:354–363

Orozco-Mosqueda MC, Velázquez-Becerra C, Macías-Rodríguez LI, Santoyo G, Flores-Cortez I, Alfaro-Cuevas R, Valencia-Cantero E (2013) *Arthrobacter agilis* UMCV2 induces iron acquisition in *Medicago truncatula* (strategy I plant) *in vitro* via dimethylhexadecylamine emission. Plant Soil 362

Parnell JJ, Berka R, Young HA, Sturino JM, Kang Y, Barnhart DM et al (2016) From the lab to the farm: an industrial perspective of plant beneficial microorganisms. Front Plant Sci 7:1110

Paulucci NS, Gallarato LA, Reguera YB, Vicario JC, Cesari AB, de Lema MBG, Dardanelli MS (2015) *Arachis hypogaea* PGPR isolated from argentine soil modifies its lipids components in response to temperature and salinity. Microbiol Res 173:1–9

Pel MJC, Van Dijken AJH, Bardoel BW, Seidl MF, Van Der Ent S, Van Strijp JAG, Pieterse CMJ (2014) *Pseudomonas syringae* evades host immunity by degrading flagellin monomers with alkaline protease AprA. Mol Plant-Microbe Interact 27:603–610

Pervaiza ZH, Contreras J, Hupp BM, Lindenberger JH, Chen D, Zhang Q, Wang C, Twigg P, Saleem M (2020) Root microbiome changes with root branching order and root chemistry in peach rhizosphere soil. Rhizosphere 16:100249

Pfeilmeier S, Caly DL, Malone JG (2016) Bacterial pathogenesis of plants: future challenges from a microbial perspective: challenges in bacterial molecular plant pathology. Mol Plant Pathol 17: 1298–1313

Pieterse CMJ, Zamioudis C, Berendsen RL, Weller DM, Van Wees SCM, Bakker PAHM (2014) Induced systemic resistance by beneficial microbes. Annu Rev Phytopathol 52:347–375

Piromyou P, Songwattana P, Greetatorn T, Okubo T, Chiba Kakizaki K, Prakamhang J, Tittabutr P, Boonkerd N, Teaumroong N, Minamisawa K (2015) The type III secretion system (T3SS) is a determinant for rice-endophyte colonization by non-photosynthetic *Bradyrhizobium*. Microbes Environ 30(4):291–300

Potshangbam M, Devi SI, Sahoo D, Strobel GA (2017) Functional characterization of endophytic fungal community associated with *Oryza sativa* L. and *Zea mays* L. Front Microbiol 8:325

Prieto-Barajas CM, Valencia-Cantero E, Santoyo G (2018) Microbial mat ecosystems: structure types, functional diversity, and biotechnological application. Electron. J Biotechnol 31

Prudent M, Salon C, Souleimanov A, Emery RJN, Smith DL (2015) Soybean is less impacted by water stress using *Bradyrhizobium japonicum* and thuricin-17 from *Bacillus thuringiensis*. Agron Sustain Dev 35:749–757

Rangjaroen C, Sungthong R, Rerkasem B, Teaumroong N, Noisangiam R, Lumyong S (2017) Untapped endophytic colonization and plant growth-promoting potential of the genus *Novosphingobium* to optimize rice cultivation. Microbes Environ 32:84–87

Raymond NS, Gómez-Muñoz B, van der Bom FJT, Nybroe O, Jensen LS, Müller-Stöver DS et al (2020) Phosphate-solubilising microorganisms for improved crop productivity: a critical assessment. New Phytol 229:1268–1277

Ricci EC (2015) Investigating the role of Pseudomonas sp. and Bacillus sp. biofilms as plant growth promoting inoculants. Master's thesis, McGill University, Montreal, QC

Rojas-Solís D, Hernandez-Pacheco CE, Santoyo G (2016) Evaluation of Bacillus and Pseudomonas to colonize the rhizosphere and their effect on growth promotion in tomato (*Physalis ixocarpa* Brot. ex Horm.). Rev Chapingo Ser Hortic 22:45–57

Rojas-Solís D, Vences-Guzmán MÁ, Sohlenkamp C, Santoyo G (2020) Antifungal and plant growth–promoting *Bacillus* under saline stress modify their membrane composition. J Soil Sci Plant Nutr 2020

Rojas-Solís D, Zetter-Salmón E, Contreras-Pérez M, Rocha-Granados MDC, Macías-Rodríguez L, Santoyo G (2018) *Pseudomonas stutzeri* E25 and *Stenotrophomonas maltophilia* CR71 endophytes produce antifungal volatile organic compounds and exhibit additive plant growth-promoting effects. Biocatal Agric Biotechnol 13:46–52

Rosier A, Medeiros FHV, Bais HP (2018) Defining plant growth promoting rhizobacteria molecular and biochemical networks in beneficial plant-microbe interactions. Plant Soil 428:35–55

Ruzzi M, Aroca R (2015) Plant growth-promoting rhizobacteria act as biostimulants in horticulture. Sci Hortic 196:124–134

Saeed Q, Xiukang W, Haider FU, Kucerik J, Mumtaz MZ, Holatko J, Naseem M, Kintl A, Ejaz M, Naveed M et al (2021) Rhizosphere bacteria in plant growth promotion, biocontrol, and bioremediation of contaminated sites: a comprehensive review of effects and mechanisms. Int J Mol Sci 22:10529

Salem SS, El-Belely EF, Niedbała G, Alnoman MM, Hassan SED, Eid AM, Shaheen TI, Elkelish A, Fouda A (2020) Bactericidal and in-vitro cytotoxic efficacy of silver nanoparticles (Ag-NPs) fabricated by endophytic actinomycetes and their use as coating for the textile fabrics. Nano 10:82

Santoyo G, del Orozco-Mosqueda MC, Govindappa M (2012) Mechanisms of biocontrol and plant growth-promoting activity in soil bacterial species of *Bacillus* and *Pseudomonas*: a review. Biocontrol Sci Tech 22:855–872

Santoyo G, Moreno-Hagelsieb G, Orozco-Mosqueda MC, Glick BR (2016) Plant growth-promoting bacterial endophytes. Microbiol Res 183:92–99

Santoyo G, Urtis-Flores CA, Loeza-Lara PD, Orozco-Mosqueda MC, Glick BR (2021) Rhizosphere colonization determinants by plant growth-promoting rhizobacteria (PGPR). Biology 10:475

Sasse J, Martinoia E, Northen T (2018) Feed your friends: do plant exudates shape the root microbiome? Trends Plant Sci 23:25–41

Schmidt JE, Gaudin A (2018) What is the agronomic potential of biofertilizers for maize? A meta-analysis. FEMS Microbiol Ecol 94:7

Sessitsch A, Mitter B (2015) 21st century agriculture: integration of plant microbiomes for improved crop production and food security. Microb Biotechnol 8:32–33

Shahzad R, Waqas M, Khan AL, Asaf S, Khan MA, Kang SM, Yun BW, Lee IJ (2016) Seed-borne endophytic *Bacillus amyloliquefaciens* RWL-1 produces gibberellins and regulates endogenous phytohormones of *Oryza sativa*. Plant Physiol Biochem 106:236–243

Shakeri E, Modarres-Sanavy SAM, Amini Dehaghi M, Tabatabaei SA, Moradi-Ghahderijani M (2016) Improvement of yield, yield components and oil quality in sesame (*Sesamum indicum* L.) by N-fixing bacteria fertilizers and urea. Arch Agron Soil Sci 62:547–560

Shanta N, Schwinghamer T, Backer R, Allaire SE, Teshler I, Vanasse A et al (2016) Biochar and plant growth promoting rhizobacteria effects on switchgrass (*Panicum virgatum* cv. Cave-in-rock) for biomass production in southern Québec depend on soil type and location. Biomass Bioenergy 95:167–173

Shivlata L, Satyanarayana T (2017) Actinobacteria in agricultural and environmental sustainability. In: Singh JS, Seneviratne G (eds) Agro-environmental sustainability. Springer, Berlin, pp 173–218

Sivasakthi S, Usharani G, Saranraj P (2014) Biocontrol potentiality of plant growth promoting bacteria (PGPR)-*Pseudomonas fluorescens* and *Bacillus subtilis*: a review. Afr J Agric Res 9: 1265–1277

Smith DL, Subramanian S, Lamont JR, Bywater-Ekegard M (2015) Signaling in the phytomicrobiome: breadth and potential. Front Plant Sci 6:709

Soliman AM, Abdel-Latif W, Shehata IH, Fouda A, Abdo AM, Ahmed YM (2021) Green approach to overcome the resistance pattern of Candida spp. using biosynthesized silver nanoparticles fabricated by *Penicillium chrysogenum* F9. Biol Trace Element Res 199:800–811

Spadaro D, Droby S (2016) Development of biocontrol products for postharvest diseases of fruit: the importance of elucidating the mechanisms of action of yeast antagonists. Trends Food Sci Technol 47:39–49

Srivastava S, Bist V, Srivastava S, Singh PC, Trivedi PK, Asif MH et al (2016) Unraveling aspects of *Bacillus amyloliquefaciens* mediated enhanced production of rice under biotic stress of *Rhizoctonia solani*. Front Plant Sci 7:587

Stokes JM, Lopatkin AJ, Lobritz MA, Collins JJ (2019) Bacterial metabolism and antibiotic efficacy. Cell Metab 30:251–259

Stringlis IA, Zamioudis C, Berendsen RL, Bakker PAHM, Pieterse CMJ (2019) Type III secretion system of beneficial rhizobacteria *Pseudomonas simiae* WCS417 and *Pseudomonas defensor* WCS374. Front Microbiol 10:2214

Subramanian P, Mageswari A, Kim K, Lee Y, Sa T (2015) Psychrotolerant endophytic *Pseudomonas sp.* strains OB155 and OS261 induced chilling resistance in tomato plants (*Solanum lycopersicum* mill.) by activation of their antioxidant capacity. Mol Plant-Microbe Interact 28: 1073–1081

Subramanian S, Ricci E, Souleimanov A, Smith DL (2016) A proteomic approach to lipo-chitooligosaccharide and thuricin 17 effects on soybean germination unstressed and salt stress. PLoS One 11:e0160660

Sulochana MB, Jayachandra SY, Kumar SA, Parameshwar AB, Reddy KM, Dayanand A (2014) Siderophore as a potential plant growth-promoting agent produced by *Pseudomonas aeruginosa* JAS-25. Appl Biochem Biotechnol 174:297–308

Thomas P, Reddy MK (2013) Microscopic elucidation of abundant endophytic bacteria colonizing the cell wall–plasma membrane peri-space in the shoot-tip tissue of banana. AoB Plants 5:plt011

Thomas P, Sekhar AC (2014) Live cell imaging reveals extensive intracellular cytoplasmic colonization of banana by normally non-cultivable endophytic bacteria. AoB Plants 6:plu002

Tiwari S, Lata C, Chauhan PS, Nautiyal CS (2016) *Pseudomonas putida* attunes morphophysiological, biochemical and molecular responses in *Cicer arietinum* L. during drought stress and recovery. Plant Physiol Biochem 99:108–117

Trdá L, Fernandez O, Boutrot F, Héloir MC, Kelloniemi J, Daire X, Adrian M, Clément C, Zipfel C, Dorey S et al (2014) The grapevine flagellin receptor VvFLS2 differentially recognizes flagellin-derived epitopes from the endophytic growth-promoting bacterium *Burkholderia phytofirmans* and plant pathogenic bacteria. New Phytol 201:1371–1384

Trivedi P, Schenk PM, Wallenstein MD, Singh BK (2017) Tiny microbes, big yields: enhancing food crop production with biological solutions. Microb Biotechnol 10:999–1003

Valentini M, Filloux A (2016) Biofilms and Cyclic di-GMP (c-di-GMP) signaling: lessons from *Pseudomonas aeruginosa* and other bacteria. J Biol Chem 291:12547–12555

Van de Mortel JE, De Vos RC, Dekkers E, Pineda A, Guillod L, Bouwmeester K, van Loon JJA, Dicke M, Raaijmakers JM (2012) Metabolic and transcriptomic changes induced in Arabidopsis by the rhizobacterium Pseudomonas fluorescens SS101. Plant Physiol 160:2173–2188

Van Oosten MJ, Pepe O, de Pascale S, Silletti S, Maggio A (2017) The role of biostimulants and bioeffectors as alleviators of abiotic stress in crop plants. Chem Biol Technol Agric 4:5

Vandana UK, Rajkumari J, Singha LP, Satish L, Alavilli H, Sudheer PDVN, Chauhan S, Ratnala R, Satturu V, Mazumder PB et al (2021) The endophytic microbiome as a hotspot of synergistic interactions, with prospects of plant growth promotion. Biology 10:101

Vargas L, Gurjao de Carvalho TL, Gomes Ferreira PC, Baldani VL, Baldani JI, Hemerly AS (2012) Early responses of rice (*Oryza sativa* L.) seedlings to inoculation with beneficial diazotrophic bacteria are dependent on plant and bacterial genotypes. Plant Soil 356:127–137

Vejan P, Abdullah R, Khadiran T, Ismail S, Nasrulhaq Boyce A (2016) Role of plant growth promoting rhizobacteria in agricultural sustainability—a review. Molecules 21:573

Wang QY, Dodd IC, Belimov AA, Jiang F (2016) Rhizosphere bacteria containing 1-aminocyclopropane-1-carboxylate deaminase increase growth and photosynthesis of pea plants under salt stress by limiting Na+ accumulation. Funct Plant Biol 43:161–172

Wu T, Xu J, Liu J, Guo WH, Li XB, Xia JB, Xie WJ, Yao ZG, Zhang YM, Wang RQ (2019) Characterization and initial application of endophytic *Bacillus safensis* strain ZY16 for improving phytoremediation of oil-contaminated saline soils. Front Microbiol 10:991

Wu T, Xu J, Xie W, Yao Z, Yang H, Sun C, Li X (2018) *Pseudomonas aeruginosa* L10: a hydrocarbon-degrading, biosurfactant-producing, and plant-growth-promoting endophytic bacterium isolated from a reed (*Phragmites australis*). Front Microbiol 9:1087

Yadong S, Li L, Alberto PM, Zhifu H, Zehan H, Cyril Z, Jian-Min Z, Jijie C (2013) Structural basis for flg22-induced activation of the Arabidopsis FLS2-BAK1 immune complex. Science 342: 624–629

Ye J, Zhang R, Nielsen S, Joseph SD, Huang D, Thomas T (2016) A combination of biochar-mineral complexes and compost improves soil bacterial processes, soil quality, and plant properties. Front Microbiol 7:372

Yi HS, Ahn YR, Song GC, Ghim SY, Lee S, Lee G, Ryu CM (2016) Impact of a bacterial volatile 2,3-butanediol on *Bacillus subtilis* rhizosphere robustness. Front Microbiol 7:993

Yu K, Pieterse CMJ, Bakker PAHM, Berendsen RL (2019) Beneficial microbes going underground of root immunity. Plant Cell Environ 42:2860–2870

Zboralski A, Filion M (2020) Genetic factors involved in rhizosphere colonization by phytobeneficial *Pseudomonas spp*. Comput Struct Biotechnol J 18:3539–3554

Zebelo S, Song Y, Kloepper JW, Fadamiro H (2016) Rhizobacteria activates (+)-delta-cadinene synthase genes and induces systemic resistance in cotton against beet armyworm (*Spodoptera exigua*). Plant Cell Environ 39:935–943

Zhou C, Ma ZY, Zhu L, Xiao X, Xie Y, Zhu J et al (2016) Rhizobacterial strain *Bacillus megaterium* bofc15 induces cellular polyamine changes that improve plant growth and drought resistance. Int J Mol Sci 17:976

Zipfel C, Oldroyd GE (2017) Plant signalling in symbiosis and immunity. Nature 543:328–336

Chapter 9
Plant Growth-Promoting Rhizobacteria (PGPR): An Indispensable Tool for Climate-Resilient Crop Production

Purabi Banerjee, Parijat Bhattacharya, Anurag Bera, and Akbar Hossain

Abstract Global climate change is exaggerating the probability of occurrences of biotic and abiotic stresses in the present scenario, drastically hampering agricultural production. Adoption of different sustainable and environmentally friendly practices is of prime importance to improve production base and resource use efficiency while maintaining soil health under climatic anomalies. Effective implementation of beneficial soil microbiota is one of those innovative strategies. The bacterial community dwelling in the rhizospheric region, responsible for enhancing the growth of the crop plants, is termed plant growth-promoting rhizobacteria (PGPR). These are involved in specific biochemical and physiological mechanisms in plants, contributing to climate-resilient crop production by alleviating different biotic and abiotic stresses. PGPR undertakes a series of strategic mechanisms to overcome stress like

---

Purabi Banerjee and Parijat Bhattacharya have contributed equally and are to be considered as the joint first authors.

P. Banerjee (✉)
Department of Agronomy, Bidhan Chandra Krishi Viswavidyalaya, Mohanpur, Nadia, West Bengal, India

Division of Crop Sciences, ICAR-Central Research Institute for Dryland Agriculture, Hyderabad, India

P. Bhattacharya
Department of Agricultural Chemistry and Soil Science, Bidhan Chandra Krishi Viswavidyalaya, Mohanpur, Nadia, West Bengal, India

School of Agriculture, Swami Vivekananda University, Barrackpore, North 24 Parganas, West Bengal, India

A. Bera
Department of Agronomy, Institute of Agricultural Sciences, Banaras Hindu University, Varanasi, Uttar Pradesh, India

A. Hossain (✉)
Division of Agronomy, Bangladesh Wheat and Maize Research Institute, Dinajpur, Bangladesh

P. Mathur et al. (eds.), *Microbial Symbionts and Plant Health: Trends and Applications for Changing Climate*, Rhizosphere Biology,
https://doi.org/10.1007/978-981-99-0030-5_9

production of biologically active metabolites, special enzymes, volatile organic compounds, biofilm, exopolysaccharides, phytohormone-like compounds and bacterial secondary metabolites, fixation of atmospheric nitrogen, solubilizing insoluble phosphatic compounds, changing the redox status and pH status of the system, etc. To intensify its application by bridging the critical research gaps, extensive research efforts on PGPR should be devoted to the enrichment of the knowledge base regarding the underlying mechanisms of soil–plant–microbe interactions and thereby boosting up redressal of challenges while focusing on the sustainable applications of PGPR. This chapter made a generous effort to compose a comprehensive but compact literature on the enormous potential of PGPR as an efficient, productive, and sustainable alternative to conventional agriculture to withstand all the environmental stresses, thereby building a strong foundation for climate-resilient sustainable crop production.

Keywords Abiotic stress · Biotic stress · Climate change · PGPR · Sustainable crop production

Abbreviations

ACC deaminase	1-Amino 1-cyclopropane carboxylase deaminase
Al	Aluminum
DNA gyrase	Deoxyribonucleic acid
Fe	Iron
HKT-1	High-affinity potassium transporter-1
IAA	Indole acetic acid
ISR	Induced systemic resistance
K	Potassium
Na	Sodium
PGPR	Plant growth-promoting rhizobacteria
pH	Potential of hydrogen
R&D	Research and development
VOCs	Volatile organic compounds

9.1 Introduction

Today's world is facing an abrupt deviation in global climate. As climate change is gearing up, the chances of occurrences of biotic and abiotic stresses are also escalating, which in turn drastically hampers agricultural production (Lipper et al. 2014; Pareek et al. 2020). One of the reasonable ways to overcome such situations is to adopt sustainable and environmentally friendly practices that can improve resource use efficiency and economic yield while maintaining the overall soil health even under a wide range of environmental extremities (Drost et al. 2020). Implementation of beneficial soil microbiota is one of those innovative strategies. In particular, some phytomicrobiome-like plant growth-promoting rhizobacteria (PGPR), categorized under beneficial microbes, enhance plant tolerance to resist different biotic as well as abiotic stresses in addition to promoting crop yield (Backer et al. 2018) through nutrient absorption, biosynthesis of phytohormones, as well as the release of antibiotics (Lyu et al. 2020; Sindhu et al. 2020). Technically, root exudates containing various organic molecules control the abundance and efficiency of plant-associated microbes living in the rhizospheric soil (Huang et al. 2019). In reciprocation, microbial signals also mediate the development of host plants to a great extent (Ortíz-Castro et al. 2009). Thus, phytomicrobiomes can be deployed to develop a sustainable and climate-resilient agricultural production system without the excessive use of extraneous chemical inputs.

The areas in soil, including rhizoplane, rhizosphere, and endosphere, nearer to the root surface are the most plant-influenced zones with maximum diversity in the microbial population (Reinhold-Hurek et al. 2015). These areas substantially regulate carbon sequestration in the soil as well as phytoremediation and contribute to a strong positive impact on crop growth and productivity (Ojuederie and Babalola 2017; Berlanas et al. 2019). More specifically, the microbes residing in the rhizospheric region efficiently contribute to carbon cycling from soil to the atmosphere and vice versa, reducing the loss of soil organic carbon by their metabolic activities (Bardgett et al. 2008; Shah et al. 2021). Basically, PGPR dwelling majorly in the plant rhizospheric region of the soil are the free-living bacteria exerting a beneficial impact on the plants (Fadiji et al. 2022). PGPR residing outside the host plant cells are directly associated with plant roots relating to the plant–microbe interacting mechanisms. However, the majority of the PGPR community colonize the plant root surface and live in between the spaces of rhizodermal layers and root hairs (Gray and Smith 2005). More or less, PGPR aid in plant growth and development in the form of biofertilizers by means of enhancing nutrient assimilation through biological nitrogen fixation, phosphate solubilization, iron acquisition, and several other diverse mechanistic principles (Sharma et al. 2013; Kuan et al. 2016). Besides, they also control pathogen infections by competition and antagonism as biocontrol agents (Beneduzi et al. 2012; Chowdhury et al. 2015). Many PGPR act as means of bioremediation and degrade various kinds of organic pollutants, thereby reducing heavy metal contamination in soils (Nie et al. 2011; Weyens et al. 2015). All these activities lead to the establishment of the climate-smart crop production system. In this chapter, a critical assessment will be carried out on the contributions of PGPR in the development of climate-resilient sustainable crop production through efficient management of biotic and abiotic stress and thereby supporting plant health.

9.2 Rhizosphere and Plant Growth-Promoting Rhizobacteria (PGPR)

The region in the immediate vicinity of plant roots is called rhizosphere. Since the early work by Hiltner (1904), the rhizosphere has been described as the soil compartment that is influenced by plant growth. This influence results from the release by the plant of organic materials, a phenomenon called rhizodeposition that consists mostly of plant metabolites (the exudates) and plant debris (dead cells, loss of mucilage, etc.) (Venturi and Keel 2016; Wang et al. 2021). Technically, there is a tripartite relationship between the plants (plant roots), soil, and microbes in the rhizosphere (Mashabela et al. 2022).

The bacterial community dwelling in the rhizospheric region that is responsible for enhancing the growth of the crop plants is termed plant growth-promoting rhizobacteria (PGPR) (de Vries et al. 2020). These microorganisms involve specific biochemical and physiological mechanisms that contribute to climate-resilient crop production by alleviating stresses (Bhattacharyya et al. 2016; Shah et al. 2021). Some of the PGPR belong to facultative endophytes invading the intercellular spaces of the host tissues to establish an association of mutual benefits (Hassan 2017; Ilangumaran and Smith 2017). The most studied genera of PGPR include *Azospirillum*, *Azotobacter*, *Gluconacetobacter*, *Arthrobacter*, *Burkholderia*, *Serratia*, *Bacillus*, *Pseudomonas,* and *Streptomyces* (Kour et al. 2020). Very recently, some genera like *Rouxiella badensis* and *Rahnella* spp. have been proposed to be promising bioinoculants with plant growth-promoting activities in addition to biocontrol properties (Ulloa-Muñoz et al. 2020; Morales-Cedeno et al. 2021).

9.3 PGPR-A Sustainable Approach against Climate Change

The green revolution necessitated the continuous use of chemical fertilizers and pesticides for upscaling agricultural productivity, which subsequently created a permanent adverse impact on the environment (Bhattacharya et al. 2019; Antar et al. 2021). On the other hand, the concurrences of different abiotic stresses in terms of thermal extremities, drought, salinity, and heavy metals stress, as well as biotic stresses concerning the prevalence of a large number of phytopathogens, are desperately triggered by global climate change, which, in consequence, considerably hamper the agricultural productivity (Ma et al. 2016). In fact, the utilization of different phytomicrobiome has been recognized by scientists as a "fresh green revolution" as a sustainable approach for the overall benefit of the agricultural production system (Lyu et al. 2020). In this scenario, plant growth-promoting rhizobacteria (PGPR) is a potential candidate for the development of an efficient, productive, and sustainable alternative to conventional agriculture against the growing use of chemical fertilizers and pesticides to deal with all the environmental stresses under changing climate (Ma et al. 2019; Khan et al. 2020a, b).

9.4 PGPR-Mediated Plant Tolerance against Abiotic Stresses

The frequency and intensity of a wide range of abiotic stresses like drought, temperature fluctuations, salinity, and heavy metal toxicity are increasing day by day under the current global scenario of changing climate, which can bring about a drastic reduction in overall production and quality of a variety of crops worldwide (Awasthi et al. 2014; Mishra et al. 2017; Vogel et al. 2019). Implementation of sustainable strategies to boost plant tolerance against such environmental abnormalities is of considerable importance to ensure global food security. The application of PGPR is considered one of the promising eco-friendly approaches to assisting crop plants to withstand the vagaries of environmental extremes (Ipek et al. 2019). At present, the importance of plant-associated rhizospheric microbiomes in crop adaptation toward abiotic stress is emerging as these microbial species are efficient enough to modulate the physiological and biochemical responses of crops and thereby facilitating improved plant tolerance to successfully thrive under adverse environments (Vacheron et al. 2013; Bhattacharyya et al. 2016).

PGPR regulates osmolytes homeostasis in the plant through the accumulation of different solutes covering sugars, proline, polyamines, polyhydric alcohols, betaines, and several other amino acids, taking a vital role in maintaining cellular osmotic balance under higher levels of moisture scarcity or salinity in soil (Vurukonda et al. 2016; Shah et al. 2021). Various PGPR strains have been often registered to take a significant role in modulating plant–water status, photosynthetic efficiency, stomatal conductance, carbohydrate metabolism, ion homeostasis, phytohormonal status, antioxidant enzymes, and water and nutrients uptake under stress (Ilangumaran and Smith 2017; Vandana et al. 2020). These physicochemical mechanisms alleviate stress through a complex signaling network interconnected with the plant–microbe interaction (Smith et al. 2017; Saritha et al. 2022). Basically, PGPR releases osmolytes and functions synergistically in combination with the osmolytes discharged by the plant to improve normal plant growth under abiotic stresses (Vardharajula et al. 2011; Shah et al. 2021). Different studies have revealed the significant role of PGPR in alleviating stress presented in Table 9.1. Inoculation with PGPR systematically induces abiotic stress tolerance via direct as well as indirect mechanisms in crop plants (Yang et al. 2009; Lata et al. 2018).

Table 9.1 Effects of PGPR on crops in ameliorating different abiotic stresses

Crop	Abiotic stress	PGPR used	Impact of PGPR	Reference
Wheat	Drought	*Streptomyces coelicolor*, *S. olivaceus*, *S. geysiriensis*	Greater phytohormones production along with effective internal water balance	Yandigeri et al. (2012)
Maize	Drought	*Burkholderia phytofirmans* & *Enterobacter* sp.	Reduced ROS-induced damage	Naveed et al. (2014)
Rice	Drought	*Pseudomonas* sp., *Bacillus cereus*, *Arthrobacter nitroguajacolicus*	Higher plant tolerance through the accumulation of proline	Gusain et al. (2015)
Pea	Low temperature	Miscellaneous PGPR strains	Growth promotion, higher resistance to DNA gyrase and inhibitor antibiotic nalidixic acid	Meena et al. (2015)
Rice	Flooding	*Pseudomonas putida*, *P. fluorescens*	Modulations in IAA and ACC deaminase activities	Etesami and Alikhani (2016)
Rice	Salinity	*Bacillus amyloliquefaciens*	Enhanced plant tolerance to salinity and upregulation of the expression of genes related to osmotic and ionic stress response mechanisms	Ilangumaran and Smith (2017)
Potato	Salinity	*Bacillus* spp.	Improved IAA production	Tahir et al. (2019)
Green gram	Drought	*Staphylococcus* sp.	Plant growth promotion through production of IAA and ACC deaminase	Jayakumar et al. (2020)
Soybean	Heat	*Bacillus cereus*	Mitigation of heat stress induced damages regarding phytohormone, amino acid and genetic expressions	Khan et al. (2020a, 2020b)
Tomato	Heat	*Bacillus cereus*	Growth improvement under high temperatures, modulations in the activities of plant growth regulators	Mukhtar et al. (2020)
Cowpea	Heavy metal	*Streptomyces acidiscabies*	Improved siderophores activity	Mellidou and Karamanoli (2022)

9.5 Broad Mechanisms of PGPR to Overcome Stress

9.5.1 PGPR Undertakes a Couple of Strategic Mechanisms to Overcome Stress

9.5.1.1 Production of Biologically Active Metabolites

Under stress, microbes produce thermostable amino acids like proline, glycine, and betaine (Vurukonda et al. 2016; Ghosh et al. 2022). These amino acids can be characterized as "compatible solutes," which, in low salt concentration, stabilizes tertiary structure of protein. Under saline environment, those maintain cellular osmotic potential as osmolytes; detoxify ROS, and thereby maintain membrane integrity (Hayat et al. 2012; Ahmed et al. 2019).

9.5.1.2 Production of Special Enzymes

PGPR is capable of modulating transcription and interferes in amino acid and/or enzyme production. Under stress conditions, ethylene is produced from ACC (1-amino 1-cyclopropane carboxylase) and it ultimately reduces root growth drastically (Husain et al. 2020). Under such condition, PGPR produces ACC (1-amino 1-cyclopropane carboxylase) deaminase enzyme to degrade ACC (del Carmen Orozco-Mosqueda et al. 2020). Moreover, microbes produce some volatile organic compounds. They lead to the production of early responsive dehydration (ERD-15) gene for conservation of moisture in plant cell. They in turn boost plant's capability to cope with salinity and drought stress (Kaushal and Wani 2016; Karnwal 2021).

9.5.1.3 Production of Volatile Organic Compounds

Microbes in the rhizosphere produce volatile organic compounds such as 3-hydroxy-2-butanone (acetoin) and (2R,3R)-butanediol (Santoro et al. 2011). These compounds provide induced systemic resistance in plants (ISR) to fight against stresses by reducing electrolyte leakage and malondialdehyde content and boosting the synthesis of photosynthetic pigments (Yasmin et al. 2019; Fig. 9.1). Also, indole acetic acid (IAA) or auxin helps increase root volume during stress and helps the plants to extract from deeper soil layers (Ghosh et al. 2013; Boivin et al. 2016).

9.5.1.4 Production of Biofilms and Exopolysaccharides

Biofilm is microbial assemblage that is irreversibly associated with a surface and acts as a "city of microbes" forming an immediate environment for their survival (Kokare et al. 2009).

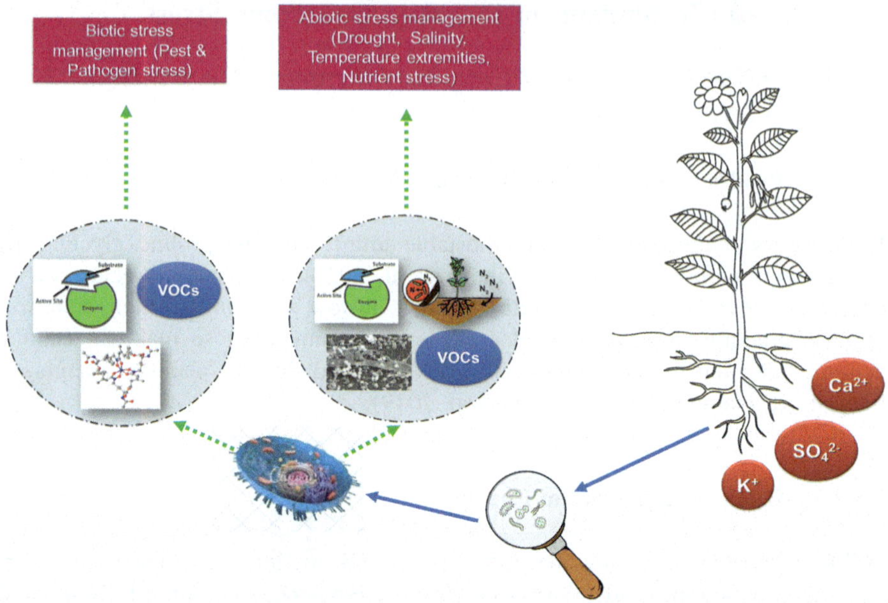

Fig. 9.1 PGPR in crop stress management

Biofilms are stabilized by extracellular polymeric substances, including polysaccharides, lipids, and proteins (Flemming and Wingender 2010; Sauer et al. 2022). These exopolysaccharides act as a protective barrier and a good cementing agent for the maintenance of soil aggregates. Better soil aggregation is particularly suitable for proper moisture conservation down the soil profile (Eze et al. 2020). Biofilms, as a polymeric network, hold a huge amount of water around its extracellular area. That is particularly suitable under moisture-stress conditions (Naseem et al. 2018; Yin et al. 2019).

9.5.1.5 Production of Bacterial Secondary Metabolites

Bacterial secondary metabolites have no direct role in the growth, development, and reproduction of PGPR, but have some important ecological functions. Those compounds originate from some of the primary metabolites through different biosynthetic pathways (Tyc et al. 2017). Secondary metabolites can be of soluble (ribosomal and nonribosomal peptides, e.g., bacteriocins and siderophores) and volatile organic compounds (terpenes) (Schmidt 2015). Iron-loving bacteria produce siderophores that contain metallic iron (Ferreira et al. 2019a, b). As a result of this, during production of siderophores, soil iron is substantially depleted (Sutak et al. 2020). It has a twofold manifestation. Firstly, it controls excess metal (iron) stress and toxicity in contaminated soils (Misslinger et al. 2021). Secondly, as iron is required for fungal germ tube formation, without metallic iron, the fungal germ tube is malformed and eventually withers away. It helps keep the soil fungal population below the harmful level (Mohanty et al. 2018).

9.5.1.6 Supply of Essential Plant Nutrients

Rhizospheric microorganisms play vital roles in enhancing the availability of essential plant nutrients. This can be mediated through two major processes, viz., fixation of atmospheric nitrogen and solubilizing insoluble phosphatic compounds. Microbes involved in atmospheric nitrogen fixation enhance nitrogen nutrition in plants. This is particularly applicable to legumes (Mehmood et al. 2018). Nitrogen-fixing organisms may be symbiotic (associated with host plant), for example, Rhizobium and Frankia, and non-symbiotic (free living, associative, and endophytes) nitrogen-fixing forms such as Anabaena, Nostoc, Azospirillum, and Azotobacter. Nitrogenase (nif) genes are required for nitrogen fixation (Bhattacharyya and Jha 2012; Nazir et al. 2018). On the other hand, PGPR has the capability of solubilizing inorganic insoluble phosphates to soluble phosphate. Bacterial genera like Azotobacter, Bacillus, Beijerinckia, Enterobacter, and Erwinia play a contributory role in solubilizing insoluble phosphate. The fixed insoluble phosphate thus becomes available for plant uptake (Zaidi et al. 2009; Nazir et al. 2018). It results in enhanced phosphorus nutrition in plants. Soils containing a higher amount of sesquioxide are inherently low in available phosphorus. This microbial process is particularly suitable in those tracts (Mulissa et al. 2015; Brahmachari et al. 2019).

9.5.1.7 Changing the Redox and Acidity/Basicity Status of the System

Microorganisms can alter the redox status of the systems through production and release of different kinds of metabolites in rhizosphere (Jia et al. 2014). This helps in the mitigation of heavy metals as a redox system can either remove the metals or can render those immobilized by changing solubility under specific situations (Borch et al. 2010; Pardue and Patrick 2018). Moreover, as microbes influence the uptake of different nutrients in rhizosphere, this is also important from the redox alteration point of view. In particular, root absorption and assimilation of ammonium and nitrate ion have a difference in Eh value; hence, when roots utilize nitrate as N source, rhizosphere redox potential gets reduced to a considerable extent (Bloom et al. 2003; Meng et al. 2019). By influencing proton pumping mechanisms, microbes can alter the pH of the entire system either partially or completely (Saberi Riseh et al. 2021). When N is available in the form of ammonium (NH_4^+), the released H^+ decreases rhizospheric pH and vice versa. Microbes release different kinds of organic acids, which in turn alter the pH. Moreover, CO_2 produced from microbial respiration eventually forms carbonic acid in the rhizosphere and thereby lowers the pH of the system (Jha and Saraf 2015). It is specifically important in enhancing nutrient uptake efficiency and blocking the uptake of heavy metals at toxic concentrations (Etesami and Adl 2020).

In the due course of growth, major stresses faced by crop plants are subdivided into abiotic and biotic stresses. Abiotic stresses include (1) drought stress, (2) salinity stress, (3) heat/thermal stress, (4) nutrient stress, (5) metal stress, and (6) acidity stress.

9.5.2 Abiotic Stresses and their Alleviation

9.5.2.1 Drought Stress

Drought stress is the most important abiotic stress affecting crop production worldwide (Kasim et al. 2013). Phytohormones such as IAA, gibberellins, ethylene, and cytokinins have been widely reported to impart greater contributory roles in the growth and development of plants (Egamberdieva 2013; Vurukonda et al. 2016). These phytohormones play important roles to escape drought and other abiotic stresses (Fahad et al. 2015). PGPRs, on the other hand, are highly capable of synthesizing phytohormones to assist the plants to cope with stressful conditions. Indole acetic acid (IAA) is one of the most dominant forms of auxins present in plant system. IAA-producing PGPRs enhance the formation of lateral root system and root hairs to help the plants in water uptake and thereby moderate the ill effects of drought (Boivin et al. 2016; Vurukonda et al. 2016). Ethylene, as a phytohormone, regulates different physiological and biochemical mechanisms (Hardoim et al. 2008; Vurukonda et al. 2016). During the biosynthesis of ethylene, S-adenosylmethionine (S-AdoMet) is converted to 1-aminocyclopropane-1-carboxylate (ACC) and 1-aminocyclopropane-1-carboxylate synthase (ACS) (Vurukonda et al. 2016). This ACC is the immediate precursor molecule of ethylene. Under stress, ethylene extensively reduces root and shoot growth. PGPRs produce ACC deaminase enzymes and thereby degrade ACC and stall excessive ethylene synthesis (Glick 2005; del Carmen Orozco-Mosqueda et al. 2020).

9.5.2.2 Salinity Stress

Soil salinity significantly affects crop production and can be considered a major abiotic stress affecting crop yield worldwide (Bouhmouch et al. 2005; Zhang et al. 2020). PGPRs such as rhizobial strains have specifically developed salinity stress tolerance mechanism that comprises accumulation of cations such as potassium. This is a biologically regulated mechanism through activation of HKT-1 or high-affinity potassium transporter-1 gene, which blocks Na^+ entry to root cells (Wangsawang et al. 2018). It also includes production of organic solutes like betaine, glycine, and proline that stabilizes protein and membrane integrity for protection against osmotic stress in saline conditions as osmoprotectants (Wdowiak-Wróbel et al. 2013; Singh et al. 2015). Prajakta et al. (2019) reported that phytohormones such as IAA also protect the symbionts from oxidative stresses.

Similar to drought, extensive production of ethylene under saline conditions causing endogenous stress in plants is curbed down through microbial production of ACC deaminase (Ali et al. 2014; Gupta and Pandey 2019).

9.5.2.3 Nutrient Stress

Sustainable production of crops is severely jeopardized with deficiency of essential nutrients (Elemike et al. 2019). PGPRs can influence the availability and uptake of plant nutrients (Mehmood et al. 2018; Nazir et al. 2018). Areas with low soil-available nitrogen curtail the quality and quantity of agricultural produce. Symbiotic and free-living rhizospheric bacteria play vital roles in fixation of atmospheric nitrogen with the help Nif gene. Similarly, in phosphorus-deficient tracts, PGPRs solubilize acid-soluble phosphate (rock phosphate) and bring it into plant-available water-soluble forms (mono calcium phosphate) (Mulissa et al. 2015) and thereby boost crop yield (Matse et al. 2020; Huang et al. 2015), reporting that co-inoculation of PGPR and Rhizobium significantly improved the growth of white clover in limited soil nitrogen condition. This was specifically accomplished through fixation of nitrogen, solubilization of phosphorus, and production of phytohormones. Similar effects were also reported in wheat (Nguyen et al. 2019), rice (Purwanto and Suharti 2021), and pulses (Andy et al. 2020).

9.5.2.4 Acidity Stress

Soil acidification is now increasingly becoming important from an agricultural point of view as it grabbed almost 70% of the potential arable lands worldwide (Kochian et al. 2004; Msimbira and Smith 2020). Plants have their own set of mechanism to overcome acidity stress. However, proper selection and inoculation of acid-soil-tolerant PGPR strain symbiotic partners can significantly improve the crop stand in susceptible areas (Zhang et al. 2020). PGPRs influence the availability and uptake of essential plant nutrients. Uptake of ammonium and nitrate forms of nitrogen significantly influence the rhizospheric pH (Saberi Riseh et al. 2021). Similar to that of drought stress management, ACC deaminase enzyme-producing strains moderate ethylene-induced stress in plant system during acidity stress (Msimbira and Smith 2020). Moreover, some of the specific genes found in rhizobia, for example, ActA (apolipoprotein N-acyl transferase) and ActR (response regulator), have been reported to code for acidity stress-tolerant protein groups (Tiwari et al. 1996; Gwin et al. 2018).

9.5.3 Biotic Stress Management by PGPR

Pathogens like bacteria, fungus, nematode, and protozoa result in diseases in crop plants. Biotic stress is also alleviated by PGPR using the following mechanisms.

9.5.3.1 Production of Protective Enzymes

Pathogenic microorganisms hamper global crop production by about 10–40% depending upon the type of crop (Savary et al. 2019). The cellular materials of pathogenic organisms are generally protected inside the cell wall that safeguards those from antibiotics. PGPR is reported to be a good source of different enzymes that act upon the cell wall of pathogens and eventually dissolve those (Gupta et al. 2021). The abovementioned group of enzyme includes chitinase, β-1,3 glucanase, protease, and cellulase (Panicker and Sayyed 2022). Similar to how chitinase degrades chitin in fungal cell wall, β-(1,3) N acetyl glucosamine of peptidoglycan of pathogenic bacterial species is degraded by β-(1,3) gluconase produced by PGPR (Gowdar et al. 2018). Brzezinska et al. (2020) isolated *Bacillus subtilis* from rhizosphere of wheat and reported that this bacterial strain effectively inhibited the growth of pathogenic fungi such as *Alternaria* and *Fusarium*. *Pseudomonas fluorescens* has also been reported to actively produce these enzymes to protect crop plants from phytopathogens (Mohammed 2020; Jamil et al. 2022).

9.5.3.2 Development of Induced Systemic Resistance

The term induced systemic resistance (ISR) can be defined as the improved state of plant resistance evoked by biological or chemical stimulants. ISR protects unexposed plants/plant parts against pests and pathogens (Pieterse et al. 2014; Romera et al. 2019). Systemic acquired resistance (SAR), on the other hand, is a mechanism of triggered defense that results in long-lasting protection against a broad spectrum of microorganisms (Tripathi et al. 2019). Plants themselves create some physical and chemical barriers before the attack of pathogens and pests. This is more of indirect nature and nonspecific to pathogens. PGPRs produce salicylic acid (SA) at the root surface and trigger SA-dependent SAR pathway. This is particularly found in willow (Salix sp.) plant (Sharma et al. 2020). In case of Arabidopsis thaliana, SA-independent ISR pathway has been reported. This mechanism uses signaling mutants of jasmonic acid (*jar1, jin1,* and *coi1*) and ethylene (*etr1, ein2, ein3,* and *eir1*) (Choudhary et al. 2007; Pieterse et al. 2014). Methylated jasmonic acid gives the signal of insect and disease incidence to neighboring plants or other parts of the same plant and thereby protect the plants from future damage (Ruan et al. 2019; Kou et al. 2021).

9.5.3.3 Production of Siderophores

Siderophores are low-molecular-weight (≤10 kDa) compounds produced by both plants and PGPRs (Johnstone and Nolan 2015). Siderophores can be classified based on the coordinating groups that are attached to ferric (Fe^{3+}) ions. The most significant groups are catecholates (Enterobactins, Agrobactins), hydroxamates (Aferrioxamines, Ferrichroms, etc.), carboxylates (Rhizoferrins, Rhizobactins, etc.), and mixed ligands (ornithine derivatives, lysine derivatives, etc.) (Sharma et al. 2018; Kramer et al. 2020). Siderophores produced by PGPR have higher affinity for soil iron (Fe^{3+}) compared with fungal siderophore. As a result, PGPR siderophores basically deplete soil iron to an extent, wherein soil fungus cannot grow the germ tube portion due to a dearth of iron, which is indispensable for infection (Ferreira et al. 2019a, b). This leads to loss in pathogenicity, and crop plants are ultimately protected from soil-borne fungi (Huo et al. 2021; Karuppiah et al. 2022). *Pseudomonas putida* and *Pseudomonas aeruginosa* have been reported to produce siderophore to inhibit soil fungal infection (Govindasamy et al. 2008; Phour and Sindhu 2022).

9.5.3.4 Production of Antibiotics and Volatile Organic Compounds

Antibiosis can be defined as inhibition of growth and infection of other microorganisms via production of low-molecular-weight bioactive compounds (antibiotics) (Kenawy et al. 2019). Antibiosis by PGPR has been reported to have greater contributory role in the management of phytopathogens (Wang et al. 2021; Lahiri et al. 2022). The genus *Bacillus* and fluorescent *Pseudomonas* spp. release a wide range of extracellular compounds that inhibit the growth of plant pathogenic fungi and bacteria (Landy and Warren 1948; Mirza et al. 2019). *Bacillus* produces bacillomycin D, iturins, mycosubtilin, zwittermicin A, etc. Antibiotic compounds produced by *Pseudomonas* include 2,4-diacetyl phloroglucinol (DAPG), pyoluteorin, and phenazines (Kenawy et al. 2019). *Pseudomonas putida* and P. *striata* produce antitumor antibiotics. These antibiotics interfere with DNA and block cancerous proliferation (Loeschcke and Thies 2015; Brehl et al. 2022). Volatile organic compounds (VOCs), on the other hand, help develop ISR and SAR in plant system and thereby help minimize crop losses by pests and pathogens. VOCs like cyclohexane and methyl isocyanide have been reported to inhibit plant pathogenic organisms such as nematodes (Deng et al. 2022).

9.6 Challenges and Prospects

In the twenty-first century, two of the massive global threats faced by human civilization are climate change and exploding population growth. As a consequence, the demand for food production is increasing in limited farmland (Zhu and Chen 2002; Bruinsma 2012; Bhattacharya et al. 2019). Till the end of the twentieth century, these problems were addressed with genetic development and chemical applications. But with the advancement of time, chemically grown foods were despised due to their ill effects on human health. Side by side, healthy organically grown foods, free from chemicals and pesticide residue, are getting more preference (Yiridoe et al. 2005; Hartman et al. 2016; Johansson et al. 2021). The emerging pandemic in recent years has also prompted us to ensure the quality of consuming food products. Following that, an increase in the usage of bio-derived materials has been observed in the past few years owing to sustainable food production (Kalra et al. 2021). Among all the bio-derived inputs, PGPR-based bio-formulations are getting highlighted in current research for their favorable impacts on crops both by using direct and indirect mechanisms (Rizvi et al. 2022). As a direct influence, PGPR enhances nutrient absorption in plants by facilitating nitrogen fixation, potassium and phosphorous solubilization, and provides stress tolerance against biotic and abiotic factors. In a broad aspect, these will enhance soil fertility, crop productivity, nutrient cycling, and disease tolerance in plants (Chandran et al. 2021).

To counter the adverse impacts of climate change while maintaining a sustainable production base of the crops, phytomicrobiome in the form of PGPR is an emerging area of research (Kashyap et al. 2018; Mitra et al. 2021). However, despite its several beneficial effects, the inclusion of PGPR as a regular management practice in the farmer's field could not be popularized yet. The transfer of technology and dissemination of proper knowledge regarding its efficiency at the field level is still not successful even though trials in controlled conditions have been performed satisfactorily (Shah et al. 2021). The major constraint in the adoption of PGPR in field conditions is associated with the uncertainty of response of the inoculated strains. Another challenge is the microbial survival rate under field conditions. In normal conditions, if the organism cannot survive and propagate properly, re-inoculation may be recommended to the farmers (Tabassum et al. 2017; Mohanty et al. 2021). The success of the PGPR application on a broad scale also depends on its marketing and commercialization. Before approaching the farmer's field, a proper study and research on appropriate microbial strains, market analysis, and survey are necessary for successful implementation (Mushtaq 2021).

To intensify its application, multidisciplinary research is needed to be conducted to gain insight into the rhizospheric chemistry and recognize suitable microbial strains for preparing efficient PGPR formulations (Riaz et al. 2021). To bridge this critical research gap, there is a substantial need to interpret the issues in terms of uncertainties, inconsistencies, and challenges with respect to the practical implementation of PGPR-based formulations in the agricultural sector (Kalyanasundaram et al. 2021; Rani et al. 2022). Various research efforts on PGPR have elucidated its great potential in the management of several climatic stresses in the agricultural context. Further research should be devoted to the enrichment of the knowledge base regarding the underlying mechanisms of soil–plant–microbe interactions and thereby boosting up redressal of challenges while focusing on the sustainable applications of PGPR (Ma et al. 2019; Gupta et al. 2021; Vocciante et al. 2022).

9.7 Conclusion

The growing demand for food production necessitates exuberant improvement in food production, which is indeed being challenged by the changing climate scenario. For the last couple of decades, chemical inputs like fertilizers and pesticides have been put into use for addressing this issue, which has not always been environmentally benign. However, in recent years, PGPR has been introduced to establish mutualistic relationships with host plants as a sustainable approach for the agricultural system. In fact, PGPR offers an eco-friendly green alternative to synthetic agrochemicals. Due to this fact, it has received considerable attention from global scientific communities. To replace conventional inorganic chemicals, large-scale production and commercialization of PGPRs are indispensable shortly. Successful and effective application of microbial formulation for better nutrient assimilation, regulation of plant physiological processes, and crop protection strongly demand meticulous dovetailing of microorganisms with the host plant as well as technological improvements in research and development (R & D) and marketing strategies. Therefore, it is prima facie impromptu to carry out an in-depth investigation on promising phytomicrobiomes and isolate them for better utilization in the farmer's field. This chapter acknowledges the momentous significance of PGPR-based microbial formulations in light of climate-resilient sustainable crop production.

References

Ahmed S, Roy Choudhury A, Chatterjee P, Samaddar S, Kim K, Jeon S, Sa T (2019) The role of plant growth-promoting rhizobacteria to modulate proline biosynthesis in plants for salt stress alleviation. In: Plant Growth Promoting Rhizobacteria for Sustainable Stress Management Springer, Singapore, pp. 1–20

Ali S, Charles TC, Glick BR (2014) Amelioration of high salinity stress damage by plant growth-promoting bacterial endophytes that contain ACC deaminase. Plant Physiol Biochem 80:160–167

Andy AK, Masih SA, Gour VS (2020) Isolation, screening and characterization of plant growth promoting rhizobacteria from rhizospheric soils of selected pulses. Biocatalys Agric Biotechnol 27:101685

Antar M, Lyu D, Nazari M, Shah A, Zhou X, Smith DL (2021) Biomass for a sustainable bioeconomy: an overview of world biomass production and utilization. Renew Sust Energ Rev 139:110691

Awasthi R, Kaushal N, Vadez V, Turner NC, Berger J, Siddique KH (2014) Individual and combined effects of transient drought and heat stress on carbon assimilation and seed filling in chickpea. Func Plant Biol 41:1148–1167

Backer R, Rokem JS, Ilangumaran G, Lamont J, Praslickova D, Ricci E (2018) Plant growth-promoting rhizobacteria: context, mechanisms of action, and roadmap to commercialization of biostimulants for sustainable agriculture. Front Plant Sci 9:1473

Bardgett RD, Freeman C, Ostle NJ (2008) Microbial contributions to climate change through carbon cycle feedbacks. ISME J 2:805–814

Beneduzi A, Ambrosini A, Passaglia LMP (2012) Plant growthpromoting rhizobacteria (PGPR): their potential as antagonists and biocontrol agents. Genet Mol Biol 35:1044–1051

Berlanas C, Berbegal M, Elena G, Laidani M, Cibriain JF, Sagües A (2019) The fungal and bacterial rhizosphere microbiome associated with grapevine rootstock genotypes in mature and young vineyards. Front Microbiol 10:1142

Bhattacharya P, Sengupta S, Halder S (2019) Chapter-5 Customized, fortified and nano enabled fertilizers-prioritizing and profiteering sustainability in agriculture. Chief Editor Dr. RK Naresh, 69

Bhattacharyya PN, Goswami MP, Bhattacharyya LH (2016) Perspective of beneficial microbes in agriculture under changing climatic scenario: a review. J Phytology 8:26–41

Bhattacharyya PN, Jha DK (2012) Plant growth-promoting rhizobacteria (PGPR): emergence in agriculture. World J Microbiol Biotechnol 28(4):1327–1350

Bloom AJ, Meyerhoff PA, Taylor AR, Rost TL (2003) Root development and absorption of ammonium and nitrate from the rhizosphere. J Plant Growth Regul 21:416–431

Boivin S, Fonouni-Farde C, Frugier F (2016) How auxin and cytokinin phytohormones modulate root microbe interactions. Front Plant Sci 7:1240

Borch T, Kretzschmar R, Kappler A, Cappellen PV, Ginder-Vogel M, Voegelin A, Campbell K (2010) Biogeochemical redox processes and their impact on contaminant dynamics. Environ Sci Technol 44(1):15–23

Bouhmouch I, Souad-Mouhsine B, Brhada F, Aurag J (2005) Influence of host cultivars and Rhizobium species on the growth and symbiotic performance of Phaseolus vulgaris under salt stress. J Plant Physiol 162(10):1103–1113

Brahmachari K, Sarkar S, Santra DK, Maitra S (2019) Millet for food and nutritional security in drought prone and red laterite region of Eastern India. Int J Plant Soil Sci 26(6):1–7

Brehl C, Brass HU, Lüchtrath C, Böckmann L, Ihling N, Classen T, Büchs J (2022) Optimized prodigiosin production with Pseudomonas putida KT2440 using parallelized noninvasive online monitoring. Biotechnol Progress e3245

Bruinsma J (2012) What are the likely developments in world agriculture towards 2050. In Global Forum on Agricultural Research (GFAR)

Brzezinska MS, Kalwasińska A, Świątczak J, Żero K, Jankiewicz U (2020) Exploring the properties of chitinolytic bacillus isolates for the pathogens biological control. Microb Pathog 148:104462

del Carmen Orozco-Mosqueda M, Glick BR, Santoyo G (2020) ACC deaminase in plant growth-promoting bacteria (PGPB): an efficient mechanism to counter salt stress in crops. Microbiol Res 235:126439

Chandran H, Meena M, Swapnil P (2021) Plant growth-promoting rhizobacteria as a green alternative for sustainable agriculture. Sustainability 13(19):10986

Choudhary DK, Prakash A, Johri BN (2007) Induced systemic resistance (ISR) in plants: mechanism of action. Indian J Microbiol 47(4):289–297

Chowdhury SP, Hartmann A, Gao XW, Borriss R (2015) Biocontrol mechanism by root-associated *Bacillus amyloliquefaciens* FZB42-a review. Front Microbiol 6:780

Deng X, Wang X, Li G (2022) Nematicidal effects of volatile organic compounds from microorganisms and plants on plant-parasitic nematodes. Microorganisms 10(6):1201. https://doi.org/10.3390/microorganisms10061201

Drost SM, Rutgers M, Wouterse M, De Boer W, Bodelier PL (2020) Decomposition of mixtures of cover crop residues increases microbial functional diversity. Geoderma 361:114060

Egamberdieva D (2013) The role of phytohormone producing bacteria in alleviating salt stress in crop plants. In *Biotechnological techniques of stress tolerance in plants*. Studium, Houston, TX, pp. 21–39

Elemike EE, Uzoh IM, Onwudiwe DC, Babalola OO (2019) The role of nanotechnology in the fortification of plant nutrients and improvement of crop production. Appl Sci 9(3):499

Etesami H, Adl, SM (2020) Plant growth-promoting rhizobacteria (PGPR) and their action mechanisms in availability of nutrients to plants. Phyto-Microbiome in stress regulation: 147–203

Etesami H, Alikhani HA (2016) Co-inoculation with endophytic and rhizosphere bacteria allows reduced application rates of N-fertilizer for rice plant. Rhizosphere 2:5–12

Eze S, Dougill AJ, Banwart SA, Hermans TD, Ligowe IS, Thierfelder C (2020) Impacts of conservation agriculture on soil structure and hydraulic properties of Malawian agricultural systems. Soil Tillage Res 201:104639

Fadiji AE, Babalola OO, Santoyo G, Perazzolli M (2022) The potential role of microbial biostimulants in the amelioration of climate change-associated abiotic stresses on crops. Front Microbiol 12:829099

Fahad S, Hussain S, Bano A, Saud S, Hassan S, Shan D, Huang J (2015) Potential role of phytohormones and plant growth-promoting rhizobacteria in abiotic stresses: consequences for changing environment. Environ Sci Pollut Res 22(7):4907–4921

Ferreira C, Vilas-Boas Â, Sousa CA, Soares H, Soares EV (2019a) Comparison of five bacterial strains producing siderophores with ability to chelate iron under alkaline conditions. AMB Express 9(1):78. https://doi.org/10.1186/s13568-019-0796-3

Ferreira MJ, Silva H, Cunha A (2019b) Siderophore-producing rhizobacteria as a promising tool for empowering plants to cope with iron limitation in saline soils: a review. Pedosphere 29(4): 409–420

Flemming HC, Wingender J (2010) The biofilm matrix. Nat Rev Microbiol 8(9):623–633

Ghosh PK, Saha P, Mayilraj S, Maiti TK (2013) Role of IAA metabolizing enzymes on production of IAA in root, nodule of *Cajanus cajan* and its PGP rhizobium sp. Biocatal Agric Biotechnol 2(3):234–239

Ghosh S, Lahiri D, Nag M, Sarkar T, Pati S, Edinur HA, Kumar M, Mohd Zain M, Ray RR (2022) Precision targeting of food biofilm-forming genes by microbial scissors: CRISPR-Cas as an effective modulator. Front Microbiol 13:964848. https://doi.org/10.3389/fmicb.2022.964848

Glick BR (2005) Modulation of plant ethylene levels by the bacterial enzyme ACC deaminase. FEMS Microbiol Lett 251(1):1–7

Govindasamy V, Senthilkumar M, Upendra-Kumar AK (2008) PGPR-biotechnology for management of abiotic and biotic stresses in crop plants. Potent Microorg Sust Agric:26–48

Gowdar SB, Deepa H, Amaresh YS (2018) A brief review on biocontrol potential and PGPR traits of Streptomyces sp. for the management of plant diseases. J Pharmacogn Phytochem 7(5):03–07

Gray E, Smith D (2005) Intracellular and extracellular PGPR: commonalities and distinctions in the plant–bacterium signaling processes. Soil Biol Biochem 37:395–412

Gupta K, Dubey NK, Singh SP, Kheni JK, Gupta S, Varshney A (2021) Plant growth-promoting rhizobacteria (PGPR): current and future prospects for crop improvement. In: Current trends in microbial biotechnology for sustainable agriculture. Springer, Singapore, pp. 203–226

Gupta S, Pandey S (2019) ACC deaminase producing bacteria with multifarious plant growth promoting traits alleviates salinity stress in French bean (Phaseolus vulgaris) plants. Front Microbiol 10:1506

Gusain YS, Singh U, Sharma A (2015) Bacterial mediated amelioration of drought stress in drought tolerant and susceptible cultivars of rice (*Oryza sativa* L.). Afr J Biotechnol 14:764–773

Gwin CM, Prakash N, Belisario JC, Haider L, Rosen ML, Martinez LR, Rigel NW (2018) The apolipoprotein N-acyl transferase Lnt is dispensable for growth in Acinetobacter species. Microbiology 164(12):1547

Hardoim PR, van Overbeek LS, van Elsas JD (2008) Properties of bacterial endophytes and their proposed role in plant growth. Trends Microbiol 16(10):463–471

Hartman GL, Pawlowski ML, Herman TK, Eastburn D (2016) Organically grown soybean production in the USA: constraints and management of pathogens and insect pests. Agronomy 6(1):16

Hassan SE (2017) Plant growth-promoting activities for bacterial and fungal endophytes isolated from medicinal plant of *Teucrium polium* L. J Adv Res 8(6):687–695. https://doi.org/10.1016/j.jare.2017.09.001

Hayat S, Hayat Q, Alyemeni MN, Wani AS, Pichtel J, Ahmad A (2012) Role of proline under changing environments: a review. Plant Signal Behav 7(11):1456–1466

Hiltner L (1904) Uber nevere Erfahrungen und Probleme auf dem Gebiet der Boden Bakteriologie und unter besonderer Beurchsichtigung der Grundungung und Broche. Arbeit Deut Landw Ges Berlin 98:59–78

Huang AC, Jiang T, Liu YX, Bai YC, Reed J, Qu B (2019) A specialized metabolic network selectively modulates Arabidopsis root microbiota. Science 364:6389

Huang XF, Zhou D, Guo J, Manter DK, Reardon KF, Vivanco JM (2015) *Bacillus* spp. from rainforest soil promote plant growth under limited nitrogen conditions. J Appl Microbiol 118(3):672–684

Huo H, Gu Y, Cao Y, Liu N, Jia P, Kong W (2021) Antifungal activity of camphor against four phytopathogens of *fusarium*. Europe PMC. https://doi.org/10.21203/rs.3.rs-274895/v1

Husain T, Fatima A, Suhel M, Singh S, Sharma A, Prasad SM, Singh VP (2020) A brief appraisal of ethylene signaling under abiotic stress in plants. Plant Signal Behav 15(9):1782051

Ilangumaran G, Smith DL (2017) Plant growth promoting rhizobacteria in amelioration of salinity stress: a systems biology perspective. Front Plant Sci 8:1768

Ipek M, Arikan S, Pirlak L, Esitken A (2019) Sustainability of crop production by PGPR under abiotic stress conditions. In: Kumar A, Meena V (eds) Plant growth promoting rhizobacteria for agricultural sustainability. Springer, Singapore, pp 293–314. https://doi.org/10.1007/978-981-13-7553-8_15

Jamil F, Mukhtar H, Fouillaud M, Dufossé L (2022) Rhizosphere signaling: insights into Plant–Rhizomicrobiome interactions for sustainable agronomy. Microorganisms 10(5):899

Jayakumar A, Padmakumar P, Nair IC, Radhakrishnan E (2020) Drought tolerant bacterial endophytes with potential plant probiotic effects from Ananas comosus. Biologia 75:1769–1778. https://doi.org/10.2478/s11756-020-00483-1

Jha CK, Saraf M (2015) Plant growth promoting rhizobacteria (PGPR). J Agric Res Dev 5:108–119

Jia Y, Huang H, Chen Z, Zhu YG (2014) Arsenic uptake by rice is influenced by microbe-mediated arsenic redox changes in the rhizosphere. Environ Sci Technol 48(2):1001–1007

Johansson E, Prieto-Linde ML, Larsson H (2021) Locally adapted and organically grown landrace and ancient spring cereals—a unique source of minerals in the human diet. Foods 10(2):393

Johnstone TC, Nolan EM (2015) Beyond iron: non-classical biological functions of bacterial siderophores. Dalton Transact 44(14):6320–6339

Kalra S, Dixit S, Lyall BW (2021) Health Halo': a new opportunity for organic food in COVID upsurge. J Univ Shanghai Sci Technol 23:113–119

Kalyanasundaram GT, Syed N, Subburamu K (2021) Recent developments in plant growth-promoting rhizobacteria (PGPR) for sustainable agriculture. Recent Dev Appl Microbiol Biochem:181–192

Karnwal A (2021) Screening and identification of abiotic stress-responsive efficient antifungal pseudomonas spp. from rice rhizospheric soil. BioTechnologia J Biotechnol Comput Biol Bionanotechnol 102(1):5–19

Karuppiah V, He A, Lu Z, Wang X, Li Y, Chen J (2022) *Trichoderma asperellum* GDFS1009-mediated maize resistance against *Fusarium graminearum* stalk rot and mycotoxin degradation. Biol Control 174:105026

Kashyap PL, Srivastava AK, Tiwari SP, Kumar S (2018) Microbes for Climate Resilient Agriculture. John Wiley & Sons, Hoboken, NJ

Kasim WA, Osman ME, Omar MN, El-Daim A, Islam A, Bejai S, Meijer J (2013) Control of drought stress in wheat using plant-growth-promoting bacteria. J Plant Growth Regul 32(1): 122–130

Kaushal M, Wani SP (2016) Rhizobacterial-plant interactions: strategies ensuring plant growth promotion under drought and salinity stress. Agric Ecosyst Environ 231:68–78

Kenawy A, Dailin DJ, Abo-Zaid GA, Malek RA, Ambehabati KK, Zakaria KHN, Enshasy HAE (2019) Biosynthesis of antibiotics by PGPR and their roles in biocontrol of plant diseases. In: Plant growth promoting rhizobacteria for sustainable stress management, Springer, Singapore, pp. 1–35

Khan MA, Asaf S, Khan AL, Jan R, Kang SM, Kim KM, Lee IJ (2020a) Thermotolerance effect of plant growth-promoting *Bacillus cereus* SA1 on soybean during heat stress. BMC Microbiol 20(1):1–14

Khan N, Bano A, Ali S, Babar MA (2020b) Crosstalk amongst phytohormones from planta and PGPR under biotic and abiotic stresses. Plant Growth Regul 90:189–203

Kochian LV, Hoekenga OA, Pineros MA (2004) How do crop plants tolerate acid soils? Mechanisms of aluminum tolerance and phosphorous efficiency. Ann Rev Plant Biol 55(1):459–493

Kokare CR, Chakraborty S, Khopade AN, Mahadik KR (2009) Biofilm: importance and applications

Kou X, Feng Y, Yuan S, Zhao X, Wu C, Wang C, Xue Z (2021) Different regulatory mechanisms of plant hormones in the ripening of climacteric and non-climacteric fruits: a review. Plant Mol Biol 107(6):477–497

Kour D, Kaur T, Devi R, Rana KL, Yadav N, Rastegari AA (2020) Biotechnological applications of beneficial microbiomes for evergreen agriculture and human health. In: Rastegari AA, Yadav AN, Yadav N (eds) New and future developments in microbial biotechnology and bioengineering. Elsevier, pp 255–279. https://doi.org/10.1016/b978-0-12-820528-0.00019-3

Kramer J, Özkaya Ö, Kümmerli R (2020) Bacterial siderophores in community and host interactions. Nat Rev Microbiol 18(3):152–163

Kuan KB, Othman R, Abdul Rahim K, Shamsuddin ZH (2016) Plant growth-promoting rhizobacteria inoculation to enhance vegetative growth, nitrogen fixation and nitrogen remobilisation of maize under greenhouse conditions. PLoS One 11:e0152478

Lahiri D, Nag M, Sayyed RZ, Gafur A, Ansari MJ, Ray RR (2022) PGPR in biofilm formation and antibiotic production. In: Antifungal metabolites of rhizobacteria for sustainable agriculture, Springer, Cham, pp. 65–82

Landy M, Warren GH (1948) Bacillomycin; an antibiotic from Bacillus subtilis active against pathogenic fungi. In: Proceedings of the Society for Experimental Biology and Medicine. Society for Experimental Biology and Medicine, New York, N.Y., 67(4):539–541

Lata R, Chowdhury S, Gond SK, White JF Jr (2018) Induction of abiotic stress tolerance in plants by endophytic microbes. Lett Appl Microbiol 66(4):268–276

Lipper L, Thornton P, Campbell BM, Baedeker T, Braimoh A, Bwalya M (2014) Climate-smart agriculture for food security. Nat Clim Chang 4:1068–1072

Loeschcke A, Thies S (2015) *Pseudomonas putida*—a versatile host for the production of natural products. Appl Microbiol Biotechnol 99(15):6197–6214

Lyu D, Backer R, Subramanian S, Smith DL (2020) Phytomicrobiome coordination signals

Ma Y, Rajkumar M, Zhang C, Freitas H (2016) Beneficial role of bacterial endophytes in heavy metal phytoremediation. J Environ Manag 174:14–25

Ma Y, Vosátka M, Freitas H (2019) Editorial: beneficial microbes alleviate climatic stresses in plants. Front Plant Sci 10:595

Mashabela MD, Tugizimana F, Steenkamp PA, Piater LA, Dubery IA, Mhlongo MI (2022) Untargeted metabolite profiling to elucidate rhizosphere and leaf metabolome changes of wheat cultivars (*Triticum aestivum* L.) treated with the plant growth-promoting rhizobacteria

Paenibacillus alvei (T22) and *Bacillus subtilis*. Front Microbiol 13:971836. https://doi.org/10.3389/fmicb.2022.971836

Matse DT, Huang CH, Huang YM, Yen MY (2020) Effects of coinoculation of rhizobium with plant growth promoting rhizobacteria on the nitrogen fixation and nutrient uptake of Trifolium repens in low phosphorus soil. J Plant Nutr 43(5):739–752

Meena RK, Singh RK, Singh NP, Meena SK, Meena VS (2015) Isolation of low temperature surviving plant growth–promoting rhizobacteria (PGPR) from pea (*Pisum sativum* L.) and documentation of their plant growth promoting traits. Biocatal Agric Biotechnol 4(4):806–811

Mehmood U, Inam-ul-Haq M, Saeed M, Altaf A, Azam F, Hayat S (2018) A brief review on plant growth promoting rhizobacteria (PGPR): a key role in plant growth promotion. Plant Prot 2(2): 77–82

Mellidou I, Karamanoli K (2022) Unlocking PGPR-mediated abiotic stress tolerance: what lies beneath. Front Sustain Food Syst 6:832896

Meng D, Li J, Liu T, Liu Y, Yan M, Hu J, Yin H (2019) Effects of redox potential on soil cadmium solubility: insight into microbial community. J Environ Sci 75:224–232

Mirza R, Azeem M, Qaisar U (2019) Influence of *Peganum harmala* peptides on the transcriptional activity of biofilm related genes in sensitive and resistant strains of Pseudomonas aeruginosa and Staphylococcus aureus. Pak J Pharm Sci 32(5):2341–2345

Mishra J, Singh R, Arora NK (2017) Alleviation of heavy metal stress in plants and remediation of soil by rhizosphere microorganisms. Front Microbiol 8:1706

Misslinger M, Hortschansky P, Brakhage AA, Haas H (2021) Fungal iron homeostasis with a focus on Aspergillus fumigatus. Biochim Biophys Acta (BBA) Mol Cell Res 1868(1):118885

Mitra D, Rodríguez AMD, Cota FIP, Khoshru B, Panneerselvam P, Moradi S, Sagarika MS, Anđelkovic S, de los Santos-Villalobos S, Mohapatra PKD (2021) Amelioration of thermal stress in crops by plant growth-promoting rhizobacteria. Physiol Mol Plant Pathol 115:101679

Mohammed AF (2020) Optimization of cellulase and chitinase enzymes production by plant growth promoting rhizobacteria. Novel Res Microbiol J 4(1):641–652

Mohanty P, Singh PK, Chakraborty D, Mishra S, Pattnaik R (2021) Insight into the role of PGPR in sustainable agriculture and environment. Front Sust Food Syst 5:667150

Mohanty SK, Gonneau C, Salamatipour A, Pietrofesa RA, Casper B, Christofidou-Solomidou M, Willenbring JK (2018) Siderophore-mediated iron removal from chrysotile: implications for asbestos toxicity reduction and bioremediation. J Hazard Mater 341:290–296

Morales-Cedeno LR, De Los Santos-Villalobos S, Santoyo G (2021) Functional and genomic analysis of *Rouxiella badensis* SER3 as a novel biocontrol agent of fungal pathogens. Front Microbiol 12:709855

Msimbira LA, Smith DL (2020) The roles of plant growth promoting microbes in enhancing plant tolerance to acidity and alkalinity stresses. Front Sust Food Syst 4:106

Mukhtar T, Smith D, Sultan T, Seleiman MF, Alsadon AA, Ali S, Saad MA (2020) Mitigation of heat stress in *Solanum lycopersicum* L. by ACC-deaminase and exopolysaccharide producing Bacillus cereus: effects on biochemical profiling. Sustainability 12(6):2159

Mulissa JM, Carolin RL, Ruth AS, Fassil A (2015) Characterization of phosphate solubilizing rhizobacteria isolated from lentil growing areas of Ethiopia. African J Microbiol Res 9(25): 1637–1648

Mushtaq Z (2021) PGPR: present role, mechanism of action and future prospects along bottlenecks in commercialization. EQA-Int J Environ Qual 41:9–15

Naseem H, Ahsan M, Shahid MA, Khan N (2018) Exopolysaccharides producing rhizobacteria and their role in plant growth and drought tolerance. J Gen Microbiol 58(12):1009–1022

Naveed M, Mitter B, Reichenauer TG, Wieczorek K, Sessitsch A (2014) Increased drought stress resilience of maize through endophytic colonization by Burkholderia phytofirmans PsJN and Enterobacter sp. FD17. Environ Exp Bot 97:30–39

Nazir N, Kamili AN, Shah D (2018) Mechanism of plant growth promoting rhizobacteria (PGPR) in enhancing plant growth—a review. Int J Manag Technol Eng 8:709–721

Nguyen ML, Spaepen S, du Jardin P, Delaplace P (2019) Biostimulant effects of rhizobacteria on wheat growth and nutrient uptake depend on nitrogen application and plant development. Arch Agron Soil Sci 65(1):58–73

Nie M, Wang YJ, Yu JY, Xiao M, Jiang LF, Yang J (2011) Understanding plant-microbe interactions for phytoremediation of petroleum polluted soil. PLoS One 6:e17961

Ojuederie OB, Babalola OO (2017) Microbial and plant-assisted bioremediation of heavy metal polluted environments: a review. Int J Environ Res Public Health 14:1504

Ortíz-Castro R, Contreras-Cornejo HA, Macías-Rodríguez L, LópezBucio J (2009) The role of microbial signals in plant growth and development. Plant Signal Behav 4(8):701–712

Panicker S, Sayyed RZ (2022) Hydrolytic enzymes from PGPR against plant fungal pathogens. In: Antifungal metabolites of rhizobacteria for sustainable agriculture, Springer, Cham, pp. 211–238

Pardue JH, Patrick WH (2018) Changes in metal speciation following alteration of sediment redox status. In Metal contaminated aquatic sediments (pp. 169–185). Routledge

Pareek A, Dhankher OP, Foyer CH (2020) Mitigating the impact of climate change on plant productivity and ecosystem sustainability. J Exp Bot 71(2):451–456

Phour M, Sindhu SS (2022) Mitigating abiotic stress: microbiome engineering for improving agricultural production and environmental sustainability. Planta 256(5):1–34

Pieterse CM, Zamioudis C, Berendsen RL, Weller DM, Van Wees SC, Bakker PA (2014) Induced systemic resistance by beneficial microbes. Annu Rev Phytopathol 52:347–375

Prajakta BM, Suvarna PP, Raghvendra SP, Alok RR (2019) Potential biocontrol and superlative plant growth promoting activity of indigenous Bacillus mojavensis PB-35 (R11) of soybean (Glycine max) rhizosphere. SN Appl Sci 1(10):1–11

Purwanto P, Suharti WS (2021) Nutrient uptake, chlorophyll content, and yield of rice (Oryza sativa L.) under the application of PGPR consortium. Biosaint J Biol Biol Educ 13:3

Rani K, Tigga P, Roy A, Das A, Trivedi A (2022) Nutrient availability and plant productivity through PGPR: mechanisms. Potential and Constraints Agric Biotechnol 75:138

Reinhold-Hurek B, Bünger W, Burbano CS, Sabale M, Hurek T (2015) Roots shaping their microbiome: global hotspots for microbial activity. Annu Rev Phytopathol 53:403–424

Riaz U, Murtaza G, Anum W, Samreen T, Sarfraz M, Nazir MZ (2021) Plant growth-promoting rhizobacteria (PGPR) as biofertilizers and biopesticides. In: Hakeem KR, Dar GH, Mehmood MA, Bhat RA (eds) Microbiota and biofertilizers. Springer, Cham, pp 181–196

Rizvi A, Ahmed B, Khan MS, El-Beltagi HS, Umar S, Lee J (2022) Bioprospecting plant growth promoting rhizobacteria for enhancing the biological properties and phytochemical composition of medicinally important crops. Molecules 27(4):1407

Romera FJ, García MJ, Lucena C, Martínez-Medina A, Aparicio MA, Ramos J, Pérez-Vicente R (2019) Induced systemic resistance (ISR) and Fe deficiency responses in dicot plants. Front Plant Sci 10:287

Ruan J, Zhou Y, Zhou M, Yan J, Khurshid M, Weng W, Zhang K (2019) Jasmonic acid signaling pathway in plants. Int J Mol Sci 20(10):2479

Saberi Riseh R, Ebrahimi-Zarandi M, Tamanadar E, Moradi Pour M, Thakur VK (2021) Salinity stress: toward sustainable plant strategies and using plant growth-promoting rhizobacteria encapsulation for reducing it. Sustainability 13(22):12758

Santoro MV, Zygadlo J, Giordano W, Banchio E (2011) Volatile organic compounds from rhizobacteria increase biosynthesis of essential oils and growth parameters in peppermint (*Mentha piperita*). Plant Physiol Biochem 49(10):1177–1182

Saritha M, Kumar P, Panwar NR, Burman U (2022) Intelligent plant–microbe interactions. Arch Agron Soil Sci 68(7):1002–1018

Sauer K, Stoodley P, Goeres DM, Hall-Stoodley L, Burmølle M, Stewart PS, Bjarnsholt T (2022) The biofilm life cycle: expanding the conceptual model of biofilm formation. Nat Rev Microbiol 20(10):608–620

Savary S, Willocquet L, Pethybridge SJ, Esker P, McRoberts N, Nelson A (2019) The global burden of pathogens and pests on major food crops. Nat Ecol Evol 3(3):430–439
Schmidt EW (2015) The secret to a successful relationship: lasting chemistry between ascidians and their symbiotic bacteria. Invert Biol 134(1):88–102
Shah A, Nazari M, Antar M, Msimbira LA, Naamala J, Lyu D, Rabileh M, Zajonc J, Smith DL (2021) PGPR in agriculture: a sustainable approach to increasing climate change resilience. Front Sustain Food Syst 5:667546. https://doi.org/10.3389/fsufs.2021.667546
Sharma A, Sidhu GPS, Araniti F, Bali AS, Shahzad B, Tripathi DK, Landi M (2020) The role of salicylic acid in plants exposed to heavy metals. Molecules 25(3):540
Sharma R, Bhardwaj R, Gautam V, Kohli SK, Kaur P, Bali RS, Vig AP (2018) Microbial siderophores in metal detoxification and therapeutics: recent prospective and applications. In: Plant microbiome: stress response, Springer, Singapore, pp. 337–350
Sharma SB, Sayyed RZ, Trivedi MH, Gobi TA (2013) Phosphate solubilizing microbes: sustainable approach for managing phosphorus deficiency in agricultural soils. Springerplus 2(1):1–14
Sindhu S, Dahiya A, Gera R, Sindhu SS (2020) Mitigation of abiotic stress in legume-nodulating rhizobia for sustainable crop production. Agribiol Res 9:444
Singh M, Kumar J, Singh S, Singh VP, Prasad SM (2015) Roles of osmoprotectants in improving salinity and drought tolerance in plants: a review. Rev Environ Sci Biotechnol 14(3):407–426
Smith DL, Gravel V, Yergeau E (2017) Signaling in the phytomicrobiome. Front Plant Sci 8:611
Sutak R, Camadro JM, Lesuisse E (2020) Iron uptake mechanisms in marine phytoplankton. Front Microbiol 11:566691
Tabassum B, Khan A, Tariq M, Ramzan M, Khan MSI, Shahid N, Aaliya K (2017) Bottlenecks in commercialisation and future prospects of PGPR. Appl Soil Ecol 121:102–117
Tahir M, Ahmad I, Shahid M, Shah GM, Farooq ABU, Akram M (2019) Regulation of antioxidant production, ion uptake and productivity in potato (*Solanum tuberosum* L.) plant inoculated with growth promoting salt tolerant *Bacillus* strains. Ecotoxicol Environ Saf 178:33–42
Tiwari RP, Reeve WG, Dilworthan MJ, Glenn AR (1996) An essential role for actA in acid tolerance of *Rhizobium melilotix*. Microbiology 142(3):601–610
Tripathi D, Raikhy G, Kumar D (2019) Chemical elicitors of systemic acquired resistance—salicylic acid and its functional analogs. Curr Plant Biol 17:48–59
Tyc O, Song C, Dickschat JS, Vos M, Garbeva P (2017) The ecological role of volatile and soluble secondary metabolites produced by soil bacteria. Trends Microbiol 25(4):280–292
Ulloa-Muñoz R, Olivera-Gonzales P, Castañeda-Barreto A, Villena GK, Tamariz-Angeles C (2020) Diversity of endophytic plant-growth microorganisms from *Gentianella weberbaueri* and *Valeriana pycnantha*, highland Peruvian medicinal plants. Microbiol Res 233:126413
Vacheron J, Desbrosses G, Bouffaud ML, Touraine B, Moënne-Loccoz Y, Muller D, Prigent-Combaret C (2013) Plant growth-promoting rhizobacteria and root system functioning. Front Plant Sci 4:356
Vandana UK, Singha B, Gulzar ABM, Mazumder PB (2020) Molecular mechanisms in plant growth promoting bacteria (PGPR) to resist environmental stress in plants. In Molecular aspects of plant beneficial microbes in agriculture Academic Press, pp. 221–233
Vardharajula S, Zulfikar Ali S, Grover M, Reddy G, Bandi V (2011) Drought-tolerant plant growth promoting *Bacillus* spp.: effect on growth, osmolytes, and antioxidant status of maize under drought stress. J Plant Interact 6:1–14
Venturi V, Keel C (2016) Signaling in the rhizosphere. Trends Plant Sci 21(3):187–198
Vocciante M, Grifoni M, Fusini D, Petruzzelli G, Franchi E (2022) The role of plant growth-promoting rhizobacteria (PGPR) in mitigating plant's environmental stresses. Appl Sci 12(3): 1231
Vogel E, Donat MG, Alexander LV, Meinshausen M, Ray DK, Karoly D (2019) The effects of climate extremes on global agricultural yields. Environ Res Lett 14:054010
de Vries FT, Griffiths RI, Knight CG, Nicolitch O, Williams A (2020) Harnessing rhizosphere microbiomes for drought-resilient crop production. Science 368(6488):270–274

Vurukonda SSKP, Vardharajula S, Shrivastava M, Skz A (2016) Enhancement of drought stress tolerance in crops by plant growth promoting rhizobacteria. Microbiol Res 184:13–24

Wang H, Liu R, You MP, Barbetti MJ, Chen Y (2021) Pathogen biocontrol using plant growth-promoting bacteria (PGPR): role of bacterial diversity. Microorganisms 9(9):1988

Wangsawang T, Chuamnakthong S, Kohnishi E, Sripichitt P, Sreewongchai T, Ueda A (2018) A salinitytolerant japonica cultivar has Na+ exclusion mechanism at leaf sheaths through the function of a Na+ transporter Os HKT 1; 4 under salinity stress. J Agron Crop Sci 204(3):274–284

Wdowiak-Wróbel S, Leszcz A, Małek W (2013) Salt tolerance in Astragalus cicer microsymbionts: the role of glycine betaine in osmoprotection. Curr Microbiol 66(5):428–436

Weyens N, Beckers B, Schellingen K, Ceulemans R, Van der Lelie D, Newman L (2015) The potential of the Ni-resistant TCEdegrading *Pseudomonas putida* W619-TCE to reduce phytotoxicity and improve phytoremediation efficiency of poplar cuttings on a Ni-TCE Cocontamination. Int J Phytoremediation 17:40–48

Yandigeri MS, Meena KK, Singh D, Malviya N, Singh DP, Solanki MK (2012) Drought-tolerant endophytic actinobacteria promote growth of wheat (*Triticum aestivum*) under water stress conditions. Plant Growth Regul 68:411–420

Yang J, Kloepper JW, Ryu CM (2009) Rhizosphere bacteria help plants tolerate abiotic stress. Trends Plant Sci 14:1–4

Yasmin H, Nosheen A, Naz R, Keyani R, Anjum S (2019) Regulatory role of rhizobacteria to induce drought and salt stress tolerance in plants. In *Field crops: sustainable management by PGPR*, Springer, Cham, pp. 279–335

Yin W, Wang Y, Liu L, He J (2019) Biofilms: the microbial "protective clothing" in extreme environments. Int J Mol Sci 20(14):3423

Yiridoe EK, Bonti-Ankomah S, Martin RC (2005) Comparison of consumer perceptions and preference toward organic versus conventionally produced foods: a review and update of the literature. Renew Agric Food Syst 20(4):193–205

Zaidi A, Khan M, Ahemad M, Oves M (2009) Plant growth promotion by phosphate solubilizing bacteria. Acta Microbiol Immunol Hung 56(3):263–284

Zhang J, Singh D, Guo C, Shang Y, Peng S (2020) Rhizobia at extremes of acidity, alkalinity, salinity, and temperature. In: Microbial versatility in varied environments, Springer, Singapore, pp. 51–65

Zhu ZL, Chen DL (2002) Nitrogen fertilizer use in China–contributions to food production, impacts on the environment and best management strategies. Nutr Cycl Agroecosyst 63(2):117–127

Chapter 10
Plant–Endophyte Interactions: A Driving Phenomenon for Boosting Plant Health under Climate Change Conditions

Saurav Anand Gurung, Arun Kumar Rai, Kiran Sunar, and Keshab Das

Abstract The endosymbiotic and ubiquitous group of microorganisms, including bacteria, fungi, and archaea that dwell inside host plant tissues without causing any harm to the plant, are termed endophytes. Over the past few decades, these microorganisms have been widely used in sustainable agriculture as biofertilizers, biocontrols, or inducers of abiotic stress tolerance. The effective colonization of the host plant by the endophyte is a necessary step in order to start these advantageous plant–microbe interactions that are engaged in boosting plant growth. Although plants naturally possess the ability to recognize potentially invasive microorganisms and launch defense mechanisms against them, these defense responses are alternatively modulated by the inoculation of beneficial endophytes, as well as pathogenic microorganisms. Furthermore, the rapidly evolving climate change conditions due to global warming and its accompanying anomalies pressurize the plant adaptation systems. Fortunately, beneficial endophytic microorganisms with potential in abiotic stress tolerance have been identified and thus can be a driving phenomenon in improving plant health under these climatic stress conditions.

Keywords Endophytes · Colonization · Plant growth promotion · Host defense · Climate change

S. A. Gurung · A. K. Rai (✉)
Department of Botany, School of Life Sciences, Sikkim University, Tadong, Sikkim, India
e-mail: akrai@cus.ac.in

K. Sunar · K. Das
Department of Botany, Balurghat Mahila Mahavidyalaya, Balurghat, Dakshin Dinajpur, West Bengal, India

P. Mathur et al. (eds.), *Microbial Symbionts and Plant Health: Trends and Applications for Changing Climate*, Rhizosphere Biology,
https://doi.org/10.1007/978-981-99-0030-5_10

Abbreviations

ABA	Abscisic acid
ACC	1-aminocyclopropane-1-carboxylic acid
AHL	N-acyl-homoserine lactones
BGL	Beta-1, 3-glucanase
CHS	Chalcone synthase
DAPG	2,4 Diacetylphloroglucinol
DRE/CRT	Dehydration-responsive element/C-repeat
EPS	Exopolysaccharides
ET	Ethylene
ETI	Effector-triggered immunity
FMO1	Flavin monooxygenase 1
GA	Gibberellins
HCN	Hydrogen cyanide
HMGR	3-Hydroxy-3-methylglutaryl-CoA reductase
IAA	Indole acetic acid
ISCs	Isochorismatases
ISR	Induced systemic resistance
JA	Jasmonic acid
LCOs	Acyl lipo-chitooligosaccharides
LEA	Late embryogenesis abundant
LOX	Lipoxygenases
LPS	Lipopolysaccharides
MAMPs	Microbe-associated molecular patterns
MAPK	Mitogen-activated protein kinase
MCPs	Methyl-accepting chemotaxis proteins
MGD	Mutant of *Gluconacetobacter diazotrophicus* strain Pal5
NPR1	Non-expressor of pathogenesis-related 1 genes
OPDA	12-Oxo-phytodienoic acid
Oxc	Oxalate decarboxylase
PAL	Phenyalanine ammonia lyase
PR 1	Pathogenesis-related 1
PRRs	Pattern recognition receptors
Pss	*Pseudomonas syringae* pv. *syringae*
PTI	Pattern-triggered immunity
QQ	Quorum quenching
QS	Quorum sensing
ROS	Reactive oxygen species
SA	Salicylic acid
SAG	SA and its glucoside
SAR	Systemic acquired resistance
SylA	SyringolinA
T3SE	Type III secreted effectors

10.1 Introduction

Numerous microbial communities interact with plants, and some of these communities enter and remain in plant tissue without causing diseases or others having detrimental effects. The endosymbiotic and ubiquitous group of microorganisms, including bacteria, fungi, and archaea that dwell inside host plant tissues without causing any harm to the plant, are termed endophytes. de Bary 1866 was the first to use the word "endophyte" to distinguish it from the epiphytes that inhabit the plant's surface. Endophytic microorganisms have been extensively utilized in sustainable agriculture over the past few decades as biofertilizers, biocontrols, or inducers of abiotic stress tolerance (Afridi et al. 2019; Gulati et al. 2008; Rungin et al. 2012). However to initiate these favorable plant–microbe interactions involved in promoting plant growth, successful endophyte colonization of the host plant is an essential step. Colonization involves a general series of phenomenon, including attachment, entry, mobilization, and establishment (Wilson et al. 2002) encompassing entry, growth, and multiplication of endophyte populations within the host plant. In addition due to their advantage of being comparatively insulated from the competitive, high-stress environment of the soil, endophytic bacteria may be of particular interest (Sturz et al. 2000). Therefore, a deeper comprehension of the patterns of endophytic microorganism's colonization as well as the persistence of these microorganisms in the rhizosphere and in plants is a crucial requirement for managing and developing efficient ways for sustainable agriculture.

Endophytes can significantly contribute to improving plant's tolerance to biotic and abiotic variables that leads to a great deal of economic loss and yield. Plants possess defense systems, which may be local owing to pattern-triggered immunity (PTI) or effector-triggered immunity (ETI), as well as systemic that includes salicylic acid (SA) signaling-based systemic acquired resistance (SAR) and ethylene (ET) and jasmonic acid (JA) signaling-based induced systemic resistance (ISR) (Arora et al. 2013; de Wit 2007; Muthukumar et al. 2017; Schuhegger et al. 2006). Modulations of these host defenses by endophytes are a necessary phenomenon to maintain symbiosis in host–endophyte interactions. Studies have also been focused on studying pathogen–microbe–plant interactions to understand the eliciting mechanisms in colonization and modulation of host defenses by these pathogenic microorganisms. Apart from these biotic influences, abiotic factors have also been increasingly affecting the plant health and its functioning. Unpredicted global climate change negatively influences plant growth and physiological activities by inducing multiple abiotic stresses and reduction in the quality of arable land and water, thus decreasing agricultural productivity (Abdelrahman et al. 2017). Additionally, the population growth forecasts of 9 billion by 2050 requiring a global food production increase of up to 70% further aggravate the problem, requiring thus an urgent need for alternative nonchemical-based biofertilizers and biopesticides. In this context, the symbiotic endophytes with beneficial properties ranging from nutrient mobilization to phytohormone release while also providing protection against phytopathogens can be a huge boost to plant health and sustainable agriculture development.

This chapter summarizes thus the diverse colonization patterns of endophyte involving several genetic and chemical signals facilitating the process, the beneficial attributes of these symbiotic endophytic microorganisms, modulation of host defense systems to maintain these interactions, as well as the functions of these beneficial microorganisms in the scenario of climate change and subsequent abiotic stress conditions.

10.2 Host–Endophyte Interactions and Molecular Signaling: Molecular and Chemical Signals for Successful Colonization

Colonization by endophytic microorganisms may follow several complex mechanisms; however, the fundamental hierarchy of colonization, viz., attachment [involving adhesins structures such as flagella, pilli, fimbrae, cell surface polysaccharides, EPS (exopolysaccharides) and LPS (lipopolysaccharides)], entry or penetration (involving celluloytic enzymes such as cellulases, endoglucanases, xylanases, and pectinases), and establishment (includes the host--endophyte-based biological interactions, viz., molecular, metabolic, and physical) remains general (Wilson et al. 2002). Attachment or adhesion of endophytes to the plant surfaces marks the initial step in the process of colonization. Studies have highlighted the role of bacterial endophytes-based exopolysaccharides (EPS) in aiding the attachment of these microbes to the root surfaces. In one such study involving rice root as colonization surface, the EPS produced by nitrogen-fixing endophytic bacterium *Gluconacetobacter diazotrophicus* strain Pal5 was reported to be an essential factor in the colonization process. The gumD gene homolog, which was predicted to be crucial for producing EPS, was mutated in the study, and the resulting mutant (MGD) of *Gluconacetobacter diazotrophicus* strain Pal5 performed normally in terms of growth and nitrogen fixation but was unable to produce EPS as a result, which prevented it from attaching to the root surface. However, EPS isolated from the culture supernatants of the wild-type strain PAL5 was successful in partially restoring wild-type biofilm production and plant colonization when added to the mutant MGD (C. H. S. G. Meneses et al. 2011). In addition, further studies also indicated the protective role of these EPS biomolecules in defending the endophytes from the host oxidative damage and elevated free radical levels (C. Meneses et al. 2017). Besides, lipopolysaccharides (LPS) have been demonstrated to be another essential component in colonization of Gram-negative bacterial endophytes. In one study, the rfbB and rfbC genes of *Herbaspirillum seropedicae*, which are involved in LPS production, were mutated. The mutant's ability to attach to maize root surface plantlets was 100-fold lower than that of the wild-type (Balsanelli et al. 2010). Later, it was postulated that maize root lectins bind to the N-acetyl glucosamine residues of the LPS O-antigen and are thus required for subsequent associative interactions and colonization (Balsanelli et al. 2013). Interestingly, a two-mode adsorption and anchoring phenomenon of *Azospirillum brasilense* strain Sp7 has been described in attachment of these plant growth-promoting microorganisms in wheat roots. Bacterial surface proteins like flagellin regulate the initial phase's weak, reversible, and unspecific binding, which creates the adsorption phase. The solid anchoring of adsorbed and free bacteria via bacterial extracellular polysaccharide constitutes the second phase (Michiels et al. 1991). Along similar lines, Rossi et al. (2016) presented the significance of flagellin-mediated adhesion to maize roots for *Azospirillum brasilense* in addition to its role in motility. Nonmotile cells were produced by *A. brasilense* mutants of the flagellin modification genes, flmA and flmB. These mutations also impacted *A. brasilense* cell's capacity to adhere to maize roots and to produce LPS and EPS.

After attachment, the next immediate step would be entry or penetration of endophytes into the host tissues. Previous investigations have revealed several potential entry points, which include apical root zone or lateral root emergence areas, stems, and leaves (Vaishnav et al. 2019; Yaron and Römling 2014; Zinniel et al. 2002). In contrast, the unfavorable conditions provided by the phyllosphere, including the fluctuations in UV light, lack of nutrients, and water availability, limit the distribution and frequency of microbial populations in the leaves (Compant et al. 2010; Hallmann 2001). Alternatively to counter this limitation, microbes may form large aggregate of leaf colonizing microorganisms, which are better suited to bear unfavorable conditions provided by the phyllosphere. A study demonstrated establishment of large aggregates of *Pseudomonas syringae* pv. *syringae* (Pss) by means of N-acyl-homoserine lactones (AHL)-dependent quorum-sensing phenomenon, which subsequently increased the fitness of these microorganisms as they were better adapted to desiccation in the leaf surface compared to solitary distributions (Monier and Lindow 2003). To date, two modes of entry in the rhizosphere compartment have been recognized, viz., active and passive penetration via thin-walled root surfaces on root hairs and lateral roots with cracks, respectively. While passive penetration occurs at cracks in the root surfaces, the active process occurs with the assistance of cellulolytic enzymes to hydrolyze exodermal cell walls of plants (Hardoim et al. 2015; Muthukumar et al. 2017). Reinhold et al. 2006 demonstrated the process of active penetration using a nitrogen-fixing endophyte *Azoarcus* sp. strain BH72 isolated from Kallar grass. Because of mutations in the endoglucanase gene eglA, *Azoarcus* sp. BH72 has altered endoglucanase activity, which resulted in significantly reduced intracellular colonization of root epidermis cells. This demonstrates an active bacterial colonization course and suggests that endoglucanase of *Azoarcus* sp. BH72 is a critical factor in successful endophytic colonization of rice roots. Additionally, previous studies have demonstrated how these hydrolytic enzymes help necrotrophic fungal endophytes to enter the host tissues (Lebeda et al. 2001; Talbot 2010).

Successful colonization of the endophytes is curtained by the compatible host–endophyte interactions involving molecular and chemical crosstalk between the two, which marks the establishment phase. Any interaction between a plant and a microbe primarily begins with a physical encounter, followed by a number of physical and chemical barriers that must be surmounted in order to successfully form an association. However, to maintain and colonize the host plant, the "balanced antagonism" hypothesis was put forth to explain how an endophyte avoids triggering the host's defenses, ensures self-resistance before being rendered ineffective by the host's toxic metabolites, and is able to develop inside its host without producing any obvious signs of infection or disease (Schulz et al. 1999). According to a preliminary study, the higher level of hydrolytic enzymes (pectinases) produced by the phytopathogen, *Erwinia carotovora* var. *carotovora* TN5, compared to non-pathogenic endophytes isolated from rice plants may account for the greater host defense system response, leading to phytopathogen biocontrol (Elbeltagy et al. 2000). However, further studies are required to explain the proposed hypothesis. In another study following the inoculation of pathogenic or non-pathogenic strains of *Fusarium oxysporum,* a differential accumulation of host's β-l, 3-glucanase and chitinase isoforms in tomato roots was observed. The accumulation of the aforementioned hydrolytic enzymes increase threefold in a pathogenic strain compared to a non-pathogenic strain, with the author highlighting the development of symptoms and vascular colonization as a key differentiating feature (Kim and Hwang 1994). The imbalance of antagonistic forces between a pathogen and a host can cause disease or the pathogen's death, depending on whether the balance is in the pathogen's favor or the host's. However, in the case of beneficial endophytes, which may also possess several virulence factors that are blocked by plant defense mechanisms, the link appears to be asymptomatic and avirulent if plant defense and endophytic virulence are in harmony (Kusari et al. 2012). Whole-genomic analysis of an endophyte *Enterobacter* sp. strain 638 isolated from poplar showed the presence of genes primarily involved in counteracting the plant defense mechanisms, including superoxide dismutase, hydroperoxide reductases, and catalases (Taghavi et al. 2010). Recent investigations using the endophyte bacterium *Bacillus subtilis* strain BSn5 revealed one key mechanism that inhibited the plant's defensive response by producing the lantibiotic subtilomycin to bind self-produced flagellin. Subtilomycin binds to flagellin and affects flg22-induced plant defense, which promotes the endophytic colonization in *Arabidopsis thaliana* and *Amorphophallus konjac* (Deng et al. 2019). ADP-ribosyltransferase activity of type III secreted effectors (T3SE) HopF2 expressed by *Pseudomonas syringae* was successful in inhibiting stomatal immunity and flg22-induced reactive oxygen species (ROS) in a different study, rendering *Arabidopsis thaliana* immunity ineffective (Hurley et al. 2014a).

Understanding colonization pattern, or the route of colonization, and the identification of terminus tissues in the host plant, can help one better comprehend the establishment and, more significantly, the colonization processes. The entrance of a pathogenic endophytic fungus, *Ralstonia solanacearum,* was described as occurring through lateral roots that reached the xylem vessels via the cortex (Álvarez et al. 2010). An interesting finding in the colonization pattern of *Burkholderia* sp. strain PsJN was observed. Utilizing gfp-marked strain of *Burkholderia* PsJN microscopic studies confirmed the entry points of this plant growth-promoting endophyte in grapevine (*Vitis vinifera* L.) to be primary and secondary roots and especially lateral root emergence sites or root tips, suggesting a passive entry of this endophyte. Further, it was also observed in rhizodermal cells, inter- and intracellular spaces of cortical cells, the endodermis, and xylem vessels. Differential existential pattern could be observed with significant accumulation in the substomatal chambers of grapevine leaves compared to the stems, thus suggesting the leaf as sink and stem as only transition in the intracellular route of colonizing microbes (Compant et al. 2005). In contrast, a study involving colonization of endophytes *Enterobacter ludwigii* strain EnVs6 and *Pantoea vagans* strain PaVv7 in grapevine (*Vitis vinifera* L.) demonstrated similar colonization pattern, including primary and secondary roots, root tips, rhizodermal cells, and cortex up to xylem vessels; however, they were not observed in systemic aerial parts of the plants. Based on the observed phenomenon, the author explained that the two endophytes utilized may not be a systemic colonizer as *Burkholderia* sp. strain PsJN or subsequently more time would be necessary for the movement to the aerial parts (Lòpez-Fernàndez et al. 2016). In another study, the colonization progression of *Paenibacillus polymyxa* in the roots of *Arabidopsis thaliana* was studied. The study corroborated previous findings of root tips as the entry or initiation sites for colonization. In addition, the study demonstrated that the colonized microbes were primarily localized in the intercellular spaces of the root tissue outside the vascular cylinder (Timmusk et al. 2005).

There are currently two contrasting theories for the origin and colonization of endophytic microorganisms in host plants. One stands for the independence of host plant in patterning the colonization and existence of these ubiquitous microorganisms, and the other believes in the role of host plants and their secondary chemistry in the colonization interaction between the endophytes and host plants. Numerous types of research using the *Arabidopsis thaliana* plant to examine the endophyte community structure have demonstrated the influence of soil on the type of endophytes that colonize *A. thaliana* roots. In one such study, the root endophytic bacterial assemblage showed significant dependence on the two soil types, viz., Cologne or Golm soils, highlighting the soil-type-dependent root colonization processes. This study thus classified soil as an important variable affecting the quantitative and qualitative composition of the *Arabidopsis* root-inhabiting microbiota, in contrast to the host genotype, which was found to have a limited impact on the root endophyte profile (Bulgarelli et al. 2012). In parallel to previous findings, endophytic community in *Arabidopsis thaliana, Arabidopsis lyrata,* and *Arabidopsis halleri* were highly influenced by environmental conditions such as soil types, soil batches, and controlled and noncontrolled climates instead of the host's genome. Interestingly, *Cardamine hirsuta* and *Arabidopsis thaliana* also presented similar community distribution of core bacterial phyla, including *Actinomycetales, Burkholderiales,* and *Flavobacteriales* based on environment-dependent establishments (Schlaeppi et al. 2014). In another study, environmental factors like drought, low nutrients, and shade directed a significant effect on the colonization of endophytic fungi *Serendipita herbamans* sp. nov on invasive knotweed (*Reynoutria* ssp.). While *S. herbamans* strongly colonized invasive knotweed in low-nutrient and shaded conditions, colonization under drought conditions was much lesser (Garnica et al. 2022). Furthermore, in one study it was observed that factors such as rainfall and phenological stage had a significant influence on the distribution of the endophytic fungi community in the leaves of sorghum (*Sorghum bicolor* L. Moench).

In contrast to these findings, a widely accepted view about the colonization of endophytic communities is that the host holds substantial impact on the endophytic community structure. In one study involving four *Psychotria* species (*Psychotria gracilenta, P. capitata, P. acuminata,* and *P. marginata*), a highly significant influence of host species on endophytic fungi community was observed. Based on Bray–Curtis dissimilarity in fungal community composition, the highest dissimilarity of 86% between *P. gracilenta* and *P. capitata* was observed. In addition, the study found associations between metabolic profile differences between the host (all the four *Psychotria* species and *P. horizontalis*) and the structure of leaf endophytic communities (Christian et al. 2020). Host plant's hormone signaling is an additional mechanism that affects the colonization of endophytic bacterial communities. A study demonstrated the negative role of plant hormone such as ethylene and salicylic acid, which activates the plant defense mechanism and reduces the colonization of endophytic bacteria. In contrast to the wild-type plant, *Klebsiella pneumonia* strain 342 was shown to be hypercolonized on an ethylene-insensitive mutant of the host *Medicago truncatula* (Iniguez et al. 2005). Along similar lines, another plant hormone jasmonic acid addition proved crucial for rice defense mechanisms and stress responses as it reduced the colonization by the diazotroph *Azoarcus* sp., indicating that plant defense mechanisms may be in charge of regulating endophytic colonization (Miché et al. 2006) (Table 10.1).

Several reports have demonstrated the influence of host exudates in promoting microbial colonization of host tissue. Plants perceive flavonoids as allowing symbiotic infection of the root by inducing transcription of the genes for the biosynthesis of the rhizobial signaling molecules known as Nod factors. In addition, flavonoids have also demonstrated their effectiveness in serving as a chemoattractant for the endophytic symbionts. Dharmatilake and Bauer 1992 demonstrated three main nod inducers (viz., 4′,7-dihydroxyflavone, 4′,7-dihydroxyflavanone, and 4,4′--dihydroxy-2-methoxychalcone) secreted in alfalfa root exudates each worked well as chemoattractants for *Rhizobium meliloti* even at low concentrations. It is interesting to note that in one study endophytic fungus *Phomopsis liquidambaris* strain B3, which increased the root exudates in plants, was used to considerably improve the symbiotic relationship between rhizobia and peanut (*Arachis hypogaea* L.). In peanut plants inoculated with *Phomopsis liquidambaris* strain B3, the root exudates were increased 50-fold, which considerably boosted chemotaxis by 11.6–36.4%, rhizobial biofilm production by 13.2–34.0%, and growth by 12.0–51.2% (Wang et al. 2021). Along similar lines, the effects of endophytic actinobacterial strains *Microbispora* sp. strain CP56, *Actinomadura* sp. strain CP84B, and *Streptomyces* spp. CP200B and CP21A on the chickpea-*Mesorhizobium* symbiosis were examined. The results demonstrated enhanced rhizobial growth, early nodule development (rhizobial chemotaxis, biofilm formation, and nod gene expression), increased secondary root formation, and upregulated expression of nif and genes involved in flavonoid synthesis (PAL: phenyalanine ammonia lyase; CHS: chalcone synthase) (Xu et al. 2022). Amino acids are another major component observed in the root

Table 10.1 Major genes involved in facilitating colonization of endophytes in different plant systems

Gene involved in colonization	Colonizing microorganism	Plant colonized	Colonizing parts	References
Outer membrane porin F (Oprf)	*Pseudomonas fluorescens*	Wheat	Roots	De Mot and Vanderleyden (1991)
Flagellin epimerase/ dehydratase and NAD (P)dependent—Amino-transferase (Flma or Flmb genes)	*Azospirillum brasilense*	Maize	Roots	Rossi et al. (2016)
Chemotaxis sensory protein (Ctaa, Ctab, and Ctac)	*Pseudomonas fluorescens* strain Pf0–1	Tomato	Roots	Oku et al. (2012)
Oxalate decarboxylase (Oxc, Bphyt_6740)	*Burkholderia phytofirmans* strain PsJN	White lupin and maize	Roots	Kost et al. (2013)
Rhamnose biosynthesis (Rfbb and Rfbc)	*Herbaspirillum seropedicae*	Maize	Roots	Balsanelli et al. (2010)
Type IV pilus retraction protein (Pilt)	*Azoarcus* sp. strain BH72	Paddy	Roots	Böhm et al. (2007)
Nodulation factor (Nodc)	*Rhizobium* sp.	Peanut	Roots	Wang et al. (2021)
Gum cluster (Gumd)	*Gluconacetobacter diazotrophicus* strain PAL5	Paddy	Roots	Meneses et al. (2011)
Flagellin (Flic3)	*Azoarcus* sp. strain BH72	Paddy (*Oryza sativa* subsp. indica cv. IR36)	Roots	Buschart et al. (2012)
Diguanylatecyclases (DGC2884)	*Pantoea* sp. strain YR343	Wheat (*Triticum aestivum*) and *Populus trichocarpa* strain BESC819	Roots	Bible et al. (2021)
Methyl-accepting chemotaxis proteins (MCPs) CtaA, CtaB, and CtaC	*Pseudomonas fluorescens* strain Pf0–1	Tomato	Roots	Oku et al. (2012)
Thpg1 (endopolygalacturonase)	*Trichoderma harzianum* strain T34	Tomato	Roots	Morán-Diez et al. (2009)
Endoglucanase (Egla)	*Azoarcus* sp. strain BH72	Kallar grass and rice	Roots	Reinhold et al. (2006)
Type III effector (Hopf2)	*Pseudomonas syringae*	*Arabidopsis thaliana* ecotype Col-0	Leaves	Hurley et al. (2014a, b)

exudates representing the chemotactic ligands for chemotaxis sensory proteins, including methyl-accepting chemotaxis proteins (MCPs)-CtaA, CtaB, and CtaC in *Pseudomonas fluorescens* strain Pf0-1. In mutation-based studies conducted by Oku et al. 2012, the triple mutants of CtaA, CtaB, and CtaC demonstrated weak chemotactic responses to cysteine and proline and no responses to the other 18 amino acids compared to the wild-type strain, suggesting its role in chemotactic sensory responses. Further research revealed that CtaA and CtaB are the principal MCPs for amino acids and root exudate, and that CtaC contributes less to chemotaxis to amino acids and root exudate, as seen by the double mutant's (including CtaA and CtaB) significantly reduced chemotactic response compared to the triple mutants. Studies on cluster root exudates highlighted fairly significant concentrations of organic acid ions such as citrate and oxalate (Dessureault-Rompré et al. 2007). Interestingly, *Burkholderia* species that are beneficial to plants showed evidence of a characteristic known as oxalotrophy, or the capacity to utilize oxalate as a carbon source. This property of oxalate metabolism was utilized in a colonization-based study of *Burkholderia phytofirmans* strain PsJN in white lupin and maize plants where the mutants (mutation in oxalate decarboxylase (oxc, Bphyt_6740) involved in oxalate degradation) displayed decreased colonization compared to wild-type strains (Kost et al. 2013). Recently, strigolactones found in root exudates of host plants have also been found to promote colonization of endophytic microorganisms, especially the mycorrhizal fungi. The specific mycorrhizal components, which are strigolactone-enhanced short-chain chitin oligomers, activate the symbiotic signaling pathway (SYM), which favorably affects colonization (López-Ráez et al. 2017). Despite numerous studies that focus on a single factor, it is safe to say that colonization is a complicated process that acknowledges a variety of factors into account. Endophyte's ability to successfully colonize a plant depends on a variety of variables, including the plant's tissue type, genotype, microbial taxon and strain type, and biotic and abiotic environmental conditions (Hardoim et al. 2015).

10.3 Endophytes and their Beneficial Plant Growth-Promoting Attributes

The production or stimulation of various types of phytohormones, supply, or facilitation of nutrients such as nitrogen, phosphorus, and necessary minerals, such as iron, as well as the regulation of plant ethylene levels through ACC deaminase activity, are the direct mechanisms used to assist plants with growth improvement, whereas the biocontrol aspect of microbial inputs and tolerance to abiotic factors is considered to be one of the indirect mechanisms (Arora et al. 2013; Shabanamol et al. 2018).

10.3.1 Direct Mechanisms of Plant Growth Promotion

10.3.1.1 Biological Fixation of the Atmospheric Nitrogen

Despite being the biggest limiting factor for plant development, atmospheric nitrogen (N_2) cannot be reduced directly by plants. Chemical N fertilization is therefore used commercially to supply nitrogen nutrition to the plant. However, decades of continuous and excessive application have had a severe impact on the soil fertility, the quality of the surface and groundwater, and the emission of greenhouse gases like carbon dioxide and nitrogen dioxide. In light of this, endophytic diazotrophic microorganisms that have biological nitrogen-fixing properties can be a sustainable alternative being both cost- and environmentally efficient (Bhattacharjee et al. 2008). Numerous studies have detected endophytic bacterial diazotrophs of several genera *Pseudomonas, Stenotrophomonas, Xanthomonas, Acinetobacter, Rhanella, Enterobacter, Pantoea, Shinella, Agrobacterium,* and *Achromobacter* (Taulé et al. 2012), *Herbaspirillum* sp. *strain B501* (Elbeltagy et al. 2001), *Novosphingobium sediminicola* and *Ochrobactrum intermedium* (Muangthong et al. 2015),and *Lysinibacillus sphaericus* (Shabanamol et al. 2018) isolated from several plant species. Yimer (2019) described nitrogen-fixing microorganisms into two major categories: symbiotic nitrogen-fixing organisms that form specialized nitrogen-fixing structures called nodules, such as members of the rhizobiaceae family, which symbiotically coexist with both non-leguminous plants and leguminous plants, including *Frankia* and *Rhizobia,* respectively, and non-symbiotic nitrogen fixers, which may be associative, free living, and endophytic in nature. These include members of cyanobacteria (*Anabaena, Nostoc*), *Azospirillum, Azotobacter, Gluconoacetobacter diazotrophicus,* and *Azocarus*. In recent investigations, however, diverse populations of non-rhizobial endophytic bacteria have been isolated, along with predominant rhizobial population. Other non-rhizobial endophytic bacteria belonging to the *Pseudomonadaceae* (19.8%), *Enterobacteriaceae* (11.8%), and *Paenibacillaceae* (5.3%) families were also found in one study, in addition to the prevalent *Bradyrhizobiaceae* present at 66% in the soybean root nodule (Sharaf et al. 2019). While several endophytic bacteria with the ability to biologically fix nitrogen in crops have been researched, the genetic and physiological mechanisms of endophytic nitrogen fixing are still poorly understood (Nag et al. 2020).

10.3.1.2 Phosphate Solubilization

Along with nitrogen, phosphorus represents an important macronutrient essential for plant growth (Ezawa et al. 2002; Takahashi and Anwar 2007). Due to the distinctive presence of phosphorus in the soil in fixed forms, such as calcium phosphates in alkaline soils and aluminum or iron phosphates in acidic soils, phosphate-solubilizing microorganisms play a crucial role in transforming these insoluble forms of phosphate into forms that plants may use for assimilation (Oteino et al.

2015; Shiferaw et al. 2013). Studies have highlighted that the population of phosphate solubilizers in the roots (endophytes) and near the roots (rhizospheric microorganisms) is always higher, especially in phosphate-deficient soils (Aranda et al. 2011; Gulati et al. 2008; Sharma et al. 2007). These phosphate-solubilizing microorganisms play a crucial role in biogeochemical cycle of phosphorus through mineralization of inorganic and organic forms of phosphates in the soil. The underlying mechanisms include production of acid phosphatases and phytases for solubilization of organic phosphates, whereas organic acids such as gluconic acid, sulfuric acid, and nitric acid have been reported to perform solubilization of inorganic phosphates (Goldstein 1994; Nannipieri et al. 2011). In one study, the endophytic bacteria *Pseudomonas* strain L321 was capable of producing medium to high levels of gluconic acids involved in solubilization of inorganic phosphates and also leads to significant growth promotion of *Pisum sativum* L. var. *Early Onward* compared to uninoculated plants under phosphate-limiting conditions (Oteino et al. 2015).

10.3.1.3 Production of Phytohormones

Apart from solubilization and acquisition of nutrients for the host plants, endophytes can also produce plant hormones with potential role to enhance plant growth (Naureen et al. 2022; Patten and Glick 2002; Sirrenberg et al. 2007). Production of these phytohormones from rhizospheric microorganisms, including rhizospheric bacteria, has been widely reported, and it is believed that up to 80% of the rhizospheric bacteria can produce indole acetic acid (Patten and Glick 1996). As a significant plant hormone, indole acetic acid (IAA) plays a variety of physiological tasks in plants, including modulating cellular signaling, regulating plant development, and inducing plant defense (Glick 2012; Gravel et al. 2007; Spaepen et al. 2007). Apart from IAA, other phytohormones produced by endophytes such as abscisic acid (ABA), cytokinins, and gibberellins (GA) have also been reported (Bhore et al. 2010; Bilal et al. 2018; Cohen et al. 2009).

10.3.1.4 ACC Deaminase Activity

Ethylene, apart from being an important plant hormone, is also a stress hormone that may have negative implications on plant growth and development (Dubois et al. 2018); however, endophytes with potential 1-aminocyclopropane-1-carboxylic acid (ACC) deaminase activity may control the levels of stress hormone by converting precursor of ethylene, ACC, into α-ketobutyrate (Gamalero and Glick 2015). In one intriguing study, the treatment of French bean seedlings with ACC-producing bacterial consortia of *Aneurinibacillus aneurinilyticus* strain ACC02 and *Paenibacillus* sp. strain ACC06 resulted in a significant reduction in salinity stress-stimulated ethylene levels (60%), as well as increases in root length (110%), root fresh weight (45%), shoot length (60%), and shoot fresh weight (255%), as well as root biomass (220%), shoot biomass (425%), and total chlorophyll content (57%) (Gupta and Pandey 2019). Another study found that two salt-tolerant endophytes with ACC deaminase activity, *Kocuria rhizophila* strain 14ASP and *Cronobacter sakazakii* strain OF115, increased wheat plant growth in terms of fresh and dry root masses, shoot length, and leaf count, as well as higher relative water contents, K+/Na + ratios, chlorophyll contents, and antioxidant activities under salt stress conditions compared to control plants. These studies thus demonstrated the role of these plant growth-promoting endophytes in influencing plant growth under stress conditions (Afridi et al. 2019).

10.3.1.5 Production of Siderophores

Iron (Fe) is the most essential microelement and also acts as a cofactor for several enzymatic reactions. In aerobic conditions, iron exists predominantly as Fe^{3+} and forms hydroxides and oxyhydroxides that are both insoluble forms of iron, which cannot be assimilated by the plant (Lindsay and Schwab 1982). Thus, in order to solubilize iron from organic compounds or minerals, chelating agents called siderophore, which are made of protein molecules with low molecular weight, are crucial. These molecules have been identified in bacteria, fungi, and graminaceous plants (Hider and Kong 2010). Interestingly, one study confirmed that the sole factor in driving plant growth promotion was siderophore activity of the endophyte *Streptomyces* sp. strain GMKU 3100 isolated from *Oryza sativa L. cv. KDML 105*. The treatment of rice and mugbean plants with mutant (desD gene mutations abolishing the siderophore production) and untreated controls produced less effective results in terms of plant growth parameters like root and shoot biomass and lengths compared to the wild-type with functional siderophore activity (Rungin et al. 2012). In addition, siderophore-producing microorganisms remove iron from the environment and restrict the accessibility of iron to phytopathogens, protecting the plant from pathogens in the process (Buysens et al. 1996). In recent studies, Arya et al. (2018) demonstrated the effectiveness of two siderophore-producing *Pseudomonas fluorescens* strains SPs9 and SPs20 in disease control of 100% and 75%, respectively, against tomato wilt pathogen *Fusarium oxysporum* f. sp. *lycopersici*.

10.3.2 Indirect Mechanisms of Plant Growth Promotion

By creating a competitive environment for getting nutrients, secreting antimicrobial substances, and generating systemic resistance in the host plants, microorganisms indirectly enhance plant health (Sangal et al. 2018). As previously discussed, siderophore production is an important mechanism for biocontrol activity by generating a competitive environment for the phytopathogen survival. In addition, production of hydrogen cyanide (HCN) can also be an important mechanism of pathogen inhibition. In one study, four unidentified endophytic bacteria MJiN9,

MJiR4, MJiR9, and MJiR13 isolated from mungbean (*Vigna radiata*) were all screened positive as HCN producers. In particular, *Fusarium oxysporium* and *Aspergillus niger*, two phytopathogenic species, were antagonistically affected by the best HCN producer, MJiR9 (Bhutani et al. 2018). Secondly, antibiosis or production of antibiotics by endophytic microorganisms is a major mechanism used by these biocontrol agents to antagonize phytopathogens. Endophytes-based antibiotics such as pyrrocidines A and B, pyrrolnitrin, and 2, 4-diacetylphloroglucinol (DAPG) have shown significant inhibitory activity against *Aspergillus flavus* and *Fusarium verticillioides, Fusarium moniliforme,* and *Ralstonia solanacearum,* respectively (Ramesh et al. 2009; Rodrigo et al. 2007; Wicklow et al. 2005). Thirdly, to defend their host plants from bacterial, viral, and fungal phytopathogens, beneficial endophytes can potentially employ a technique known as induced systemic resistance (ISR) (Alvin et al. 2014). Salicylic acid (SA), jasmonic acid (JA), and ethylene (ET)-mediated pathways, which are often a network of related signaling pathways involved in ISR induction, can be used by endophytic microorganisms to initiate ISR (Pieterse et al. 2012). Studies have shown involvement of both SA and JA/ET pathways in ISR-based defense responses (Conn et al. 2008; Niu et al. 2011). The modulation of these defense responses in plants driven by endophytes has been discussed below.

10.4 Endophytes Modulate Host Defense Mechanisms under Biotic Stress Conditions

Plants that come into contact with microorganisms may develop local defensive mechanisms based on the recognition of either microbe-associated molecular patterns (MAMPs or PAMPs), or effector proteins known as pattern-triggered immunity (PTI) or effector-triggered immunity (ETI), respectively. Through transmembrane pattern recognition receptors (PRRs), plants may identify microbial- or pathogen-associated molecular patterns (MAMPs or PAMPs), such as bacterial flagellin and fungal chitin. This action sets off the initial line of immune defense, known as pattern-triggered immunity (PTI). However, pathogens can still evade the PTI process by either suppressing PTI signaling through virulent factors or avoiding recognition of pattern recognition receptors (PRRs) present in plant membranes (Guo et al. 2009). Remarkably, type 3 secretion systems (T3SS) in bacterial pathogens have been studied for their potential to suppress PTI-based immune protection by targeting membrane-bound PRRs and their co-receptors and downregulating PAMP-triggered defense response (Macho and Zipfel 2015). In order to thus stop the spread of the infection, successive effector-triggered immunity (ETI) detects the pathogen effectors and causes a hypersensitive reaction (HR) (Jones and Dangl 2006). Furthermore, plant tissues far from infection sites may develop systemic resistance, which includes systemic acquired resistance (SAR)-resistance response linked to pathogens requiring salicylic acid (SA) signaling and induced systemic resistance (ISR)-resistance response generated due to beneficial microorganisms, which requires ethylene (ET) and jasmonic acid (JA) signaling instead (Mengistu 2020; Vandana et al. 2021).

Plants have the intrinsic capacity to detect and identify potentially invasive microbes and to launch defense mechanisms (de Wit 2007). It is noteworthy to understand that the immunological and symbiotic processes promoted by plant cell-surface receptors have similar yet distinct recognition events (Zipfel and Oldroyd 2017). Arbuscular mycorrhizal and rhizobial-based interactions with the plant have shown to need the association of SYMRK/DMI2 kinases with the LRR-receptor and subsequent interactions with the 3-hydroxy-3-methylglutaryl-CoA reductase (HMGR) involved in mevalonate pathway (Stracke et al. 2002; Venkateshwaran et al. 2015). In spite of the fact that the microbial signals, such as chitin in pathogenic fungi and peptidoglycan in pathogenic bacteria, are similar to the Myc factors signals and the Nod factors for arbuscular mycorrhizal symbioses and root–nodule symbioses, respectively, the lipid modifications on the chitin-oligosaccharide backbone of these symbiotic signals, which transform them into acyl lipo-chitooligosaccharides (LCOs), can be an important differentiating factor (Oldroyd 2013). Research studies have demonstrated that beneficial endophytes can suppress the plant defense response during rhizobium–legume symbiosis (Gourion et al. 2015). In one study involving tomato cell line Msk8, the highly sensitive perception for flagellin from phytopathogen *Pseudomonas syringae* pv *tabaci* strain 511 was demonstrated due to the opposing characteristics of the flagellin structure, where the N terminal displayed the very potent elicitor of necrotic responses while the peptides (flg15-Δ4 to flg15-Δ7) truncated at C terminus displayed an antagonistic activity against response induction. Remarkably this flagellin peptide flg15-Δ5 displayed its competitive antagonistic activity against induction of necrotic response for other elicitor-active phytopathogen preparations, including *Escherichia coli, Erwinia carotovora, Erwinia chrysanthemi, Pseudomonas syringae, P. aeruginosa,* and *P. fluorescence*. In addition, oxidative burst profiles for tomato leaves treated with flagellin peptide flg15-Δ5 of symbiotic endophytes such as *Agrobacterium tumefaciens* and *Rhizobium meliloti* were observed; however, a rapid oxidative burst was observed after treatment with flg15 from the phytopathogen (Felix et al. 1999). In a different study, MAMP signaling in *Arabidopsis* roots under the influence of the flagelin Flg22 was found to play a significant role in the production of the antibiotic camalexin, which protects against necrotrophic pathogens like *Botrytis cinerea, Alternaria brassicicola,* and *Rhizoctonia solani* and causes the deposition of callose. However, the beneficial endophyte *Pseudomonas fluorescens* strain WCS417r could suppress the MAMP signaling owing to the secretion of low molecular compounds ultimately promoting successful root colonization protecting itself from the MAMP-elicited antimicrobials and consequently inducing systemic resistance in the host plant (Millet et al. 2010) (Table 10.2).

Table 10.2 Different classes of host plant-based receptor kinase complex and involved microbe molecular pattern. Plants use receptor kinases or receptor-like proteins with extracellular domains bearing the lysin motif (LysM) to perceive GlcNAc-containing microbial compounds such as chitin and LCOs. For proteinaceous elicitors like flg22 from flagellin, there is LRR-type extracellular receptors (Zipfel and Oldroyd 2017)

Host plant	Microbe molecular pattern involved	Microbes-associated receptors types in plants
Arabidopsis thaliana	Chitin	LysM-receptor kinases-CERK1 and LYK5
Rice	Chitin	LysM-receptor-like protein-CERK1 and CEBiP
Arabidopsis thaliana	Chitin	LysM-receptor kinases-CERK1 and LYM2
Lotus	Nod-LCOs	LysM-receptor kinases-NFR1 and NFR5
Medicago	Nod-LCOs	LysM-receptor kinases-LYK3 and NFP
Parasponia andersonii	Myc-LCOs	LysM-receptor kinase- NFP and NFR5
Arabidopsis thaliana	flg22	LRR-receptor kinase-FLS2 and BAK1

10.4.1 Role of Quorum Sensing in Modulation of Host Defense Mechanisms

Quorum sensing (QS) is the name given to the phenomena whereby bacteria can detect and react to self-generated signal molecules to coordinate their behavior in response to population size (Fuqua et al. 1994). A quorum signal involving N-acyl-homoserine lactones (AHL), the most extensively researched of the quorum signals, is induced when the population of Gram-negative bacteria reaches a certain threshold. This signal has been extensively investigated for its role in gene expression changes, protein profile alterations, root development modification, symbiosis, motility, and biofilm formation (Lugtenberg 2015; Schuhegger et al. 2006). Furthermore, studies reflect the dependency of AHLs' chain length for its biological performance while short chains AHLs (4–10 carbons) were involved in plant growth-promoting activity and longer chains (12–16 carbons), on the other hand, were involved in disease resistance against numerous phytopathogens (von Rad et al. 2008; Zarkani et al. 2013). Interestingly, the roles of AHLs in host defense mechanisms have now been widely explored. In one study, *Serratia liquefaciens* strain MGl, the wild-type AHL-producing strain, was compared with AHL-negative strain *Serratia liquefaciens* strain MG44 for its systemic resistance induction in tomato plants against *Alternaria alternata*. The results demonstrated no discernible differences in the two strain's ability to colonize tomato roots, but the former showed more than 70% reduction in necrotic cell death after *A. alternata* infection of the leaves, while the latter showed no significant differences in its ability to control the pathogen growth. Furthermore, the study also validated the role of AHL-producing strain *Serratia liquefaciens* strain MGl in the stimulation of salicylic acid (SA) levels, which showed a threefold elevation compared to the AHL-negative strain MG44 (Schuhegger et al. 2006). Another study found that the application of oxo-C14-HSL (long chain AHL) led to cell wall reinforcements that included callose deposition, phenolic compounds, and lignins both locally and distantly, reflecting the role of these quorum signal molecules in influencing systemic host defense and strengthening resistance in *Arabidopsis thaliana* against bacterial pathogens. 12-oxo-phytodienoic acid (cis-OPDA) and SA were found to be the major drivers in this stress response, indicating the involvement of salicylic acid/oxylipin pathway in oxo-C14-HSL-based resistance development in *Arabidopsis thaliana* (Schenk et al. 2014). Interestingly, in one elaborative transcriptomic study, researchers found that two QS-deficient strains, viz., AHL-deficient strain (PA23–6863-lacking all the QS system) and phzR mutant (PA23phzR-lacking only PhzRI quorum sensing (QS) system) of *Pseudomonas chlororaphis* PA23 showed a differential expression pattern in the context of secondary metabolites compared to the wild PA23 strain. Exoprotease, hydrogen cyanide, and phenazine (PHZ, which are necessary for the production of antibiotics) all showed ≥ 3.42 $\log_2$fold lower expression levels than the wild PA23 strain. Similar to PA23phzR and PA23–6863, antibiotic pyrrolnitrin (PRN) biosynthesis gene expression was downregulated by 1.93 and 4.17 $\log_2$fold, respectively, compared to the wild PA23 strain. Furthermore, the study evidenced approximately 13% of the PA23's genome to be controlled by QS (Shah et al. 2020). Contrary to utilizing quorum sensing as a biocontrol tool in inhibiting phytopathogens growth in host plants, recent studies involve novel strategies to combat plant diseases, including quorum quenching (QQ) and AHL activity interference in bacterial phytopathogens. This novel mechanism interferes with the formerly useful QS systems; however, this interference and quenching are targeted for bacterial pathogens instead. *Pseudomonas segetis* strain P6 demonstrated considerable QQ activity in a recent investigation against a variety of synthetic AHLs (including C4-HSL, C6-HSL, 3-O-C6-HSL, C8-HSL, 3-O-C8-HSL, C10-HSL, 3-OH-C10-HSL, C12-HSL, and 3-O-C12-HSL), as well as important plant bacterial pathogens such *Dickeya solani, Pectobacterium atrosepticum, P. carotovorum* subsp. *carotovorum,* and *P. syringae* pv. tomato. Furthermore, co-culture experiments highlighted that P6 strain's AHL degrading activity caused reduction of virulence factors expression in *Dickeya solani, Pectobacterium atrosepticum,* and *P. carotovorum* subsp. *carotovorum* (Rodríguez et al. 2020).

10.4.2 Host Defense-Related Transcriptional Alterations Brought on by Interactions Between Plants and Microbes in Plant Cells

A shift in the expression pattern of key genes involved in plant core metabolic pathways is a defining characteristic of interactions between plants and their endophytes. According to numerous findings, the inoculation of beneficial endophytes has an impact on the host plant's defense mechanisms. In one such study involving interaction between biocontrol fungus *Trichoderma harzianum* and *Arabidopsis thaliana,* the downregulation of systemic acquired resistance (SAR)-responsive genes, including PR-1 (pathogenesis-related 1), FMO1 (flavin monooxygenase 1), WRKY transcription factor, and two glutation transferases putatively implicated in SAR response, was shown by microarray hybridization tests. Unexpectedly, the jasmonic acid-induced genes Lox1, ETR1, or PDF1.2 were similarly downregulated, despite the fact that *Trichoderma* and other helpful microbes are thought to defend plants from phytopathogens by way of an induced systemic resistance (ISR)-based defense system (Morán-Diez et al. 2012). Another study using transcriptome analysis showed that major plant defense hormones like ethylene, jasmonic acid, and salicylic acid were stimulated as a result of the interaction between PGP *Bacillus cereus* strain CGMCC 5977 and *Nicotiana tabacum.* All anticipated LOX genes, lipoxygenases (LOX), with the exception of Nitab4.5, and all MAPKs were highly elevated for JA-based signaling, as were 98.7% of the genes connected to the ethylene signaling pathway. For SA-based signaling, there was a considerable induction of the anticipated phytoalexin deficient-4 proteins pathogenesis-related protein 1 (PR1) and beta-1, 3-glucanase (BGL). In addition, upregulation of genes involved in carbohydrate metabolism pathways was observed with further validation with reduction of starch content of *B. cereus*-treated leaves (Li et al. 2020). One study demonstrated the biocontrol activity of *Bacillus altitudinis* strain JSCX-1 against *Phytophthora sojae* Kaufmann & Gerdemann in the soybean host plant (*Glycine max* (L.) Merr.). JSCX-1's biocontrol efficiency against *Phytophthora sojae* was established in in vitro greenhouse studies to be 49.28 ± 3.42%. In order to comprehend the likely pathway of the defense response in the host plant, additional investigations based on transcriptional expression analysis with regard to the regulation of plant defense-related PR1a, LOX, and ERF genes specific to *G. max* were carried out. However, the genes LOX and ERF related to induced systemic resistance (ISR) involved in jasmonic acid (JA) and/or ethylene (ET)-signaling pathways were not significantly impacted by the microbial treatment. It is interesting to note that biocontrol inoculation upregulated PR1a involved in salicylic acid-regulated systemic acquired resistance (SAR) defense response in plants. Negative crosstalk between GmPR1a caused the JA-responsive gene GmLOX to be suppressed. The study demonstrated that an SA-dependent signaling pathway may be in charge of controlling the disease resistance brought on by JSCX-1 (X. Lu et al. 2017).

To date, we have studied the various gene expression patterns based on beneficial or commensal microbe–plant interactions; however, studying also the pathogen–microbe plant interactions would be necessary to understand the eliciting mechanisms driving pathogenicity in host plants as well as drawing inferences related to host defense modulations. Salicylic acid-based signaling is a major plant hormone signaling pathway that has a crucial influence on developing resistance and defense response in plants against phytopathogens. However, pathogens have demonstrated intricate mechanisms to downregulate SA-based signaling using microbial effectors. Several studies have identified the major immune regulator non-expressor of pathogenesis-related 1 genes (NPR1), in a variety of immunological signaling pathways, including basal defense, SAR, induced systemic resistance, and ETI, as well as in regulating crosstalk between SA and other phytohormones, to be crucial for SA signaling (Pajerowska-Mukhtar et al. 2013). Following this impression, Liu et al. (2014) demonstrated the role of isochorismatases (ISCs) in suppressing salicylate-mediated innate immunity in plants. Salicylic acid is biosynthesized in chloroplasts of plants via two pathways: the isochorismate pathway, where chorismate is first converted to isochorismate and then to salicylic acid; and the phenylalanine ammonia-lyase (PAL) pathway, where phenylalanine is converted by PAL to trans-cinnamic acid, which then serves as a precursor for various routes of SA biosynthesis. The study identified two genes from phytopathogens, *Verticillium dahliae* (VdIsc1) and *Phytophthora sojae* (PsIsc1), whose putative encoded proteins showed substantial homology with known ISCs and displayed ISCs' activity, i.e., hydrolyze host isochorismate. This novel class of effectors was very influential in destroying the precursor of plant defense-signaling hormone salicylic acid. While the overexpressed transformants drastically reduced the levels of SA and SAG in soybean seedlings, the PsIsc1 gene silenced transformants of *P. sojae* raised the concentrations of both free SA and its glucoside (SAG) by threefold. Also, the relative expression of soybean PR1 transcripts was significantly decreased in overexpressed transformants. In another study, researchers demonstrated the activity of salicylate hydroxylase in degradation of SA. Shy1, a cytoplasmic salicylate hydroxylase extracted from *Ustilago maydis,* was transcriptionally upregulated during pathogenic development of the smut fungus *U. maydis*. However, compared to the wild-type CL13 solo pathogenic *U. maydis* strain, Shy1-negative mutants showed no significant reduction in the virulence symptoms using maize seedlings (Rabe et al. 2013). Interestingly, numerous studies have shown that pathogens can take advantage of the antagonistic crosstalk between the SA and JA pathways and increase plants' vulnerability to pathogenic or biotrophic infections. One study found interactions between the pathogenic *Hyaloperonospora arabidopsidis* RxLR effector protein, HaRxL44, and mediator subunit 19a (MED19a), in addition to the fact that it destabilized MED19a through proteasome-dependent degradation, shifting the balance from SA-based defense to JA/ET-based defense response in *Arabidopsis thaliana* (Caillaud et al. 2013). Apoplastic cysteine proteases, including PIP2, RCR3, CP1A, CP2, and XCP2, are positive regulators of the SA-induced plant defense response and function by inducing PR-1 gene expression. Alternatively, effectors targeting these cysteine proteases have also been reported, and these include AVR2, EPIC1, EPIC2B, Pit2, and CCP (Mueller et al. 2013; Rooney et al. 2005; Tian et al. 2007; van der Linde et al. 2012). Although there is no direct evidence of any effector interacting with PR-1 protein, a study has generated an indirect way of effector-PR-1 interaction where certain pathogenic effectors like syringolinA (SylA) and type III effector XopJ have been shown to prevent proteasome-mediated degradation of NPR1, thus influencing the immune function by compromising NPR1 turnover (Schellenberg et al. 2010; Üstün et al. 2013) (Fig. 10.1).

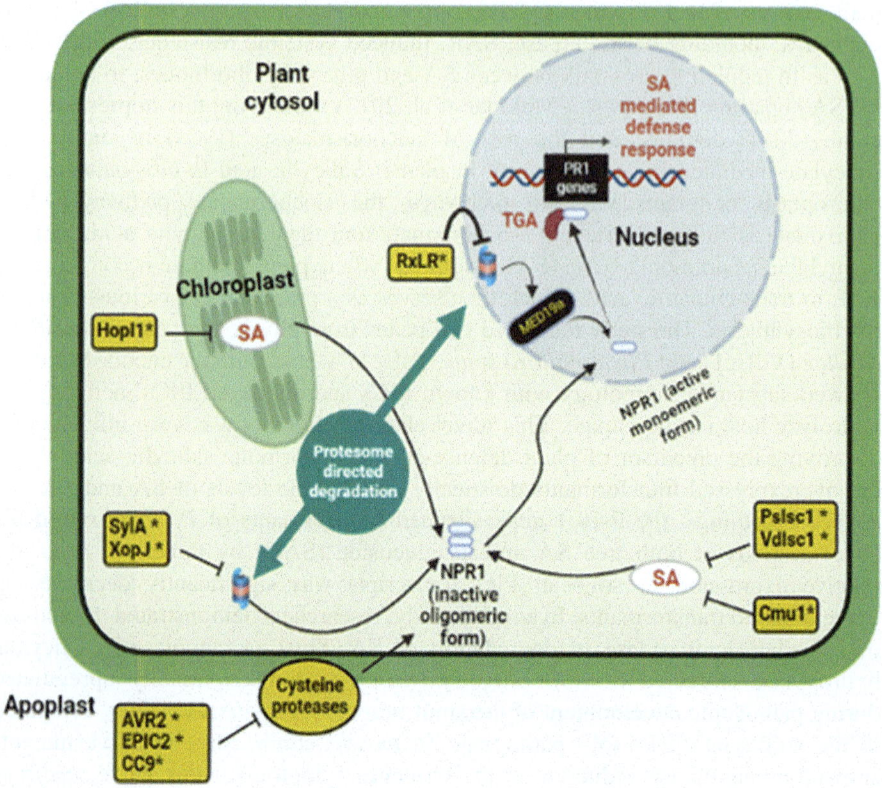

Fig. 10.1 Various pathogenic effectors involved in downregulation of salicylic acid-mediated defense response (source: Tanaka et al. 2015). Plant hormone salicylic acid (SA) is a crucial component of PAMP and ETI-triggered immunity because it triggers defense mechanisms in plants primarily through pathogenesis-related (PR) proteins. However, several pathogens-associated pathogenic effectors can downregulate SA biosynthesis and accumulation, allowing the pathogens to colonize the plant and eventually cause disease. Isochorismatase effectors, viz., PsIsc1 and VdIsc1, which are secreted by filamentous fungi like *Phytophthora sojae* and *Verticillium dahliae,* respectively, are implicated in inhibiting salicylate-mediated innate immunity by hydrolyzing isochorismate, which serves as a precursor for SA production (Liu et al. 2014). The pathogenic effector chorismate mutase Cmu1, which is secreted by *Ustilago maydis*, plays a role in reducing the availability of chorismate for SA production (Djamei et al. 2011). The J domain o***f Pseudomonas syringae***-specific HopI1 effector, which is unique to *Pseudomonas syringae*, is used to attach to heat-shock proteins (Hsp70), and the other component is a chloroplast-targeting sequence that remodels the chloroplasts and ultimately inhibits SA formation (Jelenska et al. 2007). Pathogenic effectors like syringolin A (SylA) and Xop J affect SA-dependent defense response by suppressing NPR1 turnover (Schellenberg et al. 2010; Üstün et al. 2013). RxLR, a different pathogenic effector from ***Hyaloperonospora arabidopsidis,*** interacts with MED 19a, a positive regulator of plant immunity, and permits proteasomal degradation of the latter (Caillaud et al. 2013). Cysteine proteases that play a major role in plant defense signaling can be inhibited by several pathogenic effectors AVR2, EPIC1, EPIC2B, Pit2, and CCP (Mueller et al. 2013; Rooney et al. 2005; Tian et al. 2007; van der Linde et al. 2012)

10.5 Endophytes as a Tool to Combat Climate Change

As evolution is a gradual process, the need to adapt for plants is increased by the rapid occurrence of global warming and the associated aberrations encompassing climate change circumstances (Bashir et al. 2019). There is no doubt that these extreme conditions of abiotic stresses have a serious impact on the growth and yield of plant, severely influencing crop destruction (Mittler 2006), increasing the frequency of pest attack (Coakley et al. 1999; Seherm and Coakley 2003), and enforcing competition with the easily adaptable weed species (Ziska et al. 2010). However, studies have shown that plant molecular and cellular responses display resistance against these environmental stress conditions. Transcriptomic studies have highlighted alterations in transcription factors, protein kinases, and phosphatases occurring in plant due to these abiotic stress conditions. Transcription factors belonging to CBF/DREB1family have been found to be involved in activating DRE/CRT (dehydration-responsive element/C-repeat) class of stress-responsive genes. Overexpression of these transcription factors could enhance tolerance of plants to freezing, salt, and drought stress conditions (Jaglo-Ottosen et al. 1998; Kasuga et al. 1999; Thomashow 2001). However, these late embryogenesis–abundant (LEA)-type genes (such as the DRE/CRT class of genes) are not always involved in stress-based responses in plants. For instance, stress response signals involving mitogen-activated protein kinase (MAPK)-based pathways that contribute to production of compatible osmolytes and antioxidants and SOS pathway signaling that contributes to control iron homeostasis under stress conditions do not necessarily implicate LEA-type genes signaling responses during cold, drought, and salt stress conditions (Xiong et al. 2002). In addition, numerous studies have also identified Ca^{2+} influx as a potential sensor for abiotic stress-based signals (Knight 1999; Plieth et al. 1999; Sanders et al. 1999; Sangwan et al. 2001). Early responses to low temperature, drought, and salinity stress conditions involve increase in cytosolic Ca^{2+} either through influx of the Ca^{2+} or the release of calcium through intracellular compartments possibly through the secondary signaling molecules like inositol polyphosphates, cyclic ADP ribose, and nicotinic acid adenine dinucleotide phosphate (Knight 1999; Sanders et al. 1999). Apart from these, ROS molecule stimulation under abiotic stress conditions may further induce ROS scavengers that may act to increase tolerance under environmentally stressful conditions (Kocsy et al. 2001). It is thus understood that some plant species have developed sophisticated, integrated physiological, or cellular processes that are governed by regulatory and functional genes, enabling them to thrive in harsh settings. However, there are very few plants that can thrive in these hostile environments with high levels of abiotic stress (Alpert 2000). Endophytes that confer stress tolerance in a broad range of host plants could thus promote plant community survival in highly stressed environments.

Beneficial properties of endophytes not only suffice plant growth promotion under optimal conditions, and resistance to phytopathogens but also include plant adaptation under abiotic stress settings such as unfavorable temperatures (high or low), high salinity, drought, and water stress conditions (Aly et al. 2011). In one study, the hsp70 gene isolated from widely known biocontrol endophyte *Trichoderma harzianum* was cloned and expressed in transgenic lines of *Arabidopsis thaliana* seedlings. Compared to the control transgenic lines H713, H1613 and H1651 showed higher germination percentage of 25–43% at 38 °C, 31–56% at 45°C, and 16–25% (38°C/45°C) despite showing no differences in control conditions at 22°C. Apart from increased thermotolerance, these transgenic lines also displayed higher germination under salt (NaCl), oxidative (H_2O_2), and osmotic stress (mannitol) being 5–41%, 25–42%, and 28–34%, respectively (Montero-Barrientos et al. 2010). In another study, *Fusarium culmorum* isolate FcRed1 (SaltSym) isolated from *Leymus mollis* was used to study its role in conferring resistance against salt stress. The evaluation demonstrated that percent shoot biomass and root biomass in SaltSym noninoculated plants under stress decreased by 6.59% and 15.8% drop, respectively, compared to SaltSym inoculated plants under salt stress conditions. Interestingly, paraquat-induced photobleaching (ROS activity) = based studies in the case of salt and drought stress involving SaltSym endophyte along with another endophyte *Curvularia protuberata* isolate Cp4666D (TempSym1) isolated from *Dichanthelium lanuginosum* thriving in geothermal soils showed that while rice plants inoculated with these endophytes were fairly tolerant to photobleaching (11%) the noninoculated plants were highly susceptible to photobleaching (100%). In these inoculated plants, the reduced ROS activity substantially corresponds with stress tolerance and may be essential to the process (Redman et al. 2011). Rodriguez et al. (2008) stress experiments on dunegrass (*Leymus mollis*) employing endophyte *Fusarium culmorum* (FcRed1) isolated from dunegrass growing in coastal habitats of the San Juan Island Archipelago showed that while all plants bent over with age, symbiotic dunegrass plants and unstressed control plants remain hydrated compared to nonsymbiotic or plants with no colonization of FcRed1, which wilted and lost turgor under drought and salt stress conditions. Interestingly, in the absence of stress both *F. culmorum* Fc18 isolated from agricultural environment (that does not confer resistance against salt stress conditions) and *Fusarium culmorum* (FcRed1) colonized dunegrass, rice, and tomato plants in equivalent order; however, in the presence of salt stress only FcRed1 was found to be colonizing in higher range than Fc18 and also conferred significant salt tolerance with 100% survivability. Accordingly, the aforementioned findings showed that the salt tolerance provided by FcRed1 is a habitat-specific symbiotic adaption. In a different study, desert-adapted halotolerant endophytes, viz., *Periconia macrospinosa, Neocamarosporium chichastianum,* and *Neocamarosporium goegapense,* were isolated from *Seidlitzia rosmarinus, Zygophyllum eichwaldii*, and *Haloxylon ammodendron*. The study demonstrated that endophyte-treated cucumber and tomato plants significantly promoted plant growth parameters like biomass, shoot weight, root wet weight, and leaf chlorophyll content compared to the nontreated plants (control) under salt and drought stress conditions. Furthermore, a significant increase in the chlorophyll content and antioxidant enzymes of cucumber and tomato plants compared to the control were achieved with *N. chichastianum, N. goegapense,* and *P. macrospinosa*-treated plants under these stress conditions (Hosseyni Moghaddam et al. 2021). Based on these studies, a general conclusion can be drawn about the serious and constant difficulties plants face under climate change conditions. In addition to plants being sessile organisms, the impact of these climatic changes can be dramatic and disturbing. Thus, in order to thrive in their natural habitat and be able to withstand a variety of stresses over their lifetimes, they must host beneficial endophytes. Microbial biopriming is an adaptive technique that causes a rise in resistance and/or stress tolerance as well as an aggravated defense response to stress-challenged conditions (H. Lu et al. 2021).

10.6 Conclusion

Natural techniques have received a lot of attention because agriculture is anticipated to transition to ecologically sustainable practices. In this context, biofertilizers with incorporation of beneficial endophytes can produce various biostimulatory activities in plants, including mineral uptake, enhanced plant hormone production, and modulation of induced systematic resistance in plants. Climate change under the current situations and the resulting biotic and abiotic stress events ultimately result in adverse conditions for plant growth. However, studies have demonstrated that several beneficial endophytes, viz., *Trichoderma harzianum, Fusarium culmorum* strain FcRed1, *Periconia macrospinosa, Neocamarosporium chichastianum,* and *Neocamarosporium goegapense,* can improve stress tolerance against various abiotic and biotic stress conditions. Symbiotic associations of these endophytic microorganisms also depend on the colonizing potential that involves molecular and chemical signals for attachment, penetration, and establishment in the host plant encompassing host defense modulation as well. While a great number of beneficial endophytes and their colonization process in several host plants have been studied so far, still the temporal and spatial aspects of the endophytic colonization process at the molecular level, as well as early signaling and interchange of signals between endophyte strains and host plants, are yet to be understood. The process of endophyte colonization may be better understood with further research on the molecular and biochemical underpinnings of plant–endophyte interactions.

Acknowledgments The authors sincerely acknowledge the facilities provided by the Department of Botany, Sikkim University, Sikkim, and the Department of Biotechnology for funding services.

References

Abdelrahman M, El-Sayed M, Jogaiah S, Burritt DJ, Tran LSP (2017) The "STAY-GREEN" trait and phytohormone signaling networks in plants under heat stress. Plant Cell Rep 36(7): 1009–1025

Afridi MS, Mahmood T, Salam A, Mukhtar T, Mehmood S, Ali J, Khatoon Z, Bibi M, Javed MT, Sultan T, Chaudhary HJ (2019) Induction of tolerance to salinity in wheat genotypes by plant growth promoting endophytes: involvement of ACC deaminase and antioxidant enzymes. Plant Physiol Biochem 139:569–577

Alpert P (2000) The discovery, scope, and puzzle of desiccation tolerance in plants. Plant Ecol 151(1):5–17

Álvarez B, Biosca EG, López MM (2010) On the life of *Ralstonia solanacearum*, a destructive bacterial plant pathogen. Tech and education topics in app microbio and microbial biotech (January): 267–279

Alvin A, Miller KI, Neilan BA (2014) Exploring the potential of endophytes from medicinal plants as sources of antimycobacterial compounds. Microbiol Res 169(7–8):483–495

Aly AH, Debbab A, Proksch P (2011) Fungal endophytes: unique plant inhabitants with great promises. Appl Microbiol Biotech 90(6):1829–1845

Aranda S, Montes-Borrego M, Jiménez-Díaz RM, Landa BB (2011) Microbial communities associated with the root system of wild olives (*Olea europaea* L. subsp. europaea var. *sylvestris*) are good reservoirs of bacteria with antagonistic potential against *Verticillium dahliae*. Plant Soil 343(1):329–345

Arora NK, Tewari S, Singh R (2013) In: Arora NK (ed) Multifaceted plant-associated microbes and their mechanisms diminish the concept of direct and indirect PGPRs BT- Plant microbe symbiosis: fundamentals and advances. Springer India, New Delhi, pp 411–449

Arya N, Rana A, Rajwar A, Sahgal M, Sharma AK (2018) Biocontrol efficacy of siderophore producing indigenous pseudomonas strains against fusarium wilt in tomato. Natl Acad Sci Lett 41(3):133–136

Balsanelli E, Serrato RV, De Baura VA, Sassaki G, Yates MG, Rigo LU, Pedrosa FO, De Souza EM, Monteiro RA (2010) *Herbaspirillum seropedicae* rfbB and rfbC genes are required for maize colonization. Environ Microbiol 12(8):2233–2244

Balsanelli E, Tuleski TR, de Baura VA, Yates MG, Chubatsu LS, de Oliveira PF, de Souza EM, Monteiro RA (2013) Maize root lectins mediate the interaction with *Herbaspirillum seropedicae* via N-acetyl glucosamine residues of lipopolysaccharides. PLoS One 8(10):e77001

Bashir K, Matsui A, Rasheed S, Seki M (2019) Recent advances in the characterization of plant transcriptomes in response to drought, salinity, heat, and cold stress. F1000 Res 8

Bhattacharjee RB, Singh A, Mukhopadhyay SN (2008) Use of nitrogen-fixing bacteria as biofertiliser for non-legumes: prospects and challenges. Appl Microbiol Biotechnol 80(2): 199–209

Bhore SJ, Ravichantar N, Loh CY (2010) Screening of endophytic bacteria isolated from leaves of *Sambung Nyawa* [*Gynura procumbens* (Lour.) Merr.] for cytokinin-like compounds. Bioinformation 5(5):191

Bhutani N, Maheshwari R, Suneja P (2018) Isolation and characterization of plant growth promoting endophytic bacteria isolated from *Vigna radiata*. Indian J Agric Res 52:6

Bible AN, Chang M, Morrell Falvey JL (2021) Identification of a diguanylate cyclase expressed in the presence of plants and its application for discovering candidate gene products involved in plant colonization by *Pantoea* sp. YR343. PLoS One 16(7):e0248607

Bilal L, Asaf S, Hamayun M, Gul H, Iqbal A, Ullah I, Lee IJ, Hussain A (2018) Plant growth promoting endophytic fungi *Asprgillus fumigatus* TS1 and *Fusarium proliferatum* BRL1 produce gibberellins and regulates plant endogenous hormones. Symbiosis 76(2):117–127

Böhm M, Hurek T, Reinhold-Hurek B (2007) Twitching motility is essential for endophytic rice colonization by the N2-fixing endophyte *Azoarcus* sp. strain BH72. Mol Plant Microbe Interact 20(5):526–533

Bulgarelli D, Rott M, Schlaeppi K, Loren V, van Themaat E, Ahmadinejad N, Assenza F, Rauf P, Huettel B, Reinhardt R, Schmelzer E, Peplies J (2012) Revealing structure and assembly cues for *Arabidopsis* root-inhabiting bacterial microbiota. Nature 488(7409):91–95

Buschart A, Sachs S, Chen X, Herglotz J, Krause A, Reinhold-Hurek B (2012) Flagella mediate endophytic competence rather than act as MAMPs in rice-*Azoarcus* sp. strain BH72 interactions. Mol Plant-Microbe Interact 25(2):191–199

Buysens S, Heungens K, Poppe J, Höfte M (1996) Involvement of pyochelin and Pyoverdin in suppression of *Pythium*-induced damping-off of tomato by *Pseudomonas aeruginosa* 7NSK2. Appl Environ Microbiol 62(3):865–871

Caillaud MC, Asai S, Rallapalli G, Piquerez S, Fabro G, Jones JD (2013) A downy mildew effector attenuates salicylic acid–triggered immunity in *Arabidopsis* by interacting with the host mediator complex. PLoS Biol 11(12):e1001732

Christian N, Sedio BE, Florez-Buitrago X, Ramírez-Camejo LA, Rojas EI, Mejía LC, Palmedo S, Rose A, Schroeder JW, Herre EA (2020) Host affinity of endophytic fungi and the potential for reciprocal interactions involving host secondary chemistry. Am J Bot 107(2):219–228

Coakley SM, Scherm H, Chakraborty S (1999) Climate change and plant disease management. Annu Rev Phytopathol 37:399–426

Cohen AC, Travaglia CN, Bottini R, Piccoli PN (2009) Participation of abscisic acid and gibberellins produced by endophytic *Azospirillum* in the alleviation of drought effects in maize. Botany 87(5):455–462

Compant S, Clément C, Sessitsch A (2010) Plant growth-promoting bacteria in the rhizo- and endosphere of plants: their role, colonization, mechanisms involved and prospects for utilization. Soil Biol Biochem 42(5):669–678

Compant S, Reiter B, Sessitsch A, Nowak J, Clément C, Ait Barka E (2005) Endophytic colonization of *Vitis vinifera* L. by plant growth-promoting bacterium *Burkholderia* sp. strain PsJN. Appl and environmental. Microbiology 71(4):1685–1693

Conn VM, Walker AR, Franco CMM (2008) Endophytic actinobacteria induce defense pathways in *Arabidopsis thaliana*. Mol Plant-Microbe Interact 21(2):208–218

de Bary A (1866) Morphologie und physiologie der pilze, flechten und myxomyceten. Engelmann

De Mot R, Vanderleyden J (1991) Purification of a root-adhesive outer membrane protein of root-colonizing *Pseudomonas fluorescens*. FEMS Microbiol Lett 81(3):323–327

Deng Y, Chen H, Li C, Xu J, Qi Q, Xu Y, Zhu Y, Zheng J, Peng D, Ruan L, Sun M (2019) Endophyte *Bacillus subtilis* evade plant defense by producing lantibiotic subtilomycin to mask self-produced flagellin. Commun Biol 2(1):1–2

Dessureault-Rompré J, Nowack B, Schulin R, Luster J (2007) Spatial and temporal variation in organic acid anion exudation and nutrient anion uptake in the rhizosphere of *Lupinus albus* L. Plant Soil 301(1–2):123–134

Dharmatilake AJ, Bauer WD (1992) Chemotaxis of *Rhizobium meliloti* towards nodulation gene-inducing compounds from alfalfa roots. Appl Environ Microbiol 58(4):1153–1158

Djamei A, Schipper K, Rabe F, Ghosh A, Vincon V, Kahnt J, Osorio S, Tohge T, Fernie AR, Feussner I, Feussner K (2011) Metabolic priming by a secreted fungal effector. Nature 478(7369):395–398

Dubois M, Van den Broeck L, Inzé D (2018) The pivotal role of ethylene in plant growth. Trends Plant Sci 23(4):311–323

Elbeltagy A, Nishioka K, Sato T, Suzuki H, Ye B, Hamada T, Isawa T, Mitsui H, Minamisawa K (2001) Endophytic colonization and in planta nitrogen fixation by a *Herbaspirillum* sp. isolated from wild rice species. Appl Environ Microbiol 67(3–12):5285–5293

Elbeltagy A, Nishioka K, Suzuki H, Sato T, Sato YI, Morisaki H, Mitsui H, Minamisawa K (2000) Isolation and characterization of endophytic bacteria from wild and traditionally cultivated rice varieties. Soil Sci Plant Nutr 46(3):617–629

Ezawa T, Smith SE, Smith FA (2002) P metabolism and transport in AM fungi. Plant Soil 244(1): 221–230

Felix G, Duran JD, Volko S, Boller T (1999) Plants have a sensitive perception system for the most conserved domain of bacterial flagellin. Plant J For Cell Mol Biol 18(3):265–276

Fuqua WC, Winans SC, Greenberg EP (1994) Quorum sensing in bacteria: the LuxR-LuxI family of cell density-responsive transcriptional regulators. J Bacteriol 176(2):269–275

Gamalero E, Glick BR (2015) Bacterial modulation of plant ethylene levels. Plant Physiol 169(1): 13–22

Garnica S, Liao Z, Hamard S, Waller F, Parepa M, Bossdorf O (2022) Environmental stress determines the colonization and impact of an endophytic fungus on invasive knotweed. Biol Invasions 24(6):1785–1795

Glick BR (2012) Plant growth-promoting bacteria: mechanisms and applications. Scientifica 2012

Goldstein AH (1994) Involvement of the quinoprotein glucose dehydrogenase in the solubilization of exogenous phosphates by gram-negative bacteria. Phosphate in microorganisms: cellular and molecular biology. ASM Press, Washington, DC, pp 197–203

Gourion B, Berrabah F, Ratet P, Stacey G (2015) Rhizobium-legume symbioses: the crucial role of plant immunity. Trends Plant Sci 20(3):186–194

Gravel V, Antoun H, Tweddell RJ (2007) Growth stimulation and fruit yield improvement of greenhouse tomato plants by inoculation with *Pseudomonas putida* or *Trichoderma atroviride*: possible role of indole acetic acid (IAA). Soil Bio Biochem 39(8):1968–1977

Gulati A, Rahi P, Vyas P (2008) Characterization of phosphate-solubilizing fluorescent *pseudomonads* from the rhizosphere of seabuckthorn growing in the cold deserts of Himalayas. Curr Microbiol 56(1):73–79

Guo M, Tian F, Wamboldt Y, Alfano JR (2009) The majority of the type III effector inventory of *Pseudomonas syringae* pv. tomato DC3000 can suppress plant immunity. Mol Plant Microbe Interact 22(9):1069–1080

Gupta S, Pandey S (2019) ACC deaminase producing bacteria with multifarious plant growth promoting traits alleviates salinity stress in French bean (*Phaseolus vulgaris*) plants. Front Microbiol 1506

Hallmann J (2001) Plant interactions with endophytic bacteria. Biotic interactions in plant-pathogen associations: 87–119

Hardoim PR, Van Overbeek LS, Berg G, Pirttilä AM, Compant S, Campisano A, Döring M, Sessitsch A (2015) The hidden world within plants: ecological and evolutionary considerations for defining functioning of microbial endophytes. Microbiol Mol Biol Rev 79(3):293–320

Hider RC, Kong X (2010) Chemistry and biology of siderophores. Nat Prod Rep 27(5):637–657

Hurley B, Lee D, Mott A, Wilton M, Liu J, Liu YC (2014a) The *pseudomonas syringae* type III effector HopF2 suppresses *arabidopsis* stomatal immunity. PLoS One 9(12):1–20

Hurley B, Lee D, Mott A, Wilton M, Liu J, Liu Yulu C (2014b) The *pseudomonas syringae* type III effector HopF2 suppresses *Arabidopsis* stomatal immunity. PLoS One 9(12):e114921

Iniguez AL, Dong Y, Carter HD, Ahmer BM, Stone JM, Triplett EW (2005) Regulation of enteric endophytic bacterial colonization by plant defenses. Mol Plant-Microbe Interact 18(2):169–178

Jaglo-Ottosen KR, Gilmour SJ, Zarka DG, Schabenberger O, Thomashow MF (1998) *Arabidopsis* CBF1 overexpression induces COR genes and enhances freezing tolerance. Science 280(5360): 104–106

Jelenska J, Yao N, Vinatzer BA, Wright CM, Brodsky JL, Greenberg JT (2007) AJ domain virulence effector of pseudomonas syringae remodels host chloroplasts and suppresses defenses. Curr Biol 6:499–508

Jones JDG, Dangl JL (2006) The plant immune system. Nature 444(7117):323–329

Kasuga M, Liu Q, Miura S, Yamaguchi-Shinozaki K, Shinozaki K (1999) Improving plant drought, salt, and freezing tolerance by gene transfer of a single stress-inducible transcription factor. Nat Biotechnol 17(3):287–291

Kim YJ, Hwang BK (1994) Differential accumulation of β-1, 3-glucanase and chitinase isoforms in pepper stems infected by compatible and incompatible isolates of Phytophthora capsici. Physiol Mol Plant Pathol 45(3):195–209

Knight H (1999) Calcium signaling during abiotic stress in plants. In International review of cytology, 269–324. Elsevier

Kocsy G, Galiba G, Brunold C (2001) Role of glutathione in adaptation and signalling during chilling and cold acclimation in plants. Physiol Plant 113(2):158–164

Kost T, Stopnisek N, Agnoli K, Eberl L, Weisskopf L (2013) Oxalotrophy, a widespread trait of plant-associated *Burkholderia* species, is involved in successful root colonization of lupin and maize by *Burkholderia phytofirmans*. Front Microbiol 4(JAN):1–9

Kusari S, Hertweck C, Spiteller M (2012) Perspective chemical ecology of endophytic fungi : origins of secondary metabolites. CHBIOL 19(7):792–798

Lebeda A, Luhová L, Sedlářová M, Jančová D (2001) The role of enzymes in plant-fungal pathogens interactions/Die Rolle der Enzyme in den Beziehungen zwischen Pflanzen und pilzlichen Erregern. J Plant Dis Protect:89–111

Li Y, Zhao M, Chen W, Du H, Xie X, Wang D, Dai Y, Xia Q, Wang G (2020) Comparative transcriptomic analysis reveals that multiple hormone signal transduction and carbohydrate metabolic pathways are affected by *Bacillus cereus* in *Nicotiana tabacum*. Genomics 112(6): 4254–4267

van der Linde K, Hemetsberger C, Kastner C, Kaschani F, van der Hoorn RA, Kumlehn J, Doehlemann G (2012) A maize cystatin suppresses host immunity by inhibiting apoplastic cysteine proteases. Plant Cell 24(3):1285–1300

Lindsay WL, Schwab AP (1982) The chemistry of iron in soils and its availability to plants. J Plant Nutr 5(4–7):821–840

Liu T, Song T, Zhang X, Yuan H, Su L, Li W, Xu J, Liu S, Chen L, Chen T, Zhang M (2014) Unconventionally secreted effectors of two filamentous pathogens target plant salicylate biosynthesis. Nat Commun 5(1):1

Lòpez-Fernàndez S, Compant S, Vrhovsek U, Bianchedi PL, Sessitsch A, Pertot I, Campisano A (2016) Grapevine colonization by endophytic bacteria shifts secondary metabolism and suggests activation of defense pathways. Plant Soil 405(1–2):155–175

López-Ráez JA, Shirasu K, Foo E (2017) Strigolactones in plant interactions with beneficial and detrimental organisms: the Yin and Yang. Trends Plant Sci 22(6):527–537

Lu H, Wei T, Lou H, Shu X, Chen Q (2021) A critical review on communication mechanism within plant-endophytic fungi interactions to cope with biotic and abiotic stresses. J Fungi 7:9

Lu X, Zhou D, Chen X, Zhang J, Huang H, Wei L (2017) Isolation and characterization of *Bacillus altitudinis* JSCX-1 as a new potential biocontrol agent against *Phytophthora sojae* in soybean [*Glycine max* (L.) Merr.]. Plant Soil 416(1):53–66

Lugtenberg B (2015) Principles of plant-microbe interactions: Microbes for sustainable agriculture. Principles of Plant-Microbe Interact: Microbes for Sustainable Agriculture: 1–448

Macho AP, Zipfel C (2015) Targeting of plant pattern recognition receptor-triggered immunity by bacterial type-III secretion system effectors. Curr Opin Microbiol 23:14–22

Meneses C, Gonçalves T, Alquéres S, Rouws L, Serrato R, Vidal M, Baldani JI (2017) *Gluconacetobacter diazotrophicus* exopolysaccharide protects bacterial cells against oxidative stress in vitro and during rice plant colonization. Plant Soil 416(1):133–147

Meneses CH, Rouws LF, Simões-Araújo JL, Vidal MS, Baldani JI (2011) Exopolysaccharide production is required for biofilm formation and plant colonization by the nitrogen-fixing endophyte Gluconacetobacter diazotrophicus. Mol Plant Microbe Interact 24(12):1448–1458

Mengistu AA (2020) Endophytes: colonization, behaviour, and their role in defense mechanism. Int J Microbiol 2020

Miché L, Battistoni F, Gemmer S, Belghazi M, Reinhold-Hurek B (2006) Upregulation of jasmonate-inducible defense proteins and differential colonization of roots of *Oryza sativa* cultivars with the endophyte *Azoarcus* sp. Mol Plant Microbe Interact 19(5):502–511

Michiels KW, Croes CL, Vanderleyden J (1991) Two different modes of attachment of *Azospirillum brasilense* Sp7 to wheat roots. J Gen Microbiol 137(9):2241–2246

Millet YA, Danna CH, Clay NK, Songnuan W, Simon MD, Werck-Reichhart D, Ausubel FM (2010) Innate immune responses activated in *Arabidopsis* roots by microbe-associated molecular patterns. Plant Cell 22(3):973–990

Mittler R (2006) Abiotic stress, the field environment and stress combination. Trends Plant Sci 11(1):15–19

Moghaddam MS, Safaie N, Soltani J, Hagh-Doust N (2021) Desert-adapted fungal endophytes induce salinity and drought stress resistance in model crops. Plant Physiol Biochem 160:225–238

Monier J-M, Lindow SE (2003) Differential survival of solitary and aggregated bacterial cells promotes aggregate formation on leaf surfaces. Proc Natl Acad Sci 100(26):15977–15982

Montero-Barrientos M, Hermosa R, Cardoza RE, Gutierrez S, Nicolas C, Monte E (2010) Transgenic expression of the *Trichoderma harzianum* hsp70 gene increases *Arabidopsis* resistance to heat and other abiotic stresses. J Plant Physiol 167(8):659–665

Morán-Diez E, Hermosa R, Ambrosino P, Cardoza RE, Gutiérrez S, Lorito M, Monte E (2009) The ThPG1 endopolygalacturonase is required for the *trichoderma harzianum-* plant beneficial interaction. Mol Plant Microbe Interact 22(8):1021–1031

Morán-Diez E, Rubio B, Domínguez S, Hermosa R, Monte E, Nicolás C (2012) Transcriptomic response of *Arabidopsis thaliana* after 24 h incubation with the biocontrol fungus *Trichoderma harzianum*. J Plant Physiol 169(6):614–620

Muangthong A, Youpensuk S, Rerkasem B (2015) Isolation and characterisation of endophytic nitrogen fixing bacteria in sugarcane. Trop Life Sci Res 26(1):41–51

Mueller AN, Ziemann S, Treitschke S, Aßmann D, Doehlemann G (2013) Compatibility in the *Ustilago maydis*–maize interaction requires inhibition of host cysteine proteases by the fungal effector Pit2. PLoS Pathog 9(2):e1003177

Muthukumar A, Udhayakumar R, Naveenkumar R (2017) Role of bacterial endophytes in plant disease control. (November):133–161

Nag P, Shriti S, Das S (2020) Microbiological strategies for enhancing biological nitrogen fixation in nonlegumes. J Appl Microbiol 129(2):186–198

Nannipieri P, Giagnoni L, Landi L, Renella G (2011). Role of phosphatase enzymes in soil. In Phosphorus in action (pp. 215–243). Springer, Berlin, Heidelberg

Naureen A, Nasim FU, Choudhary MS, Ashraf M, Grundler FM, Schleker AS (2022) A new endophytic fungus CJAN1179 isolated from the Cholistan desert promotes lateral root growth in Arabidopsis and produces IAA through tryptophan-dependent pathway. Arch Microbiol 204(3): 1–13

Niu DD, Liu HX, Jiang CH, Wang YP, Wang QY, Jin HL, Guo JH (2011) The plant growth–promoting rhizobacterium *Bacillus cereus* AR156 induces systemic resistance in Arabidopsis thaliana by simultaneously activating salicylate-and jasmonate/ethylene-dependent signaling pathways. Mol Plant-Microbe Interact 24(5):533–542

Oku S, Komatsu A, Tajima T, Nakashimada Y, Kato J (2012) Identification of chemotaxis sensory proteins for amino acids in *Pseudomonas fluorescens* Pf0-1 and their involvement in chemotaxis to tomato root exudate and root colonization. Microbes Environ 27(4):462–469

Oldroyd GE (2013) Speak, friend, and enter: signalling systems that promote beneficial symbiotic associations in plants. Nat Rev Microbiol 11(4):252–263

Oteino N, Lally RD, Kiwanuka S, Lloyd A, Ryan D, Germaine KJ, Dowling DN (2015) Plant growth promotion induced by phosphate solubilizing endophytic *Pseudomonas* isolates. Front Microbiol 6

Pajerowska-Mukhtar KM, Emerine DK, Mukhtar MS (2013) Tell me more: roles of NPRs in plant immunity. Trends Plant Sci 18(7):402–411

Patten CL, Glick BR (1996) Bacterial biosynthesis of indole-3-acetic acid. Can J Microbiol 42(3): 207–220

Patten CL, Glick BR (2002) Role of *Pseudomonas putida* indoleacetic acid in development of the host plant root system. Appl Environ Microbiol 68(8):3795–3801

Pieterse CM, Van der Does D, Zamioudis C, Leon-Reyes A, Van Wees SC (2012) Hormonal modulation of plant immunity. Annu Rev Cell Dev Biol 28:489–521

Plieth C, Hansen U, Knight H, Knight MR (1999) Temperature sensing by plants: the primary characteristics of signal perception and calcium response. Plant J 18(5):491–497

Rabe F, Ajami-Rashidi Z, Doehlemann G, Kahmann R, Djamei A (2013) Degradation of the plant defence hormone salicylic acid by the biotrophic fungus *Ustilago maydis*. Mol Microbiol 89(1): 179–188

von Rad U, Klein I, Dobrev PI, Kottova J, Zazimalova E, Fekete A, Hartmann A, Schmitt-Kopplin P, Durner J (2008) Response of *Arabidopsis thaliana* to N-hexanoyl-DL-homoserine-lactone, a bacterial quorum sensing molecule produced in the rhizosphere. Planta 229(1):73–85

Ramesh R, Joshi AA, Ghanekar MP (2009) Pseudomonads: major antagonistic endophytic bacteria to suppress bacterial wilt pathogen, *Ralstonia solanacearum* in the eggplant (*Solanum melongena* L.). World J Microbiol Biotechnol 25(1):47–55

Redman RS, Kim YO, Woodward CJ, Greer C, Espino L, Doty SL, Rodriguez RJ (2011) Increased fitness of rice plants to abiotic stress via habitat adapted Symbiosis: a strategy for mitigating impacts of climate change. PLoS One 6(7):e14823

Reinhold HB, Maes T, Gemmer S, Van Montagu M, Hurek T (2006) An endoglucanase is involved in infection of rice roots by the not-cellulose-metabolizing endophyte *Azoarcus* sp. strain BH72. Mol Plant Microbe Interact 19(2):181–188

Rodrigo M, Pizzirani-Kleiner AA, Araujo WL, Jose MR (2007) Diversity of cultivated endophytic bacteria from sugarcane: genetic and biochemical characterization of *Burkholderia cepacia* complex isolates. Appl Environ Microbiol 73(22):7259–7267

Rodríguez M, Torres M, Blanco L, Béjar V, Sampedro I, Llamas I (2020) Plant growth-promoting activity and quorum quenching-mediated biocontrol of bacterial phytopathogens by *Pseudomonas segetis* strain P6. Sci Rep 10(1):4121

Rodriguez RJ, Henson J, Van Volkenburgh E, Hoy M, Wright L, Beckwith F, Kim YO, Redman RS (2008) Stress tolerance in plants via habitat-adapted symbiosis. ISME J 2(4):404–416

Rooney HC, Van't Klooster JW, van der Hoorn RA, Joosten MH, Jones JD, de Wit PJ (2005) *Cladosporium* Avr2 inhibits tomato Rcr3 protease required for Cf-2-dependent disease resistance. Science (New York, NY) 308(5729):1783–1786

Rossi FA, Medeot DB, Liaudat JP, Pistorio M, Jofré E (2016) In *Azospirillum brasilense*, mutations in flmA or flmB genes affect polar flagellum assembly, surface polysaccharides, and attachment to maize roots. Microbiol Res 190:55–62

Rungin S, Indananda C, Suttiviriya P, Kruasuwan W, Jaemsaeng R, Thamchaipenet A (2012) Plant growth enhancing effects by a siderophore-producing endophytic *streptomycete* isolated from a Thai jasmine rice plant (*Oryza sativa* L. cv. KDML105). Antonie Van Leeuwenhoek 102(3): 463–472

Sanders D, Brownlee C, Harper JF (1999) Communicating with calcium. Plant Cell 11(4):691–706

Sangal V, Goodfellow M, Blom J, Tan GY, Klenk HP, Sutcliffe IC (2018) Revisiting the taxonomic status of the biomedically and industrially important genus *Amycolatopsis*, using a phylogenomic approach. Front Microbiol 9:22–81

Sangwan V, Foulds I, Singh J, Dhindsa RS (2001) Cold-activation of *Brassica napus* BN115 promoter is mediated by structural changes in membranes and cytoskeleton, and requires Ca2+ influx. Plant J 27(1):1–12

Schellenberg B, Ramel C, Dudler R (2010) *Pseudomonas syringae* virulence factor syringolin a counteracts stomatal immunity by proteasome inhibition. Mol Plant-Microbe Interact 23(10): 1287–1293

Schenk ST, Hernández-Reyes C, Samans B, Stein E, Neumann C, Schikora M, Reichelt M, Mithöfer A, Becker A, Kogel KH, Schikora A (2014) N-acyl-homoserine lactone primes plants for cell wall reinforcement and induces resistance to bacterial pathogens via the salicylic acid/ oxylipin pathway. Plant Cell 26(6):2708–2723

Schlaeppi K, Dombrowski N, Oter RG, Van Themaat EVL, Schulze-Lefert P (2014) Quantitative divergence of the bacterial root microbiota in *Arabidopsis thaliana* relatives. Proc Natl Acad Sci U S A 111(2):585–592

Schuhegger R, Ihring A, Gantner S, Bahnweg G, Knappe C, Vogg G, Hutzler P, Schmid M, Van Breusegem F, Eberl LE, Hartmann A (2006) Induction of systemic resistance in tomato by

N-acyl-L-homoserine lactone-producing rhizosphere bacteria. Plant Cell Environ 29(5): 909–918

Schulz B, Römmert AK, Dammann U, Aust H, Jür SD (1999) The endophyte-host interaction: a balanced antagonism? Mycological Res 103(10):1275–1283

Seherm H, Coakley SM (2003) Plant pathogens in a changing world. Australas Plant Pathol 32(2): 157–165

Shabanamol S et al (2018) Characterization and in planta nitrogen fixation of plant growth promoting endophytic diazotrophic Lysinibacillus sphaericus isolated from rice (Oryza sativa). Physiol Mol Plant Pathol 102:46–54

Shah N, Gislason AS, Becker M, Belmonte MF, Fernando WD, de Kievit TR (2020) Investigation of the quorum-sensing regulon of the biocontrol bacterium *Pseudomonas chlororaphis* strain PA23. PLoS One 15(2):e0226232

Sharaf H, Rodrigues RR, Moon J, Zhang B, Mills K, Williams MA (2019) Unprecedented bacterial community richness in soybean nodules vary with cultivar and water status. Microbiome 7(1): 1–18

Sharma K, Dak G, Agrawal A, Bhatnagar M, Sharma R (2007) Effect of phosphate solubilizing bacteria on the germination of *Cicer arietinum* seeds and seedling growth. J Herb Med Toxicol 1(1):61–63

Shiferaw D, Diriba M, Gezahegn B (2013) Effect of phosphate solubilizing bacteria on seed germination and seedling growth of faba bean (*Vicia faba* L.). Int J Agric Res 8(3):123–136

Sirrenberg A, Göbel C, Grond S, Cempinski N, Feussner I, Pawlowski K (2007) *Piriformospora indica* induces increased root branching in Arabidopsis through IAA production. In Abstract in Meeting Report of Research Group (Vol. 546)

Spaepen S, Vanderleyden J, Remans R (2007) Indole-3-acetic acid in microbial and microorganism-plant signaling. FEMS Microbiol Rev 31(4):425–448

Stracke S, Kistner C, Yoshida S, Mulder L, Sato S, Kaneko T, Tabata S, Sandal N, Stougaard J, Szczyglowski K, Parniske M (2002) A plant receptor-like kinase required for both bacterial and fungal symbiosis. Nature 417(6892):959–962

Sturz AV, Christie BR, Nowak J (2000) Bacterial endophytes: potential role in developing sustainable systems of crop production. Crit Rev Plant Sci 19(1):1–30

Taghavi S et al (2010) Genome sequence of the plant growth promoting endophytic bacterium Enterobacter sp. 638. PLoS Genet 6(5):19

Takahashi S, Anwar MR (2007) Wheat grain yield, phosphorus uptake and soil phosphorus fraction after 23 years of annual fertilizer application to an andosol. Field Crop Res 101(2):160–171

Talbot NJ (2010) Living the sweet life: how does a plant pathogenic fungus acquire sugar from plants? PLoS Biol 8(2):e1000308

Tanaka S, Han X, Kahmann R (2015) Microbial effectors target multiple steps in the salicylic acid production and signaling pathway. Front Plant Sci 6:349

Taulé C, Mareque C, Barlocco C, Hackembruch F, Reis VM, Sicardi M, Battistoni F (2012) The contribution of nitrogen fixation to sugarcane (*Saccharum officinarum* L.), and the identification and characterization of part of the associated diazotrophic bacterial community. Plant Soil 356(1–2):35–49

Thomashow MF (2001) So what's new in the field of plant cold acclimation? Lots! Plant Physiol 125(1):89–93

Tian M et al (2007) A *Phytophthora infestans* cystatin-like protein targets a novel tomato papain-like apoplastic protease. Plant Physiol 143(1):364–377

Timmusk S, Grantcharova N, Wagner EGH (2005) *Paenibacillus polymyxa* invades plant roots and forms biofilms. 71(11):7292–7300

Üstün S, Bartetzko V, Börnke F (2013) The Xanthomonas campestris type III effector XopJ targets the host cell proteasome to suppress salicylic-acid mediated plant defence. PLoS Pathog 9(6): e1003427

Vaishnav A, Shukla AK, Sharma A, Kumar R, Choudhary DK (2019) Endophytic bacteria in plant salt stress tolerance: current and future prospects. J Plant Growth Regulat 38(2):650–668

Vandana UK, Rajkumari J, Singha LP, Satish L, Alavilli H, Sudheer PD, Chauhan S, Ratnala R, Satturu V, Mazumder PB, Pandey P (2021) The endophytic microbiome as a hotspot of synergistic interactions, with prospects of plant growth promotion. Biology 10(2):1–29

Venkateshwaran M, Jayaraman D, Chabaud M, Genre A, Balloon AJ, Maeda J, Forshey K, den Os D, Kwiecien NW, Coon JJ, Barker DG (2015) A role for the mevalonate pathway in early plant symbiotic signaling. Proc Natl Acad Sci 112(31):781–9786

Wang HW, Ma CY, Xu FJ, Lu F, Zhang W, Dai CC (2021) Root endophyte-enhanced peanut-rhizobia interaction is associated with regulation of root exudates. Microbiol Res 250(November 2020):126765

Wicklow DT, Roth S, Deyrup ST, Gloer JB (2005) A protective endophyte of maize: *acremonium zeae* antibiotics inhibitory to *aspergillus flavus* and *fusarium verticillioides*. Mycological Res 109(5):610–618

Wilson M, MacNab R, Henderson B (2002) Bacterial adhesion as a virulence mechanism. Bacterial disease mechanisms: an introduction to cellular microbiology 1: 353–404

de Wit PJGM (2007) How plants recognize pathogens and defend themselves. Cell Mol Life Sci 64(21):2726–2732

Xiong L, Schumaker KS, Zhu JK (2002) Cell signaling during cold, drought, and salt stress. Plant Cell 14(Suppl):165–183

Xu T, Vo QA, Barnett SJ, Ballard RA, Zhu Y, Franco CM (2022) Revealing the underlying mechanisms mediated by endophytic actinobacteria to enhance the rhizobia—chickpea (*Cicer arietinum* L.) symbiosis. Plant Soil 474(1–2):299–318

Yaron S, Römling U (2014) Biofilm formation by enteric pathogens and its role in plant colonization and persistence. Microbial Biotech 7(6):496–516

Yimer D (2019) Components, mechanisms of action, success under greenhouse and field condition, market availability, formulation and inoculants development on biofertilizer. Biomed J Sci Technical Res 12(4):9366–9372

Zarkani AA, Stein E, Röhrich CR, Schikora M, Evguenieva-Hackenberg E, Degenkolb T, Vilcinskas A, Klug G, Kogel KH, Schikora A (2013) Homoserine lactones influence the reaction of plants to *Rhizobia*. Int J Mol Sci 14(8):17122–17146

Zinniel ZDK, Denise K, Pat L, Harris NB, Zhengyu F, Daniel K, Phyllis H, Carol AI, Alahari A, Raúl GB, Anne KV (2002) Isolation and characterization of endophytic colonizing bacteria from agronomic crops and prairie plants. Appl Environ Microbiol 68(5):2198–2208

Zipfel C, Oldroyd GED (2017) Plant signalling in symbiosis and immunity. Nature 543(7645): 328–336

Ziska L, Tomecek M, Gealy D (2010) Competitive interactions between cultivated and red rice as a function of recent and projected increases in atmospheric carbon dioxide. Agron J 102

Chapter 11
Deciphering the Role of Growth-Promoting Bacterial Endophytes in Harmonizing Plant Health

L. Rajendran, D. Durgadevi, R. Kavitha, T. Archana, S. Harish, V. Sendhilvel, T. Raguchander, and G. Karthikeyan

Abstract Bacterial endophytes invade plants similarly to phytopathogens, making them suitable as biological agents without causing any harmful effects. Endophytic bacteria colonize a broad spectrum of plant species, and it has often been stated that all plants in nature harbor endophytic bacteria. The main source of endophytic colonization is believed to be the rhizosphere and phyllosphere, where high similarity was found between the bacteria present internally in the root and in the rhizosphere or phyllosphere. They have the ability to manage plant diseases, nematodes, insects, and other organisms through the innovative or novel production of antifungal/bacterial chemicals and unique metabolites. Several authors also reported that endophytic bacteria promote plant growth by means of IAA, gibberellin, and cytokinin, and possess various antibiotic genes. Some endophytes offer increased resistance to pathogens, thus making them ideal candidates for biological control. More recently, bacterial formulations using endophytes have been developed and used to combat various diseases of economically important crops in some developing countries. In China, for instance, a liquid formulation of *B. subtilis* has been extensively tested to control sheath blight disease in rice under field conditions. Although this type of formulation can be produced in large quantity using a simple fermentation and formulation processes, it may be difficult to store and may have a

L. Rajendran (✉)
HRS, Tamil Nadu Agricultural University, Ooty, India
e-mail: rajendran.l@tnau.ac.in

D. Durgadevi · R. Kavitha · T. Archana · T. Raguchander · G. Karthikeyan
Department of Plant Pathology, Tamil Nadu Agricultural University, Coimbatore, Tamil Nadu, India

S. Harish
Department of Oilseeds, Tamil Nadu Agricultural University, Coimbatore, Tamil Nadu, India
e-mail: harish.s@tnau.ac.in

V. Sendhilvel
ICAR-KVK, Virinchipuram, TNAU, Vellore, India

P. Mathur et al. (eds.), *Microbial Symbionts and Plant Health: Trends and Applications for Changing Climate*, Rhizosphere Biology,
https://doi.org/10.1007/978-981-99-0030-5_11

relatively short shelf life. Hence, improved formulations and delivery systems are important for implementation and commercialization of effective biological control.

Keywords Biocontrol · Endophytes · Functional role · Isolation · Omics · Plant growth-promoting rhizobacteria

Abbreviations

ARDRA	Amplified rDNA restriction analysis
ARISA	Automated ribosomal intergenic spacer analysis
CMV	Cucumber mosaic virus
CTAB	Cetyltrimethylammonium bromide
DGGE	Denaturing gradient gel electrophoresis
FAME	fatty acid methyl ester analysis
ISR	Induced systemic resistance
LB	Luria broth
PAL	Phenylalanine ammonia lyase
PGPE	Plant growth-promoting endophyte
POX	Peroxidase
PPO	Polyphenol oxidase
TGGE	Temperature gradient gel electrophoresis
T-RFLP	Terminal restriction fragment length polymorphism
TSA	Tryptic soy agar
TSWV	Tomato spotted wilt virus

11.1 Introduction

Plants are naturally inhabited by diverse microorganisms like fungi, bacteria, and viruses, and the interactions of microorganisms with plants may be symbiotic or pathogenic. Similarly, another type of interaction that needs to be overlooked is an endophytic association, especially with the bacteria. It is defined as the bacteria that colonize the intercellular and intracellular spaces of all recognized plant compartments throughout all or a portion of their existence without any harm to the plant. Endophytic bacteria were categorized into more than 200 genera, with the majority belonging to the three phyla Firmicutes, Actinobacteria, and Proteobacteria. It could be Gram-negative (*Achromobacter*, *Acinetobacter*, *Pseudomonas,* etc.) or Gram-positive (most commonly *Bacillus* spp.). Further, transmission routes those bacterial endophytes can take, including horizontal and vertical transmission in that vertical transmission, may occur by pollen and seeds, whereas horizontal transmission occurs through the atmosphere, soil, and insects (Adeleke et al. 2022). Seed bacterial endophytes that infect the seed's embryo and endosperm after being transported vertically can protect plants from biotic and abiotic stress (Archana et al. 2020). However, endophytes are divided into two categories based on where they live in the host: obligatory (unable to grow outside of plants and more likely to spread by seed than through rhizospheric soil) and facultative (mostly free living in soil and causes infection when the favorable condition prevails). Some endophytes move through their host's root xylem and transpiration current with the help of flagella, while others use intracellular spaces. The rhizosphere has a wider variety of microbial communities than the endosphere. According to Santoyo et al. (2016), bacterial endophytes have an advantage over rhizobacteria because they live within a plant's tissues, giving them a chance to constantly interact with plant cells and their metabolites. They may also be more resilient to biotic and abiotic stresses than rhizospheric bacteria. A few endophytic microbes fix nitrogen, solubilize phosphate, or boost nutrient intake by forming symbiotic relationships. These microorganisms also produce siderophores, detoxify contaminants, and increase resilience to biotic and abiotic challenges, which promote plant growth and development. Several techniques have been developed to assess the structure and makeup of the host-associated microbial community. The best techniques for studying the functions of native species are those that depend on culture and select particular populations that can thrive in solitary environments (Lebeis 2014). In contrast, technologies that can analyze the makeup of bacterial communities can identify uncultivable, slowly developing, or sparsely numerous microorganisms. These two approaches determine endophytic communities, each having its limitations in the identification of endophyte composition.

11.2 Culture-Dependent Techniques

The most widely used method for the isolation of endophytic bacteria is culture-dependent techniques, which include isolation from the surface-sterilized plant tissues and cultivation on an appropriate medium. Characterization is also done utilizing several methods, such as morphological, enzymatic, or fatty acid analysis like fatty acid methyl ester (FAME) analysis. These culture-dependent approaches are simple, inexpensive, and have the potential for additional phylogenetic, physiological, and biochemical characterization (Puri et al. 2018). Importantly, culturable isolates that have been characterized can be used as inoculants in the future. Utilizing culture-dependent approaches, significant knowledge regarding the microbial diversity and functionality of microbial communities associated with plants was assessed (Pei et al. 2017). These cultured bacterial endophytes displayed a variety of characteristics that aid in the promotion of plant growth, including the uptake of water and nutrients, including nitrogen, phosphorus, zinc, and other nutritional components. Culture-dependent techniques mostly depend on the manipulation of plant tissue, cultivation material, and growth conditions. Plant tissues were surface sterilized with 1% sodium hypochlorite, 70% ethanol, and three rinses in sterilized distilled water for culture-dependent analysis. Using phosphate buffer, 1 gram of plant tissues was

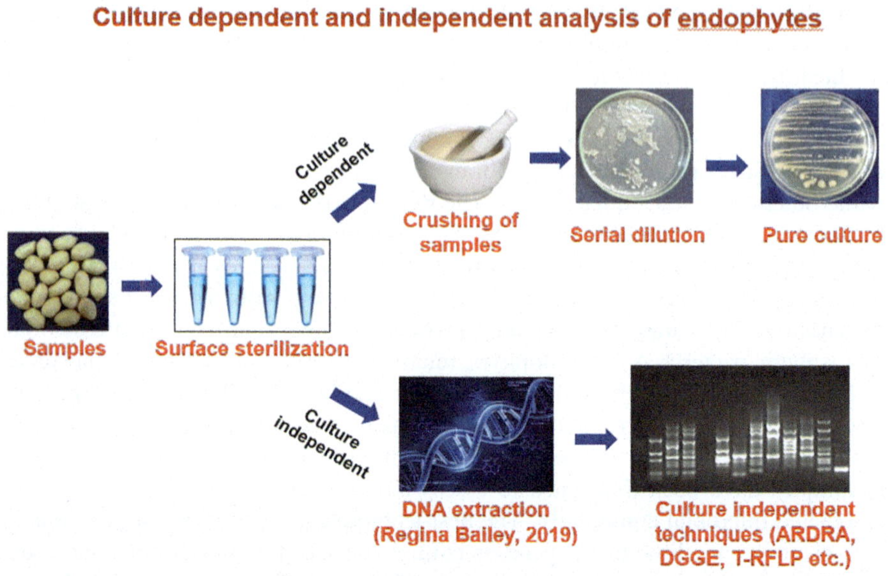

Fig. 11.1 Culture-dependent and -independent analysis of endophytes

crushed and 1 ml of the resulting suspension was then put on Petri dishes containing Luria broth (LB) or tryptic soy agar (TSA), where it was incubated at room temperature. One milliliter of the most recent rinse water was put into Petri dishes containing LB/TSA medium as a check (Rath et al. 2018). After the incubation period, colonies that appeared on the majority of the plates were chosen and characterized using morphological traits like colony color, shape, and margin, as well as biochemical processes like starch hydrolysis, catalase test, methyl red test, KOH test, and citrate utilization test. To make conclusions about the readily isolated endophytes, culture-dependent techniques are widely used (Eevers et al. 2016) (Fig. 11.1), which provide the raw material for genome sequence and to determine plant phenotype, including bacterial endophyte colonization, even though it lacks the sensitivity to detect the shift in endophytic community composition.

11.3 Culture-Independent Techniques

Most prokaryotes are recalcitrant to isolate under laboratory conditions; hence, culture-independent techniques, namely, molecular methods, have been widely used to characterize the endophytic communities in vitro. Compared to culture-dependent procedures, these culture-independent methods give accurate and quick taxonomy of bacterial endophytes (Pereira et al. 2011). Various finger printing techniques depend on gDNA to amplify the specific marker of endophytic bacteria. The 16S rRNA gene, which is frequently used to identify bacteria, is based on a comparative sequencing analysis of protein-coding genes (Wang et al. 2019). The cetyltrimethylammonium bromide (CTAB) method was used to extract the genomic DNA from the chosen bacterial endophytes. The 16S rRNA gene from the chosen efficient endophytic bacteria was amplified using the universal primers 27F and 1492R. 50 mL of 10X PCR buffer, 5 mL of sample, 10 mM dNTP, 75 mM MgCl2, 1.25 U of Taq polymerase, and 15 pmol of each primer were used in the process. The PCR procedure was carried out under cyclic conditions, which includes initial denaturation at 94 °C for 5 min, denaturation at 94 °C for 1 min, annealing at 55 ° C for 40 sec, extension at 72 °C for 1 min, and final extension at 72 °C for 10 min (40 cycles).

Analysis of root-associated mRNA for the expression of nitrogenase genes implied that these bacteria were nevertheless active. It might be the reason why most of the nitrogen-fixing bacteria were found low *in planta.* Several 16S rRNA gene fragments were amplified using a variety of finger printing techniques, including denaturing gradient gel electrophoresis (DGGE), automated ribosomal intergenic spacer analysis (ARISA), terminal restriction fragment length polymorphism (T-RFLP), amplified rDNA restriction analysis (ARDRA), and temperature gradient gel electrophoresis (TGGE) (Ali et al. 2021). Studies of the intricate endophytic communities are conducted directly from the plant using metagenomics, also known as environmental genomics. For culture-independent techniques, it is not necessary to perform bacteria isolation or culturing, and cloning steps are also not required. The probing techniques usage was based on the hybridization of synthetically created oligonucleotides to certain target sequences in bacterial DNA. These probes have an advantage over traditional PCR, where they are connected to a radioactive or fluorescent label that allows visual detection of the target following hybridization. The target sequence only determines the specificity of the probe; apart from that, stringency and washing conditions are also needed. The most popular technique for counting labeled bacteria among probe-labeled techniques, fluorescent in situ hybridization (FISH), uses fluorescence microscopy. Through the use of flow cytometric analysis, endophytic bacteria are analyzed both quantitatively and qualitatively (Janatiningrum et al. 2018). Endophytic bacteria present in a sample or suspension are fluorescently labeled using one or more specialized probes and, then, the flow cytometer is used to analyze the labeled fluid to identify and quantify the bacteria. The use of probes may limit the number of bacteria that can be targeted, which is a potential drawback of probe-based approaches. RT-qPCR is a different method of amplifying DNA that employs fluorescent dyes for the detection and amplification of PCR reactions in a single-tube format. The increase in fluorescent signal, which is proportional to the amount of DNA generated after each cycle, is used to assess the accumulation of PCR product after all PCR cycles. This method is paired with RT, a method for calculating the number of mRNA transcripts. The transcript demonstrates live cells' metabolic activity (Beltran-Garcia et al. 2021). The most suitable and widely applicable method for studying the complex endophytic bacterial communities was denaturing gradient gel electrophoresis (DGGE) analysis. This technique is based on the sequence-dependent separation of a mixture of amplified DNA fragments on an acrylamide gel identical in size containing a gradient of denaturing components. In an increasingly denaturing environment, stranded DNA (dsDNA) becomes (partially) single-stranded (ssDNA). This denaturation, instead of discrete portions of the fragment, will gradually denature through called melting domains. The passage of the DNA fragment through the acrylamide gel is drastically slowed and halted due to the conformational change. The nucleotide sequence and % G + C content of the fragment determine the position of the gel in electrophoresis (Fig. 11.1).

11.4 Plant Growth-Promoting Traits (PGPs)

11.4.1 Phytohormone Regulation

Plant hormones, as key chemical constituents, can provide tolerance to a variety of abiotic and biotic stresses. They function as effector molecules that are known to be involved in signal perception, transduction, cellular homeostasis, and gene expression. As a result, they play a major role in disease resistance (Bari and Jones 2009). The hormones, namely, auxin, IAA, and ABA, are involved in regulating a wide range of biological functions, including plant responses to biotic and abiotic stressors. For example, *P. fluorescens*-mediated control of Fusarium head blight disease of barley involves both IAA and ABA (Petti et al. 2012). Auxin is the most common phytohormone secreted by plant endophytic bacteria to stimulate root growth, which results in the formation of lateral, adventitious roots and root hairs, thereby increasing the surface area of the root (Hilbert et al. 2012). *Streptomyces atrovirens*, *Streptomyces olivaceoviridis*, *Streptomyces rimosus*, *Streptomyces rochei,* and *Streptomyces viridis* are all IAA-producing endophytes that improve seed germination, root elongation, and growth. Beneficial bacteria such as *Bacillus amyloliquefaciens, Azospirillum,* and *Rhizobia* produce IAA via both the tryptophan (trp)-dependent pathway (indole-3-pyruvic acid-IPyA) and the trp-independent pathway (Abd-alla et al. 2013). Cytokinins are related to the growth of plants, which aid in impeding senescence. The endophytes that produce cytokinin-like compounds could be used to extend the shelf life of cut flowers, leafy vegetables, and fruits (Bhore et al. 2010). When compared to other endophytes such as *Azotobacter beijerinckii*, *Pseudomonas fluorescens,* and *Pseudomonas putida*, *Azotobacter chroococcum* was the most important producer of cytokinin (Dudeja et al. 2012). Endophytic bacteria can activate the ACC synthase, allowing higher plants to produce ACC. In many leguminous plants, the conversion of ACC to ethylene by ACC oxidase suppresses nodule formation and plant growth. Endophytic bacteria can reduce ethylene levels by producing ACC deaminase, which cleaves ACC and promotes plant growth by removing *in planta* ethylene inhibition (Fig. 11.2).

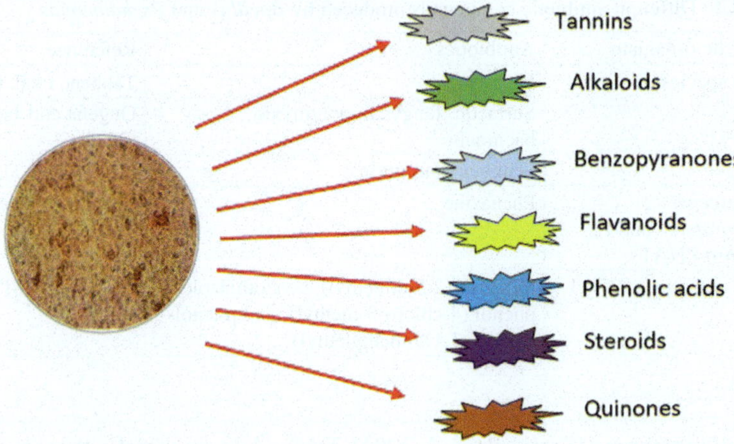

Fig. 11.2 Secondary metabolites production by the plant growth-promoting endophytic bacteria

11.4.2 Antibiotic Synthesis

Antibiotic production is an effective mechanism used by many endophytes to control soil plant pathogens. Because of the presence of biosynthetic genes, this ability can be correlated to the induction of systemic resistance (ISR) in plants due to the underlying antimicrobial mechanism as antibiotics can act as determinants, triggering the response (Xu et al. 2019). *Bacillus* and *Streptomyces* species are the most common antibiotic-producing Gram-positive bacterial endophytes found in a variety of environments (Ek-Ramos et al. 2019). Endophytic *Bacillus* can also produce a variety of compounds such as fengycins, iturins, and surfactins (Hardoim 2019).

Ramyabharathi and Raguchander (2014) identified the phthalic acid-producing *Bacillus subtilis* EPCO16 strain, which has antifungal activity against *Fusarium oxysporum* f. sp. *lycopersici.* A study by D'Alessandro et al. (2014) showed the presence of antibiotic compound 2,3-BD in the endophytic bacteria *Enterobacter aerogenes,* which showed resistance against Northern corn leaf blight, *Setosphaeria turcica.* A list of a few compounds produced by *Bacillus* and *Pseudomonas* is provided in Table 11.1.

11.4.3 Siderophores

Many endophytes can produce siderophores, a mechanism that is crucial for their survival in low levels of iron. It provides a viable edge for chelating Fe^{3+} and gaining ferric iron can deny the pathogenic organisms and other competing microbes. This ferric-siderophore complex will be inaccessible to other harmful organisms, but siderophore-producing microbes can use it via specialized receptors in their

Table 11.1 Different antibiotic compounds produced by *Bacillus* and *Pseudomonas*

Endophytic organism	Antibiotics	Reference
Bacillus species	Peptide	Tabbene et al. (2009)
	Surfactin, fengycin, polymyxin, bacitracin	Ongena and Jacques (2008)
	Fengycin and iturin	Savadogo et al. (2011)
P. fluorescens, *P. chlororaphis,* and P. *aeruginosa* PNA1	Phenazine	Shanmugaiah et al. (2010)
Bacillus velezensis ZSY-1	Pyrazine (2,5-dimethyl), benzothiazole, phenol (4-chloro-3-methyl), and phenol-2,4-bis (1,1-dimethylethyl)	Gao et al. (2017)

outer-cell membrane (Sessitsch et al. 2004). Depending on moiety, these molecules can be categorized into three groups, which donate oxygen ligands for Fe^{3+}: (1) catecholates or phenolates, (2) hydroxamates or carboxylates, and (3) the mixed types. Siderophore molecules containing catecholate groups are enterobactin (*Streptomyces*), pyochelin (*Pseudomonas aeruginosa*), and vibriobactin (*Vibrio cholerae*). The siderophores with hydroxamate groups include alcaligin (*Alcaligenes denitrificans*) and staphyloferrin (*Staphylococcus* spp.), whereas examples of mixed types are mycobactin (*Mycobacterium tuberculosis*) and petrobactin (*B. anthrasis*). Many endophytes are reported as siderophore producers, whereas *Pseudomonas* sp. are identified as one of the major groups (Jacob et al. 2020). Pseudobactin, a siderophore produced by the endophyte *Pseudomonas fluorescens,* restricted the growth of *Erwinia caratovora* (Sayyed et al. 2013). Siderophore producing seven endophytic strains from ginseng *(Panax ginseng* C.A. Meyer), namely, *Bacillus cereus*, *B. flexus*, *B. megaterium*, *Lysinibacillus fusiformis*, *L. sphaericus*, *Microbacterium phyllosphaerae,* and *Micrococcus luteus,* were identified by Vendan et al. (2010). According to Li et al. (2009), *Pantoea* spp. from cotton roots exhibited antagonistic activity by producing siderophores. Similarly, *Pseudomonas* and *Burkholderia* and two species of *Pantoea* (*P. ananatis* and *P. agglomerans*) were proven to be an excellent producer of siderophores (Yang et al. 2008). Verma et al. (2011) testified that three endophytic bacteria from the roots of *Azadirachta indica* unveiled their potential in suppressing the growth of *Alternaria alternata,* the causal agent of early blight disease in tomatoes. Spore suspension of these strains of endophytic *Streptomyces* pointedly encouraged the growth of plants besides inhibiting the growth of *A. alternata*. Further, siderophore-producing *Pseudomonas* strain GRP3 application enhanced plant health and physiology like chlorophyll a and chlorophyll b of *Vigna radiata* (Sharma et al. 2003). Siderophore-mediated iron acquisition is associated with biocontrol properties of *Trichoderma* spp., which is accomplished by competition for iron and, ultimately, suppression of *Fusarium lycopersici* causing wilt in tomatoes (Segarra et al. 2009).

11.4.4 Phosphate Solubilization

Phosphorus is an imperative element for plant growth, which promotes root development, early flowering, tillering, and metabolic activities like protein synthesis (War and Shaktawat 2003). The association and colonization of rhizospheric bacteria on roots involve direct competition while endophytes may confer more benefit due to less competition. Through the metabolic processes, a variety of soil microbes can solubilize the inaccessible P by releasing organic acids that directly dissolve the rock phosphate or chelating calcium ions that release P into the solution (Walia et al. 2017). Forty endophytic bacterial isolates were isolated from banana roots and tested for solubilization of phosphorus by amending different organic and inorganic sources of phosphorus, in which 67.5% of isolates solubilized the phosphorus from tricalcium phosphate and 7.5% from soy lecithin. Significant differences were found in an average colony and zone of solubilization diameters. The majority of isolates belonged to *Bacillus* spp., among which EB 12 (*Bacillus pumilus*) unveiled the maximum colony diameter at an average of 29.3 mm and EB. 18 (*Bacillus* sp.) exhibited the largest solubilization zone diameter (45.43 mm) (Amanda et al. 2017). Different degrees of phosphate-solubilizing activity and gluconic acid synthesis were seen in endophytic PSB isolated from maize grown in Brazilian Cerrado soil. All of the PSB under investigation produced gluconic acid, and *Bacillus* species were the most effective phosphate solubilizers (De Abreu et al. 2017). *Pseudomonas* strains L111, L228, and L321 possessed virtuous phosphate solubilization activity and particularly in L321 this trait was expressed under phosphate-limiting conditions, which resulted in enhanced plant growth promotion of *Pisum sativum* L. plants (Oteino et al. 2015). Similarly, Jasim et al. 2013 isolated PSB belonging to *Klebsiella* sp., *Bacillu*s sp., and *Enterobacter* sp. from *Piper nigrum,* which significantly enhanced the growth-promoting ability of *Vigna radiata* seedlings. Vendan et al. (2010) scrutinized endophytic bacterial isolates from ginseng (*Panax ginseng*) for their phosphate-solubilizing ability by detecting extracellular solubilization of precipitated tricalcium phosphate with glucose as the sole source of carbon. Half of the endophytic isolates tested showed phosphate-solubilizing ability. Based on the solubilization zone, *Lysinibacillus fusiformis* documented maximum solubilization of mineral phosphate. In the same study, endophytic isolates of *Bacillus cereus* and *B. megaterium* also showed notable solubilization activity.

11.4.5 Induced Systemic Resistance

Endophytes reveal ISR by altering the host's physiological and biochemical reactions, resulting in the synthesis of defense chemicals against pathogens. ISR is normally activated by pathogen-related proteins (PR proteins), the synthesis of polyphenols, flavonoids, phytoalexins, or the induction of signal transduction pathways triggered by jasmonate/SA or ethylene. The PR proteins inhibit pathogen progression and aid in pathogen defense by inducing biotic stress tolerance and reactive oxygen species, which in turn regulate the function of important signaling molecules. Jha (2019) found that the paddy plants were inoculated with *P. aeruginosa* and *P. pseudoalcaligenes* and showed phenolics and flavonoid production along with the induction of PR proteins like enzymes β-1,3-glucanase and catalase in plants even in the absence of pathogen *Pyricularia grisea*, causing blast. The role of defense-related enzymes such as l-phenylalanine ammonia lyase (PAL), peroxidase (POX), polyphenol oxidase (PPO), and (chitinase and β-1,3 glucanase) in disease resistance has been reported by different researchers. A study testified a novel endophytic bacterium *B. oryzicola* YC7007 with antagonistic activity against *F. fujikuroi*, responsible for bakanae disease in rice via ISR and antibiotic production (Hossain et al. 2016). Likewise, Abdallah et al. (2017) isolated five *Bacillus* spp. with antagonistic activities against fusarial wilt and were found to produce SA, with the maximum production recorded for *B. subtilis* SV41. Accordingly, SV41-based treatments showed a slight upregulation in acidic PR-1 and PR-3 expression genes in tomato plants inoculated or not with *F. oxysporum* f. sp. *lycopersici*. Romero-Cortes et al. (2019) conveyed two endophytic bacteria *B. amyloliquefaciens* and *P. fluorescens* individually and in combination investigated the modulation of defense-related genes and systemic resistance against *Alternaria alternata* in *Withania somnifera* plants and enhanced the expression of SA and jasmonic acid-responsive genes in the stressed plants. Endophytic *B. cereuses* S42, from *Nicotiana glauca* stem, possess antagonistic activity against *F. oxysporum* f. sp. *lycopersici*. Culture supernatant and whole-cell suspensions of this endophytic had drastically reduced the wilt by 87–96% and also enhanced the plant growth by 39–79% when compared to inoculated and uninoculated control by phthalic acids and dibutyl esters production (Abdallah et al. 2016) (Fig. 11.3).

11.5 Functional Role in Biocontrol

11.5.1 Endophytic Bacteria in Disease Management

Two strategies, viz., (a) preemptive colonization of the internal tissue and (b) stimulating plant defense mechanisms or producing metabolites that are suppressive to the pathogen have been reported to be involved in plant–pathogen management. In case of preemptive colonization, endophytic bacteria that colonize the same ecological niche as the pathogen does and the closely related species, namely, *P. fluorescens* to control *P. syringae, Agrobacterium radiobacter* to manage *A. tumefaciens,* and *Enterobacter* spp. to manage *Erwinia amylovora*. Endophytic bacteria have been shown to control *Verticillium alboatrum* and *R. solani* on potato (Pleban et al. 1995; Nowak et al. 1995) and rice (Krishnamoorthy and Gnanamanickam 1997), *Fusarium oxysporum* f.sp. *vasinfectum* on cotton (Chen et al. 1995), *Sclerotium rolfsii* on bean (Pleban et al. 1995), *F. oxysporum* f.sp. *pisi*

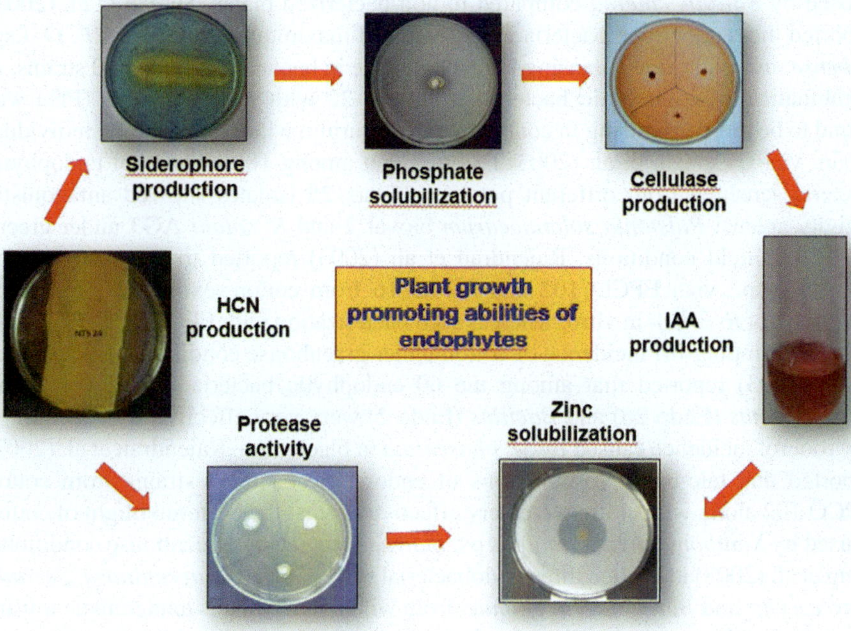

Fig. 11.3 Plant growth-promoting traits (PGP) of bacterial endophytes in plant health

on pea (Benhamou et al. 1996a), and *Urocystis fagacearum* on oak (Brooks et al. 1994). Endophytic bacteria from potato was found to be antagonistic against *Clavibacter michiganensis* subsp. *sepidonicus*, causing bacterial ring rot (Van Buren et al. 1993), and against *Erwinia carotovora* var. *atroseptica* causing soft rot (Sturz and Matheson 1996) in potato. Antibiosis has been shown for endophytic bacteria isolated from maize against *F. moniliforme* (Hinton and Bacon 1995) and potato against *Fusarium sambuctnum*, *F. avenaceum*, *F. oxysporum*, and *R. solani*. Some endophytic bacteria such as *P. fluorescens* 89B-61 induce systemic resistance (ISR) against *P. syringae* pv. *lachrymans* (Liu et al. 1995). An unclarified mode of action is represented by *Kluyvera ascorbata* and *Alcaligene splechaudii* controlling *Xanthomonas campestris* pv. *campestris* (Assis et al. 1998).

Bacillus species are among the most common bacteria found to colonize plants endophytically (Lilley et al. 1996; Mahaffee and Kloepper 1997), and their endophytic ability could play a role in the biocontrol of vascular plant pathogens. Rangeshwaran et al. (2002) reported that endophytic bacteria play a significant role in plant protection against soil-borne plant pathogens and the overall productivity of agricultural ecosystems. Reiter et al. (2002) reported that screening of endophytic bacteria against *E. carotovora* sub sp. *atroseptica* revealed that 38% of the endophytes protected tissue culture potato plants from black leg disease. Barka et al. (2002) reported that in vitro bacterization of grapevine with plant growth-promoting rhizobacterium, *Pseudomonas* PSJN, reduced the incidence of gray mold caused by *Botrytis cinerea* compared to non-bacterized plants. Harish et al. (2003) reported that endophytic bacteria EB 22 showed maximum inhibition of *F. O.* f.sp. *cubense* mycelial growth causing Panama disease in banana among the 40 strains. A combination of endophytic bacterial strain EP 22 with *P. fluorescens* (Pf1) was found to be more promising in controlling the fusarium wilt pathogen than individual strain. van Overbeek et al. (2003) reported that among 180 isolates of endophytic bacteria screened from different potato cultivars 28 isolates showed antagonistic activity against *Ralstonia solanacearum* biovar 2 and *R. solani* AG3 under greenhouse and field conditions. Rajendran et al. (2003) reported that two isolates of *Bacillus* spp., viz., EPCO 102 and EPCO 16 from cotton, exhibited the highest inhibition of *R. solani* in vitro, whereas seed bacterization with EPCO 102 effectively reduced damping-off incidence of cotton under greenhouse conditions. Karthikeyan et al. (2003) reported that among the 60 endophytic bacteria from black gram, *Pseudomonas* (Endo 26) and *Bacillus* (Endo 2) were very effective in reducing the dry root rot incidence caused by *M. phaseolina* in blackgram. Rajendran et al. (2003) reported that talc-based formulations of endophytic bacterial strains from cotton EPCO 102 along with chitin were very effective against the bacterial blight of cotton caused by *Xanthomonas axonopodis* pv. *malvacearum* under greenhouse conditions. Ting et al. (2003) identified three endobacterial isolates, viz., *P. aeruginosa*, *Serratia marcescens,* and *Burkholderia glumae,* from wild banana that is antagonistic toward *F. O.* f.sp. *cubense* causing Panama disease of banana recording high percentage of inhibition of radial growth (PIRG) value.

11.5.2 Endophytes in Insect Pest Management

Most of the research carried out so far have been dealt with the utilization of endophytic fungi in the management of insect pests (Webber 1981; Funk et al. 1983; Gaynor and Hunt 1983; Claydon et al. 1985; Johnson et al. 1985), whereas very few reports have been dealt by the use of endophytic bacteria. Generally, fluorescent pseudomonads influence the growth and development of insects at all stages, including egg, larva, pupa, and adult. *Pseudomonas maltophila* affects the growth of *Helicoverpa zeae,* the corn earworm, leading to 60% reduction in adult emergence. Pupae and adults emerging from the bacteria-infected larvae remained smaller (Bong and Sikorowski 1991). Qingwen et al. (1998) reported that the relative growth rate, consumption rate, and digestibility of feed by larvae of *H. armigera* have been affected when they feed on cotton plants treated with *Pseudomonas gladioli* due to increased polyphenol and terpenoid content in treated plants. As fluorescent pseudomonads are effective rhizosphere colonizers and are endophytic in nature in the plant, attempts have been made to transfer the insecticidal crystal protein from *Bacillus thuringiensis* to *P. fluorescens.* Thus, genetically engineered *P. fluorescens* has been found to be effective against lepidopteran insect pest. Transgenic *P. cepacia* 526 with the crystal protein gene has consistently shown insecticidal activity against tobacco hornworm (Stock et al. 1990). Induction of systemic resistance (ISR) by PGPR against various diseases and pests was considered the most desirable approach in crop protection. *Pseudomonas*-treated rice leaves altered the feeding behavior of leaf folder and reduced larval and pupal weight, increasing the larval mortality and incidence of malformed adults in vitro (Radjacommare et al. 2002). Rajendran et al. (2003) reported that the application of endophytic bacterial strain, Pf1, reduced the aphid and bollworm incidence within 5 days of spraying. Altered feeding behavior of boll worm larvae, reduction in larval as well as pupal weight, and increase in larval mortality were observed in in vitro boll worm bioassay.

11.6 Mechanism of Biocontrol

11.6.1 Growth Promotion Activity

Rhizobacteria have been found to increase plant growth after inoculation onto seeds as a seed treatment or priming are therefore called "plant growth-promoting rhizobacteria" (PGPR) (Kloepper et al. 1980). The mechanisms of growth promotion are complex and appear to comprise both changes in the microbial balance in the rhizosphere and alterations in host plant physiology (Glick et al. 1999). van Peer and Schippers (1988) documented the increased root and shoot fresh weight of tomato, cucumber, lettuce, and potato as a result of bacterization with *Pseudomonas* strains. Many bacteria can produce auxins, gibberellins, cytokinins, and ethylene (Frankenberger and Arshad 1995), and they have often been inferred that rhizobacteria-produced auxins are responsible for growth promotion. Fluorescent pseudomonads increased the plant growth of rice and cotton by 27 and 40 per cent, respectively, when applied to the seed (Sakthivel and Gnanamanickam 1987). In potato plantlets grown in vitro, PsJN increased cytokinin content by inducing synthesis in the early stages of plant growth and development (Lazarovits and Nowak, 1997). Various fluorescent *Pseudomonas* strains improved vegetative seed germination, plant height, cane diameter, Brix values, and cane weight in sugarcane (Viswanathan and Samiyappan 1999). Similarly, the application of *P. fluorescens* Pf1 as seed treatment followed by soil application against root rot effectively supported the enhanced plant growth, better native rhizobium nodulation, and grain yield (Jayashree et al. 2000). *Pseudomonas* PsJN induced plant growth in parallel with an antagonistic effect against *Botrytis cinerea* in grapes grown in vitro (Barka et al. 2002). Bacterization of peanut seeds with fluorescent pseudomonads GRC_2 resulted in increased seed germination, early seedling growth, fresh nodule weight, grain yield, and reduced charcoal rot incidence (Gupta et al. 2002). Direct stimulation may include the enzyme ACC deaminase secreted by PGPR organisms that can lower plant ethylene levels, which are produced during stress conditions (Glick et al. 1999). The suggested role of ACC deaminase is to ensure that the concentration of ethylene does not become elevated to the point that growth (especially of roots) is retarded (Glick, 1995; Glick et al. 1998; Glick et al. 1999), and the significance of ACC deaminase gene has been documented in many of the crop plants in terms of improving plant growth (Mayak et al. 2004).

11.6.2 Induced Systemic Resistance (ISR)

Induced resistance is a state of enhanced defensive capacity developed by a plant when appropriately stimulated (van Loon et al. 1998). In 1991, two research groups independently described induced systemic resistance (ISR) as the mode of action of disease suppression by non-pathogenic rhizobacteria (Van Peer et al. 1991; Wei et al. 1991). Since then, the involvement of ISR in disease suppression has been studied in a wide range of biological control microorganisms, and, in many cases, ISR was found to be involved. Rhizobacteria-mediated induced systemic resistance (ISR) is phenotypically similar to the better-known systemic acquired resistance (SAR), the induced state that develops when plants successfully activate their defense mechanism in response to primary infection by a pathogen, notably when the latter induces a hypersensitive reaction through which it becomes limited in a local necrotic lesion of brown, desiccated tissue (Ryals et al. 1996; Sticher et al. 1997). SAR is a generally occurring phenomenon that confers an enhanced defensive capacity against all types of pathogens. Under the influence of the primary infection, a signal – the nature of which is still unclear – is generated and transported throughout the plant, thereby establishing the induced state. SAR is characterized by a requirement for salicylic acid (SA) as a signal and by the SA-mediated accumulation of several families of pathogenesis-related proteins (PRs), among which are chitinases and glucanases with potential antifungal activity (Durner et al. 1997).

11.6.3 Peroxidase (PO)

Peroxidases have been found to play a major role in the regulation of plant cell elongation, phenol oxidation, polysaccharide crosslinking, IAA oxidation, crosslinking of extensin monomers, oxidation of hydroxyl–cinnamyl alcohols into free radical intermediates and wound healing (Vidhyasekaran et al. 1997), and catalyzes the last step in the biosynthesis of lignin and other oxidative phenols (Bruce and West 1989). PO is associated with disease resistance in plants (Hammerschmidt and Kuć 1982). Viswanathan and Samiyappan (1999) reported that enhanced levels of PO were noticed in fluorescent pseudomonads-treated sugarcane in response to infection by *C. falcatum.* In cucumber, increased activity of PO was noticed in plants treated with *P. corrugata* strain 13 and induction of isoperoxidases by this bacterium may play an important role in ISR (Chen et al. 2000). In groundnut, increased activity of PO was observed due to the application of *P. fluorescens* and PO isoforms were expressed at higher levels (Meena et al. 2000). In tomato, peroxidase isoform was significantly increased in PGPR-treated plants inoculated with TSWV compared to the control plants (Kandan et al. 2002). Bharathi (2001) reported that chilli plants treated with mixtures of strains of PGPR, viz., Pf1 + *B. subtilis* + Neem + chitin, showed enhanced PO activity against CMV. The biochemical analysis of plant defense-related enzymes showed the induction of new isoforms of superoxide dismutase in tomato plants treated with arbuscular mycorrhizal fungi against *Phytophthora* infection (Pozo et al. 2001). Two peroxidase isoforms have been induced in the fluorescent pseudomonads-treated rice plants challenged with *R. solani* (Nandakumar et al. 2001). High-level expression of PO was reported in *P. fluorescens* Pf1-treated tomato plants challenged with *F. oxysporum* f.sp. *lycopersici* (Ramamoorthy et al. 2002). Sible (2003) reported that three isoforms of PO were expressed on fruit peel tissue upon treatment with the bioformulation containing *P. fluorescens*, FP7, amended with chitin. Harish et al. (2003) reported more induction of PO isoforms in the endophytic bacteria-treated plants challenged with viruliferous aphids than in untreated plants against BBTV.

11.6.4 Polyphenol Oxidase (PPO)

PPO usually accumulates upon wounding in plants and can be induced via the octadecanoid defense signal pathway (Constabel et al. 1995). PGPR-treated sugarcane after pathogen inoculation showed comparatively enhanced induction of PPO isoforms than the untreated plants (Viswanathan and Samiyappan 1999). The activity of PPO was stimulated by the root application of *P. corrugata* 13 *and P. aureofaciens* in cucumber roots in response to infection by *P. aphanidermatum* and correlated with disease resistance (Chen et al. 2000). Application of a mixture of PGPR strains to chillies expressed more isoforms of PPO when compared to control plants against CMV (Bharathi 2001). Expression of new PPO isoform was observed in *P. fluorescens* Pf1-treated tomato plants challenged with *F. oxysporum* f.sp. *lycopersici* (Ramamoorthy et al. 2002). Similarly, increased activity of PPO was observed in tomato by fluorescent pseudomonads in response to infection by *tomato spotted wilt virus* (Kandan et al. 2002). Reported that *P. fluorescens* induced a twofold increase in PPO in banana roots against fusarium wilt disease.

11.6.5 Phenylalanine Ammonia Lyase (PAL)

Phenylalanine ammonia lyase is the first enzyme involved in the phenylpropanoid pathway and plays a key role in the biosynthesis of phenolics and phytoalexins (Bashan et al. 1985). Induction of enzymes such as PAL and PO leading to the accumulation of phenolics and lignin can occur in response to insect and pathogen attacks, exposure to oxidizing pollutants, and mechanical stimulation, and is thought to function in the resistance of plants to damage by these stresses. PAL is the key enzyme in inducing the synthesis of salicylic acid (SA) which induces systemic resistance in many plants. An increase in the level of mRNAs encoding for PAL was recorded in the early stage of interaction between bean roots and various rhizobacteria (Zdor and Anderson 1992). Plants treated with *Pseudomonas* strains initially showed higher levels of PAL compared to the control (Chen et al. 2000). Similarly, in tomato and chilli increased activity of PAL was observed in *P. fluorescens* strain Pf1-treated plants challenged with *F. oxysporum* f.sp. *lycopersici* and *C. capsici,* respectively (Ramamoorthy et al. 2001). Radjacommare et al. (2004) reported that seedling dip with talc-based formulation of *P. fluorescens* induced the activity of PAL in finger millet leaves against blast disease. Seed treatment and seedling root dipping with beneficial bacteria induced early and enhanced levels of PAL in rice plants (Prabhukarthikeyan et al. 2022).

11.6.6 Scavengers of Active Oxygen Species (AOS)

The oxidative burst contributes to the defense of the host plant in several ways, i.e., by means of crosslinking of cell wall proteins, which render the cell wall more resistant to attack by fungal enzymes (Bradley et al. 1992), act directly as toxins against the pathogens (Mehdy 1994), and act as second messengers for the activation of a variety of defense genes (Lamb and Dixon 1997). Scavengers of active oxygen species like catalase (which catalyzes the decomposition of H_2O_2) (Scandalios 1994), superoxide dismutase (which scavenges $\cdot O_2^-$) (Bowler et al. 1992), and peroxidase (which scavenges H_2O_2) suppress the oxidative burst (Vera-Estrell et al. 1993) and inhibit tissue necrotization. Plants produce active oxygen species such as superoxide anion ($\cdot O_2^-$), hydrogen peroxide (H_2O_2), and hydroxyl radical ($\cdot OH$) are one of the earliest responses of the plants to attempted infection by pathogens (Grant and Loake 2000). Catalase and peroxidase are of particular interest because of their role in binding SA, which plays an important role in induced resistance (Anderson et al. 1998). β-glucosidase produces "free" SA, which can enhance plant disease resistance. In plants, β-glucosidases are reported to function in phytohormone metabolism (Smith and van Staden 1978) and defense against phytopathogens and herbivores (Poulton 1990). Perhaps the well-documented function of β-glucosidase in plants is the release of HCN from cyanogenic glucosides in tissues damaged by herbivores and pathogens (Poulton 1990).

Phenolics are fungitoxic in nature and increase the physical and mechanical strength of the host cell wall. Lignin is a phenolic polymer that is difficult to be breached by pathogens and has been implicated in plant defense against pests and diseases (Nicholson and Hammerschmidt 1992). Treatment of pea seeds with *P. fluorescens* strain 63–28 resulted in the formation of structural barriers, viz., cell wall apposition (papillae) and deposition of newly formed callose and accumulation of phenolic compounds at the site of penetration of invading hyphae of *P. ultimum* and *F. oxysporum* f. sp. *pisi* (Benhamou et al. 1996a, 1996b). M'Piga et al. (1997) reported that in tomatoes, bacterization with the same bacterial strain has brought about cell wall thickening, deposition of phenolic compounds, and formation of callose, resulting in restricted growth of *F. oxysporum* f.sp. *radicis-lycopersici*. Thus, the rapid strengthening of reaction sites of fungal and insect entry delays the infection process and allows sufficient time for the host to build up other defense reactions.

11.6.7 Pathogenesis-Related Proteins (PRs)

Induced resistance by PGPR is associated with the accumulation of PR proteins. Pathogenesis-related proteins are designated as PRs and are defined as proteins coded by the host plant but induced specifically in pathological or related situations. They are not only accumulated locally in the infected leaves but also induced systemically associated with the development of systemic-induced resistance against further infection by pathogens (van Loon et al. 1994). Originally, five main classes of PRs (PR1-5) were characterized by both biochemical and molecular techniques in tobacco (van Loon et al. 1987). In 1994, nomenclature for PRs was proposed based on their grouping into families sharing amino acid sequences, serological relationships, and/or biological functions (van Loon et al. 1994). Later, van Loon and van Strien (1999) classified 14 classes of PRs (PR1–14).

Chitinases are PR-proteins that hydrolyze chitin, a major cell wall component (3–10%) of higher fungi. Chitinases cleave a bond between C_1 and C_4 of two consecutive *N*-acetyl glucosamine (GlcNAc) either by endolytic or exolytic mechanisms. Enhanced accumulation of chitinase in tobacco and bean leaves was observed in response to the application of *Pseudomonas* spp. to roots (Zdor and Anderson 1992). Maurhofer et al. (1994) reported that ISR by *P. fluorescens* CHA0 against tobacco necrosis virus (TNV) in tobacco was associated with the accumulation of PR proteins, namely, β-1, 3 glucanases and endochitinases. They also established that ineffective strain P3 did not accumulate PR proteins, indicating the involvement of PR proteins in the induction of resistance. In pea, seed bacterization with *P. fluorescens* 63–28 has signaled the host to produce hydrolytic enzymes such as chitinases and β-1, 3 glucanases (Benhamou et al. 1996b). Mahendran (1996) found that seed treatment with *P. fluorescens* induced chitinase isoforms with molecular weights of 31, 25, and 23 kDa in rice. Botha et al. (1998) observed that differential expression of chitinase isozymes in near–isogenic wheat lines challenged with Russian wheat aphid. Isoelectrofocusing (IEF) gels revealed the existence of multiple isoforms of chitinase with an additional band in the R. *solani*-challenged plants. Ramamoorthy et al. (2002) reported the induction of 46 kDa chitinase due to *P. fluorescens* Pf1 treatment when challenge inoculated with *F. oxysporum* f.sp. *lycopersici* in tomato. McKenzie et al. (2002) studied the expression of PR proteins, viz., chitinase, β 1,3glucanase, and peroxidase, due to Gemini virus infection and *Bemisia tabaci* infestation in tomato. The results suggest that a significant increase in the enzymatic activity of all measured PR proteins compared to control plants was only seen in viruliferous whitefly-infested plants. It is concluded that PR protein response is much more intense when it is attacked by whiteflies carrying *tomato mottle virus* than by white fly alone. Accumulation of chitinases and peroxidases with the onset of ISR by endophytic bacteria has also been observed in some plants. The transgenic plants expressing the insect chitinase appear to be more resistant to feeding damage by tobacco hornworms and tobacco budworms (Ding et al. 1997; Ding 1995). PR-6 is a proteinase inhibitor implicated in defense against insects and other herbivores, microorganisms, and nematodes (Ryan 1990). PR-9 family of peroxidases is likely to function in strengthening plant cell walls by catalyzing lignin deposition in reaction to microbial attack (van Loon and van Strien 1999). PR-10 type is structurally related to ribonucleases and is effective against viruses (Moiseyev et al. 1997). The collective function of several PRs may be effective in inhibiting pathogen growth, multiplication, and/or spread of the pathogen and be responsible for the state of induced resistance (Ryals et al. 1996). Durgadevi et al. (2021) reported the induction of chitinase gene in *Bacillus*-treated rice plants challenge inoculated with *R. solani*.

11.6.8 Interactions between Signaling Molecules Involved in Plant Defense

In response to the attack by pathogens or pests, plants activate signaling cascades, leading to the accumulation of endogenous hormones that trigger the defense responses. Salicyclic acid (SA), jasmonic acid (JA), and ethylene (E) are plant-specific hormones involved in communicating the attack by many pathogens and pests in a broad range of plant species. SA, JA, and E signaling cascades do not activate defenses independently, but they rather establish complex interactions that determine the response mounted in each condition. The interaction among SA, JA, and ET defense signaling pathways can be antagonistic, cooperative, or synergistic, depending on the plant species, the combination of organisms attacking the plant and the developmental and physiological state of the plant (Rojo et al. 2003). Malamy et al. (1990) reported that in tobacco infected with TMV endogenous salicylic acid levels rose before the onset of enhanced resistance to TMV and the expression of PR-1. The resistance to many pathogens like *Erysiphe cichoracearum*, *E. orontii*, *Oidium lycopersicum*, *Botrytis cinerea,* and *Plectosphaerella cucumerina* (Thomma et al. 1999; Ellis and Turner 2001) requires both JA and E for the activation of defense mechanisms. In tomato, salicylic acid induces resistance to the systemic movement of *cucumber mosaic virus* but does not affect its replication or cell-to-cell movement. However, in the case of *tobacco mosaic virus* in tobacco, salicylic acid appears to induce interference with the synthesis of viral DNA (Murphy et al. 2001). The signaling steps, like the perception of the cue to the regulation of gene expression required for defense, have been studied extensively (Glazebrook 2005; Dong 2001; León et al. 2001; Asai et al. 2002, Holt et al. 2003).

11.7 Role of Omics in Biocontrol

Endophytes, including endophytic fungi, endophytic bacteria, and endophytic actinomycetes, exist in diverse organs, tissues, and intercellular spaces of plants without any detrimental impact on the host plant (Wilson 1995; Jia et al. 2016). Entophytes developed a symbiotic beneficial relationship with host plants during long-term co-evolution. Compared with soil microorganisms, endophytes within plants possessed direct impacts on plants.

11.7.1 Metagenomics

The main actors in the biogeochemical processes of the plant environment are endophytic bacteria. The plant contains millions of different prokaryotic and eukaryotic species, including bacteria, archaea, viruses, phages, fungi, and protozoa. There are two categories of techniques used to determine the taxonomic diversity and phylogenetic relationships of microorganisms: culture-dependent and -independent techniques. In culture-dependent procedures, only bacteria that can grow in a lab setting can be identified and examined, such as (1) the plate count method and (2) the BIOLOG plate method (Biolog Inc., Hayward, CA, USA). The soil-found unculturable microorganisms were not amenable to this approach. There are four different types of culture-independent methods: (1) microbial lipid-based [fatty acid methyl esters, DNA re-association, and phospholipid fatty acid analysis (PLFA)]; (2) non-PCR-methods (DNA and, GC based); (3) PCR-associated (RFLP, RAPD, DGGE, FISH, etc.); and (4) sequence-associated (clonal library sequencing, marker-based, and whole metagenome shotgun sequencing). Shotgun sequencing and marker gene-based sequencing are two of the most popular sequencing methods now being employed because of their simplicity, efficiency, and affordability. Metagenomics and transcriptomics, two recent cutting-edge approaches, can reveal details about the functional traits of different endophyte species (Kaul et al. 2016). Although closely related strains of bacteria have 16S rRNA sequences that are $>98\%$ identical, some bacterial characteristics are less conserved across evolutionary lineages and differ in these strains (Martiny et al. 2015). The identification of specific species, strains, or putative functional units in a bacterial community that are essential for interacting with host plants and subsequently promoting development may be made possible with further advancements in these techniques.

11.7.2 Plant–Endophyte Interactions in Genomic and Post-Genomic Era

Research on plant–microbe interactions is becoming more and more dependent on multi-omics, which includes genomes, transcriptomics, proteomics, and metabolomics, as a result of recent advancements in high-throughput sequencing and systems biology technologies. While transcriptomics examines RNA and gene expression patterns, genomics works with complete genome sequences. While metabolomics endeavors to unravel the biosynthesis, disintegration, and conversion of metabolites in a specific organism, proteomics analyzes the architectures, functionalities, and interconnections of dynamic proteins. It has unveiled the genes that code for their plant-colonization preferences, as well as the generation of a range of bioactive substances, such as genes for the synthesis of hormones and antibiotics in plants, fixing nitrogen, and acquiring nutrition (K, P, Fe, etc.). *Bacillus subtilis* Bbv57's potential as a bioactive agent for phytopathogens was identified by genome analysis (Raguchander et al. 2022). By using a proteomic analysis based on mass spectrometry, Lery et al. (2011) found 78 differentially expressed proteins connected to the endophytic Glucono acetobacter–sugarcane relationship. Untargeted metabolomics involves a thorough examination of the sample's many known and undiscovered compounds, and targeted metabolomics, which involves the measurement of the sample's particular metabolites (Roberts et al. 2012). The results of untargeted metabolomics can be validated by targeted metabolomics, and untargeted metabolomics can be used to find novel compounds. While investigating the causal linkages between genes and metabolites, integrated analysis of transcriptomics and metabolomics could assess both the genes and the metabolites that are differentially expressed along a time curve. Through bioinformatic methods like functional annotation and pathway enrichment, it is possible to identify the important genes, metabolites, and signaling and metabolic pathways, which improves understanding of the physiological and molecular mechanisms. Recent studies on rice, potatoes, tobacco, and other plants have made extensive use of combined transcriptomics–metabolomics analysis.

Several genes included in iron uptake and metabolism were found by studying the endophytic genomes. Compared to *E. coli* K12, Enterobacter sp. 638 possesses nine ABC transporters for siderophore complexes in its genome (Taghavi et al. 2010). With 22 iron TonB receptor genes, Azoarcus sp. BH72 has twice as many as its free-living soil buddy EbN1 (Krause et al. 2006). These findings imply that endophytic bacteria can effectively compete with other microbes, particularly phytopathogens, for iron in low-iron environments. The endophytic genomes have been found to contain several genes involved in quorum sensing (QS), in addition to the favorable features for plants discussed above. For instance, the genome of *Serratia proteamaculans* 568 has 24 luxR QS genes. Significant colonization-related attributes, notably swimming motility or biofilm formation, are controlled by QS in the similar endophytic bacterium *S. plymuthica* G3 (Liu et al. 2011). These features are QS-independent in certain free-living Serratia spp., thus raising the intriguing possibility that the exact function of QS relies on the bacterium's ecology. Further genome sequencing uncovered genes that may be crucial for the endophytic way of life. In order to defend it against plant reactive oxygen species (ROS), diazotrophic *K. pneumoniae* (Kp) 342 possesses genes for superoxide dismutases, catalases, peroxidases, and reductases (Fouts et al. 2008). Metaproteogenomics links the proteome and the genome of the environmental sample. This allows the identification of more proteins (functions) than proteomics alone. Rice's phyllosphere and rhizosphere's microbial communities were studied using a metaproteogenomic method (Wang et al. 2019). A novel approach for closely examining endophytism and revealing the characteristics needed to harbor plants as a habitat has been made available through endophyte genome analysis. It has revealed the genes crucial for the endophytic lifestyle that are found frequently in endophyte genomes, including those for nitrogen fixation, the production of phytohormones (IAA, GA, etc.), the acquisition of minerals (Fe, P, etc.), stress tolerance, adhesion, and other genes related to colonization (Firrincieli et al. 2015). These PGP traits explain the role of microbes in nutrient cycling as well as their ability to colonize plant endosphere.

The genetic characteristics that directly or indirectly determine the various bioactivities, as well as colonization preferences, have been identified by whole-genome analysis of endophytic bacteria. It helps with the identification of specific genes involved in metabolic processes like transport, endophytic secretory system, surface attachment and insertion elements, plant growth stimulation, and antibiotic mechanisms. These investigations have improved our knowledge of the ecology and evolution of endophytes. The endophytic bacteria *Pantoea ananatis* has the potential to be used in biotechnology since it contains genes for N-acyl homoserine lactone synthases and hydrolases, hyperadherence factors, and fusaric acid resistance proteins (Megías et al. 2016). Studies on comparative genomics have indicated that the selection of dissimilar habitats is mostly influenced by differences in the metabolic, secretory, transport, and surface attachment proteins. As potential molecular determinants underlying the divergent phenotypic behavior of closely related species during host plant colonization as symbiotic endophytes or as phytopathogen, Monteiro et al. (2012) identified lipopolysaccharide and adhesins. Analyzing the genomes of endophytic isolates and their non-endophytic counterparts exposes characteristics that are probably necessary for initiating and sustaining plant–microbial interactions. Realizing the genetic influences on niche acclimation is also useful (Lòpez-Fernàndez et al. 2015). Genomes of endophytic species that are directly connected but have diverse operational roles in the host plant can also be examined to learn how adaptable they are and how their evolution has progressed. Three strains of *P. ananatis* have had their genomes compared by Sheibani-Tezerji et al. (2015). The three strains have various ways of interacting with the host plant despite being separated as maize seed endophytes. Comparing the strains' genomes revealed changes in the genes associated with secretory proteins, integrases, transposons, and phages. Thus, understanding the metabolic diversity among the constituents of a microbiome can be achieved through comparative genome analysis of endophytes for various metabolite gene clusters, and this insight can then be utilized in metabolic engineering.

Understanding the potential of endophytes, which are uncultured microbial populations, requires a metagenomics approach (Dinsdale et al. 2008). In this method, DNA is taken from the entire population and its gene content is examined. It has been discovered that transcriptomics is a practical method for examining the microbial populations connected to various plants (Sheibani-Tezerji et al. 2015). By contrasting the transcriptomes of groups of interdependent species, it helps us

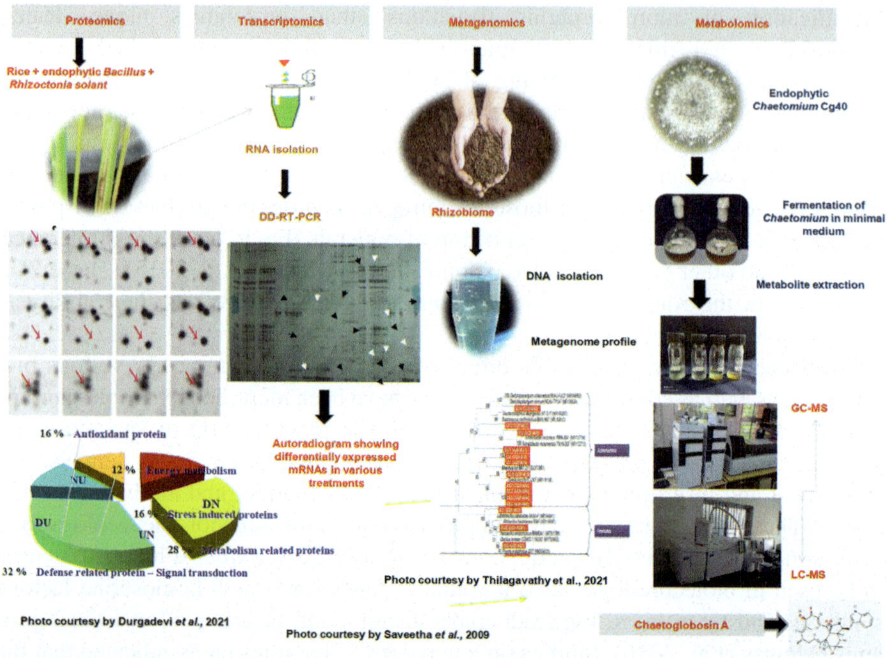

Fig. 11.4 Plant–endophytic bacteria interactions in genomic era

understand how microbial communities respond to changing environments. A thorough examination of the genes that are differentially expressed in microbes and plants would shed light on the structure and operation of their mutualistic relationships. A more accurate picture of the simultaneous gene expression in both symbiotic partners is given by dual RNA-seq transcriptional profiling. Discovered that nutrient absorption was increased when *Azospirillum brasilense* colonized the wheat roots. By contrasting the transcriptomes of plants with and without endophytes, we can gain a better understanding of the mechanisms underpinning endophyte-mediated disease resistance and plant growth promotion capacities (Fig. 11.4).

11.7.3 Proteomics and Metaproteomics Study

Post-genomic investigations of microbial communities are becoming more and more frequent as a result of recent advances in technology as genomics-based analyses are unable to reveal the actual function of microbial communities in situ. Proteomics is the comprehensive examination of the numerous proteins that an organism generates (Wilkins et al. 1996), whereas in metaproteomics, the functional expression of the metagenome is identified, and the metabolic processes occurring within a community during the time of sampling are elucidated. Another name for it is whole community proteomics. Using a proteomic analysis based on mass spectrometry, Lery et al. (2011) identified 78 proteins that were differently expressed between the sugarcane–Glucono acetobacter interaction model and the control. Studies based on the proteome are inadequate without genetic data. Due to the existence of interfering compounds such as alkaloids, polyphenols, thick polysaccharides, lipids, organic acids, and other secondary metabolites, protein extraction and sample preparation are also complex steps in proteomic investigations. In their 2012 research on the microbial communities in the phyllosphere and rhizosphere of rice, Knief et al. (2012) used a metaproteogenomic technique. Despite the nifH genes being present in both microenvironments, the results demonstrated that only the rhizosphere had expression. New targets for functional studies include proteins involved in plant–endophyte interactions that could not be examined in farmed isolates. It is possible to successfully determine plant–bacterial relationships by using the plant-associated bacterial protein secretion system (Downie 2010). During tripartite interactions of fluorescent pseudomonas (Fp 7) with turmeric rhizome rot, Prabhukarthikeyan et al. (2017) reported 12 differentially expressed proteins, including beta subunit-like tryptophan synthase, phosphoglycerate kinase, cellulose synthase, ATP synthase, cysteine-rich peptide, and ribosomal protein. The inability to identify proteins using the metaproteomic technique due to the lack of closely comparable reference genomes would be overcome by combining the metagenome and metaproteome analyses. Endophytic *B. subtilis* EBPBS4 of rice differentially expressed 15 resistant proteins, among which disease resistance protein RGA1, ascorbate peroxidases, chitinase, and serine-threonine protein kinase were involved in sheath blight disease resistance. These proteins were found to be differentially regulated during the tripartite interaction of host, pathogen, and biocontrol agent (Durgadevi et al. 2021).

11.7.4 Volatilomics in Plant Growth Regulation

A collection of beneficial bacteria known as PGPR can increase plant development through a variety of processes, including fixing nitrogen, solubilizing phosphorus, disease resistance, enhanced nutrient uptake, altered plant hormone concentrations, and the production of volatile organic compounds (VOCs). Microbial VOCs play a significant antagonistic and growth-promoting role in plants. In contrast to plant growth stimulants, microorganisms can influence plant development indirectly through the production of volatile organic compounds (VOCs). By controlling plant hormone synthesis and transport, as well as fostering systemic tolerance and resistance, microbial VOCs can aid in plant growth. VOCs generated by PGPR control the synthesis and transportation of plant hormones. According to *Bacillus subtilis* GB03 VOCs improved the stimulation of plant development through the cytokinin pathway. After screening PGPR from the soil of a tropical rainforest, Huang et al. (2015) determined that it was *Bacillus subtilis*. Studies have shown that these strains' volatile compounds can control the expression of genes associated with 3-indoleacetic acid (IAA) to support the growth of *A. thaliana*. According to Zhang et al. (2008), soil bacteria can enhance the photosynthesis of *A. thaliana* by reducing the levels of ABA and glucose induction in the plant. After being subjected to GB03 (*Bacillus subtilis*), *A. thaliana* produced fewer downstream metabolites and transcripts involved in the production of ABA. Xie et al. (2020) found that VOCs produced by *Bacillus subtilis* DZSY21 strongly inhibited the growth of *Curvularia lunata*. In addition, the volatiles increased the buildup of intracellular ROS in conidia while suppressing the germination of conidia and the upregulation of pathogenicity genes in *C. lunata* mycelium. The effects of VOCs produced by *Bacillus subtilis* SYST2 on the control of plant hormones in tomato (*Lycopersicum esculentum*) plants were demonstrated by Tahir et al. (2017). The growth-related genes were differentially expressed as a result of SYST2 volatiles, and tomato (*Lycopersicum esculentum*) growth was aided by higher auxin, gibberellin, and cytokinin concentrations. Sridharan et al. (2020) reported *T. longibrachiatum* EF5 endophytic fungal antagonist of rice exerts 138 mVOCs, when *T. longibrachiatum* EF5 interacted with *Sclerotium rolfsii* and *Macrophomina phaseolina*. *T. longibrachiatum* EF5 controls the levels of a variety of substances during these interactions, including longifolene, caryophyllene, 1-butanol 2-methyl, cedrene, and cuprenene. These substances play a role in the biosynthesis of sesquiterpenoid and alkane as well as the hydrolysis of trimethylamine. We shed light on the several ways *T. longibrachiatum* EF5 engages in antagonistic behavior, including hyperparasitism, antagonism, and biocontrol via mVOCs. In addition to their ability to combat pathogens, these metabolites may also help plants proliferate.

11.7.5 Practical Applications

The awareness of the use of biocontrol agents as seed treatment/soil application in integrated disease management (IDM) in cereals, pulses and oilseeds, other crops, etc., has been strengthened over the decades for the benefit of the farming community as an alternative to intensive chemical farming. Further, bioformulation has been slowly developed and evaluated from talc to liquid for ease of application through a drip, which in turn reduces dependency on labors.

The success of biological control of plant diseases depends on the availability of effective formulation of biocontrol agents, their survival during storage, rapid multiplication, and colonization after inoculation (Becker and Schwinn 1993). Different formulations are being developed; however, formulations should be easily applicable, should have longer shelf life, and cost-effective. Mass multiplication of PGPR in a suitable medium and development of a powder formulation was first carried out in the 1980s. A dried powder formulation of PGPR is especially important for seed treatment and soil application. The survival and the effectiveness of methylcellulose in a powder formulation for coating sugar beet seed has been well documented. A talc-based powder formulation of PGPR has been developed for inoculation of potato seed pieces (Kloepper and Schroth 1981) and found that potato seed treatment with powder formulation produced a higher level of root colonization by PGPR than aqueous preparations. Similarly, the suitability of different carriers for the development of stable formulation has been tested and found that in talc-based and peat-based formulations, the bacterial population has been stable up to 240 days of storage period (Vidhyasekaran and Muthamilan 1995).

Further, several systems have been reported for endophytic bacteria. The use of LS213, a commercial formulation containing formulated spores of *B. subtilis* strain GB03, *B. amyloliquefaciens* strain IN 937a, and chitosan as a carrier, significantly increased the growth of tomato, cucumber, tobacco, and pepper in addition to protection against bacterial spot and late blight of tomato, angular leaf spot of cucumber, and blue mold of tobacco. Different microbial formulations have been developed using liquid or solid materials as carriers. The solid formulations are based on inorganic or organic carriers, prepared in solid, granular, or powdery forms and classified according to their particle sizes or application mode (Malusa and Vassilev 2014; Stamenković et al. 2018). Liquid inoculants are microbial cultures modified with water, oil, or polymers (i.e., additives) that improve cell-suspension viscosity, stability, and dispersion capacity (Bashan et al. 2016; Malusá et al. 2016). Further, Manikandan et al. (2010) reported that liquid formulation with glycerol amendment maintained higher population of *P. fluorescens,* (Pf1) up to 6 months of storage. Combination of seed treatment, seedling dip, and soil drenching of liquid formulation of *P. fluorescens,* Pf1, recorded minimum disease incidence of *Fusarium* wilt of tomato under glasshouse (17.33%) and field (4.81%) conditions (Table 11.2).

11.8 Conclusion and Future Thrusts

To conclude, the use of biocontrol agents is going to be mandatory in plant disease management, which is the need of the hour and essential to safeguard the environment and soil health. In addition, it is inseparable from organic farming and in future proper identity with accession number and liquid formulation with high cell count will mitigate plant diseases in a timely manner. The comprehensive study of endophytes and their di or tri-partite way of interactions in the induction of plant responses is highly essential for developing new strategies for enhanced plant growth activity with simultaneous disease control. The present era is focusing on the development of bioproducts with more feasibility, accessibility, and availability, and the biological product either as microbes or microbial secondary product-based formulation would be effective and an eco-friendly measure in the twentieth century. The major constraints in the biological-originated product are quality and shelf life of the product. Intensive research is required to develop the commercial product with extended shelf life with stable organic compound to suppress the plant disease development. The exploration of phylloplane microbial engineering to develop a novel product would be an innovative strategy to simplify the disease management. As mentioned, all these examples add to a growing number of findings that are paving the way to strategies that use bacterial endophytes to boost plant immunity.

Table 11.2 Effects of different bioformulations of epi–endophytes against major crop diseases

Formulation	Epi/endophytes	Crop	Disease	References
Talc based	*P. fluorescens*	Rice	Sheath blight	Radjacommare et al. (2002)
		Tomato	*TSWV*	Kandan et al. (2005)
		Mung bean	*Macrophomina* root rot	Saravanakumar et al. (2007)
		Rice	Sheath rot	Saravanakumar et al. (2007)
		Tea	Blister blight	Saravanakumar et al. (2009)
		Sugarcane	Red rot	
		Mango	Anthracnose	
		Rice	Sheath blight	Vidhyasekaran and Muthamilan (1999)
Talc based	*B. subtilis* (EPC 5) as bioconsortia	Coconut	Basal stem rot	Rajendran et al. (2018)
Lignite based	*P. fluorescens*	Rice	Sheath blight	Vidhyasekaran and Muthamilan (1999)
Peat based	*P. fluorescens*	Rice	Sheath blight	Vidhyasekaran and Muthamilan (1999)
	P. chlororaphis, *B. subtilis*	Turmeric	Rhizome rot	Nakkeeran et al. (2004)
Liquid based	*P. fluorescens*	Tomato	*Fusarium* wilt	Manikandan et al. (2010)
		Mango	Stem end rot	Parthasarathy et al. (2016)
	P. fluorescens (FP 7)	Turmeric	Rhizome rot	Prabhukarthikeyan et al. (2017)
	B. Subtilis (Bbv 57)	Cut flower	*Fusarium* wilt	Ramyabharathi et al. (2020)
Water in oil based	*Fluorescent pseudomonads* (FP7)	Banana	Anthracnose	Faisal et al. (2014)

Further, to understand the endophytic lifestyle of symbionts, combined genomic and transcriptome analysis is more helpful. With this, efficient delivery system has to be strengthened for the benefits of farming community at the field level while using the technology developed at the doorstep of laboratory.

References

Abd-Alla MH, El-Sayed ESA, Rasmey AHM (2013) Indole-3-acetic acid (IAA) production by Streptomyces atrovirens isolated from rhizospheric soil in Egypt. J Biol Earth Sci 3(2):182–193

Abdallah RAB, Mokni-Tlili S, Nefzi A, Jabnoun-Khiareddine H, Daami-Remadi M (2016) Biocontrol of Fusarium wilt and growth promotion of tomato plants using endophytic bacteria isolated from Nicotiana glauca organs. Biol Control 97:80–88

Abdallah RAB, Stedel C, Garagounis C, Nefzi A, Jabnoun-Khiareddine H, Papadopoulou KK, Daami-Remadi M (2017) Involvement of lipopeptide antibiotics and chitinase genes and induction of host defense in suppression of Fusarium wilt by endophytic Bacillus spp. in tomato. Crop Prot 99:45–58

Adeleke BS, Fadiji AE, Ayilara MS, Igiehon ON, Nwachukwu BC, Babalola OO (2022) Strategies to enhance the use of endophytes as bioinoculants in agriculture. Horticulturae 8(6):498

Ali M, Ali Q, Sohail MA, Ashraf MF, Saleem MH, Hussain S, Zhou L (2021) Diversity and taxonomic distribution of endophytic bacterial community in the rice plant and its prospective. Int J Mol Sci 22(18):10165

Amanda DMM, Izabela CPG, Silvia N, Adelica AX, Wellington SG, Jose AD, Marlon CTP (2017) Phosphate solubilization by endophytic bacteria isolated from banana trees. Ann Braz Acad Sci 89(4):2945–2954

Anderson MD, Chen Z, Klessig D (1998) Possible involvement of lipid peroxidation in salicylic acid-mediated induction of PR-1 gene expression. Phytochemistry 47:555–566

Archana T, Rajendran L, Manoranjitham SK, Santhana Krishnan VP, Paramasivan M, Karthikeyan G (2020) Culture-dependent analysis of seed bacterial endophyte, Pseudomonas spp. EGN 1 against the stem rot disease (Sclerotium rolfsii Sacc.) in groundnut. Egypt J Biol Pest Control 30(1):1–13

Asai T, Tena G, Plotnikova J, Willmann MR, Chiu W-L, Gomez-Gomez L, Boller T, Ausubel FM, Sheen J (2002) MAP kinase signaling cascade in Arabidopsis innate immunity. Nature 415:977–983

Assis SMP, de Silveira EB, Mariano LR, Menezes D (1998) Endophytic bacteria—method for isolation and antagoniststic potential against cabbage black rot. Summer Phytopathologica 24: 216–220

Bari R, Jones JDG (2009) Role of plant hormones in plant defence responses. Plant Mol Biol 69: 473–488

Barka EA, Gognies S, Nowak J, Audran J, Belarbi A (2002) Inhibitory effect of endophyte bacteria on *Botrytis cinerea* and its influence to promote the grapevine growth. Biol Control 24:135–142

Bashan Y, Okon Y, Henis Y (1985) Peroxidase, polyphenol oxidase and phenols in relation to resistance against *Pseudomonas syringae* pv. *Tomato* in tomato. Canadian J Bot 65:366–372

Bashan Y, de-Bashan LE, Prabhu SR (2016) Superior polymeric formulations and emerging innovative products of bacterial inoculants for sustainable agricultura and the environment. In: Singh HB, Sarma BK, Keswani C (eds) Agriculturally important microorganisms – commercialization and regulatory requirements in Asia. Springer, Berlin, pp 15–46. https://doi.org/10.1007/978-981-10-2576-1_2

Becker JO, Schwinn FJ (1993) Control of soil-borne pathogens with living bacteria and fungi: status and outlook. Pestic Sci 37(4):355–363

Beltran-Garcia MJ, Martinez-Rodriguez A, Olmos-Arriaga I, Valdez-Salas B, Chavez-Castrillon YY, Di Mascio P, White JF (2021) Probiotic endophytes for more sustainable banana production. Microorganisms 9(9):1805

Benhamou N, Belanger RR, Paulitz TC (1996a) Induction of differential host responses by *Pseudomonas fluorescens* in Ri T-DNA-transformed pea roots after challenge with *Fusarium oxysporum* f. sp. *pisi* and *Pythium ultimum*. Phytopathology 86:114–178

Benhamou N, Kloepper JW, Quadt-Hallman A, Tuzun S (1996b) Induction of defense-related ultrastructural modifications in pea root tissues inoculated with endophytic bacteria. Plant Physiol 112(3):919–929

Bharathi R (2001) Development of a rhizobacteria based bio-formulation for the management of major pests and diseases in chillies. *M.Sc. (Ag.) thesis*, Tamil Nadu Agricultural University, Coimbatore-3, India

Bhore SJ, Ravichantar N, Loh CY (2010) Screening of endophytic bacteria isolated from leaves of Sambung Nyawa [Gynura procumbens (Lour.) Merr.] for cytokinin-like compounds. Bioinformation 5(5):191

Bong CFJ, Sikorowski PP (1991) Effects of cytoplasmic polyhedrosis virus and bacterial contamination on growth and development of the corn earworm, *Heliothis zeae* (Boddie). J Invertebr Pathol 57:406–412

Botha AM, Nagel MAC, Westhuizen AJV, Botha FC (1998) Chitinase isoenzymes in near-isogenic wheat lines challenged with Russian wheat aphid, exogenous ethylene and mechanical wounding. Bot Bull Acad Sin 39:99–106

Bowler C, Van Montagu M, Inzé D (1992) Superoxide-dismutase and stress tolerance. Annu Rev Plant Physiol Plant Mol Biol 43:83–116

Bradley DJ, Kjellbom P, Lamb CJ (1992) Elicitor and wound-induced oxidative cross-linking of a praline-rich plant cell wall protein: a novel, rapid defense response. Cell 70:21–30

Brooks DS, Gonzales CF, Appel DN, Filer TH (1994) Evaluation of endophytic bacteria as potential biocontrol agents for oak wilt. Biol Control 4:373–381

Bruce RJ, West CA (1989) Elicitation of lignin biosynthesis and isoperoxidase activity by pectic fragments in suspension cultures of castor bean. Plant Physiol 91(3):889–897

Chen C, Bauske EM, Musson G, Rodriguez-Kabana R, Kloepper JW (1995) Biological control of *Fusarium* wilt on cotton by use of endophytic bacteria. Biol Control 5:83–91

Chen C, Belanger RR, Benhamou N, Paulitz TC (2000) Defense enzymes induced in cucumber roots by treatment with plant growth-promoting rhizobacteria (PGPR) and Pythium aphanidermatum. Physiol Mol Plant Pathol 56(1):13–23

Claydon N, Grove JF, Pople M (1985) Elm bark beetle boring and feeding deterrents from *Phomopsis oblonga*. Phytochemistry 24:937–943

Constabel CP, Bergey DR, Ryan CA (1995) Systemin activates synthesis of wound-inducible tomato leaf polyphenol oxidase via the octadecanoid defense signaling pathway. Proc Natl Acad Sci 92(2):407–411

D'Alessandro MARCO, Erb M, Ton J, Brandenburg A, Karlen D, Zopfi J, Turlings TC (2014) Volatiles produced by soil-borne endophytic bacteria increase plant pathogen resistance and affect tritrophic interactions. Plant Cell Environ 37(4):813–826

De Abreu CS, Figueiredo JEF, Oliveira-Paiva CA, Dos Santos VL, Gomes EA, Ribeiro VP et al (2017) Maize endophytic bacteria as mineral phosphate solubilizers

Ding X (1995) *Manduca* chitinase mediated resistance to tobacco bud worm (*Heliothis virescens*) and tobacco horn worm (*Manduca sexta*) larvae in transgenic tobacco plants. *Ph.D. Dissertation*. Kansas State University, Manhattan, USA

Ding X, Gopalakrishnan B, Johnson LB, White FF, Wang X, Morgan TD, Kramer KJ, Muthukrishnan S (1997) Insect resistance of transgenic tobacco expressing an insect chitinase gene. Transgenic Res 53:654–659

Dinsdale EA, Edwards RA, Hall D, Angly F, Breitbart M, Brulc JM, Furlan M, Desnues C, Haynes M, Li L, McDaniel L (2008) Functional metagenomic profiling of nine biomes. Nature 452(7187):629–632

Dong X (2001) Genetic dissection of systemic acquired resistance. Curr Opin Plant Biol 4(4):309–314

Downie JA (2010) The roles of extracellular proteins, polysaccharides and signals in the interactions of rhizobia with legume roots. FEMS Microbiol Rev 34(2):150–170

Dudeja SS, Giri R, Saini R, Suneja-Madan P, Kothe E (2012) Interaction of endophytic microbes with legumes. J Basic Microbiol 52(3):248–260

Durgadevi D, Harish S, Manikandan R, Prabhukarthikeyan SR, Alice D, Raguchander T (2021) Proteomic profiling of defense/resistant genes induced during the tripartite interaction of *Oryza sativa*, *Rhizoctonia solani* AG1-1A, and *Bacillus subtilis* against rice sheath blight. Physiol Mol Plant Pathol 115:101669

Durner J, Shah J, Klessig DF (1997) Salicylic acid and disease resistance in plants. Trends Plant Sci 2(7):266–274

Eevers N, Hawthorne JR, White JC, Vangronsveld J, Weyens N (2016) Exposure of Cucurbita pepo to DDE-contamination alters the endophytic community: a cultivation dependent vs a cultivation independent approach. Environ Pollut 209:147–154

Ek-Ramos MJ, Gomez-Flores R, Orozco-Flores AA, Rodríguez-Padilla C, González-Ochoa G, Tamez-Guerra P (2019) Bioactive products from plant-endophytic Gram-positive bacteria. Front Microbiol 10:463

Ellis C, Turner JG (2001) The *Arabidopsis* mutant Cev 1 has constitutively active jasmonate and ethylene signal pathways and enhanced resistance to pathogens. Plant Cell 13:1025–1033

Faisal MP, Nagendran K, Karthikeyan G, Raguchander T, Prabakar K (2014) Water in oil based PGPR formulation of *Pseudomonas fluorescens* (FP7) showed enhanced resistance against *Colletotrichum musae*. Crop Prot 65:186–193

Firrincieli A, Otillar R, Salamov A, Schmutz J, Khan Z, Redman RS, Fleck ND, Lindquist E, Grigoriev IV, Doty SL (2015) Genome sequence of the plant growth promoting endophytic yeast *Rhodotorula graminis* WP1. Front Microbiol 6:978

Fouts DE, Tyler HL, DeBoy RT, Daugherty S, Ren Q, Badger JH, Durkin AS, Huot H, Shrivastava S, Kothari S, Dodson RJ (2008) Complete genome sequence of the N2-fixing broad host range endophyte Klebsiella pneumoniae 342 and virulence predictions verified in mice. PLoS Genet 4(7):e1000141

Frankenberger WT, Arshad M (1995) Phytohormones in soils microbial production and function. Marcel Dekker, New York

Funk CR, Halisky PM, Johnson MC, Siegel MR, Stewart AV, Ahmad S, Hurley RH, Harvey IC (1983) An endophytic fungus and resistance to sod webworms: association in *Lolium perenne*. Bio/Technology 1:189–191

Gao Z, Zhang B, Liu H, Han J, Zhang Y (2017) Identification of endophytic Bacillus velezensis ZSY-1 strain and antifungal activity of its volatile compounds against Alternaria solani and Botrytis cinerea. Biol Control 105:27–39

Gaynor DL, Hunt WF (1983) The relationship between nitrogen supply, endophytic fungus and Argentine stem weevil resistance in ryegrass. Proc NewZealand Grassland Association 44:257–263

Glazebrook J (2005) Contrasting mechanisms of defense against biotrophic and necrotrophic pathogens. Annu Rev Phytopathol 43:205

Glick BR (1995) The enhancement of plant growth by free-living bacteria. Can J Microbiol 41 (2):109–117

Glick BR, Penrose DM, Li J (1998) A model for the lowering of plant ethylene concentrations by plant growth-promoting bacteria. J Theor Biol 190(1):63–68

Glick BR, Patten CL, Holguin G, Penrose DM (1999) Biochemical and genetic mechanisms used by plant growth promoting bacteria. Imperial College Press, London

Grant JJ, Loake GJ (2000) Role of reactive oxygen intermediates and cognate redox signalling in disease resistance. Plant Physiol 124:21–29

Gupta C, Dubey R, Maheshwari D (2002) Plant growth enhancement and suppression of Macrophomina phaseolina causing charcoal rot of peanut by fluorescent Pseudomonas. Biol Fertil Soils 35(6):399–405

Hammerschmidt R, Kuć J (1982) Lignification as a mechanism for induced systemic resistance in cucumber. Physiol Plant Pathol 20(1):61–71

Hardoim PR (2019) Biologically active compounds from bacterial endophytes. In: Jha S (ed) Endophytes and secondary metabolites. Springer, Cham, pp 3–31

Harish S, Rajendran L, Suganthi C, Radjacommare R, Kavino M, Lavanya N, Ramanathan A, Kumar N Samiyappan R (2003) Antagonism of endophytic bacteria against *Fusarium oxysporum* f.sp. *cubense* causing panama disease of banana. Proceedings on the 6th International Plant Growth Promoting Rhizobacteria (PGPR) Workshop held at Calicut, Kerala, India, 5-–10 October, 2003. Abstracts and Short Papers. 132–139

Hilbert M, Voll LM, Ding Y, Hofmann J, Sharma M, Zuccaro A (2012) Indole derivative production by the root endophyte Piriformospora indica is not required for growth promotion but for biotrophic colonization of barley roots. New Phytol 196:520–534

Hinton DM, Bacon CW (1995) *Enterobacter clocae* is an endophytic symbiont of corn. Mycopathology 129:117–125

Holt JA, Luo G, Billin AN, Bisi J, McNeill YY, Kozarsky KF, Jones SA (2003) Definition of a novel growth factor-dependent signal cascade for the suppression of bile acid biosynthesis. Genes Dev 17(13):1581–1591

Hossain MT, Khan A, Chung EJ, Rashid MHO, Chung YR (2016) Biological control of rice bakanae by an endophytic Bacillus oryzicola YC7007. Plant Pathol J 32(3):228

Huang XF, Zhou D, Guo J, Manter DK, Reardon KF, Vivanco JM (2015) Bacillus spp. from rainforest soil promote plant growth under limited nitrogen conditions. J Appl Microbiol 118 (3):672–684

Jacob J, Krishnan GV, Thankappan D, Amma DKBNS (2020). Endophytic bacterial strains induced systemic resistance in agriculturally important crop plants. In: *Microbial Endophytes* 75–105. Woodhead Publishing

Janatiningrum I, Solihin DD, Meryandini A, Lestari Y (2018) Comparative study on the diversity of endophytic actinobacteria communities from Ficus deltoidea using metagenomic and culture-dependent approaches. Biodiversitas J Biol Divers 19(4):1514–1520

Jasim B, John Jimtha C, Jyothis M, Radhakrishnan EK (2013) Plant growth promoting potential of endophytic bacteria isolated from Piper nigrum. Plant Growth Regul 71(1):1–11

Jayashree K, Shanmugam V, Raguchander T, Ramanathan A, Samiyappan R (2000) Evaluation of Pseudomonas fluorescens (Pf-1) against blackgram and sesame root-rot disease. J Biol Control:55–61

Jha Y (2019) The importance of zinc-mobilizing rhizosphere bacteria to the enhancement of physiology and growth parameters for paddy under salt-stress conditions. Jordan J Biol Sci 12:2

Jia M, Chen L, Xin HL, Zheng CJ, Rahman K, Han T, Qin LP (2016) A friendly relationship between endophytic fungi and medicinal plants: a systematic review. Front Microbiol 7:906

Johnson MC, Dahlman LD, Siegel MR, Bush LP, Latch GCM, Potter DA, Varney DR (1985) Insect feeding deterrents in endophyte-infected tall fescue. Appl Environ Microbiol 49:568–571

Kandan A, Commare RR, Nandakumar R, Ramiah M, Raguchander T, Samiyappan R (2002) Induction of phenyl propanoid metabolism by Pseudomonas fluorescens against tomato spotted wilt virus in tomato. Folia Microbiol 47(2):121–129

Kandan A, Ramiah M, Vasanthi VJ, Radjacommare R, Nandakumar R, Ramanathan A, Samiyappan R (2005) Use of Pseudomonas fluorescens-based formulations for management of tomato spotted wilt virus (TSWV) and enhanced yield in tomato. Biocontrol Sci Tech 15 (6):553–569

Karthikeyan M, Bhaskaran R, Kandan A, Radjacommare R, Ramanathan A Samiyappan R (2003) Endophytic bacteria for biological control of dry root rot (*Macrophomina phaseolina*) in black gram. Proceedings on the 6th International Plant Growth Promoting Rhizobacteria (PGPR) Workshop held at Calicut, kerala, India, 5–10 October, 2003. Abstracts and Short Papers. 146–152

Kaul S, Sharma TK, Dhar M (2016) "Omics" tools for better understanding the plant–endophyte interactions. Front Plant Sci 7:955

Kloepper JW, Leong J, Teintze M, Schroth MN (1980) *Pseudomonas* siderophores: a mechanism explaining disease suppressive soils. Curr Microbiol 4:317–320

Kloepper JW, Schroth MN (1981) Development of a powder formulation of rhizobacteria for inoculation of potato seed pieces. Phytopathology 71(6):590–592

Knief C, Delmotte N, Chaffron S, Stark M, Innerebner G, Wassmann R, Von Mering C, Vorholt JA (2012) Metaproteogenomic analysis of microbial communities in the phyllosphere and rhizosphere of rice. ISME J 6(7):1378–1390

Krause A, Ramakumar A, Bartels D, Battistoni F, Bekel T, Boch J, Böhm M, Friedrich F, Hurek T, Krause L, Linke B (2006) Complete genome of the mutualistic, N2-fixing grass endophyte *Azoarcus* sp. strain BH72. Nat Biotechnol 24(11):1384–1390

Krishnamoorthy K, Gnanamanickam SS (1997) Enumeration, location and characterization of endophytic bacteria within sugar beet roots. Can J Bot 63:1262–1265

Lamb C, Dixon RA (1997) The oxidative burst in plant disease resistance. Annu Rev Plant Physiol Plant Mol Biol 48:251–275

Lazarovits G, Nowak J (1997) Rhizobacteria for improvement of plant growth and establishment. HortScience 32(2):188–192

Lebeis SL (2014) The potential for give and take in plant–microbiome relationships. Front Plant Sci 5:287

León J, Rojo E, Sánchez-Serrano JJ (2001) Wound signalling in plants. J Exp Bot 52(354):1–9

Lery LM, Hemerly AS, Nogueira EM, von Krüger WM, Bisch PM (2011) Quantitative proteomic analysis of the interaction between the endophytic plant-growth-promoting bacterium Gluconacetobacter diazotrophicus and sugarcane. Mol Plant Microbe Interact 24(5):562–576

Li CH, Zhao MW, Tang CM, Li SP (2009) Population dynamics and identification of endophytic fungi antagonistic toward plant-pathogenic fungi in cotton root. Microb Ecol 59:344–356

Lilley AK, Fry JC, Bailey MJ, Day MJ (1996) Comparison of aerobic heterotrophic taxa isolated from four root domains of mature sugar beet (*Beta vulgaris*). FEMS Microbiol Ecol 21(3): 231–242

Liu L, Kloepper JW, Tuzun S (1995) Induction of systemic resistance in cucumber against bacterial leaf spot by plant growth promoting rhizhobacteria. Phytopathology 85:843–847

Liu X, Jia J, Popat R, Ortori CA, Li J, Diggle SP, Gao K, Cámara M (2011) Characterisation of two quorum sensing systems in the endophytic Serratia plymuthicastrain G3: differential control of motility and biofilm formation according to life-style. BMC Microbiol 11(1):1–12

van Loon LC, Bakker PAHM, Pieterse CMJ (1998) Systemic resistance induced by rhizosphere bacteria. Annu Rev Phytopathol 36:453–483

van Loon LC, Gerristen YAM, Ritter CE (1987) Identification, purification and characterization of pathogenesis-related proteins from virus-infected Samsun NN tobacco leaves. Plant Mol Biol 9: 593–609

van Loon LC, Pierpoint WS, Boller T, Conejero V (1994) Recommendations for naming plant pathogenesis-related proteins. Plant Mol Biol Rep 12:245–264

van Loon LC, van Strien EA (1999) The families of pathogenesis-related proteins, their activities and comparative analysis of PR-1 type proteins. Physiol Mol Plant Pathol 49:359–376

Lòpez-Fernàndez S, Sonego P, Moretto M, Pancher M, Engelen K, Pertot I, Campisano A (2015) Whole-genome comparative analysis of virulence genes unveils similarities and differences between endophytes and other symbiotic bacteria. Front Microbiol 6:419

M'piga P, Bélanger RR, Paulitz TC, Benhamou N (1997) Increased resistance to *fusarium oxysporum* f. sp. *radicislycopersici* in tomato plants treated with the endophytic bacterium *Pseudomonas fluorescens* strain 63–28. Physiol Mol Plant Pathol 50:301–320

Mahaffee WF, Kloepper JW (1997) Temporal changes in the bacterial communities of soil, rhizosphere, and endorhiza associated with field-grown cucumber (Cucumis sativus L.). Microb Ecol 34(3):210–223

Mahendran R (1996) Induction of Chitinases in Rice (*Oryza sativa* L.) seeds for the management of Brown spot Pathogen. *M.Sc. (Ag.,) Thesis*, TNAU, Coimbatore-3, India. 78p

Malamy J, Carr JP, Klessig DF, Raskin I (1990) Salicylic acid: a likely endogenous signal in the resistance response of tobacco to viral infection. Science 250:1002–1004

Malusa E, Vassilev N (2014) A contribution to set a legal framework for biofertilisers. Appl Microbiol Biotechnol 98:6599–6607

Malusà E, Pinzari F, Canfora L (2016) Efficacy of biofertilizers: challenges to improve crop production. In: Microbial inoculants in sustainable agricultural productivity. Springer, New Delhi, pp 17–40

Manikandan R, Saravanakumar D, Rajendran L, Raguchander T, Samiyappan R (2010) Standardization of liquid formulation of *Pseudomonas fluorescens* Pf1 for its efficacy against *Fusarium* wilt of tomato. Biol Control 54:83–89

Martiny JB, Jones SE, Lennon JT, Martiny AC (2015) Microbiomes in light of traits: a phylogenetic perspective. Science 350(6261):aac9323

Maurhofer M, Hase C, Meuwly P, Métraux JP, Défago G (1994) Induction of systemic resistance of tobacco to *tobacco necrosis virus* by the root-colonizing *Pseudomonas fluorescens* strain CHA0: influence of the *gac*A gene and of pyoverdine production. Phytopathology 84:139–146

Mayak S, Tirosh T, Glick BR (2004) Plant growth-promoting bacteria that confer resistance to water stress in tomatoes and peppers. Plant Sci 166(2):525–530

McKenzie CL, Shatters RG, Doostdar H, Lee SD, Inbar M, Mayer RT (2002) Effect of *Geminivirus* infection and *Bemisia* infestation on accumulation of pathogenesis-related proteins in tomato. Archiv Insect Biochem Physiol 49:203–214

Meena B, Ramamoorthy V, Marimuthu T, Velazhahan R (2000) Pseudomonas fluorescens mediated systemic resistance against late leaf spot of groundnut. J Mycol Plant Pathol 30(2):151–158

Megías E, Megías M, Ollero FJ, Hungria M (2016) Draft genome sequence of Pantoea ananatis strain AMG521, a rice plant growth-promoting bacterial endophyte isolated from the Guadalquivir marshes in southern Spain. Genome Announc 4(1):e01681–e01615

Mehdy MC (1994) Active oxygen species in plant defense against pathogens. Plant Physiol Mol Plant Pathol 49:143–158

Moiseyev GP, Fedoreyena LI, Zhuravlev YN, Yasnetskaya E, Jekel PA, Beintema JJ (1997) Primary structures of two ribonucleases from ginseng calluses. New members of the PR-10 family of intracellular pathogenesis-related plant proteins. FEBS Lett 407:207–210

Monteiro RA, Balsanelli E, Tuleski T, Faoro H, Cruz LM, Wassem R, de Baura VA, Tadra-Sfeir MZ, Weiss V, DaRocha WD, Muller-Santos M (2012) Genomic comparison of the endophyte Herbaspirillum seropedicae SmR1 and the phytopathogen Herbaspirillum rubrisubalbicans M1 by suppressive subtractive hybridization and partial genome sequencing. FEMS Microbiol Ecol 80(2):441–451

Murphy AM, Gilliland A, Wong CE, West J, Singh DP, Carr JP (2001) Signal transduction in resistance to plant viruses. Eur J Plant Pathol 107:121–128

Nakkeeran S, Kavitha K, Mathiyazhagan S, Fernando WGD, Chandrasekar G, Renukadevi P (2004) Induced systemic resistance and plant growth promotion by *Pseudomonas chlororaphis* strain PA-23 and *Bacillus subtilis* strain CBE4 against rhizome rot of turmeric (*Curcuma longa* L.). Can J Plant Pathol 26:417–418

Nandakumar R, Babu S, Viswanathan R, Raguchander T, Samiyappan R (2001) Induction of systemic resistance in rice against sheath blight disease by Pseudomonas fluorescens. Soil Biol Biochem 33(4–5):603–612

Nicholson RL, Hammerschmidt R (1992) Phenolic compound and their role in disease resistance. Annu Rev Phytopathol 30:369–389

Nowak J, Asiadu SK, Lazarovits G, Pillay V, Stewart A, Smith C, Liu Z (1995) Enhancement of *in vitro* growth and transplant stress tolerance of potato and vegetable plantlets co-cultured with a plant growth promoting pseudomonad bacterium. In: Carre F, Chagvardieff P (eds) Ecophysiology and photosynthetic *in vitro* cultures. Commissariat l'energie atomique, France, pp 173–179

Ongena M, Jacques P (2008) Bacillus lipopeptides: versatile weapons for plant disease biocontrol. Trends Microbiol 16(3):115–125

Oteino N, Lally RD, Kiwanuka S, Lloyd A, Ryan D, Germaine KJ, Dowling DN (2015) Plant growth promotion induced by phosphate solubilizing endophytic Pseudomonas isolates. Front Microbiol 6:745

van Overbeek L, van Elsas D, van Vuurde J, Colon L (2003) Disease control in potato by endophytes. Proceedings on the 6th International Plant Growth Promoting Rhizobacteria (PGPR) Workshop held at Calicut, Kerala, India, 5–10 October, 2003. Abstracts and Short Papers. 507–509

Parthasarathy S, Gurudevan T, Subramanian KS, Kuppusamy P (2016) Bacterial antagonists and hexanal-induced systemic resistance of mango fruits against *Lasiodiplodia theobromae* causing stem-end rot. J Plant Interact 11(1):158–166. https://doi.org/10.1080/17429145.2016.1252068

van Peer R, Schippers B (1988) Plant growth response in bacterization with selected *Pseudomonas* spp. strains and rhizosphere microbial development in hydroponic cultures. Can J Microbiol 35: 456–463

Pei C, Mi C, Sun L, Liu W, Li O, Hu X (2017) Diversity of endophytic bacteria of Dendrobium officinale based on culture-dependent and culture-independent methods. Biotechnol Biotechnol Equip 31(1):112–119

Pereira P, Ibáñez F, Rosenblueth M, Etcheverry M, Martínez-Romero E (2011) Analysis of the bacterial diversity associated with the roots of maize (*Zea mays* L.) through culture-dependent and culture-independent methods. Int Schol Res Notices:2011

Petti C, Reiber K, Ali SS, Berney M, Doohan FM (2012) Auxin as a player in the biocontrol of Fusarium head blight disease of barley and its potential as a disease control agent. BMC Plant Biol 12:224

Pleban S, Ingel F, Chet I (1995) Control ofRhizoctonia solani andSclerotium rolfsii in the greenhouse using endophyticBacillus spp. Eur J Plant Pathol 101(6):665–672

Poulton EJ (1990) Cyanogenesis in plants. Plant Physiol 94:401–405

Pozo MJ, Cordier C, Dumas-Gaudot E, Gianinazzi S, Barea JM, Azcón-Aguilar C (2001) Localized versus systemic effect of arbuscular mycorrhizal fungi on defence responses to Phytophthora infection in tomato plants. J Exp Bot 53(368):525–534

Prabhukarthikeyan SR, Manikandan R, Durgadevi D, Keerthana U, Harish S, Karthikeyan G, Raguchander T (2017) Bio-suppression of turmeric rhizome rot disease and understanding the molecular basis of tripartite interaction among *Curcuma longa*, *Pythium aphanidermatum* and *Pseudomonas fluorescens*. Biol Control 111:23–31

Prabhukarthikeyan SR, Keerthana U, Baite MS, Panneerselvam P, Mitra D, Kumar RN, Rath PC (2022) Bacillus rhizobacteria: a versatile biostimulant for sustainable agriculture. In: New and future developments in microbial biotechnology and bioengineering. Elsevier, pp 33–44

Puri RR, Adachi F, Omichi M, Saeki Y, Yamamoto A, Hayashi S, Itoh K (2018) Culture-dependent analysis of endophytic bacterial community of sweet potato (*Ipomoea batatas*) in different soils and climates. J Adv Microbiol 13:1–12

Qingwen Z, Ping L, Gang W, Qingnian C (1998) On the biochemical mechanism of induced resistance of cotton to cotton bollworm by cutting off young seedling at plumular axis. Acta Phytophylacica Sinica 25:209–212

Radjacommare R, Nandakumar R, Kandan A, Suresh S, Bharathi M, Raguchander T, Samiyappan R (2002) *Pseudomonas fluorescens* based bioformulation for the management of sheath blight disease and leaf folder insect in rice. Crop Prot 21:671–677

Radjacommare R, Ramanathan A, Kandan A, Harish S, Thambidurai G, Sible GV, Ragupathi T, Samiyappan R (2004) PGPR mediates induction of pathogenesis-related (PR) proteins against the infection of blast pathogen in resistant and susceptible ragi [*Eleusine coracana* (L.) Gaertner] cultivars. Plant Soil 266:165–176

Raguchander T, Karthikeyan G, Sendhilvel V, Harish S, Karthiba L, Rajendran L, Veera RR, Raveendran M, Jonathan EI, Prabakar K (2022) Complete genome sequence analysis of Bacillus subtilis Bbv57, a promising biocontrol agent against Phytopathogens. Int J Mol Sci 23 (17):9732. https://doi.org/10.3390/ijms23179732

Rajendran L, Harish S, Radjacommare R, Kandan A, Sible GV, Lavanya, N, Durairaj, C., Ramanathan A, Samiyappan R (2003) Bacterial endophytes as potential biocontrol agents for important pest and disease of cotton. Proceedings on the 6th International Plant Growth Promoting Rhizobacteria (PGPR) Workshop held at Calicut, Kerala, India , 5–10 October, 2003. Abstracts and Short Papers. 202–207

Rajendran L, Karthikeyan G, Raguchander T, Samiyappan R (2018) Effect of endophytic bacteria Bacillus subtilis EPC 5 on basal stem rot disease spread and nut yield in coconut plantations. J Appl Hortic 20(3):207–212

Ramamoorthy V, Raguchander T, Samiyappan R (2002) Enhancing resistance of tomato and hot pepper to *pythium* diseases by seed treatment with fluorescent pseudomonads. Eur J Plant Pathol 108:429–441

Ramamoorthy V, Viswanathan R, Raguchander T, Prakasam V, Samiyappan R (2001) Induction of systemic resistance by plant growth promoting rhizobacteria in crop plants against pest and diseases. Crop Protect 20:1–11

Ramyabharathi SA, Raguchander T (2014) Efficacy of secondary metabolites produced by EPCO16 against tomato wilt pathogen f. sp. lycopercisi. J Mycol Plant Pathol 44(2):148

Ramyabharathi SA, Sankari Meena K, Rajendran L, Raguchander T, Jonathan EI (2020) Potential of a rhizobacterium *Bacillus subtilis* (Bbv 57) on *Fusarium oxysporum* f. sp. *gerberae* and *Meloidogyne incognita* infecting Gerbera grown in protected cultivation. Eur J Plant Pathol (online)

Rangeshwaran R, Wasnikar AR, Prasad RD, Anjula N, Sunanda CR (2002) Isolation of endophytic bacteria for biological control of wilt pathogens. J Biol Control 16(2):125–134

Rath M, Mitchell TR, Gold SE (2018) Volatiles produced by *Bacillus mojavensis* RRC101 act as plant growth modulators and are strongly culture-dependent. Microbiol Res 208:76–84

Reiter B, Pfeifer U, Schwab H, Sessitsch A (2002) Response of endophytic bacterial communities in potato plants to infection with *Erwinia carotovora* subsp. *atroseptica*. Appl Environ Microbiol 68(5):2261–2268

Roberts LD, Souza AL, Gerszten RE, Clish CB (2012) Targeted metabolomics. Curr Protoc Mol Biol 98(1):30–32

Rojo E, Solano R, Sanchez-Serrano JJ (2003) Interactions between signaling compounds involved in plant defense. J Plant Growth Regulator 22:82–98

Romero-Cortes T, Pérez España VH, López Pérez PA, Rodríguez-Jimenes GDC, Robles-Olvera VJ, Aparicio Burgos JE, Cuervo-Parra JA (2019) Antifungal activity of vanilla juice and vanillin against *Alternaria alternata*. CyTA-J Food 17(1):375–383

Ryals JA, Neuenschwander UH, Willits MG, Molina A, Steiner HY, Hunt MD (1996) Systemic acquired resistance. Plant Cell 8(10):1809

Ryan CA (1990) Protease inhibitors in plants: genes for improving defenses against insects and pathogens. Annu Rev Phytopathol 28:425–449

Sakthivel N, Gnanamanickam SS (1987) Evaluation of Pseudomonas fluorescens for suppression of sheath rot disease and for enhancement of grain yields in rice (Oryza sativa L.). Appl Environ Microbiol 53(9):2056–2059

Santoyo G, Moreno-Hagelsieb G, del Carmen Orozco-Mosqueda M, Glick BR (2016) Plant growth-promoting bacterial endophytes. Microbiol Res 183:92–99

Saravanakumar D, Harish S, Loganathan M, Vivekananthan R, Rajendran L, Raguchander T, Samiyappan R (2007) Rhizobacterial bioformulation for the effective management of *Macrophomina* root rot in mungbean. Archiv Phytopathol Plant Protect 40:323–337

Saravanakumar D, Lavanya N, Muthumeena K, Raguchander T, Samiyappan R (2009) Fluorescent pseudomonad mixtures mediate disease resistance in rice plants against sheath rot (*Sarocladium oryzae*) disease. BioControl 54:273–286

Savadogo A, Tapi A, Chollet M, Wathelet B, Traoré AS, Jacques PH (2011) Identification of surfactin producing strains in Soumbala and Bikalga fermented condiments using polymerase chain reaction and matrix assisted laser desorption/ionization-mass spectrometry methods. Int J Food Microbiol 151(3):299–306

Sayyed RZ, Chincholkar SB, Reddy MS, Gangurde NS, Patel PR (2013) Siderophore producing PGPR for crop nutrition and phytopathogen suppression. In: Bacteria in agrobiology: disease management (449–471). Springer, Berlin, Heidelberg

Scandalios JG (1994) Regulation and properties of plant catalases. In: Foyer CH, Mullineaux PM (eds) Causes of photo-oxidative stress and amelioration of defense systems in plants. CRC Press, Boca Raton, FL, pp 275–316

Segarra G, Van der Ent S, Trillas I, Pieterse CMJ (2009) MYB72 a node of convergence in induced systemic resistance triggered by a fungal and a bacterial beneficial microbe. Plant Biol:1190–1196

Sessitsch A, Reiter B, Berg G (2004) Endophytic bacterial communities of field-grown potato plants and their plant-growth-promoting and antagonistic abilities. Can J Microbiol 50(4): 239–249

Shanmugaiah V, Mathivanan N, Varghese B (2010) Purification, crystal structure and antimicrobial activity of phenazine-1-carboxamide produced by a growth-promoting biocontrol bacterium, Pseudomonas aeruginosa MML2212. J Appl Microbiol 108(2):703–711

Sharma A, Johri BN, Sharma AK, Glick BR (2003) Plant growth-promoting bacterium pseudomonas sp. strain GRP3 influences iron acquisition in mung bean (Vigna radiata L. Wilzeck). Soil Biol Biochem 35(7):887–894

Sheibani-Tezerji R, Rattei T, Sessitsch A, Trognitz F, Mitter B (2015) Transcriptome profiling of the endophyte Burkholderia phytofirmans PsJN indicates sensing of the plant environment and drought stress. MBio 6(5):e00621–e00615

Sible GV (2003) Biotechnology based approaches for the early detection and management of post harvest anthracnose in banana. Ph. D. Thesis. Tamil Nadu Agricultural University, Coimbatore

Smith AR, van Staden J (1978) Changes in endogenous cytokinin levels in kernels of *Zea mays* L. during imbibition and germination. J Exp Bot 29:1067–1073

Sridharan AP, Thankappan S, Karthikeyan G, Uthandi S (2020) Comprehensive profiling of the VOCs of *Trichoderma longibrachiatum* EF5 while interacting with *Sclerotium rolfsii* and *Macrophomina phaseolina*. Microbiol Res 236:126436

Stamenković S, Beškoski V, Karabegović I, Lazić M, Nikolić N (2018) Microbial fertilizers: a comprehensive review of current findings and future perspectives. Span J Agric Res 16:1

Sticher L, Mauch-Mani B, Métraux AJ (1997) Systemic acquired resistance. Annu Rev Phytopathol 35(1):235–270

Stock CA, Mcloughlin TJ, Klein JA, Adang M (1990) Expression of a *bacillus thuringiensis* crystal protein gene in *Pseudomonas cepacia* 526. Can J Microbiol 36:879–884

Sturz AV, Matheson BG (1996) Populations of endophytic bacteria which influence host-resistance to *Erwinia* induced bacterial soft rot in potato tuber. Plant Soil 184:265–271

Tabbene O, Slimene IB, Djebali K, Mangoni ML, Urdaci MC, Limam F (2009) Optimization of medium composition for the production of antimicrobial activity by Bacillus subtilis B38. Biotechnol Prog 25(5):1267–1274

Taghavi S, Van Der Lelie D, Hoffman A, Zhang YB, Walla MD, Vangronsveld J, Newman L, Monchy S (2010) Genome sequence of the plant growth promoting endophytic bacterium Enterobacter sp. 638. PLoS Genet 6(5):1000943

Tahir HA, Gu Q, Wu H, Raza W, Hanif A, Wu L, Colman MV, Gao X (2017) Plant growth promotion by volatile organic compounds produced by *Bacillus subtilis* SYST2. Front Microbiol 8:171

Thomma BPHJ, Eggermont K, Koenraad FMJ, Broekaert WF (1999) Requirement of functional ethylene-insensitive 2 gene for efficient resistance of *Arabidopsis thaliana* to infection by *Botrytis cinerea*. Plant Physiol 121:1093–1101

Ting ASY, Meon S, Kadir J, Son R Gurmit S (2003) Potential role of endophytes in the biocontrol of *Fusarium* wilt. 2nd International symposium on *Fusarium* wilt on banana held at Salvador de Bahia, Brazil on 22–26 September, 2003.32

Van Buren AM, Andren C, Isbimaru CA (1993) Biological control of bacterial ring rot pathogen by endophytic bacteria isolated from potato. Phytopathology 83:1406

Van Peer R, Niemann GJ, Schippers B (1991) Induced resistance and phytoalexin accumulation in biological control of fusarium wilt of carnation by Pseudomonas sp. strain WCS417r. Phytopathology 91:728–734

Vendan RT, Yu YJ, Lee SH, Rhee YH (2010) Diversity of endophytic bacteria in ginseng and their potential for plant growth promotion. J Microbiol 48(5):559–565

Vera-Estrell R, Blumwald E, Higgins VJ (1993) Non-specific glycopeptide elicitors of *Cladosporium fulvum*: evidence for involvement of active oxygen species in elicitor-induced effects on tomato cell suspensions. Physiol Mol Plant Pathol 42:9–22

Verma VC, Singh SK, Prakash S (2011) Bio-control and plant growth promotion potential of siderophore producing endophytic streptomyces from Azadirachta indica a. Juss J Basic Microbiol 51(5):550–556

Vidhyasekaran P, Muthamilan M (1995) Development of formulations of *Pseudomonas fluorescens* for control of chickpea wilt. Plant Dis 79:782–786

Vidhyasekaran P, Muthamilan R (1999) Evaluation of powder formulation of *Pseudomonas fluorescens* Pf1 for control of rice sheath blight. Biocontrol Sci Tech 9:67–74

Vidhyasekaran P, Rabindran R, Muthamilan M, Nayar K, Rajappan K, Subramanian N, Vasumathi K (1997) Development of powder formulation of Pseudomonas fluorescens for control of rice blast. Plant Pathol 46:291–297

Viswanathan R, Samiyappan R (1999) Induction of systemic resistance by plant growth promoting rhizobacteria against red rot disease caused by Colletotrichum falcatum Went. in sugarcane. Proc Sugar Technol Assoc India 61:24–39

Wang SS, Liu JM, Sun J, Sun YF, Liu JN, Jia N, Dai XF (2019) Diversity of culture-independent bacteria and antimicrobial activity of culturable endophytic bacteria isolated from different dendrobium stems. Sci Rep 9(1):1–12

War S, Shaktawat MS (2003) Influence of phosphorus sources, levels and solubilizers on yield, quality and nutrient uptake of soybean (Glycine max)—wheat (Triticum aestivum) cropping system in southern Rajasthan. Indian J Agric Sci 73(1):3–7

Walia A, Guleria S, Chauhan A, Mehta P (2017) Endophytic bacteria: role in phosphate solubilization. In: Endophytes: crop productivity and protection, pp 61–93

Webber J (1981) A natural control of Dutch elm disease. Nature (London) 292:449–451

Wei G, Kloepper JW, Tuzun S (1991) Induction of systemic resistance of cucumber to Colletotrichum orbiculare by select strains of plant growth-promoting rhizobacteria. Phytopathology 81(11):1508–1512

Wilkins MR, Sanchez JC, Gooley AA, Appel RD, Humphery-Smith I, Hochstrasser DF, Williams KL (1996) Progress with proteome projects: why all proteins expressed by a genome should be identified and how to do it. Biotechnol Genet Eng Rev 13(1):19–50

Wilson D (1995) Endophyte: the evolution of a term, and clarification of its use and definition. Oikos:274–276

Xie S, Liu J, Gu S, Chen X, Jiang H, Ding T (2020) Antifungal activity of volatile compounds produced by endophytic *Bacillus subtilis* DZSY21 against *Curvularia lunata*. Ann Microbiol 70(1):1–10

Xu W, Wang F, Zhang M, Ou T, Wang R, Strobel G, Xie J (2019) Diversity of cultivable endophytic bacteria in mulberry and their potential for antimicrobial and plant growth-promoting activities. Microbiol Res 229:126328

Yang JH, Liu HX, Zhu GM, Pan YL, Xu LP, Guo JH (2008) Diversity analysis of antagonists from rice-associated bacteria and their application in biocontrol of rice diseases. J Appl Microbiol 104(1):91–104

Zdor RE, Anderson AJ (1992) Influence of root colonizing bacteria on the defense responses in bean. Plant Soil 140:99–107

Zhang H, Xie X, Kim MS, Kornyeyev DA, Holaday S, Paré PW (2008) Soil bacteria augment Arabidopsis photosynthesis by decreasing glucose sensing and abscisic acid levels in planta. Plant J 56(2):264–273

Chapter 12
Endophytic Microbes and Their Role in Plant Health

Charishma Krishnappa, B. S. Kavya, H. M. Akshay Kumar, Priya Reddy, G. Rajeshwar Rao, and K. Darshan

Abstract Endophytes represent an important group of organisms having an intimate relationship with their hosts which mostly colonize intercellular spaces and form a range of different associations with them. They can be symbionts, mutualists, commensals, and trophobiotic. Endophytes are believed to enhance the growth and development of the plants and also a storehouse of secondary metabolites. Important mechanisms such as phytostimulation and biofertilization involved in growth promotion and biocontrol activity against disease suppression have been reported. Asymptomatic association and beneficial traits coveted on plants attracted interest to study them as a potential source of plant benefiters. Recent advances in omics have revealed rich diversity of endophytes from various crops. There is an increasing opportunity to utilize these endophytes for plant growth promotion and disease suppression. The common bacterial endophytes belong to *Pseudomonas*, *Bacillus*, and *Burkholderia*, while the major fungal endophytes are *Fusarium*, *Aspergillus*, *Colletotrichum*, *Diaporthe*, *Gliocladium*, *Lecanicillium*, *Phyllosticta*, and *Trichoderma*. These endophytes provide a large scope for their utilization as plant probiotics and biostimulants.

Charishma Krishnappa, B. S. Kavya, H. M. Akshay Kumar, Priya Reddy, G. Rajeshwar Rao and K. Darshan contributed equally with all other contributors.

C. Krishnappa · H. M. Akshay Kumar
Division of Plant Pathology, ICAR-Indian Agricultural Research Institute, New Delhi, India

B. S. Kavya
Department of Plant Pathology, University of Agricultural Sciences, Bangalore, Karnataka, India

P. Reddy
Faculty of Biological Science, Friedrich Schiller University, Jena, Germany
e-mail: preddy@ice.mpg.de

G. Rajeshwar Rao · K. Darshan (✉)
Forest Protection Division, ICFRE-Tropical Forest Research Institute, Jabalpur, Madhya Pradesh, India

P. Mathur et al. (eds.), *Microbial Symbionts and Plant Health: Trends and Applications for Changing Climate*, Rhizosphere Biology,
https://doi.org/10.1007/978-981-99-0030-5_12

Keywords Biofertilization · Biocontrol · Diversity · Endophytes · Phytosimulation

12.1 Introduction

Microorganisms are ubiquitous. Life forms on earth are majorly represented by various groups of microorganisms. The several groups majorly include bacteria, fungi, actinomycetes, and algae. Numerous efforts have been made to characterize the microbiome associated with a few essential niches like humans, animals, and plants. Globally, the Human Microbiome Project (HMP) and the Earth Microbiome Project (EMP) have been launched to tackle certain microbiome-related areas, viz., their structural composition, their utility, and their interaction behavior with the surrounding environment and modifications of adaptation to benefit them (Hood 2012). Recent advancements in the human microbiome studies have also paved the the way for its advanced studies in plant system. The potential of the microbiome is understood in its association with various beneficial activities.

The plant's evolutionary and ecological dynamics have been under the continuous influence of the plant and its associated microbiome. Plant and associated beneficial organisms share a balanced antagonism so that the resistance mechanisms and virulence factors posed by each other will be nullified. Plant and harmful microorganisms share a robust antagonistic activity which leads to disease. Plant and associated beneficial organisms share a synergistic antagonism against pathogens; all these conditions are acted upon under the same environmental condition (Kusari et al. 2012). The plant-associated microbiome can be categorized as beneficial organisms, harmful microorganisms, and neutral organisms. Further, they can be grouped into endophytes, epiphytes, commensals, antagonists, and synergists (Kusari et al. 2012). Pesticides can be substituted by biopesticides (Rani et al. 2021; Archana et al. 2022). A variety of biologically derived compounds had pesticidal action against insect pests and diseases (Shivakumara et al. 2022). Many reports of the plant and its associated microbiome have been published. This is one such effort made to gather information on plant endophytes.

Endophytes are a group of organisms that share an intimate relationship with their host. The mutual relationships between plant and endophytes are evident from their intra-/inter-cell colonization. Various study groups have reported that plant health and survival depend on these organisms (Khare et al. 2018). Different groups have tried to define endophytes. Bulgarelli and his group defined endophytes as the set of microbial genomes located inside plant organs (Bulgarelli et al. 2013).

12.2 History

The word endophyte was believed to be coined by De Bary (1866). Initially, an endophyte was defined as any organism which grows within plant tissues (Khare et al. 2018). The word endophyte takes its roots from Greek. Endo means within; phyton means plant. The spectrum of hosts is very broad. One typical example of this association is legume and rhizobium symbiosis. Sixty million years of evolutionary history between plants and endophytes has been reported (Sprent 2008).

Vogl, in 1898, first reported an endophyte as a mycelium within a grass seed, *Lolium temulentum* (Mano and Morisaki 2008). Later, in 1926, Perotti found the association between bacterial flora in root tissues. He assumed them to be non-pathogenic organisms (Perotti 1926).

There is a growing interest in the role of the endophytes in ameliorating various stresses on plants, including biotic stress (such as pathogenic microbes) and abiotic stress (such as drought and salt stress), because endophytes can produce phytohormones, fixing nitrogen, antagonistic substances, and enzymes, which play a vital role in plants' response to biotic and abiotic stress (Yu et al. 2019).

12.3 Methods to Detect and Identify Endophytes in Plant Tissues

Establishing an endophytic relation of an organism within a plant is often tricky, as epiphytic contamination is a problem, always. There are a few techniques for establishing this relation, as reviewed by Schulz and Boyle (2006):

1. Location of an endophyte within plant tissue, histological observation. Generally, surface-sterilized plant tissue is incubated on a suitable nutrient medium and observed for the emergence of growth. Fungi can be visualized as mycelium or spores; bacteria can be visualized as distinct cells.
2. Isolation of an endophyte after complete surface sterilization. This is often quite a practice to isolate an endophyte. A sterility check (a control plate where there is no growth of any organism) is an essential step in this method.
3. Immunological methods (ELISA, dot blot assay, tissue printing, or immunogold labeling). Polyclonal and monoclonal antibodies are sensitive and specific to detecting endophytes.
4. The molecular method, wherein the endophytic DNA (whether a bacteria or fungi) is amplified directly from the colonized plant tissues.
5. Microscopic observation. Tagging an organism with a reporter (e.g., Gfp) is essential for this method. The organism is visualized based on the fluorescence emitted.

12.4 Diversity of Endophytes

Species diversity, richness, and abundance are significant measures of an endophyte community structure. This community structure within a plant is not rigid. This dynamic relation is driven by abiotic and biotic factors such as soil conditions, biogeography, plant species, microbe-microbe interactions, and plant-microbe interactions, both at local and larger scales (Gaiero et al. 2013). Sometimes, it is decided by the genetic dispositions of both developmental stages and nutritional status. Genetic disposition means a change in the hereditary nature of plants or organisms. Loss or gain of a trait leads to endophytism. In *Colletotrichum magna*, a single mutation transformed it from a pathogenic fungus into an endophyte (Freeman and Rodriguez 1993).

12.5 Nature of an Endophyte

Hardoim and his group explained two groups of endophytes: obligate and facultative endophytes. The former cannot find a life outside of plants. Seed transmission is considered a significant source of dissemination. The latter are biphasic. Facultative endophytes have a free-living life in the soil, but when an opportunity arises, they will colonize the plants. Gaiero and his co-workers defined a third group called passive endophytes. This group does not have the cellular machinery required to colonize the plant but does so accidentally due to stochastic events like open wounds at the entry sites (Hardoim et al. 2008; Gaiero et al. 2013).

12.6 Differences Between an Endophyte and Pathogen Colonization of a Plant

Any organism to colonize a plant should avoid recognition from the plant. Here lies the difference between an endophyte and a pathogen. The main biological question is to what extent the recognition is avoided to overcome the plant resistance. Different factors are responsible for the maintenance of balance in a continuum of life inside the plant. Many study groups have tried to address this issue. A few major revelations made so far are discussed here. One primary consideration is the presence of organisms. Their concentration in number is the reason not to activate the plant's immune response. It is believed that endophytes are often present in lower quantities than pathogens.

A balance between nutrient requirements and plant defense was crucial in determining whether an organism turns endophytic or pathogenic. Thus, an imbalance would result in plant resistance (arresting the growth of the pathogen) or disease initiation (circumvention of plant defense). Effector-triggered immunity (ETI) is the common race-off phenomenon between plants and pathogens. Several groups have revealed that endophytes or any plant growth-promoting organism would not trigger the defense above the threshold line required for effector-triggered susceptibility (ETS). Only potential pathogens can breach ETI and cause ETS. Effectorome studies from pathogens and non-pathogenic organisms revealed several similarities and uncommonness. The difference in the number of effectors, sequence differences, and other characteristics ultimately decide whether a relation is mutualistic or parasitic (pathogenic). The mutualistic relationship between an endophyte and a plant does not necessarily stimulate high-order resistance, which can affect endophyte colonization. However, detailed studies on such aspects are yet to address this concept (Kogel et al. 2006; Collinge et al. 2022).

12.7 Endophyte Biodiversity

To develop an efficient bioformulation for growth promotion and disease management, we need to study the differences in the constituent microbes which occur naturally colonizing the phyllosphere, spermosphere, endosphere, rhizosphere, and mycorrhizal associations of crop plants. Compared to epiphytes, understanding the endosphere microbiome composition is the most important as they are most beneficial for growth and development; the microbiome in the root endosphere is substantially less diverse compared to the rhizosphere or the bulk soil (Liu et al. 2017). This shows that when habitats shift from soil to roots, roots act as efficient habitat filters (Bulgarelli et al. 2013). There is proof these root endophytes end up moving up to the stems and leaves via the apoplast in xylem vessels (Chi et al. 2005). Therefore, it is expected that the microbiomes of plant leaves and shoots have substantial taxonomic and functional similarities to those of roots (Bodenhausen et al. 2013; Bulgari et al. 2014; Bai et al. 2015). Most of the studies related to endophytic diversity were limited to cultivated crops, some biomes have been particularly understudied, and most studies on cultivated crops due to their high biodiversity, sampling difficulties, or the least attention (i.e., tropical rainforests, mangrove forests, flooded grasslands, coastal dunes, etc.). Endophyte richness also differs with climatic variations; for example, in southwestern Arizona, Bowman and Arnold (2018) discovered that *Pinus ponderosa* hosted more diverse foliar fungal endophyte communities at mid-to-high elevations in comparison to lower elevations, and Zimmerman and Vitousek (2012) reported greater fungal endophyte richness at wetter, low elevation sites on a Hawaiian mountainside. With data on hosts analyzed for endophytes, Poaceae was by far the most thoroughly investigated family, followed by Fabaceae, Pinaceae, and Asteraceae. Foliar endophytes were the subject of the vast majority of investigations, followed by endophytes of the root and stem. Comparatively, floral tissues and plant propagules were understudied for endophyte diversity. The number of studies characterizing endophyte fungal biodiversity is much higher than that of endophytic bacterial diversity (Harrison and Griffin 2020). Higher biological diversity ultimately implies higher chemical diversity. This raises the possibility that new chemical structures and physiologically active substances might be found in biodiverse regions. Endophytic microbiomes of invasive plants in both their native and invaded habitats differ significantly. For instance, compared to the native range of *Centaurea solstitialis*, Lu-Irving et al. (2019) observed lower richness in the phyllosphere and endophytic root bacteria. Similar to this, Shipunov et al. (2008) described a significant change in the fungal endophyte assemblage found in *Centaurea stoebe* leaves between native and invaded regions of its range. Thus, the history of host invasion has an impact on endophyte biodiversity.

12.8 Fungal Endophytes

Mycorrhizal and fungal endophytes differ in the manner that the latter are totally contained inside plant tissues and they can develop within roots, stems, and leaves. They only emerge out of the plant to sporulate at the time, but the former initially colonize plant roots and spread into the rhizosphere (Sherwood and Carroll 1974; Carroll 1998; Stone et al. 2004).

Earlier fungal endophytes were of two major kinds, the non-clavicipitaceous endophytes (NC endophytes), which may be collected from asymptomatic tissues of nonvascular plants, ferns and allies, conifers, and angiosperms, and the clavicipitaceous endophytes (C endophytes), which infect certain grasses (Clay and Schardl 2002; Arnold 2007). C endophytes belong to a single class with a narrow host range (class I), and NC endophytes were comprised of three different classes with a broad host range (class II, class III, and class IV) based on a variety of life history traits and the ecological importance of those traits (host range, tissue colonization and the extent of colonization, in planta diversity, and transmission patterns) (Rodriguez et al. 2009). They can also be classified as symptomatic or asymptomatic endophytes; asymptomatic endophytes can become pathogenic (Delaye et al. 2013).

Depending on how they reproduce, endophytes are classified as either sexual or asexual endophytes; for instance, the sexually and asexually reproducing genera *Epichloe* and *Neotyphodium* (formerly *Acremonium*) have been established within the *Epichloe* endophytes (Moon et al. 1999; Leuchtmann et al. 2000). And based on transmission mode in the host, there are two kinds of endophytes: true endophytes, which are transferred from parent to progeny and are called vertically transmitted endophytes, e.g., *Neotyphodium*, which are seed transmitted (Saikkonen et al. 2013), and horizontally transmitted endophytes which are passed among various members of a population (Hartley and Gange 2009).

Endophytes can also be categorized as generalized and specialized endophytes based on their host range, generalized endophytes being most common (Cook et al. 2014), seed-borne *Epichloe* fungi specialized to the members of the Poaceae (Clay and Schardl 2002; Rudgers et al. 2009), and swainsonine-producing *Alternaria* fungi to specific Fabaceous taxa (Cook et al. 2014; Panaccione et al. 2014).

Fungal endophytes are ubiquitous in nature; in the rice ecosystem, *Fusarium*, *Aspergillus*, *Curvularia*, *Penicillium*, *Gilmaniella*, and *Arthrobotrys foliicola* were the most often found endophytic fungal genera. The spermosphere had more *Fusarium* and *Curvularia* than the other fungus combined. *Penicillium* and *Aspergillus* were primarily found in the leaf sheath and leaf blade, respectively. *Gilmaniella* and *A. foliicola* were found associated only with the roots and leaf blades, respectively (Zakaria et al. 2010). In the mangrove ecosystem, the most dominating fungal endophytes were found to be from the genera *Pestalotiopsis* followed by *Alternaria* and *Cladosporium* (Hamzah et al. 2018). The genus *Alternaria* is the most predominant endophyte associated with salt tolerance, followed by *Cladosporium* and *Fusarium* (Khalmuratova et al. 2020). *Phomopsis archeri, Alternaria alternata*, and *Pestalotiopsis besseyi* in the stem and *Leptostroma* sp. and *P. lingam* in needle tissues were found as dominating endophytes in pine trees of coniferous forests in China (Wang and Guo 2007). *Phomopsis cassiae* is a major phyllosphere endophytic fungus isolated from *Cassia spectabilis* (Silva et al. 2006). Botryosphaeriales (30.5%) and Diaporthales (34.4%), which mainly were represented by the species *Neofusicoccum parvum* (23.4%) and *Phomopsis* sp. (24.7%), were found to be most predominant in the waterlogging species *Distylium chinense* (Duan et al. 2019). *Sclerotinia sclerotiorum, Rhizoctonia solani, Botrytis cinerea*, and *Fusarium oxysporum* were discovered to be the core of *Ocimum sanctum* phyllosphere mycoendophytes (Chowdhary and Kaushik 2015). Coffee's (*Coffea arabica*) most prevalent endophytic genera were *Colletotrichum*, *Fusarium*, *Penicillium*, and *Xylaria* (Oliveira et al. 2014; Vega et al. 2010).

12.9 Bacterial Endophytes

When compared to bulk soil, which contains between 10^6 and 10^9 bacteria per gram of soil, root endophytic habitats have only around 10^4 to 10^8 bacteria per gram of root tissues; this demonstrates that roots serve as effective habitat filters when habitats transition from soil to roots (Bulgarelli et al. 2013). Proteobacteria in relative abundance (50%), Actinobacteria (10%), Firmicutes (10%), and Bacteroidetes (10%) often make up the majority of the root endophytic bacterial communities (Liu et al. 2017).

Research on constituents of endophyte has found that a small number of bacterial types may be dominant in the plant root endosphere, adding more support for the active and practical selection of bacteria by the plants. Gammaproteobacteria (genera *Enterobacter*, *Pseudomonas*, and Stenotrophomonas) make up the core of bacterial endophytes in the root of rice as well as sweet potatoes (Marques et al. 2015; Ferrando and Fernandez Scavino 2015). The endosphere of plant tissues, in certain instances, was shown to be dominated by just one or two bacterial species, particularly the genus *Pseudomonas* (34%) in *Populus deltoides* roots (Gottel et al. 2011), and *Pseudomonas* (52%) and *Enterobacter* (35.5%) in sugarcane stems dominate the rhizosphere (Magnani et al. 2013).

As reviewed by Rana et al. in 2020a, b, it was shown that the most common bacterial endophyte genus is *Bacillus* on wheat (Zhao et al. 2015), maize (Gond et al. 2015), rice (Ji et al. 2014), sugarcane (Manoel da Silva et al. 2015), chickpea (Joseph and Priya 2011), peanut (Wang et al. 2013), soybean, and bean. The most prevalent is *Bacillus amyloliquefaciens*, whereas *Bacillus oryzicola* YC7010T and *Bacillus endocardicis* CCBAU 05776T came from rice and soybean (Chung et al. 2015; Zhang et al. 2012). *Pseudomonas*, a member of the phylum Proteobacteria, is the second most common genus of endophytic bacteria. *Pseudomonas aeruginosa*, *Pseudomonas fulva*, *Pseudomonas lini*, *Pseudomonas monteilii*, *Pseudomonas putida*, and *Pseudomonas thivervalensis* were among the several species of *Pseudomonas* reported from wheat (Verma et al. 2019), maize (Montanez et al. 2009), rice (Mbai et al. 2013), sugarcane (Govindarajan et al. 2007), peanut (Wang et al. 2013), soybean (Zhao et al. 2018), and bean (Dinic et al. 2014). The following most common genera are *Burkholderia*; several species of *Burkholderia*, which include *Burkholderia vietnamiensis*, *Burkholderia silvatlantica*, and *Burkholderia tropica*, were reported from maize (Estrada et al. 2002), rice (Rangjaroen et al. 2015), soybean (Perin et al. 2006), and sugarcane (Govindarajan et al. 2008).

Numerous studies have shown that from bulk soil to the root endosphere, Acidobacteria, Gemmatimonadetes, and Archaea all successively decline, while Proteobacteria, notably Gammaproteobacteria, considerably rise in relative abundance (Edwards et al. 2015; Liu et al. 2017). Generally, the number and species of endophytic actinomycetes in plant roots are more than those in other parts of plants. *Streptomyces* and *Micromonospora* are the dominant genera and have been acknowledged as valuable resources for antibiotics. Proteobacteria, Actinobacteria, and Firmicutes were found by Oliveira et al. (2014) to be the main groups in the leaf microbiome of common bean plants (*Phaseolus vulgaris*). *Microbispora*, *Micromonospora*, *Nocardioides*, and *Streptomyces* were the most common actinomycete endophytes of wheat (*Triticum aestivum*) (Coombs and Franco 2003).

While Trichosphaeriales, ascomycetes, have been reported from wheat plants with evidence of being niche-specific (Ofek-Lalzar et al. 2016), *Fusarium*, one of the filamentous fungi genera, has been reported as being predominant in wheat (Bacon and Hinton 2007), maize (Bacon and Hinton 1996), rice (Tian et al. 2004), and soybean (Garcia-Romera et al. 1998). Nitrogen-fixing bacteria *Rhizobium* is a symbiotic endophyte of leguminous plants (Sprent et al. 2013); while some *Azospirillum* species have been observed to be connected with cereal grasses with the potential to fix nitrogen, *Frankia* has been discovered to be associated with higher plants (Vessey et al. 2005).

12.10 Endophytes and Plant Growth Promotion

Endophytes are found in all or most plant species. An alliance between endophytes and host tissues has direct benefits for plant growth promotion by aiding enzyme activity and nutrient exchange (Murphy et al. 2018; Khan et al. 2015).

Plant growth-promoting endophytes (PGPE) augment plant growth by modulating phytohormones (gibberellin, auxin, cytokinin, and ethylene) and nutrient availability (phosphorus solubilization, siderophore production, nitrogen fixation, etc.) and are involved in bioremediation of heavy metals (Vardharajula et al. 2017; Afzal et al. 2019). Thus, they act as biostimulants and biofertilizers (Baron and Rigobelo 2022).

12.10.1 Underlying Mechanisms in Plant Growth Promotion

PGPE displays three disparate mechanisms in PGP activity, viz., phytostimulation, biofertilization (directly), and biocontrol (indirectly) (Santos et al. 2018; Baron and Rigobelo 2022; Wu et al. 2021) (Fig 12.1). The implication of each of the mechanisms in PGP, which are directly involved, is discussed below with some

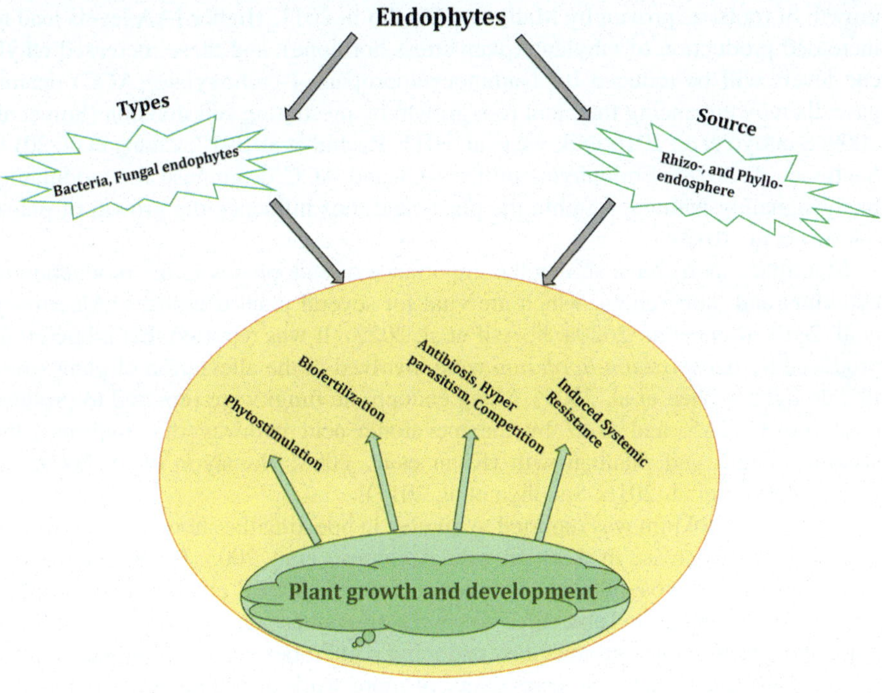

Fig. 12.1 Underlying mechanisms involved in plant growth and development. Diverse endophytes such as bacterial and fungal obtained from rhizosphere and phyllosphere contribute toward plant growth and development. The important mechanisms involved are believed to be phytostimulation which further helps plant in the production of important phytohormones and their control; biofertilization which helps plants in nutrient fixation, solubilization, mobilization, and production; and disease suppression by means of antibiosis, hyper-parasitism, competition, and induced systemic resistance

illustrations. However, the biocontrol activity of endophytes will be discussed under mechanisms of disease management.

12.10.1.1 Phytostimulation

The mechanism of plant growth regulation by stimulating phytohormones' production is phytostimulation (Beneduzi et al. 2012). PGPE is capable of modulating several phytohormone levels, viz., gibberellin, auxin, cytokinin, and ethylene. These phytohormones are the signal molecules that play a significant role in various physiological and biological processes in plants (Asif et al. 2022).

The very first auxin produced by endophytic fungi and bacteria is indole-3-acetic acid (IAA), which is involved in the regulation of cell division, thereby controlling the root and shoot development, and involved in vascular tissue differentiation (Kuklinsky-Sobral et al. 2004; Tsavkelova et al. 2007; Dias et al. 2009; Taghavi et al. 2009; Jaroszuk-Ściseł et al. 2014; Asif et al. 2022). The lower level of IAA has a positive effect on plant growth, whereas the higher level of IAA causes stunted growth of roots, as proven by Malik and Sindhu in 2011. Higher IAA levels lead to increased production of ethylene (plant stress hormone), and these increased ethylene levels will be reduced by 1-aminocyclopropane-1-carboxylate (ACC) deaminase, thereby enhancing the plant root growth by promoting cell division (Sun et al. 2009; Santoyo et al. 2016; Nikolic et al. 2011; Rashid et al. 2012; Zhang et al. 2011; Santos et al. 2018). Endophytes utilize IAA and ACC deaminase for modulating many signaling pathways within the plants and thus influence the growth of plants (Gaiero et al. 2013).

Manifold reports have shown the importance of endophytes in the production of cytokinin and gibberellins, which are vital for several plant responses (Albermann et al. 2013; Rana et al. 2020a, b; Asif et al. 2022). It was reported that gibberellins produced by *Azospirillum lipoferum* were involved in the alleviation of plant stress like drought (Cohen et al. 2009). Many endophytic fungi were reported to produce GA1, GA3, GA5, and GA7 by the mevalonic acid pathway that improves the seedling length and plant growth (Khan et al. 2008; Hamayun et al. 2009a, b, 2010; Shahzad et al. 2016; Sandhya et al. 2017).

Similarly, cytokinin was reported to involve in boosting the shoot and root weight and the overall increase in shoot growth (Arkhipova et al. 2005; Bhore et al. 2010).

Illustrations of biosynthesis of phytohormones by PGPE and its effect on plant growth are presented in Table 12.1. From the abovementioned studies, it is evident that the phytohormones produced by endophytic microbes positively influence plant growth. Also, it displays the essentiality of more work on elucidating the targeted signaling pathways for the synthesis of phytohormones by endophytes and the mechanisms involved in the growth promotion.

Table 12.1 Some illustrations of endophyte-mediated biosynthesis of phytohormones

Endophyte	Host	Hormone	PGP activity	References
Endophytic fungi				
Penicillium citrinum IR-3-3	*Ixeris repens*	GA	Increased the length of seedling	Khan et al. (2008)
Aspergillus fumigatus	Soybean	GA	Enhanced the length and fresh and dry weight of plant	Hamayun et al. (2009a)
Cladosporium sphaerospermum	Soybean	GA	Plant growth-promoting	Hamayun et al. (2009b)
Cladosporium sp. MH-6	Cucumber	GA	Increased shoot length and other growth attributes	Hamayun et al. (2010)
Paecilomyces formosus strain LHL10	Cucumber	GA	Enhanced shoot length and allied growth	Khan et al. (2012)
Phoma glomerata and *Penicillium* sp.	Cucumber	IAA and GA	Increased plant biomass and plant growth parameters	Waqas et al. (2012)
Colletotrichum fructicola CMU-A109	*Coffea arabica*	IAA	Coleoptile stimulation elongation of maize, rye, and rice	Numponsak et al. (2018)
Endophytic bacteria				
Bacillus subtilis	Lettuce	Cytokinin	30 % increase in shoot and root weight	Arkhipova et al. (2005)
Bacillus subtilis strain NA-108	*Fragaria ananassa* Duchesne	IAA	Enhanced shoot length and number of leaves and increased dry weight of both shoot and root	de Melo Pereira et al. (2012)
Bacillus subtilis and *B. cereus*	*Capsicum annuum* L.	Auxin (indole acetic acid)	Increase the total biomass of plants, stem diameter, number of flower buds, and root volume	Pena-Yam et al. (2016)
Bacillus amyloliquefaciens strain RWL-1	*Oryza sativa* L.	Gibberellins	Prolonged seed growth and plant development	Shahzad et al. (2016)
Acinetobacter baumannii, *Enterobacter asburiae*, many *Pseudomonas* spp., and *Sinorhizobium meliloti*	Maize	IAA, gibberellins, and cytokinin	Plant growth-promoting, tolerance to drought stress	Sandhya et al. (2017)
Bacillus amyloliquefaciens	*Capsicum annuum* L.	IAA	Enhanced seed germination and seedling vigor	Gowtham et al. (2018)
B. amyloliquefaciens strain B14 and *Bacillus* sp. strains B19 and P12	*Phaseolus vulgaris*	Auxins	Increased germination potential and stem and root length	Sabate et al. (2018)
Pseudomonas aeruginosa, *Bradyrhizobium* sp.	Soybean	IAA	Improved growth parameters	Kumawat et al. (2019)

(continued)

Table 12.1 (continued)

Endophyte	Host	Hormone	PGP activity	References
Pseudomonas spp.	*Nicotiana tabacum*	IAA, ACC deaminase	Promoted root growth	Li et al. (2019)
Acinetobacter guillouiae	Wheat	IAA	Plant growth-promoting	Rana et al. (2020a, b)

12.10.1.2 Biofertilization

Another way of promoting plant growth by endophytes is nutrient acquisition. They are capable enough to solubilize and facilitate both macro- and micronutrients from soil to host plants and increase crop productivity. Thus, another alternative for chemical fertilizers is called biofertilizers (Rana et al. 2020a, b).

12.10.1.3 Nitrogen Fixation

One of the primary macronutrients required by plants is nitrogen, which increases the growth and yield of the crop. Endophytic microbes are more efficient biological nitrogen fixers than rhizospheric microbes because of their capacity to colonize inside the plant tissues where excess carbon content and less or lack of oxygen can be seen in the endosphere (James and Olivares 1998). More interestingly, unlike rhizospheric microbes, they never form nodule-like structures in roots (Ladha and Reddy 2003). This associative symbiosis enables some endophytic bacteria to produce nitrogen-fixing genes that are able to convert atmospheric nitrogen to nitrate and ammonia, which are further utilized by plants (Bhattacharjee et al. 2008; Santi et al. 2013).

Several authors reported the enhanced biomass production in different crops after endophytic bacterial inoculation that provided the plant with atmospheric nitrogen by nitrogenase activity (Table 12.2).

12.10.1.4 Phosphorus Solubilization

Phosphorus is an essential macronutrient after nitrogen. Most of the phosphorus in soil is unavailable to the plants because of its insoluble and immobilized nature (Gouda et al. 2018). Some of the endophytes possess phosphorus-transporting ability by dissolving the precipitated insoluble phosphorus and thus increasing the availability of phosphorus to the plants (Gaiero et al. 2013). Hence, endophytic microbes are considered phosphate solubilizers.

Various mechanisms are involved in phosphorus solubilization by endophytic microbes, viz., chelation; lowering pH of the medium; secretion of organic acids (acetic acid, glycolic acid, malonic acid, oxalic acid, etc.), exopolysaccharides, and phosphorus-solubilizing enzymes (phosphatase, C-P lyase and phytase); and ion

Table 12.2 Some illustrations of endophyte-mediated nutrient acquisition

Endophyte	Host	PGP activity	References
Herbaspirillum seropedicae and *Acetobacter diazotrophicus*	Sugarcane	N_2 fixation	Triplett (1996)
Serratia marcescens	Rice	N_2 fixation	Gyaneshwar et al. (2001)
Burkholderia	Maize	N_2 fixation	Estrada et al. (2002)
Klebsiella sp. GR9	Sugarcane	N_2 fixation (*nif-D* genes)	Govindarajan et al. (2007)
Methylobacterium spp.	Citrus	Siderophore production	Lacava et al. (2008)
Bacillus and *Sphingopyxis*	Strawberry	Phosphate solubilization and enhanced shoot and root growth	Dias et al. (2009)
Pantoea, Pseudomonas, Rahnella, Herbaspirillum, Azospirillum, Rhizobium (*Agrobacterium*), and *Brevundimonas*	Maize	N_2 fixation (*nifH* genes)	Montanez et al. (2009)
Burkholderia vietnamiensis	*Populus trichocarpa*	Enhanced nitrogen content (37%) by N_2 fixation	Xin et al. (2009)
Bacillus megaterium, Micrococcus luteus, B. cereus, Lysinibacillus fusiformis	Ginseng	N_2 fixation (*nifH* gene), phosphate solubilization	Vendan et al. (2010)
Pantoea J49	Peanut	Phosphate solubilizations, siderophore production	Taurian et al. (2010)
Bacillus spp.	–	Siderophore production	Han et al. (2011)
Enterobacter spp. and *Klebsiella* spp.	Sugarcane	N_2 fixation, phosphate solubilizations, and siderophore production	Lin et al. (2012)
Pseudomonas spp. and *Bacillus* spp.	Sugarcane	Phosphorus solubilization and siderophore production	Chauhan et al. (2013)
Bacillus sp. strain EUCB 10	*Eucalyptus*	Phosphorus solubilization and biological nitrogen fixation	Paz et al. (2012)
Pseudomonas strains L111, L228, and L321	*Miscanthus giganteus*	P solubilization activity	Oteino et al. (2015)
Klebsiella variicola strain DX120E	Sugarcane	Promoted the uptake of N, P, and K	Wei et al. (2014)
Bacillus subtilis (EB-126) and *Bacillus* sp. (EB-47)	Banana	N_2 fixation, phosphate solubilization	Andrade et al. (2014)
Pantoea dispersa	Cassava	Promoted P uptake	Chen et al. (2014)
Bacillus megaterium and *B. amyloliquefaciens*	Wheat	N_2 fixation; solubilization of P, K, and Zn; production of siderophore and ammonia	Verma et al. (2015)

(continued)

Table 12.2 (continued)

Endophyte	Host	PGP activity	References
Acinetobacter sp. and *Bacillus* sp.	*Phyllanthus amarus*	P solubilization activity	Joe et al. (2016)
Enterobacter sp. (strain MCR1.48)	*Acacia polyphylla*	High phosphorus solubilization index	Castro et al. (2018)
Enterobacter, *Acinetobacter*, *Pseudomonas*, and *Bacillus*	Soybean	N_2 fixation	Zhao et al. (2018)
Trichoderma koningii ST-KKU1, *Macrophomina phaseolina* SS1L10, and *M. phaseolina* SS1R10	Sunchoke	P solubilization and siderophore production	Suebrasri et al. (2020)
Trichoderma harzianum	*Phaseolus vulgaris*	Siderophore production	Eslahi et al. (2020)

exchange (Van Der Heijden et al. 2008; Yi et al. 2008; Ahmed and Shahab 2009; Singh et al. 2020). The mechanism opted for P solubilization depends on the species of endophyte. The role of endophyte in P solubilization is summarized in Table 12.2.

12.10.1.5 Siderophore Production

Iron is the principal micronutrient of life required by plants as a co-factor of most essential enzymes in their metabolism. Like phosphorus, iron present in the soil is also not available to plants because of its low solubility. Insoluble iron predominantly exists in ferric (Fe3+) forms of hydroxides, oxides, carbonates, and phosphates in soil (Ma et al. 2016). Under such iron deficiency conditions, microbes produce low molecular weight iron-chelating agents called siderophores that bind Fe3+. These bound siderophores help the plants to acquire soluble iron through ligand exchange or chelate degradation (Rajkumar et al. 2009; Ma et al. 2016).

Endophytic microbes are reported to produce these siderophores, help the plant uptake of iron, and influence plant metabolism, as shown in Table 12.2. Along with this, siderophores produced by endophytes also bind other essential elements and make them unavailable to phytopathogens and, thus, outcompete them by preventing their propagation (Lacava et al. 2008; Han et al. 2011).

12.10.2 Defense Mechanism

Endophytes are ubiquitously present in the universe and have shown their presence in combating plant diseases. Endophytes colonize the plants and protect the plants against pathogens, viz., root colonization by *Fusarium oxysporum* endophytes reduces wilt disease symptoms caused by pathogenic *Fusarium* strains in tomato (Constantin et al. 2020), *Trichoderma* parasitize *S. sclerotiorum* and *Rhizoctonia solani* hyphae (Harman et al. 2004), and cell wall extract of *Piriformospora indica* reduced the nematode infection of *Heterodera schachtii* in *Arabidopsis* (Daneshkhah et al. 2013). Thus, endophytes put forth different mechanisms to protect the plant against invaders. It might use the direct or indirect mechanism or a combination of both to fight the pathogens; a few of them are discussed below (Table 12.3) (Fig. 12.1).

12.10.2.1 Direct Mechanism

Endophyte colonizes the host plant, providing protection to the plant through antagonism mechanisms like the production of chemical compounds and antibiotics against the pathogen (antibiosis), competing for the space and nutrients with the invading pathogens, or causing the death of the pathogen through hyper-parasitism.

Antibiosis

Antibiosis is a mechanism in which the endophytic fungi/microorganism suppresses the physiology and growth of the disease-causing pathogen. It is evident that the host plant species harbored many endophytes, which protect the plants against pathogenic organisms. For instance, the endophytic fungi, WF4 isolated from finger millet have the control of *Fusarium graminearum* that causes fungal diseases by the production of antifungal compounds (Mousa et al. 2015), *Trichoderma harzianum* strain (M10) produces secondary metabolite harzianic acid and protect the tomato plant against *Rhizoctonia solani* (Manganiello et al. 2018) and in a dual culture test conducted, showed 51% inhibition in the mycelial growth of *R. solani* due to *B. subtilis* SR22 indicating potent antagonistic behavior (Rashad et al. 2022). Antibiosis was the most prevalent antagonistic mechanism of *C. globosum* against *B. sorokiniana* in agreement with studies by Aggarwal et al. (2004) and Darshan et al. (2020a, b).

Hyper-parasitism

Hyper-parasitism involves the growth of a biocontrol agent toward the target organism and dissolution of the target pathogens' cell wall or membrane by the activity of lytic enzymes or PR proteins, finally collapsing the pathogen (Fadiji and Babalola 2020). Antagonistic mechanisms of *C. globosum* fungus are exerted through different studies. *C. globosum* mycoparasitizes and produces antifungal metabolites that help in the lysis of the cell wall of the pathogen (Darshan et al. 2020a, b). *Burkholderia gladioli* strain E39CS3 (BG-E39), a saffron endophyte, portrayed antifungal activity and induced cell death in the mycelia of *Fusarium oxysporum*. It produced chitinases as well as β-1,3-glucanase, which are involved in the pathogen cell wall degradation (Ahmad et al. 2022). Endophytes *Streptomyces*

Table 12.3 Some of the illustrations of endophyte-mediated disease suppression

Mechanism	Endophyte	Pathogen	References
Antibiosis	*Bacillus subtilis* SR22	*Rhizoctonia solani*	Rashad et al. (2022)
	Pseudomonas protegens Pf-5 and *Pseudomonas fluorescens* 2P24	*Aphanomyces euteiches*	Lai et al. (2022)
	Trichoderma harzianum strain M10	*Rhizoctonia solani* (secondary metabolite harzianic acid)	Manganiello et al. (2018)
	Bacillus spp.	*Fusarium oxysporum* f. sp. *lycopersici*	Abdallah et al. (2022)
Competition	*Trichoderma harzianum* isolate (TRIC8)	*A. alternata*, *B. cynodontis*, and *F. culmorum*	Guclu and Ozer (2022)
	F. oxysporum	*V. dahliae*	Mulero-Aparicio et al. (2020)
	Pseudomonas fluorescens	*P. viridiflava*	Karablieh et al. (2017)
Hyper-parasitism	*Trichoderma harzianum* isolate (TRIC8)	*B. cynodontis* and *F. culmorum*	Guclu and Ozer (2022)
	Burkholderia gladioli strain E39CS3 (BG-E39)	*Fusarium oxysporum*	Ahmad et al. (2022)
	Streptomyces hygroscopicus OsiSh-2	*Magnaporthe oryzae*	Xu et al. (2019)
	Bacillus halotolerans Y6	*Verticillium dahliae*	Zhang et al. (2019)
Systemic response (ISR)	*Piriformospora indica*	*Rhizoctonia cerealis*, *Fusarium graminearum* (wheat)	Li et al. (2022a, b)
	Streptomyces	*Fusarium oxysporum* f. sp. *lycopersici Race 3*	Abbasi et al. (2019)
	T. harzianum (ThM9) and *T. virens* (TvA1)	*P. cubensis*	Rauf et al. (2022)
	Burkholderia sp. SSG	Broad spectrum of pathogens	Kong et al. (2022)
	Bacillus subtilis 26D and *B. subtilis* Ttl2	Potato virus X and potato virus Y	Veselova et al. (2022)
	Bacillus velezensis JRX-YG39	*Botrytis cinerea*	Feng et al. (2022)
Secondary compounds	*D. foliorum*	*C. albicans*	Dos Santos et al. (2021)
	T. hamatum FB10	*M. incognita*	Baazeem et al. (2021)
	Aspergillus flavus, *A. fumigatus*, and *Fusarium lateritium*	*Eupenicillium brefeldianum*, *Penicillium echinulatum*, *Alternaria phragmospora*, *Fusarium oxysporum*, *Fusarium verticillioides*, and *Alternaria alternata*	Kamel et al. (2020)

(continued)

Table 12.3 (continued)

Mechanism	Endophyte	Pathogen	References
Growth promotion	*Actinomycetes* sp.	*Sclerotinia sclerotiorum*	Elshafie and Camele (2022)
	Penicillium simplicissimum CN7, *Talaromyces flavus* BC1, and *Trichoderma konilangbra* DL3	*F. oxysporum*, fungus *L. theobromae* N13, and *N. dimidiatum*	Huong et al. (2022)
	B. ambifaria XN08	*R. cerealis*	An et al. (2022)
	Piriformospora indica	*Stemphylium vesicarium*	Roylawar et al. (2021)
	Pseudochrobactrum kiredjianiae A4	*Fusarium graminearum*	Fu et al. (2017)

hygroscopicus OsiSh-2 showed strong direct suppression of mycelial growth, conidial germination, and appressorial formation of *Magnaporthe oryzae* (Xu et al. 2019). Endophytic bacterium *Bacillus halotolerans* Y6 showed strong antagonistic abilities that inhibit *Verticillium dahliae* spore germination and mycelial growth. The results of the enzyme activity assay, heterologous expression, and gene knockdown showed that the key virulence factor of Y6 for antagonizing *V. dahliae* was β-glucanase Bgy6 (Zhang et al. 2019).

Competition

Competition is the mechanism through which endophytes proliferate and colonize the plants, thus inhibiting the space for other pathogenic fungi to colonize. They also take up the nutrients and lead the pathogenic organisms to starvation and then death. *Streptomyces sporocinereus* have antagonistic activity on *Magnaporthe oryzae* that causes the blast, hence protecting the rice plant against blast disease (Zeng et al. 2018). It was found that *F. oxysporum* f. sp. *lycopersici* colonized the melon root surface and inhibited conidial germination and hyphal elongation of pathogenic *F. oxysporum*, competing for nutrients rather than disease resistance (Iida et al. 2022). Competition by most endophytes usually works in combination with other endophytic mechanisms like improving the plant fitness by nutrient absorption and by warding off the pathogens by triggering the production of secondary metabolites, volatiles, and antibiotics, which are detrimental for the pathogen. Competition might not always work, especially when the pathogenic fungi dominate the space in the plant, but it is interesting to see how the plant differentiates between the pathogenic and non-pathogenic fungi and promote the beneficial one to colonize (Fadiji and Babalola 2020).

12.10.2.2 Indirect Mechanism

Endophytes, along with directly defending the pathogen, also perform through indirect mechanisms like better growth and development of a plant, inducing systemic response, and changing the transcriptional and metabolic profile of the host plant. A few of the mechanisms are briefed below.

Endophytes induce the host plant to produce metabolites to prevent the disease or kill the pathogen, along with improving the growth and development of the plants. Endophytes trigger the process of host metabolism to make their hosts produce specific metabolites (Li et al. 2022a, b), viz., antifungal compound pyrrolnitrin by *Burkholderia ambifaria* XN08 (An et al. 2022). A wide array of secondary compounds, viz., phytoalexins, sugars, and amino acids are alkaloids, terpenoids, flavonoids, benzopyranones, tannins, phenolic acids, quinones, steroids, tetralones, and chinones are biosynthesized by endophytes and provide the plants with better resistance against phytopathogens (Joseph and Priya 2011). Some elicitors like glycoprotein, polysaccharides, and lipopolysaccharides trigger plant defense mechanisms and increase secretion of plant secondary metabolites, which effectively reduce attack by pathogens (Gao et al. 2010). Benzoxazinone glucosides play a major role in defending against pathogens in cereals like wheat, rye, and barley. Phytoalexins like camalexin help fight *A. brassicicola* and phenylalanine and terpenoids against *M. oryzae*. Few metabolites like 3-deoxyanthocyanidin and 2-hydroxy-octadecatrienoic acid are produced in the *Arabidopsis* leaves against *Colletotrichum* (Piasecka et al. 2015) and avenacins in the disease resistance of oats (Mugford et al. 2009). Recent "omics" studies have proved that these endophytes, in interaction with the host plant, alter the transcriptomic and metabolomic profile of the plant and, in turn, promote growth and development and also render disease resistance (Kundu et al. 2022).

12.11 Conclusion

Recent chemical pollution owing to the use of fertilizers and pesticides warrants an alternate approach to manage the environment by keeping food production at required scale. Endophytes are the potential candidates for sustainable agriculture in order to reduce the use of synthetic chemicals. Their contribution to enhancing plant biomass, shoot and root growth, seed germination, and nutrient acquisition is indispensable. Forthcoming reports should focus on the exact mechanism behind the nutrient acquisition and signaling pathways targeted by particular phytohormones produced by endophytes that remain elusive. Further, exploring novel endophytes, their characterization, and their mode of action in promoting plant growth is crucial.

References

Abbasi S, Safaie N, Sadeghi A, Shamsbakhsh M (2019) *Streptomyces* strains induce resistance to *Fusarium oxysporum* f. sp. *lycopersici* race 3 in tomato through different molecular mechanisms. Front Microbiol 10:1505

Abdallah RAB, Catalina S, Constantine G, Ahlem N, Hayfa JK, Kalliope KP, Mejda DR (2022) Involvement of lipopeptide antibiotics and chitinase genes and induction of host defense in suppression of *Fusarium* wilt by endophytic *Bacillus* spp. in tomato. Crop Prot 99:45–48

Afzal I, Shinwari ZK, Sikandar S, Shahzad S (2019) Plant beneficial endophytic bacteria: mechanisms, diversity, host range and genetic determinants. Microbiol Res 221:36–49

Aggarwal R, Tewari AK, Srivastava KD, Singh DV (2004) Role of antibiosis in the biological control of spot blotch (*Cochliobolus sativus*) of wheat by *Chaetomium globosum*. Mycopathologia 157(4):369–377

Ahmad T, Bashir A, Farooq S, Riyaz-Ul-Hassan S (2022) *Burkholderia gladioli* E39CS3, an endophyte of *Crocus sativus* Linn., induces host resistance against corm-rot caused by *Fusarium oxysporum*. J Appl Microbiol 132(1):495–508

Ahmed N, Shahab S (2009) Phosphate solubilization: their mechanism genetics and application. Internet J Microbiol 9(1):1–19

Albermann S, Linnemannstons P, Tudzynski B (2013) Strategies for strain improvement in *Fusarium fujikuroi*: overexpression and localization of key enzymes of the isoprenoid pathway and their impact on gibberellin biosynthesis. Appl Microbiol Biotechnol 97(7):2979–2995

An C, Ma S, Liu C, Ding H, Xue W (2022) *Burkholderia ambifa*ria XN08: a plant growth-promoting endophytic bacterium with biocontrol potential against sharp eyespot in wheat. Front Microbiol 13:906724

Andrade LF, de Souza GLOD, Nietsche S, Xavier AA, Costa MR, Cardoso AMS, Pereira MCT, Pereira DFGS (2014) Analysis of the abilities of endophytic bacteria associated with banana tree roots to promote plant growth. J Microbiol 52(1):27–34

Archana HR, Darshan K, Lakshmi MA, Ghoshal T, Bashayal BM, Aggarwal R (2022) Biopesticides: a key player in agro-environmental sustainability. In: Trends of applied microbiology for sustainable economy, Academic, pp 613–653

Arkhipova TN, Veselov SU, Melentiev AI, Martynenko EV, Kudoyarova GR (2005) Ability of bacterium *Bacillus subtilis* to produce cytokinins and to influence the growth and endogenous hormone content of lettuce plants. Plant Soil 272(1):201–209

Arnold AE (2007) Understanding the diversity of foliar endophytic fungi: progress, challenges, and frontiers. Fungal Biol Rev 21(2-3):51–66

Asif R, Yasmin R, Mustafa M, Ambreen A, Mazhar M, Rehman A, Umbreen S, Ahmad M (2022) Phytohormones as plant growth regulators and safe protectors against biotic and abiotic stress. Plant hormones: recent advances, new perspectives and applications, pp 115

Baazeem A, Almanea A, Manikandan P, Alorabi M, Vijayaraghavan P, Abdel-Hadi A (2021) *In vitro* antibacterial, antifungal, nematocidal and growth promoting activities of *Trichoderma hamatum* FB10 and its secondary metabolites. J Fungus 7(5):331

Bacon CW, Hinton DM (1996) Symptomless endophytic colonization of maize by *Fusarium moniliforme*. Can J Bot 74(8):1195–1202

Bacon CW, Hinton DM (2007) Bacterial endophytes: the endophytic niche, its occupants, and its utility. In: Plant-associated bacteria. Springer, Dordrecht, pp 155–194

Bai ZQ, Lin X, Wang J, Zhou X, Liu J, Yang B, Yang X, Liao S, Wang L, Liu Y (2015) New meroterpenoids from the endophytic fungus *Aspergillus flavipes* AIL8 derived from the mangrove plant *Acanthus ilicifolius*. Mar Drugs 13(1):237–248

Baron NC, Rigobelo EC (2022) Endophytic fungi: a tool for plant growth promotion and sustainable agriculture. Mycology 13(1):39–55

Beneduzi A, Ambrosini A, Passaglia LM (2012) Plant growth-promoting rhizobacteria (PGPR): their potential as antagonists and biocontrol agents. Genet Mol Biol 35(4):1044–1051

Bhattacharjee RB, Singh A, Mukhopadhyay SN (2008) Use of nitrogen-fixing bacteria as biofertilizer for non-legumes: prospects and challenges. Appl Microbiol Biotechnol 80(2): 199–209

Bhore SJ, Ravichantar N, Loh CY (2010) Screening of endophytic bacteria isolated from leaves of Sambung Nyawa [*Gynura procumbens* (lour.) Merr.] for cytokinin-like compounds. Bioinformation 5(5):191–197

Bodenhausen N, Horton MW, Bergelson J (2013) Bacterial communities associated with the leaves and the roots of *Arabidopsis thaliana*. PLoS One 8(2):e56329

Bowman EA, Arnold AE (2018) Distributions of ectomycorrhizal and foliar endophytic fungal communities associated with Pinus ponderosa along a spatially constrained elevation gradient. *Am J Bot* 105(4):687–699

Bulgarelli D, Schlaeppi K, Spaepen S, Van Themaat EVL, Schulze-Lefert P (2013) Structure and functions of the bacterial microbiota of plants. Annu Rev Plant Biol 64:807–838

Bulgari D, Casati P, Quaglino F, Bianco PA (2014) Endophytic bacterial community of grapevine leaves influenced by sampling date and phytoplasma infection process. BMC Microbiol 14(1): 1–11

Carroll G (1998) Fungal endophytes in stems and leaves: from latent pathogen to mutualistic symbiont. Ecology 69(1):2–9

Castro RA, Dourado MN, Almeida JRD, Lacava PT, Nave A, Melo ISD, de Azevedo JL, Quecine MC (2018) Mangrove endophyte promotes reforestation tree (*Acacia polyphylla*) growth. Braz J Microbiol 49(1):59–66

Chauhan H, Bagyaraj DJ, Sharma A (2013) Plant growth-promoting bacterial endophytes from sugarcane and their potential in promoting growth of the host under field conditions. Exp Agric 49(1):43–52

Chen Y, Fan JB, Du L, Xu H, Zhang QH, He YQ (2014) The application of phosphate solubilizing endophyte *Pantoea dispersa* triggers the microbial community in red acidic soil. Appl Soil Ecol 84:235–244

Chi F, Shen SH, Cheng HP, Jing YX, Yanni YG, Dazzo FB (2005) Ascending migration of endophytic rhizobia, from roots to leaves, inside rice plants and assessment of benefits to rice growth physiology. Appl Environ Microbiol 71(11):7271–7278

Chowdhary K, Kaushik N (2015) Fungal endophyte diversity and bioactivity in the Indian medicinal plant *Ocimum sanctum* Linn. PLoS One 10(11):e0141444

Chung EJ, Hossain MT, Khan A, Kim KH, Jeon CO, Chung YR (2015) *Bacillus oryzicola* sp. nov., an endophytic bacterium isolated from the roots of rice with antimicrobial, plant growth promoting, and systemic resistance inducing activities in rice. Plant Pathol J 31(2):152–164

Clay K, Schardl C (2002) Evolutionary origins and ecological consequences of endophyte symbiosis with grasses. Am Nat 160(S4):S99–S127

Cohen AC, Travaglia CN, Bottini R, Piccoli PN (2009) Participation of abscisic acid and gibberellins produced by endophytic *Azospirillum* in the alleviation of drought effects in maize. Botany 87(5):455–462

Collinge DB, Jensen B, Jorgensen HJ (2022) Fungal endophytes in plants and their relationship to plant disease. Curr Opin Microbiol 69:102177

Constantin ME, Vlieger BV, Takken FLW, Rep M (2020) Diminished pathogen and enhanced endophyte colonization upon co-inoculation of endophytic and pathogenic fusarium strains. Microorganisms 8(4):544

Cook D, Gardner DR, Pfister JA (2014) Swainsonine-containing plants and their relationship to endophytic fungi. Agric Food Chem 62(30):7326–7334

Coombs JT, Franco CM (2003) Visualization of an endophytic *Streptomyces* species in wheat seed. Appl Environ Microbiol 69(7):4260–4262

Daneshkhah R, Cabello S, Rozanska E, Sobczak M, Grundler FMW, Wieczorek K, Hofmann J (2013) *Piriformospora indica* antagonizes cyst nematode infection and development in arabidopsis roots. J Exp Bot 64(12):3763–3774

Darshan K, Aggarwal R, Bashyal BM, Haritha Mohan M (2020a) Deciphering the network of interconnected pathways of *Chaetomium globosum* antagonistic related genes against *Bipolaris sorokiniana* using RNA seq approach. Biol Control 34(4):258–269

Darshan K, Aggarwal R, Bashyal BM, Singh J, Shanmugam V, Gurjar MS, Solanke AU (2020b) Transcriptome profiling provides insights into potential antagonistic mechanisms involved in *Chaetomium globosum* against *Bipolaris sorokiniana*. Front Microbiol 11:578115

De Bary A (1866) Morphologie und Physiologie der Pilze, Flechten und Myxomyceten Hofmeister's handbook of physiological botany. Wilhelm Engelmann, Leipzig

De Melo Pereira GV, Magalhaes KT, Lorenzetii ER, Souza TP, Schwan RF (2012) A multiphasic approach for the identification of endophytic bacterial in strawberry fruit and their potential for plant growth promotion. Microb Ecol 63(2):405–417

Delaye L, Garcia-Guzman G, Heil M (2013) Endophytes versus biotrophic and necrotrophic pathogens-are fungal lifestyles evolutionarily stable traits? Fungal Divers 60(1):125–135

Dias AC, Costa FE, Andreote FD, Lacava PT, Teixeira MA, Assumpcao LC, Araujo WL, Azevedo LL, Melo IS (2009) Isolation of micropropagated strawberry endophytic bacteria and assessment of their potential for plant growth promotion. World J Microbiol Biotechnol 25(2): 189–195

Dinic Z, Ugrinovic M, Bosnic P, Mijatovic M, Zdravkovic J, Miladinovic M, Josic D (2014) Solubilization of inorganic phosphate by endophytic *Pseudomonas* sp. from French bean nodules. Ratarstvoi povrtarstvo 51(2):100–105

Dos Santos GD, Gomes RR, Goncalves R, Fornari G, Maia BHLNS, Schmidt-Dannert C, Gaascht F, Glienke C, Schneider GX, Colombo IR, Degenhardt-Goldba J, Pietsch JLM, Costa-Ribeiro MCV, Vicente VA (2021) Molecular identification and antimicrobial activity of foliar endophytic fungi on the Brazilian pepper tree (*Schinus terebinthifolius)* reveal new species of *Diaporthe*. Curr Microbiol 78(8):3218–3229

Duan X, Xu F, Qin D, Gao T, Shen W, Zuo S, Dong J (2019) Diversity and bioactivities of fungal endophytes from *Distylium chinense*, a rare waterlogging tolerant plant endemic to the three gorges reservoir. BMC Microbiol 19(1):1–14

Edwards J, Johnson C, Santos-Medellin C, Lurie E, Podishetty NK, Bhatnagar S, Sundaresan V (2015) Structure, variation, and assembly of the root-associated microbiomes of rice. Proc Natl Acad Sci 112(8):E911–E920

Elshafie HS, Camele I (2022) Rhizospheric actinomycetes revealed antifungal and plant-growth promoting activities under a controlled environment. Plan Theory 11(14):1872

Eslahi N, Kowsari M, Motallebi M, Zamani MR, Moghadasi Z (2020) Influence of recombinant *Trichoderma* strains on the growth of bean (*Phaseolus vulgaris* L) by increased root colonization and induction of root growth-related genes. Sci Hortic 261:108932

Estrada P, Mavingui P, Cournoyer B, Fontaine F, Balandreau J, Caballero-Mellado J (2002) A N_2-fixing endophytic *Burkholderia* sp. associated with maize plants cultivated in Mexico. Can J Microbiol 48(4):285–294

Fadiji AE, Babalola OO (2020) Elucidating mechanisms of endophytes used in plant protection and other bioactivities with multifunctional prospects. Front Bioeng Biotechnol 8:467

Feng B, Chen D, Jin R, Li E, Li P (2022) Bioactivities evaluation of an endophytic bacterial strain *Bacillus velezensis* JRX-YG39 inhabiting wild grape. BMC Microbiol 22:170

Ferrando L, Fernandez Scavino A (2015) Strong shift in the diazotrophic endophytic bacterial community inhabiting rice (*Oryza sativa*) plants after flooding. FEMS Microbiol Ecol 91(9): fiv104

Freeman S, Rodriguez RJ (1993) Genetic conversion of a fungal plant pathogen to a nonpathogenic, endophytic mutualist. Science 260(5104):75–78

Fu Y, Gao H, Li H, Qin Y, Tang W, Lu J, Li M, Shao L, Liu H (2017) Change of growth promotion and disease resistant of wheat seedling by application of biocontrol bacterium *Pseudochrobactrum kiredjianiae* A4 under simulated microgravity. Acta Astronaut 139:222–227

Gaiero JR, McCall CA, Thompson KA, Day NJ, Best AS, Dunfield KE (2013) Inside the root microbiome: bacterial root endophytes and plant growth promotion. Am J Bot 100(9): 1738–1750

Gao FK, Dai CC, Liu XZ (2010) Mechanisms of fungal endophytes in plant protection against pathogens. Afr J Microbiol Res 4(13):1346–1351

Garcia-Romera I, Garcia-Garrido JM, Martin J, Fracchia S, Mujica MT, Godeas A, Ocampo JA (1998) Interactions between saprotrophic *Fusarium* strains and arbuscular mycorrhizas of soybean plants. Symbiosis 24:235–246

Gond SK, Bergen MS, Torres MS, White JF Jr (2015) Endophytic *bacillus* spp. produce antifungal lipopeptides and induce host defence gene expression in maize. Microbiol Res 172:79–87

Gottel NR, Castro HF, Kerley M, Yang Z, Pelletier DA, Podar M, Schadt CW (2011) Distinct microbial communities within the endosphere and rhizosphere of *Populus deltoides* roots across contrasting soil types. Appl Environ Microbiol 77(17):5934–5944

Gouda S, Kerry RG, Das G, Paramithiotis S, Shin HS, Patra JK (2018) Revitalization of plant growth promoting rhizobacteria for sustainable development in agriculture. Microbiol Res 206: 131–140

Govindarajan M, Kwon SW, Weon HY (2007) Isolation, molecular characterization and growth-promoting activities of endophytic sugarcane diazotroph *Klebsiella* sp. GR9. World J Microbiol Biotechnol 23(7):997–1006

Govindarajan M, Balandreau J, Kwon SW, Weon HY, Lakshminarasimhan C (2008) Effects of the inoculation of *Burkholderia vietnamensis* and related endophytic diazotrophic bacteria on grain yield of rice. Microb Ecol 55(1):21–37

Gowtham HG, Murali M, Singh SB, Lakshmeesha TR, Murthy KN, Amruthesh KN, Niranjana SR (2018) Plant growth promoting rhizobacteria-*Bacillus amyloliquefaciens* improves plant growth and induces resistance in chilli against anthracnose disease. Biol Control 126:209–217

Guclu T, Ozer N (2022) *Trichoderma harzianum* antagonistic activity and competition for seed colonization against seedborne pathogenic fungi of sunflower. Lett Appl Microbiol 74(6): 1027–1035

Gyaneshwar P, James EK, Mathan N, Reddy PM, Reinhold-Hurek B, Ladha JK (2001) Endophytic colonization of rice by a diazotrophic strain of *Serratia marcescens*. J Bacteriol 183(8): 2634–2645

Hamayun M, Khan SA, Khan MA, Khan AL, Kang SM, Kim SK, Joo GJ, Lee IJ (2009a) Gibberellin production by pure cultures of a new strain *of Aspergillus fumigatus*. World J Microbiol Biotechnol 25(10):1785–1792

Hamayun M, Afzal Khan S, Ahmad N, Tang DS, Kang SM, Na CI, Sohn EY, Hwang YH, Shin DH, Lee BH, Kim IJ, Lee IJ (2009b) *Cladosporium sphaerospermum* is a new plant growth-promoting endophyte from the roots of *Glycine max* (L.) Merr. World J Microbiol Biotechnol 25(4):627–632

Hamayun M, Khan SA, Khan AL, Rehman G, Kim YH, Iqbal I, Husssain J, Sohn EY, Lee IJ (2010) Gibberellin production and plant growth promotion from pure cultures of *Cladosporium* sp. MH-6 isolated from cucumber (*Cucumis sativus* L.). Mycologia 102(5):989–995

Hamzah TNT, Lee SY, Hidayat A, Terhem R, Faridah-Hanum I, Mohamed R (2018) Diversity and characterization of endophytic fungi isolated from the tropical mangrove species, *Rhizophora mucronata*, and identification of potential antagonists against the soil-borne fungus, *Fusarium solani*. Front Microbiol 9:1707

Han S, Zhang S, Lin T, Gong M (2011) Screening for siderophore-producing endophytic bacteria against *Fusarium oxysporum*. Hunan Agric Sci Technol 12(7):994–996

Hardoim PR, van Overbeek LS, van Elsas JD (2008) Properties of bacterial endophytes and their proposed role in plant growth. Trends Microbiol 16(10):463–471

Harman G, Howell C, Viterbo A, Chet I, Lorito M (2004) *Trichoderma* species-opportunistic, avirulent plant symbionts. Nat Rev Microbiol 2:43–56

Harrison JG, Griffin EA (2020) The diversity and distribution of endophytes across biomes, plant phylogeny and host tissues: how far have we come and where do we go from here? Environ Microbiol 22(6):2107–2123

Hartley SE, Gange AC (2009) Impacts of plant symbiotic fungi on insect herbivores: mutualism in a multitrophic context. Annu Rev Entomol 54(1):323–342

Hood L (2012) Tackling the microbiome. Science 336(6086):1209

Huong NTM, Hoai PTT, Thao PTH, Huong TT, Chinh VD (2022) Growth stimulation, phosphate resolution, and resistance to fungal pathogens of some endogenous fungal strains in the rhizospheres of medicinal plants in Vietnam. Molecules 27(16):5051

Iida Y, Ogata A, Kanda H, Nishi O, Sushida H, Higashi Y, Tsuge T (2022) Biocontrol activity of nonpathogenic strains of *Fusarium oxysporum*: colonization on the root surface to overcome nutritional competition. Front Microbiol 13:826677

James EK, Olivares FL (1998) Infection and colonization of sugar cane and other graminaceous plants by endophytic diazotrophs. Crit Rev Plant Sci 17(1):77–119

Jaroszuk-Scisel J, Kurek E, Trytek M (2014) Efficiency of indoleacetic acid, gibberellic acid and ethylene synthesized in vitro by *Fusarium culmorum* strains with different effects on cereal growth. Biologia 69(3):281–292

Ji SH, Gururani MA, Chun SC (2014) Isolation and characterization of plant growth promoting endophytic diazotrophic bacteria from Korean rice cultivars. Microbiol Res 169(1):83–98

Joe MM, Devaraj S, Benson A, Sa T (2016) Isolation of phosphate solubilizing endophytic bacteria from *Phyllanthus amarus* Schum & Thonn: evaluation of plant growth promotion and antioxidant activity under salt stress. J Appl Res Med Aromat Plants 3(2):71–77

Joseph B, Priya RM (2011) Bioactive compounds from endophytes and their potential in pharmaceutical effect: a review. Am J Biochem Mol Biol 1(3):291–309

Kamel NM, Abdel-Motaal FF, El-Zayat SA (2020) Endophytic fungi from the medicinal herb *Euphorbia geniculata* as a potential source for bioactive metabolites. Arch Microbiol 202:247–255

Karablieh NA, Dokh AA, Mutlak I, Abdulhadi Z (2017) *In vitro* biological control of *Pseudomonas viridiflava* by *Pseudomonas fluorescens* via siderophore competition. Jordan J Agric Sci 13(3): 629–643

Khalmuratova I, Choi DH, Woo JR, Jeong MJ, Oh Y, Kim YG, Kim JG (2020) Diversity and plant growth-promoting effects of fungal endophytes isolated from salt-tolerant plants. Microb Biotechnol 30(11):1680–1687

Khan SA, Hamayun M, Yoon H, Kim HY, Suh SJ, Hwang SK, Kim JM, Lee IJ, Choo YS, Yoon UH, Kong WS, Lee BM, Kim JG (2008) Plant growth promotion and *Penicillium citrinum*. BMC Microbiol 8(1):1–10

Khan AL, Hamayun M, Kang SM, Kim YH, Jung HY, Lee JH, Lee IJ (2012) Endophytic fungal association via gibberellins and indole acetic acid can improve plant growth under abiotic stress: an example of *Paecilomyces formosus* LHL10. BMC Microbiol 12(1):1–14

Khan AL, Hussain J, Al-Harrasi A, Al-Rawahi A, Lee IJ (2015) Endophytic fungi: resource for gibberellins and crop abiotic stress resistance. Crit Rev Biotechnol 35(1):62–74

Khare E, Mishra J, Arora NK (2018) Multifaceted interactions between endophytes and plant: developments and prospects. Front Microbiol 9:2732

Kogel KH, Franken P, Huckelhoven R (2006) Endophyte or parasite–what decides? Curr Opin Plant Biol 9(4):358–363

Kong P, Li X, Gouker F, Hong C (2022) cDNA transcriptome of Arabidopsis reveals various defense priming induced by a broad-spectrum biocontrol agent *Burkholderia* sp. SSG Int J Mol Sci 23(6):3151

Kuklinsky-Sobral J, Araujo WL, Mendes R, Geraldi IO, Pizzirani-Kleiner AA, Azevedo JL (2004) Isolation and characterization of soybean-associated bacteria and their potential for plant growth promotion. Environ Microbiol 6(12):1244–1251

Kumawat KC, Sharma P, Sirari A, Singh I, Gill BS, Singh U, Saharan K (2019) Synergism of *Pseudomonas aeruginosa* (LSE-2) nodule endophyte with *Bradyrhizobium* sp. (LSBR-3) for improving plant growth, nutrient acquisition and soil health in soybean. World J Microbiol Biotechnol 35(3):1–17

Kundu A, Mishra S, Kundu P, Jogawat A, Vadassery J (2022) *Piriformospora indica* recruits host-derived putrescine for growth promotion in plants. Plant Physiol 188:2289–2307

Kusari S, Hertweck C, Spiteller M (2012) Chemical ecology of endophytic fungi: origins of secondary metabolites. Chem Biol 19(7):792–798

Lacava PT, Silva-Stenico ME, Araujo WL, Simionato AVC, Carrilho E, Tsai SM, Azevedo JL (2008) Detection of siderophores in endophytic bacteria *Methylobacterium* spp. associated with *Xylella fastidiosa* subsp. *pauca*. Pesqui Agropecu Bras 43(4):521–528

Ladha JK, Reddy PM (2003) Nitrogen fixation in rice systems: state of knowledge and future prospects. Plant Soil 252(1):151–167

Lai X, Niroula D, Burrows M, Wu X, Yan Q (2022) Identification and characterization of bacteria-derived antibiotics for the biological control of pea aphanomyces root rot. Microorganisms 10: 1596

Leuchtmann A, Schmidt D, Bush LP (2000) Different levels of protective alkaloids in grasses with stroma-forming and seed-transmitted *Epichloë/Neotyphodium* endophytes. J Chem Ecol 26(4): 1025–1036

Li J, Zheng B, Hu R, Liu Y, Jing Y, Xiao Y, Sun M, Chen W, Zhou Q (2019) *Pseudomonas* species isolated from tobacco seed promote root growth and reduce lead contents in *Nicotiana tobacum* K326. Can J Microbiol 65(3):214–223

Li L, Guo N, Feng Y, Duan M, Li C (2022a) Effect of *Piriformospora indica*-induced systemic resistance and basal immunity against *Rhizoctonia cerealis* and *Fusarium graminearum* in wheat. Front Plant Sci 13:836940

Li Z, Wen W, Qin M, He Y, Xu D, Li L (2022b) Biosynthetic mechanisms of secondary metabolites promoted by the interaction between endophytes and plant hosts. Front Microbiol 13:928967

Lin L, Li Z, Hu C, Zhang X, Chang S, Yang L, Li Y, An Q (2012) Plant growth-promoting nitrogen-fixing enterobacteria are in association with sugarcane plants growing in Guangxi, China. Microbes Environ 27(4):391–398

Liu H, Carvalhais LC, Crawford M, Singh E, Dennis PG, Pieterse CM, Schenk PM (2017) Inner plant values: diversity, colonization and benefits from endophytic bacteria. Front Microbiol 8: 2552

Lu-Irving P, Harenčár JG, Sounart H, Welles SR, Swope SM, Baltrus DA, Dlugosch KM (2019) Native and invading yellow starthistle (*Centaurea solstitialis*) microbiomes differ in composition and diversity of bacteria. Msphere 4(2):e00088–e00019

Ma Y, Rajkumar M, Zhang C, Freitas H (2016) Beneficial role of bacterial endophytes in heavy metal phytoremediation. J Environ Manag 174:14–25

Magnani GS, Cruz LM, Weber H, Bespalhok JC, Daros E, Baura V, Souza EM (2013) Culture-independent analysis of endophytic bacterial communities associated with Brazilian sugarcane. Genet Mol Res 12:4549–4558

Malik DK, Sindhu SS (2011) Production of indole acetic acid by *Pseudomonas* sp.: effect of coinoculation with *Mesorhizobium* sp. Cicer on nodulation and plant growth of chickpea (*Cicer arietinum*). Physiol Mol Biol 17(1):25–32

Manganiello G, Sacco A, Ercolano MR, Vinale F, Lanzuise S, Pascale A, Napolitano M, Lombardi N, Lorito M, Woo SL (2018) Modulation of tomato response to *Rhizoctonia solani* by *Trichoderma harzianum* and its secondary metabolite Harzianic acid. Front Microbiol 9: 1966

Mano H, Morisaki H (2008) Endophytic bacteria in the rice plant. Microbes Environ 23(2):109–117

Manoel da Silva J, Carvalho dos Santos TM, Santos de Albuquerque L, Coentro Montaldo Y, Lima U, de Oliveira J, Mesquita da Silva SG, da Rocha Oliveira Teixeira R (2015) Potential of the endophytic bacteria (*Herbaspirillum* spp. and *Bacillus* spp.) to promote sugarcane growth. Aust J Crop Sci 9(8):754–760

Marques JM, da Silva TF, Vollu RE, de Lacerda JRM, Blank AF, Smalla K, Seldin L (2015) Bacterial endophytes of sweet potato tuberous roots affected by the plant genotype and growth stage. Appl Soil Ecol 96:273–281

Mbai FN, Magiri EN, Matiru VN, Nganga J, Nyambati VCS (2013) Isolation and characterization of bacterial root endophytes with potential to enhance plant growth from Kenyan Basmati rice. Am Int J Contemp Res 3(4):25–40

Montanez A, Abreu C, Gill PR, Hardarson G, Sicardi M (2009) Biological nitrogen fixation in maize (*Zea mays* L.) by 15N isotope-dilution and identification of associated culturable diazotrophs. Biol Fertil Soils 45(3):253–263

Moon CD, Tapper BA, Scott B (1999) Identification of *Epichloë* endophytes in planta by a microsatellite-based PCR fingerprinting assay with automated analysis. Appl Environ Microbiol 65(3):1268–1279

Mousa WK, Schwan A, Davidson J, Strange P, Liu H, Zhou T, Auzanneau FI, Raizada MN (2015) An endophytic fungus isolated from finger millet (*Eleusine coracana*) produces anti-fungal natural products. Front Microbiol 6:1157

Mugford ST, Qi X, Bakht S, Hill L, Wegel E, Hughes RK, Papadopoulou K, Melton R, Philo M, Sainsbury F (2009) A serine carboxypeptidase-like acyltransferase is required for synthesis of antimicrobial compounds and disease resistance in oats. Plant Cell 21(8):2473–2484

Mulero-Aparicio A, Varo A, Agustí-Brisach C, López-Escudero FJ, Trapero A (2020) Biological control of Verticillium wilt of olive in field. Crop Prot 128:104993

Murphy BR, Doohan FM, Hodkinson TR (2018) From concept to commerce: developing a successful fungal endophyte inoculant for agricultural crops. J Fungi 4(1):24

Nikolic B, Schwab H, Sessitsch A (2011) Metagenomic analysis of the 1-aminocyclopropane-1-carboxylate deaminase gene (*acdS*) operon of an uncultured bacterial endophyte colonizing *Solanum tuberosum* L. Arch Microbiol 193(9):665–676

Numponsak T, Kumla SN, Matsui K, Lumyong S (2018) Biosynthetic pathway and optimal conditions for the production of indole-3-acetic acid by an endophytic fungus, *Colletotrichum fructicola* CMU-A109. PLoS One 13(10):e0205070

Ofek-Lalzar M, Gur Y, Ben-Moshe S, Sharon O, Kosman E, Mochli E, Sharon A (2016) Diversity of fungal endophytes in recent and ancient wheat ancestors *Triticum dicoccoides* and *Aegilops sharonensis*. FEMS Microbiol Ecol 92(10):fiw152

Oliveira R, Souza R, Lima T, Cavalcanti M (2014) Endophytic fungal diversity in coffee leaves (*Coffea arabica*) cultivated using organic and conventional crop management systems. Mycosphere 5(4):523–530

Oteino N, Lally RD, Kiwanuka S, Lloyd A, Ryan D, Germaine KJ, Dowling DN (2015) Plant growth promotion induced by phosphate solubilizing endophytic *Pseudomonas* isolates. Front Microbiol 6:745

Panaccione DG, Beaulieu WT, Cook D (2014) Bioactive alkaloids in vertically transmitted fungal endophytes. Funct Ecol 28(2):299–314

Paz ICP, Santin RCM, Guimaraes AM, Rosa OPP, Dias ACF, Quecine MC, Azevedo JL, Matsumura ATS (2012) Eucalyptus growth promotion by endophytic *Bacillus* spp. Genet Mol Res 11(4):3711–3720

Pena-Yam LP, Ruiz-Sanchez E, Barboza-Corona JE, Reyes-Ramirez A (2016) Isolation of Mexican *Bacillus* species and their effects in promoting growth of chili pepper (*Capsicum annuum* L. cv Jalapeno). Indian J Microbiol 56(3):375–378

Perin L, Martínez-Aguilar L, Paredes-Valdez G, Baldani JI, Estrada-De Los Santos P, Reis VM, Caballero-Mellado J (2006) *Burkholderia silvatlantica* sp. *nov*., a diazotrophic bacterium associated with sugar cane and maize. Int J Syst Evol Microbiol 56(8):1931–1937

Perotti R (1926) On the limits of biological enquiry in soil science. Proc Int Soc Soil Sci 2:146–161

Piasecka A, Jedrzejczak-Rey N, Bednarek P (2015) Secondary metabolites in plant innate immunity: conserved function of divergent chemicals. New Phytol 206(3):948–964

Rajkumar M, Ae N, Freitas H (2009) Endophytic bacteria and their potential to enhance heavy metal phytoextraction. Chemosphere 77(2):153–160

Rana KL, Kour D, Kaur T, Devi R, Yadav AN, Yadav N, Dhaliwal HS, Saxena AK (2020a) Endophytic microbes: biodiversity, plant growth-promoting mechanisms and potential applications for agricultural sustainability. Antonie Leeuwenhoek 113(8):1075–1107

Rana KL, Kour D, Kaur T, Sheikh I, Yadav AN, Kumar V, Suman A, Dhaliwal HS (2020b) Endophytic microbes from diverse wheat genotypes and their potential biotechnological applications in plant growth promotion and nutrient uptake. Proc Natl Acad Sci India Sect B: Biol Sci 90(5):969–979

Rangjaroen C, Rerkasem B, Teaumroong N, Noisangiam R, Lumyong S (2015) Promoting plant growth in a commercial rice cultivar by endophytic diazotrophic bacteria isolated from rice landraces. Ann Microbiol 65(1):253–266

Rani AT, Kammar V, Keerthi MC, Rani V, Majumder S, Pandey KK, Singh J (2021) Biopesticides: an alternative to synthetic insecticides. In: Microbial technology for sustainable environment. Springer, Singapore, pp 439–466

Rashad YM, Abdalla SA, Sleem MM (2022) Endophytic *Bacillus subtilis* SR22 triggers defense responses in tomato against rhizoctonia root rot. Plan Theory 11:2051

Rashid S, Charle TC, Glick BR (2012) Isolation and characterization of new plant growth-promoting bacterial endophytes. Appl Soil Ecol 61:217–224

Rauf M, Ur-Rahman A, Arif M, Gul H, Ud-Din A, Hamayun M, Lee IJ (2022) Immunomodulatory molecular mechanisms of *Luffa cylindrica* for downy mildews resistance induced by growth-promoting endophytic fungi. J Fungi 8:689

Rodriguez RJ, White JF Jr, Arnold AE, Redman ARA (2009) Fungal endophytes: diversity and functional roles. New Phytol 182(2):314–330

Roylawar P, Khandagale K, Randive P, Shinde B, Murumkar C, Ade A, Singh M, Gawande S, Morelli M (2021) *Piriformospora indica* primes onion response against *Stemphylium* leaf bight disease. Pathogens 10(9):1085

Rudgers JA, Afkhami ME, Rúa MA, Davitt AJ, Hammer S, Huguet VM (2009) A fungus among us: broad patterns of endophyte distribution in the grasses. Ecology 90(6):1531–1539

Sabate DC, Brandan CP, Petroselli G, Erra-Balsells R, Audisio MC (2018) Biocontrol of *Sclerotinia sclerotiorum* (Lib.) de Bary on common bean by native lipopeptide-producer *Bacillus* strains. Microbiol Res 11:21–30

Saikkonen K, Gundel PE, Helander M (2013) Chemical ecology mediated by fungal endophytes in grasses. J Chem Ecol 39(7):962–968

Sandhya V, Shrivastava M, Ali SZ, Sai Shiva Krishna Prasad V (2017) Endophytes from maize with plant growth promotion and biocontrol activity under drought stress. Russ Agric Sci 43(1): 22–34

Santi C, Bogusz D, Franche C (2013) Biological nitrogen fixation in non-legume plants. Ann Bot 111(5):743–767

Santos MLD, Berlitz DL, Wiest SLF, Schunemann R, Knaak N, Fiuza LM (2018) Benefits associated with the interaction of endophytic bacteria and plants. Braz Arch Biol Technol 61: e18160431

Santoyo G, Moreno-Hagelsieb G, del Carmen O-MM, Glick BR (2016) Plant growth-promoting bacterial endophytes. Microbiol Res 183:92–99

Schulz B, Boyle C (2006) What are endophytes? In: Microbial root endophytes. Springer, Berlin, pp 1–13

Shahzad R, Waqas M, Khan AL, Asaf S, Khan MA, Kang SM, Yun BW, Lee IJ (2016) Seed-borne endophytic *Bacillus amyloliquefaciens* RWL-1 produces gibberellins and regulates endogenous phytohormones of *Oryza sativa*. Plant Physiol Biochem 106:236–243

Sherwood M, Carroll G (1974) Fungal succession on needles and young twigs of old-growth Douglas fir. Mycologia 66:499–506

Shipunov A, Newcombe G, Raghavendra AK, Anderson CL (2008) Hidden diversity of endophytic fungi in an invasive plant. *Am J Bot* 95(9):1096–1108

Shivakumara KT, Keerthi MC and Polaiah AC (2022) Efficacy of different biorational insecticides against *Aphis nerii* Boyer de Fonscolombe (Hemiptera. Aphididae) on *Gymnema sylvestre* (R. Br) under laboratory and field conditions. J Appl Res Med Aromat. Plants 28:100358

Silva GH, Teles HL, Zanardi LM, Young MCM, Eberlin MN, Hadad R, Araújo ÂR (2006) Cadinane sesquiterpenoids of *Phomopsis cassiae*, an endophytic fungus associated with *Cassia spectabilis* (Leguminosae). Phytochemistry 67(17):1964–1969

Singh B, Boukhris I, Kumar V, Yadav AN, Farhat-Khemakhem A, Kumar A, Singh D, Blibech M, Chouayekh H, Alghamdi OA (2020) Contribution of microbial phytases to the improvement of plant growth and nutrition: a review. Pedosphere 30(3):295–313

Sprent JI (2008) 60Ma of legume nodulation. What's new? What's changing? J Exp Bot 59(5): 1081–1084

Sprent JI, Ardley JK, James EK (2013) From north to south: a latitudinal look at legume nodulation processes. S Afr J Bot 89:31–41

Stone JK, Polishook JD, White JF (2004) Endophytic fungi. In: Biodiversity of fungi: inventory and monitoring methods, vol 241, p 270

Suebrasri T, Harada H, Jogloy S, Ekprasert J, Boonlue S (2020) Auxin-producing fungal endophytes promote growth of sunchoke. Rhizosphere 16:100271

Sun Y, Cheng Z, Glick BR (2009) The presence of a 1-aminocyclopropane-1-carboxylate (ACC) deaminase deletion mutation alters the physiology of the endophytic plant growth-promoting bacterium *Burkholderia phytofirmans* PsJN. FEMS Microbiol Lett 296(1):131–136

Taghavi S, Garafola C, Monchy S, Newman L, Hoffman A, Weyens N, Barac T, Vangronsveld J, van der Lelie D (2009) Genome survey and characterization of endophytic bacteria exhibiting a beneficial effect on growth and development of poplar trees. Appl Environ Microbiol 75(3): 748–757

Taurian T, Anzuay MS, Angelini JG, Tonelli ML, Ludueña L, Pena D, Ibáñez F, Fabra A (2010) Phosphate-solubilizing peanut associated bacteria: screening for plant growth-promoting activities. Plant Soil 329(1):421–431

Tian XL, Cao LX, Tan HM, Zeng QG, Jia YY, Han WQ, Zhou SN (2004) Study on the communities of endophytic fungi and endophytic actinomycetes from rice and their antipathogenic activities in vitro. World J Microbiol Biotechnol 20(3):303–309

Triplett EW (1996) Diazotrophic endophytes: progress and prospects for nitrogen fixation in monocots. Plant Soil 186(1):29–38

Tsavkelova EA, Cherdyntseva TA, Botina SG, Netrusov AI (2007) Bacteria associated with orchid roots and microbial production of auxin. Microbiol Res 162(1):69–76

Van Der Heijden MG, Bardgett RD, Van Straalen NM (2008) The unseen majority: soil microbes as drivers of plant diversity and productivity in terrestrial ecosystems. Ecol Lett 11(3):296–310

Vardharajula S, Shaik ZA, Vurukonda SKP, Shrivastava M (2017) Plant growth promoting endophytes and their interaction with plants to alleviate abiotic stress. Curr Biotechnol 6(3): 252–263

Vega FE, Simpkins A, Aime MC, Posada F, Peterson SW, Rehner SA, Arnold AE (2010) Fungal endophyte diversity in coffee plants from Colombia, Hawai'i, Mexico and Puerto Rico. Fungal Ecol 3(3):122–138

Vendan RT, Yu YJ, Lee SH, Rhee YH (2010) Diversity of endophytic bacteria in ginseng and their potential for plant growth promotion. J Microbiol 48(5):559–565

Verma P, Yadav AN, Khannam KS, Panjiar N, Kumar S, Saxena AK, Suman A (2015) Assessment of genetic diversity and plant growth promoting attributes of psychrotolerant bacteria allied with wheat (*Triticum aestivum*) from the northern hills zone of India. Ann Microbiol 65(4): 1885–1899

Verma SK, Kharwar RN, White JF (2019) The role of seed-vectored endophytes in seedling development and establishment. Symbiosis 78(2):107–113

Veselova SV, Sorokan AV, Burkhanova GF, Rumyantsev SD, Cherepanova EA, Alekseev VY, Sarvarova ER, Kasimova AR, Maksimov IV (2022) By modulating the hormonal balance and ribonuclease activity of tomato plants *Bacillus subtilis* induces defense response against potato virus X and potato virus Y. Biomol Ther 12(2):288

Vessey JK, Pawlowski K, Bergman B (2005) Root-based N2-fixing symbioses: legumes, actinorhizal plants, *Parasponia* sp. and cycads. Plant Soil 274:51–57

Wang Y, Guo LD (2007) A comparative study of endophytic fungi in needles, bark, and xylem of *Pinus tabulaeformis*. Botany 85(10):911–917

Wang S, Wang W, Jin Z, Du B, Ding Y, Ni T, Jiao F (2013) Screening and diversity of plant growth promoting endophytic bacteria from peanut. Afr J Microbiol Res 7(10):875–884

Waqas M, Khan AL, Kamran M, Hamayun M, Kang SM, Kim YH, Lee IJ (2012) Endophytic fungi produce gibberellins and indoleacetic acid and promote host-plant growth during stress. Molecules 17(9):10754–10773

Wei CY, Lin L, Luo LJ, Xing YX, Hu CJ, Yang LT, Li YR, An Q (2014) Endophytic nitrogen-fixing *Klebsiella variicola* strain DX120E promotes sugarcane growth. Biol Fertil Soils 50(4):657–666

Wu W, Chen W, Liu S, Wu J, Zhu Y, Qin L, Zhu B (2021) Beneficial relationships between endophytic bacteria and medicinal plants. Front Plant Sci 12:646146

Xin G, Zhang G, Kang JW, Staley JT, Doty SL (2009) A diazotrophic, indole-3-acetic acid-producing endophyte from wild cottonwood. Biol Fertil Soils 45(6):669–674

Xu T, Cao L, Zeng J, Franco CMM, Yang Y, Hu X, Liu Y, Wang X, Gao Y, Bu Z, Shi L, Zhou G, Zhou Q, Liu X, Zhu Y (2019) The antifungal action mode of the rice endophyte Streptomyces hygroscopicus OsiSh-2 as a potential biocontrol agent against the rice blast pathogen. Pestic Biochem Physiol 160:58–69

Yi Y, Huang W, Ge Y (2008) Exopolysaccharide: a novel important factor in the microbial dissolution of tricalcium phosphate. World J Microbiol Biotechnol 24(7):1059–1065

Yu X, Zhang W, Lang D, Zhang X, Cui G, Zhang X (2019) Interactions between endophytes and plants: beneficial effect of endophytes to ameliorate biotic and abiotic stresses in plants. J Plant Biol 62(1):1–13

Zakaria L, Yaakop AS, Salleh B, Zakaria M (2010) Endophytic fungi from paddy. Trop Life Sci Res 21(1):101

Zeng J, Xu T, Cao L, Tong C, Zhang X, Luo D, Han S, Pang P, Fu W, Yan J (2018) The role of iron competition in the antagonistic action of the rice endophyte *Streptomyces sporocinereus* OsiSh-2 against the pathogen *Magnaporthe oryzae*. Microb Ecol 76:1021–1029

Zhang YF, He LY, Chen ZJ, Wang QY, Qian M, Sheng XF (2011) Characterization of ACC deaminase-producing endophytic bacteria isolated from copper-tolerant plants and their potential in promoting the growth and copper accumulation of *Brassica napus*. Chemosphere 83(1):57–62

Zhang YZ, Chen WF, Li M, Sui XH, Liu HC, Zhang XX, Chen WX (2012) *Bacillus endoradicis* sp. *nov.*, an endophytic bacterium isolated from soybean root. Int J Syst Evol Microbiol 62(2):359–363

Zhang L, Li W, Tao Y, Zhao S, Yao L, Cai Y, Niu Q (2019) Overexpression of the key virulence factor 1,3–1,4-β-d-Glucanase in the endophytic bacterium *Bacillus halotolerans* Y6 to improve verticillium resistance in cotton. J Agric Food Chem 67(24):6828–6836

Zhao L, Xu Y, Lai XH, Shan C, Deng Z, Ji Y (2015) Screening and characterization of endophytic *Bacillus* and *Paenibacillus* strains from medicinal plant *Lonicera japonica* for use as potential plant growth promoters. Braz J Microbiol 46:977–989

Zhao L, Xu Y, Lai X (2018) Antagonistic endophytic bacteria associated with nodules of soybean (*Glycine max* L.) and plant growth-promoting properties. Braz J Microbiol 49:269–278

Zimmerman NB, Vitousek PM (2012) Fungal endophyte communities reflect environmental structuring across a Hawaiian landscape. Proc Natl Acad Sci 109(32):13022–13027

Chapter 13
Multitrophic Reciprocity of AMF with Plants and Other Soil Microbes in Relation to Biotic Stress

Supriya Sharma, V. Bhuvaneswari, Bandana Saikia, R. Karthik, B. Rajeshwaran, P. Shree Naveena, and Mateti Gayithri

Abstract The plant roots are invariably colonized by an extent of tiny microorganisms including arbuscular mycorrhizal fungi (AMF) which support plant health in multiple ways. The symbiotic relationship with host plants is achieved through various signaling mechanisms and metabolic pathways which guide their mutual survival. Besides their plant growth-promoting activities, they communicate with other soil microbes to improve plant fitness during unfavorable circumstances. Rhizobium and mycorrhization helper bacteria (MHB) interaction with AM fungi provides us with new insight for enhanced protection of plants against abiotic factors. In addition to that, the alteration of host nutrition, modifications in root physiology and morphology, increase plant defense. Thus, remodeling of the rhizospheric microbial community by AM fungi activates their multifaceted mechanisms in host plants against phytopathogens insect pests, nematodes, and phanerogamic parasites. It modifies the host physiology which ultimately benefits plants. The current knowledge about AM fungi with their potentiality in crop production and protection is discussed in detail.

Keywords AM fungi · Biotic stress · Plant immunity · Symbiosis

Abbreviations

ENOD11	Early nodulin 11 gene
IPD3	Interacting protein of does not make infections 3

S. Sharma (✉) · V. Bhuvaneswari · B. Saikia · R. Karthik · M. Gayithri
Assam Agricultural University, Jorhat, India
e-mail: supriya.sharma@aau.ac.in; bhuvaneswari.v.adj19@aau.ac.in; bandana.saikia.adj18@aau.ac.in

B. Rajeshwaran · P. Shree Naveena
Kerala Agricultural University, Thiruvananthapuram, India

P. Mathur et al. (eds.), *Microbial Symbionts and Plant Health: Trends and Applications for Changing Climate*, Rhizosphere Biology,
https://doi.org/10.1007/978-981-99-0030-5_13

LOX	Lipoxygenase
MDA	Malondialdehyde
NPR1	Non-expressor of pathogenesis-related genes
PAL	Phenylalanine ammonia lyase
POD	Peroxidase
SOD	Superoxide dismutase
SYMRK	Symbiosis receptor-like kinase

13.1 Paleobiology of Glomerales

Arbuscular mycorrhizal (AM) symbiosis is the most extensive mutualistic association. It concerns 80% of land plants and involves fungi belonging to the phylum Glomeromycota. Glomeromycota is one of the eight recognized phyla under Kingdom Fungi according to the classification given by Kirk (2001), based on the molecular phylogeny of their 18S ribosomal ribonucleic acid (rRNA) genes and the type of mitochondrial cristae present. The rRNA data also depicts their close relationships with Ascomycota and Basidiomycota (Schussler et al. 2001), unlike the earlier artificial classification which placed them with Zygomycota. The biochemical and genetic characterization of the organisms was not possible up to a certain period due to their obligate symbiotic nature. Still, with recent advancements in molecular biology, the creation of new taxa based on combining phenotypic and genetic data and the integration of fossil taxa have led to systematic reevaluations and descriptions of new species (Sturmer 2012). The first mycorrhizal gene obtained from the spores of several genera from root cultures was sequenced which was the highly conserved single subunit (SSU) rRNA gene, and these fungi were found to be 350 million years old based on the molecular substitution rates of SSU sequences. Due to their widespread abundance, importance in modern ecosystems, and fossils of plant roots 450 million years old with hyphae and spores of Glomeromycota fungi, it is widely hypothesized that fungi from different lineages (including the glomeromycetes) played a crucial role in facilitating the evolution of land plants (Pirozynsky and Malloch 1975; Field et al. 2019).

The phylum Glomeromycota is divided into four orders and nine families (Schussler et al. 2001; Kirk 2001). These orders include Glomerales, Diversisporales, Archaeosporales, and Paraglomerales. The order Glomerales has only one family, viz., Glomeraceae, with one genus *Glomus*. Glomeromycete fungi produce several characteristic structures that have been conserved through time: mycelia consisting of mostly coenocytic hyphae; specific hyphal configurations within the host plant cells that maximize plant-fungal nutrient exchange, such as hyphal coils and arbuscules; thin-walled vesicles used as storage organs within the plant host; and spores (Schussler and Walker 2011). The spores are multinucleate, cover a broad range of sizes (40–800 μm), and can be found singly, in small clusters, or inside sporocarps (Walker et al. 2018). Spores can have variable shapes, with globose, ovoid, or irregular outlines, and can have walls with complex organization comprising multiple distinct layers (Schussler et al. 2001). However, depending on the specific taphonomic and diagenetic history of specimens, in some cases, only some of these layers are preserved in fossil spores (Walker et al. 2018), which present a thicker outer layer and a thinner inner layer, with the space between these two layers thought to correspond to another layer that is rarely preserved (Dotzler et al. 2006).

13.2 Metabolic Pathways Involved in Symbiotic Association with Plants

13.2.1 Pre-symbiosis

The process of the growth of AM fungi prior to root colonization is called pre-symbiosis. It comprises the following steps: spore germination, hyphal growth, host recognition, and appressorium formation. As spores have been germinated both in vitro and under field condition, it indicates that the germination does not rely only on the host root exudates, which mostly enhance the rate of germination. Arbuscular mycorrhizal fungal spores germinate under amiable conditions of CO_2 concentration, soil matrix, pH, and phosphorus level.

The phosphorus concentration in the soil and strigolactones induce the development of AM hyphae. Apart from strigolactones, plants release N-acetyl glucosamine derivatives which play a dominant role in triggering fungal reprogramming at the pre-contact stage. Sulfated and nonsulfated lipochitooligosaccharides (s/nsMyc-LCOs) and fungal short-chain chitin oligomers generate pre-symbiosis reactions in the host. Phenol and flavonoid derivatives take part in a crucial role for generating hyphal branching. Even though flavonoid derivatives impact the earlier phases of the fungal life cycle, the results of the observations with flavonoid-deficient mutants of maize expressed that they are not dispensable for symbiosis (Becard et al. 1995).

A reduced level of phosphorus concentration in the soil enhances hyphal growth and branching intensity. Under the low-phosphorus condition, the root exudates stimulate and intensify the host root colonization. Mitochondrial activity of the AMF spore is induced by the host root exudates as the oxygen consumption rate is elevated by 30% following exposure to root exudates. In no time, perceptions of fungal signals bring about nuclear and perinuclear Ca^{2+} changes in rhizodermal cells. Various components of a common symbiosis signaling pathway (CSSP) regulate the transduction of Ca^{2+} signals and even regulate rhizobial infection during nodulation (Chabaud et al. 2011).

The LRR (leucine-rich repeat) receptor kinase (DMI2, SYMRK) along with an interacting transcription factors (IPD3, Cyclops) are highly essential for Ca^{2+} spiking which decode and transduce Ca^{2+} signals in host plants in response to AM fungi signals. CSSP components do not always play an important part in plant responses during AM because alternative pathways also exist (Miller et al. 2013). Saito et al. (2007) observed that in the common symbiosis pathway, mutants of the rice chitin receptor Chitin Elicitor Receptor Kinase 1 (CERK1) and mutants of genes exhibited the same aberrant swollen hyphopodia.

For the formation of appressoria, an AM fungus does not depend on chemical signals produced by the plant. However, for further hyphal penetration and growth into the root cortex, signaling between symbionts is essential. Carbohydrate molecules are signals for a majority of plant/fungal interactions and are appropriate candidates for appressorial formation.

13.2.2 Symbiosis

After penetration into the parenchyma, the fungus establishes profusely branched distinguishable structures called arbuscules for nutrient exchange with the plant. The arbuscules constitute the place of active bidirectional transfer of nutrients among plants and fungi. Plant exudates might attract symbionts as well as pathogens or saprotrophs. Plants have to identify probable symbionts for their colonization and should elicit a defense response against pathogens. So, the symbiosis signal is called a "door opener" in symbiotic interactions.

Fungal hyphae release the prime signal of unspecified nature; it switches the promoter of the symbiosis-related gene ENOD11 in roots. Direct physical contact is not necessary amidst hyphae and roots, designating that it is mediated by a diffusible material (Kosuta et al. 2003). In addition, the AM symbiosis has a profound effect, particularly in the case of phosphorus metabolism which in succession controls the production of strigolactones (Smith et al. 2011).

Right now there is an ample experimental evidence to advise that GRAS TFs are efficient in initiating intra- and interspecific interactions that form a transcriptional regulatory complex that controls crucial processes in the AM symbiosis (Ho-Plagaro and García-Garrido 2022). Takeda et al. (2015) proved that gibberellic acids are necessary in the case of *Lotus japonicus* (legume) for AM development. The alternating GA concentrations in the host roots impact the expression of AM-induced genes positively or negatively that mediate AM fungal infection and colonization. Crosstalk between SLs and GA is surging: an SL-dependent interaction between D14, GA signaling repressor, and the SL receptor SLR1. Lately, GA signaling appeared to manage the biogenesis of SLs, by downregulation of interrelated genes (Ito et al. 2017).

13.2.3 Post-symbiosis

Arbuscular mycorrhizal fungi are obligate symbionts. The AM fungi consume the photosynthesis products in the form of hexoses. Arbuscules play a major role in transferring carbon from plants to fungi. The conversion of hexose to trehalose and glycogen takes place inside the mycelium. By the pentose phosphate pathway, the pentose is produced from the intraradical hexose (Campos et al. 2018).

Lipid biosynthesis takes place in the intraradical mycelium. They are transported to extraradical hyphae, the place where they are metabolized or stored. The disintegration of lipids into hexoses is termed gluconeogenesis. This process occurs in the extraradical mycelium. As the plants supply more carbon to the AM fungi accordingly, the fungi enhance the uptake and transfer of phosphorus to the plant. They may vary in their ability to provide phosphorus to plants. Phosphorus uptake is six times more in the presence of mycorrhiza compared to root hairs (Ma et al. 2022).

13.3 Interaction Between Mycorrhizae with Other Beneficial Microbes

In nature, in the mycorrhizosphere around the root system, there are numerous arrays of microorganisms that communicate with each other for their survival and to perform various metabolic functions in the plant ecosystem. Plants frequently engage in multiple mutualistic associations with various microorganisms in their rhizosphere. The mutualistic interactions between plant-microbe consequently enhances the plant growth and development (Jia and Gray 2008), susceptibility to various diseases (Gao et al. 2012), and response to herbivory (Barber 2012) and abiotic stress (Redman et al. 2011). An enormous number of tiny living propagules of AM fungi, plant growth-promoting rhizobacteria, soil bacteria, root nodule bacteria, and root endophytes compete for host plant root colonization to benefit themself. In general, there was a belief for a long time that the mycorrhizae symbiotic association is a bipartite interaction (fungus-plant interaction). But, the belief is broken down with the studies of mycorrhizae interaction with other microbes that help in maintaining plant health in climate regimes under various stresses in tripartite relations.

13.3.1 AMF with Nitrogen-Fixing Rhizobium

Legume crops and nitrogen-fixing rhizobium bacteria co-evolved in a symbiotic manner over 80 million years ago (Sprent 2007). In this interaction, the plant imparts its beneficial endosymbiont with photosynthate, besides other nutrients, in exchange for valuable fixed nitrogen in the form of ammonium and amino acids (Udvardi and Day 1997). This process is achieved by *Rhizobium* perceiving the plants' signals like flavonoids which further activate the Nod factor secretion leading to the fixation of atmospheric nitrogen into nitrogen forms that can be used by the host. Legume-rhizobium interaction is not the only interaction that benefits the host plant. Back in

1987, Newman and Reddell reported that 90% of legumes constitutes to form mycorrhizal symbiosis. AMF colonizes plant roots and develops various intraradical structures (hypha, vesicles, arbuscules, and spores) and intensifies plant nutrient and mineral acquisition (especially phosphorus) (Smith and Read 2008). In legumes, this integral tripartite (plant-rhizobium-AMF) interaction has been widely studied throughout the years. Each symbiont plays an essential part in natural ecosystems by impacting plant productivity, community structure, and nutrition (Bauer et al. 2012). However, co-inoculation of both symbioses brings about a synergistic effect on host biomass, growth, and fitness, combinedly increasing plant fitness more than would be estimated by their individual effects (Jia et al. 2004; Rydlova et al. 2011; Meena et al. 2018). In *Vicia faba*, the combined application of AMF and *Rhizobium* remarkably enhances production compared to individual treatments of the same (Jia et al. 2004). Perennial legumes such as *Acacia* spp., *Anadenanthera* spp., and *Cajanus cajan* also showed synergistic effects during legume-AMF-rhizobia interactions (Primieri et al. 2022). The benefit of annual/perennial legume-rhizobia-AMF interactions is highly synergistic manner as rhizobia and AMF help plants obtain complementary resources (N and P, respectively). N and P are possibly co-limiting, because the movement of the acquired P into the plant requires more N and vice versa. Besides, rhizobia nodules have more P requirements (Sulieman and Tran 2015), and AMF inflicts more demand on N, along with plant-derived lipids and carbohydrates (Johnson 2010). Hence, coupling with one symbiont may maximize the plant's ability.

AMF and rhizobia follow similar sequences of initial signaling, entry, and development during interaction with legumes. The host roots release strigolactones (a group of sesquiterpenes) that act as an inducer for hyphal branching (Akiyama and Hayashi 2006). Moreover, rising flavonoid biosynthesis was identified in plant roots upon AMF infection (Abdel-Lateif et al. 2012). Spore germination to hyphal branching and secondary spore formation were increased due to flavonoids (Abdel-Lateif et al. 2012). Similarly, in rhizobium symbiosis, flavonoids play a crucial role in *nod* gene expressions and nodule development (Abdel-Lateif et al. 2012). Early signaling in both types of microbial interactions is determined by the plant flavonoids.

Attachment and entry of AM fungi or rhizobia is a multi-step procedure that involves diversified degrees of host participation. In order to enter into the host cell, AM fungi exert mechanical pressure moreover, degradative enzymes such as exo- and endoglucanases, xyloglucanases, cellulases, pectolytic enzymes (Harrison 1999). Hydrolytic enzymes are also released by rhizobium which act on the plant cell wall (Martinez-Molina et al. 1979). Considering the apparent similarities in signal perception and entry of the microbes in both mycorrhizal and nodule symbioses, an overlap in the activation of gene expression can be expected. Molecular and genetic studies indicated that infection processes in both symbioses and the microsymbionts which are extremely similar give rise to a common signaling cascade for initiating root nodules and AMF (Cullimore and Denarie 2003). The signaling cascade for nodule formation is the *Nod* signals, whereas arbuscular mycorrhizal fungi produce analogous *Myc* signals, whose perception is essential to initiate infection.

13.3.2 AMF with Mycorrhiza Helper Bacteria (MHB)

The background of mycorrhization helper bacteria started in the 1990s with the inoculation of ectomycorrhizae fungi *Laccaria laccata* in the Douglas fir nursery. It was found that sporocarps and ectomycorrhizas of *Laccaria laccata* harbor some bacteria which stimulate the symbiotic association of AMF with roots. In addition, these bacteria were always localized in the sporocarps which indicate that the AM fungi are highly dependent upon their association to complete their life cycle (Garbaye et al. 1990; Duponnois and Garbaye 1991). The stimulator (helper) bacteria *Pseudomonas fluorescens* BBc6 isolated from the sporocarps were inoculated with *L. laccata* to improve the percent of mycorrhizal short roots; thereby, the bacteria are called mycorrhiza helper bacteria (Duponnois and Garbaye 1991). Later, Frey-Klett et al. (2007) categorized the MHB into two broad groups based on their function, viz., mycorrhiza helper bacteria and mycorrhization helper bacteria. The former signifies the effect of functional changes in already well-established AMF symbiosis, and the latter deals with the stimulation of bacteria in the establishment of a symbiotic association of AMF with their host. Bharadwaj et al. (2008) reported that *Pseudomonas*, *Stenotrophomonas*, and *Arthrobacter* were the dominant bacterial genera out of the 385 bacteria isolated from the surface-decontaminated spores of *Glomus intraradices* and *Glomus mosseae.* The electron microscopic studies revealed that the cytoplasm of the *Gigaspora margarita* harbors thousands of bacterial endosymbionts, and further identifications of them with rRNA exhibited they belong to pseudomonads group II, genus *Burkholderia* (Bianciotto et al. 1996). A high abundance of these bacterial genera with AM fungal spores is favored through modification of the micro-environment by AM fungi in the plant rhizosphere (Artursson et al. 2006). The AM fungi recruit the bacteria from the mycorrhizosphere niche *via* the secretion of exudates from mycelium, spores, and plant roots. However, the nature, composition, and quantity of exudates from different AM fungal mycelium and various plant roots decide the bacterial population in the rhizosphere (Lynch and Whipps 1990; Bais et al. 2006). *Pseudomonas* and *Arthrobacter* were highly attracted to the exudates of mycorrhizae-colonized roots of tomatoes compared to the exudates of non-mycorrhizal tomato roots (Sood 2003).

Several researchers reported that the MHB plays a major role in the receptivity of AM fungi signaling to short roots, guides the fungus to recognize the host root, and benefits the mycorrhizae by changing the physio-chemical properties of the rhizosphere. The prominent characteristic of MHB is to induce short root development (Schrey et al. 2005; Labbe et al. 2014), thereby stimulates the ectomycorrhiza to infect maximum number of lateral roots (Poole et al. 2001). This host root differentiation can be directly initiated by the bacteria as a majority of these MHB can also perform as plant growth-promoting rhizobacteria (PGPR) (Bending et al. 2002). Auxin- (Spaepen et al. 2014) and ethylene-producing bacteria (Brunetti et al. 2021) frequently regulate the root architecture; consequently, auxin equilibrium in the rhizosphere stimulates the ectomycorrhizal formation (Felten et al. 2010).

13.4 Increased Fitness of Plants Colonized with AMF Against Biotic Stress

Plant pathogens and insect pests are a major limiting factor for food production which accounts for about 20–40 percent yield losses. Every year, plant diseases and invasive insects cost the global economy about US$220 billion and US$70 billion, respectively (FAO 2019). To combat this pest impact, agriculture fields are dumped with chemical pesticides which in other ways have caused negative consequences on public health and the environment. To offset these harmful effects, the use of benign microbes seems to be a viable option. Microbes based on their mode of action generally called as mycoparasites (*Trichoderma* spp.), entomopathogens like *Bacillus* spp. and Nucleopolyhedrosis virus (NPV) etc. However, recent past reports emphasize that AMFs are also capable of alleviating the severity of plant pathogens and insect pests in addition to their plant growth-promoting attributes. Upon AMF colonization, the plant tends to undergo various physiological changes ranging from enhanced nutrient acquisition to protection against various biotic stressors which include pathogenic fungi, bacteria, virus, insects, nematodes, parasitic plants, etc. that cost host plant fitness and productivity (Dowarah et al. 2021).

13.4.1 Effect on Plant Pathogens

As plants are a sessile entity, they are continually confronted with various pathogenic microbes that take a toll on plant growth and health. AMF symbiosis proves to be a better option for managing plant pathogens that were briefly given in various tables, viz., pathogenic fungi (Table 13.1), bacteria (Table 13.2), and viruses (Table 13.3). Mycorrhiza-colonized plants have evolved multifaceted mechanisms to combat harmful microbes, including alteration in host plant nutrition, competition for niche and photosynthates, altered root morphology and physiology, enhanced plant defense, and change in rhizosphere microbiota which, when worked independently or together, can collectively grant plant protection. These mechanisms are elaborated below.

Table 13.1 Effects of AM fungi against plant pathogenic fungi

Host plant	AMF inoculants	Pathogenic fungus/ disease	Effect of AMF on pathogen	Association effect	References
Tomato	*Glomus mosseae*	*Erwinia carotovora* pv. *carotovora*	–	Competition for host nutrients Mycorrhiza inoculated plant expressed normal growth in the presence of pathogen	García-Garrido et al. (1992)
Brinjal	*Glomus etunicatum*, *Gigaspora margarita*	*Verticillium dahliae*	–	Accumulation of lignin in the first- and second-order roots delayed disease occurrence	Matsubara et al. (1995)
Tomato	*Glomus mosseae*, *Glomus intraradices*	*Phytophthora parasitica*	–	Activation of hydrolytic and antioxidant enzymes (chitinase, superoxide dismutase, chitosanase, β-1,3-glucanase)	Cordier et al. (1998)
Pea	*Glomus intraradices*	*Aphanomyces euteiches*	–	Maximized phosphorus uptake Decreased the disease severity	Bodker et al. (1998)
Potato	*Glomus etunicatum*	*Rhizoctonia solani*	–	Phytoalexin accumulation (rishitin and solavetivone)	Yao et al. (2003)
Cucumber	*Glomus intraradices*	*Colletotrichum orbiculare*	–	Reduced pathogenicity due to callose formation in leaves	Lee and George (2005)
Cucumber	*Glomus mosseae*	*Colletotrichum orbiculare*	–	Mycorrhiza-colonized plants inhibited anthracnose disease in combination with bioprotective isolates and PGP microbiota in soil	Chandanie et al. (2006)
Tomato	AMF	*Alternaria solani*	–	Reduced the disease symptoms Induced systemic resistance	Fritz et al. (2006)
Barley	*Glomus mosseae*	*Gaeumannomyces graminis* var. *tritici*	–	Reduced root lesions with increased SA content around roots	Khaosaad et al. (2007)
Date palm	AMF		–		Jaiti et al. (2008)

(continued)

Table 13.1 (continued)

Host plant	AMF inoculants	Pathogenic fungus/disease	Effect of AMF on pathogen	Association effect	References
		Fusarium oxysporum f. sp. *albedinis*		Seedling disease resistance due to an increase in phenolics (hydroxycinnamic acid)	
Tomato	*Glomus macrocarpum* *Glomus fasciculatum*	*F. oxysporum* f. sp. *lycopersici*	–	Increased JA, PAL activity, phenol and foliar trichome density	Kapoor (2008)
Tomato	*Glomus intraradices*	*Rhizoctonia solani*	–	Reduced tomato seedlings damping-off has driven association with other PGP microbiota in soil	Amer and Abou-El-Seoud (2008)
Cotton	*Glomus etunicatum*, *Glomus intraradices*, *Glomus versiforme*	*Verticillium dahliae*	–	Reduced disease index	Kobra et al. (2009)
Common bean	*Glomus mosseae* + *Glomus intraradices* + *Glomus clarum* + *Gigaspora gigantea* + *Gigaspora margarita*	*Rhizoctonia solani*	–	Decreased disease severity and index with increased plant growth and nutrition, cell wall thickening, cytoplasmic granulation, and antimicrobial compounds	Abdel-Fattah et al. (2010)
Potato	*Glomus* sp.	*Phytophthora infestans*	–	Systemic defense priming with enhanced expression of pathogenesis-related genes	Gallou et al. (2011)
Rice	*Glomus intraradices*	*Magnaporthe oryzae*	–	Reduced disease severity Systemic induction of host defense genes (OsNPR1, OsAP2, OsEREBP, and OsJAmyb)	Campos-Soriano et al. (2012)
Barley	*Glomus intraradices*, *Glomus mosseae*, and *Gigaspora rosea*	*Gaeumannomyces graminis* var. *tritici*	–	Reduced number of infected root lesions	Castellanos-Morales et al. (2012)
Tomato	*Glomus fasciculatum*	*Fusarium oxysporum* f. sp. *lycopersici*	–	Resistance with delayed progression of the root vascular pathogen to the shoot due to enhanced JA-mediated defense	Nair et al. (2015a)

Tomato	*Glomus fasciculatum*	*Alternaria alternata*	–	Enhanced LOX activity correlated with fourfold higher MeJA in leaves	Nair et al. (2015b)
Tomato	*Funneliformis mosseae*	*Alternaria solani*	–	Increased JA activity and expression of defense genes and enzymes (PAL, LOX, chitinase, β-1,3-glucanase)	Song et al. (2015)
Grapevine	*Rhizophagus irregularis*	*Plasmopara viticola* or *Botrytis cinerea*	–	Elevated levels of active forms of stilbenoids (i.e., d-viniferins, trans-form of resveratrol, and pterostilbene)	Bruisson et al. (2016)
Wheat	*Rhizophagus irregularis*	*Pseudomonas putida* KT2440	–	Enhanced systemic priming of chitosan-induced callose	Perez-de-Luque et al. (2017)
Tomato	*Glomus mosseae*	*Cladosporium fulvum*	–	Increase in antioxidant activities of SOD and POD with decreased levels of H_2O_2 and MDA	Wang et al. (2018)
Tomato	*Glomus fasciculatum*, *Acaulospora laevis*	*Fusarium oxysporum* f. sp. *lycopersici*	–	Increased amino acid contents (arginine and proline) conferred disease resistance	Manila and Nelson (2017)
Soybean	*Rhizophagus irregularis*	*Fusarium virguliforme*	–	Upregulation of defense-related genes	Marquez et al. (2019)
Cowpea	*Glomus versiforme*	*Erysiphe flexuosa*	–	Improved host tolerance with increased P uptake	Omomowo et al. (2018)
Tomato	*Glomus* sp.	*Fusarium oxysporum* f. sp. *lycopersici*	–	Production of antimicrobial compounds Reduced disease incidence with arrested mycelia growth	Kumari and Prabina (2019)
Chili	*Glomus* sp.	*Pythium aphanidermatum*	–	Increased antimicrobial compounds including fatty acids, dibutyl phthalate, and terpenoid, shahamin-B, hindering pathogen establishment	Kumari and Srimeena (2019)
Banana	*Glomus* spp.	*Fusarium oxysporum* f. sp. *cubense*	–	Absence of wilting symptom	Rodríguez-Castillo et al. (2019)

(continued)

Table 13.1 (continued)

Host plant	AMF inoculants	Pathogenic fungus/ disease	Effect of AMF on pathogen	Association effect	References
Peach	*Acaulospora scrobiculata*	Peach replant disease	–	Increased antioxidant enzymes and upregulation of plant defense genes	Gao et al. (2020)
Cucumber	*Claroideoglomus claroideum*	*Pythium ultimum*	+	Decreased shoot dry weight	Ravnskov et al. (2020)
Rice variety Maratelli	*Funneliformis mosseae*; *Rhizophagus irregularis*	*Magnaporthe oryzae*	+	Increased disease susceptibility	Campo et al. (2020)
Tomato	*Rhizophagus irregularis*	*Botrytis cinerea*	–	Lignin and yatein accumulation from root to shoot provide enhanced resistance	Sanmartín et al. (2020a)
Tomato	*Rhizophagus irregularis*	*Botrytis cinerea*	–	Increased callose accumulation	Sanmartín et al. (2020a)
Pea	*Claroideoglomus etunicatum*, *Claroideoglomus claroideum*, *Rhizophagus irregularis*, *Funneliformis geosporus*, and *Funneliformis mosseae*	*Didymella pinodes*	–	High level of defense-related proteins and secondary metabolites Induced systemic resistance	Sistani et al. (2020)
Grapevine	*Rhizophagus irregularis*	Root pathogens	–	Stimulation of fatty acid (palmitoleic, linoleic, and linolenic acids) and PR protein expression in roots and enhanced defense hormone contents in leaves	Goddard et al. (2021)
Tomato	Consortium: 26 species *Claroideoglomus* sp., *Rhizophagus* sp., *Funneliformis* sp., *Septoglomus*, *Diversispora* sp., *Glomus* sp., *Acaulospora* sp., *Entrophospora* sp., *Gigaspora* sp., and *Scutellospora* sp.	*Verticillium dahliae*	–	Decreased lipid peroxidation and high antioxidant enzyme activity in roots (SOD, CAT, POX, PPO)	Rahou et al. (2021)
Tomato	*Rhizoglomus intraradices*, *Funneliformis mosseae*, and *Claroideoglomus etunicatum*	*Fusarium oxysporum* f. sp. *lycopersici*	–	Less fusaric acid accumulation in roots Increased chlorophyll, phosphate-metabolizing enzyme, antioxidant	Hashem et al. (2021)

				enzymatic systems Reduced disease incidence	
Apple	*Glomus* sp. and *Paraglomus* sp.	Apple replant disease	–	Increased antioxidant- and defense-related enzymes (SOD, POD, CAT, PAL, PPO, APX, chitinase, and β-1,3-glucanase)	Wang et al. (2021)
Wheat	*Rhizoglomus intraradices*	*Fusarium pseudograminearum*	–	Increased antioxidant defense (SOD, CAT, GPX, and GSH) and decreased lipid peroxidation level Decreased disease severity	Spagnoletti et al. (2021)
Tomato	*Rhizoglomus intraradices*	*Fusarium oxysporum* f. sp. *lycopersici*	–	Decreased disease severity Higher root lengths and volume and surface areas	Raza et al. (2022)
Orange	*Funneliformis mosseae*	*Phytophthora parasitica*	-	Increases SA content in roots Higher expression of plant defense gene (*PtPR1*, *PtPR4*, and *PtPR5*)	Tian et al. (2021)
Watermelon	Consortium: *Claroideoglomus* sp., *Funneliformis* sp., *Diversispora* sp., *Glomus* sp., and *Rhizophagus* sp.	*Rhizoctonia solani*	–	Reduced disease incidence Increased root activity and antioxidant enzymes	Wu et al. (2021)
Artichoke	*Glomus viscosum*	*Verticillium dahliae*	–	Increased antioxidant levels (APX, MDHAR, and SOD activities, higher content of ASC and GSH) Decreased levels of lipid peroxidation and H_2O_2	Villani et al. (2021)
Tomato	*Gigaspora margarita*	*Botrytis cinerea*	–	Priming effect of SA- and JA-related defense genes	Fujita et al. (2022)
Tomato	*Rhizophagus irregularis*	*Fusarium oxysporum*	–	Induced expression of the jasmonic acid-related defense genes (AOC and LOXD and increased activities of PPO and PAL)	Wang et al. (2022)
Tomato	*Funneliformis mosseae* + *Glomus fasciculatum*	*Fusarium oxysporum* f. sp. *lycopersici*	–	Reduced the wilt severity to 77.44 %, under both pot and field trials compared to other treatments	Devi et al. (2022)

Table 13.2 Effects of AM fungi against plant pathogenic bacteria

Host plant	AM fungi	Pathogenic bacteria	Effect of AMF on pathogen	Association effect	References
Tomato	*Glomus mosseae*	*Pseudomonas syringae*	–	Neither growth of tomato nor percentage of vesicular arbuscular infection was negatively affected by pathogenic bacteria	Garcia-Garrido and Ocampo (1989)
Mulberry	*Glomus fasciculatum* *Glomus mosseae*	*P. syringae* pv. *mori*	–	Reduced blight incidence	Sharma et al. (1995)
Eucalyptus	AMF	*Pseudomonas solanacearum*	–	Improved disease resistance with reduced disease percentage of 10.0% to 17.5%	Mingqin (2004)
Medicago	*Glomus intraradices* *Glomus gigantea* *Glomus versiforme*	*Xanthomonas campestris*	–	Both local and systemic defense responses in root and shoot reduced pathogen development	Liu et al. (2007)
Tobacco	*Glomus mosseae*	*Ralstonia solanacearum*	–	Reduced disease incidence Increased activity of antioxidant enzymes (PPO, PAL, and POD)	Yuan et al. (2022)
Wheat	*Funneliformis mosseae*	*Xanthomonas translucens*	–	Enhanced disease resistance	Fiorilli et al. (2018)
Orange	*Paraglomus occultum*	*Xanthomonas axonopodis* pv. *citri*	–	Enhanced disease resistance Higher levels of superoxide radical quenching enzymes and pathogen defense gene expressions (PtEPS1 and PtPR4)	Miao-Miao et al. (2019)
Tomato	*Gigaspora margarita*	*Pseudomonas syringae* pv. *tomato* DC3000	–	Priming effect of salicylic acid (SA)- and jasmonic acid (JA)-related defense genes	Fujita et al. (2022)

Table 13.3 Effects of AM fungi against viral diseases

Host plant	AM fungi	Pathogenic virus	Effect of AMF on pathogen	Association effect	References
Citrus	*G. etunicatum*	Citrus tristeza virus and citrus rugose virus	+	Increased leaf shock symptom	Nemec and Myhre (1984)
Mung bean	*Gigaspora gilmorei* *Acaulospora marrowae* *Glomus fasciculatum* *Glomus constrictum*	Yellow mosaic virus	+	Mycorrhizal colonization and spore formation were reduced	Jayaraman and Kumar (1995)
Tobacco	*Glomus intraradices*	Tobacco mosaic virus	–	Higher disease incidence with the severity of necrotic lesions	Shaul et al. (1999)
Tomato	*Glomus mosseae*	Yellow disease	–	Reduced disease symptoms	Lingua et al. (2002)
Potato	*Glomus intraradices*	Potato virus Y	+	Reduced plant growth and root development Increased disease severity with high virus multiplication	Sipahioglu et al. (2009)
Tomato	*Funneliformis mosseae*	Tomato spotted wilt virus	–	Increases virus titer showing delay in disease recovery	Miozzi et al. (2011)
Cucumber	*Glomus mosseae*	Cucumber mosaic virus	–	Induction of JA defense gene reduced disease severity	Elsharkawy et al. (2012)
Tomato	*Funneliformis mosseae*	Tomato yellow leaf curl Sardinia virus	–	Possibility of higher JA levels	Maffei et al. (2014)
Grapevine	*R. intraradices*	Grapevine fanleaf virus	–	Reduced gall formation on roots	Hao et al. (2018)
Tomato	*Funneliformis mosseae* + *Rhizoglomus clarum* + *Rhizophagus aggregatus*	Tomato mosaic virus	–	Upregulation of defense genes (PAL1 and HQT) involved in antiviral polyphenolic compounds (flavonoids and chlorogenic acid) Reduced disease severity and	Aseel et al. (2019)

(continued)

Table 13.3 (continued)

Host plant	AM fungi	Pathogenic virus	Effect of AMF on pathogen	Association effect	References
				incidence and virus accumulation level	
Tomato	*Rhizoglomus irregulare*	Tomato bushy stunt virus	–	Disease resistance due to enhanced expression of methylation-related genes	Khoshkhatti et al. (2020a)
Tomato	*Rhizoglomus irregulare*	Tomato mosaic virus	+	Enhanced level of virus accumulation and a decreased expression of pathogenesis-related genes	Khoshkhatti et al. (2020b)
Tomato	*Funneliformis mosseae*	Cucumber mosaic virus	–	Increased SA level and less ROS accumulation	Miozzi et al. (2020)
Tomato	*Funneliformis mosseae*	*Beet curly top Iran virus* (BCTIV)	+	Enhanced disease symptom and viral accumulation	Ebrahimi et al. (2020)

13.4.1.1 Altered Nutrient Uptake

Increased nutrient uptake facilitated by AMF colonization has major implications on plant health and fitness resulting in increased tolerance or resistance upon pathogen attack. Several nutrients, especially phosphorus, are acquired by plants as a result of symbiosis. A study by Garmendia et al. (2004) provided evidence that *Glomus deserticola* mycorrhized pepper plants when challenged against *Verticillium* wilt (*V. dahlia*) exhibited greater yield than non-mycorrhizal ones with increased phosphorus uptake rate providing enhanced tolerance under limited phosphorus. However, it is not yet clear that the increased nutrient uptake would solely be ascribed to pathogen tolerance, and there are contradictory reports as well. For example, in a mycorrhized plant, reduced disease severity was observed with no change in P uptake suggesting a mechanism independent of phosphorus status (Fritz et al. 2006). Moreover, this improved nutrient uptake facilitated by extraradical hyphae of mycorrhizal fungi will help plant development by overcoming a pathogen-incurred loss by facilitating regrowth of the infected root tissues (Singh et al. 2000). Up to date, there are several reviews quoting improved nutrient uptake as the possible mechanism in guarding plants against pathogens (Azcon-Aguilar and Barea 1996; Huang et al. 2003; Xavier and Boyetchko 2004; Wehner et al. 2010; Schouteden et al. 2015; Bora and Lokhandwala 2016; Comby et al. 2017).

13.4.1.2 Competition for Niche and Photosynthates

Competition for common resources by AMF can be beneficial when they can directly halt the pathogenic infection of stressors by physical exclusion when competing for colonization site and host nutrients like reduced carbon at the plant-AMF interface for survival and multiplications of microbes (Xavier and Boyetchko 2004). Several reports have speculated this as a direct possible mechanism in the downregulation of pathogen attacks (Kobra et al. 2009; Spagnoletti et al. 2017). There is no strong evidence that could conclude that competition for nutrients is the generalized mechanism against pathogen attack.

13.4.1.3 Alteration of Root Morphology and Physiology

Remodeling of plant root architecture by AMF is considered to be one of the direct mechanisms helping to reduce pathogen infection through the processes of lignification (Matsubara et al. 1995; Sanmartín et al. 2020a) or callose deposition (Lee and George 2005; Perez-de-Luque et al. 2017; Sanmartín et al. 2020b) in plant cells which acts as penetration barriers to pathogen proliferation (Cordier et al. 1998; Vigo et al. 2000; Balestrini and Bonfante 2014; Weng et al. 2022) or indirectly compensating the pathogen-incurred loss by a profuse growth of higher-order roots (Matsubara et al. 1995). Change in host metabolism will hold obvious alterations in root exudation pattern with a change in its pH and/or chemical composition with fewer stimulatory compounds or inhibitory compounds affecting the pathogen (Vierheilig et al. 2008). Root exudation dynamics of AMF (*R. intraradices*)-colonized tomato plants were seen to attenuate microconidia germination of *Verticillium* wilt pathogen (*Fusarium oxysporum* f. sp. *lycopersici*), when tested in vitro (Raza et al. 2022). Contrarily, Castellanos-Morales et al. (2012) reported *Glomus* spp. and *G. rosea* colonization in barley reduces take-all disease (*Gaeumannomyces graminis* var. *tritici*) without much alteration in root growth. However, to date, there is no solid proof for advocating this mechanism.

13.4.1.4 Alteration of Plant Defense

The plant also tends to develop an enhanced defense strategy known as mycorrhiza-induced resistance (MIR) as a response to AMF colonization which shares defense of both systemic acquired resistance (SAR) which is elicited upon pathogen infection and induced systemic resistance (ISR) often by non-pathogenic rhizobacteria on root colonization phytohormones as defense signaling agents, priming plants to quickly combat several plants' pathogenic microbe attack (Cameron et al. 2013; Nishad et al. 2020). Studies by researchers flaunting systemic priming against pathogens concerning mycorrhized plants were often found to display upregulation of defense-related genes through omic data (Aseel et al. 2019; Gao et al. 2020; Tian et al. 2021; Wang et al. 2022), jasmonic acid (Kapoor 2008; Fujita et al. 2022), salicylic acid (Khaosaad et al. 2007), defense hydrolytic enzymes such as Β-1,3-glucanase and chitinase (Cordier et al. 1998; Song et al. 2015; Wang et al. 2021), and antioxidant enzymes (Wang et al. 2017; Hashem et al. 2021; Wu et al. 2021; Spagnoletti et al. 2021; Rahou et al. 2021). Metabolite profiling studies reported enhanced production of several secondary metabolites encompassing various compounds including flavonoids (Aseel et al. 2019), stilbenoids (Bruisson et al. 2016), terpenoids (Yao et al. 2003), and phenolics (Jaiti et al. 2008; Kapoor 2008; Abdel-Fattah et al. 2010) which ruled out a plethora of plant pathogens. Besides existing local/systemic defense, the mycorrhizal hyphal network connects the neighboring plants in the community which serves as an interplant defense signaling in addition to a nutrient allocation by forming common mycelial networks (CMN) (Ferlian et al. 2018).

13.4.1.5 Alteration of Rhizosphere

Changes in the root exudation pattern will directly influence the microbial composition of micro-niche and myco-rhizosphere both quantitatively and qualitatively by recruiting beneficial microorganisms like phosphate solubilizers and mycorrhiza helper bacteria surrounding the host root (Tahat et al. 2010). These microbiota-mediated changes were reported for their positive effect on the host plant with the altered microbial communities stimulating antagonistic compounds which defend against plant pathogens (Azcon-Aguilar and Barea 1996; Johansson et al. 2004; Siasou et al. 2009).

As reported from numerous works of literature, mycorrhizal symbiosis in plants is often perceived as beneficial, though a large literature highlights the context dependency of this interaction. The colonization of AMF itself is affected by many abiotic and biotic factors; if all the colonization is successful, a positive effect is not always assured, i.e., mycorrhiza-induced susceptibility (MIS).

The biotic factor which governs the colonization includes host genotype specificity, AMF species, and pathogen life cycle. The differential mycorrhiza-induced resistance by *R. irregularis* in three cultivars of common bean and two tomato cultivars, where, despite the same colonization rate, resistance was induced in one cultivar of tomato and two cultivars of beans (Mora-Romero et al. 2015). Hayek et al. (2012) demonstrated the AMF species specificity against petunia root pathogen, *Thielaviopsis basicola*, where *Glomus mosseae* induced its resistance but *G. intraradices* and *Gigaspora rosea* failed to do so. Mycorrhizal bio-protection not only seems to rely on the host plant genotype and root-colonizing AMF species but also on the degree of AM root colonization, with some AMF species protecting with a high colonization rate while some can act even with a low colonization rate (Castellanos-Morales et al. 2012).

Abiotic factors like soil moisture, temperature, and soil phosphorus content affect the colonization rate of AMF. Temperature influences the rate of colonization of AMF and its synchrony with pathogens. For example, the optimum temperature for

Rhizoctonia solani is 22–24 °C, while the AMF needs 20–22 °C for colonization. In such conditions, the mycorrhizal plant will not be effective against pathogens (Jain and Sethi 1988). Soil moisture is one of the factors which favor disease development; on the other hand, it negatively impacts the colonization of AMF. Iqbal et al. (1988) reported that higher moisture content of 25% was unfavorable for AMF, while the same moisture content enhanced the *R. solani* incidence. The soil's P content is primarily responsible for mycorrhizal colonization since an increase in soluble phosphate content negatively affects the AMF colonization rate (Mosse 1973; Sparling and Tinker 1978).

The stimulatory effect of AMF on crop plants is hardly observed, and inhibitory effects on plant growth also prevail (Jacott et al. 2017). Plants are affected when they regulate both AMF and pathogenic microbes *via* common molecular pathways like in *Medicago truncatula* where the *Glomus* sp. signaling factor gene (RAM2) is being shared by an oomycete pathogen for appressoria formation and colonization (Wang et al. 2012). Plants showing detrimental effects upon AMF colonization results in "mycorrhiza-induced susceptibility" (MIS) which is very well documented in AMF interaction with viral pathogens than with pathogenic fungi and bacteria, altering host plant defenses supervening pathogen susceptibility (Miozzi et al. 2019) (Table 13.3). However, there are reports which suggest that microbial consortia with several AMF species (Liu et al. 2007; Abdel-Fattah et al. 2010; Castellanos-Morales et al. 2012; Aseel et al. 2019; Rahou et al. 2021; Hashem et al. 2021; Wu et al. 2021; Devi et al. 2022) or AMF with other beneficial microbes including beneficial fungi (Elsharkawy et al. 2012) and bacteria (Zhang et al. 2012; Perez-de-Luque et al. 2017; Devi et al. 2022) have the pooled benefit of complementary effects of each when they orchestrate synergistically. Undeniably, consortia engineering is a prerequisite, or otherwise attaining plant protection becomes elusive when they behave antagonistically (Rodríguez-Castillo et al. 2019). Emerging research evidence also suggests that the combined application of chemicals such as melatonin, biochar, polyamine, strigolactone, and dopamine and also soil microorganisms like mycorrhiza helper bacteria and plant rhizosphere bacteria can also steer the beneficial effects of this extremely fine rapport of AMF and host plants (Zhu et al. 2022). Therefore, biocontrol by AMF should be overlooked in an integrated approach rather than devised as a single tool to attain its overall potential.

13.4.2 Effects of AMF Against Herbivorous Insects

Plant tolerance and resistance are two variable defense mechanisms, through which plants respond against various biotic (pathogens, pests, nematodes, etc.) and abiotic stressors (drought, temperature, etc.) (Gull et al. 2019). In the context of herbivory, plant resistance is the trait that negatively affects the performance of herbivores through structural or chemical modifications, whereas plant tolerances can compensate for the negative effects of herbivores without causing any harm to them (Mitchell et al. 2016). The presence of root symbiotic microbes, especially AMF, induces plant defense against herbivores through various mechanisms, which have been reported over the past three decades (Rabin and Pacovsky 1985; Gange et al. 1994; Gange 2001). However, the effects of AMF on plant defensive traits range from beneficial to detrimental, depending on the context of the feeding guild and specificity of insect, AMF species, host plant, and environmental conditions (Barber et al. 2013; Tao et al. 2016).

13.4.2.1 AMF-Induced Plant Resistance Against Herbivores

AMF colonization sensitizes the plant immune system, like plant beneficial microbes (PGPR, endophytes) through defense priming (Jung et al. 2012). Priming keeps the plant defense system in a highly vigilant state without activating any costly defense mechanisms but mounts rapid and efficient defense activities against various attackers including herbivores upon future attack, which is highly mediated by various phytohormones like JA, SA, ABA, and ET (Hartley and Gange 2009; Pieterse et al. 2014). AMF-triggered plant defense is called mycorrhiza-induced resistance (MIR) which is analogous to induced systemic resistance (IST) in plants (Kadam et al. 2020). In response to MIR, the plant produces various secondary metabolites belonging to alkaloids (Andrade et al. 2013), terpenoids (Sharma et al. 2017), and phenols (Frew and Wilson 2021) which may negatively affect the performance of herbivorous insects and mites either altering the plant attractiveness (Babikova et al. 2014) or directly antagonizing by toxins and anti-digestive metabolites (Rivero et al. 2021) and also attract natural enemies via changes in plant volatile blends in mycorrhized plants (Kaur et al. 2020).

However, mycorrhization negatively affects herbivore performance; it is not always true for all insect species. The magnitude and direction of AMF effects on herbivores change through the lifestyle and feeding guild of the insects (Koricheva et al. 2009; Hartley and Gange 2009). Generalist herbivores which are highly susceptible to innate plant defense are adversely affected by the mycorrhization of host plants (Song et al. 2013), whereas specialist herbivores which possess the ability to thrive against plant defense perform well in mycorrhized plants because of improved nutrient availability to the plants (Gehring and Bennett 2009). Similarly, the chewing and mining insects which activate JA-mediated defense response in plants while feeding are negatively affected by mycorrhization because of the AMF priming effect (Selvaraj et al. 2020), but sap-sucking insects which elevate SA-mediated responses perform well in the mycorrhized plants (Babikova et al. 2014). It is highly imperative to consider the notion that SA and JA are mutually antagonistic to each other (Huot et al. 2014).

Furthermore, the common mycorrhizal network (CMN) shared between different plant communities acts as a conduit for transferring various plant defense signals (JA, MeSA) and nutrients to neighboring plants which may also assist in priming the plant defense against future herbivore attacks (Babikova et al. 2013). The effects of AMF on plant defense traits against herbivores are summarized in Table 13.4.

Table 13.4 Effects of AMF on plant defense traits against herbivores

Insect pest	AMF	Host plant	Effect of AMF on insect	Possible mechanism	References
		AMF effects on plant resistance traits			
Tetranychus urticae	*Glomus mosseae*	*Phaseolus vulgaris*	–	Mycorrhized plant changes herbivore-induced plant volatiles which attracts predatory mites	Schausberger et al. (2012)
Helicoverpa armigera	*Glomus mosseae*	Tomato	–	AMF induces jasmonic acid-mediated defense	Song et al. (2013)
Dermolepida albohirum	*Glomus etunicatum* *Glomus intraradices* *Glomus coronatum* *Glomus mosseae*	Sugarcane	–	Increased silicone concentration imparts resistance	Frew et al. (2017)
Aphis nerii *Danaus plexippus*	*Rhizophagus intraradices*, *Funneliformis mosseae*, *G. aggregatum*, *Claroideoglomus etunicatum*	*Asclepias curassavica*, *A. latifolia*, *A. syriaca*, and *A. incarnata*	±	Changes in cardenolides in relation to AMF percent colonization affect herbivore performance	Meier and Hunter (2018)
Acyrthosiphon pisum	*Rhizophagus irregularis*	*Medicago truncatula*	±	The difference in AMF colonization level changes aphid performance	Maurya et al. (2018)
Spodoptera littoralis	*Rhizophagus irregularis*	Tomato	–	AMF colonizations work complementary to JA-mediated resistance and improve trichome density	Formenti and Rasmann (2019)
Trichoplusia ni	*Rhizophagus irregularis*	*Solanum tuberosum*	–	Insect weight gain is decreased by systemic upregulation of *12-oxo-phytodienoic reductase 3*, *allene oxide synthase*, *protease inhibitor type I*, and PAL	Schoenherr et al. (2019)
Acyrthosiphon pisum	*Rhizophagus irregularis*	*Medicago truncatula*	–	Reduced foliar N content and MIR	Garzo et al. (2020)

(continued)

Table 13.4 (continued)

Insect pest	AMF	Host plant	Effect of AMF on insect	Possible mechanism	References
Spodoptera frugiperda	Commercial AMF mix	Sorghum-sudangrass	–	Mycorrhized plants attract more hymenopteran and dipteran enemies	Kaur et al. (2020)
Spodoptera litura	*Glomus intraradices*	Black gram	–	Prime the plant defense to produce phenolics and various superoxide radical quenching enzymes	Selvaraj et al. (2020)
Helicoverpa punctigera	*Claroideoglomus etunicatum*, *Funneliformis coronatum*, *Funneliformis mosseae*, *Rhizophagus irregularis*	*Triticum aestivum*	–	AMF communities affect plant phenolic concentration which imparts resistance to herbivory	Frew and Wilson (2021)
Spodoptera exigua	*Funneliformis mosseae*	Tomato	–	Primed production of defensive compounds like alkaloids, fatty acid derivatives, and PP polyamine conjugates imparts resistance	Rivero et al. (2021)
Gypsy moth	*Glomus mosseae*	*Populus alba* × *P. berolinensis*	–	Upregulation of JA biosynthesis pathway and enhanced production of insect toxic secondary metabolites	Jiang et al. (2021)
Tetranychus urticae	*Rhizophagus irregularis*	*Citrus aurantium*	–	Upregulated expression of JA-mediated genes and metabolites	Manresa-Grao et al. (2022)
Locusta migratoria manilensis	*Funneliformis mosseae*	*Elymus* sp.	–	Enhanced production of defense enzymes in both donor and receiver plants which connected through CMN	Yu et al. (2022)
Lymantria dispar	*Funneliformis mosseae*, *Rhizophagus intraradices*	*Rosa rugosa*	–	Primed accumulation of JA- and ABA-mediated secondary metabolites in leaf	Lv et al. (2022)

13.4.2.2 AMF-Induced Plant Tolerance Against Herbivores

Plant tolerances are yet another defense strategy plants used to thrive under stressful conditions. It is highly specific to certain genotypes of plant species (Mitchell et al. 2016). The nutrient status of soil, innate growth rate of plant genotype, and resource allocations are the mechanisms through which plants can withstand or compensate for the loss incurred by pests without affecting them (Tao et al. 2016). Unlike plant-resistant traits, plant-tolerant traits will not impose any selection pressure on insects for resistance development against the plant defense system, and there will be no trade-offs between plant defense and growth, which is why tolerant traits will be considered as highly advantageous than the resistant traits (Peterson et al. 2017). Despite the benefits of tolerance, very few studies have concentrated on the effects of AMF on the tolerant traits of plants against herbivores. However, the research in this area is considerably increasing in the recent past.

Improving nutrient status is considered one of the strategies through which AMF plants can withstand insect herbivory. According to the research conducted by Tao et al. (2016) and Real-Santillan et al. (2019) in two variable mycorrhized plants milkweed and maize, an increase in foliar P concentration imparts tolerance against *Danaus plexippus* and *Spodoptera frugiperda*, respectively. Contrary to those, Bernaola et al. (2018) and Bernaola and Stout (2021) reported that there was insignificance that exists between plant nutrient status and insect tolerance traits in mycorrhized rice against various pests. The differences that arise in plant tolerance may be due to the difference in AMF species, insects, and plants involved. However, in all the cases, there was no yield reduction despite increases in herbivory in mycorrhized plants. To date, there were no reports on the mycorrhized plant tolerance against disease-causing insect vectors and sap-sucking insects, which may hold potential for future endeavors in this area. Table 13.5 summarizes the effects of AMF on plant-resistant traits against herbivores.

13.4.3 Effect of AMF on Plant Parasitic Nematodes

Plant parasitic nematodes (PPN) are dominant lower invertebrate organisms, threaten the world's agriculture, and cause significant yield loss to various crops (Bernard et al. 2017; Abd-Elgawad 2020). AMF can be used as an effective biocontrol agent and an alternative to nematicides for nematode control as they do not harm the environment (Wani et al. 2017). Much research conducted on AMF against PPN proves that AMF provides effective control against PPN. The AMF controls PPN through various mechanisms (Schouteden et al. 2015; Wani et al. 2017). The colonization of AMF which competes with PPN for resources and controls PPN, modification in root growth due to AMF which affects nematode penetration, and more biomass production due to more uptakes of phosphorus and other nutrients which confers resistance to plants and compensates loss caused by

Table 13.5 Effects of AMF on plant tolerance trait against herbivore

Insect pest	AMF	Host plant	Effect of AMF on insect	Possible mechanism	References
AMF effects on plant tolerance traits					
Danaus plexippus	*Rhizophagus irregularis*, *Funneliformis mosseae*, *Glomus aggregatum*, *Claroideoglomus etunicatum*	*Asclepias* sp.	+	Increases in foliar P concentration impart tolerance	Tao et al. (2016)
Lissorhoptrus oryzophilus Spodoptera frugiperda	*Rhizophagus irregularis*, *Funneliformis mosseae*, *Glomus deserticola*, *Rhagium fasciculatum*, *Sclerocystis dussii*, *Glomus microaggregatum*	*Oryza sativa*	+	Increase in plant biomass—nutritional quantity hypothesis	Bernaola et al. (2018)
Spodoptera frugiperda	*Glomus* spp., *Acaulospora* spp., *Gigaspora* spp., *Intraspora* spp.	*Zea mays*	+	Mycorrhization increases plant growth rate and foliar P concentration	Real-Santillan et al. (2019)
Dermolepida albohirtum	*Claroideoglomus etunicatum*, *Funneliformis coronatum*, *Funneliformis mosseae*, *Rhizophagus irregularis*	*Saccharum* spp.	+	Through resource reallocation mechanism, mycorrhized plant increases shoot biomass	Frew et al. (2020)
Lissorhoptrus oryzophilus	*Rhizophagus irregularis*, *Funneliformis mosseae*, *Glomus aggregatum*, *Claroideoglomus etunicatum*	*Oryza sativa*	+	Increase in plant biomass and yield without any major changes in plants' nutrient constitutions	Bernaola and Stout (2021)
Lymantria dispar	*Glomus intraradices*	*Populus alba* × *P. berolinensis*	+	Altered C/N ratio and reduced expression of flavonoid biosynthesis gene reduce plant resistance	Jiang et al. (2022)

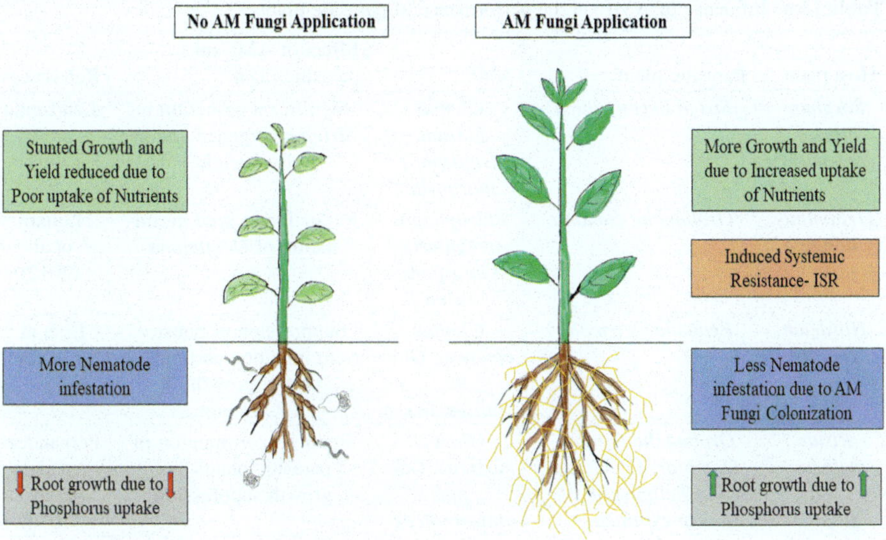

Fig. 13.1 Possible effect of AMF on PPN and plant

PPN and by providing induced systemic resistance (Fig. 13.1). The PPN takes carbon (C) as an important source for their growth (Rodriguez-Kabana and King 1980). Possibly, the application of AMF prevents nematode growth by mainly competing for C source. The AMF controls endoparasitic nematodes by directly competing for resources, whereas, because of the delay in the colonization of AMF in root tips (Smith et al. 1992), the indirect control (by modifying plant physiology) is achieved for ectoparasitic nematodes (Hol and Cook 2005).

The application of AMF potentially reduces various plant parasitic nematodes. For example, root-knot nematode (RKN) causes more damage among all the nematode species, and the various species of RKN can be controlled by the application of AMF (Calvet et al. 2001; Labeena et al. 2002; Waceke et al. 2002; Rao et al. 2003). The ectoparasitic nematodes such as *Tylenchorhynchus* sp. and *Criconemella* sp. can also be effectively managed by *Glomus* sp. (Jain et al. 1998; Kassab and Taha 1990). The population of the most important sedentary endoparasitic cyst nematode species such as *Heterodera glycines* and *H. cajani* is reduced when the AMF, viz., *Glomus mosseae* and *G. fasciculatum*, are applied, respectively (Todd et al. 2001 and Siddiqui and Mahmood 1995). *Radopholus similis* is an important migratory endoparasitic nematode in bananas, and its infestation is reduced with the application of *G. mosseae* (Elsen et al. 2001). Likewise, the application of AMF substantially reduces the population of many other PPN and not only controls PPN infestation but also improves plant growth and yield which benefits mankind.

Table 13.6 Influence of AMF on the performance of parasitic plants

Host plant	Parasitic plant	AMF	Effect of AMF on parasitic plant	References
Sorghum bicolor, *Zea mays*	*Striga hermonthica*	*Glomus clarum*, *Gigaspora margarita*	Significant reduction in *Striga* emergence and its dry weight	Lendzemo et al. (2005)
Helianthus annuus	*Orobanche cumana*	*Rhizophagus irregularis*, *Gigaspora rosea*	Reduction in seed germination of *O. cumana*	Louarn et al. (2012)
Hordeum vulgare	*Pedicularis tricolor*	*Glomus mosseae*, *Glomus intraradices*	Suppression of haustorium initiation and reduction of growth of *P. tricolor*	Li et al. (2012)
Pisum sativum	*Orobanche crenata*, *O. minor*, *O. foetida*, *Phelipanche aegyptiaca*	*Glomus mosseae*, *Glomus intraradices*	Reduced germination of parasitic plants and growth suppression	Fernandez-Aparicio et al. (2010)
Elymus nutans	*Pedicularis kansuensis*	*Glomus mosseae*	Prevent haustorium formation and significant reduction in survival of *P. kansuensis*	Sui et al. (2014)
Poa annua	*Odontites vulgaris*	*Glomus lamellosum*, *G. mosseae*	AMF did not affect the growth of the host as well as parasitic plant	Salonen et al. (2001)
Trifolium pratense	*Rhinanthus serotinus*	*Glomus claroideum*	Increased biomass of host plant but no effect on the parasitic plant	Salonen et al. (2001)

13.4.4 Effect of AMF on Parasitic Plants

Parasitic plants are one of the most important and noxious pests for agricultural crops (Erdogan 2022), and *Striga*, *Orobanche*, and *Cuscuta* cause more damage (Fernandez-Aparicio et al. 2020). The AMF acts as a biocontrol agent for these plant parasites. Possibly, the AMF prevents parasitic plant growth by the formation of haustorium and reducing seed germination (Li et al. 2012; Louarn et al. 2012). The interaction between the host plant, the parasitic plants, and the AMF is a highly evolved mechanism. Interestingly, the signaling compounds secreted by plants for the germination of parasitic plants also induce the hyphal germination of AMF. For example, the strigolactone produced by the host plant induces the germination of *Striga* as well as the AMF (Bouwmeester et al. 2007). However, various studies show that the AMF can be a potential alternative to other control measures. For example, the application of AMF substantially reduces the growth of various parasitic plants which are depicted in Table 13.6. The AMF helps to improve soil nutrient status (Jeffries et al. 2003), and possibly it helps to prevent the growth of some parasitic plants. For instance, *Striga* is an important root hemiparasite, and it is well-grown in poor soils (Parker and Riches 1993), and it may be controlled by the application of AMF as it improves soil fertility.

13.5 Conclusion

The glomeromycete fungus which explicitly exhibits symbiotic relationships with plants in the form of AM associations and nutrient cycling is a natural ecological asset to humans and the farming community in particular. As per the fossil-generated data, the AM fungi colonization on plant roots is 450 million years old, and its contribution toward facilitating the evolution of land plants needs to be reminded quite frequently to all stakeholders to unravel its potentiality for application in all biological fields. The amount of research work done in this line is encouraging up to now, but to fully realize its effects and benefits, an extra mile of work ought to be undertaken by the school of scientists. Thus, AM fungi can play a crucial role in mitigating many plant diseases in different dimensions, viz., boosting the overall health of the plants and thereby augmenting their capacity to combat a broad spectrum of biotic or abiotic stress.

Acknowledgments The authors sincerely acknowledge the facilities provided by the Department of Plant Pathology, Assam Agricultural University, Jorhat-13, Assam.

References

Abdel-Fattah GM, El-Dohlob SM, El-Haddad SA, Hafez EE, Rashad YM (2010) An ecological view of arbuscular mycorrhizal status in some Egyptian plants. J Environ Sci 37:123–136

Abd-Elgawad MM (2020) Plant-parasitic nematodes and their biocontrol agents: current status and future vistas. In: Management of phytonematodes: recent advances and future challenges, pp 171–203

Abdel-Lateif K, Bogusz D, Hocher V (2012) The role of flavonoids in the establishment of plant roots endosymbioses with arbuscular mycorrhiza fungi, rhizobia and Frankia bacteria. Plant Signal Behav 7(6):636–641

Akiyama K, Hayashi H (2006) Strigolactones: chemical signals for fungal symbionts and parasitic weeds in plant roots. Ann Bot 97(6):925–931

Amer MA, Abou-El-Seoud II (2008) Mycorrhizal fungi and *Trichoderma harzianum* as biocontrol agents for suppression of Rhizoctonia solani damping-off disease of tomato. Commun Agric Appl Biol Sci 73(2):217–232

Andrade SAL, Malik S, Sawaya ACHF, Bottcher A, Mazzafera P (2013) Association with arbuscular mycorrhizal fungi influences alkaloid synthesis and accumulation in *Catharanthus roseus* and *Nicotiana tabacum* plants. Acta Physiol Plant 35(3):867–880

Artursson V, Finlay RD, Jansson JK (2006) Interactions between arbuscular mycorrhizal fungi and bacteria and their potential for stimulating plant growth. Environ Microbiol 8:1–10

Aseel DG, Rashad YM, Hammad SM (2019) Arbuscular mycorrhizal fungi trigger transcriptional expression of flavonoid and chlorogenic acid biosynthetic pathways genes in tomato against tomato mosaic virus. Sci Rep 9(1):1–10

Azcon-Aguilar C, Barea JM (1996) Arbuscular mycorrhizas and biological control of soil-borne plant pathogens-an overview of the mechanisms involved. Mycorrhiza 6:457–464

Babikova Z, Gilbert L, Bruce TJ, Birkett M, Caulfield JC, Woodcock C, Pickett JA, Johnson D (2013) Underground signals carried through common mycelial networks warn neighboring plants of aphid attacks. Ecol Lett 16(7):835–843

Babikova Z, Gilbert L, Bruce T, Dewhirst SY, Pickett JA, Johnson D (2014) Arbuscular mycorrhizal fungi and aphids interact by changing host plant quality and volatile emission. Funct Ecol 28(2):375–385

Bais HP, Weir TL, Perry LG, Gilroy S, Vivanco JM (2006) The role of root exudates in rhizosphere interactions with plants and other organisms. Annu Rev Plant Biol 57:233–266

Balestrini R, Bonfante P (2014) Cell wall remodeling in mycorrhizal symbiosis: a way towards biotrophism. Front Plant Sci 5:237

Barber NA (2012) Arbuscular mycorrhizal fungi are necessary for the induced response to herbivores by Cucumis sativus. J Plant Ecol 6:171–176

Barber NA, Kiers ET, Hazzard RV, Adler LS (2013) Context-dependency of arbuscular mycorrhizal fungi on plant-insect interactions in an agroecosystem. Front Plant Sci 4:338

Bauer J, Kleczewski N, Bever J, Clay K, Reynolds H (2012) Nitrogen-fixing bacteria, arbuscular mycorrhizal fungi, and the productivity and structure of prairie grassland communities. Oecologia 170:1089–1098

Becard G, Taylor LP, Douds DD, Pfeffer PE, Doner LW (1995) Flavonoids are not necessary to plant signal compounds in arbuscular mycorrhizal symbioses. Mol Plant-Microbe Interact 8(2): 252–258

Bending GD, Poole EJ, Whipps JM, Read DJ (2002) Characterization of bacteria from *Pinus sylvestris-Suillus luteus* mycorrhizas and their effects on root-fungus interactions and plant growth. FEMS Microbiol Ecol 39:219–227

Bernaola L, Stout MJ (2021) The effect of mycorrhizal seed treatments on rice growth, yield, and tolerance to insect herbivores. J Pest Sci 94(2):375–392

Bernaola L, Cosme M, Schneider RW, Stout M (2018) Belowground inoculation with arbuscular mycorrhizal fungi increases the local and systemic susceptibility of rice plants to different pest organisms. Front Plant Sci 9:747

Bernard GC, Egnin M, Bonsi C (2017) The impact of plant-parasitic nematodes on agriculture and methods of control. Nematology 10:121–151

Bharadwaj DP, Lundquist PO, Persson P, Alstrom S (2008) Evidence for specificity of cultivable bacteria associated with arbuscular mycorrhizal fungal spores. FEMS Microbiol Ecol 65:310–322

Bianciotto V, Bandi CD, Minerdi M, Sironi H, Tichy V, Bonfante P (1996) An obligately endosymbiotic mycorrhizal fungus itself harbors obligately intracellular bacteria. Appl Environ Microbiol 62:3005–3010

Bodker L, Kjoller R, Rosendahl S (1998) Effect of phosphate and the arbuscular mycorrhizal fungus *Glomus intraradices* on disease severity of root rot of peas (*Pisum sativum*) caused by *Aphanomyces euteiches*. Mycorrhiza 8(3):169–174

Bora M, Lokhandwala A (2016) Mycorrhizal association: a safeguard for plant pathogen. In: Plant, soil and microbes, pp 253–275

Bouwmeester HJ, Roux C, Lopez-Raez JA, Becard G (2007) Rhizosphere communication of plants, parasitic plants and AM fungi. Trends Plant Sci 12(5):224–230

Bruisson S, Maillot P, Schellenbaum P, Walter B, Gindro K, Deglene-Benbrahim L (2016) Arbuscular mycorrhizal symbiosis stimulates key genes of the phenylpropanoid biosynthesis and stilbenoid production in grapevine leaves in response to downy mildew and grey mold infection. Phytochemistry 131:92–99

Brunetti C, Saleem AR, Della Rocca G, Emiliani G, De Carlo A, Balestrini R, Centritto M (2021) Effects of plant growth-promoting rhizobacteria strain to produce ACC deaminase on photosynthesis, isoprene emission, ethylene formation, and growth of *Mucuna pruriens* (L.) DC. in response to water deficit. J Biotechnol 331:53–62

Calvet C, Pinochet J, Hernandez-Dorrego A, Estaun V, Camprubí A (2001) Field microplot performance of the peach-almond hybrid GF-677 after inoculation with arbuscular mycorrhizal fungi in a replant soil infested with root-knot nematodes. Mycorrhiza 10(6):295–300

Cameron DD, Neal AL, Van Wees SC, Ton J (2013) Mycorrhiza-induced resistance: more than the sum of its parts? Trends Plant Sci 18(10):539–545

Campo S, Martín-Cardoso H, Olive M, Pla E, Catala-Forner M, Martínez-Eixarch M, San Segundo B (2020) Effect of root colonization by arbuscular mycorrhizal fungi on growth, productivity and blast resistance in rice. Rice 13(1):1–14

Campos P, Borie F, Cornejo P, Lopez-Raez JA, Lopez-García A, Seguel A (2018) Phosphorus acquisition efficiency related to root traits: is mycorrhizal symbiosis a key factor to wheat and barley cropping? Front Plant Sci 9:752

Campos-Soriano LIDIA, Garcia-Martinez J, Segundo BS (2012) The arbuscular mycorrhizal symbiosis promotes the systemic induction of regulatory defense-related genes in rice leaves and confers resistance to pathogen infection. Mol Plant Pathol 13(6):579–592

Castellanos-Morales V, Cardenas-Navarro R, Garcia-Garrido JM, Illana A, Ocampo JA, Steinkellner S, Vierheilig H (2012) Bioprotection against *Gaeumannomyces graminis* in barley a comparison between arbuscular mycorrhizal fungi. Plant Soil Environ 58(6):256–261

Chabaud M, Genre A, Sieberer BJ, Faccio A, Fournier J, Novero M, Bonfante P (2011) Arbuscular mycorrhizal hyphopodia and germinated spore exudates trigger Ca^{2+} spiking in the legume and nonlegume root epidermis. New Phytol 189(1):347–355

Chandanie WA, Kubota M, Hyakumachi M (2006) Interactions between plant growth promoting fungi and arbuscular mycorrhizal fungus *Glomus mosseae* and induction of systemic resistance to anthracnose disease in cucumber. Plant Soil 286(1):209–217

Comby M, Gacoin M, Robineau M, Rabenoelina F, Ptas S, Dupont J, Baillieul F (2017) Screening of wheat endophytes as biological control agents against Fusarium head blight using two different in vitro tests. Microbiol Res 202:11–20

Cordier C, Pozo MJ, Barea JM, Gianinazzi S, Gianinazzi-Pearson V (1998) Cell defense responses associated with localized and systemic resistance to *Phytophthora parasitica* induced in tomato by an arbuscular mycorrhizal fungus. Mol Plant-Microbe Interact 11(10):1017–1028

Cullimore J, Denarie J (2003) How legumes select their sweet talking symbionts. Sci 302(5645): 575–578

Devi NO, Tombisana Devi RK, Debbarma M, Hajong M, Thokchom S (2022) Effect of endophytic Bacillus and arbuscular mycorrhiza fungi (AMF) against Fusarium wilt of tomato caused by *Fusarium oxysporum* f. sp. *lycopersici*. Egypt J Biol. Pest Control 32(1):1–14

Dotzler N, Krings M, Taylor TN, Agerer R (2006) Germination shields in *Scutellospora* (Glomeromycota: Diversisporales, Gigasporaceae) from the 400 million-year-old Rhynie chert. Mycol Prog 5:178–184

Dowarah B, Gill SS, Agarwala N (2021) Arbuscular mycorrhizal fungi in conferring tolerance to biotic stresses in plants. J Plant Growth Regul 1–16

Duponnois R, Garbaye J (1991) Mycorrhization helper bacteria associated with the Douglas fir-*Laccaria laccata* symbiosis: effects in aseptic and in glasshouse conditions. Ann Sci For 48(3):239–251

Ebrahimi S, Eini O, Koolivand D (2020) Arbuscular mycorrhizal symbiosis enhances virus accumulation and attenuates resistance-related gene expression in tomato plants infected with beet curly top Iran virus. J Plant Dis Prot 127(3):341–348

Elsen A, Declerck S, De Waele D (2001) Effects of *Glomus intraradices* on the reproduction of the burrowing nematode (*Radopholus similis*) in dixenic culture. Mycorrhiza 11(1):49–51

Elsharkawy MM, Shimizu M, Takahash H, Hyakumachi M (2012) The plant growth-promoting fungus *Fusarium equiseti* and the arbuscular mycorrhizal fungus *Glomus mosseae* induce systemic resistance against cucumber mosaic virus in cucumber plants. Plant Soil 361(1): 397–409

Erdogan P (2022) Parasitic plants in agriculture and management. Parasitic Plants 1

FAO (2019). https://www.fao.org/news/story/en/item/1187738/icode/

Felten J, Legue V, Ditengou FA (2010) Lateral root stimulation in the early interaction between *Arabidopsis thaliana* and the ectomycorrhizal fungus *Laccaria bicolor*: is fungal auxin the trigger? Plant Signal Behav 7:864–867

Ferlian O, Biere A, Bonfante P, Buscot F, Eisenhauer N, Fernandez I, Martinez-Medina A (2018) Growing research networks on mycorrhizae for mutual benefits. Trends Plant Sci 23(11): 975–984

Fernandez-Aparicio M, Garcia-Garrido JM, Ocampo JA, Rubiales D (2010) Colonization of field pea roots by arbuscular mycorrhizal fungi reduces Orobanche and Phelipanche species seed germination. Weed Res 50(3):262–268

Fernandez-Aparicio M, Delavault P, Timko MP (2020) Management of infection by parasitic weeds: a review. Plan Theory 9(9):1184

Field KJ, Bidartondo MI, Rimington WR, Hoysted GA, Beerling DJ, Cameron DD, Duckett JG, Leake JR, Pressel S (2019) Functional complementarity of ancient plant-fungal mutualisms: contrasting nitrogen, phosphorus and carbon exchanges between Mucoromycotina and Glomeromycotina fungal symbionts of liverworts. New Phytol 223:908–921

Fiorilli V, Vannini C, Ortolani F, Garcia-Seco D, Chiapello M, Novero M, Domingo G, Terzi V, Morcia C, Bagnaresi P, Moulin L, Bracale M, Bonfante P (2018) Omics approaches revealed how arbuscular mycorrhizal symbiosis enhances yield and resistance to leaf pathogens in wheat. Sci Rep 8(1):1–18

Formenti L, Rasmann S (2019) Mycorrhizal fungi enhance resistance to herbivores in tomato plants with reduced jasmonic acid production. Agronomy 9(3):131

Frew A, Wilson BA (2021) Different mycorrhizal fungal communities differentially affect plant phenolic-based resistance to insect herbivory. Rhizosphere 19:100365

Frew A, Powell JR, Allsopp PG, Sallam N, Johnson SN (2017) Arbuscular mycorrhizal fungi promote silicon accumulation in plant roots, reducing the impacts of root herbivory. Plant Soil 419(1):423–433

Frew A, Powell JR, Johnson SN (2020) Aboveground resource allocation in response to root herbivory as affected by the arbuscular mycorrhizal symbiosis. Plant Soil 447(1):463–473

Frey-Klett P, Garbaye J, Tarkka M (2007) The mycorrhiza helper bacteria revisited. New Phytol 176:22–36

Fritz M, Jakobsen I, Lyngkjaer MF, Thordal-Christensen H, Pons-Kühnemann J (2006) Arbuscular mycorrhiza reduces the susceptibility of tomatoes to *Alternaria solani*. Mycorrhiza 16(6): 413–419

Fujita M, Kusajima M, Fukagawa M, Okumura Y, Nakajima M, Akiyama K, Nakashita H (2022) Response of tomatoes primed by mycorrhizal colonization to virulent and avirulent bacterial pathogens. Sci Rep 12(1):1–12

Gallou A, Mosquera HPL, Cranenbrouck S, Suarez JP, Declerck S (2011) Mycorrhiza-induced resistance in potato plantlets challenged by *Phytophthora infestans*. Physiol Mol Plant Pathol 76(1):20–26

Gange AC (2001) Species-specific responses of a root-and shoot-feeding insect to arbuscular mycorrhizal colonization of its host plant. New Phytol 150(3):611–618

Gange AC, Brown VK, Sinclair GS (1994) Reduction of black vine weevil larval growth by vesicular-arbuscular mycorrhizal infection. Entomol Exp Appl 70(2):115–119

Gao XX, Lu M, Wu H, Zhang R, Pan J, Tian S, Li H, Liao (2012) Co-inoculation with rhizobia and AMF inhibited soybean red crown rot: from field study to plant defense-related gene expression analysis. PLoS One 7(3):e33977

Gao WQ, Lu LH, Srivastava AK, Wu QS, Kuca K (2020) Effects of mycorrhizae on physiological responses and relevant gene expression of peach affected by replant disease. Agronomy 10(2): 186

Garbaye J, Duponnois R, Wahl JL (1990) The bacteria associated with *Laccaria laccata* ectomycorrhizas or sporocarps: effect on symbiosis establishment on Douglas Fir. Symbiosis 9:267–273

Garcia-Garrido JM, Ocampo JA (1989) Effect of VA mycorrhizal infection of tomato on damage caused by Pseudomonas syringae. Soil Biol Biochem 21(1):165–167

García-Garrido JM, Garcia-Romera I, Ocampo JA (1992) Cellulase production by the vesicular–arbuscular mycorrhizal fungus *Glomus mosseae* (Nicol. and Gerd.) Gerd. and Trappe. New Phytol 121(2):221–226

Garmendia I, Goicoechea N, Aguirreolea J (2004) Antioxidant metabolism in asymptomatic leaves of verticillium-infected pepper associated with an arbuscular mycorrhizal fungus. J Phytopathol 152(11–12):593–599

Garzo E, Rizzo E, Fereres A, Gomez SK (2020) High levels of arbuscular mycorrhizal fungus colonization on *Medicago truncatula* reduces plant suitability as a host for pea aphids (*Acyrthosiphon pisum*). Insect Sci 27(1):99–112

Gehring C, Bennett A (2009) Mycorrhizal fungal–plant–insect interactions: the importance of a community approach. Environ Entomol 38(1):93–102

Goddard ML, Belval L, Martin IR, Roth L, Laloue H, Deglene-Benbrahim L, Chong J (2021) Arbuscular mycorrhizal symbiosis triggers major changes in primary metabolism together with modification of defense responses and signaling in both roots and leaves of *Vitis vinifera*. Front Plant Sci 1675

Gull A, Lone AA, Wani NUI (2019) Biotic and abiotic stresses in plants. In: Abiotic and biotic stress in plants, pp 1–19

Hao Z, Tuinen D, Fayolle L, Chatagnier O, Li X, Chen B, Gianinazzi S, Gianinazzi-Pearson V (2018) Arbuscular mycorrhiza affects grapevine fanleaf virus transmission by the nematode vector *Xiphinema index*. Appl Soil Ecol 129:107–111

Harrison MJ (1999) Molecular and cellular aspects of the arbuscular mycorrhizal symbiosis. Annu Rev Plant Biol 50(1):361–389

Hartley SE, Gange AC (2009) Impacts of plant symbiotic fungi on insect herbivores: mutualism in a multitrophic context. Annu Rev Entomol 54(1):323–342

Hashem A, Akhter A, Alqarawi AA, Singh G, Almutairi KF, AbdAllah EF (2021) Mycorrhizal fungi induced activation of tomato defense system mitigates Fusarium wilt stress. Saudi J Biol Sci 28(10):5442–5450

Hayek S, Grosch R, Gianinazzi-Pearson V, Franken P (2012) Bioprotection and alternative fertilisation of petunia using mycorrhiza in a soilless production system. Agron Sustain Dev 32(3):765–771

Hol WG, Cook R (2005) An overview of arbuscular mycorrhizal fungi–nematode interactions. Basic Appl Ecol 6(6):489–503

Ho-Plagaro T, García-Garrido JM (2022) Multifarious and interactive roles of GRAS transcription factors during arbuscular mycorrhiza development. Front Plant Sci 13

Huang J, Luo S, Zeng R (2003) Mechanisms of plant disease resistance induced by arbuscular mycorrhizal fungi. J Appl Ecol 14(5):819–822

Huot B, Yao J, Montgomery BL, He SY (2014) Growth–defense tradeoffs in plants: a balancing act to optimize fitness. Mol Plant 7(8):1267–1287

Iqbal SH, Nasim G, Niaz M (1988) Role of vesicular-arbuscular mycorrhiza as a deterrent to damping off caused by *Rhizoctonia solani* in *Brassica oleracea*. Biologia

Ito S, Yamagami D, Umehara M, Hanada A, Yoshida S, Sasaki Y, Asami T (2017) Regulation of strigolactone biosynthesis by gibberellin signaling. Plant Physiol 174(2):1250–1259

Jacott CN, Murray JD, Ridout CJ (2017) Trade-offs in arbuscular mycorrhizal symbiosis: disease resistance, growth responses and perspectives for crop breeding. Agronomy 7(4):75

Jain RK, Sethi C (1988) Influence of endomycorrhizal fungi *Glomus fasciculatum* and *G. eplgaeus* on penetration and development of *Heterodera cajani* on cowpea. Indian J Nematol 18(1): 89–93

Jain RK, Hasan N, Singh RK, Pandey PN (1998) Interaction between plant parasitic nematodes, *Meloidogyne incognita*, *Tylenchorhynchus vulgaris* and VAM fungi, *Glomus fasciculatum* on lucerne. Ann Plant Protect Sci 6(1):37–40

Jaiti F, Kassami M, Meddich A, El Hadrami I (2008) Effect of arbuscular mycorrhization on the accumulation of hydroxycinnamic acid derivatives in date palm seedlings challenged with *Fusarium oxysporum* f. sp. *albedinis*. J Phytopathol 156(11–12):641–646

Jayaraman J, Kumar D (1995) Influence of mungbean yellow mosaic virus on mycorrhizal fungi associated with *Vigna radiata* var. PS16. Indian Phytopathol 48(1):108–110

Jeffries P, Gianinazzi S, Perotto S, Turnau K, Barea JM (2003) The contribution of arbuscular mycorrhizal fungi in sustainable maintenance of plant health and soil fertility. Biol Fertil Soils 37(1):1–16

Jia Y, Gray VM (2008) Growth yield of *Vicia faba* L in response to microbial symbiotic associations. S Afr J Bot 74:25–32

Jia Y, Gray VM, Straker CJ (2004) The influence of rhizobium and arbuscular mycorrhizal fungi on nitrogen and phosphorus accumulation by *Vicia faba*. Ann Bot 94:251–258

Jiang D, Tan M, Wu S, Zheng L, Wang Q, Wang G, Yan S (2021) Defense responses of arbuscular mycorrhizal fungus-colonized poplar seedlings against gypsy moth larvae: a multiomics study. Hortic Res 8

Jiang D, Lin R, Tan M, Yan J, Yan S (2022) The mycorrhizal-induced growth promotion and insect resistance reduction in *Populus alba*× *P. berolinensis* seedlings: a multi-omics study. Tree Physiol 42(5):1059–1069

Johansson JF, Paul LR, Finlay RD (2004) Microbial interactions in the mycorrhizosphere and their significance for sustainable agriculture. FEMS Microbiol Ecol 48(1):1–13

Johnson NC (2010) Resource stoichiometry elucidates the structure and function of arbuscular mycorrhizas across scales. New Phytol 185:631–647

Jung SC, Martinez-Medina A, Lopez-Raez JA, Pozo MJ (2012) Mycorrhiza-induced resistance and priming of plant defences. J Chem Ecol 38(6):651–664

Kadam SB, Pable AA, Barvkar VT (2020) Mycorrhiza induced resistance (MIR): a defence developed through synergistic engagement of phytohormones, metabolites and rhizosphere. Funct Plant Biol 47(10):880–890

Kapoor R (2008) Induced resistance in mycorrhizal tomato is correlated to concentration of jasmonic acid. J Biol Sci 8(3):49–56

Kassab AS, Taha AHY (1990) Aspects of the host-parasite relationships of nematodes and sweet potato 1. Population dynamics and interaction of *Criconemella* spp., *Rotylenchulus reniformis*, *Tylenchorhynchus* spp. and endomycorrhiza. Ann Agric Sci (Cairo) 35(1):497–508

Kaur J, Chavana J, Soti P, Racelis A, Kariyat R (2020) Arbuscular mycorrhizal fungi (AMF) influences growth and insect community dynamics in Sorghum-sudangrass (Sorghum x drummondii). Arthropod Plant Interact 14(3):301–315

Khaosaad T, Garcia-Garrido JM, Steinkellner S, Vierheilig H (2007) Take-all disease is systemically reduced in roots of mycorrhizal barley plants. Soil Biol Biochem 39(3):727–734

Khoshkhatti N, Eini O, Koolivand D (2020a) Effect of *Rhizoglomus irregulare* symbiosis on reducing disease severity in tomato plants infected with tomato bushy stunt virus with emphasis on the expression of methylation-related genes. Crop Biotech 10(30):43–55

Khoshkhatti N, Eini O, Koolivand D, Pogiatzis A, Klironomos JN, Pakpour S (2020b) Differential response of mycorrhizal plants to tomato bushy stunt virus and tomato mosaic virus infection. Microorganisms 8(12):2038

Kirk PM (2001) Ainsworth and Bisby's dictionary of the fungi, 9th edn. C.A.B International, Oxon

Kobra N, Jalil K, Youbert G (2009) Effects of three Glomus species as biocontrol agents against Verticillium-induced wilt in cotton. J Plant Prot Res

Koricheva J, Gange AC, Jones T (2009) Effects of mycorrhizal fungi on insect herbivores: a meta-analysis. Ecology 90(8):2088–2097

Kosuta S, Chabaud M, Lougnon G, Gough C, Denarie J, Barker DG, Becard G (2003) A diffusible factor from arbuscular mycorrhizal fungi induces symbiosis-specific MtENOD11 expression in roots of *Medicago truncatula*. Plant Physiol 131(3):952–962

Kumari SMP, Prabina BJ (2019) Protection of tomato, Lycopersicon esculentum from wilt pathogen, *Fusarium oxysporum* f. sp. *lycopersici* by arbuscular mycorrhizal fungi, *Glomus* sp. Int J Curr Microbiol Appl Sci 8:1368–1378

Kumari SMP, Srimeena N (2019) Arbuscular mycorrhizal fungi (AMF) induced defense factors against the damping-off disease pathogen, *Pythium aphanidermatum* in chilli (*Capsicum annum*). Int J Curr Microbiol Appl Sci 8:2243–2248

Labbe JL, Weston DJ, Dunkirk N (2014) Newly identified helper bacteria stimulate ectomycorrhizal formation in Populus. Front Plant Sci 5:1–10

Labeena P, Sreenivasa MN, Lingaraju S (2002) Interaction effects between arbuscular mycorrhizal fungi and root-knot nematode *Meloidogyne incognita* on tomato. Indian J Nematol 32(2): 118–120

Lee YJ, George E (2005) Development of a nutrient film technique culture system for arbuscular mycorrhizal plants. Hort Sci 40(2):378–380

Lendzemo VW, Kuyper TW, Kropff MJ, van Ast AV (2005) Field inoculation with arbuscular mycorrhizal fungi reduces *Striga hermonthica* performance on cereal crops and has the potential to contribute to integrated Striga management. Field Crops Res 91(1):51–61

Li AR, Smith SE, Smith FA, Guan KY (2012) Inoculation with arbuscular mycorrhizal fungi suppresses initiation of haustoria in the root hemiparasite *Pedicularis tricolor*. Ann Bot 109(6): 1075–1080

Lingua G, DAgostino G, Massa N, Antosiano M, Berta G (2002) Mycorrhiza-induced differential response to a yellows disease in tomato. Mycorrhiza 12(4):191–198

Liu J, Maldonado-Mendoza I, Lopez-Meyer M, Cheung F, Town CD, Harrison MJ (2007) Arbuscular mycorrhizal symbiosis is accompanied by local and systemic alterations in gene expression and an increase in disease resistance in the shoots. Plant J 50(3):529–544

Louarn J, Carbonne F, Delavault P, Becard G, Rochange S (2012) Reduced germination of *Orobanche cumana* seeds in the presence of arbuscular mycorrhizal fungi or their exudates. PLoS One 7(11):e49273

Lv Y, Wang F, Chen H, Zhang T, Yan J, Hu Y (2022) Unraveling arbuscular mycorrhizal fungi-induced resistance of purple branch rose (*Rosa rugosa* 'Zizhi') *to Lymantria dispar* based on metabolomics. Biol Control 172:104971

Lynch JM, Whipps JM (1990) Substrate flow in the rhizosphere. Plant Soil 129:1–10

Ma J, Wang W, Yang J, Qin S, Yang Y, Sun C, Huang J (2022) Mycorrhizal symbiosis promotes nutrient content accumulation and affects the root exudates in maize. BMC Plant Biol 22(1): 1–13

Maffei G, Miozzi L, Fiorilli V, Novero M, Lanfranco L, Accotto GP (2014) The arbuscular mycorrhizal symbiosis attenuates symptom severity and reduces virus concentration in tomato infected by tomato yellow leaf curl Sardinia virus (TYLCSV). Mycorrhiza 24(3):179–186

Manila R, Nelson R (2017) Nutrient uptake and promotion of growth by arbuscular mycorrhizal fungi in tomato and their role in bio-protection against the tomato wilt pathogen. J Microbiol Biotechnol 3(4):4246

Manresa-Grao M, Pastor-Fernández J, Sanchez-Bel P, Jaques JA, Pastor V, Flors V (2022) Mycorrhizal symbiosis triggers local resistance in citrus plants against spider mites. Front Plant Sci 13

Marquez N, Giachero ML, Gallou A, Debat HJ, Declerck S, Ducasse DA (2019) Transcriptome analysis of mycorrhizal and nonmycorrhizal soybean plantlets upon infection with *Fusarium virguliforme*, one causal agent of sudden death syndrome. Plant Pathol 68(3):470–480

Martinez-Molina E, Morales VM, Hubbell DH (1979) Hydrolytic enzyme production by Rhizobium. Appl Environ Microbiol 38(6):1186–1188

Matsubara Y, Tamura H, Harada T (1995) Growth enhancement and Verticillium wilt control by vesicular-arbuscular mycorrhizal fungus inoculation in eggplant. J Jpn Soc Hortic Sci

Maurya AK, Kelly MP, Mahaney SM, Gomez SK (2018) Arbuscular mycorrhizal symbiosis alters plant gene expression and aphid weight in a tripartite interaction. J Plant Interact 13(1):294–305

Meena RS, Vijayakumar V, Yadav GS, Mitran T (2018) Response and interaction of *Bradyrhizobium japonicum* and arbuscular mycorrhizal fungi in the soybean rhizosphere. Plant Growth Regul 84(2):207–223

Meier AR, Hunter MD (2018) Mycorrhizae alter toxin sequestration and performance of two specialist herbivores. Front Ecol Evol 6:33

Miao-Miao XIE, Zhang YC, Li-Ping LIU, Ying-Ning ZOU, Qiang-Sheng WU, Kamil KUCA (2019) Mycorrhiza regulates signal substance levels and pathogen defense gene expression to resist citrus canker. Not Bot Horti Agrobot Cluj Napoca 47(4):1161–1167

Miller JB, Pratap A, Miyahara A, Zhou L, Bornemann S, Morris RJ, Oldroyd GE (2013) Calcium/Calmodulin-dependent protein kinase is negatively and positively regulated by calcium, providing a mechanism for decoding calcium responses during symbiosis signaling. Plant Cell 25(12):5053–5066

Mingqin G, Yu C, Fengzhen W (2004) Resistance of the AM fungus Eucalyptus seedlings against *Pseudomonas solanacearum*. Forest Res 17(4):441–446

Miozzi L, Catoni M, Fiorilli V, Mullineaux PM, Accotto GP, Lanfranco L (2011) Arbuscular mycorrhizal symbiosis limits foliar transcriptional responses to viral infection and favors long-term virus accumulation. Mol Plant-Microbe Interact 24(12):1562–1572

Miozzi L, Vaira AM, Catoni M, Fiorilli V, Accotto GP, Lanfranco L (2019) Arbuscular mycorrhizal symbiosis: plant friend or foe in the fight against viruses? Front Microbiol 10:1238

Miozzi L, Vaira AM, Brilli F, Casarin V, Berti M, Ferrandino A, Nerva L, Accotto GP, Lanfranco L (2020) Arbuscular mycorrhizal symbiosis primes tolerance to cucumber mosaic virus in tomato. Viruses 12(6):675

Mitchell C, Brennan RM, Graham J, Karley AJ (2016) Plant defense against herbivorous pests: exploiting resistance and tolerance traits for sustainable crop protection. Front Plant Sci 7:1132

Mora-Romero GA, Cervantes-Gamez RG, Galindo-Flores H, González-Ortíz MA, Felix-Gastelum-R, Maldonado-Mendoza IE, Lopez-Meyer M (2015) Mycorrhiza-induced protection against pathogens is both genotype-specific and graft-transmissible. Symbiosis 66(2):55–64

Mosse B (1973) Advances in the study of vesicular-arbuscular mycorrhiza. Annu Rev Phytopathol 11(1):171–196

Nair A, Kolet SP, Thulasiram HV, Bhargava S (2015a) Role of methyl jasmonate in the expression of mycorrhizal induced resistance against *Fusarium oxysporum* in tomato plants. Physiol Molecul Plant Pathol 92:139–145

Nair A, Kolet SP, Thulasiram HV, Bhargava S (2015b) Systemic jasmonic acid modulation in mycorrhizal tomato plants and its role in induced resistance against *Alternaria alternata*. Plant Biol 17(3):625–631

Nemec S, Myhre D (1984) Virus-*Glomus etunicatum* interactions in citrus rootstocks. Plant Dis 68: 311–314

Newman EI, Reddell P (1987) The distribution of mycorrhizas among families of vascular plants. New Phytol 106:745–751

Nishad R, Ahmed T, Rahman VJ, Kareem A (2020) Modulation of plant defense system in response to microbial interactions. Front Microbiol 11:1298

Omomowo IO, Fadiji AE, Omomowo OI (2018) Assessment of bio-efficacy of *Glomus versiforme* and *Trichoderma harzianum* in inhibiting powdery mildew disease and enhancing the growth of cowpea. Ann Agric Sci 63:917

Parker C, Riches CR (1993) Parasitic weeds of the world: biology and control. CAB international

Perez-de-Luque A, Tille S, Johnson L, Pascual-Pardo D, Ton J, Cameron DD (2017) The interactive effects of arbuscular mycorrhiza and plant growth-promoting rhizobacteria synergistically enhance host plant defences against pathogens. Sci Rep 7(1):1–10

Peterson RK, Varella AC, Higley LG (2017) Tolerance: the forgotten child of plant resistance. Peer J 5:e3934

Pieterse CM, Zamioudis C, Berendsen RL, Weller DM, Van Wees SC, Bakker PA (2014) Induced systemic resistance by beneficial microbes. Annu Rev Phytopathol 52:347–375

Pirozynsky KA, Malloch DW (1975) The origin of land plants: a matter of mycotrophism. Biosystems 6:153–164

Poole EJ, Bending GD, Whipps JM, Read DJ (2001) Bacteria associated with *Pinus sylvestris-Lactarius rufus* ectomycorrhizas and their effects on mycorrhiza formation in vitro. New Phytol 151:743–751

Primieri S, Magnoli SM, Koffel T, Sturmer SL, Bever JD (2022) Perennial, but not annual legumes synergistically benefit from infection with arbuscular mycorrhizal fungi and rhizobia: a meta-analysis. New Phytol 233(1):505–514

Rabin LB, Pacovsky RS (1985) Reduced larva growth of two lepidoptera (Noctuidae) on excised leaves of soybean infected with a mycorrhizal fungus. J Econ Entomol 78(6):1358–1363

Rahou YA, Boutaj H, Boutasknit A, Douira A, Benkirane R, El Modafar Cherkaoui, Meddich A (2021) Colonization of tomato roots with arbuscular mycorrhizal fungi changes of antioxidative activity and improves tolerance to *Verticillium dahliae*. Plant Cell Biotechnol Molecul Biol 65–81

Rao MS, Naik D, Shylaja M, Reddy PP (2003) Management of *Meloidogyne incognita* on eggplant by integrating endomycorrhiza, Glomus fasciculatum with bio-agent *Verticillium chlamydosporium* under field conditions. Indian J Nematol 33(1):29–32

Ravnskov S, Cabral C, Larsen J (2020) Mycorrhiza induced tolerance in *Cucumis sativus* against root rot caused by *Pythium ultimum* depends on fungal species in the arbuscular mycorrhizal symbiosis. Biol Control 141:104133

Raza S, Akhter A, Wahid F, Hashem A, AbdAllah EF (2022) *Rhizophagus intraradices* and tomato-basil companionship affect root morphology and root exudate dynamics in tomato under fusarium wilt disease stress. Appl Ecol Environ Res 20(1):235–249

Real-Santillan RO, Del-Val E, Cruz-Ortega R, Contreras-Cornejo HA, Gonzalez-Esquivel CE, Larsen J (2019) Increased maize growth and P uptake promoted by arbuscular mycorrhizal fungi coincide with higher foliar herbivory and larval biomass of the fall armyworm *Spodoptera frugiperda*. Mycorrhiza 29(6):615–622

Redman RSYO, Kim CJDA, Woodward C, Greer L, Espino SL, Doty RJ, Rodriguez (2011) Increased fitness of rice plants to abiotic stress via habitat adapted symbiosis: a strategy for mitigating impacts of climate change. PLoS One 6:e14823

Rivero J, Lidoy J, Llopis-Gimenez A, Herrero S, Flors V, Pozo MJ (2021) Mycorrhizal symbiosis primes the accumulation of antiherbivore compounds and enhances herbivore mortality in tomato. J Exp Bot 72(13):5038–5050

Rodríguez-Castillo G, Molina-Rodríguez M, Cambronero-Heinrichs JC, Quiros-Fournier JP, Lizano-Fallas V, Jimenez-Rojas C, Rodriguez-Rodriguez CE (2019) Simultaneous removal of neonicotinoid insecticides by a microbial degrading consortium: detoxification at reactor scale. Chemosphere 235:1097–1106

Rodriguez-Kabana R, King PS (1980) Use of mixtures of urea and blackstrap molasses for control of root-knot nematodes in soil. Nematropica 38–44

Rydlova J, Puschel D, Sudova R, Gryndler M, Mikanova O, Vosatka M (2011) Interaction of arbuscular mycorrhizal fungi and rhizobia: effects on flax yield in spoil-bank clay. J Plant Nutr Soil Sci 174(1):128–134

Saito K, Yoshikawa M, Yano K, Miwa H, Uchida H, Asamizu E, Kawaguchi M (2007) NUCLEOPORIN85 is required for calcium spiking, fungal and bacterial symbioses, and seed production in Lotus japonicus. Plant Cell 19(2):610–624

Salonen V, Vestberg M, Vauhkonen M (2001) The effect of host mycorrhizal status on host plant–parasitic plant interactions. Mycorrhiza 11(2):95–100

Sanmartín N, Pastor V, Pastor-Fernández J, Flors V, Pozo MJ, Sanchez-Bel P (2020a) Role and mechanisms of callose priming in mycorrhiza-induced resistance. J Exp Bot 71(9):2769–2781

Sanmartín N, Sanchez-Bel P, Pastor V, Pastor-Fernández J, Mateu D, Pozo MJ, Flors V (2020b) Root-to-shoot signalling in mycorrhizal tomato plants upon *Botrytis cinerea* infection. Plant Sci 298:110595

Schausberger P, Peneder S, Jurschik S, Hoffmann D (2012) Mycorrhiza changes plant volatiles to attract spider mite enemies. Funct Ecol 26(2):441–449

Schoenherr AP, Rizzo E, Jackson N, Manosalva P, Gomez SK (2019) Mycorrhiza-induced resistance in potato involves priming of defense responses against cabbage looper (Noctuidae: Lepidoptera). Environ Entomol 48(2):370–381

Schouteden N, De Waele D, Panis B, Vos CM (2015) Arbuscular mycorrhizal fungi for the biocontrol of plant-parasitic nematodes: a review of the mechanisms involved. Front Microbiol 6:1280

Schrey SD, Schellhammer M, Ecke M, Hampp R, Tarkka M (2005) Mycorrhiza helper bacterium Streptomyces AcH 505 induces differential gene expression in the ectomycorrhizal fungus Amanita muscaria. New Phytol 168:205–216

Schussler A, Walker C (2011) Evolution of the 'plant-symbiotic' fungal phylum, Glomeromycota. In: Wöstemeyer J, Pöggeler S (eds) The Mycota. Evolution of fungi and fungal-like organisms, vol XIV. Springer, pp 163–185

Schussler A, Schwarzott D, Walker C (2001) A new fungal phylum, the Glomeromycota: phylogeny and evolution. Mycol Res 105:1413–1421

Selvaraj A, Thangavel K, Uthandi S (2020) Arbuscular mycorrhizal fungi (*Glomus intraradices*) and diazotrophic bacterium (Rhizobium BMBS) primed defense in blackgram against herbivorous insect (*Spodoptera litura*) infestation. Microbiol Res 231:126355

Sharma DD, Govindaiah S, Katiyar RS, Das PK, Janardhan L, Bajpai AK, Choudhry PC, Janardhan L (1995) Effect of VA-mycorrhizal fungi on the incidence of major mulberry diseases. Indian J Seric 34:34–37

Sharma E, Anand G, Kapoor R (2017) Terpenoids in plant and arbuscular mycorrhiza-reinforced defence against herbivorous insects. Ann Bot 119(5):791–801

Shaul O, Galili S, Volpin H, Ginzberg I, Elad Y, Chet I, Kapulnik Y (1999) Mycorrhiza-induced changes in disease severity and PR protein expression in tobacco leaves. Mol Plant-Microbe Interact 12(11):1000–1007

Siasou E, Standing D, Killham K, Johnson D (2009) Mycorrhizal fungi increase biocontrol potential of *Pseudomonas fluorescens*. Soil Biol Biochem 41:1341–1343

Siddiqui ZA, Mahmood I (1995) Biological control of *Heterdora cajani* and *Fusarium udum* by *Bacillus subtilis*, *Bradhyrhizobium japonicum* and *Glomus fasciculatum* on pigeon pea. Fundam Appl Nematol 18(6):559–566

Singh R, Adholeya A, Mukerji KG (2000) Mycorrhiza in control of soil-borne pathogens. In: Mukerji KG, Chamola BP, Singh J (eds) Mycorrhizal biology. Kluwer, New York, pp 173–196

Sipahioglu MH, Demir S, Usta M, Akkopru A (2009) Biological relationship of potato virus Y and arbuscular mycorrhizal fungus *Glomus intraradices* in potato. Pest Tech 3:63–66

Sistani RN, Desalegn G, Kaul HP, Wienkoop S (2020) Seed metabolism and pathogen resistance enhancement in *Pisum sativum* during colonization of arbuscular mycorrhizal fungi: an integrative metabolomics-proteomics approach. Front Plant Sci 11:872

Smith S, Read D (2008) Mycorrhizal symbiosis, 3rd edn. Academic, New York

Smith SE, Dickson S, Walker NA (1992) Distribution of VA mycorrhizal entry points near the root apex: is there an un-infectible zone at the root tip of leek or clover. New Phytol 122(3): 469–477

Smith SE, Jakobsen I, Gronlund M, Smith FA (2011) Roles of arbuscular mycorrhizas in plant phosphorus nutrition: interactions between pathways of phosphorus uptake in arbuscular mycorrhizal roots have important implications for understanding and manipulating plant phosphorus acquisition. Plant Physiol 156(3):1050–1057

Song YY, Ye M, Li CY, Wang RL, Wei XC, Luo SM, Zeng RS (2013) Priming of anti-herbivore defense in tomato by arbuscular mycorrhizal fungus and involvement of the jasmonate pathway. J Chem Ecol 39(7):1036–1044

Song Y, Chen D, Lu K, Sun Z, Zeng R (2015) Enhanced tomato disease resistance primed by arbuscular mycorrhizal fungus. Front Plant Sci 6:786

Sood SG (2003) Chemotactic response of plant-growth-promoting bacteria towards roots of vesicular-arbuscular mycorrhizal tomato plants. FEMS Microbiol Ecol 45:219–227

Spaepen S, Bossuyt S, Engelen K, Marchal K, Vanderleyden J (2014) Phenotypical and molecular responses of *Arabidopsis thaliana* roots as a result of inoculation with the auxin-producing bacterium *Azospirillum brasilense*. New Phytol 201(3):850–861

Spagnoletti F, Carmona M, Gómez NET, Chiocchio V, Lavado RS (2017) Arbuscular mycorrhiza reduces the negative effects of *M. phaseolina* on soybean plants in arsenic-contaminated soils. Appl Soil Ecol 121:41–47

Spagnoletti FN, Carmona M, Balestrasse K, Chiocchio V, Giacometti R, Lavado RS (2021) The arbuscular mycorrhizal fungus *Rhizophagus intraradices* reduces the root rot caused by *Fusarium pseudograminearum* in wheat. Rhizosphere 19:100369

Sparling GP, Tinker PB (1978) Mycorrhizal infection in Pennine grassland. 1. Levels of infection in the field [England]. J Appl Ecol (UK)

Sprent JI (2007) Evolving ideas of legume evolution and diversity: a taxonomic perspective on the occurrence of nodulation. New Phytol 174:11–25

Sturmer S (2012) A history of the taxonomy and systematics of arbuscular mycorrhizal fungi belonging to the phylum Glomeromycota. Mycorrhiza 22:247–258

Sui XL, Li AR, Chen Y, Guan KY, Zhuo L, Liu YY (2014) Arbuscular mycorrhizal fungi: potential biocontrol agents against the damaging root hemiparasite *Pedicularis kansuensis*. Mycorrhiza 24(3):187–195

Sulieman S, Tran LSP (2015) Phosphorus homeostasis in legume nodules as an adaptive strategy to phosphorus deficiency. Plant Sci 239:36–43

Tahat MM, Kamaruzaman S, Othman R (2010) Mycorrhizal fungi as a biocontrol agent. Plant Pathol J 9(4):198–207

Takeda N, Handa Y, Tsuzuki S, Kojima M, Sakakibara H, Kawaguchi M (2015) Gibberellin regulates infection and colonization of host roots by arbuscular mycorrhizal fungi. Plant Signal Behav 10(6):e1028706

Tao L, Ahmad A, de Roode JC, Hunter MD (2016) Arbuscular mycorrhizal fungi affect plant tolerance and chemical defences to herbivory through different mechanisms. J Ecol 104(2): 561–571

Tian L, Zou YN, Wu QS, Kuca K (2021) Mycorrhiza-induced plant defence responses in trifoliate orange infected by *Phytophthora parasitica*. Acta Physiol Planta 43(3):1–8

Todd TC, Winkler HE, Wilson GWT (2001) Interaction of *Heterodera glycines* and *Glomus mosseae* on soybean. J Nematol 33(4S):306

Udvardi MK, Day DA (1997) Metabolite transport across symbiotic membranes of legume nodules. Ann Rev Plant Physiol Plant Mol Biol 48:493–523

Vierheilig H, Steinkellner S, Khaosaad T, Garcia-Garrido JM (2008) The biocontrol effect of mycorrhization on soilborne fungal pathogens and the autoregulation of the AM symbiosis: one mechanism, two effects?. In: Mycorrhiza. Springer, Berlin, pp 307–320

Vigo C, Norman JR, Hooker JE (2000) Biocontrol of the pathogen *Phytophthora parasitica* by arbuscular mycorrhizal fungi is a consequence of effects on infection loci. Plant Pathol 49(4): 509–514

Villani A, Tommasi F, Paciolla C (2021) The arbuscular mycorrhizal fungus *Glomus viscosum* improves the tolerance to verticillium wilt in artichoke by modulating the antioxidant defense systems. Cell 10(8):1944

Waceke JW, Waudo SW, Sikora R (2002) Effect of inorganic phosphatic fertilizers on the efficacy of an arbuscular mycorrhiza fungus against a root-knot nematode on pyrethrum. Int J Pest Manag 48(4):307–313

Walker C, Harper CJ, Brundrett MC, Krings M (2018) Looking for arbuscular mycorrhizal fungi in the fossil record: an illustrated guide. In: Krings M, Cuneo N, Harper CJ, Rothwell GW (eds) Transformative paleobotany. Academic Press/Elsevier, Papers to Commemorate the Life and Legacy of Thomas N. Taylor, pp 481–517

Wang Y, Huang J, Gao Y (2012) Arbuscular mycorrhizal colonization alters subcellular distribution and chemical forms of cadmium in *Medicago sativa* L. and resists cadmium toxicity. PLoS One 7(11):e48669

Wang S, He G, Chen M, Zuo T, Xu W, Liu X (2017) The role of antioxidant enzymes in the ovaries. Oxidative medicine and cellular longevity

Wang YY, Yin QS, Qu Y, Li GZ, Hao L (2018) Arbuscular mycorrhiza-mediated resistance in tomato against *Cladosporium fulvum*-induced mould disease. J Phytopathol 166(1):67–74

Wang M, Zhang R, Zhao L, Wang H, Chen X, Mao Z, Yin C (2021) Indigenous arbuscular mycorrhizal fungi enhance resistance of apple rootstock 'M9T337' to apple replant disease. Physiol Mol Plant Pathol 116:101717

Wang L, Yang D, Ma F, Wang G, You Y (2022) Recent advances in responses of arbuscular mycorrhizal fungi-plant symbiosis to engineered nanoparticles. Chemosphere 286:131644

Wani KA, Manzoor J, Shuab R, Lone R (2017) Arbuscular mycorrhizal fungi as biocontrol agents for parasitic nematodes in plants. In: Mycorrhiza-nutrient uptake, biocontrol, ecorestoration, pp 195–210

Wehner J, Antunes PM, Powell JR, Mazukatow J, Rillig MC (2010) Plant pathogen protection by arbuscular mycorrhizas: a role for fungal diversity? Pedobiologia 53(3):197–201

Weng W, Yan J, Zhou M, Yao X, Gao A, Ma C, Ruan J (2022) Roles of arbuscular mycorrhizal fungi as a biocontrol agent in the control of plant diseases. Microorganisms 10(7):1266

Wu M, Yan Y, Wang Y, Mao Q, Fu Y, Peng X, Ahammed GJ (2021) Arbuscular mycorrhizal fungi for vegetable (VT) enhance resistance to *Rhizoctonia solani* in watermelon by alleviating oxidative stress. Biol Control 152:104433

Xavier LJ, Boyetchko SM (2004) Arbuscular mycorrhizal fungi in plant disease control. In: Fungal biotechnology in agricultural, food, and environmental applications, pp 183–194

Yao Q, Li X, Ai W, Christie P (2003) Bi-directional transfer of phosphorus between red clover and perennial ryegrass via arbuscular mycorrhizal hyphal links. Eur J Soil Biol 39(1):47–54

Yu L, Zhang W, Geng Y, Liu K, Shao X (2022) Cooperation with arbuscular mycorrhizal fungi increases plant nutrient uptake and improves defenses against insects. Front Ecol Evol 10

Yuan W, Ruan S, Qi G, Wang R, Zhao X (2022) Plant growth-promoting and antibacterial activities of cultivable bacteria alive in tobacco field against *Ralstonia solanacearum*. Environ Microbiol 24(3):1411–1429

Zhu B, Gao T, Zhang D, Ding K, Li C, Ma F (2022) Functions of arbuscular mycorrhizal fungi in horticultural crops. Sci Hortic 303:111219

Chapter 14
Effect of Temperature and Defense Response on the Severity of Dry Root Rot Disease in Chickpea Caused by *Macrophomina phaseolina*

Preeti, Dinesh Panwar, Poonam Saini, and Jitendra Kumar Vats

Abstract Severe outbreaks of diseases are the consequences of interactions between pathogenic microorganisms and cultivated crops. Plants are recurrently exposed to various biotic stresses in their innate environment. To endure such unfavorable circumstances, they have acquired complex mechanisms to react and acclimatize to biotic stresses. Plants are host to many contagious diseases caused by both living and nonliving agencies like viruses, bacteria, fungi, nematodes, and stress conditions. Comparatively small proportions of pathogens effectively assault the host plant which results in diseases. Plant disease is a physiological disorder or structural irregularity that is destructive to the plant and its products that suppress its monetary value. Plants are sessile and responsive organisms that encounter a variety of environmental strains. Throughout their life cycle, plants respond to diverse intimidations arising from their external environment. Fungal pathogens have a meticulous impact causing a major damage and yield losses in agriculture. Due to their sessile nature plants develops a broad range of strategies that cooperatively work in defense against biotic and abiotic stresses. Plant diseases can be viewed as obstinate roadblocks to the rapid advancement in agricultural yield. Interpreting the molecular basis of plant–pathogen interaction facilitates us to select plant resistance genes. The present work is focused on the consequence of fungus *Macrophomina phaseolina* on the growth of chickpea plants via monitoring various biochemical parameters which play important role in defense response upon infection with the pathogen. Comparative studies on the effect of *M. phaseolina* toxin in chickpea as

Preeti (✉) · D. Panwar
Chaudhary Charan Singh University, Meerut, Uttar Pradesh, India

P. Saini
Rakesh P.G College, Pilani, Rajasthan, India

J. K. Vats
Jai Prakash University, Chapra, Bihar, India

P. Mathur et al. (eds.), *Microbial Symbionts and Plant Health: Trends and Applications for Changing Climate*, Rhizosphere Biology,
https://doi.org/10.1007/978-981-99-0030-5_14

well as other host plants like *Brassica oleracea, Brassica juncea, Helianthus annus*, and *Vigna mungo* via detached leaf bioassay method are also planned.

Keywords Chickpea · Chitinase gene · Resistant

Abbreviations

Avr gene	Avirulence gene
F. solani	*Fusarium solani*
FAOSTAT	Food and agriculture organization corporate statistical database
H_2O_2	Hydrogen peroxide
HR	Hypersensitive response
IAR	Induced systemic resistance
ICARDA	International Center for Agricultural Research in the Dry Areas
ICRISAT	International Crops Research Institute for Semi-Arid Tropics
M.P	*Macrophomina phaseolina*
PAL	Phenylalanine
PAMPs	Pathogen-associated molecular patterns
PR proteins	Pathogenesis-related proteins
R gene	Resistant gene
ROS	Reactive oxygen species
SA	Salicylic acid
SAR	Systemic acquired resistance
UV	Ultraviolet

14.1 Introduction

Microorganisms and plants live in the vicinity of each other hence; they affect each other in both harmful and useful manners (Ortíz-Castro et al. 2009). Plants have developed a variety of mechanisms to counter the various stresses of both abiotic and biotic origins (Gull et al. 2019). Plants have the mechanism to identify stress, when sensed; they activated different signaling pathways to combat the stress. Different genes activated and started their expression and gene products are essentially required for defense. For successful control, plants must recognize the stress-causing organism and should trigger the best possible defense process to encounter the pathogen (Kaur et al. 2022). To develop a successful plant defense mechanism, the pathogen is the central part of the process. Plants have a high risk of diseases of biotic origin like bacterial, fungal, and viral pathogens and this results in keen loss to plant production (Hussain and Usman 2019). A number of diverse options are accessible for farmers to prevent and cure infections of crops. Some alternatives comprise development of disease-resistant varieties, biological check, rotation of crops, tillage, and use of insecticides and pesticides, etc. (Barzman et al. 2015). Substantial attempts have been achieved to devise environmentally friendly approaches for preventing and curing diseases of plants and to save mankind from health-related risks (El-Gamal Nadia et al. 2007; Food and Agriculture Organization of the United Nations Rome 2017).

Severe outbreaks of diseases are the consequences of interactions between pathogenic microorganisms and cultivated crops (Lamichhane and Venturi 2015). Plants are recurrently exposed to various biotic stresses in their innate environment (Iqbal et al. 2021). To endure in such unfavorable circumstances, they have acquired complex mechanisms to react and acclimatize to biotic stresses (Fujita et al. 2006; Isah 2019). Plants are host to many contagious diseases caused by both living and nonliving agencies like viruses, bacteria, fungi, nematodes, and stress conditions. Comparatively small proportions of pathogens effectively assault the host plant which results in diseases (Mahajan and Tuteja 2005; Guiñazú et al. 2012; Nazarov et al. 2020).

Plants are sessile and responsive organisms that encounter a variety of environmental stresses. During their life cycle, plants respond to diverse intimidations arising from their external environment. Fungal pathogens have a meticulous impact causing a major damage and yield losses in agriculture (Fujita et al. 2006; Shuping and Eloff 2017). Due to their sessile nature plants develops a broad range of strategies that cooperatively work in defense against biotic and abiotic stresses (Agrios 2005). Plant diseases can be viewed as obstinate roadblocks to the rapid advancement in agricultural yield. Interpreting the molecular basis of plant–pathogen interaction facilitates us to select plant resistance genes (Prabhukarthikeyan et al. 2020).

Fungal infection employs different pathways for colonizing host plants. Fungal pathogen either directly penetrates the cell wall or enters through the natural openings like stomata to invade the host (Zabka et al. 2008; Pawlowski and Hartman 2016). Like other plant interacting microbes, fungal pathogen secretes effectors proteins that interacts with host and invade plant defense mechanisms (Choudhary et al. 2012). Nonetheless, the plant developed diverse defense mechanisms that mark the effector proteins, which results in innate immunity that is often linked with localized cell death (Dodds et al. 2009; Selin et al. 2016). The molecular mechanisms involved in the perception, signaling, and reaction in plant–pathogen interactions are major elements in the study of accurate resistance or susceptibility (Peyraud et al. 2017). The effector proteins called elicitors are responsible for activating resistance in plants against fungal infection (Ito et al. 1997). Highlighted plant signaling pathways contribute to finding natural and man-made compounds known as elicitors that are accountable for immunity against pathogen infection (Gomez et al. 2004; AbdulMalik et al. 2020).

Plant opposes pathogen attacks by activating a broad variety of mechanisms, either constitutive or inducible, which adds to the immunity of plants against pathogens. Constitutive defenses are preexisting resistance mechanisms which control infection immediately after revelation of host to pathogen (Andersen et al. 2018). The inducible defenses are associated with quick and efficient commencement of cellular defense responses, which are induced only after contact with a pathogen (Canesi and Pruzzo 2016).

On detection of pathogen, plant activates hypersensitive response against infection. The recognition of pathogens normally occurs through the binding of avirulence (Avr) genes and R genes secreted by pathogen and plant, respectively (Balint-Kurti 2019). If the R gene of host and Avr gene of pathogen is muted or absent, no detection is possible and results in disease (Grover and Gowthaman 2003). As a consequence of putative adherence of these two host–pathogen proteins, a signaling cascade is triggered and results in a variety of plant defense mechanisms (Tyler 2002). The immune resistances are concerned with confinement of pathogens spreading out. Plants have several polymorphic R genes and their products which enable them to function as a receptor for Avr proteins secreted by several pathogens (Health 2000; Petit-Houdenot and Fudal 2017).

Plant produces many low molecular weight antibiotic phytoalexin in response to infection (Loser and Weltring 1998). Phytoalexins hold the development of bacterial and fungal pathogens in vivo and in vitro, and accumulation of these antibiotics throughout an infection can induce resistance against pathogen (Grayer and Kokubun 2001; Jeandet et al. 2014).

On the whole, structurally related phytoalexins are produced by plants of the same species and these phytoalexins enable a plant to introduce a toxic cocktail against encroaching pathogen (Ortega et al. 2005). These phytoalexins are produced in the cell adjacent to infected cell and delivered at the site of infection in the form of lipid vesicles, creating a noxious microenvironment and forbidding ecesic of pathogen. Accruement of phytoalexin is frequently linked with hypersensitive response, although phytoalexins are secreted by only living cells (Pusztahelyi et al. 2015).

Secondary metabolites like phenolic compounds having antimicrobial activity are mainly concerned with resistance against pathogens. Polyphenols are commonly found in plants and they have been reported to have numerous biological effects including antioxidant activity (Othman et al. 2019). Phenolic compounds function as phytoalexins, phytoanticipins, structural roadblocks, and regulators of pathogenicity or activators of plant defense genes against fungal bacterial and viral infection in plants (Tripathi et al. 2022). At the initial stage of defense, these phenolic compounds quickly accumulate at the site of infection, which limits or deliberate the pathogen growth and plays an essential part during resistance (Pratyusha 2022).

One of the important enzymes employed in the phenylpropanoid pathway that regulates stress by governing the synthesis of phenolic substances is Phenylalanine ammonia lyase (PAL; EC 4.3.1.5) (Wen et al. 2005; Zhang and Liu 2015). Lignin, phenolic substances, and phytoalexins add inflexibility and durability to cell walls and render barriers against pathogen (Conèeica et al. 2006; Lee et al. 2019). Involution of an antioxidant immune system in plants defends them against oxidative damage caused by ROS (reactive oxygen species) production or the cleaning of already produced reactive oxygen species (ROS) (Ali et al. 2018; Cavelcanti et al. 2007; Nita and Grzybowski 2016).

As, a result of environmental stimuli, the phenylpropanoid pathway signaling is a far-flung biochemical stress response (Chatterjee and Ghosh 2008). L-Phenylalanine is deaminated to *trans*-cinnamic acid by PAL (Barberan and Espin 2001), which is the initial step in the synthesis and accumulation of plants' secondary metabolites like lignin and phytoalexins during pathogen attack (Ozyigit et al. 2007; Ramírez-Gómez et al. 2019). PAL also induces the synthesis of salicylic acid (SA) and it can regulate the defense mechanism by activating the synthesis of phenolic compounds which is responsible for inducing systemic resistance in many plants (Wen et al. 2005; Lefevere et al. 2020).

Therefore, several antioxidant catalysts such as peroxidase (POX; EC 1.11.1.7) involve in ROS production during the pathogenesis. POX is one of the key oxidoreductive enzymes that take part in cell wall synthesis like lignifications and suberization of host plant cells, oxidation of phenols during infection (Chittoor et al. 1999; Quiroga et al. 2005; Minatel et al. 2017).

Plants have developed a variety of defense mechanisms against fungi which make up the assembly of secondary metabolites, proteins, and peptides bearing antifungal activity (Keller 2019). Plants exhibit an innate immune system by developing resistance like alteration of cell wall composition and oxidative burst of cells that cure infection and biosynthesis of chemicals like phytoalexin and pathogenesis-related proteins (Nürnberger and Scheel 2001; Wan et al. 2021). Such resistance can be sparked by exposing the plant to pathogenic microbes, or synthetically by salicylic acid and jasmonic acid (Wu and Bradford 2003; Betsuyaku et al. 2018).

This type of resistance is called as Induced Systemic Resistance (IAR) or Systemic Acquired Resistance (SAR). Among all induced defense mechanisms, accumulation of "Pathogenesis Related (PR) proteins" is the most vital immunity of plants against pathogenic infection (Adrienne and Barbara 2006; Boccardo et al. 2019). For instance, degradation of chitin by chitinase is one of the key components of many pathogenic fungi cell walls (Bartnick 1968). These enzymes restrict fungal growth by inducing dissolution of hyphal tips (Mauch et al. 1988). The transfer of chitinase genes to other plants may reduce or hinder the disease symptoms (Grover and Gowthaman 2003).

For the development of resistant cultivars, quick methods are the need of the hour for the successful evaluation of host resistance and variability of pathogenic micro-organisms (Chen et al. 2019). Alternative methods like detached leaf assay are often employed in place of whole plant evaluation and have several advantages also like precise quantification of disease development along with determination of pathogen reproduction (Aregbesola et al. 2020).

Chickpea (*Cicer arietinum* L.) is the second most important legume crop world-wide (Madurapperumage et al. 2021) (Fig. 14.1). Chickpea have high protein content, it is utilized as a vegetable and dry bean (Wallace et al. 2016). As, a supplement of high proteins cereals, legume crops are used as one of the most beneficial answers to protein malnutrition in developing countries (Maphosa and Jideani 2017). Fungus is major mutualistic symbionts and parasite of plants; hence, their analysis is an essential component of plant pathology and plant science (Vågsholm et al. 2020; Choudhary et al. 2012). The worldwide production of chickpea is approximately threefold of lentils and consumption is second most to dry beans among pluses (FAOSTAT 2017; ICARDA 1993). Chickpea is a

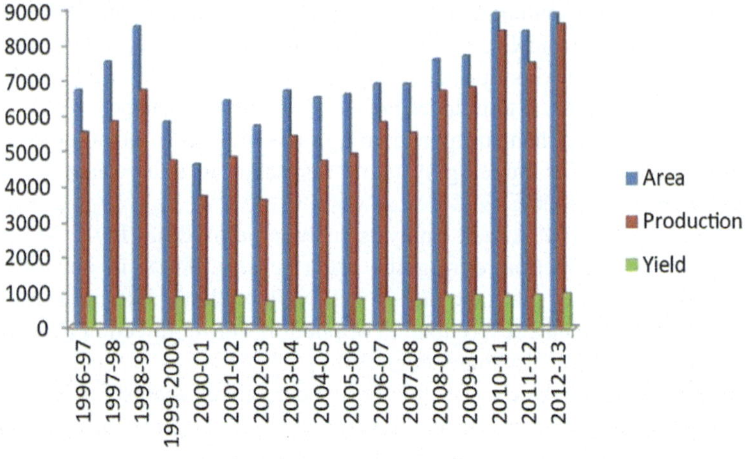

Fig. 14.1 All India area, production, and yield of chana

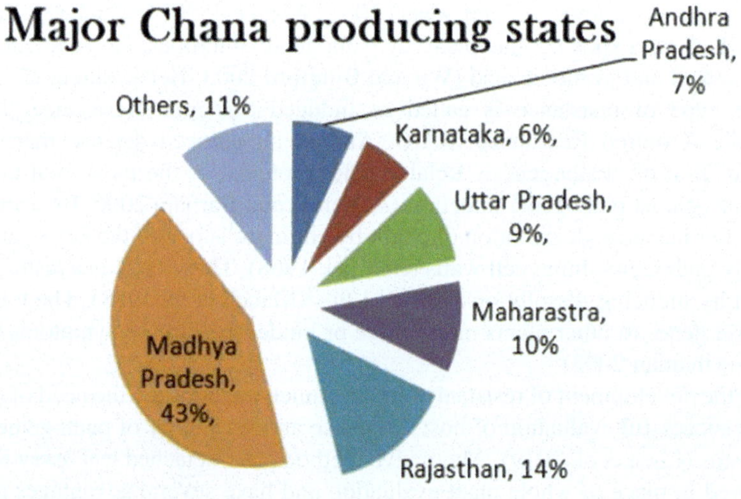

Fig. 14.2 The state-wise production of chickpea

self-pollinating crop and rarely cross-pollinated; only 0–1% is reported (ICRISAT 2014, 2019). The state-wise production of chickpeas is mentioned in Fig. 14.2.

Pathogenic fungi secrete phytotoxins causing damage to plant tissue having symptoms including necrosis, chlorosis, wilting, water soaking, and eventually death of plants. In various host–pathogen interactions phytotoxins have constituted as virulence factors which could be used for screening out resistant cells (Pontes et al. 2020). This soilborne fungal infection is among the most crucial biotic factors confining the yield of chickpea, leading to severe monetary loss (Tarafdar et al. 2018).

Macrophomina phaseolina (Tassi) goid a soil inhabiting fungal phytopathogen, causes charcoal rot in over 500 plant species including chickpea causing a major loss in its yield (Ghosh et al. 2016). It continues to be a disputing task in managing the soilborne disease in natural environment. It is dispersed universally and is predominant in arid, tropical, and subtropical regions, especially areas having very high temperatures and low rainfall. It is a key phytopathogenic fungus infecting a huge number of plant species and existing for up to 15 years in the soil as a saprophyte (Kaur et al. 2022). It causes severe yield losses in epidemic years (Degani et al. 2022). Climate conditions of the desert support the growth and survival of *M. phaseolina* hence; disease is quite common in Rajasthan and leads to heavy losses of this crop (Marquez et al. 2021).

M. phaseolina is considered as one of the soilborne fungi that infect large numbers of host plants through roots, stems, and crown regions, near the soil surface (Sinha et al. 2016). It causes several plant diseases as charcoal rot. *M. phaseolina* is considered as one of the soilborne fungi that infect a large number of host plants through roots, stems, and crown regions near the soil surface (Arora and Dhurwe 2019). It causes several plant diseases such as charcoal rot, leaf blight, and damping off. So, the fungus caused great economic losses. Some physiological factors that affect the growth and pathogenicity of *M. phaseolina* were indicated. Root infections due to soilborne pathogens are often primary restraints in the production of legume crops (Ijaz et al. 2012; Wolfgang et al. 2019). Screening of germplasm to obtain resistance or tolerant cultivar may be vital and can be used as a basis of resistance indicator in plant breeding programmes (Zhou et al. 2018).

M. phaseolina releases a numeral of phytotoxins, such as asperlin, isoasperlin, phaseolinic acid, phaseolinone, phomalactone, and phomenon (Mahato et al. 1987; Nallasivam and Fernandes 2017). Phaseolinone and phaseolinic acid seem most significant of others that cause symptoms of disease in plants comparable to those caused by the disease-causing microbes (Mathur 1968; Dhingra and Sinclair 1985; Siddiqui et al. 1979; Bhattacharya et al. 1992). The release of phaseolinone phytotoxin during the infection caused by *M. phaseolina* and defended by the crucial role of cell wall degrading enzymes in the disease unveiling and procession (Chan and Sackston 1973).

Many new emerging plant diseases are surfaced due to the increasing temperature globally at an alarming rate (Velásquez et al. 2018; Chobe et al. 2020; Palit et al. 2020) and one of them is root rot disease of chickpea and the etiological agent is *Macrophomina phaseolina*, a soilborne necrotrophic fungus (Marquez et al. 2021; Dell'Olmo et al. 2022). Poor apical growth, yellowish or dull green leaf, reduced vigor, and twig dieback are the symptoms shown by the plant infected with dry root rot disease (Chobe et al. 2019). Further, leaves suddenly start wilting and dried on the trees, if the extent of disease is severe (Garrett et al. 2016). Recent studies identified the southern and central states of India with more than 35% disease incidence as dry root rot hotspots (Abbas et al. 2020). There are many assumptions and one assumption suggests that climate change is the principal cause. One of the reasons identified for disease spread is rising temperature (30–35 °C) and dry soil conditions (less than 60%) which are suitable for disease progression (Ghosh et al.

2017; Adhikary et al. 2019). Scientific reports suggested that the pathogen can tolerate harsh environmental conditions such as extremities pH, temperature, and drought (Almomani et al. 2013; Burdon and Zhan 2020). Chickpea is more infection prone during flowering and podding season coupled with drought and higher temperature coincidently (Devi et al. 2022). The focus of scientists is to develop more disease-resistant varieties with better management methodology (Hatmi et al. 2015).

Chickpea Plants Get Diseases in These Conditions
They are in search of more disease favorable factors and strategies to counter the disease at the molecular level. Recently, a few genes were identified from chickpea genome which encodes for defense enzymes like chitinase and endochitinase (Iriti and Faoro 2009). Both enzymes are suitable tools for the management of dry root rot disease (Veliz et al. 2017). As stated earlier, climate change and global warming tilt the balance of hosts and pathogens toward pathogens seriously (Karthik et al. 2021).

The epidemiology of plant diseases is depended upon several environmental and climatic factors (Sharma et al. 2019), as life of pathogens are profoundly affected by the amount of greenhouse gases, humidity, rainfall, soil moisture and temperature (Sharma et al. 2015). Hence, to develop an insight of knowledge of host–pathogen interaction, studies should be taken in multivariant scenarios at cellular, physiological, and molecular levels and should also include both abiotic and biotic stress factors (Singh et al. 2021).

In many economically important plant species, the biochemical basis of disease resistance remains unexplored as yet. Such studies are not available in many common and economically important crops of Rajasthan including Chickpeas. Further, there are no reports as yet on the biochemical basis of disease resistance of the rot disease caused by *M. phaseolina* in *Cicer arietinum* plants.

The present work is focused on the consequence of fungus *M. phaseolina* on the growth of chickpea plants via monitoring various biochemical parameters which play important role in defense response upon infection with the pathogen (Ramegowda and Senthil-Kumar 2015). Comparative studies on the effect of *M. phaseolina* toxin in chickpea as well as other host plants like *Brassica oleracea*, *Brassica juncea*, *Helianthus annus*, and *Vigna mungovia* detached leaf bioassay method are also planned.

14.2 Historical Backgrounds

Physiology of disease resistance is one of the most attractive areas of research. Rather, they depend upon the natural defense response of individual cells and receive the systemic signals from the zone of infection (Amil-Ruiz et al. 2011; Nishad et al. 2020). Pathogenic microorganisms employ different life strategies that can be generally classified into obliterate the host and feed upon the substances (necrotrophs) and those that utilize a surviving host to finish their life cycle (biotrophs) (Moënne-Loccoz et al. 2015). Pathogenic necrosis is frequently caused by accumulation of poisonous substances (Abu Qamar et al. 2006). Pathogenic fungus multiply in apoplast (intercellular spaces) either inserting through stomata or hydathodes or lesions (Melotto et al. 2008; Bos et al. 2010) whereas fungi directly enter epidermal cells or enlarge hyphae on top among plant cells (Craig et al. 2009; Ojha et al. 2008; Anand et al. 2007; MikuliPetkovšek et al. 2009; Amil-Ruiz et al. 2011).

Among the antimicrobial substances produced by the plants, phenolics are the most important one. Synthesis of PAL is increased in many plant species as they detect any sort of stress such as herbicide intervention, injury, nutrient deficiency, radiation, and microbial attack (Donkor et al. 2019).

Induced disease resistance can be defined as the progression of active resistance dependent on the host plant's chemical or physical barriers started by abiotic or biotic agents, it sensitizes the host to act in response rapidly after detecting infection in the form of accumulation of phytoalexin, phenols, lignifications, pathogenesis-related (PR) proteins and elicitation of enzymes like chitinase, peroxidase, and polyphenol oxidase (Aal et al. 2019). Induced disease resistance in plants have positive effects on plant growth and yield all of which are important characteristics for disease management strategies (Lawton et al. 1995). Several publish reports by many investigators that resistance could be systemically induced by the interaction of both biotic and abiotic stress like chemical compounds with plants. Many chemical compounds are known to induce resistance by activating defense networks in plants. Several investigators studied the effectiveness of these chemical inducers on charcoal rot disease incidence caused by *M. phaseolina* (Mohammad et al. 2007; El-Fiki et al. 2004). Inducer of systemic resistance sensitizes the plant to respond rapidly after infection (Glazebrook et al. 1997).

14.2.1 Host–Pathogen Interaction

There are two classes of plant resistance, i.e., host specific and nonhost specific. An interaction between a specific genotype of the host and the pathogen is categorized into host-specific resistance, also called pathogen race-specific resistance. The resistance exhibited by the whole plant species toward a specific pathogen or parasite is categorized into nonhost-specific resistance. Resistance of plant species against common pathogens is included in nonhost-specific resistance (Health 2000). However, the biochemical changes during the course of infection are almost similar in both host and nonhost-resistant plants (Somssica and Hahlbrock 1998).

14.2.1.1 Systemic Acquired Resistance

Host plants have come through from previous infection by plant pathogens then it can automatically protect against further pathogen infections. It seems that disease-causing pathogens protect the plant from further infections against homologous pathogens, although a plant may or may not carry the resistance-specific gene. The presence of these resistance-specific genes forces the plant to protect itself from subsequent microbial infection. This kind of reaction is termed Systemic Acquired Resistance (SAR). The establishment of SAR in a plant is linked with many defense reactions, like pathogenesis-related proteins and many defense-associated enzymes (Neuenschwander et al. 1996). SAR refers to a distinct signal transduction pathway that plays an important role in the ability of plants to defend themselves against the pathogen (Ryals et al. 1996; Gozzo 2003). Salicylic acid (SA) is widely distributed in plants (Raskin 1990) and plays a key role in SAR. In addition, certain natural and synthetic compounds can trigger similar plant responses (Kessmann et al. 1994; Metraux et al. 1990). The phenylalanine pathway is the main route of SA biosynthesis (Ogawa et al. 2006).

14.2.1.2 Salicylic Acid

Salicylic acid, present naturally in very low quantities in plants, is an endogenous growth regulator (Dempsey and Klessig 2017). It is the precursor of aspirin normally found in plants. It reacts with two proteins of the cell membrane. One of the proteins has catalase activity and this catalase activity is stopped when reacts with salicylic acid, resulting in the accumulation of hydrogen peroxide at the binding site locally (Ribnicky et al. 1998) (Fig. 14.3). Such form of accumulation of hydrogen peroxide brings about many reactions in cells that finally augment their resistance to pathogens (Durrant and Dong 2004). Another protein that has a high affinity to salicylic acid directly starts the expression of the gene for resistance (Zhang et al. 2007; Heidel et al. 2004).

14.2.2 Elicitors and Their Functions

Initially, the elicitor molecules only comprise phytoalexin but at present elicitors are a group of molecules generally used against different pathogen infections (Nürnberger 1999). Elicitors are divided into different classes on the basis of physical or chemical properties, types of biotic and abiotic stresses, and on the basis of their evolution and molecular organization (Table 14.1). Elicitors are classified into two main classes, "general elicitors" and "race specific elicitors" (Bektas and Eulgem 2015). General elicitors are involved in the resistant mechanism of specific and nonspecific host plants, while race-specific elicitors are able to activate pathogen resistance in particular host cultivars (Radman et al. 2003).

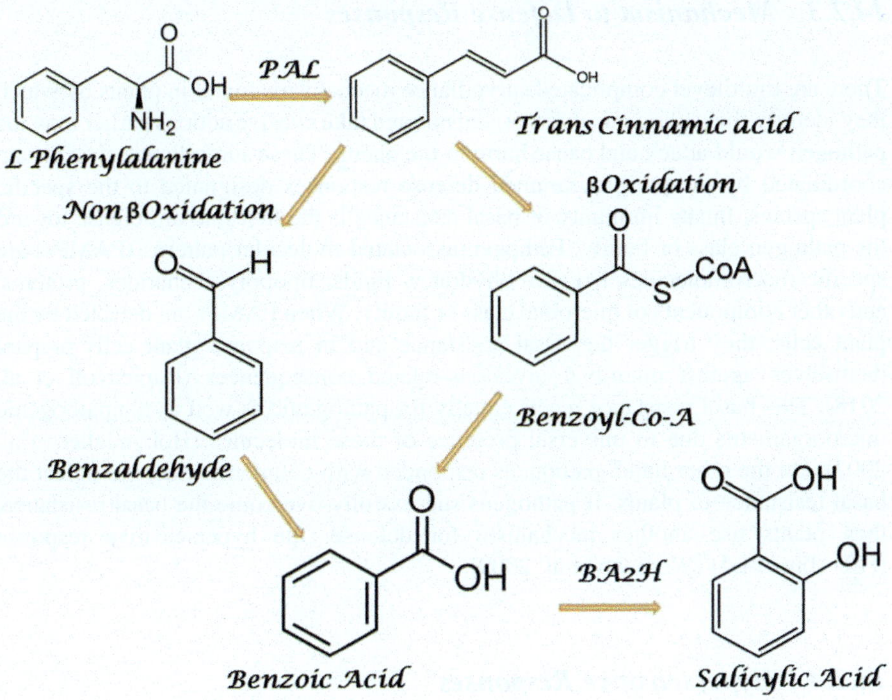

Fig. 14.3 Proposed pathway of SA biosynthesis in Tobacco

Table 14.1 List of various kinds of plant elicitors

Type of elicitors and their examples			
	Chemical elicitors		
Physical elicitors	Abiotic elicitors	Biotic elicitors	
Injury	Metal ions	(1) Complex composition Yeast cell wall Mycelia cell wall, and fungal spores	(2) Defined composition (2.1) Carbohydrates Oligosaccharides: Mannuronate, guluronate, mannan, and galacturonides Polysaccharides: Alginate, pectin, and chitosan (2.2) Proteins Peptides: Glutathione Proteins: Cellulase and oligandrin (2.3) Lipids Lipopolysaccharides (2.4) Glycoproteins Not characterized (2.5) Volatiles C_6–C_{10} compounds

14.2.3 Mechanism to Defense Responses

There are multilevel complicated surveillance mechanisms found in plants by which they identify the pathogenic microorganisms and take evasive actions earlier than the pathogen would attack and cause harm to the plants. These surveillance systems are coordinated by the pre-programmed defense responses designated to the specific plant species. Innate immunity or basal resistance is the first line of defense toward the pathogen class or family. Pathogen-associated molecular patterns (PAMPs) are specific macromolecules like carbohydrates, lipids, lipopolysaccharides, proteins, and other components of microbial class or family. When PAMPs are detected by the plant cells, they trigger the basal resistance and in response plant cells prepare themselves against microbial invasions-related consequences (Rapicavoli et al. 2018). This basal resistance reacts equally for pathogenic as well as nonpathogenic microorganisms due to universal presence of these molecules (Holzmueller et al. 2007). On the other hand, pathogens responded with countermeasures to defeat the basal resistance of plants. If pathogens successfully overcome the basal resistance, then plants use another mechanism for defense: the hypersensitive response (HR) (Fig. 14.4) (Wenjuan et al. 2009).

14.2.4 Hypersensitive Responses

To activate a hypersensitive response, physical or physiological contact is necessary between pathogen and host (Health 2000). The phenomenon of HR was described in wheat plants infected with rust fungi by Stakman 1915. Hypersensitivity is a

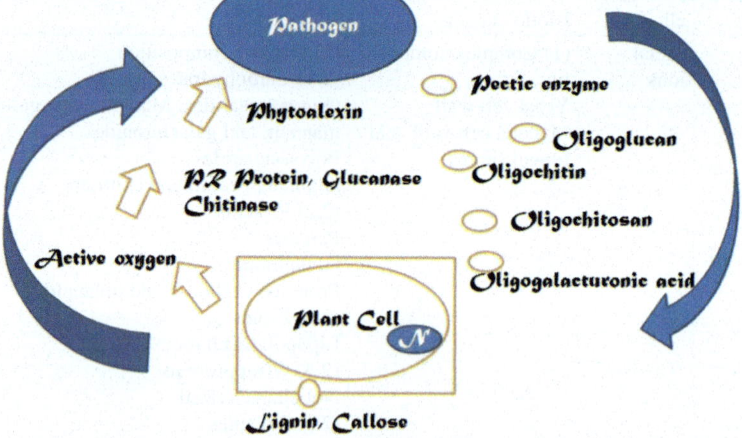

Fig. 14.4 Biosynthesis from shikimate and phenylalanine derivatives of the phenylpropanoid pathway

hyperimmune response shown by the host activated by the pathogen after entering the host cell (Kortekamp and zyprian 2003), and this sometimes causes death of host cell and confined the pathogen at the site. Researchers reported that apoptosis in animals is very much similar to HR in plants (Morel and Dangl 1997). Now, the above definition is made broader by the inclusion of defense gene expression (Iakimova et al. 2005). A plant is able to survive if it could identify the possible threat of different pathogens and respond appropriately through proper defense mechanisms (Nishad et al. 2020), and this event is the result of R gene of host plant and Avr gene of pathogenic microorganisms (Flor 1971). This Avr-R gene binding starts the appropriate host plant defense and results in the death of host cell also termed as HR (Levine et al. 1996). The product of R gene either works as a receptor for the Avr gene products (Yang et al. 1996) or recognizes the Avr factor indirectly through a co-receptor as supported by accumulating evidence (Dixon et al. 1994). Although resistance responses are often associated with HR, in some cases, they can occur without or with very little cell death (Bendahmane et al. 1999; Ehrenfield et al. 2005).

14.2.5 Phytoalexin

Plants produces and accumulate some antimicrobial compounds in plants when attacked by bacterial or fungal pathogen called phytoalexin (Wilkens et al. 2010). Phytoalexins are also produced in response to non-biological factors like experimental exposure to the plant with ions of heavy metals (Cu or Hg) and with UV light of a short wavelength (Grayer and Kokubun 2001). Phytoalexins are well documented in the field of plant defense. Much research has been conducted on the elicitation process and specific elicitors have been discovered (Dakora and Phillips 1996). Phytoalexins are thought to be synthesized in a cell adjacent to the infection site, in response to a signal produced either by the invading pathogen or by infected host cells (N'Guyen et al. 2021). The infected cell becomes a toxic microenvironment for the invading pathogen (Wharton and Nicholson 2000).

14.2.6 Phenylalanine Ammonia Lyase

PAL (E.C.4.3.1.5) is the enzyme that catalyze core reactions in phenylpropanoid metabolism leading to the production of plant phenolics including those that are produced under stress (Kliebenstein 2004; Jones 1984). PAL has been generally recognized as a marker of environmental stress in different plant species (MacDonald and D'Cunha 2007). It contributes to deamination of L-phenylalanine to cinnamic acid. Since PAL activity is known to affect the accumulation of polyphenols any stress such as fungal pathogen inoculation or mechanical wall injury will alter its level and consequently its products. Thus, activation of defense

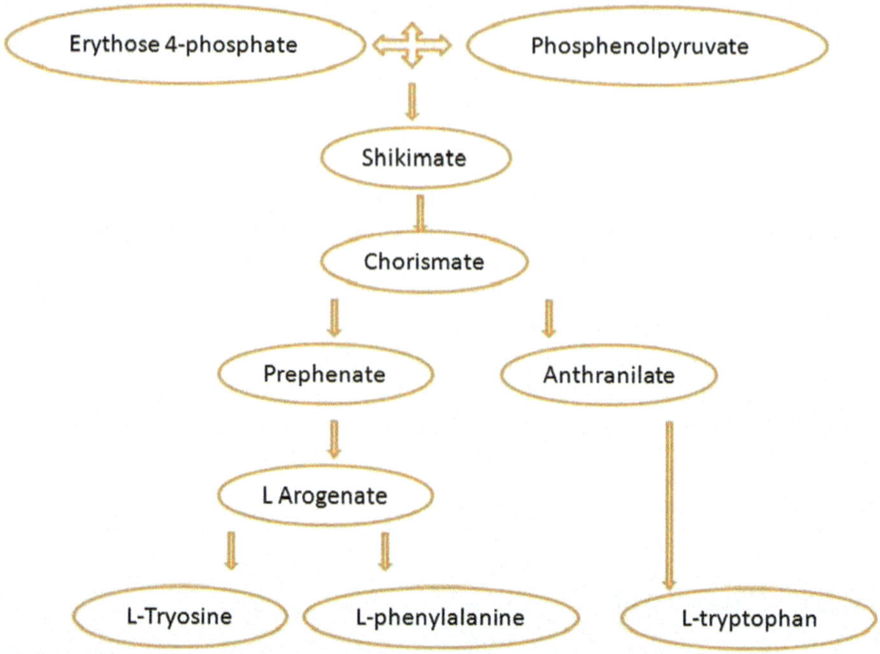

Fig. 14.5 Innate immune response of plants in plant–pathogen interaction

response pathway can be detected by measuring the PAL activity after pathogen inoculation (Lavanya et al. 2022) (Fig. 14.5).

14.2.7 Oxidative Burst

The oxidative burst is generally defined as a rapid production of high levels of ROS in response to external stimuli (Dangl and Jones 2001; Mittler et al. 2004). Hydroxyl radicals, hydrogen peroxide, and superoxide anions are the most important members of ROS (Apel and Hirt 2004). All of these species are extremely reactive and toxic which can lead to the oxidative demolition of host cells (Gechev et al. 2006). The first evidence of production of ROS came into light in 1983 when potato tuber disks were inoculated with an incompatible race of *Phytophthora infestans*, and superoxide radicals were generated and then transformed into hydrogen peroxide (Doke 1983). Similar findings were reported by the interaction of avirulent strain of *Pseudomonas syringae* strain DC300 with *Arabidopsis thaliana* (Alvarez et al. 1998). These investigations confirm the fact that the production of ROS in plants is thought to play an important signaling role in the activation and establishment of plant defense (Huang et al. 2019; Levine et al. 1994).

14.2.8 *Peroxidase*

Peroxidase is a member of the oxidoreductases (E.C.1.11.1.7) family, catalyses the oxidation of various inorganic and organic substrates in the presence of H_2O_2 (Koksal and Gulcin 2008). They were also reported to mediate plant defense against biological stresses (Prasannath 2017). Peroxidases worked both in favor of plants as well as against them also, some fractions of peroxidases promote growth by their indole acetic acid and oxidase activity (Takahama and Yoshitama 1998), while on the other hand, certain fractions cause cell wall rigidification and tissue lignifications resulted in retardation of growth (Passardi et al. 2005). Peroxidases studied in date palms considering their relationships with cultivar resistance to Bayoud disease (Gay and Tuzun 2000). For example, a positive correlation was established between decline in occurrence of *Fusarium* wilt in banana and induction of defense-related enzymes peroxidase and polyphenol oxidase (Sauerborn et al. 2002).

14.2.9 *Polyphenols*

Phenolic compounds are a large class of plant secondary metabolites distributed widely in the plant kingdom. Plants synthesize a variety of aromatic substances and a majority of them are either phenols or their oxygen-substituted derivatives (Othman et al. 2019; Ozyigit et al. 2007). Most of them are produced as secondary metabolites as simple phenols and phenolic acids, alkaloids, essential oils, flavones, flavonoids, flavonols, quinones, tannins, and terpenoids (Tungmunnithum et al. 2018). All of these compounds contain one or more than one benzene rings with attached hydroxyl groups that may be variously elaborated with methyl, methoxyl, amino, or glycosyl groups (Kumar and Goel 2019). Various researchers reported that a substantial increase in the production of phenolics is observed in plants in response to fungal attacks (Li and Steffens 2002).

14.2.10 *Toxins of* M. phaseolina

M. phaseolina was grown in vitro and it was first reported that the fungus produces phytotoxic substances and the symptoms were alike of that pathogen produces in plants (Mathur 1968). A variety of phytotoxin specifically phaseolinone, phomenon, phaseolinic acid, phomalactone, isoasperlin and asperlin are reported to be produced by *M. phaseolina* (Dhar et al. 1982; Mahato et al. 1987; Bhattacharya et al. 1992). Several plant pathogenic fungi produce phytotoxins related to disease progression, which interact with the cell machinery. Later one toxin was named phaseolinone (Siddiqui et al. 1979) (Fig. 14.6), later this toxin become the most studied toxin which persuades disease symptoms in host plant (Junior et al. 2021). It is a nonspecific exotoxin that is highly stable and nonbiodegradable even at high temperatures. The toxin affects germination of seeds, growth of seedlings, and tissue culture necrosis during regeneration and callusing (Abbas et al. 2018). It also inhibits the germination of seeds and causes wilting of seedlings (Vishal and Kumar 2018).

The isolation and structure of another toxin, phaseolinic acid were reported from the culture of filtrate of *M. phaseolina* (Mahato et al. 1987). This toxin did not affect germination of seeds or seedling growth but was able to induce nonspecific leaf necrosis in several plants (Rather et al. 2020).

14.2.11 Pathogenic-Related Protein

Pathogenesis-related proteins were first described by Van Loon, who observed the accumulation of various novel proteins in leaves of tobacco by tobacco mosaic virus (van Loon and van Strien 1999). Plant and pathogen interaction results in infection or plant avert the infection (Abdulkhair and Alghuthaymi 2016). Like animals, plants do not have circulating phagocytes, however, to prevent microbial invasion they have a complex thick cell wall (Tyagi et al. 2014). Plants develop various defense mechanisms of innate pathogen-specific resistance such as varying composition of cell wall, synthesis of phytoalexins, oxidative burst, and pathogenesis-related proteins (Andersen et al. 2018). All of these responses are produced by plants equally whether the microorganism is virulent, nonvirulent, or nonpathogenic and can be induced artificially by low molecular weight chemical compounds and by using volatile molecules like jasmonate and salicylic acid (Wu and Bradford 2003), 2,6-Dichloro-isonicotinic acid or benzo (1,2,3) thiadiazole-7-carbothioic acid Smethyl ester (BTH) (Vallad and Goodman 2003). Such a type of response by plants is termed as Induced Systemic Resistance (IAR) or Systemic Acquired Resistance (SAR) (Oostendorp et al. 2001).

Among the responses exhibited by the host plant, production of "Pathogenesis Related (PR) proteins" is considered most important as these proteins provide systemic resistance to whole plant upon microbial invasion (Adrienne and Barbara 2006). These PR proteins are synthesized by a number of organisms throughout all kingdoms which are low molecular weight, small, cysteine rich, and basic in nature. PR proteins vary in their mechanism of action, species specificity, and primary structure (Leiter et al. 2005).

Fig. 14.6 Chemical structure of phaseolinone

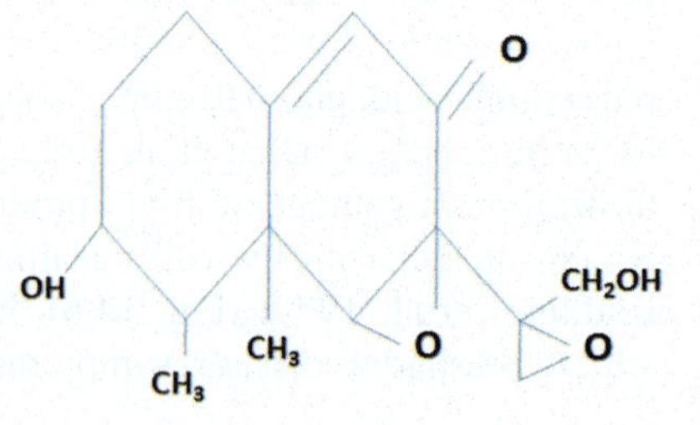

Table 14.2 Main properties of classified families of PR proteins

Family	Type member	Typical size (kDa)	Properties	Proposed microbial target
PR-1	Tobacco PR-1a	15	Antifungal	Unknown
PR-2	Tobacco PR-2	30	β-1,3Glucanase	β-1,3Glucanase
PR-3	Tobacco P,Q	25–30	Chitinase (class I, II, IV, V, VI)	Chitin
PR-4	Tobacco "R"	15–20	Chitinase class I, II	Chitin
PR-5	Tobacco S	25	Thaumatin like	Membrane
PR-6	Tomato Inhibitor 1	8	Proteinase inhibitors	-a
PR-7	Tomato P69	75	Endoproteinase	-a
PR-8	Cucumber chitinase	28	Chitinase class III	Chitin
PR-9	Tobacco "lignin forming peroxidase"	35	Peroxidase	-a
PR-10	Parsley "PR1"	17	Ribonuclease like	-a
PR-11	Tobacco "class V Chitinase"	40	Chitinase class I	Chitin
PR-12	Radish Rs-AFP3	5	Defensin	Membrane
PR-13	Arabidopsis THI2.1	5	Thionin	Membrane
PR-14	Barley L TP4	9	Lipid transfer protein	Membrane
PR-15	Barley OxOa (germin)	20	Oxalate oxidase	-a
PR-16	Barley OxOLP	20	Oxalate oxidase like	-a
PR-17	Tobacco PRp27	27	Unknown	-a

At present, 17 families have been assigned to PR proteins as per their structure, size, and properties are given in Table 14.2 (Van Loon et al. 2006).

14.2.12 Chitinase and Function in the Plant

Fungal cell wall is made by a polymer chitin (Garcia-Rubio et al. 2020) and enzyme chitinases (E.C. 3.2.1.14) catalyze the cleavage of bond between C1 and C4 of two successive N-acetyl-D-glucosamine units of chitin (Zeyen et al. 2002; Krolicka et al. 2018). Chitinases are produced by animals, plants, bacteria, and fungi. Plant produces endo-chitinases and works on fungal hyphae after invasion and hamper growth of pathogenic fungi (Sharma et al. 2016). Previous investigators investigated in vitro growth inhibition of fungi with plant chitinases (Mauch et al. 1988; Pusztahelyi 2018). The first report of chitinase induction was reported against *Fusarium solani* by pea plant. The protein extracted from plant was tested against 18 fungi and was found effective toward 15 fungi (Bettini et al. 1998).

One purified chitinase may inhibit only one or the same cell wall compositional fungal pathogen while a consortium of chitinases, β-1, 3-glucanase, and PR protein inhibit most of the tested fungi and establish the fact that synergistic activity is must for fungal growth inhibition (Veliz et al. 2017). A number of studies verified these results in tobacco, grapes, and chickpea (Giri et al. 1998). These investigations provide insight that a specific chitinase enzyme is produced by the host against a specified pathogen and this may not affect other pathogens. Tobacco plants produce the enzyme chitinase class I in response to *F. solani* (Jach et al. 1995). Several reports indicated that expression of chitinases in response to fungal pathogens is more powerful in resistant varieties when compared with susceptible varieties of tomato wheat and sugar beet (Lawrence et al. 2000; Manghwar et al. 2021).

Plants produce elicitor compounds which result in disease resistance through PR gene expression quickly and affect the cell wall synthesis of germinating fungal spores (Lavanya et al. 2022). On the basis of the primary structure of chitinase, plant chitinases have been divided into seven classes, from class I to class VII (González et al. 2015). Class III chitinases are distinctive in structure and have no similarity with other classes of plant chitinases (Gomez et al. 2004). These chitinases belong to the PR-8 family and family 18 of glycosyl hydrolases (Chen et al. 2019, 2020).

14.3 Conclusion

Chickpea (*Cicer arietinum*) is a self-pollinated, cool season, and rabi legume. India is the largest producer of chickpea which is the second most essential cool seasonal crop in the world. The major production areas of chickpeas are located in Maharashtra, Uttar Pradesh, Madhya Pradesh, Punjab, Rajasthan, and Haryana. Chickpea in semiarid regions suffers from charcoal root rot diseases induced by *Macrophomina phaseolina* (Tassi) Goid. *Macrophomina phaseolina* is a soil inhabiting root fungus so it is difficult to remove the pathogen from spreading the disease. This pathogen-induced disease is a global problem worldwide and is predominant in the area having scanty rainfall and high temperature in regions like arid, subtropical, and tropical areas. Hence, the present work has attempted to screen varieties of chickpeas. A positive correlation was found between peroxidase activity and resistant response of the varieties. The screening of germplasm to obtain resistant varieties is important because these can be used in plant breeding programmes, breeding lines, and mapping population for *M. phaseolina.*

The estimation of phenylalanine ammonia lyase, phenols, salicylic acid, and total protein activity of both control and inoculated plants were determined for its correlation with resistant response.

The observed value of polyphenols content reveals that when chickpea plant has been infected with pathogen by foliar suspension of mycelium, its phenolics concentration rises as consequence of resistance to disease. The polyphenol content has been observed to increase in inoculated plants in comparison to the control one.

PAL is the main enzyme in inducing the synthesis of salicylic acid which induces systemic acquired resistance in the host. In the present study, the resistant genotype inoculated with fungus showed increased PAL activity. The susceptible genotype showed decreased PAL activity.

Throughout pathogen infection, the secretion of salicylic acid soon increases in host with the increase in PAL catalysis, in order to combat increasing infection in plants by activating SAR.

Detached leaf assay is a rapid and economical way to do pathogenicity tests and the procedure has been used to study many host–pathogen interaction systems in dicots. The detached leaf assay is a rapid and reliable method to evaluate germplasm, breeding lines, and mapping population for *M. phaseolina* resistance. The method can be also used to monitor symptom development and infection parameters over time.

To expand insight into the molecular basis of resistance, defense-related genes were analyzed for their expression levels in response to pathogen attack. The gene selected for analysis was chickpea chitinase. It was observed that expression levels of chitinase genes were significantly higher in the resistant variety as compared to the susceptible one. A good correlation was observed between chitinase activity and fungal pathogen resistance.

From the results presented here, it can be inferred that the inoculation of chickpea with *M. phaseolina* results in activation of host defense system. Further investigations about factors that impart in induction and regulation of this gene expression may help to provide a better comprehension of plant–microbe interactions and mechanisms of plant responses to infection. Moreover, the results of these studies can be useful for constructing transgenic-resistant plants. Overall findings may suggest the mechanism of induced resistance in chickpea plants and together with the study of toxins and mechanism of gene regulation will open new and fascinating ways for plant genetic engineering for crop improvement.

References

Aal AEA, Abd-El-Kader DA, Khalifa MMA, Ali MAS (2019) Management of peanut cercospora leaf spot using resistant cultivars and inducer resistance chemicals. J Agric Res 46(3):665–684

Abbas MS, El-Shabrawi HM, Soliman AS, Selim AM (2018) Optimization of germination, callus induction, and cell suspension culture of African locust beans *Parkia biglobosa* (Jacq.). Benth J Genet Eng Biotechnol 16(1):191–201. https://doi.org/10.1016/j.jgeb.2017.10.012

Abbas HK, Bellaloui N, Butler AM, Nelson JL, Abou-Karam M, Shier WT (2020) Phytotoxic responses of soybean (*Glycine max* L.) to Botryodiplodin, a toxin produced by the charcoal rot disease fungus. Toxins (Basel) 12:25. https://doi.org/10.3390/toxins12010025

Abdulkhair WM, Alghuthaymi MA (2016) Plant pathogens. In: Plant growth. IntechOpen. https://doi.org/10.5772/65325

AbdulMalik NA, Kumar IS, Nadarajah K (2020) Elicitor and receptor molecules: orchestrators of plant defense and immunity. Int J Mol Sci 21(3):963. https://doi.org/10.3390/ijms21030963

Abu Qamar S, Chen X, Dhawan R, Bluhm B, Sameron LS, Dietrich RA, Mengiste T (2006) Expression profiling and mutant analysis reveals complex regulatory networks involved in *Arabidopsis* response to botrytis infection. Plant J 48:28–44

Adhikary NK, Chowdhury MR, Begum T, Mallick R (2019) Integrated management of stem and root rot of sesame (*Sesamum indicum* L.) caused by (Tassi) Goid. Int J Curr Microbiol Appl Sci 8:804–808. https://doi.org/10.20546/ijcmas.2019.804.089

Adrienne CS, Barbara JH (2006) Parallels in fungal pathogenesis on plant and animal hosts. Eukaryot Cell 5:1941–1949

Agrios GN (2005) (2005) Plant disease caused by nematodes. In: Agrios GN (ed) Plant pathology, 5th edn. Elsevier Academic, London, p 608

Ali M, Cheng Z, Ahmad H, Hayat S (2018) Reactive oxygen species (ROS) as defenses against a broad range of plant fungal infections and case study on ROS employed by crops against *Verticillium dahliae* wilts. J Plant Interact 13(1):353–363. https://doi.org/10.1080/17429145.2018.1484188

Almomani F, Alhawatema M, Hameed K (2013) Detection, identification and morphological characteristic of: the charcoal rot disease pathogens isolated from infected plants in Northern Jordan. Arch Phytopathol Plant Prot 46:1005–1014. https://doi.org/10.1080/03235408.2012.756174

Alvarez ME, Pennell RI, Meijer PJ, Ishikawa A, Dixon RA, Lamb C (1998) Reactive oxygen intermediates mediate a systemic signal network in establishment of plant immunity. Cell 92:773–784

Amil-Ruiz F, Blanco-Portales R, Muñoz-Blanco J, Caballero JL (2011) The strawberry plant defense mechanism: a molecular review. Plant Cell Physiol 52:1873–1903

Anand T, Thiruvergadam R, Gandhi K, Velapan P, Ramasamy S (2007) Chemically and biologically mediated systemic resistance in cucumber (*Cucumis sativus* L.) against *Pseudoperonospora cubensis* and *Erysiphe cichoracearum*. Phytopathol Meditrr 46:259–271

Andersen EJ, Ali S, Byamukama E, Yen Y, Nepal MP (2018) Disease resistance mechanisms in plants. Genes 9(7):339. https://doi.org/10.3390/genes9070339

Apel K, Hirt H (2004) Reactive oxygen species: metabolism, oxidative stress, and signal transduction. Annu Rev Plant Biol 55:373–399

Aregbesola E, Ortega-Beltran A, Falade T, Jonathen G, Hearne S, Bandopadhyay R (2020) A detached leaf assay to rapidly screen for resistance of maize to *Bipolaris maydis*, the causal agent of southern corn leaf blight. Eur J Plant Pathol 156:133–145. https://doi.org/10.1007/s10658-019-01870-4

Arora MK, Dhurwe U (2019) causal organism of charcoal rot: review article. East Afr Scholars J Agric Life Sci 2(2). https://doi.org/10.36349/easjals.2019.v02i02.006

Balint-Kurti P (2019) The plant hypersensitive response: concepts, control and consequences. Mol Plant Pathol 20(8):1163–1178. https://doi.org/10.1111/mpp.12821

Barberan TF, Espin JC (2001) Phenolics compounds and related enzymes as determinants of quality of fruits and vegetables. J Sci Food Agric 81:853–876

Bartnick GS (1968) Cell wall chemistry, morphogenesis, and taxonomy of fungi. Annu Rev Microbiol 22:87–108

Barzman M, Bàrberi P, Birch ANE, Boonekamp P, Graf B, Hommel B, Jensen JE, Kiss J, Kudsk P, Lamichhane JR, Messéan A, Moonen A, Ratnadass A, Ricci P, Sarah J, Sattin M (2015) Eight principles of integrated pest management. Agron Sustain Dev 35:1199–1215. https://doi.org/10.1007/s13593-015-0327-9

Bektas Y, Eulgem T (2015) Synthetic plant defense elicitors. Front Plant Sci 5:804. https://doi.org/10.3389/fpls.2014.00804

Bendahmane A, Kanyuka K, Baulcombe DC (1999) The Rx gene from potato controls separate virus resistance and cell death response. Plant Cell 11:781–792

Betsuyaku S, Katou S, Takebayashi Y, Sakakibara H, Nomura N, Fukuda H (2018) Salicylic acid and jasmonic acid pathways are activated in spatially different domains around the infection site during effector-triggered immunity in *Arabidopsis thaliana*. Plant Cell Physiol 59(1):8–16. https://doi.org/10.1093/pcp/pcx181

Bettini P, Cosi E, Pellegrini MG, Turbanti L, Vendramin G, Buiatti M (1998) Modification of competence for in vitro response to *Fusarium oxysporum* in tomato cells. III. PR-protein gene expression and ethylene evolution in tomato cell lines transgenic for phytohormone-related bacterial genes. Theor Appl Genet 97:575–583

Bhattacharya D, Dhar TK, Ali E (1992) An enzyme immunoassay of phaseolinone and its application in estimation of the amount of toxin in -infected seeds. Appl Environ Microbiol 58:1970–1974

Boccardo NA, Segretin ME, Hernandez I, Mirkin FG, Chacon O, Lopez Y, Borrás-Hidalgo O, Bravo-Almonacid FF (2019) Expression of pathogenesis-related proteins in transplastomic tobacco plants confers resistance to filamentous pathogens under field trials. Sci Rep 9:279. https://doi.org/10.1038/s41598-019-39568-6

Bos JIB, Prince D, Pitino M, Maffei ME, Win J, Hogenhout SA (2010) A functional genomics approach identifies candidate effectors from the aphid species *Myzuspersicae* (Green Peach Aphid). PLoS Gen 6:e1001216

Burdon JJ, Zhan J (2020) Climate change and disease in plant communities. PLoS Biol 18(11): e3000949. https://doi.org/10.1371/journal.pbio.3000949

Canesi L, Pruzzo C (2016) Specificity of innate immunity in bivalves: a lesson from bacteria. Environ Sci. https://doi.org/10.1016/B978-0-12-803252-7.00006-0

Cavelcanti FR, Resende ML, Lima SP, Silveira JA, Oliveira JT (2007) Activities of antioxidant enzymes and photosynthetic responses in tomato pre-treated by plant activators and inoculated by *Xanthomonas vesicatoria*. Physiol Mol Plant Pathol 68:198–208

Chan YH, Sackston WE (1973) Non-specificity of the necrosis inducing toxin of *Sclerotium bataticola*. Can J Bot 51:690–692

Chatterjee A, Ghosh SK (2008) Alteration in biochemical components in mesta plants infected with yellow vein mosaic disease. Brazilian J Plant Physiol 20:267–275

Chen A, Sun J, Matthews A, Armas-Egas L, Chen N, Hamill S, Mintoff S, Tran-Nguyen LTT, Batley J, Aitken EAB (2019) Assessing variations in host resistance to *Fusarium oxysporum* f sp. *cubense* Race 4 in *Musa* species, with a focus on the subtropical race 4. Front Microbiol 10: 1062. https://doi.org/10.3389/fmicb.2019.01062

Chen W, Jiang X, Yang Q (2020) Glycoside hydrolase family 18 chitinases: the known and the unknown. Biotechnol Adv 43:107553. https://doi.org/10.1016/j.biotechadv.2020.107553

Chickpea: An ancient crop for the modern world. http://exploreit.icrisat.org/page/chickpea/685/60, ICRISAT, Downloaded 26 January 2014

Chittoor JM, Leach JE, White FF (1999) Induction of peroxidase during defense against pathogens. In: Datta SK, Muthukrishnan S (eds) Pathogenesis related proteins in plants. CRC Press, Boca Raton, pp 171–193

Chobe DR, Singh R, Tarafdar A, Chandran US, Ghosh R, Sharma M (2019) Deciphering the response of putative mutants against *Rhizoctonia bataticola* [(Taub.) Butler] causing dry root rot of chickpea. J Mycol Plant Pathol 49(4):358–367

Chobe DR, Tarafdar A, Chandran USS, Sudharani SM (2020) First report of *Ectophoma multirostrata* causing root rot in chickpea. Plant Dis 104:1866–1866. https://doi.org/10.1094/PDIS-01-19-0121-PDN

Choudhary MI, Zafar S, Khan NT, Ahmad S, Noreen S, Marasini BP, Al-Khedhairy AA, Rahman AU (2012) Biotransformation of dehydroepiandrosterone with and β-glucuronidase inhibitory activity of transformed products. J Enzyme Inhib Med Chem 27:348–355

Conèeica LF, Ferreres F, Tavores RM, Dios AC (2006) Induction of phenolic compounds in *Hypericum pertoralum* L. cells by *Colletotricohum gloeosprioides* elicitation. Phytochemistry 67:149–155

Craig A, Ewan R, Mesmar J, Gudipati V, Sadanandom A (2009) E3 ubiquitin ligases and plant innate immunity. J Exp Bot 60:1123–1132

Dakora FD, Phillips DA (1996) Diverse functions of isoflavonoids in legumes transcend anti-microbial definitions of phytoalexins. Physiol Mol Plant Pathol 49:1–20

Dangl JL, Jones JD (2001) Plant pathogens and integrated defense responses to infection. Nature 411:826–833

Degani O, Becher P, Gordani A (2022) Pathogenic interactions between *Macrophomina phaseolina* and *Magnaporthiopsis maydis* in mutually infected cotton sprouts. Agriculture 12:255. https://doi.org/10.3390/agriculture12020255

Dell'Olmo E, Tripodi P, Zaccardelli M, Sigillo L (2022) Occurrence of on chickpea in Italy: pathogen identification and characterization. Pathogens. 11(8):842. https://doi.org/10.3390/pathogens11080842

Dempsey DA, Klessig DF (2017) How does the multifaceted plant hormone salicylic acid combat disease in plants and are similar mechanisms utilized in humans? BMC Biol 15:23. https://doi.org/10.1186/s12915-017-0364-8

Devi P, Jha UC, Prakash V, Kumar S, Parida SK, Paul PJ, Prasad PVV, Sharma KD, Siddique KHM, Nayyar H (2022) Response of physiological, reproductive function and yield traits in cultivated chickpea (*Cicer arietinum* L.) under heat stress. Front Plant Sci 13:880519. https://doi.org/10.3389/fpls.2022.880519

Dhar TK, Siddiqui KAI, Ali E (1982) Structure of phaseolinone, a novel phytotoxin from Macrophomina phaseolina. Tetrahedron Lett 23:5459–5462

Dhingra OD, Sinclair JB (1985) Basic plant pathology methods. CBS Publishers and Distributors Delhi, 435 p

Dixon RA, Harrison MJ, Lamb CJ (1994) Early events in the activation of plant defense responses. Annu Rev Phytopathol 32:479–501

Dodds PN, Rafiqi M, Gan PH, Hardham AR, Jones DA, Ellis JG (2009) Effectors obiotrophic fungi and oomycetes: pathogenicity factors and triggers of host resistance. New Phytol 183:993–1000

Doke N (1983) Involvement of superoxide anion generation in the hypersensitive response of potato tuber to infection with an incompatible race of *Phytophthora infestans* and to the hypal wall components. Physiol Plant Pathol 23:345–357

Donkor D, Mirzahosseini Z, Bede J, Bauce E, Despland E (2019) Detoxification of host plant phenolic aglycones by the spruce budworm. PLoS One 14(5):e0208288. https://doi.org/10.1371/journal.pone.0208288

Durrant WE, Dong X (2004) Systemic acquired resistance. Annu Rev Phytopathol 42:185–209

Ehrenfield N, Canon P, Stance C, Medina C, Arce-Johnson P (2005) Tobamovirus coat protein CPCg induces an HR-like response in sensitive tobacco plants. Mol Cells 19:418–427

El-Fiki AII, Mohamed FG, El-Deeb AA, Khalifa MMA (2004) Some applicable methods for controlling sesame charcoal rot disease () under greenhouse conditions. Egypt J Phytopathol 32:87–101

El-Gamal Nadia G, Abd-El-Kareem F, Fotouh YO, El-Mougy NS (2007) Induction of systemic resistance in potato plants against late and early blight diseases using chemical inducers under greenhouse and field conditions. Res J Agric Biol Sci 3:73–81

FAOSTAT (2017) Available online at: www.fao.org/faostat/en/#home

Flor HH (1971) Current status of the gene–for-gene concept. Annu Rev Phytopathol 9:275–296

Fujita K, Komatsu K, Tanaka K (2006) An *in vitro* model for studying vascular injury after laser microdissection. Histochem Cell Biol 125:509–514

Garcia-Rubio R, de Oliveira HC, Rivera J, Trevijano-Contador N (2020) The fungal cell wall: *Candida*, *Cryptococcus*, and *Aspergillus* species. Front Microbiol 10:2993. https://doi.org/10.3389/fmicb.2019.02993

Garrett KA, Nita M, De Wolf ED, Esker PD, Gomez-Montano L, Sparks AH (2016) Plant pathogens as indicators of climate change. In: Letcher TM (ed) Climate change. Elsevier, Amsterdam, pp 325–338. https://doi.org/10.1016/B978-0-444-63524-2.00021-X

Gay PA, Tuzun S (2000) Involvement of a novel peroxidase isozyme and lignifications in hydathodes in resistance to black rot disease in cabbage. Can J Bot 78:1144–1149

Gechev TS, Van Breusegem F, Stone JM, Denev I, Laloi C (2006) Reactive oxygen species as signals that modulate plant stress responses and programmed cell death. Bioassays 28:1091–1101

Ghosh R, Tarafdar A, Sharma M (2016) Rapid detection of *Fusarium oxysporum* f. sp. ciceris from disease infested chickpea fields by loop-mediated isothermal amplification. Indian Phytopathol 69:47–50

Ghosh R, Tarafdar A, Sharma M (2017) Rapid and sensitive diagnoses of dry root rot pathogen of chickpea (*Rhizoctonia bataticola* (Taub.) Butler) using loop-mediated isothermal amplification assay. Sci Rep 7:42737. https://doi.org/10.1038/srep42737

Giri AP, Harsulkar AH, Deshpande VV, Sainan MN, Gupta VS, Ranjekar PK (1998) Chickpea defensive proteinase inhibitors can be inactivated by Podborer gut proteinases. Plant Physiol 116:393–401

Glazebrook J, Rogers EE, Ausubel FM (1997) Use of *Arabidopsis* for genetic dissection of plant defense responses. Annu Rev Genet 31:547–569

Gomez LD, Noctor G, Knight MR, Foyer CH (2004) Regulation of calcium signalling and gene expression by glutathione. J Exp Bot 55:1851–1856

González LMG, El Kayal W, Morris JS, Cooke JEK (2015) Diverse chitinases are invoked during the activity-dormancy transition in spruce. Tree Genetics Genomes 11:41. https://doi.org/10.1007/s11295-015-0871-0

Gozzo F (2003) Systemic acquired resistance in crop protection: from nature to a chemical approach. J Agric Food Chem 51:4487–4503

Grayer RJ, Kokubun T (2001) Plant-fungal interactions: the search for phytoalexins and other antifungal compounds from higher plants. Phytochemistry 56:253–263

Grover A, Gowthaman R (2003) Strategies for development of fungus resistant transgenic plants. Curr Sci 84:330–340

Guiñazú LB, Andrés JA, Rovera M, Rosas SB (2012) Isolation and characterization of rhizobacteria antagonistic to (Tassi) Goid, causal agent of Alfalfa damping-off. In: Malik A, Grohmann E (eds) Environmental protection strategies for sustainable development strategies for sustainability, pp 329–339

Gull A, Lone AA, Wani NUI (2019) Biotic and abiotic stresses in plants. In: Abiotic and biotic stress in plants. IntechOpen. https://doi.org/10.5772/intechopen.85832

Hatmi S, Gruau C, Trotel-Aziz P, Villaume S, Rabenoelina F, Baillieul F, Eullaffroy P, Clement C, Ferchichi A, Aziz A (2015) Drought stress tolerance in grapevine involves activation of polyamine oxidation contributing to improved immune response and low susceptibility to *Botrytis cinerea*. J Exp Bot 66:775–787. https://doi.org/10.1093/jxb/eru436

Health M (2000) Non host resistance and nonspecific plant defense. Curr Opin Plant Biol 3:315–319

Heidel AJ, Clarke JD, Antonovics J, Dong X (2004) Fitness costs of mutations affecting the systemic acquired resistance pathway in *Arabidopsis thaliana*. Genetics 168:2197–2206

Holzmueller EJ, Jose S, Jenkins MA (2007) Influence of calcium, potassium and magnesium on *Cornus florida* L. density and resistance to dogwood anthracnose. Plant Soil 290:189–199

Huang H, Ullah F, Zhou D-X, Yi M, Zhao Y (2019) Mechanisms of ROS regulation of plant development and stress responses. Front Plant Sci 10:800. https://doi.org/10.3389/fpls.2019.00800

Hussain F, Usman F (2019) Fungal biotic stresses in plants and its control strategy. In: Abiotic and biotic stress in plants. IntechOpen. https://doi.org/10.5772/intechopen.83406

Iakimova E, Atanassov A, Woltering E (2005) Chemical and pathogen-induced programmed cell death in plants. Biotechnol Eq 19:124–138

ICARDA (1993) Descriptors for Chickpea (*Cicer arietinum* L.), International Board for Plant Genetic Resources, Rome, Italy; International Crops Research Institute for the Semi-Arid Tropics, Patancheru, India and International Center for Agricultural Research in the Dry Areas, Aleppo, Syria. ISBN: 92-9043-137-7

ICRISAT Happenings (2019) Real-time surveillance decrypts dry root rot spread in Central India. https://www.icrisat.org/real-time-surveillance-decrypts-dry-root-rot-spread-in-central-india/. Accessed 26 Apr 2019

Ijaz S, Sadaqat HA, Khan MN (2012) A review of the impact of charcoal rot (). J Agric Sci:1–6

Iqbal Z, Iqbal MS, Hashem A, Abd Allah EF, Ansari MI (2021) Plant defense responses to biotic stress and its interplay with fluctuating dark/light conditions. Front Plant Sci 12:631810. https://doi.org/10.3389/fpls.2021.631810

Iriti M, Faoro F (2009) Chemical diversity and defense metabolism: how plants cope with pathogens and ozone pollution. Int J Mol Sci 10:3371–3399

Isah T (2019) Stress and defense responses in plant secondary metabolites production. Biol Res 52: 39. https://doi.org/10.1186/s40659-019-0246-3

Ito Y, Kaku H, Shibuya N (1997) Identification of a high-affinity binding protein for *N*-acetylchitooligosaccharide elicitor in the plasma membrane of suspension-cultured rice cells by affinity labelling. Plant J 12:347–356. https://doi.org/10.1046/j.1365-313X.1997.12020347.x

Jach G, Görnhardt B, Mundy J, Logemann J, Pinsdorf E, Leah R, Schell J, Maas C (1995) Enhanced quantitative resistance against fungal disease by combinatorial expression of different barley antifungal proteins in transgenic tobacco. Plant J 8:97–109

Jeandet P, Hébrard C, Deville M-A, Cordelier S, Dorey S, Aziz A, Crouzet J (2014) Deciphering the role of phytoalexins in plant-microorganism interactions and human health. Molecules 19(11): 18033–18056. https://doi.org/10.3390/molecules191118033

Jones DH (1984) Phenylalanine ammonia lyase: regulation of its induction and role in plant development. Phytochemistry 23:1349–1359

Junior MCP, Pandini JA, Hein DPR, da Silveira JA, Stangarlin JR, Portz RL, Bonett LP (2021) Phaseolin induction on common-bean cultivars and biological control of *Colletotrichum lindemuthianum* 89 race by *Baccharis trimera* (Less.) Dc. Braz Arch Biol Technol e21200816. https://doi.org/10.1590/1678-4324-75years-2021200816

Karthik S, Reddy MS, Yashaswini G (2021) Climate change and its potential impacts on insect-plant interactions. In: The nature, causes, effects and mitigation of climate change on the environment. IntechOpen. https://doi.org/10.5772/intechopen.98203

Kaur S, Samota MK, Choudhary M, Choudhary M, Pandey AK, Sharma A, Thakur J (2022) How do plants defend themselves against pathogens-biochemical mechanisms and genetic interventions. Physiol Mol Biol Plants 28:485–504. https://doi.org/10.1007/s12298-022-01146-y

Keller NP (2019) Fungal secondary metabolism: regulation, function and drug discovery. Nat Rev Microbiol 17:167–180. https://doi.org/10.1038/s41579-018-0121-1

Kessmann H, Staub T, Hofmann C, Maetzke T, Herzog J, Ward E, Uknes S, Ryals J (1994) Induction of systemic acquired disease resistant in plants by chemicals. Annu Rev Phytopathol 32:439–459

Kliebenstein DJ (2004) Secondary metabolites and plant/environment interactions: a view through *Arabidopsis thaliana* tinged glasses. Plant Cell Environ 27:675–684

Koksal E, Gulcin I (2008) Purification and characterization of peroxidase from cauliflower (*Brassica oleracea* L. var. *botrytis*) buds. Protein Pept Lett 15:320–326

Kortekamp A, Zyprian E (2003) Characterization of plasmopara-resistance in grapevine using *in vitro* plants. J Plant Physiol 160:1393–1400

Krolicka M, Hinz SWA, Koetsier MJ, Joosten R, Eggink G, van den Broek LAM, Boeriu CG (2018) Chitinase Chi1 from *Myceliophthora thermophila* C1, a thermostable enzyme for chitin and chitosan depolymerization. J Agric Food Chem 66(7):1658–1669. https://doi.org/10.1021/acs.jafc.7b04032

Kumar N, Goel N (2019) Phenolic acids: natural versatile molecules with promising therapeutic applications. Biotechnol Rep 24:e00370

Lamichhane JR, Venturi V (2015) Synergisms between microbial pathogens in plant disease complexes: a growing trend. Front Plant Sci 6:385. https://doi.org/10.3389/fpls.2015.00385

Lavanya SN, Niranjan-Raj S, Jadimurthy R, Sudarsan S, Srivastava R, Tarasatyavati C, Rajashekara H, Gupta VK, Nayaka SC (2022) Immunity elicitors for induced resistance against the downy mildew pathogen in pearl millet. Sci Rep 12:4078. https://doi.org/10.1038/s41598-022-07839-4

Lawrence CB, Singh NP, Qui J, Gardner RG, Tuzun S (2000) Constitutive hydrolytic enzymes are associated with polygenic resistance of tomato to *Alternaria solani* and may function as an elicitor release mechanism. Physiol Mol Plant Pathol 57:211–220

Lawton KA, Weymann K, Friedrich L, Vernooij B, Uknes S, Ryals J (1995) Systemic acquired resistance in *Arabidopsis* requires salicylic acid but not ethylene. Mol Plant-Microbe Interact 8: 863–870

Lee M, Jeon HS, Kim SH, Chung JH, Roppolo D, Lee H, Cho HJ, Tobimatsu Y, Ralph J, Park OK (2019) Lignin-based barrier restricts pathogens to the infection site and confers resistance in plants. EMBO J 38:e101948. https://doi.org/10.15252/embj.2019101948

Lefevere H, Bauters L, Gheysen G (2020) Salicylic acid biosynthesis in plants. Front Plant Sci 11: 338. https://doi.org/10.3389/fpls.2020.00338

Leiter E, Szappanos H, Oberparleiter C, Kaiserer L, Csernoch L, Pusztahelyi T, Emri T, Pócsi I, Salvenmoser W, Marx F (2005) Antifungal protein PAF severely affects the integrity of the plasma membrane of *Aspergillus nidulans* and induces an apoptosis-like phenotype. Antimicrob Agents Chemother 49:2445–2453

Levine A, Tenhaken R, Lamb C (1994) H_2O_2 from the oxidative burst orchestrates the plant hypersensitive disease resistance response. Cell 79:583–593

Levine A, Pennell RI, Alvarez ME, Palmer R, Lamb C (1996) Calcium-mediated apoptosis in a plant hypersensitive disease resistance response. Curr Boil 6:427–437

Li L, Steffens JC (2002) Over expression of polyphenol oxidase in transgenic tomato plants results in enhanced bacterial disease resistance. Plants 215:239–247

Loser K, Weltring KM (1998) Induction of a polyubiquitin gene (*ubi1*) by potato phytoalexins and heat shock in *Gibberella pulicaris*. Curr Genet 34:404–409

MacDonald MJ, D'Cunha GB (2007) A modern view of phenylalanine ammonia-lyase. Biochem Cell Biol 85:273–282

Madurapperumage A, Tang L, Thavarajah P, Bridges W, Shipe E, Vandemark G, Thavarajah D (2021) Chickpea (*Cicer arietinum* L.) as a source of essential fatty acids – a biofortification approach. Front Plant Sci 12:734980. https://doi.org/10.3389/fpls.2021.734980

Mahajan S, Tuteja N (2005) Mini review cold, salinity and drought stresses: an overview. Archiv Biochem Biophys 444:139–158

Mahato SB, Siddiqui KAI, Bhattacharya G, Ghosal T, Miyahara K, Sholichin M, Kawasaki T (1987) Structure and stereochemistry of phaseolinic acid: a new acid from. J Nat Prod 50:245–247

Manghwar H, Hussain A, Ali Q, Saleem MH, Abualreesh MH, Alatawi A, Ali S, Munis MFH (2021) Disease severity, resistance analysis, and expression profiling of pathogenesis-related protein genes after the inoculation of *Fusarium equiseti* in wheat. Agronomy 11(11):2124. https://doi.org/10.3390/agronomy11112124

Maphosa Y, Jideani VA (2017) The role of legumes in human nutrition. In: Functional food-improve health through adequate food. IntechOpen. https://doi.org/10.5772/intechopen.69127

Marquez N, Giachero ML, Declerck S, Ducasse DA (2021) General characteristics of pathogenicity and methods of control. Front Plant Sci 12:634397. https://doi.org/10.3389/fpls.2021.634397

Mathur SB (1968) Production of toxins and pecteolytic enzymes by two isolates of *Sclerotium bataticola* Taub. and their role in pathogenesis. Phytopathology 62:327–333

Mauch F, Mauch-maini B, Boller T (1988) Antifungal hydrolases in pea tissue ii. Inhibition of fungal growth by combination of chitinase and β- 1,3 glucanases. Plant Physiol 88:936–942

Melotto M, Mecey C, Niu Y, Chung HS, Katsir L, Yao J, Zeng W, Thines B, Staswick P, Browse J, Howe GA, He SY (2008) A critical role of two positively charged amino acids in the Jas motif of Arabidopsis JAZ proteins in mediating coronatine- and jasmonoyl isoleucine-dependent interactions with the COI1 F-box protein. Plant J 55:979–988

Metraux JP, Signer H, Ryals J, Ward E, Bem MW, Gaudin J, Raschdorf K, Schmid E, Blum W, Inverardi B (1990) Increase in salicylic acid at the onset of systemic acquired resistance in cucumber. Science 250:1004–1006

MikuliPetkovšek M, Stampar F, Veberic R (2009) Accumulation of phenolic compounds in apple in response to infection by the scab pathogen, *Venturia inaequalis*. Physiol Mol Plant Pathol 74: 60–67

Minatel IO, Borges CV, Ferreira MI, Gomez HAG, Chen CO, Lima GPP (2017) Phenolic compounds: functional properties, impact of processing and bioavailability. In: Soto-Hernandez M, Palma-Tenango M, Garcia-Mateos MR (eds) Phenolic compounds - biological activity. IntechOpen. https://doi.org/10.5772/66368

Mittler R, Vnderauwera S, Gollery M, Breusegen FV (2004) Reactive oxygen gene network of plants. Trends Plant Sci 9:490–498

Moënne-Loccoz Y, Mavingui P, Combes C, Normand P, Steinberg C (2015) Microorganisms and biotic interactions. In: Bertrand JC, Caumette P, Lebaron P, Matheron R, Normand P, Sime-Ngando T (eds) Environmental microbiology: fundamentals and applications. Springer, Dordrecht. https://doi.org/10.1007/978-94-017-9118-2_11

Mohammad R, Thomas W, Hamed K, Abbas JL, Richard ET, Gabriel LB, Sciumbato (2007) Soybean charcoal rot disease fungus *Macrophomina phaseolina* in Mississippi produces the phytotoxin (-) botryodiplodin but no detectable phaseolinone. J Nat Prod 70:128–129

Morel JB, Dangl JL (1997) The hypersensitive response and the induction of cell death in plants. Cell Death Differ 4:671–683

N'Guyen GQ, Raulo R, Porquier A, Iacomi B, Pelletier S, Renou J-P, Bataillé-Simoneau N, Campion C, Hamon B, Kwasiborski A, Colou J, Benamar A, Hudhomme P, Macherel D, Simoneau P, Guillemette T (2021) Responses of the necrotrophic fungus *Alternaria brassisicola* to the Indolic Phytoalexin Brassinin. Front Plant Sci 11:611643. https://doi.org/10.3389/fpls.2020.611643

Nallasivam JL, Fernandes RA (2017) A protecting-group-free synthesis of (+)-nephrosteranic, (+)-protolichesterinic, (+)- nephrosterinic, (+)- phaseolinic, (+)-rocellaric acids and (+)-methylenolactocin. Org Biomol Chem 15:708–716. https://doi.org/10.1039/C6OB02398C

Nazarov PA, Baleev DN, Ivanova MI, Sokolova LM, Karakozava MV (2020) Infectious plant diseases: etiology, current status, problems and prospects in plant protection. Acta Naturae 12(3):46–59. https://doi.org/10.32607/actanaturae.11026

Neuenschwander U, Lawton K, Ryals J (1996) Systemic acquired resistance. In: Stacey G, Keen NT (eds) Plant microbe interaction. Chapman and Hall, New York, pp 81–106

Nishad R, Ahmed T, Rahman VJ, Kareem A (2020) Modulation of plant defense system in response to microbial interactions. Front Microbiol 11:1298. https://doi.org/10.3389/fmicb.2020.01298

Nita M, Grzybowski A (2016) The role of the reactive oxygen species and oxidative stress in the pathomechanism of the age-related ocular diseases and other pathologies of the anterior and posterior eye segments in adults. Oxid Med Cell Longev 3164734. https://doi.org/10.1155/2016/3164734

Nürnberger T (1999) Signal perception in plant pathogen defense. Cell Mol Life Sci 55:167–182

Nürnberger T, Scheel D (2001) Signal transmission in plant immune response. Trends Plant Sci 6: 372–374

Ogawa D, Nakajima N, Seo S, Mitsuhara S, Kamad H, Ohashi Y (2006) The phenylalanine pathway is the main route of salicylic acid biosynthesis in Tobacco mosaic virus-infected tobacco leaves. Plant Biotechnol 23:392–398

Ojha S, Chakraborty MR, Dutta S, Chatterjee NC (2008) Influence of VAM on nutrient uptake and growth of custard-apple. Asian J Exp Sci 22:221–224

Oostendorp M, Kuz W, Dietrich B, Staub T (2001) Induced resistance in plants by chemicals. Eur J Plant Pathol 107:19–28

Ortega X, Velasquez JC, Pérez LM (2005) IP3 production in the hypersensitive response of lemon seedlings against *Alternaría alternate* involves active protein tyrosine kinases but not a G-protein. Biol Res 38:89–99

Ortíz-Castro R, Contreras-Cornejo HA, Macías-Rodríguez L, López-Bucio J (2009) The role of microbial signals in plant growth and development. Plant Signal Behav 4(8):701–712. https://doi.org/10.4161/psb.4.8.9047

Othman L, Sleiman A, Abdel-Massih RM (2019) Antimicrobial activity of polyphenols and alkaloids in Middle Eastern plants. Front Microbiol 10:911. https://doi.org/10.3389/fmicb.2019.00911

Ozyigit II, Kahraman MV, Ercan O (2007) Relation between explant age, total phenols and regeneration response in tissue cultured cotton (*Gossypium hirsutum* L.). Afr J Biotechnol 6: 3–8

Palit P, Kudapa H, Zougmore R, Kholova J, Whitbread A, Sharma M, Varshney RK (2020) An integrated research framework combining genomics, systems biology, physiology, modelling and breeding for legume improvement in response to elevated CO_2 under climate change scenario. Curr Plant Biol 22:100149. https://doi.org/10.1016/j.cpb.2020.100149

Passardi F, Cosio C, Penel C, Dunand C (2005) Peroxidases have more functions than a Swiss army knife. Plant Cell Rep 24:255–265

Pawlowski ML, Hartman GL (2016) Infection mechanisms and colonization patterns of fungi associated with soybean. In: Fungal pathogenicity. IntechOpen. https://doi.org/10.5772/62305

Petit-Houdenot Y, Fudal I (2017) Complex interactions between fungal avirulence genes and their corresponding plant resistance genes and consequences for disease resistance management. Front Plant Sci 8:1072. https://doi.org/10.3389/fpls.2017.01072

Peyraud R, Dubiella U, Barbacci A, Genin S, Raffaele S, Roby D (2017) Advances on plant-pathogen interactions from molecular toward systems biology perspectives. Plant J 90(4): 720–737. https://doi.org/10.1111/tpj.13429

Pontes JGM, Fernandes LS, dos Santos RV, Tasic L, Fill TP (2020) Virulence factors in the phytopathogen–host interactions: an overview. J Agric Food Chem 68(29):7555–7570. https://doi.org/10.1021/acs.jafc.0c02389

Prabhukarthikeyan SR, Parameswaran C, Keerthana U, Teli B, Jag PTK, Cayalvizhi B, Panneerselvam P, Senapati A, Nagendran K, Kumari S, Yadav MK, Aravindan S, Sanghamitra S (2020) Understanding the plant-microbe interactions in CRISPR/CAS9 era: indeed a sprinting start in marathon. Curr Genomics 21(6):429–443. https://doi.org/10.2174/1389202921999200716110853

Prasannath K (2017) Plant defense-related enzymes against pathogens: a review. AGRIEAST. J Agric Sci 11(1):38–48. https://doi.org/10.4038/agrieast.v11i1.33

Pratyusha S (2022) Phenolic compounds in the plant development and defense: an overview. In: Hasanuzzaman M, Nahar K (eds) Plant stress physiology- perspectives in agriculture. IntechOpen. https://doi.org/10.5772/intechopen.102873

Pusztahelyi T (2018) Chitin and chitin-related compounds in plant–fungal interactions. Mycology 9(3):189–201. https://doi.org/10.1080/21501203.2018.1473299

Pusztahelyi T, Holb IJ, Pócsi I (2015) Secondary metabolites in fungus-plant interactions. Front Plant Sci 6:573. https://doi.org/10.3389/fpls.2015.00573

Quiroga M, Guerrero C, Botella MA, Barcelo A, Amaya I, Medina MI, Alonso FJ, de Forchetti SM, Tigier H, Castro VV, Fontes MSW (2005) Plant defense and antimicrobial peptides. Prot Pept Lett 12(1):11–16. https://doi.org/10.2174/0929866053405832

Radman R, Saez T, Bucke C, Keshavarz T (2003) Elicitation of plants and microbial cell system. Biotechnol App Biochem 37:91–102

Ramegowda V, Senthil-Kumar M (2015) The interactive effects of simultaneous biotic and abiotic stresses on plants: mechanistic understanding from drought and pathogen combination. J Plant Physiol 176:47–54. https://doi.org/10.1016/j.jplph.2014.11.008

Ramírez-Gómez XS, Jiménez-García SN, Campos VB, Campos MLG (2019) Plant metabolites in plant defense against pathogens. In: Plant diseases - current threats and management trends. IntechOpen. https://doi.org/10.5772/intechopen.87958

Rapicavoli JN, Blanco-Ulate B, Muszyński A, Figueroa-Balderas R, Morales-Cruz AM, Azadi P, Dobruchowska JM, Castro C, Cantu D, Roper MC (2018) Lipopolysaccharide O-antigen delays plant innate immune recognition of *Xylella fastidiosa*. Nat Commun 9:390. https://doi.org/10.1038/s41467-018-02861-5

Raskin I (1990) Role of salicylic acid in plants. Annu Rev Plant Physiol Plant Mol Biol 43:439–463

Rather BA, Mir IR, Masood A, Anjum NA, Khan NA (2020) Nitric oxide pre-treatment advances seed germination and alleviates copper-induced photosynthetic inhibition in Indian Mustard. Plants 9(6):776. https://doi.org/10.3390/plants9060776

Ribnicky DM, Shulaev V, Raskin I (1998) Intermediates of salicylic acid biosynthesis in tobacco. Plant Physiol 118:565–572

Ryals J, Neuenschwander UH, Willits MG, Molina A, Steiner HY, Hunt M (1996) Systemic acquired resistance. Plant Cell 8:1809–1819

Sauerborn J, Buschmann H, Ghiasi GK, Kogel KH (2002) Benzothiadiazole activates resistance in sunflower (*Helianthus annuus*) to the root parasitic weed *Orobanche Cumana*. Phytopathology 92:59–64

Selin C, de Kievit TR, Belmonte MF, Fernando WGD (2016) Elucidating the role of effectors in plant-fungal interactions: progress and challenges. Front Microbiol 7:600. https://doi.org/10.3389/fmicb.2016.00600

Sharma M, Ghosh R, Pande S (2015) Dry root rot (*Rhizoctonia bataticola* (Taub.) Butler): an emerging disease of chickpea- where do we stand? Arch Phytopathol Plant Prot 48:13–16, 797–812. https://doi.org/10.1080/03235408.2016.1140564

Sharma M, Sengupta A, Ghosh R, Agarwal G, Tarafdar A, Nagavardhini A, Pande S, Varshney RK (2016) Genome wide transcriptome profiling of *Fusarium oxysporum* f sp. *ciceris* conidial germination reveals new insights into infection-related genes. Sci Rep 6:37353. https://doi.org/10.1038/srep37353

Sharma M, Ghosh R, Tarafdar A, Rathore A, Chobe DR, Kumar AV, Gaur PM, Samineni S, Gupta O, Singh NP, Saxena DR, Saifulla M, Pithia MS, Ghante PH, Mahalinga DM, Upadhyay JB, Harer PN (2019) Exploring the genetic cipher of chickpea (*Cicer arietinum* L.) through identification and multi-environment validation of resistant sources against *Fusarium* wilt (*Fusarium oxysporum* f. sp. *ciceris*). Front Sustain Food Syst 3:78. https://doi.org/10.3389/fsufs.2019.00078

Shuping DSS, Eloff JN (2017) Use of plants to protect plants and food against fungal pathogens: a review. Afr J Tradit Complement Altern Med 14(4):120–127. https://doi.org/10.21010/ajtcam.v14i4.14

Siddiqui KAI, Gupta AK, Paul AK, Banerjee AK (1979) Purification and properties of a heat-resistant exotoxin produced by *Macrophomina phaseolina* (Tassi) Goid in culture. Experientia 35:1222–1223

Singh M, Singh J, Maurya S, Kumar S, Meena AK, Sharma P, Lakhran L (2021) Root rot disease incited by in arid legumes and their management: a review. Legume Res LR-4714:1–7. https://doi.org/10.18805/LR-4714

Sinha R, Gupta A, Senthil-Kumar M (2016) Understanding the impact of drought on foliar and xylem invading bacterial pathogen stress in chickpea. Front Plant Sci 7:902. https://doi.org/10.3389/fpls.2016.00902

Somssica IE, Hahlbrock K (1998) Pathogen defense in plants-a paradigm of biological complexity. Trends Plant Sci 3:86–90

Stakman EC (1915) Relation between *Puccina graminis* and plants highly resistant to its attack. J Agric Res 4:193–199

Takahama U, Yoshitama K (1998) Hydoxycinnamic acid esters enhance peroxidase-dependent oxidation of differences in the enhancement among the esters. J Plant Res 111:97–100

Tarafdar A, Rani TS, Chandran USS, Ghosh R, Chobe DR, Sharma M (2018) Exploring combined effect of abiotic (soil moisture) and biotic (*Sclerotium rolfsii* Sacc.) stress on collar rot development in chickpea. Front Plant Sci 9:1154. https://doi.org/10.3389/fpls.2018.01154

Tripathi R, Tewari R, Singh KP, Keswani C, Minkina T, Srivastava AK, De Corato U, Sansinenea E (2022) Plant mineral nutrition and disease resistance: a significant linkage for sustainable crop protection. Front Plant Sci 13:883970. https://doi.org/10.3389/fpls.2022.883970

Tungmunnithum D, Thongboonyou A, Pholboon A, Yangsabai A (2018) Flavonoids and other phenolic compounds from medicinal plants for pharmaceutical and medical aspects: an overview. Medicines 5(3):93. https://doi.org/10.3390/medicines5030093

Tyagi H, Jha S, Sharma M, Giri J, Tyagi AK (2014) Rice SAPs are responsive to multiple biotic stresses and overexpression of OsSAP1, an A20/AN1 zinc-finger protein, enhances the basal resistance against pathogen infection in tobacco. Plant Sci 225:68–76

Tyler BM (2002) Molecular basis of recognition between *Phytophthora* pathogens and their hosts. Annu Rev Phytopathol 40:137–167

Vågsholm I, Arzoomand NS, Boqvist S (2020) Food security, safety, and sustainability- getting the trade-offs right. Front Sustain Food Syst 4:16. https://doi.org/10.3389/fsufs.2020.00016

Vallad GE, Goodman RM (2003) Systemic acquired resistance and induced systemic resistance in conventional agriculture. Alliance Crop Soil Environ Sci Soc 44:1920–1934

van Loon LC, van Strien EA (1999) Families of pathogenesis-related proteins, their activities, and comparative analysis of PR-1 type proteins. Physiol Mol Plant Pathol 55:85–97

van Loon LC, Rep M, Pieterse CMJ (2006) Significance of inducible defense-related proteins in infected plants. Annu Rev Phytopathol 44:135–162

Velásquez AC, Castroverde CDM, He SY (2018) Plant and pathogen warfare under changing climate conditions. Curr Biol 28(10):PR619-R634. https://doi.org/10.1016/j.cub.2018.03.054

Veliz EA, Martínez-Hidalgo P, Hirsch AM (2017) Chitinase-producing bacteria and their role in biocontrol. AIMS Microbiol 4(3):689–705. https://doi.org/10.3934/microbiol.2017.3.689

Vishal B, Kumar PP (2018) Regulation of seed germination and abiotic stresses by gibberellins and abscisic acid. Front Plant Sci 9:838. https://doi.org/10.3389/fpls.2018.00838

Wallace TC, Murray R, Zelman KM (2016) The nutritional value and health benefits of chickpeas and hummus. Nutrients 8(12):766. https://doi.org/10.3390/nu8120766

Wan J, He M, Hou Q, Zou L, Yang Y, Wei Y, Chen X (2021) Cell wall associated immunity in plants. Stress Biol 1:3. https://doi.org/10.1007/s44154-021-00003-4

Wen Z, Liao W, Chen S (2005) Production of cellulase/β-glucosidase by the mixed fungi culture of *Trichoderma reesei* and *Aspergillus phoenicis* on dairy manure. Appl Biochem Biotechnol:93–104

Wenjuan L, Ping H, Jiyun J (2009) Potassium influenced phenylalanine ammonia lyase, peroxidases and polyphenol oxidases in *Fusarium graminearum* infected maize (*Zea mays* L.). In: Proceedings of international plant nutrition colloquium XVI

Wharton PS, Nicholson RL (2000) Temporal synthesis and radiolabelling of the sorghum 3-deoxyanthocyannidin phytoalexins and the antocyanin, cyaniding 3-dimaloynl glucoside. New Phytol 145:457–469

Wilkens A, Paulsen J, Wray V, Winterhalter P (2010) Structures of two novel trimeric stilbenes obtained by horseradish peroxidase catalyzed biotransformation of trans-resveratrol and viniferin. J Agric Food Chem 58:6754–6761

Wolfgang A, Taffner J, Guimarães RA, Coyne D, Berg G (2019) Novel strategies for soil-borne diseases: exploiting the microbiome and volatile-based mechanisms toward controlling *Meloidogyne*-based disease complexes. Front Microbiol 10:1296. https://doi.org/10.3389/fmicb.2019.01296

Wu C, Bradford KJ (2003) Class I chitinase and â-1,3-glucanase are differentially regulated by wounding, methyl jasmonate, ethylene and gibberellin in tomato seeds and leaves. Plant Physiol 133:263–273

Yang Y, Yuan Q, Gabriel DW (1996) Water soaking Function(s) of xcmh1005 are redundantly encoded by members of the *Xanthomonas avr/pth* gene family. Mol Plant Microbe Interact 9: 105–113

Zabka M, Lesniak W, Prus W, Kuznicki J, Filipek A (2008) Sgt1 has co-chaperone properties and is up-regulated by heat shock. Biochem Biophys Res Commun 370:179–183

Zeyen RJ, Kruger WM, Lyngkjaer MF, Carver TLW (2002) Differential effects of D-mannose and 2-deoxy-D-glucose on attempted powdery mildew fungal infection of inappropriate and appropriate Gramineae. Physiol Mol Plant Pathol 61:315–323

Zhang X, Liu C (2015) Multifaceted regulations of gateway enzyme phenylalanine ammonia-lyase in the biosynthesis of phenylpropanoids. Mol Plant 8(1):17–27. https://doi.org/10.1016/j.molp.2014.11.001

Zhang X, Dai Y, Xiong Y, Defraia C, Li J, Dong X, Mou Z (2007) Overexpression of arabidopsis map kinase 7 leads to activation of plant basal and systemic acquired resistance. Plant J 52: 1066–1079

Zhou M, Tang Y, Deng X, Ruan C, Ding M, Shao J, Tang Y, Wu Y, (2018) Description of cultivated common buckwheat. In: Zhou M, Suvorova G, Woo SH (eds) Buckwheat germplasm in the world, pp 53–60. https://doi.org/10.1016/B978-0-12-811006-5.00006-9

Chapter 15
Emerging Roles of Plant Growth Promoting Rhizobacteria in Salt Stress Alleviation: Applications in Sustainable Agriculture

Varsha Venugopalan, Dinakar Challabathula, and Kavya Bakka

Abstract Increase in soil salinity is a serious concern as it adversely impacts the growth, development and yield of agricultural crops. The disruption in the ionic composition thereby affecting the fertility of the soil is observed in salt-affected soils warranting the development of technologies for improved crop growth under salinity conditions. Whilst generation of transgenic plants harbouring genes important for imparting tolerance to salinity stress conditions is considered a feasible biotechnological approach, the usage of transgenic crops for human consumption is not well received. Further, breeding approaches for generation of salt tolerant varieties is time consuming. In this context, the usage of halotolerant Plant Growth Promoting Rhizobacteria (PGPR) for improvement of salinity tolerance in crop plants is considered as an eco-friendly and cost-effective approach. This approach will also maintain the soil microbial diversity thereby protecting it from salinity-induced degradation. Few PGPR such as *Pseudomonas* and *Bacillus* species although have been effectively used for imparting plant growth promotion under control conditions, isolation of halotolerant PGPR, their effective utilization and optimization for agronomical application in salt-affected soils are key factors for improvement of salinity tolerance in crop plants. Further, whilst many of the isolated halotolerant PGPR show plant-growth promoting traits under laboratory conditions, their efficiency is questionable under field conditions. In the current chapter, the methodologies employed for inoculation of isolated halotolerant PGPR either singly or as a consortium to agricultural crops and the changes observed in the plants towards improvement in salinity tolerance will be discussed. The efficient delivery methods with carriers like nanoparticles, biopolymers and organic carriers along with

---

V. Venugopalan · K. Bakka (✉)
Department of Microbiology, School of Life Sciences, Central University of Tamil Nadu, Thiruvarur, India
e-mail: kavyabakka@cutn.ac.in

D. Challabathula
Department of Life Sciences, School of Life Sciences, Central University of Tamil Nadu, Thiruvarur, India
e-mail: dinakarc@cutn.ac.in

P. Mathur et al. (eds.), *Microbial Symbionts and Plant Health: Trends and Applications for Changing Climate*, Rhizosphere Biology,
https://doi.org/10.1007/978-981-99-0030-5_15

preparation of formulations, the route of administration for improvement of salinity tolerance in crop plants and their survival in saline soil will be elaborately discussed.

Keywords Carriers · Consortium · PGPR · Plant–microbe interaction · Salinity stress · Seed treatment

Abbreviations

ABA	Abscisic acid
ACC deaminase	1-Aminocyclopropane-1-carboxylate deaminase
AHL	Acyl homoserine lactone
APX	Ascorbate peroxidase
CAT	Catalase
CMC	Carboxymethyl cellulose
DHAR	Dihydroascorbate reductase
GR	Glutathione reductase
H_2O_2	Hydrogen peroxide
IAA	Indole-3-acetic acid
PGPB	Plant growth promoting bacteria
PGPR	Plant growth promoting rhizobacteria
ROS	Reactive oxygen species
SOD	Superoxide dismutase

15.1 Introduction

Plants being sessile are exposed to different environmental stress conditions out of which salinity is known to seriously impact the irrigated land resulting in impairment of plant growth, development and yield (Qadir et al. 2014). Soil salinity is caused due to the accumulation of various kinds of ions such as chlorides of sodium, calcium, magnesium, sulphates or carbonates which accumulate due to improper irrigation (Bartels and Dinakar 2013; Vaishnav et al. 2015; Pandey and Gupta 2020). Although plants absorb essential nutrients in the form of water-soluble salts, excessive intracellular accumulation negatively impacts plant growth and development. Plants respond to salinity stress in an ion-dependent manner initially causing the stomatal closure, causing a reduction in transpiration and later slowing down the metabolic processes (Rajendran et al. 2009). Further, prolonged exposure to high salinity stress leads to oxidative damage to intracellular biomolecules such as proteins, nucleic acids and lipids ultimately causing plant cell death. The majority of agricultural crops like *Oryza sativa* and *Phaseolus vulgaris* L. are glycophytes that are known to be adversely affected by salinity stress (Horie et al. 2012; Gupta and Pandey 2019). Whilst halophytic plants are adapted to grow, reproduce and complete their life cycle in high salinity conditions (500 mM NaCl or more), the glycophytic plants are extremely sensitive to salt in the soil and thus stress mitigation measures are required (Avinash and Bhakthi 2017). The plants' innate ability to tolerate salinity stress is determined by the genetic architecture and existence of numerous pathways that function to protect the cellular metabolism. However, since majority of the crop species are sensitive to salinity stress, these kinds of mechanisms do not exist in them.

The excess Na^+ and Cl^- ions in the saline soils delays seed germination and affect the viability of the embryos due to reduced water uptake (Munns and Tester 2008; Zhao et al. 2021). Although excess salinity is known to cause detrimental effects in glycophytic plants, moderate salinity was known to delay the transition phase of flowering without affecting the vegetative growth (Li et al. 2017). The osmotic stress caused by salinity augments the protein glycation in *Arabidopsis*, particularly the photosynthetic enzymes, thereby seriously affecting photosynthesis (Gagan et al. 2012). Salinity-induced stomatal closure limiting the CO_2 uptake, direct effect of salinity on the activity of photosynthetic enzymes and increased oxidative damage to the photosynthetic apparatus are other reasons for decreased photosynthetic activity in glycophytic plants during salinity stress (Analin et al. 2020).

Employment of halotolerant PGPR is one of the simple and feasible ways to induce salinity stress tolerance in agricultural crops. Compared to generation of transgenic plants and breeding varieties for induction of salinity stress tolerance, inoculating halotolerant PGPR to crop plants is less time consuming and feasible approach (Ilangumaran and Smith 2017). Out of the many important plant growth-promoting traits possessed by the halotolerant PGPR, phosphate solubilization potential, nitrogen fixing ability, production of phytohormones and siderophores are considered very important due to their involvement in the direct stimulation of plant growth and development under stress conditions (Dodd and Perez-Alfocea 2012; Bakka and Challabathula 2020; Bakka et al. 2022). The halotolerant PGPR strains *Pseudomonas fluorescens* isolated from the rhizosphere of plants grown in salty regions enhanced the growth of *Zea mays* plants upon inoculation (Zerrouk et al. 2016). The existing literature strongly suggests usage of halotolerant PGPR as one of the best approaches to mitigate salinity stress responses in plants. Further, the PGPR which improves soil fertility along with plant growth can also be used as a replacement for agrochemicals which are known to affect the quality of the soil (Vejan et al. 2016). In the current chapter, the methodologies that are employed for inoculation of isolated halotolerant PGPR to agricultural crops and the changes observed in the plants towards improvement in salinity tolerance will be discussed. Further, in order to increase the shelf life and activity of the bacteria in the soil, the data available on efficient delivery methods with carriers like nanoparticles, biopolymers, and organic carriers will be discussed. The strategies for the preparation of formulations and route of administration of the isolated PGPR will be elaborated.

15.2 Halotolerant PGPR

Halotolerant bacteria isolated from saline environments with PGPR traits are used as bioinoculants since they have the potential to enhance plant growth under saline stress conditions through direct or indirect mechanisms. These bacteria tolerate salt conditions by maintaining their membrane stability through the modification of fatty acid composition. The bacteria isolated from the rhizosphere or from the interior surfaces of the roots of halophilic plants when used as inoculants colonize the interior or exterior of the host plant roots and aid in mitigating the salinity stress responses in host plants. Reports strongly suggest that these bacteria can improve soil fertility, thereby replacing the usage of chemical fertilizers. The halotolerant PGPR has the ability to reside outside (rhizosphere) as well as inside the root cells (endophytic). *Klebsiella* sp. mediates the stress tolerance in plants and enhances the growth of host plants (Singh et al. 2015). Whilst PGPR such as *Agrobacterium, Serratia, Azotobacter, Chromobacterium* and *Erwinia* are extracellular and reside outside the root surface, PGPR such as *Rhizobium, Allorhizobium, Mesorhizobium* and *Bradyrhizobium* are known to be intracellular, existing inside the root cells. Additionally, a few PGPRs also produce hydrolytic enzymes, biosurfactants, siderophores and antibiotics to prevent growth and survival of pathogenic microorganisms (Bomle et al. 2021). The inoculation of *Lactuca sativa* (Lettuce) plants with the combination of PGPR *Pseudomonas mendocina* and arbuscular mycorrhizal *Glomnus mosseae* led to better hydration and reduction of Na^+ uptake under salinity stress conditions (Kohler et al. 2009). The rhizobacterial strains such as *Pseudomonas, Bacillus, Mucilaginibacter* and *Rhizobium* sp. improved the growth and salinity tolerance of *Arabidopsis thaliana* plants (Fan et al. 2020). In another study, inoculation of salt tolerant isolate KH-2 followed by the treatment with *Azotobacter* increased the yields of *Oryza sativa* (Javed et al. 2020). Whilst inoculation of a PGPR as a single entity or as a consortium was shown to have beneficial effect on host plants in ameliorating salinity stress responses, reports indicated better performance of the plants when inoculated with the bacterial consortium. In agreement with this, co-inoculation of salt tolerant bacteria like *Bacillus subtilis* and *Arthrobacter* sp. provided more salinity tolerance to *Triticum aestivum* than single-species inoculation (Upadhyay and Singh 2015). Compared to single bacterial inoculations, reports suggested the combinations of PGPR have a considerable effect on germination and plant growth. Furthermore, the application of a consortium of rhizobacteria from the same species was more effective than different species. The number of leaves, leaf area and leaf biomass were higher in the maize plants inoculated with the same species such as *Pseudomonas fluorescens* and *Pseudomonas putida* (Noumavo et al. 2013).

The PGPR isolated from the rhizosphere or roots of halophytes were found to be tolerant to salinity. Upon inoculation of these halotolerant PGPR to glycophytic host plants, the improvement in salinity tolerance along with increased growth and yield was observed. The changes observed in host plants upon inoculation with halotolerant PGPR are shown in Fig. 15.1. Halotolerant PGPRs like *Bacillus*

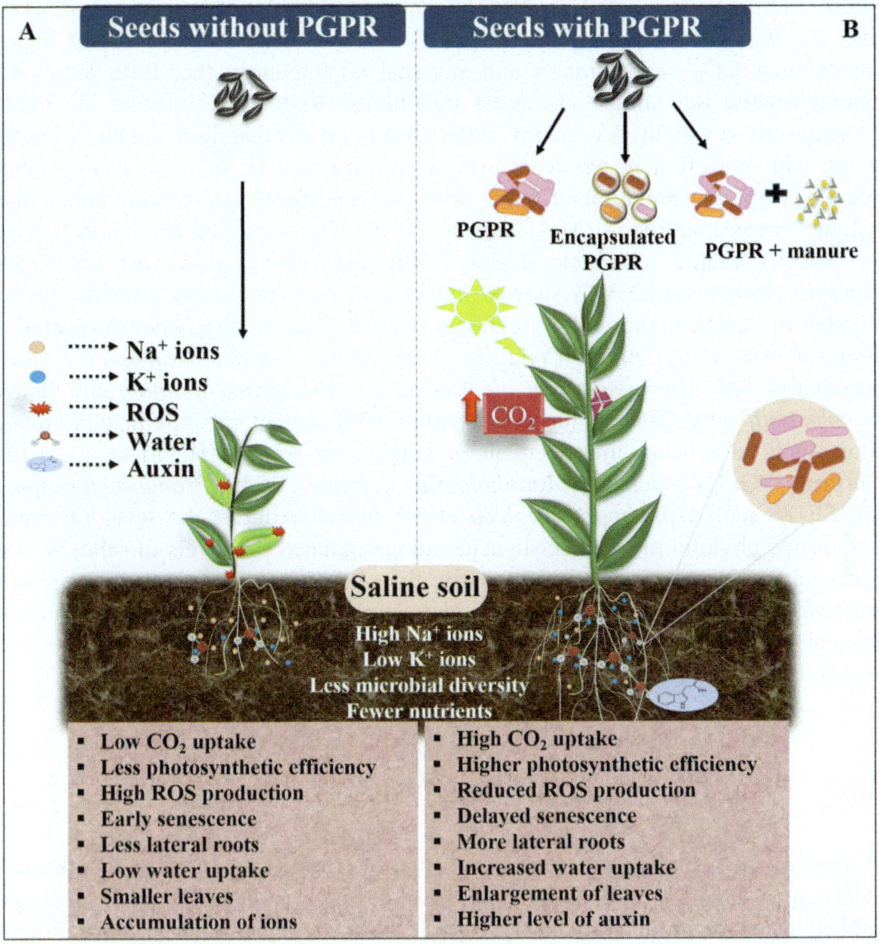

Fig. 15.1 Effect of PGPR on plant growth traits under salinity stress (**a**) Untreated seeds (**b**) Seeds treated with PGPR

endophyticus, Bacillus tequilensis, Planococcus rifietoensis, Variovorax paradoxus and *Arthrobacter agilis* isolated from the roots of halophyte *Salicornia europaea* were tolerant to salinity and showed more colonization to the roots of host plants aiding in increased seed germination, increased growth and better protection from salinity stress-induced oxidative damage (Zhao et al. 2016). The inoculation of halotolerant *Bacillus subtilis* GOT9 to *Arabidopsis thaliana* led to increased formation of lateral roots in host plants and upregulation of *SOS1* gene, a Na^+/H^+ antiporter for mitigating the salinity stress. Further, the upregulation of several stress-responsive genes such as *RD29B* and *RAB18* was also observed. The *Brassica campestris* plants inoculated with *B. subtilis* showed increased expression of stress-responsive *BrDREB1D, BrWRKY7* and *BrCSD3* transcript levels to augment the stress conditions (Woo et al. 2020). The consortium of *Pseudarthrobacter oxydans* and *Bacillus zhangzhouensis* inoculated *Beta vulgaris* L. seeds exhibited higher intercellular CO_2 concentration and stomatal conductance, therefore, increasing photosynthetic rate under salt stress conditions (Redondo-Gómez et al. 2022). Increased fresh weight, dry weight, better absorption of potassium and phosphorous along with reduction in ethylene levels were observed in *Solanum lycopersicum* seedlings upon inoculation of seeds with *Achromobacter piechaudii* under high salinity stress conditions (Mayak et al. 2004). The seeds of *Raphanus sativus* L. (radish) treated with halotolerant PGPR *Staphylococcus kloosii* EY37 and *Kocuria erythromyxa* EY43 showed higher root biomass, shoot biomass, higher number of leaves, higher chlorophyll and relative water content when compared to uninoculated control plants (Yildirim et al. 2008). *Capsicum annum* L. plants inoculated with *Microbacterium oleivorans, Brevibacterium iodinum* and *Rhizobium massiliae* exhibited an increase in chlorophyll content, increase in antioxidants along with significant improvement in height and weight (Hahm et al. 2017). Further, PGPRs such as *Burkholderia cepacia* SE4, *Promicromonospora* sp. SE188 and *Acinetobacter calcoaceticus* SE370 reduced the level of stress-responsive phytohormone abscisic acid and upregulated the levels of salicylic acid and gibberellin in host plants. Similarly, the plant growth promotion along with tolerance to salinity upon inoculation with halotolerant PGPB was observed in many crop plants such as *Oryza sativa*, *Lycopersicum esculentum*, *Triticum aestivum*, and *Raphanus sativus* (Kang et al. 2014).

15.3 Plant Growth Promoting Traits

A very low percentage of soil microbiota is known to possess plant growth promoting traits such as production of phytohormones like indole-3-acetic acid and gibberellic acid, production of siderophores for chelating iron, solubilization of inorganic phosphate to a usable form for the plants, fixation of atmospheric nitrogen and supplementation of micro- and macronutrients from the soil to the plants (Glick 1995). PGPRs are also known to produce enzymes that stop ethylene production thereby preventing senescence in leaves. Production of ACC (1-aminocyclopropane-1-carboxylate) deaminase enzyme is one of the key features that is observed in majority of the PGPR. ACC is the precursor of ethylene which plays an important role in the defence mechanism under stress conditions. It accumulates and inhibits the growth of plants in case of continuous stress. The ACC deaminase-producing bacteria reduce the ethylene levels in plants by converting the ACC, a precursor of ethylene biosynthesis to α-ketobutyrate and ammonia thereby facilitating the growth and survival of host plants in saline stress conditions (Glick 2014; Mahmood et al. 2017). The inoculation of ACC deaminase-producing bacteria *Stenotrophomonas maltophilia* SBP-19 to *Triticum aestivum* showed improvement in plant growth as it utilizes ACC (Singh and Jha 2017). Inoculation of ACC deaminase-producing bacteria such as *Aneurinibacillus aneurinilyticus* and *Paenibacillus* sp. to *Phaseolus vulgaris* (French bean) has led to increased IAA production, phosphate solubilization and production of ammonia to support the growth of seedlings under salinity stress conditions (Gupta and Pandey 2019). Seeds treated with ACC deaminase-producing bacterial strains isolated from rice rhizosphere exhibited a higher rate of seed germination as well as salinity tolerance and reduction in ethylene production (Bal et al. 2013).

Besides ACC deaminase-producing bacteria, acyl-homoserine lactone (AHL) producing bacteria has the ability to exhibit tolerance towards salinity through quorum sensing. Exogenous application of signal molecules like AHL produced by the consortium of *Aeromonas* sp. showed higher chlorophyll content, stomatal conductance and transpiration rate in both salt sensitive and salt tolerant *Triticum aestivum* over control plants (Nawaz et al. 2020a). N-3-oxo-hexanoyl-homoserine lactone (3OC6-HSL) is a signalling molecule belonging to AHLs family that is responsible for the population density of Gram-negative bacteria. The external application of 3OC6-HSL in *Arabidopsis thaliana* and *Triticum aestivum* enhanced physiological parameters like root elongation and shoot length. The increased total chlorophyll content and decreased level of malondialdehyde and Na^+ uptake were observed in 3OC6-HSL treated plants. The expression of genes involved in ABA-dependent pathways including *COR15a, RD22* and *ERD1* were upregulated in 3OC6-HSL treated plants and mitigated the stress conditions (Zhao et al. 2020). *Pseudomonas oryzihabitans* capable of producing ACC deaminase exerted a difference in their transcriptomic and metabolic profiles and detoxified the excess ions generated in saline conditions. Metabolites such as xylose, mannose, sedoheptulose and rhamnose have been accumulated under salinity and were decreased when *S. lycopersicum* is treated with *P. oryzihabitans*. Likewise, the transcription of stress responsive genes is also regulated with PGPR treatment (Mellidou et al. 2021). The inoculation of rhizospheric bacteria *Bacillus oumilus* and an endophytic bacteria *Pseudomonas pseudoalcaligenes* to *Oryza sativa* plants stimulated various kinds of osmoprotectants such as glycine betaine-like quaternary ammonium compounds and antioxidant proteins at a higher level in the early stages of plant growth (Jha et al. 2011).

Biofilm forming as well as ACC deaminase-producing PGPRs like *Bacillus licheniformis* and *Pseudomonas plecoglossicida*, are found to be effective in mitigating salt stress by increasing antioxidant enzymes. Biofilm formation of these PGPRs confers strong rhizospheric adherence and colonization which provide better survival strength in *Helianthus annus* (sunflower) against salinity (Yasmeen et al. 2020). Similarly, the treatment of salt tolerant isolate *Bacillus pumilis* in *Triticum aestivum* showed an excellent biofilm formation colonization behaviour under saline conditions. The elevated levels of antioxidant enzymes like catalase (CAT) and superoxide dismutase (SOD) showed that the bacterial treatment is significantly alleviating the stress condition (Ansari et al. 2019). *Burkholderia* sp. treatment to *Oryza sativa* showed better colonization on the root surface and significantly reduced reactive oxygen species (ROS) through the activation of antioxidant enzymes. Besides, it reduced the level of ethylene under salt treatment when compared to control (Sarkar et al. 2018).

15.4 Halotolerant PGPR-Mediated Salinity Stress Tolerance

It is undoubtedly clear that the inoculation of halotolerant PGPR to glycophytic plants improves the salinity tolerance levels. Improved photosynthesis, increased growth and development, increase in antioxidants, decreased oxidative damage and increased yield are some of the key features that are observed in PGPR inoculated plants (Fig. 15.1). Inoculated plants show modulated metabolism leading to improved growth and increased stress tolerance. Salinity stress causes a decrease in the availability of Fe due to decreased release of Fe-chelating compounds thereby leading to Fe deficiency and interveinal chlorosis in salinity stress exposed plants (Abbas et al. 2015; Ferchichi et al. 2016; Li et al. 2016). The halotolerant PGPR can survive in high saline environments due to the presence of chelators for Fe called as siderophores. Presence of high-affinity siderophores in halotolerant PGPR ensures removal of Fe from Fe^{3+} complexes in the soil aiding in supply of Fe to the host plants. This bacterial-mediated Fe supplementation to host plants is observed in many agricultural crops such as corn, wheat, rice and mungbean helping them to sustain growth under salinity stress conditions (Kavamura et al. 2013; Ramadoss et al. 2013; Singh et al. 2015; Mahmood et al. 2017). The halotolerant PGPR are also endowed with the production of certain enzymes that have the capacity to hydrolyze the organic and inorganic insoluble and unusable phosphorous compounds from the soil and convert them into soluble forms for plant assimilation facilitating growth and development. *Bacillus amyloliquefaciens* CZ-6 strain isolated from saline environments has the ability to produce IAA and ACC deaminase for increased development of shoot and root. It also possesses antifungal activity with the secretion of chitinase and protease enzymes (Zhou et al. 2021). As organic phosphorus occurs in the form of phytic acid, the enzyme phytase is necessary for the solubilization of phytate to increase phosphorus in the soil. The halotolerant PGPR *Klebsiella variicola* produces maximum phytase enzyme and has the highest phytate solubilization index (Kusale et al. 2021). Many reports have shown the phosphate solubilization potential of isolated halotolerant PGPB as a plant growth-promoting trait for increased growth, development and salinity stress tolerance (Dodd et al. 2010; Costa-Gutierrez et al. 2020; Shultana et al. 2020; Tchakounté et al. 2020). *Bacillus magaterium* inoculation influences the auxin redistribution in *Arabidopsis thaliana* and maintains the root architecture under salinity (Wang et al. 2021). *Waito C* rice, deficient with gibberellic acid showed an increase in germination, length and biomass when it is inoculated with *Leifsonia soli* SE134 through GA synthesis (Lucke et al. 2020*). Plants inoculated with Pseudomonas fluorescens strains showed better morphology due to increased production of cytokinins* (García de Salamone et al. 2001).

The regulation of ion homeostasis which is critical for the development of plants is observed in many of the PGPR inoculated plants. The efficient regulation of osmolytes together with the activation of antioxidant enzymes keeps a check on oxidative stress parameters thereby protecting the plants from salinity-induced damage (Jha and Subramanian 2014). Exopolysaccharides produced by halotolerant PGPR play an important role in binding to the root surfaces and limiting the uptake of cations to alleviate salinity stress in plants (Morcillo and Manzanera 2021). A variety of exopolysaccharides and osmoprotectants are produced by the halotolerant PGPR facilitating the growth and development of plants under salinity stress conditions (Dar et al. 2021; Morcillo and Manzanera 2021). Co-inoculation of *Bradirhizobium japonicum* and *Bacillus subtilis* to *Glycine max* showed higher colonization and exopolysaccharides production, protecting the roots of host plants from salinity stress-induced damages and Na^+ accumulation (Han and Lee 2005). Inoculation of *A. thaliana* plants with halotolerant PGPR *Bacillus megateriumn* ZS-3 caused Na^+ phagocytosis and an increase in K^+/Na^+ ratio (Shi et al. 2022). Upregulation of Vegetative Storage Protein (VSP) regulating the sodium transporters thereby reducing the accumulation of Na^+ ions is observed in soybean plants inoculated with halotolerant PGPR (Vaishnav et al. 2015). In another study, *Capsicum annum* inoculated with *Bacillus amyloliquefaciens* increased the concentration of phosphorous and potassium and reduced the concentration of Ca, Na^+ and Cd under salinity stress, leading to increased chlorophyll content along with increased growth and development of plants (Kazerooni et al. 2021). Plants alleviate oxidative stress during salinity by the exclusion of Na^+ and influx of K^+ ions. A significant reduction in shoot and root Na^+ by 32.8% and 24.5%, respectively, was observed in inoculated plants exposed to 200 mM NaCl. Additionally, significant increases in K^+ content of shoot and root by 32.5% and 35.8%, respectively, were observed in inoculated plants subjected to 200 mM NaCl (Singh and Jha 2017). The inoculation of *Bacillus cereus, Serratia marcescens* and *Pseudomonas aeruginosa* to *Triticum aestivum* markedly increased K^+ and Ca^{2+} contents, antioxidant enzyme activities and membrane stability during salinity stress (Desoky et al. 2020).

15.5 Effects of Inoculation of Halotolerant PGPR on Plants Under Salinity Stress

Bacillus megaterium is known to be one of the excellent PGPR for providing tolerance towards salinity stress. *A. thaliana* inoculated with *B. megaterium* ZS-3, *Serratia marcescens* 90-166 and *Pseudomonas fluorescens* WCS417 showed increased lateral root formation due to redistribution of auxin in the root meristems (Shi et al. 2010; Zamioudis et al. 2013; Wang et al. 2021; Shi et al. 2022). Further, whilst the *aux1-7, etr1* and *ein2* mutants of *Arabidopsis thaliana* defective in auxin and ethylene signalling pathways have shown limited root hairs and lateral root formation, inoculation of endophytic *Bacillus megaterium* to these mutant plants have led to an increased number of root hairs and lateral roots suggesting the modulation of auxin and ethylene signalling in plants by bacteria leading to enhanced growth (Lopez-Bucio et al. 2007). The rhizobacteria isolated from the halotolerant grass *Distichlis spicata* improved the root architecture in *Arabidopsis thaliana, Cucumis sativus* (Cucumber) and *Citrullus lanatus* (Watermelon) under salinity (Palacio-Rodriguez et al. 2017). Although, nutrients have been found to be high with the incorporation of both chemical fertilizers as well as PGPR bacteria in *Solanum lycopersicum*, there is a reduction of water uptake in case of chemical fertilizer under salinity but not with the treatment of PGPR as it improves the root development (Cordero et al. 2018).

Bacterial colonization is very essential to exhibit their effect on the plants during salinity stress conditions. SAL-17 and SAL-21 strains of *Aeromonas* sp. were found to have better colonization efficiency leading to promotion in root growth under salinity stress conditions (Nawaz et al. 2020b). Although, photosynthetic processes are found to be very sensitive to salinity stress, inoculation of isolated halotolerant PGPR has shown increased photosynthesis under salinity stress conditions (Bakka and Challabathula 2020; Taj and Challabathula 2021). The salinity stress-induced decrease in photosynthesis is due to decreased chlorophyll content, reduced activity of PSII leading to decreased efficiency, reduced electron transport rate and increased oxidative damage to photosynthetic processes (Acosta-Motos et al. 2015). Inoculation of two strains of *Alcaligenes faecalis* in *Triticum aestivum* has led to increased PSII efficiency during salinity stress conditions emphasizing on the beneficial role of halotolerant PGPR for protecting photosynthesis (Babar et al. 2021). The compounds secreted by the PGPR are known to directly influence the physiology, growth and development of plants thereby enhancing the salinity stress tolerance (Bakka et al. 2022).

Essential oils, flavonoids, phenolic compounds and rosmarinic acid produced from *Rosmarinus officinalis* are widely used in therapeutics and cosmetics. However, the production of these compounds has been affected severely due to salinity stress. The utilization of PGPR instead of chemical fertilizers positively influenced the growth and yield of essential oils from *R. officinalis* (Dehghani et al. 2019). The pigments that are present in the PGPR-treated plants have been estimated using biochemical methods. Chlorophyll ‘a’, ‘b’ and carotenoids essential for photosynthetic activity and assimilation were found higher in 75 days old plants after consortium treatment of *Rhizobium, Pseudomonas* and *Bacillus* (Robert et al. 2017). The spilling of petroleum in the soil during transportation will co-contaminate with salt and inorganic compounds. Microbial inoculation in plants has beneficial role in phytoremediation for the removal of contaminants in the soil using plants. The pooled inoculation of PGPR and arbuscular mycorrhizal fungi in oats relieved saline-alkali soil contaminated by petroleum which is evaluated by measuring the total petroleum hydrocarbon present in the soil (Xun et al. 2015). Usually, ridge sowing provides better cultivation as excess water flows through the furrows between them. Hereby, the inoculation of *Pseudomonas putida* in wheat seeds with ridge sowing showed better efficiency in their growth and nutrient uptake when compared to the other methods (Ullah et al. 2022). Various applications of PGPR for improvement of salinity tolerance along with higher yield in agricultural crops are shown in Fig. 15.2.

The lipid membranes undergo lipid peroxidation and produce malondialdehyde under stress conditions which are reduced by PGPR inoculation (Singh et al. 2016;

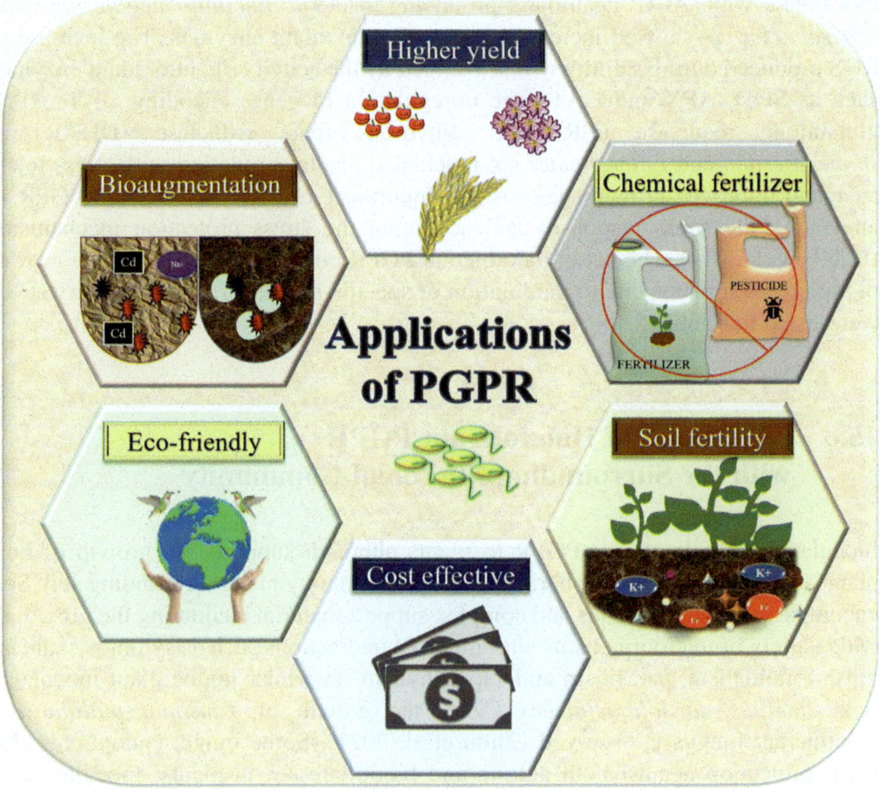

Fig. 15.2 Applications of PGPR in sustainable agriculture

Alexander et al. 2020). Some PGPRs change the composition of lipid bilayers to cope with stress. *Ochrobactrum* strain L115 accumulates more phosphatidylcholine to maintain structural integrity under the combination of high temperature and salinity. *Arachis hypogaea* inoculated with this strain were found to show increased root and shoot biomass and higher yield (Paulucci et al. 2015).

Priming of *Oryza sativa* seeds with *Bacillus* sp. L81 and *Aeromonas* sp. AMG272 showed increased levels of ascorbate peroxidase (APX) and superoxide dismutase (SOD) correlating with improved salinity tolerance and better growth (García-Cristobal et al. 2015). Inoculation of *Pseudomonas mendocina* to the glycophytic plant *Lactuca sativa* resulted in high salinity tolerance with increased shoot biomass and higher activity of antioxidant enzymes (Kohler et al. 2009). Salinity stress-induced cell death through the activation of the caspase-like protease enzyme activity is reduced in *Oryza sativa* plants inoculated with halotolerant PGPR *Pseudomonas pseudoalcaligenes* and *Bacillus pumilus*. The inoculated plants showed twofold reduction in caspase activity and better survival rate under high saline environments (Jha and Subramanian 2014). *Abelmoschus esculentus* L. plants inoculated with ACC deaminase-producing bacteria *Bacillus megaterium* and *Enterobacter* sp. showed increased levels of antioxidant enzymes. The increase in ROS produced during salinity was scavenged by the activity of antioxidant enzymes such as SOD, APX and CAT. The upregulation of genes encoding CAT, APX, glutathione reductase (GR) and dihydroascorbate reductase (DHAR) in *B. megaterium* and *Enterobacter* sp. inoculated plants correlating with stress tolerance and improved growth suggests the importance of these halotolerant PGPR in alleviating the stress responses and activating the stress protection mechanisms (Habib et al. 2016). The list of halotolerant PGPR positively influencing the growth of various host plants due to modulation of specific parameters under salinity stress were listed in Table 15.1.

15.6 Interaction of Halotolerant PGPR with the Surrounding Microbial Community

Inoculation of halotolerant PGPR to plants although supports the growth of host plants, it modifies the rhizospheric microbial community in the surrounding soil. Soil encompasses a heterogenous and complex support material facilitating the growth of wide variety of microorganisms with multiple interactions such as symbiosis, antagonism, mutualism, parasitism and saprophytism. In winter jujube plant inoculated with *Bacillus amyloliquefaciens* CZ-6, the growth of *Tausonia pullulans*, a halotolerant fungus is observed (Zhou et al. 2021). Some fungal endophytes also show protection against both abiotic and biotic stresses in plants. Inoculation of *Solanum lycopersicum* plants with *Pseudomonas* sp. S3 showed better resistance to salinity stress along with antifungal activity against *Rhizoctonia solani, Fusarium oxysporum, Botrytis cinerea, Alternaria alternata* and *Aspergillus niger* (Pandey and Gupta 2020). Few PGPRs such as Rhizobium and Mesorhizobium, synthesizing ACC deaminase secrete hydrolytic enzymes to disrupt the fungal cell walls thereby affording protection to the plants (Bomle et al. 2021). Endophytic halotolerant PGPR isolated from *Cakile maritima, Matthiola tricuspidate* and *Crithmum maritimum* showed plant growth promotion under salinity stress and possessed the ability to inhibit the growth and development of phytopathogens like *Verticillium dahliae, Ralstonia solanacearum* and *Clavibacter michiganensis* (Christakis et al. 2021). *Pseudomonas stutzeeri* and *Bacillus toyonenes* isolated from *Prunus domestica* showed antifungal activity against *Verticillium dahliae* and *Fusarium oxysporum.* Correspondingly, they improved the physiological and biochemical parameters of *Solanum lycopersicum* (Essalimi et al. 2022).

Table 15.1 PGPR with their respective host and influence on plant growth under salinity stress conditions

S. no	PGPR	Host	Increase	Decrease	References
1	*Pseudomonas fluorescens*	*Zea mays*	Primary, lateral and seminal root length	Na^+ uptake	Zerrouk et al. (2016)
2	*Pseudomonas mendocina* + *Glomnus mosseae*	*Lactuca sativa* (Lettuce)	Root and shoot biomass, hydration, peroxidase, catalase	Na^+ uptake	Kohler et al. (2009)
3	*Bacillus pumilis*	*Triticum aestivum* L. (salt tolerant-Aas-11)	Fresh and dry weight of root, potassium uptake	Na^+ uptake	Nawaz et al. (2020a, b)
4	*Exiguobacterium aurantiacum*	*Triticum aestivum* L. (salt tolerant-Aas-11)	Dry weight, potassium and calcium uptake of shoot		
5	*Pseudomonas fluorescens*	*Triticum aestivum L.* (salt sensitive -Galaxy -13)	High water content, shoot potassium and root calcium uptake		
6	*Exiguobacterium aurantiacum* + *Bacillus pumilis* + *Pseudomonas fluorescens*	*Triticum aestivum L.* (both salt tolerant and salt sensitive)	Root length and biomass, shoot length and biomass, glycine betaine, calcium and potassium uptake		
7	*Pseudomonas-n, Bacillus* -L, *Mucilaginibacter*-K *and Rhizobium*-L	*Arabidopsis thaliana*	Colonization density and growth	Na^+ uptake and H_2O_2 generation	Fan et al. (2020)
8	*Salt tolerant isolate* KH-2 + *Azotobacter*	*Oryza sativa*	Yield of rice grain	H_2O_2 generation	Javed et al. (2020)
9	*Bacilluis subtilis* + *Arthrobacter sp.*	*Triticum aestivum L.*	Plant growth	Ascorbate Peroxidase (APX), CAT, Glutathione reductase	Upadhyay et al. (2012)
10	*Pseudomonas fluorescens* + *Pseudomonas putida*	*Zea mays*	Germination rate, plant dry weight, number of leaves and leaf area	Na^+ uptake and H_2O_2 generation	Noumavo et al. (2013)
11	*Stenotrophomonas maltophilia*	*Triticum aestivum L.*	Shoot and root length, chlorophyll content, SOD, CAT and Peroxidase (POX)	Ethylene, Proline, Malondialdehyde (MDA)	Singh and Jha (2017)

(continued)

Table 15.1 (continued)

S. no	PGPR	Host	Increase	Decrease	References
12	*Pseudomonas putida* UW4	Canola plants	Shoot, fresh biomass	ROS generation	Cheng et al. (2007)
13	*Consortium of Aeromonas sp.*	Saltsensitive and salt tolerant *Triticum aestivum* L.	Chlorophyll content, stomatal conductance, transpiration rate, grain yield, biomass, nitrate reductase activity	Na^+ uptake and H_2O_2 generation	Nawaz et al. (2020b)
14	*Bacillus licheniformis, Pseudomonas plecoglossicida*	*Helianthus annus*	Biofilm formation, rhizospheric adherence, colonization of bacteria, CAT, SOD and guaiacol peroxidase	Na^+ uptake	Yasmeen et al. (2020)
15	*Bacillus megaterium* ZS-3	*Arabidopsis thaliana*	K^+/Na^+ ratio	Na^+ uptake	Shi et al. (2022)
16	*Bacillus amyloliquefaciens*	*Capsicum annum* cv.	Seedling growth, chlorophyll content, potassium and phosphate	Concentration of calcium, sodium and cadmium	Kazerooni et al. (2021)
17	*Bacillus cereus, Serratia marcescens and Pseudomonas aeruginosa*	*Triticum aestivum*	Membrane stability, potassium, calcium and antioxidant enzymes	Na^+ uptake	Desoky et al. (2020)
18	*Bacillus megaterium* ZS-3, *Serratia marcescens 90-166 and Pseudomonas fluorescens* WCS417	*Arabidopsis thaliana*	Lateral root formation and redistribution of auxin	Primary root formation, Na^+ uptake and ROS	Shi et al. (2022) and Wang et al. (2021)
19	*Bacillus megaterium*	*Arabidopsis thaliana*	Number of root hairs and lateral roots, auxin and ethylene	Generation of ROS	Lopez-Bucio et al. (2007)
20	*Bacillus flexus*	*Limonium sinense*	Colonization, growth and chemotaxis	Na^+ uptake	Xiong et al. (2020)
21	*Alcaligenes faecalis*	*Triticum aestivum*	Efficiency of PSII	Na^+ uptake	Babar et al. (2021)
22	*Bacillus magaterium*	*Arabidopsis thaliana*	Root architecture and redistribution of auxin	Na^+ uptake	Wang et al. (2021)

23	*Leifsonia soli* SE134	Waito C rice	Gibberellic acid	ROS generation	Lucke et al. (2020)
24	*Ochrobactrum intermedium* strain L115	*Arachis hypogaea*	Root and shoot length, fatty acid unsaturation, phosphatidylcholine, structural integrity, root and shoot biomass	Generation of ROS and Na^+ uptake	Paulucci et al. (2015)
25	*Stenotrophomonas maltophilia* BJ01 and *Bradyrhizobium japonicum*	*Arachis hypogaea*	Tolerance to salt stress	Proline concentration	Alexander et al. (2020)
26	*Bacillus* sp. L81 *and Aeromonas* sp. AMG272	*Oryza sativa*	Plant growth and Salt stress tolerance	Ascorbate, peroxidase and superoxide dismutase	García-Cristobal et al. (2015)
27	*Pseudomonas pseudoalcaligenes and Bacillus pumilus*	*Oryza sativa*	Shoot and root length	Na^+ uptake and H_2O_2 generation	Jha and Subramanian (2014)
28	*Bacillus amyloliquefaciens* CZ-6	Winter Jujube	Population of halotolerant fungus *Tausonia pullulans*	Na^+ uptake	Zhou et al. (2021)
29	*Pseudomonas stutzeeri and Bacillus toyonenes*	*Solanum lycopersicum*	Root and shoot growth, Antioxidant enzymes	Reduced ROS	Essalimi et al. (2022)
30	*Pseudomonas oryzihabitans*	*Solanum lycopersicum*	Plant growth, upregulation of stress-responsive genes	Xylose, mannose, sedoheptulose and rhamnose	Mellidou et al. (2021)
31	*Mesorhizobium loti* MAFF303099	*Lotus japonicus*	*nifA and acdS*	Na^+ uptake	Nukui et al. (2006)
32	*Arthrobacter protophormiae* (SA3), *Rhizobium 15 leguminosarum* (R) *and Glomus mosseae* (G)	*Pisum sativum*	Plant growth and biomass	Proline and Lipid peroxidation	Barnawal et al. (2014)
33	*Enterobacter cloacae*	*Zea mays*	Plant growth and salt stress tolerance	Antioxidant enzymes	Ali et al. (2022)
34	*Dietzia natronolimnaea*	*Triticum aestivum*	Root architecture, upregulation of SOS, TaMYB and WRKY TFs	Na^+ uptake	Bharti et al. (2016b)

(continued)

Table 15.1 (continued)

S. no	PGPR	Host	Increase	Decrease	References
35	*Bacillus tequilensis*	*Triticum aestivum*	Phosphate solubilization, chlorophyll content, SOS1 and SOS2 upregulation	Na^+ uptake	Haroon et al. (2022)
36	*Arthrobacter protophormiae* (SA3), *Bacillus subtilis* (LDR2) *and Dietzia natronolimnaea* (STR1)	*Triticum aestivum*	Stomatal conductance, Efficiency of photosystem II, TaCTRI, TaDREB2	AUX/IAA1 gene	Barnawal et al. (2017)
37	*Bacillus megaterium*	*Zea mays*	PIP gene for aquaporins	Na^+ uptake	Marulanda et al. (2010)
38	*Bacillus pumilus* + *Pseudomonas pseudoalcaligenes*	*Oryza sativa*	Glycine betaine and antioxidant enzymes	Stress	Jha et al. (2011)
39	*Pseudomonas putida*	*Triticum aestivum*	Growth and nutrient uptake	ROS accumulation	Ullah et al. (2022)
40	*Bacillus pumilis*	*Triticum aestivum*	Biofilm formation, Improved colonization	Na^+ uptake	Ansari et al. (2019)
41	*Bacillus endophyticus, Bacillus tequilensis, Planococcus rifietoensis, Variovorax paradoxus and Arthrobacter agilis*	*Salicornia europea*	Colonization, Germination rate and Seedling growth	Na^+ uptake and H_2O_2 generation	Wu et al. (2011)
42	*Achromobacter piechaudii*	*Solanum lycopersicum*	Weight of seedlings, water uptake, potassium and phosphorus absorption	Ethylene production	Mayak et al. (2004)
43	*Burkholderia cepacia* SE4, *Promicromonospora* sp. SE188 *and Acinetobacter calcoaceticus* SE370	*Cucumis sativus* L.	Water potential, salicylic acid and Gibberellin	Electrolyte leakage, abscisic acid	Kang et al. (2014)
44	*Bacillus subtilis* GOT9	*Arabidopsis thaliana* and *Brassica campestris*	Lateral growth and SOS1 gene upregulation	ROS generation	Woo et al. (2020)
45	*Bacillus fortis* SSB_{21} *and Pseudomonas aeruginosa* SSB_{13}.	*Capsicum annum* L.	Chlorophyll content, relative water content, potassium ions, *CAPIP2, CaKR1, CaOSM1* and CAChi2 upregulation	Na^+ uptake	Yasin et al. (2018)

46	*Lactobacillus sp., Pseudomonas putida*	*Lactuva sativa* and *Raphanus sativus*	Radicle length	Na^+ uptake	Hussein and Joo (2018)
47	*Azotobacter chroococcum*	*Solanum lycopersicum*	Radicle and plumule length	Na^+ uptake	Mayak et al. (2004)
48	*Bacillus megaterium and Enterobacter sp.* UPMR18	*Abelmoschus esculentus* L. (Okra)	Germination percentage, growth parameters, chlorophyll content, SOD, glutathione reductase, monohydroascorbate reductase, ascorbate peroxidase and CAT	Na^+ uptake and H_2O_2 generation	Habib et al. (2016)
49	*Pseudarthrobacter oxydans and Bacillus zhangzhouensis*	*Beta vulgaris* L.	Chlorophyll, carotenoid, stomatal conductance, intercellular CO_2	Ion toxicity	Redondo-Gómez et al. (2022)
50	*Pseudomonas putida* UW4	*Camelina sativa* (camelina)	Seed and oil quality	Na^+ uptake	Heydarian et al. (2016)
51	*Aneurinibacillus aneurinilyticus and Paenibacillus sp.*	*Phaseolus vulgaris* (French bean) *Zea mays* L.	Seedling and plant growth, IAA production, phosphate solubilization, ammonia production	ROS generation	Gupta and Pandey (2019) Zhou et al. (2019)
52	*Staphylococcus kloosii* EY37 *and Kocuria erythromyxa* EY43	*Raphanus sativus* L. (radish)	Root and shoot biomass, number of leaves, chlorophyll and relative water content	Na^+ uptake and H_2O_2 generation	Yildirim et al. (2008)
53	*Serratia and Klebsiella*	*Triticum aestivum*	Shoot biomass and SOD	Na^+ uptake	Acuna et al. (2019)
54	*Microbacterium oleivorans, Brevibacterium iodinum and Rhizobium massiliae*	*Capsicum annum* L.	Height and weight of plants, chlorophyll content, ascorbate peroxidase, guaiacol peroxidase and CAT	Ion uptake	Hahm et al. (2017)
55	*Kocuria rhizophilia* Y1	*Zea mays*	Upregulated antioxidant and salt tolerance genes	Na^+ uptake and H_2O_2 generation	Li et al. (2020)
56	*Ochrobactrum pseudogrignonense*	*Triticum aestivum L.*	Germin-like proteins, SOD, peroxidase and proline	Na^+ uptake and H_2O_2 generation	Chakraborty et al. (2019)

15.7 Gene Expression Profiles in Plants Inoculated with Halotolerant PGPR

The genes involved in Salicylic acid, Jasmonic acid and Ethylene metabolic pathways such as *NPR1, PR1, ERF1, LOX2* and *AOS* were found to be upregulated at different stages of growth in ZS-3 inoculated *A. thaliana* to develop Induced Systemic Tolerance (IST) against salinity stress. The inoculated plants have also shown increased expression of *HKT1, NHX1, AVP1* and SOS pathway genes involved in constraining Na^+ uptake (Shi et al. 2022). During salinity stress, activation of SOS pathway results in the efflux of Na^+ (Bomle et al. 2021). The higher expression of SOS1 and SOS2 genes were observed in *Triticum aestivum* plants inoculated with *B. tequilensis*. Further, seed priming of *Triticum aestivum* with this PGPR positively influenced phosphate solubilization, chlorophyll content and stress tolerance (Haroon et al. 2022). The activation of *nifA* genes responsible for nitrogen fixation regulates the expression of *acdS* gene which produces ACC deaminase when *Lotus japonica* is inoculated with bacterial strains (Nukui et al. 2006). Inoculation of PGPR to plants subjected to salinity stress led to a decrease in the expression of stress-responsive WRKY2 transcripts (Kazerooni et al. 2021). The inoculation of *Arthrobacter protophormiae*, *Rhizobium leguminosarum* and *Glomus mosseae* showed better improvement in growth and biomass of *Pisum sativum* with reduction in proline and lipid peroxidation. These kinds of studies paved the way for gaining new insights into co-inoculation of bacteria to mitigate stress conditions (Barnawal et al. 2014).

Whilst increased expression of stress responsive genes such as Antibiotic-resistant Iturin C (*ItuC*), biosurfactant-producing genes (*sfp*) and the genes responsible for the production of antioxidant enzymes are upregulated upon exposure of PGPR *Enterobacter cloacae* to salinity, inoculation of this bacteria to maize plants led to increased expression of genes encoding antioxidant enzymes and osmoprotectants along with decrease in stress-responsive genes (Ali et al. 2022). *TaMYB* genes responsible for regulating proline and malondialdehyde and WRKY transcription factors that are crucial for the expression of stress-related genes and ROS scavenging were upregulated in *Triticum aestivum* plants inoculated with *Dietzia natronolimnaea* (Bharti et al. 2016a, b). The upregulation of *TaCTRI* and *TaDREB2* genes responsible for stress tolerance and downregulation of *AUX/IAA1* gene encoding for transcriptional repression of auxin-responsive genes upon inoculation of *Arthrobacter protophormiae* (SA3), *Bacillus subtilis* (LDR2) and *Dietzia natronolimnaea* (STR1) to *Triticum aestivum* plants were observed. Further, the inoculated plants have also shown improved stomatal conductance and increased efficiency of photosystem II activity (Barnawal et al. 2017). The seeds of *Zea mays* inoculated with *Kocuria rhizophilia* Y1 showed enhanced growth and biomass of plants, increased relative water content and photosynthetic performance by maintaining the levels of phytohormones such as IAA and ABA. Moreover, alleviation of salt stress through the regulation of antioxidant genes like *ZmGR1* and *ZmAPX1* and salt tolerant genes including *ZmNHX1, ZmNHX2, ZmNHX3,*

ZmWRKY58 and *ZmDREB2A* were observed (Li et al. 2020). Priming of seeds of *Arabidopsis thaliana* with *Burkholderia phytofirmans* exhibited changes in the expression of genes encoding proteins involved in ion homeostasis, ROS scavenging and ABA-dependent pathways resulting in increased tolerance towards salinity stress (Pinedo et al. 2015).

An adequate number of aquaporins present in the plasma membrane are very crucial to increasing transcellular water exchange in plants. The inoculation of *Bacillus megaterium* to *Zea mays* upregulated the PIP genes thereby facilitating the increased water transport in roots (Marulanda et al. 2010). Stress-responsive genes such as *CaPIP2, CaKR1*, *CaOSM1* and *CaChi2* are abundantly expressed in *Capsicum annum* L. treated with rhizobacterial strain *Bacillus fortis* SSB21 and *Pseudomonas aeruginosa* SSB13. The inoculated plants showed a reduction of hydrogen peroxide, malondialdehyde and proline signifying the mitigation of osmotic stress by rhizobacteria (Yasin et al. 2018).

15.8 Methods for PGPR Inoculation

The different inoculation methods employed to deliver PGPR have variable effects on the growth parameters of the plant. The PGPR inoculum can be administered as a single strain or as a consortium of multiple strains. PGPR when delivered in consortium have enhanced stress alleviation and better growth promotion. Also, the favourable responses observed in the laboratory-based studies are hard to replicate under field conditions primarily due to variations in inoculation methods and external factors like temperature, humidity and light (Lopes et al. 2021). Since several crucial growth factors like soil pH, water content, soil organic matter composition, carbon-nitrogen content, geographical factors and climate are difficult to predict, standardized inoculation methods should be practiced for successful colonization of the plant rhizosphere region (Bossio et al. 1998; Drenovsky et al. 2004; Kristin and Miranda 2013). The moisture and oxygen availability of the soil directly influences the microbial diversity of the rhizosphere which rapidly changes during flooding and drought conditions thereby reducing the microbial load of the soil (Gouda et al. 2018). Therefore, measures to facilitate the colonization of bacteria to plant material by seed treatments, continuous supply of inoculum through foliar spray, root dipping or irrigation methods and a combination of these can be administered to achieve PGPR-mediated salinity stress alleviation and consistent growth promoting effect in plants. During salinity stress, the fluctuations in environmental conditions pose an increased threat to plant well-being, thereby making it essential to employ several inoculation strategies to amend the soil microbial diversity (Mahmood et al. 2016a, b). Different methods of PGPR administration to the plant material and soil substratum have been standardized and the variations of these methods have been customized based on the survival efficacy, colonization factors and plant–microbe signalling properties. Figures 15.3 and 15.4 show different methods that are employed for halotolerant PGPR inoculation.

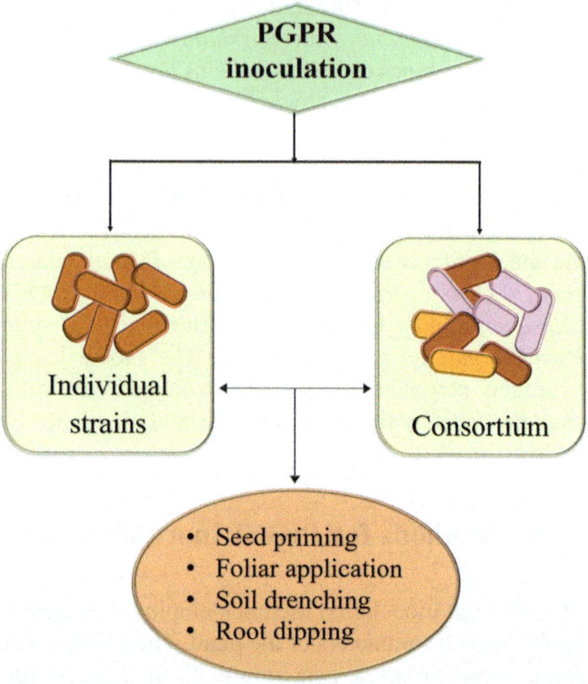

Fig. 15.3 Isolation, characterization, large-scale production and different formulations of PGPR with various routes of administration

Seed priming is an efficient method used for PGPR inoculation that involves incubation of seeds with the bacterial culture for a definite period of time before sowing onto the soil thereby facilitating rhizosphere colonization by bacteria during the seed germination process (Taylor and Harman 1990; Philippot et al. 2013). Previous studies have demonstrated that by priming the seeds of *Oryza sativa* and *Cicer arietinum* L., increased biomass and yield can be obtained (Ullah et al. 2017; Khan et al. 2019). Kernels from *Zea mays* L. upon biopriming with *Azotobacter* and *Azospirillum* sp. PGPRs resulted in higher growth, improved yield and faster development (Sharifi 2011).

Foliar application method has been proven to enhance interaction of PGPR with the plant shoot tissues thereby providing higher chances of colonization (Efthimiadou et al. 2020). PGPR is inoculated to the phyllosphere regions of the plant in the form of solution, either as a foliar spray or foliar injection (Basha et al. 2006). Soil drenching is a method where a surplus of PGPR inoculum is supplied to the soil either as single or multiple applications throughout the lifespan of the plant. Since this method involves a rapid increase in PGPR density in the rhizosphere soil, it is proven to offer improved chances of PGPR colonization (Hernández-Montiel et al. 2017). For better root colonization of PGPR, root inoculation is practiced wherein roots in the seedling stage are submerged in PGPR solution for a definite period before transplantation onto the soil. The *Oryza sativa* seedlings when subjected to root submersion showed an increase in plant height and panicle length

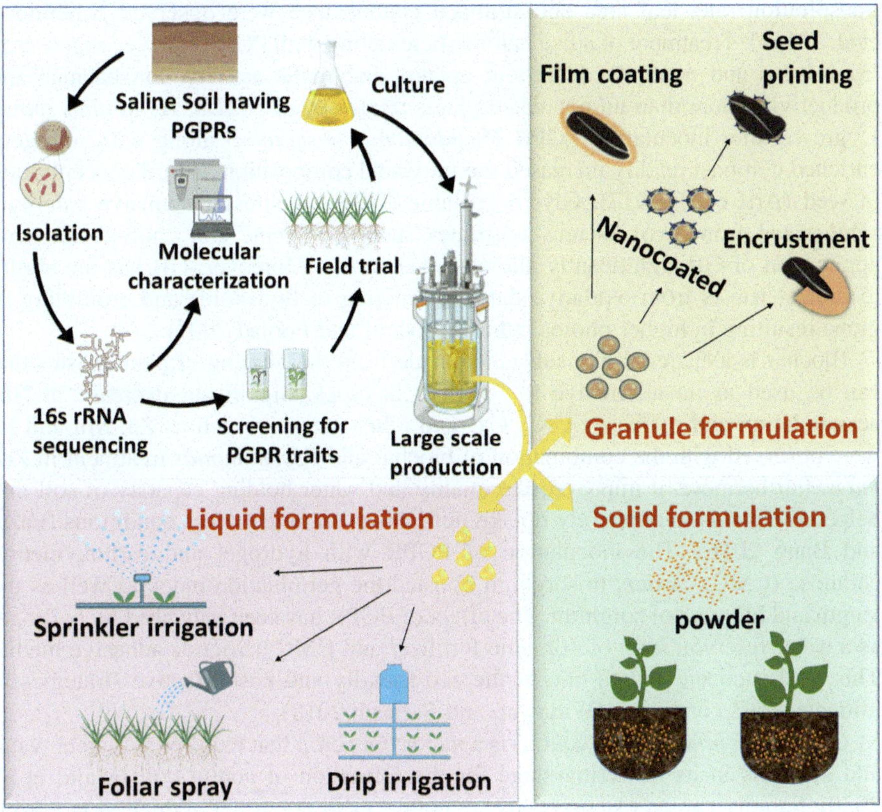

Fig. 15.4 Methods for PGPR inoculation to plants

(Ullah et al. 2017). Root inoculation using hydroponics with PGPR suspended in a nutrient medium showed increased yield in tomato and cucumber plants (Gül et al. 2013).

15.9 Increasing the Efficiency of Halotolerant PGPR

Increased leaf thickness with a reduction in leaf area along with downregulation of photosynthesis and metabolism is a strategy employed by plants to decrease salinity stress-induced damage (Acosta-Motos et al. 2015). Silicon deposition in the cell wall improves plant growth and tolerance by maintaining the apoplastic barrier (Lux et al. 2002). The application of silicon spray along with PGPRs such as *Enterobacter cloacae* and *Bacillus drentensis* in Mung beans significantly increased the physiological and biochemical parameters thereby mitigating salinity stress. Further, higher seed yield under high salinity in addition to the increase in plant growth, weight, transpiration rate, leaf area and stomatal conductance were observed (Mahmood et al. 2016b). Treatment of silica nanoparticles along with PGPR relieved maize crop from stress and maximized nutrient uptake, enzymatic activity, ion balance and productivity more than uninoculated plants (Hafez et al. 2021a). *Helianthus annus* L. grown after inoculating PGPR *Pseudomonas aeruginosa* along with nitrogen-enriched compost readily increased the fatty acid composition and oil concentration in seed (Arif et al. 2017). Glycine betaine (GB) is a stress responsive osmolyte synthesized from two discrete substrates such as glycine and choline. External application of GB significantly alleviates salt stress. Moreover, GB has the ability to protect leaves from oxidative damage, bulging of thylakoids and crumbling of grana resulting in higher photosynthesis (Ashraf and Foolad 2007).

Biochar is a charcoal-like substance made from the burning of plant wastes that can be used as an alternative to fertilizer in crops. Significant decrease in Na^+ accumulation and increase of K^+, Ca^+ and other nutrients such as Zn, Mn and Fe were observed with the combination of biochar and *Pseudomonas* treatment in *Zea mays*. Furthermore, it improved the quality and water holding capacity of soil that helped the plants to efficiently uptake nutrients even under stress conditions (Fazal and Bano 2016). The inoculation of PGPR with hydrogel and carboxymethyl cellulose (CMC) carriers in sorghum boosted the germination index as well as the length and biomass of sorghum. The effect of PGPR has been enhanced by hydrogel as a water reservoir with bio-organic fertilizer and CMC carrier as adhesive media. This is also considered as one of the eco-friendly and cost-effective strategies to mitigate stress conditions (Widawati and Suliasih 2018).

Gossypium hirsutum L. (cotton) is a major fibre crop that requires sufficient water and nutrients in its flowering stage for the formation of cotton boll (Majid et al. 2020). Salinity causes extensive loss in cotton yield due to the reduction in germination rate, dry weight of the cotton seedlings and length of root and shoot (Wu et al. 2012). The direct addition of bacterial cells in liquid form into the soil had an adverse effect on the survival of cells in the soil. Encapsulating the microbes using carrier molecules helps for long-term survival by protecting them from external stress and controlled release of microbes into the soil. The cells sheltered inside the beads have the capability to move towards the rhizosphere region for better colonization and promote the growth of the cotton plants when compared to unencapsulated cells (Trevors et al. 1993). The use of carrier immobilized PGPR showed better tolerance without genetic manipulations in plants (Chanratana et al. 2019). Whilst untreated *Klebsiella oxytoca* cells were struggling for initiation of colonization, microencapsulation of the cells with sodium alginate effectively increased the colonization potential of the bacteria within the roots and rhizosphere, thereby resulting in higher germination rate and higher cotton growth. Since the bacteria are released in a controlled manner, the retention time of microbes was prolonged and were shown to retain even after 28 days (Wu et al. 2014a).

Chitosan, a derivative of chitin, a polycationic biodegradable polymer is routinely used for encapsulating PGPR for sustainable and eco-friendly crop production (Xing et al. 2015). *Pseudomonas putida* encapsulated with chitosan derived from the exoskeletons of *Callinectes aminicola* and *Cardisoma armatus* enhanced the height and diameter of the crown, leaf area and height of *Zea mays*. It sturdily maintained the growth and biomass of plant and crown even in the presence of higher salt concentrations up to 400 mM (Agbodjato et al. 2021). Seeds of *Solanum lycopersicum* inoculated with chitosan immobilized aggregated *Methylobacterium oryzae* limited the electrolyte leakage and uptake of Na^+ ions and improved the physiological response of plants under salinity stress conditions (Chanratana et al. 2019).

Foliar application of halotolerant PGPR is also one of the efficient ways to boost the growth and development of plants. Seeds of *Capsicum annum* L. treated with *Bacillus thuringiensis* followed by foliar application of chitosan on grown plants has led to increased efficiency of PSII under salinity stress conditions correlating with activity of antioxidant enzymes and increased fresh weight and yield (Kahtani et al. 2020). The combination of three PGPRs, *Azospirillum lipoferum, Pseudomonas fluorescens* and *Pseudomonas putida* along with chitosan caused a significant increase in the growth of *Zea mays* (Agbodjato et al. 2016). Further, *Zea mays* inoculated using a consortium of *Pseudomonas taiwanensis* and *Pantoea agglomerans* with nano chitosan displayed better phosphate solubilization, germination rate, chlorophyll and carotenoid content (Agri et al. 2021).

Alginate is one of the potential and biodegradable carriers that are used in agriculture for controlled release of molecules. Bentonite is an absorbent clay that increases water holding capacity in the soil thereby enhancing the soil moisture by up to 12% along with increased productivity of millet grains (Mi et al. 2020). Similarly, starch is also a biodegradable agent which can be used for controlled release of molecules by PGPR. The mixing of alginate, starch and bentonite with the PGPRs significantly increased and retained the microbial colonies around the roots even after 64 days when compared to the free cell treatment. *Pseudomonas putida* encapsulated with the mixture of alginate, starch and bentonite significantly increased tolerance against salinity and showed increased production of IAA and gibberellin resulting in the higher biomass of cotton. The root colonization was observed to be higher in the middle region of the root as it contains more root hairs and exudates than the tip and base of the roots (He et al. 2016). The swelling of beads after addition of soil positively affects the release of *Raoultella planticola* from the alginate–starch–bentonite beads (Wu et al. 2012). The diameter of the microcapsule and releasing property of bacteria is increased with increasing concentrations of starch and sodium alginate. However, high yield was obtained with a higher concentration of sodium alginate and the biodegradability of the microcapsule is directly proportional to the concentration of starch (Wu et al. 2014b).

Polyvinyl alcohol (PVA) is a nontoxic biopolymer having high mechanical strength and chemical resistance. As PVA has good compatibility, its combination with alginate shows an effective outcome in an agricultural field. Immobilization of *Bacillus pumilis* in PVA/Sodium alginate (1:1) beads incorporated into the seedlings of *Pharbitis nil* displayed increased growth and diameter of root and shoot providing tolerance to salt stress. It also improved the phosphorus and nitrogen in the rhizospheric region which is crucial for higher crop yield (Ansari et al. 2019). Seed coating is a method of concealing microbial inoculants onto the surface of seeds with biopolymers having porous nature. It assists in the delivery of targets to their respective site and provides a suitable environment for seed germination (Rocha et al. 2019). Seeds of *Zea mays* L. were coated with *Azospirillum brasilense* using pure polymers and mixed polymers of starch, gelatin and PVA. Results from the study evidenced that the mixture of triple polymer for seed coating is less efficient than starch–gelatin mixture and single polymer. The starch–gelatin mixture maintained the structural integrity and uniform distribution of seeds. It improved the viability of cells and protection against stress conditions. When compared to uncoated, single coated and triple mixture coated; the starch–gelatin coated seeds exhibited a higher germination rate (Vercelheze et al. 2019).

The use of organic carriers like compost, peat, biogas slurry and crushed corn cob also increases the activity and shelf life of PGPR. Carriers support and maintain microbial growth by protecting them from harsh environmental conditions. Assessment of the effect of compost, biogas slurry and peat on the bacterial consortium like *Bacillus cereus* strain Y5, *Bacillus* sp. Y14 and *Bacillus subtilis* Y16 inoculated to *Triticum aestivum* under saline conditions resulted in increased efficiency of photosynthesis, increased uptake of water, reduced Na^+ uptake, increased K^+ uptake, increased chlorophyll a and b contents. It had a considerable effect on the growth, yield, height of the plant, length of the spikes, total grains in a spike and quality of straw.

Gelatin is a cost-effective and biodegradable compound that is made up of different amino acids. The blending of gelatin with alginate encapsulation of microbes effectively supports the development and tolerance of plants to abiotic stresses (Pour et al. 2019). The encapsulation of *Pantoea agglomerans* strain KL to sodium alginate and inoculation to *Oryza sativa* plants led to improvement in salinity tolerance (Bhise and Dandge 2019). The significance of zinc oxide nanoparticles in seed germination and protection from plant pathogens is well studied (Faizan et al. 2020). Salicylic acid synthesized through the isochorismate pathway is important for providing disease resistance in plants (Lefevere et al. 2020). The encapsulation of halotolerant bacteria *Pseudomonas* sp. using alginate along with salicylic acid and zinc oxide nanoparticles treatment improved the resistance of *Oryza sativa* plants to salinity and showed antifungal activity against *Sclerotium rolfsii* (Panichikkal et al. 2021). The maize seeds soaked in *Pseudomonas fluorescens* and *Enterobacter aerogenes* incorporated with biogas slurry and chemical nitrogen fertilizers showed reduced Na^+ uptake and proline content and enhanced K^+ uptake and improved quality of seed, plants and soil (Ahmed et al. 2014). Encapsulation of *Raoultella planticola* using sodium bentonite and alginate has been done to develop a cost-effective and long durability biofertilizer formulation. The smaller amount of sodium alginate needed for microcapsule is effective when it is blended with bentonite. Moreover, swelling, biodegradability and rate of release were positively correlated with the increasing concentration of sodium alginate (He et al. 2015). Hydrocarbon degrading halotolerant endophytic bacteria *Bacillus safensis* strain ZY16 isolated from *Chloris virgata* showed the ability to degrade aromatic hydrocarbons like naphthalene, phenanthrene, n-hexadecane and octacosane. Further, *B. safensis* has the ability to tolerate salt stress, produce biosurfactants and promote the growth and development of plants (Wu et al. 2019). Calcium-alginate is another combination that is formed by the interaction between Na and Ca^{2+} ions (Vena et al. 2016). The calcium-alginate microcapsule has been prepared by a pore-coagulation bath whose swelling and viability were found to be higher with increased sodium alginate concentrations. It is one of the efficient materials for bacterial encapsulation as it provides a better environment for the microbes to grow and release for a long time even under saline conditions. The germination rate of the cotton was significantly higher under salinity in the case of calcium-alginate encapsulation than in the free cell-treated plants and control. Similarly, plant height, root length and dry weight of the seedlings were greater in comparison with control plants (Wu et al. 2011).

Seeds of *Helianthus annus* L. soaked in *Azospirillum* and *Pseudomonas* led to plants with higher Relative water content, carotenoid and leaf protein contents (Naz and Bano 2015). Zeolites are crystalline solids of aluminium silicate which are used as a carrier of nutrients in plants to promote their growth and development (Ramesh and Reddy 2011). Zeolite behaves as a shelter for microorganisms as they harbour inside the clay until experiencing a suitable environment to colonize. Chabazite is one of the zeolites used for agricultural purposes. *Ranunculus asiaticus* inoculated with rhizobacteria along with chabazite and fertilized substrate was efficient in enhancing the growth and floral characteristics; explicitly the height and biomass of plants, number of leaves, number of flowers and diameter of flowers. Chlorophyll content was also found to be higher in PGPR with chabazite and fertilized substrate than individual application and control (Prisa 2020). Humic acid containing nitrogen, oxygen, hydrogen and sulphur generated from the decomposition of plants and animals supported the growth of soil microorganisms and maintains soil pH, improves soil texture, fertility and water holding capacity thereby positively affecting crop productivity. Individual application of zeolite, humic acid and PGPR on *Atriplex canescens* L. increased the height and diameter of stem and root biomass (Abdi 2019). Treatment of *Phaseolus vulgaris* L. by PGPR along with silicon fertilizer improved the growth, yield and osmotic tolerance under high saline conditions (Kumar et al. 2020a).

Silver nanoparticles became a good composite to stimulate the growth and yield of many crops. They are known to increase the production of catalase and peroxidase to scavenge free radicals generated during stress (Krishnaraj et al. 2012). Application of silver nanoparticles with *Bacillus pumilus* and *Pseudomonas moraviensis* onto *Allium cepa* L. (onion) increased the fresh weight of root, leaves and bulb. Individual inoculation of PGPRs and silver nanoparticles alone showed higher flavonoid content in leaves. However, the combination of PGPRs with silver nanoparticles increased the flavonoid content in the bulb but reduced it in the leaves. The carotenoid contents were increased and positively regulated by PGPR. These effects are being promoted by *Pseudomonas moraviensis* under stress conditions and *Bacillus pumilus* under control conditions (Jahangir et al. 2020). Silver nanoparticles synthesized from *Tagetes erecta* combined with *Bacillus* sp. had a profound effect on seed germination, phosphate solubilization and IAA production in the presence of salt in *Zea Mays* (Kumar et al. 2020b). Although irrigation with municipal wastewater provides a profound number of organic compounds and nutrients required for plant growth, the substantial amount of salt and heavy metals interfere with the development of plants. Maize seeds treated with PGPRs, like *Pseudomonas fluorescens, Bacillus cereus, Planomicrobium chinese* as well as with silver nanoparticles prior to sowing showed better root and shoot weight, length, leaf area, potassium and chlorophyll content (Khan and Bano 2016). The treatment of *Comamonas testosteroni* and silver nanoparticles to *Linum usitatissimum* L. (linseed) plants under salinity stress led to the increase of photosynthetic pigments and increase in the activity of antioxidant enzymes such as SOD, CAT, APX and peroxidase. The decrease of H_2O_2 and malondialdehyde content in the bacteria and silver nanoparticles inoculated plants indicated the alleviation of salt stress (Khalofah et al. 2021). Cucumber seeds inoculated with *Pseudomonas putida* and *Pseudomonas stutzeri* followed by spraying of silver nanoparticles had shown increased chlorophyll content, increased flavonoids, increased activity of Phenylalanine ammonia lyase and enhanced root length (Nawaz and Bano 2020).

Seed priming with silicon nanoparticles followed by inoculation of the seeds with PGPRs including *Pseudomonas fluorescens* and *Pseudomonas putida* provided tolerance against salinity stress (Hatami et al. 2021). Seeds of *Vicia faba* (faba bean) inoculated with the mixture of *Rhizobium leguminosarum* and *Bacillus circulans* and sprayed with potassium silicate showed reduced uptake of Na^+ by the plants. Improved nutrient uptake from the soil and its flow from roots to leaves and increased number of nodules and root hairs are observed in inoculated plants (Hafez et al. 2021b). Addition of *Pseudomonas pseudoalcaligenes* and *Pseudomonas putida* bacterial cultures to the rhizospheric region of *Coriandrum sativum* plants followed by foliar spraying of silicon fertilizer has shown increased plant biomass, increased antioxidant activity and salinity stress tolerance (Al-Garni et al. 2019). Nanosilica is a nontoxic metal oxide and acts as a better candidate to carry and supply nutrients because of its mesoporous nature. The seeds of *Zea mays* treated with the consortium of *Pseudomonas taiwanensis* and *Pantoea agglomerans* along with nano silicon dioxide showed increased total chlorophyll and carotenoid content along with increased activity of antioxidant enzymes. Apart from these plant parameters, it also had a beneficial effect on soil quality through the enhanced activities of fluorescein diacetate, dehydrogenase and alkaline phosphatase responsible for the recycling of nutrients (Kukreti et al. 2020). Exopolysaccharide-producing bacteria *Citrobacter freundii* along with silicon nanoparticles is employed for inoculating the seeds of *Solanum lycopersicum.* There is a significant improvement in root and shoot weight and a decrease in proline and antioxidant enzymes. The lowering of CAT, SOD and peroxidase revealed that the combination of *C. freundii* and silicon nanoparticle effectually mitigates the salt stress (Moshabaki et al. 2018).

Sodicity arises due to the presence of high amount of Na^+ salts relative to other cations in the irrigated water which negatively affects the nature of soil. Sodicity is toxic to plants and soil than salinity. Biochar produced from rice husks and corn stalks were combined with PGPRs to assess their effect on *Zea mays* L. plants exposed to sodicity. The combination of PGPR with biochar effectively improved the soil ecosystem by increasing the activity of urease and dehydrogenase enzymes. It enhanced photosynthetic pigments, relative water content and leaf area in plants.

Furthermore, higher yields were obtained in the PGPR and biochar applied plants than in untreated plants (Nehela et al. 2021). Similar observations were made in *Brassica napus* plants treated with biochar mixed with *Pseudomonas putida* and *Azotobacter chroococcum* (Ghassemi-Golezani and Abdoli 2022). The combined effect of *Pseudomonas psychrotolerans* CS51 and silicon in *Zea mays* led to increased flavonoid and polyphenol content and lower proline and sodium uptake resulting in considerable amelioration of salt stress (Kubi et al. 2021). Seeds of *Zea mays* soaked in *Pseudomonas* sp. culture were sowed with biochar and chemical fertilizer. In these combinations, biochar and *Pseudomonas* sp. reduced Na^+ uptake and increased the activity of peroxidase and electrolyte leakage similar to chemical fertilizer (Fazal and Bano 2016).

Vermicompost is the breaking down of organic matter by microorganisms that improves the quality of soil and supports microbial growth. As vermicompost houses many microbes it is beneficial for soil quality and plant growth. Treatment of *Mentha arvensis* plants with the bacterial strain *Exiguobacterium oxidotolerans* and a fungal strain *Glomus fasciculatum* together with vermicompost exhibited better growth and herb yield than individual inoculations under salinity stress conditions. The combination of bacterial and fungal cultures along with the vermicompost improved the quality of soil and led to higher tolerance to salinity stress (Bharti et al. 2016a).

Inoculation of bacterial strains *Stenotrophomonas, Raoultella* sp. and *Klebsiella* sp. isolated and identified from vermicompost to *Zea mays* L. under high saline conditions stimulated the uptake of magnesium and the activity of alpha-amylase that helps in the utilization of food reserves during germination (Zhou et al. 2019). Treatment of lettuce with the combination of vermicompost and *Pseudomonas fluorescens* showed a considerable improvement in the dry weight of the shoot and the uptake of nutrients such as phosphate, nitrogen, magnesium, iron and manganese. Similar results were obtained in the uptake of nutrients when phosphate sources are provided. These results suggested that organic vermicompost and PGPR can be used as alternatives to chemical fertilizers for sustainable agriculture (Khosravi et al. 2018).

DNA methylation is one of the protective mechanisms through which plants can mitigate stress. Whenever plants are exposed to stress conditions, cytosine hypermethylation of DNA help them to express stress-responsive genes thereby it can withstand adverse environment (Peng and Zhang 2009). The inoculation of PGPRs *Bacillus subtilis, Lactobacillus casei* and *Bacillus pumilus* along with zinc nanoparticles into *Solanum lycopersicum* improved tolerance to salinity through DNA hypomethylation (Hosseinpour et al. 2020). Extracts from seaweed *Macrocystis pyrifera* have the potential to support the growth of plants and relieve the damages caused by many biotic and abiotic stresses. Alginate and laminarin present in seaweed promote resistance to plants from salt stress (Vera et al. 2011). Mannitol regulates carbon and energy homeostasis, coenzymes and scavenges free radicals (Stoop et al. 1996; Sharma et al. 2014). The combined inoculation of PGPR *Azospirillum brasilense* and algal extracts from *Macrocystis pyrifera* considerably increased germination rate, seedling growth and plant development (Julia et al. 2020). The list of different reported PGPR formulations effective in alleviating the salinity stress responses in different host plants were indicated in Table 15.2.

Table 15.2 List of different PGPR formulations to alleviate the salinity stress in plants

S. no	PGPR	Carrier	Host	Increase	References
1	*Enterobacter cloacae* and *Bacillus drentensis*	Silicon spray	*Vigna radiata*	Plant growth, leaf area, transpiration rate, stomatal conductance and seed yield	Mahmood et al. (2016a)
2	*Pseudomonas*	Biochar	*Zea Mays*	Uptake of potassium, calcium, zinc, manganese and iron	Fazal and Bano (2016)
3	*Azospirillum lipoferum* 1103D + *Azotobacter chroococcum* 4103D + *Bacillus* sp. 3D	Hydrogel + carboxymethyl cellulose	*Sorghum bicolor*	Germination index, length and biomass of plants	Widawati and Suliasih (2018)
4	*Pseudomonas aeruginosa*	Nitrogen-enriched compost	*Helianthus annus* L.	Seed oil concentration, fatty acid composition and harvest index	Arif et al. (2017)
5	*Pseudomonas putida*	Chitosan	*Zea mays* L.	Increase in height, crown diameter and leaf area	Agbodjato et al. (2021)
6	*Bacillus thuringiensis*	Chitosan	*Capsicum annum* L.	Improved relative water content, chlorophylls a and b and fruit yield	Kahtani et al. (2020)
7	*Methylobacterium oryzae* CBMb20	Chitosan	*Solanum lycopersicum* Mill.	Decreased influx of Na^+, increased nutrient uptake and plant dry weight	Chanratana et al. (2019)
8	*A. lipoferum, P. fluorescens*	Chitosan	*Zea mays L.*	Seed vigour	Agbodjato et al. (2016)
9	*P. putida*	Chitosan		Shoot weight	
10	*P. fluorescens*	Chitosan		Number of leaves, aerial and underground biomass	
11	*A. lipoferum*	Chitosan		Leaf area	
12	*A. lipoferum, P. fluorescens, P. putida.*	Chitosan		Aerial and underground dry biomass, Nitrogen content	
13	*Pseudomonas putida*	Alginate-starch-bentonite	Cotton	Production of IAA, gibberellin, cotton biomass	He et al. (2016)
14	*Bacillus pumilus*	Polyvinylalcohol/ sodium alginate	*Pharbitis nil*	Length and diameter of plant, dry weight, improved soil properties	Ansari et al. (2019)

15	*Bacillus cereus* strain Y5, *Bacillus* sp. Y14 and *Bacillus subtilis* Y16	Compost, press mud, biogas slurry and peat	*Triticum aestivum*	Photosynthetic rate, uptake of water, transpiration rate, K^+ uptake, chlorophylls a and b	Salman et al. (2017)
16	*Pantoea agglomerans*	Sodium alginate	*Oryza sativa*	Root colonization ability, root and shoot length, fresh and dry weight, chlorophyll content	Bhise and Dandge (2019)
17	*Klebsiella oxytoca* Rs-5	Calcium-alginate	Cotton	Higher germination rate, plant height, root length and dry weight of the seedlings	Wu et al. (2011)
18	*Bacillus lumilis, Pseudomonas moraviensis*	Silver nanoparticles	*Allium cepa*	Fresh weight of leaves, roots and bulbs, increased sugar and flavonoid content of bulb	Jahangir et al. (2020)
19	*Bacillus* sp.	Silver nanoparticles	*Zea mays L.*	Seed germination, Phosphate solubilization, IAA production	Kumar et al. (2020b)
20	*Pseudomonas fluorescens, Bacillus cereus*	Silver nanoparticles	*Zea mays L.*	Root and shoot weight, length, leaf area, potassium and chlorophyll content	Khan and Bano (2016)
21	*Comamonas officinalis*	Silver nanoparticles	*Linum usitatissimum* L. (linseed)	Increased photosynthetic pigments, antioxidant enzymes, decrease in H_2O_2 and malondialdehyde	Khalofah et al. (2021)
22	*Pseudomonas fluorescens, Pseudomonas putida*	Silicon nanoparticle	*Melissa officinalis* L.	Stomatal opening, leaf water content, plant biomass and photosynthetic pigments	Hatami et al. (2021)
23	*Rhizobium leguminosarum* + *Bacillus circulans*	Potassium silicate	*Vicia faba* L. (faba bean)	Number of leaves, leaf area, stomatal conductance, vegetative growth, CAT, POD, SOD, photosynthetic machinery, number and dry weight of nodules, root hair formation, nutrient uptake	Hafez et al. (2021b)
24	*Pseudomonas taiwanensis* and *Pantoea agglomerans*	Nanosilicon dioxide	*Zea mays* L.	Chlorophyll and carotenoid content, CAT, POD, soil fertility, dehydrogenase and alkaline phosphatase	Kukreti et al. (2020)
25	*Pseudomonas putida* and *Pseudomonas stutzeri*	Silver nanoparticles	*Cucumis sativus*	Root length, chlorophylls a and b, flavonoids, leaf protein, phenylalanine ammonia lyase	Nawaz and Bano (2020)

(continued)

Table 15.2 (continued)

S. no	PGPR	Carrier	Host	Increase	References
26	*Azotobacter chroococcum* and *Pseudomonas koreensis*	Biochar	*Zea mays* L.	Soil ecosystem, leaf area, relative water content, total chlorophyll and carotenoid content, grain and stover yield	Nehela et al. (2021)
27	*Pseudomonas taiwanensis* and *Pantoea agglomerans*	Nanochitosan	*Zea mays L.*	Phosphate solubilization, siderophore production, enhanced seed germination, chlorophyll and carotenoid content, plant height	Agri et al. (2021)
28	*Pseudomonas pseudoalcaligenes* (KB-10) and *P. putida* (KB-25)	Silicon	*Coriandrum sativum*	Relative water content, photosynthetic pigments, peroxidase activity, plant biomass and salt tolerance	Al-Garni et al. (2019)
29	*Pseudomonas putida* Rs-198 and *Azotobacter chroococcum* Rs-106	Biochar	*Brassica napus* L.	Seed and oil yield, antioxidants, root length and diameter, shoot length and biomass, leaf area, chlorophyll content	Ghassemi-Golezani and Abdoli (2022)
30	*Azospirillum brasilense* Ab-V5	Starch, Gelatin	*Zea mays L.*	Bacterial viability and germination rate	Vercelheze et al. (2019)
31	*Exiguobacterium oxidotolerans* + *Glomus fasciculatum*	Vermicompost	*Mentha arvensis*	Better growth and herb yield	Bharti et al. (2016a)
32	*Pseudomonas fluorescens*	Vermicompost and phosphate sources	Lettuce	Dry weight of shoot, uptake of nutrients like P, N, Mg, Fe and Mn	Khosravi et al. (2018)
33	*Pseudomonas* sp.	Biochar	*Zea mays* L.	Increased activity of POD, electrolyte leakage	Fazal and Bano (2016)
34	*Pseudomonas psychrotolerans* CS51	Silicon	*Zea mays* L.	Increased flavonoid and polyphenol content	Kubi et al. (2021)
35	*Azospirillum* and *Pseudomonas*	Salicylic acid	*Helianthus annus L.*	Relative water content, carotenoids, phytohormone content, protein content	Naz and Bano (2015)
36	*Curtobacterium albidum* SRV4	Farmyard manure	*Oryza sativa*	Mineral restoration, plant height, tillers, panicles, yield, CAT, SOD, POX, APX	Vimal and Singh (2019)
37	*Azospirillum brasilense*	*Macrocystis pyrifera* algal extract	*Lactuca sativa*	Germination rate, plant growth, root biomass, seedling growth	Julia et al. (2020)

15.10 Conclusions and Future Prospects

Soils with high salinity pose a threat to agricultural productivity worldwide. Due to constantly changing climatic conditions, efforts are to be made to increase the salinity tolerance in agricultural crops. Development of salinity tolerant agricultural crops by genetic engineering or by breeding approaches is essential but is time consuming and expensive process. At this juncture, it is essential to gain deeper insights into cost-effective and environmentally friendly options of PGPR-mediated salinity tolerance in plants for sustainable agriculture. A detailed understanding of the mechanisms employed by the PGPR to facilitate the growth and development of plants and restoration of soil fertility is needed. Isolation of new halotolerant PGPR with novel traits important for inducing salinity tolerance in agricultural crops is required. Since majority of the isolated halotolerant PGPR show plant growth promotion in pot experiments conducted under laboratory conditions, field experiments with the single potent bacteria or bacterial consortium with the addition of stabilizing agents are required for evaluating the efficacy of the bacteria. Experiments emphasizing the stability of the PGPR in the soil and effective colonization potential in the rhizosphere or in the roots along with the commercialization of them are necessary.

References

Abbas G, Saqib M, Akhtar J (2015) Interactive effects of salinity and iron deficiency on different rice genotypes. J Plant Nutr Soil Sci 178:306–311

Abdi N (2019) Effects of the zeolite, plant growth promoting rhizobacteria (PGPR) and humic acid on early growth of *Atriplex canescens* L seedlings in Arak. Arid Biome 8(2):27–35

Acosta-Motos JR, Diaz-Vivancos P, Alvarez S, Fernandez-Garcia N, Sanchez-Blanco MJ, Hernandez JA (2015) Physiological and biochemical mechanisms of the ornamental *Eugenia myrtifolia* L plants for coping with NaCl stress and recovery. Planta 242(4):829–846

Acuna JJ, Campos M, Mora M, Jaisi DP, Jorquera MA (2019) ACCD-producing rhizobacteria from an Andean Altiplano native plant (*Parastrephia quadrangularis*) and their potential to alleviate salt stress in wheat seedlings. Appl Soil Ecol 136:84–190

Agbodjato NA, Noumavo PA, Adjanohoun A, Agbessi L, Baba-Moussa L (2016) Synergistic effects of plant growth promoting rhizobacteria and chitosan on in vitro seeds germination, greenhouse growth, and nutrient uptake of maize (*Zea mays* L). Biotechnol Res Int 1–11

Agbodjato N, Mikpon T, Babalola O, Dah-Nouvlessounon D, Amogou O, Lehmane H, Adoko M, Adjanohoun A, Baba-Moussa L (2021) Use of plant growth promoting rhizobacteria in combination with chitosan on maize crop: promising prospects for sustainable, environmentally friendly agriculture and against abiotic stress. Agron 11(11):2205

Agri U, Chaudhary P, Sharma A (2021) In vitro compatibility evaluation of agriusable nanochitosan on beneficial plant growth-promoting rhizobacteria and maize plant. Natl Acad Ssi Lett 44(6): 555–559

Ahmed M, Ahmad ZZ, Jamil M, Nazli F, Mohammad L, Zaman AMFU (2014) Integrated use of plant growth promoting Rhizobacteria, biogas slurry and chemical nitrogen for sustainable production of maize under salt-affected conditions. Pak J Bot 46(1):375–382

Alexander A, Singh VK, Mishra A (2020) Halotolerant PGPR *Stenotrophomonas maltophilia* BJ01 induces salt tolerance by modulating physiology and biochemical activities of *Arachis hypogaea*. Front Microbiol 11

Al-Garni SMS, Khan MMA, Bahieldin A (2019) Plant growth-promoting bacteria and silicon fertilizer enhance plant growth and salinity tolerance in *Coriandrum sativum*. J Plant Interact 14(1):386–396

Ali B, Wang X, Saleem MH, Sumaira Hafeez A, Afridi MS, Khan S, Zaib-Un-Nisa UI, do Amaral Júnior AT, Alataw A, Ali S (2022) PGPR-mediated salt tolerance in maize by modulating plant physiology, antioxidant defense, compatible solutes accumulation and bio-surfactant producing genes. Plants 11(3):345

Analin B, Mohanan A, Bakka K, Challabathula D (2020) Cytochrome oxidase and alternative oxidase pathways of mitochondrial electron transport chain are important for the photosynthetic performance of pea plants under salinity stress conditions. Plant Physiol Biochem 154:248–259

Ansari FA, Ahmad I, Pichtel J (2019) Growth stimulation and alleviation of salinity stress to wheat by the biofilm forming *Bacillus pumilus* strain FAB10. Appl Soil Ecol 143:45–54

Arif MS, Shahzad SM, Riaz M, Yasmeen T, Shahzad T, Akhtar MJ, Bragazza L, Buttler A (2017) Nitrogen-enriched compost application combined with plant growth-promoting rhizobacteria (PGPR) improves seed quality and nutrient use efficiency of sunflower. J Plant Nutr Soil 180(4): 464–473

Ashraf M, Foolad MR (2007) Roles of glycine betaine and proline in improving plant abiotic stress resistance. Environ Exp Bot 59(2):206–216

Avinash M, Bhakthi T (2017) Halophytes: potential resources for salt stress tolerance genes and promoters. Front Plant Sci 18(8):829

Babar M, Saif-ur-Rehman RS, Aslam K, Abbas R, Athar HR, Manzoor I, Kashif Hanif M, Naqqash T (2021) Mining of halo-tolerant plant growth promoting rhizobacteria and their impact on wheat (*Triticum aestivum* L.) under saline conditions. J King Saud Univ Sci 33(3):101372

Bakka K, Challabathula D (2020) Amelioration of salt stress tolerance in plants by plant growth-promoting rhizobacteria: insights from "Omics" approaches. In: Varma A, Tripathi S, Prasad R (eds) Plant microbe symbiosis. Springer, Cham

Bakka K, Gopika PV, Sreelakshmi H, Challabathula D (2022) Halotolerant plant growth promoting rhizobacteria: a futuristic direction to salt stress tolerance. In: Roy S, Mathur P, Chakraborty AP, Saha SP (eds) Plant stress: challenges and management in the new decade. Advances in science, technology & innovation. Springer, Cham

Bal HB, Nayak L, Das S, Adhya TK (2013) Isolation of ACC deaminase producing PGPR from rice rhizosphere and evaluating their plant growth promoting activity under salt stress. Plant and Soil 366(1–2):93–105

Barnawal D, Bharti N, Maji D, Chanotiya CS, Kalra A (2014) ACC deaminase-containing *Arthrobacter protophormiae* induces NaCl stress tolerance through reduced ACC oxidase activity and ethylene production resulting in improved nodulation and mycorrhization in *Pisum sativum*. J Plant Physiol 171(11):884–894

Barnawal D, Bharti N, Pandey SS, Pandey A, Chanotiya CS, Kalra A (2017) Plant growth-promoting rhizobacteria enhance wheat salt and drought stress tolerance by altering endogenous phytohormone levels and TaCTR1/TaDREB2 expression. Physiol Plant 161(4):502–514

Bartels D, Dinakar C (2013) Balancing salinity stress responses in halophytes and non-halophytes: a comparison between *Thellungiella* and *Arabidopsis thaliana*. Funct Plant Biol 40(9):819–831

Basha SA, Sarma BK, Singh DP, Annapurna K, Singh UP (2006) Differential methods of inoculation of plant growth-promoting rhizobacteria induce synthesis of phenylalanine ammonia-lyase and phenolic compounds differentially in chickpea. Folia Microbiol (Praha) 51:463–468

Bharti N, Barnawal D, Shukla S, Tewari SK, Katiyar RS, Kalra A (2016a) Integrated application of *Exiguobacterium oxidotolerans, Glomus fasciculatum* and vermicompost improves growth, yield and quality of *Mentha arvensis* in salt-stressed soils. Ind Crops Prod 83:717–728

Bharti N, Pandey SS, Barnawal D, Patel VK, Kalra A (2016b) Plant growth promoting rhizobacteria *Dietzia natronolimnaea* modulates the expression of stress responsive genes providing protection of wheat from salinity stress. Sci Rep 6(1):34768

Bhise KK, Dandge PB (2019) Alleviation of salinity stress in rice plant by encapsulated salt tolerant plant growth promoting bacteria *Pantoea agglomerans* strain KL and its root colonization ability. Arch Agron Soil Sci 65(14):1955–1968

Bomle DV, Kiran A, Kumar JK, Nagaraj LS, Pradeep CK, Ansari MA, Alghamdi S, Kabrah A, Assaggaf H, Dablool AS, Murali M, Amruthesh KN, Udayashankar AC, Niranjana SR (2021) Plants saline environment in perception with rhizosphere bacteria containing 1-Aminocyclopropane-1-Carboxylate Deaminase. Int J Mol Sci *22*(21):11461

Bossio DA, Scow KM, Gunapala N, Graham KJ (1998) Determinants of soil microbial communities: effects of agricultural management, season, and soil type on phospholipid fatty acid profiles. Microb Ecol 36:1–12

Chakraborty U, Chakraborty BN, Dey PL, Chakraborty AP, Sarkar J (2019) Biochemical responses of wheat plants primed with *Ochrobactrum pseudogrignonense* and subjected to salinity stress. Aric Res 8(4):427–440

Chanratana M, Joe MM, Roy Choudhury A, Anandham R, Krishnamoorthy R, Kim K, Jeon S, Choi J, Choi J, Sa T (2019) Physiological response of tomato plant to chitosan-immobilized aggregated *Methylobacterium oryzae* CBMB20 inoculation under salinity stress. 3 Biotech 9(11):397

Cheng Z, Park E, Glick B (2007) 1-Aminocyclopropane-1-carboxylate deaminase from Pseudomonas putida UW4 facilitates the growth of canola in the presence of salt. Can J Microbiol 53 (7):912–918

Christakis CA, Daskalogiannis G, Chatzaki A, Markakis EA, Mermigka G, Sagia A, Rizzo GF, Catara V, Lagkouvardos I, Studholme DJ, Sarris PF (2021) Endophytic bacterial isolates from halophytes demonstrate phytopathogen biocontrol and plant growth promotion under high salinity. Front Microbiol 12

Cordero I, Balaguer L, Rincon A, Pueyo JJ (2018) Inoculation of tomato plants with selected PGPR represents a feasible alternative to chemical fertilization under salt stress. J Plant Nutr Soil Sci 181(5):694–703

Costa-Gutierrez SB, Raimondo EE, Lami MJ, Vincent PA, Espinosa-Urgel M, de Cristóbal RE (2020) Inoculation of Pseudomonas mutant strains can improve growth of soybean and corn plants in soils under salt stress. Rhizosphere 16:100255

Dar A, Zahir ZA, Iqbal M, Mehmood A, Javed A, Hussain A, Bushra AM (2021) Efficacy of rhizobacterial exopolysaccharides in improving plant growth, physiology and soil properties. Environ Monit Assess 193:515

Dehghani BR, Azarnezhad N, Akhbari M, Ghorbani M (2019) Salinity stress and PGPR effects on essential oil changes in *Rosmarinus officinalis* L. Agric Food Secur 8(1):2

Desoky ESM, Saad AM, El-Saadony MT, Merwad ARM, Rady MM (2020) Plant growth-promoting rhizobacteria: potential improvement in antioxidant defence system and suppression of oxidative stress for alleviating salinity stress in *Triticum aestivum* (L) plants. Biocatal Aric 30:101878

Dodd IC, Perez-Alfocea F (2012) Microbial amelioration of crop salinity stress. J Exp Bot 63(9): 3415–3428

Dodd I, Zinovkina N, Safronova V, Belimov A (2010) Rhizobacterial mediation of plant hormone status. Ann Appl Biol 157:361–379

Drenovsky RE, Vo D, Graham KJ, Scow KM (2004) Soil water content and organic carbon availability are major determinants of soil microbial community composition. Microb Ecol 48:424–430

Efthimiadou A, Katsenios N, Chanioti S, Giannoglou M, Djordjevic N, Katsaros G (2020) Effect of foliar and soil application of plant growth promoting bacteria on growth, physiology, yield and seed quality of maize under Mediterranean conditions. Sci Rep 10:21060

Essalimi B, Esserti S, Rifai LA, Koussa T, Makroum K, Belfaiza M, Rifai S, Venisse JS, Faize L, Alburquerque N, Burgos L, el Jadoumi S, Faize M (2022) Enhancement of plant growth, acclimatization, salt stress tolerance and verticillium wilt disease resistance using plant growth-promoting rhizobacteria (PGPR) associated with plum trees (*Prunus domestica*). Sci Hortic 291:110621

Faizan M, Hayat S, Pichtel J (2020) Effects of zinc oxide nanoparticles on crop plants: a perspective. Analysis:83–99

Fan D, Subramanian S, Smith DL (2020) Plant endophytes promote growth and alleviate salt stress in *Arabidopsis thaliana*. Sci Rep 10(1):12740

Fazal A, Bano A (2016) Role of plant growth-promoting rhizobacteria (PGPR), biochar, and chemical fertilizer under salinity stress. Commun Soil Sci Plant 47(17):1985–1993

Ferchichi S, El Khouni A, Zorrig W, Atia A, Rabhi M, Gharsalli M, Abdelly C (2016) Nutrient uptake and use efficiencies in *Medicago ciliaris* under salinity. J Plant Nutr 39:932–941

Gagan P, Tatiana B, Rico S, Uta G, Robert B, Elena T, Stefanie S, Gerd UB, Klaus H, Wolfgang B, Andrea S, Thomas V, Claudia B, Ludger W, Andrej F (2012) Osmotic stress is accompanied by protein glycation in *Arabidopsis thaliana*. J Exp B 67(22):6283–6295

García de Salamone IE, Hynes RK, Nelson LM (2001) Cytokinin production by plant growth promoting rhizobacteria and selected mutants. Can J Microbiol 47(5):404–411

García-Cristobal J, García-Villaraco A, Ramos B, Gutierrez-Mañero J, Lucas JA (2015) Priming of pathogenesis related-proteins and enzymes related to oxidative stress by plant growth promoting rhizobacteria on rice plants upon abiotic and biotic stress challenge. J Plant Physiol 188:72–79

Ghassemi-Golezani K, Abdoli S (2022) Alleviation of salt stress in rapeseed (*Brassica napus* L) plants by biochar-based rhizobacteria: new insights into the mechanisms regulating nutrient uptake, antioxidant activity, root growth and productivity. Arch Agron Soil Sci 1–18

Glick BR (1995) The enhancement of plant growth by free living bacteria. Can J Microbiol 41:109–117

Glick BR (2014) Bacteria with ACC deaminase can promote plant growth and help to feed the world. Microbiol Res 169(1):30–39

Gouda S, Kerry RG, Das G, Paramithiotis S, Shin H, Patra JK (2018) Revitalization of plant growth promoting rhizobacteria for sustainable development in agriculture. Microbiol Res 206:131–140

Gül A, Özaktan H, Kidoğlu F, Tüzel Y (2013) Rhizobacteria promoted yield of cucumber plants in perlite under fusarium wilt stress. Scientia Horticul 153:22–25

Gupta S, Pandey S (2019) ACC deaminase producing bacteria with multifarious plant growth promoting traits alleviates salinity stress in French bean (*Phaseolus vulgaris*) plants. Front Microbiol 10

Habib SH, Kausar H, Saud HM (2016) Plant growth-promoting rhizobacteria enhance salinity stress tolerance in okra through ROS-scavenging enzymes. BioMed Res Int 1–10

Hafez EM, Osman HS, Gowayed SM, Okasha SA, Omara AED, Sami R, AbdEl-Monem AM, Abd El-Razek UA (2021a) Minimizing the adversely impacts of water deficit and soil salinity on maize growth and productivity in response to the application of plant growth-promoting rhizobacteria and silica nanoparticles. Agron 11(4):676

Hafez EM, Osman HS, El-Razek UAA, Elbagory M, Omara AED, Eid MA, Gowayed SM (2021b) Foliar-applied potassium silicate coupled with plant growth-promoting rhizobacteria improves growth, physiology, nutrient uptake and productivity of Faba Bean (*Vicia faba* L) irrigated with saline water in salt-affected soil. Plants 10(5):894

Hahm MS, Son JS, Hwang YJ, Kwon DK, Ghim SY (2017) Alleviation of salt tress in pepper (*Capsicum annum* L.) plants by plant growth-promoting rhizobacteria. J Microbiol Biotechnol 27(10):1790–1797

Han HS, Lee K (2005) Physiological responses of soybean-inoculation of *Bradyrhizobium japonicum* with PGPR in saline soil conditions. Res J Agric Biol Sci 1:216–222

Haroon U, Khizar M, Liaquat F, Ali M, Akbar M, Tahir K, Batool SS, Kamal A, Chaudhary HJ, Munis MFH (2022) Halotolerant plant growth-promoting rhizobacteria induce salinity tolerance in wheat by enhancing the expression of SOS genes. J Plant Growth Regul 41(6):2435–2448

Hatami M, Khanizadeh P, Bovand F, Aghaee A (2021) Silicon nanoparticle-mediated seed priming and Pseudomonas spp inoculation augment growth, physiology and antioxidant metabolic status in *Melissa officinalis* L plants. Ind Crops Prod 162:113238

He Y, Wu Z, Tu L, Han Y, Zhang G, Li C (2015) Encapsulation and characterization of slow-release microbial fertilizer from the composites of bentonite and alginate. Appl Clay Sci:68–75

He Y, Wu Z, Ye BC, Wang J, Guan X, Zhang J (2016) Viability evaluation of alginate-encapsulated *Pseudomonas putida* Rs-198 under simulated salt-stress conditions and its effect on cotton growth. Eur J Soil Biol 75:135–141

Hernández-Montiel LG, Chiquito-Contreras CJ, Murillo-Amador B, Vidal-Hernández L, Quiñones-Aguilar EE, Chiquito-Contreras RG (2017) Efficiency of two inoculation methods of *Pseudomonas putida* on growth and yield of tomato plants. Soil Sci Plant Nutr 17:1003–1012

Heydarian Z, Yu M, Gruber M, Glick BR, Zhou R, Hegedus DD (2016) Inoculation of soil with plant growth promoting bacteria producing 1-Aminocyclopropane-1-Carboxylate Deaminase or expression of the corresponding *acdS* gene in transgenic plants increases salinity tolerance in *Camelina sativa*. Front Microbiol 7

Horie T, Karahara I, Katsuhara M (2012) Salinity tolerance mechanisms in glycophytes: an overview with the central focus on rice plants. Rice 5:11

Hosseinpour A, Haliloglu K, Tolga CK, Ozkan G, Ozturk HI, Pour-Aboughadareh A, Poczai P (2020) Application of zinc oxide nanoparticles and plant growth promoting bacteria reduces genetic impairment under salt stress in tomato (*Solanum lycopersicum* L 'Linda'). Agriculture 10(11):521

Hussein KA, Joo JH (2018) Plant growth-promoting rhizobacteria improved salinity tolerance of *Lactuca sativa* and *Raphanus sativus*. J Microbiol Biotechnol 28(6):938–945

Ilangumaran G, Smith DL (2017) Plant growth promoting rhizobacteria in amelioration of salinity stress: a systems Biology perspective. Front Plant Sci 8:1768

Jahangir S, Javed K, Bano A (2020) Nanoparticles and plant growth promoting rhizobacteria (PGPR) modulate the physiology of onion plant under salt stress. Pak J Bot 52(4)

Javed H, Riaz A, Qureshi A, Javed K, Mujeeb F, Ijaz F, Akhtar MS, Ali MA, Gul R, Aftab M (2020) Isolation, characterization and screening of PGPR capable of providing relief in salinity stress. Eurasian J Soil Sci 9(2):85–91

Jha Y, Subramanian RB (2014) PGPR regulate caspase-like activity, programmed cell death, and antioxidant enzyme activity in paddy under salinity. Physiol Mol Biol Plants 20(2):201–207

Jha Y, Subramanian RB, Patel S (2011) Combination of endophytic and rhizospheric plant growth promoting rhizobacteria in *Oryza sativa* shows higher accumulation of osmoprotectant against saline stress. Acta Physiol Plant 33(3):797–802

Julia I, Oscar M, Analía L, Zocolo Guilherme J, Virginia L (2020) Biofertilization with *Macrocystis pyrifera* algae extracts combined with PGPR-enhanced growth in *Lactuca sativa* seedlings. J Appl Phycol 32(6):4361–4371

Kahtani MDF, Attia KA, Hafez YM, Khan N, Eid AM, Ali MAM, Abdelaal KAA (2020) Chlorophyll fluorescence parameters and antioxidant defence system can display salt tolerance of salt acclimated sweet pepper plants treated with chitosan and plant growth promoting rhizobacteria. Agris 10(8):1180

Kang SM, Khan AL, Waqas M, You YH, Kim JH, Kim JG, Hamayun M, Lee IJ (2014) Plant growth-promoting rhizobacteria reduce adverse effects of salinity and osmotic stress by regulating phytohormones and antioxidants in *Cucumis sativus*. J Plant Interact 9(1):673–682

Kavamura VN, Santos SN, da Silva JL, Parma MM, Avila LA, Visconti A, Zucchi TD, Taketani RG, Andreote FD, de Melo IS (2013) Screening of *Brazilian cacti* rhizobacteria for plant growth promotion under drought. Microbiol Res 168:183–191

Kazerooni EA, Maharachchikumbura SSN, Adhikari A, Al Sadi AM, Kang SM, Kim LR, Lee IJ (2021) Rhizospheric *Bacillus amyloliquefaciens* protects *Capsicum annuum* cv

Geumsugangsan from multiple abiotic stresses via multifarious plant growth-promoting attributes. Front Plant Sci 12:669693

Khalofah A, Kilany M, Migdadi H (2021) Phytostimulatory influence of *Comamonas testosteroni* and silver nanoparticles on *Linum usitatissimum* L under salinity stress. Plants 10(4):790

Khan N, Bano A (2016) Modulation of phytoremediation and plant growth by the treatment with PGPR, Ag nanoparticle and untreated municipal wastewater. Int J Phytoremediation 18(12): 1258–1269

Khan N, Bano A, Rahman MA, Guo J, Kang Z, Babar MA (2019) Comparative physiological and metabolic analysis reveals a complex mechanism involved in drought tolerance in chickpea (*Cicer arietinum* L.) induced by PGPR and PGRs. Sci Rep 9:2097

Khosravi A, Zarei M, Ronaghi A (2018) Effect of PGPR, Phosphate sources and vermicompost on growth and nutrients uptake by lettuce in a calcareous soil. J Plant Nutr 41(1):80–89

Kohler J, Hernández JA, Caravaca F, Roldán A (2009) Induction of antioxidant enzymes is involved in the greater effectiveness of a PGPR versus AM fungi with respect to increasing the tolerance of lettuce to severe salt stress. Environ Exp Bot *65*(2–3):245–252

Krishnaraj C, Jagan EG, Ramachandran R, Abirami SM, Mohan N, Kalaichelvan PT (2012) Effect of biologically synthesized silver nanoparticles on *Bacopa monnieri* (Linn) wettest plant growth metabolism. Process Biochem 47(4):651–658

Kristin A, Miranda H (2013) The root microbiota-a fingerprint in the soil? Plant Soil 370:671–686

Kubi HAA, Khan MA, Adhikari A, Imran M, Kang SM, Hamayun M, Lee IJ (2021) Silicon and plant growth-promoting rhizobacteria *Pseudomonas psychrotolerans* CS51 mitigates salt stress in *Zea mays* L. Agriculture 11(3):272

Kukreti B, Sharma A, Chaudhary P, Agri U, Maithani D (2020) Influence of nanosilicon dioxide along with bioinoculants on *Zea mays* and its rhizospheric soil. 3 Biotech 10(8):345

Kumar V, Kumar P, Khan A (2020a) Optimization of PGPR and silicon fertilization using response surface methodology for enhanced growth, yield and biochemical parameters of French bean (*Phaseolus vulgaris* L) under saline stress. Biocatal Agric Biotechnol 23:101463

Kumar P, Pahal V, Gupta A, Vadhan R, Chandra H, Dubey RC (2020b) Effect of silver nanoparticles and *Bacillus cereus* LPR2 on the growth of *Zea mays*. Sci Rep 10(1):20409

Kusale SP, Attar YC, Sayyed RZ, Malek RA, Ilyas N, Suriani NL, Khan N, Enshasy HA (2021, 1894) Production of plant beneficial and antioxidants metabolites by *Klebsiella variicola* under salinity stress. Molecules 26(7)

Lefevere H, Bauters L, Gheysen G (2020) Salicylic acid biosynthesis in plants. Front. Plant Sci 11

Li Z, Zhang J, Li J, Li H, Zhan G (2016) The functional and regulatory mechanisms of the *Thellungiella salsuginea* ascorbate peroxidase 6 (TsAPX6) in response to salinity and water deficit stresses. PLoS One 11

Li K, Wang Y, Han C, Zhang W, Jia H, Li X (2017) GA signaling and CO/FT regulatory module mediate salt-induced late flowering in *Arabidopsis thaliana*. Plant Growth Regu 53:195–206

Li X, Sun P, Zhang Y, Jin C, Guan C (2020) A novel PGPR strain *Kocuria rhizophila* Y1 enhances salt stress tolerance in maize by regulating phytohormone levels, nutrient acquisition, redox potential, ion homeostasis, photosynthetic capacity and stress-responsive genes expression. Environ Exp Bot 174

Lopes MJDS, Dias-Filho MB, Simone E, Gurgel C (2021) Microbes: inoculation methods and abiotic factors. Front Sustain Food Syst 5:606454

Lopez-Bucio J, Campos-Cuevas JC, Hernandez-Calderon E, Velasquez-Becerra C, Farias-Rodriguez R, Macias-Rodriguez LI, Valencia-Cantero E (2007) *Bacillus megaterium* rhizobacteria promote growth and alter root-system architecture through an auxin- and ethylene-independent signalling mechanism in *Arabidopsis thaliana*. Mol Plant Microbe Interact 20(2):207–217

Lucke M, Correa MG, Levy A (2020) the role of secretion systems, effectors, and secondary metabolites of beneficial rhizobacteria in interactions with plants and microbes. Front. Plant Sci 11

Lux A, Luxova M, Hattori T, Inanaga S, Sugimoto Y (2002) Silicification in sorghum (Sorghum bicolor) cultivars with different drought tolerance. Physiol Plant 115(1):87–92

Mahmood A, Turgay OC, Farooq M, Hayat R (2016a) Seed biopriming with plant growth promoting rhizobacteria: a review. FEMS Microbiol Ecol 92:112

Mahmood S, Daur I, Al-Solaimani SG, Ahmad S, Madkour MH, Yasir M, Hirt H, Ali S, Ali Z (2016b) Plant Growth Promoting Rhizobacteria and silicon synergistically enhance salinity tolerance of mung bean. Front Plant Sci 7

Mahmood F, Shahid M, Hussain S, Shahzad T, Tahir M, Ijaz M, Hussain A, Mahmood K, Imran M, Babar SAK (2017) Potential plant growth-promoting strain *Bacillus* sp. SR-2-1/1 decolorized azo dyes through NADH-ubiquinone, oxidoreductase activity. Biores Tech 235:176–184

Majid M, Ali M, Shahzad K, Ahmad F, Ikram RM, Ishtiaq M, Alaraidh IA, Al-hashimi A, Ali HM, Zarei T, Datta R, Fahad S, el Sabagh A, Hussain GS, Salem MZM, Habib-ur-Rahman M, Danish S (2020) Mitigation of osmotic stress in cotton for the improvement in growth and yield through inoculation of rhizobacteria and phosphate solubilizing bacteria coated Diammonium phosphate. Sustainability 12(24):10456

Marulanda A, Azcón R, Chaumont F, Ruiz-Lozano JM, Aroca R (2010) Regulation of plasma membrane aquaporins by inoculation with a *Bacillus megaterium* strain in maize (*Zea mays* L) plants under unstressed and salt-stressed conditions. Planta 232(2):533–543

Mayak S, Tirosh T, Glick BR (2004) Plant growth-promoting bacteria confer resistance in tomato plants to salt stress. Plant Physiol Biochem 42(6):565–572

Mellidou I, Ainalidou A, Papadopoulou A, Leontidou K, Genitsaris S, Karagiannis E, van de Poel B, Karamanoli K (2021) Comparative transcriptomics and metabolomics reveal an intricate priming mechanism involved in PGPR-mediated salt tolerance in Tomato. Front Plant Sci 12

Mi J, Gregorich EG, Xu S, McLaughlin NB, Liu J (2020) Effect of bentonite as a soil amendment on field water-holding capacity, and millet photosynthesis and grain quality. Sci Rep 10(1): 18282

Morcillo R, Manzanera M (2021) The effects of plant-associated bacterial exopolysaccharides on plant abiotic stress tolerance. Metabolites 11(6):337

Moshabaki IF, Tahmourespour A, Hoodaji M, Ataabadi M, Mohammadi A (2018) Influence of exopolysaccharide-producing bacteria and SiO2 nanoparticles on proline content and antioxidant enzyme activities of tomato seedlings (*Solanum lycopersicum* L) under salinity stress. Pol J Environ Stud 28(1):153–163

Munns R, Tester M (2008) Mechanisms of salinity tolerance. Annu Rev Plant Biol 59:651–658.1

Nawaz S, Bano A (2020) Effects of PGPR (*Pseudomonas* sp) and Ag-nanoparticles on enzymatic activity and physiology of cucumber. Recent Pat Food Nutr Agric 11(2):124–136

Nawaz A, Shahbaz M, Asadullah IA, Marghoob MU, Imtiaz M, Mubeen F (2020a) Potential of salt tolerant PGPR in growth and yield augmentation of wheat (*Triticum aestivum* L.) under saline conditions. Front Microbiol 11

Nawaz MS, Arshad A, Rajput L, Fatima K, Ullah S, Ahmad M, Imran A (2020b) Growth-stimulatory effect of quorum sensing signal molecule N-acyl-homoserine lactone-producing multi-trait *Aeromonas* sp. on wheat genotypes under salt stress. Front Microbiol 11

Naz R, Bano A (2015) Molecular and physiological responses of sunflower (*Helianthus Annuus* L) to PGPR and SA under salt stress. Pak J Bot 47(1):35–42

Nehela Y, Mazrou YSA, Alshaal T, Rady AMS, El-Sherif AMA, Omara AED, AbdEl-Monem AM, Hafez EM (2021, 1960) The integrated amendment of sodic-saline soils using biochar and plant growth-promoting rhizobacteria enhances maize (*Zea mays* L) resilience to water salinity. Plants 10(9)

Noumavo PA, Kochoni E, Didagbe YO, Adjanohoun A, Allagbe M, Sikirou R, Gachomo EW, Kotchoni SO, Baba-Moussa L (2013) Effect of different plant growth promoting rhizobacteria on maize seed germination and seedling development. Am J Plant Sci 4(5):1013–1021

Nukui N, Minamisawa K, Ayabe SI, Aoki T (2006) Expression of the 1-aminocyclopropane-1-carboxylic acid deaminase gene requires symbiotic nitrogen-fixing regulator gene nifa2 in *Mesorhizobium loti* MAFF303099. Appl Environ Microbiol 72(7):4964–4969

Palacio-Rodriguez R, Coria-Arellano JL, Lopez-Bucio J, Sanchez-Salas J, Muro-Perez G, Castaneda-Gaytan G, Saenz-Mata J (2017) Halophilic rhizobacteria from *Distichlis spicata* promote growth and improve salt tolerance in heterologous plant hosts. Symbiosis 73(3): 179–189

Pandey S, Gupta S (2020) Evaluation of Pseudomonas sp. for its multifarious plant growth promoting potential and its ability to alleviate biotic and abiotic stress in tomato (*Solanum lycopersicum*) plants. Sci Rep 10(1):20951

Panichikkal J, Prathap G, Nair RA, Krishnankutty RE (2021) Evaluation of plant probiotic performance of *Pseudomonas* sp encapsulated in alginate supplemented with salicylic acid and zinc oxide nanoparticles. Int J Biol Macromol 166:138–143

Paulucci NS, Gallarato LA, Reguera YB, Vicario JC, Cesari AB, Garcia de Lema MB, Dardanelli MS (2015) *Arachis hypogaea* PGPR isolated from Argentine soil modifies its lipids components in response to temperature and salinity. Microbiol Res 173:1–9

Peng H, Zhang J (2009) Plant genomic DNA methylation in response to stresses: potential applications and challenges in plant breeding. Prog Nat Sci 19(9):1037–1045

Philippot L, Raaijmakers JM, Lemanceau P, van der Putten WH (2013) Going back to the roots: the microbial ecology of the rhizosphere. Nat Rev Microbiol 11:789–799

Pinedo I, Ledger T, Greve M, Poupin MJ (2015) *Burkholderia phytofirmans* PsJN induces long-term metabolic and transcriptional changes involved in *Arabidopsis thaliana* salt tolerance. Front Plant Sci 6

Pour MM, Saberi-Riseh R, Mohammadinejad R, Hosseini A (2019) Investigating the formulation of alginate- gelatin encapsulated *Pseudomonas fluorescens* (VUPF5 and T17-4 strains) for controlling *Fusarium solani* on potato. Int J Biol Macromol 133:603–613

Prisa D (2020) Optimised fertilisation with zeolitites containing plant growth promoting rhizobacteria (PGPR) in *Ranunculus asiaticus*. GSC Biol Pharm 10(1):096–102

Qadir M, Quillérou E, Nangia V, Murtaza G, Singh M, Thomas R, Drechsel P, Noble A (2014) Economics of salt-induced land degradation and restoration. Nat Resour Forum 38:282–295

Rajendran K, Tester M, Roy SJ (2009) Quantifying the three main components of salinity tolerance in cereals. Plant Cell Environ 32:237–249

Ramadoss D, Lakkineni VK, Bose P, Ali S, Annapurna K (2013) Mitigation of salt stress in wheat seedlings by halotolerant bacteria isolated from saline habitats. Springer Plus 2(6)

Ramesh K, Reddy DD (2011) Zeolites and their potential uses in agriculture. Agris:219–241

Redondo-Gómez S, Romano-Rodríguez E, Mesa-Marín J, Sola-Elías C, Mateos-Naranjo E (2022) Consortia of plant-growth-promoting rhizobacteria isolated from halophytes improve the response of swiss chard to soil salinization. Agronomy 12(2):468

Robert AG, Kalaikandhan R, Mathivanan S, Chidambaram ALA (2017) Impact of PGPR inoculation on photosynthetic pigment and protein contents in *Arachis hypogaea* L. J Sci Agric 1(29)

Rocha I, Ma Y, Souza-Alonso P, Vosátka M, Freitas H, Oliveira RS (2019) Seed coating: a tool for delivering beneficial microbes to agricultural crops. Front Plant Sci 10

Salman S, Khan MY, Zahir ZA, Asghar HN, Chaudhry UK (2017) Comparative effectiveness of different carriers to improve the efficacy of bacterial consortium for enhancing wheat production under salt affected field conditions. Pak J Bot 49(4):1523–1530

Sarkar A, Pramanik K, Mitra S, Soren T, Maiti TK (2018) Enhancement of growth and salt tolerance of rice seedlings by ACC deaminase-producing *Burkholderia* sp. MTCC 12259. J Plant Physiol 231:434–442

Sharifi RS (2011) Study of grain yield and some of physiological growth indices in maize (*Zea mays* L.) hybrids under seed biopriming with plant growth promoting rhizobacteria (PGPR). J Food Agric Environ 189:3–4

Sharma HSS, Fleming C, Selby C, Rao JR, Martin T (2014) Plant biostimulants: a review on the processing of macroalgae and use of extracts for crop management to reduce abiotic and biotic stresses. J Appl Phycol 26(1):465–490

Shi CL, Park HB, Lee JS, Ryu S, Ryu CM (2010) Inhibition of primary roots stimulation of lateral root development in *Arabidopsis thaliana* by the rhizobacterium *Serratia marcescens* 90–166 is through both auxin-dependent and -independent signalling pathways. Mol Cells 29(3):251–258

Shi LN, Lu LX, Ye JR, Shi HM (2022) The endophytic strain ZS-3 enhances salt tolerance in *Arabidopsis thaliana* by regulating photosynthesis, osmotic stress, and ion homeostasis and inducing systemic tolerance. Front Plant Sci 13

Shultana R, Kee Zuan AT, Yusop MR, Saud HM (2020) Characterization of salt- tolerant plant growth-promoting rhizobacteria and the effect on growth and yield of saline- affected rice. PLoS One 15

Singh RP, Jha PN (2017) The PGPR *Stenotrophomonas maltophilia* SBP-9 augments resistance against biotic and abiotic stress in wheat plants. Front in Microbiol 8

Singh RP, Jha P, Jha PN (2015) The plant-growth-promoting bacterium *Klebsiella* sp. SBP-8 confers induced systemic tolerance in wheat (*Triticum aestivum*) under salt stress. J Plant Physiol 184:57–67

Singh SK, Pathak R, Choudhary V (2016) Plant growth-promoting rhizobacteria-mediated acquired systemic resistance in plants against pests and diseases. In: Choudhary D, Varma A (eds) Microbial-mediated induced systemic resistance in plants. Springer

Stoop J, Williamson J, Masonpharr D (1996) Mannitol metabolism in plants: a method for coping with stress. Trends Plant Sci 1(5):139–144

Taj Z, Challabathula D (2021) Protection of Photosynthesis by Halotolerant *Staphylococcus sciuri* ET101 in tomato (*Lycoperiscon esculentum*) and rice (*Oryza sativa*) plants during salinity stress: possible interplay between carboxylation and oxygenation in stress mitigation. Front Microbiol 11:547750

Taylor AG, Harman GE (1990) Concept and technologies of selected seed treatments. Ann Rev Phytopathol 28:321–339

Tchakounté G, Berger B, Patz S, Becker M, Fankem H, Taffouo VD, Ruppel S (2020) Selected rhizosphere bacteria help tomato plants cope with combined phosphorus and salt stresses. Microorganisms 8:1844

Trevors JT, van Elsas JD, Lee H, Wolters AC (1993) Survival of alginate-encapsulated *Pseudomonas fluorescens* cells in soil. Appl Microbiol Biotechnol 39(4–5):637–643

Ullah MA, Mahmood IA, Ali A, Nawaz Q, Sultan T, Zaman BU (2017) Effect of inoculation methods of biozote-max (plant growth promoting rhizobacteria-PGPR) on growth and yield of rice under naturally salt-affected soil. Res Plant Bio 7:24–26

Ullah S, Bano A, Ullah A, Shahid MA, Khan N (2022) A comparative study of plant growth promoting rhizobacteria (PGPR) and sowing methods on nutrient availability in wheat and rhizosphere soil under salinity stress. Rhizosphere 23

Upadhyay S, Singh DP (2015) Effect of salt-tolerant plant growth-promoting rhizobacteria on wheat plants and soil health in a saline environment. Plant Biol 17(1):288–293

Upadhyay SK, Singh JS, Saxena AK, Singh DP (2012) Impact of PGPR inoculation on growth and antioxidant status of wheat under saline conditions. Plant Biol 14(4):605–611

Vaishnav A, Kumari S, Jain S, Varma A, Choudhary DK (2015) Putative bacterial volatile-mediated growth in soybean (*Glycine max* L. Merrill) and expression of induced proteins under salt stress. J Appl Microbiol 119(2):539–551

Vejan P, Abdullah R, Khadiran T, Ismail S, Nasrulhaq BA (2016) Role of plant growth promoting rhizobacteria in agricultural sustainability-a review. Molecules 21(5):573

Vena MP, Jobbagy M, Bilmes SA (2016) Microorganism mediated biosynthesis of metal chalcogenides; a powerful tool to transform toxic effluents into functional nanomaterials. Sci Total Environ 565:804–810

Vera J, Castro J, Gonzalez A, Moenne A (2011) Seaweed polysaccharides and derived oligosaccharides stimulate defence responses and protection against pathogens in plants. Mar Drugs 9(12):2514–2525

Vercelheze AES, Marim BM, Oliveira ALM, Mali S (2019) Development of biodegradable coatings for maize seeds and their application for *Azospirillum brasilense* immobilization. Appl Microbiol Biotechnol 103(5):2193–2203

Vimal SR, Singh JS (2019) Salt tolerant PGPR and FYM application in saline soil paddy agriculture sustainability. Clim Chang Environ Sustain 7(1):61

Wang S, Na X, Yang L, Liang C, He L, Jin J, Liu Z, Qin J, Li J, Wang X, Bi Y (2021) *Bacillus megaterium* strain WW1211 promotes plant growth and lateral root initiation via regulation of auxin biosynthesis and redistribution. Plant Soil 466(1–2):491–504

Widawati S, Suliasih (2018) The effect of plant growth promoting rhizobacteria (PGPR) on germination and seedling growth of *Sorghum bicolor* L Moench. IOP Confer Ser Earth Environ Sci 166:012022

Woo OG, Kim H, Kim J, Keum HL, Lee KC, Sul WJ, Lee JH (2020) *Bacillus subtilis* strain GOT9 confers enhanced tolerance to drought and salt stresses in *Arabidopsis thaliana* and *Brassica campestris*. Plant Physiol Biochem 148:359–367

Wu Z, Zhao Y, Kaleem I, Li C (2011) Preparation of calcium–alginate microcapsuled microbial fertilizer coating *Klebsiella oxytoca* Rs-5 and its performance under salinity stress. Eur J Soil Biol 47(2):152–159

Wu Z, Guo L, Qin S, Li C (2012) Encapsulation of *R. planticola* Rs-2 from alginate-starch-bentonite and its controlled release and swelling behaviour under simulated soil conditions. J Ind Microbiol Biotechnol 39(2):317–327

Wu Z, Peng Y, Guo L, Li C (2014a) Root colonization of encapsulated *Klebsiella oxytoca* Rs-5 on cotton plants and its promoting growth performance under salinity stress. Eur J Soil Biol 60:81–87

Wu Z, He Y, Chen L, Han Y, Li C (2014b) Characterization of *Raoultella planticola* Rs-2 microcapsule prepared with a blend of alginate and starch and its release behaviour. Carbohydr polym 110:259–267

Wu T, Xu J, Liu J, Guo WH, Li XB, Xia JB, Xie WJ, Yao ZG, Zhang YM, Wang RQ (2019) Characterization and initial application of endophytic *Bacillus safensis* Strain ZY16 for improving phytoremediation of oil-contaminated saline soils. Front Microbiol 10

Xing K, Zhu X, Peng X, Qin S (2015) Chitosan antimicrobial and eliciting properties for pest control in agriculture: a review. Agron Sustain Dev 35(2):569–588

Xiong YW, Li XW, Wang TT, Gong Y, Zhang CM, Xing K, Qin S (2020) Root exudates-driven rhizosphere recruitment of the plant growth-promoting rhizobacterium *Bacillus flexus* KLBMP 4941 and its growth-promoting effect on the coastal halophyte *Limonium sinense* under salt stress. Ecotoxicol Environ Saf 194:110374

Xun F, Xie B, Liu S, Guo C (2015) Effect of plant growth-promoting bacteria (PGPR) and arbuscular mycorrhizal fungi (AMF) inoculation on oats in saline-alkali soil contaminated by petroleum to enhance phytoremediation. Environ Sci Pollut Res 22(1):598–608

Yasin NA, Akram W, Khan WU, Ahmad SR, Ahmad A, Ali A (2018) Halotolerant plant-growth promoting rhizobacteria modulate gene expression and osmolyte production to improve salinity tolerance and growth in *Capsicum annum* L. Environ Sci Pollut Res 25(23):23236–23250

Yasmeen T, Ahmad A, Arif MS, Mubin M, Rehman K, Shahzad SM, Iqbal S, Rizwan M, Ali S, Alyemeni MN, Wijaya L (2020) Biofilm forming rhizobacteria enhance growth and salt tolerance in sunflower plants by stimulating antioxidant enzymes activity. Plant Physiol Biochem 156:242–256

Yildirim E, Turan M, Donmez MF (2008) Mitigation of salt stress in radish (*Raphanus sativus*) by plant growth promoting rhizobacteria. Rom Biotechnol Lett 13(5):3933–3943

Zamioudis C, Mastranesti P, Dhonukshe P, Blilou I, Pieterse CMJ (2013) Unravelling root developmental programs initiated by beneficial *Pseudomonas* spp bacteria. Plant 162(1): 304–318

Zerrouk IZ, Benchabane M, Khelifi L, Yokawa K, Ludwig-Müller J, Baluska F (2016) A *Pseudomonas* strain isolated from date-palm rhizospheres improves root growth and promotes root formation in maize exposed to salt and aluminium stress. J Plant Physiol 191:111–119

Zhao S, Zhou N, Zhao ZY, Zhang K, Wu GH, Tian CY (2016) Isolation of endophytic plant growth-promoting bacteria associated with the halophyte *Salicornia europaea* and evaluation of their promoting activity under salt stress. Curr Microbiol 73(4):574–581

Zhao Q, Yang XY, Li Y, Liu F, Cao XY, Jia ZH, Song SS (2020) N-3-oxo-hexanoyl-homoserine lactone, a bacterial quorum sensing signal, enhances salt tolerance in Arabidopsis and wheat. Bot Stud 61(1):8

Zhao S, Zhang Q, Liu M, Zhou H, Ma C, Wang P (2021) Regulation of Plant Responses to Salt Stress. Int Mol Sci 22:4609

Zhou J, Ahmed N, Cheng Y, Qin C, Chen P, Zhang C, Zhang L (2019) Effect of inoculation of strains with ACC deaminase isolated from vermicompost on seed germination and some physiological attributes in maize (*Zea mays* L) exposed to salt stress. Pak J Bot 51(4)

Zhou Y, Hao L, Ji C, Zhou Q, Song X, Liu Y, Li H, Li C, Gao Q, Li J, Zhang P, Liu X (2021) The effect of salt-tolerant antagonistic bacteria CZ-6 on the rhizosphere microbial community of Winter Jujube (*Ziziphus jujuba* Mill "Dongzao") in saline-alkali land. BioMed Res Int 1–13

Chapter 16
Studies on Orchidoid Mycorrhizae and Mycobionts, Associated with Orchid Plants as Plant Growth Promoters and Stimulators in Seed Germination

Oindrila Chakraborty, Dinesh Kumar Agrawala, and Arka Pratim Chakraborty

Abstract Orchidaceae is one of the largest family of angiosperm with more than 28,000 accepted names Worldwide distributed from sea level at coastal plains up to high mountain ranges of the alpine Himalayas. They are known to mankind for the last several centuries for their beautiful attractive flowers and as medicinal plants. The orchids are highly interesting since their discovery due to its horticultural, medicinal, and aesthetic importance. The mycorrhizal association is essential for survival at seedling stage. The seeds of orchid are very tiny and lightweight, the embryo inside is ill-developed and there is no endosperm which restricts direct nutrient supply to the embryo. They contain a very small amount of high-value proteins and lipids but a small amount of sugar, though some species also contain some starch granules. Orchids established a mycorrhizal symbiosis with fungi in their natural habitat affects seed germination, protocorm growth, and nutrition. Several studies indicate how orchid gain mineral nutrients and even organic compounds after interaction with orchid mycorrhizal fungi (OMF). OMF plays an important role in the life cycle of orchids. Mycorrhiza also helps in increasing phytohormone secretion and activities of enzyme which helps in the germination of seeds and growth of orchids.

Keywords Orchidaceae · Orchid mycorrhizal fungi (OMF) · Protocorms

---

O. Chakraborty
Botanical Survey of India, Central National Herbarium, Howrah, West Bengal, India

D. K. Agrawala
Botanical Survey of India, CGO Complex, Salt Lake, Kolkata, West Bengal, India

A. P. Chakraborty (✉)
Department of Botany, Raiganj University, Raiganj, India

P. Mathur et al. (eds.), *Microbial Symbionts and Plant Health: Trends and Applications for Changing Climate*, Rhizosphere Biology,
https://doi.org/10.1007/978-981-99-0030-5_16

Abbreviations

AM	Arbuscular mycorrhiza
ECM	Ectomycorrhizae
GA	Gibberellic acid
HM	Heavy metal
IAA	Indole acetic acid
NAA	Naphthalene acetic acid
OMF	Orchid mycorrhizal fungi
ROS	Reactive oxygen species
VAM	Vesicular arbuscular fungi

16.1 Introduction

Orchidaceae is one of the largest family of flowering plants with nearly 28,349 accepted names worldwide, distributed throughout except the Arctic and Antarctic areas. In India, orchids are represented by c.1260 taxa under 160 genera and distributed from sea level at coastal plains up to high mountain range of alpine Himalayas (Singh et al. 2019). It is one of the highly evolved group of plants with enormous diversity, extremely beautiful flowers and highly specialized structures to ensure successful pollination and reproduction.

Orchids evinced two types of growth forms, in sympodial forms show horizontal growth forms representing mainly in the Epidendroid and Orchidoid and in monopodial ones with continuous apical growth representing the tribe Vandeae (Fig. 16.1). The high relative humidity in the temperature is suitable for growth in terrestrial and epiphytic orchids.

Orchids are magnificent with distinct shapes and ornamentation of flowers. They are wonderful creation of nature for their efficiency in functioning various floral parts. The flowers of orchids have evolved complex floral structure facilitating biotic cross-pollination which makes them superior from other plant groups in evolutionary process. The exchange of genetic materials by cross-pollination leads to development of extreme genotypic variability which can be observed through their phenotypic expressions (Singh et al. 2019).

The orchid seeds are very minute, dust-like consisting of a small embryo and the seed is enclosed by a very thin layer of transparent cells and have a limited food reserve. Germination and development of seedling are very slow in orchid plants as they are devoid of endosperm so the nutrient reserve is absent. The mycorrhizal fungi play an important role in the seed germination (Misra 2007). They depend on certain specific mycorrhiza during seed germination stage for nutrition, water, carbon, minerals, etc. through fungal hyphae to embryo. In some orchid species, partial photosynthesis occur but they depend on some fungi for carbon intake, this nutritional strategy is known as “mixotropy” or “partial mycoheterotrophy” whereas

Fig. 16.1 (**a**) Sympodial growth. (**b**) Monopodial growth. (Source: Live photographs O. Chakraborty and D.K. Agrawala)

in some orchid plants photosynthesis process never takes place and they fully depend on fungus for the nutrient supply (Gebauer and Meyer 2003; Julou et al. 2005; Selosse and Roy 2009; Jacquemyn et al. 2017). The orchid plants take their food and water through the root hairs and velamen from moisture in the air and the host plants (Sathiyadash et al. 2020).

16.2 Historical Background of Orchids

16.2.1 Pre-linnaean

The history of Indian Orchids known from Charaka in his literature "Charaka Samhita," described medicinal value of Vandaceous taxon.

16.2.2 Linnaean and Post-linnaean

William Roxburgh (1832) in his book "Flora Indica" described 57 species of orchid under five genera and proposed many new taxa. The pioneering work of Indian Orchids by John Lindley (1830–1859) described all the known genera and species collected earlier in his significant books "Genera and species of Orchidaceous plants" (1830–1840), "Folia Orchidacea" (1852–1859) and contribution to the

Orchidology of India 1857–1859. Considerable work on orchids was done in India during the pre-independence era.

J. D. Hooker described 1600 species in his book "Flora of British India" (1888–1890). This was the new dawn in the modern study of Indian orchids. He has also come up with descriptions and illustrations of hundred species in "A century of Indian Orchids" in the Annals of the Royal Botanic Garden, Calcutta (1895). G. King and R. Pantling provided descriptions with illustrations in the Orchids of Sikkim Himalaya (1898).

After the changed geo-political boundary of India in the post-independence period, documentation of Indian Orchids has caught speed.

- Pradhan (1976, 1979) covered most of the orchid species in India known by that time in his two volumes "Guide to identification and culture of Indian Orchids."
- More recently, Misra (2007) has reported 1331 Indian Orchid species under 185 genera.

Checklist/field guide for different states were published the mentionable ones are H. J. Chowdhery and D. K. Agrawala (2013) etc. for western Himalayan region, Lucksom S. Z. (2007). *The Orchids of Sikkim and North East Himalaya*. CONCEPT, Siliguri.

Of late in the recent centuries and recent time revisionary studies on many important genera have been completed in Indian perspective by the Botanical Survey of India on significant groups. An account of orchids for the entire country has been addressed recently by Singh et al. (2019).

16.3 Morphology of Orchids

The flowers of orchids are trimerous and zygomorphic with a wide range of variation in size, shape, form, and coloration among all other angiosperm families. They comprise three sepals, dorsal one is usually different than lateral two. Out of the three petals, the inner two look alike; the third petal is modified in various shapes, sizes, postures, and sometimes color from the other petals known as "labellum" or "lip." The base of the lip saccate or having sac-like to tubular extension—a hollow appendage called spur and usually secretes nectar to attract pollinators. Gynostemium or column formed by the fusion of the stamens and style from base to apex, showing wide range of variation in size, texture, position, etc. The anther consists of a cap-like structure known as rostellum which helps to prevent pollinia of the same flower from reaching the stigma, thus acting as an obstacle for self-pollination. The pollen grains in orchids stick to each other by a viscid or waxy substance forming a mass called pollinarium (Singh et al. 2019). Fruits are capsules, vary in shape, size, and texture in different groups of orchids. A single capsule carries a million of seeds that are very light and dispersed by wind. The seeds of orchids are tiniest, dust-like among all the flowering plants. They are devoid of endosperms, consist of a central spot of densely grown protoplasmic cells, an undifferentiated embryo and a reticulate, papery, one-cell layer seed coat that extends at either ends into tiny wings (Singh et al. 2019; Misra 2007) (Fig. 16.2). Often among the seeds, the fruit is filled with hygroscopic hairs which on maturity become dry and loose that help in dehiscence of the capsules (Singh et al. 2019).

16.4 Mycorrhiza: Mycorrhiza and Its Types

The term mycorrhiza derived from the Greek word "*mykes*" means mushroom or fungi and "*rhizo*" means root. Fungal roots were discovered by the German botanist A. B. Frank in the last century in forest trees such as pine. Mycorrhiza is a mutualistic symbiotic association between the roots of higher plants and fungus. Mycorrhizas are symbiotic associations of specialized soil fungi and plants that are involved in nutrient transfer (Brundrett 2002). The phosphorous can be accumulated in the fungal hyphae and transport it to the root tissue in plant cells. Mycorrhiza is classified into three main categories: Ectomycorrhiza, Endomycorrhiza, and Ectendomycorrhiza. Endomycorrhiza is further divided into five types: Vesicular Arbuscular Mycorrhiza (VAM), Orchidoid Mycorrhiza, Monotropoid Mycorrhiza, Ericoid Mycorrhiza, and Arbutoid Mycorrhiza (Table 16.1). Endomycorrhiza (Vesicular arbuscular mycorrhiza (VAM) now known as arbuscular mycorrhiza, AM) plays a very important role on enhancing the plant growth and yield due to an increased supply of phosphorus to the host plant. Mycorrhizal plants can absorb and accumulate several times more phosphate from the soil or solution than non-mycorrhizal plants. Plants inoculated with endomycorrhiza are more resistant to some root diseases. Arbuscular Mycorrhizal (AM) fungi (or Vesicular Arbuscular Mycorrhizal, VAM fungi), belonging to the Phylum Glomeromycota are symbionts with terrestrial plant roots. It is now generally recognized that they improve not only the phosphorus nutrition of the host plant but also its growth, which may result in an increase in resistance to drought stress and some diseases. Plants with Vesicular Arbuscular Mycorrhiza are omnipresent, while those with ectomycorrhizae are important in ecosystems. There are some mycorrhizae that occurs in a particular family of plants and some plants also have roots that remains non-mycorrhizal. Orchid mycorrhizae are differing from other types of mycorrhizae as they occur in stems as well as roots and fungi may recolonize older cells (Hadley 1982; Smith and Read 1997). The orchid mycorrhizae are morphologically alike to other mycorrhizae and they associate with the phylogenetically distinct group of soil fungi so that they have limited specializations as mycorrhizal fungi showing less evidence of co-evolution with their hosts (Rasmussen 1995; Currah et al. 1997; Brundrett 2002). In 1886, Wahrlich surveyed more than 50 cultivated orchid species and found fungal associations in them. In successive surveys found that in terrestrial orchids, mycorrhizae are ubiquitous and also present in some epiphytic orchid species.

In orchids, mycorrhiza that has been found mainly falls under endomycorrhizal group popularly known as Orchidoid Mycorrhiza. Mostly reported endomycorrhiza

Fig. 16.2 (**a**–**j**) *Smitinandia micrantha* (Lindl.) Holttum (**a**), Habit (**b**), Inflorescence (**c**), Flower (**d**), Sepals and Petals (**e**), Column with ovary and labellum (**f**), Labellum with spur (**g**), Labellum (dorsal view and ventral view) (**h**), Clinandrium (**i**), Pollinia with stipe and viscidium (**j**), anther cap and pollinia (Source: O. Chakraborty and D.K. Agrawala)

Table 16.1 Classification of mycorrhizae on the basis of morphological and anatomical features

Mycorrhiza			
Ectomycorrhizae	Ectomycorrhizae are the association where fungi form a mantle around roots and there is no hyphal penetration of cells. Fungal hyphae is generally separate and the distinct Hartig's net is present between the cells.		
Endomycorrhizae	Mycorrhizal association in which the fungal hyphae present on the surface of the root as individual thread directly penetrates into the root hairs, other epidermal cells and into cortical cells.	**VAM fungi**	VAM fungi are obligate biotropic as they are completely depending on plants for their survival.
		Orchidoid mycorrhizae	Fungi belong to basidiomycotina and colonize only members of family Orchidaceae and their association is pseudomycorrhizal, plays an important role in establishment of seedling.
		Ericoid mycorrhizae	This is found in roots of plants belonging to order ericales and the rootless are covered by a loosely woven mesh of dark brown septate hyphae from which branches penetrate the cortical cells.
		Monotropid mycorrhizae	The fungi colonized achlorophyllous members of angiosperms belonging to family monotropaceae. Fungal sheath present.
		Arbutoid mycorrhizae	The associations are variants of ectomycorrhizae found in certain plants in Ericaceae characterized by hyphae coils in epidermal cells.
Ectendomycorrhizae	The formation of Ectendomycorrhizae starts with the formation of Hartig's net which grow behind the apical meristem of the growing root. The Hartig net penetrates between the epidermal and outer cortical cells and later extends up to inner cortex.		

found in association with orchids is different species of *Rhizoctonia* (Burgeff 1959; Warcup and Talbot 1967, 1980) like *R. solanii* and *R. repens*. Symbiotic association with orchids is also found in fungal genera like *Sebacina*, *Armillaria*, *Corticium*, *Marasamius*, *Thanetophora*, and *Glomus*. Other than this, Saha and Rao (2006) reported fungal isolates like *Cochliobolus spicifer*, and *Trichoderma viride* associated with epiphytes and ground orchids. Downie (1957, 1959) and Masuhara et al. (1993) reported the isolation of *Rhizoctonia goodyrae-repentis* fungi from terrestrial orchid *Goodyera repens* and *Rhizoctonia solani* from *Dactylorchis purpurella*. Aggarwal and Zettler (2010) reported *Ceratobasidium* with *Dactylorhiza hatagirea* and *Rhizoctonia solanii* with *Zeuxine strateumatica* by Rajkumar and Kaushik (2007). The occurrence of orchidaceous mycorrhiza in 83 orchid taxa has been enumerated by Wang and Qiu (2006). Mycorrhizae previously reported in some species like *Spathoglottis plicata* (Hadley and Williamson 1972; Senthilkumar and Krishnamurthy 1996, 1998a, b); *Vanilla planifolia* (Madhaiyan et al. 2003); *Calanthe triplicate* (Kaliamoorthy 2007); *Rhynchostylis retusa* (Radhika and Rodrigues 2007); *Oberoniaensiformis*, *Oberoniaplatycaulon*; and *Malaxis versicolor* (Bagyalakshmi et al. 2010). Saprophytic mycorrhizal fungi like *Mycena* sp. found in some orchids like *Anoctochillus roxburghii* (Guo et al. 1997) and *Cymbidium sinense* (Fan et al. 1996). Kaushik (1983) reported mycorrhizal association in 53 taxa and a pack of elongated cells with mycorrhizal pelotons and septate hyphae in photographs of longitudinal sections through microscope taken by microtomy of roots in *Calanthe gracilis*. The mycorrhizal endophytes and their association with North Indian Orchidaceae have been studied by Vij and Sharma (1983, 1988), Vij et al. (1985). Mycorrhizal endophytes documented elaborately by Vij et al. (1985) in terrestrial orchid *Spiranthes lancea*. Senthilkumar et al. (2000) confirmed the presence of mycorrhizal association in the epiphytic orchid *Acampe praemorsa* in which mycorrhiza colonized in the roots. Bukhari 2020 observed the presence of a loose hyphal network in the cortical region of *Porpax reticulata* and Bukhari and Herlekar (2015) observed a similar in other epiphytic orchids of western ghats. The present status of endophytic mycorrhiza has been reported by Sharma et al. (1988). An account of orchidaceous mycorrhizal fungi has been presented by Singh and Varma (2000). The status of Orchidoid mycorrhizal technique to invest them in a depth manner was reviewed by Vij et al. (2002). Varma (1994) reported the important role of mycorrhizal fungi for substantiating orchid seedlings by tissue culture techniques. Mycorrhizal association in terrestrial orchids has been reported by Katiyar et al. (1986). Kumar and Krishnamurthy (1998) conducted a cytochemical study on the mycorrhizae in the terrestrial orchid *Spathoglottis plicata*. Kaushik and Pal (2011) described the identification features, classification, and cultural characteristics of various species of *Rhizoctonia*. Rajkumar and Kaushik (2007) isolated a cellulose producing mycorrhizal fungus *Rhizoctonia solani* from *Zeuxine strateumatica*. Colonization of *M. dendrobii* on *D. candidum* helps growth with phytohormone secretion (Kaur 2020).

The presence of a hyphal structure in the cells of an orchid seed was first time recorded by Link and the presence of fungi in an orchid was first described by Reissek. The fungal root discovered by the German botanist A.B. Frank and the symbiotic relationship between orchid nutrition and the presence of endophytic hyphae was first suggested by Frank. The pioneering work of Bernard depicting the infection of suitable fungus for germination of orchid seeds and the importance of fungi in seed germination of orchids was realized. In their entire life cycle, the orchids depend on mycorrhizal fungi or at least one till the initial establishment of seedlings started (Weber and Webster 2001). Zettler (1997) has depicted orchids as mycotrophs (fungus-fedders). Comprehensive data has been published on the infection that occurs naturally in orchids but work on tropical species is still anticipated (Hadley 1984).

16.5 Protocorms

The French botanist Bernard coined the term "protocorm" means a stage of development of the orchid embryo. Protocorms are the mass of cells forming temporary tuberoid structures developed by the swelling of embryo after germination of orchid seeds. The fungus digests the testa and breaks the starch and cellulose into sugar for the mass of cells (embryo) helping in germination. After the protocorm grows into a suitable size, the new shoots form the structure known as mycorrhizome, thus helping in growth of seedlings (Manoharachary 2019) (Fig. 16.3).

16.6 Orchid Fungi

Dearnaley (2007) describes orchid mycorrhizas (OM) are symbiotic associations between fungi and terrestrial, epiphytic, or lithophytic species of the Orchidaceae. In this association, fungal hyphae enter into the parenchyma cells of germinating seeds,

Fig. 16.3 Different stages in the lifecycle of an orchid. Mycorrhizal fungi are mandatory for seed germination and seedling development. Adult green photosynthetic orchid can acquire nutrients through mycorrhizal fungi and even supplement carbon through mixotrophy (Source: Sathiyadash et al. 2020)

Table 16.2 Orchid mycobiont genera with "*Rhizoctonia*" anamorph stages, their systematic status and ecological preferences (Source: Rasmussen and Rasmussen 2014)

Teleomorph	Family	Life forms
Sebacina	Sebacinaceae, probably polyphyletic	Ectomycorrhiza Ericoid mycorrhiza Saprotrophs
Tulasnella	Tulasnellaceae	Saprotrophs Ectomycorrhiza
Ceratobasidium	Ceratobasidiaceae	Saprotrophs Ectomycorrhiza
Thanatephorus	Ceratobasidiaceae	Plant pathogens Parasites on fungi

protocorms, seedlings, or in the roots of adult plants and form intracellular hyphal coils (Smith and Read 2008). The highly coiled hyphal structures formed by fungi within the root cortex known as pelotons (Sathiyadash et al. 2012). Pelotons (from the French word for "small round hank") coined by Noel Bernard, a pioneer researcher on orchid mycorrhiza. Orchid mycorrhizas are characterized by the presence of coils of undifferentiated hyphae in cortical cells of the root, stem, or protocorms of orchid species and it is also able to recolonize older cells (Hadley 1982; Masuhara and Katsuya 1992; Smith and Read 1997; Peterson et al. 1998). Orchid mycorrhizas are morphologically different from other mycorrhizas and involve a phylogenetically distinct group of soil fungi (Rasmussen 1995; Currah et al. 1997). The considerable variations are found in distribution of mycorrhizas within the roots or stem of orchids. The sporadic information on mycorrhizal colonization has been found in epiphytic orchid species whereas more widespread in terrestrial orchids (Burgeff 1959; Rasmussen 1995). There are two types of orchid mycorrhizas recognized by early researchers (1) tolypophagy, which occurs in most of the species (2) phytophagy, found in myco-heterotrophic species (e.g., *Gastrodia*) (Rasmussen 1995). Mostly the endophytic mycorrhizae known to form orchid mycorrhizae are Basidiomycetes and belong to the genus *Rhizoctonia*. In some members of the Basidiomycete "rust" lineage (Atractiellomycetes, Pucciniomycotina) are mycobionts of orchids (Kottke et al. 2009).

The *Rhizoctonia* like fungi is characterized by certain hyphal morphological traits, including lack of conidia; the hyphal branch at a right angle; a septum located a short distance from a constricted ramification and the presence of monilioid cells (a special type of cell with rounder shape) mainly in sclerotic structures. The young hyphae either binucleate or multinucleate and rarely produce fruiting bodies, also difficult to distinguish the differentiation of species in anamorphic stage. The teleomorph of *Rhizoctonia* can be *Ceratobasidium*, *Thanatephorus*, *Tulasnella*, and *Sebacina* differs morphologically (Table 16.2).

16.6.1 Phenology

Rasmussen (1995) observed some data about phenology of the fungi associated with the orchids or the impact of low winter temperatures or dry summer weather on the activity of external mycelium. In some regions of the world growing season is delimited by periods of cold or dry conditions but some mycorrhizal fungi are active in a season other than the growing season and possess a network of hyphae and propagules in soils (Brundrett and Abbott 1994; Braunberger et al. 1997). Batty et al. (2000) assumed that the orchid fungus behaves in a similar way to other mycorrhizal fungi and there will be a particular season in which the condition of soil and activity of fungi will be conducive for germination of the orchid seedling. Alexandar and Hadley (1985) observed the terrestrial orchids *Bletilla striata* and *Goodyera repens* showed signs of fungal infection throughout the year with a peak from December to May. The main limiting factor of endophytic activity is the lack of soil moistures and not low temperature.

16.6.2 Entry and Colonization of Fungi in Orchids

Mycorrhizal colonization in epiphytic orchids is reported sporadically but it is widespread in terrestrial orchids. In different genera of terrestrial orchids colonization is distinct, within their roots or stems, and the pattern of infection in the whole plant (i.e., in root, collar, stem, and rhizome) associated with particular fungal types. The pattern of colonization of mycorrhizal fungi within plants is determined by host cell properties but mycorrhizal morphological features can be correlated with the presence of certain fungi (Brundrett 2002) (Fig. 16.4).

In the life cycle of many orchids two stages of infection are observed: (1) During the germinating of seedling primary infection occurs and (2) The new roots of adult plants were reinfected. The main area of infection is living epidermal hair cells and basal cells of seed (Burgeff 1959) followed by formation of hyphal clusters thinly enclosed by the host cytoplasm and surrounded by the covered layer.

The work on *Acanthephippium bicolor* showed the entry of fungi through the deformed part of root of epidermal cells. The exodermal region of fungal hyphae punctures the velamen tissue before entry. The fungal hyphae enter into the cortical region as the passage cells of the cortical region are not joined like the other cells of the exodermis. Fungal hyphae can also enter through the root hairs which is reported by Senthilkumar and Krishnamurthy (1998a, b) in *Spathoglottis plicata* and also by Weber and Webster (2001) in certain species of *Dactylorhiza*. The fungal colonization is observed to begin in the inner cortical cells.

Successful and stable establishment of symbiosis is reflected by the formation of pelotons, tight interwoven coils, consider as the distinctive character of orchidoid mycorrhiza (Currah and Zelmer 1992). The colonization of fungal hyphae usually began in the inner cortical cells, even sometimes in outer cortical region and at times

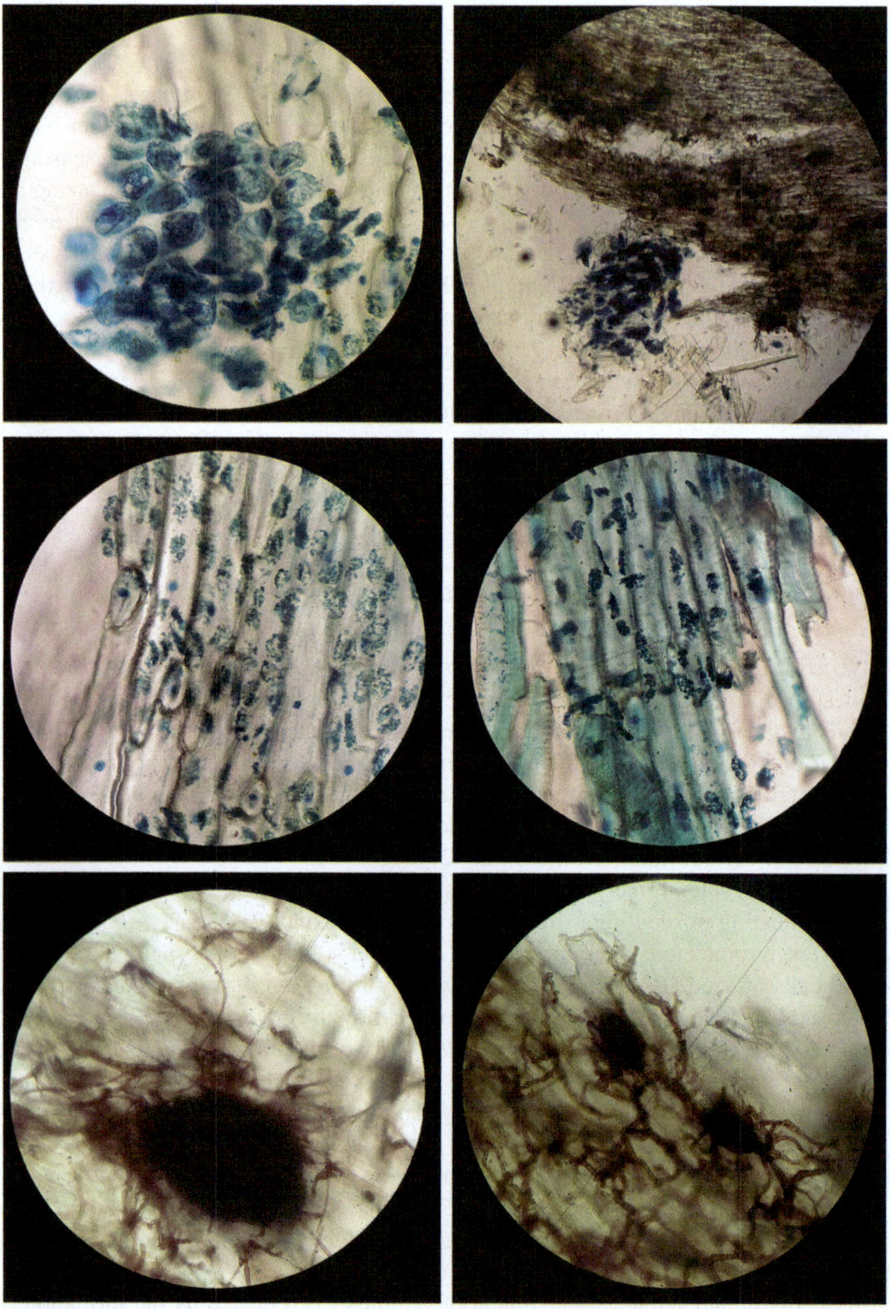

Fig. 16.4 Mycorrhizal association in epiphytic monopodial orchid root (Source: O. Chakraborty and A.P. Chakraborty)

in a few passage cells of exodermis. The roots of *Rhizoctonia* sp. colonizes which increasing plant height, biomass, formation of root and length of the root in *Cymbidium kanran* (Lee et al. 2003). The plant height, biomass, and the number of new buds has been reported to increase many a times after being colonized by *Mycena* sp. In *Paphiopedilum villosum* and *Acanthephippium bicolor*, the pelotons are observed to occupy the entire cell cavities, particularly in the inner cortex region. The cortical zone with mature pelotons forms the digestion layer and the fungal hyphae extending from these mature pelotons are finer. Before attaining maturity, early stages of fungal colonization are found to be in the form of loose aggregates of hyphae (Manoharachary 2019). After the digestion of pelotons in the cortical cells release lipid droplets in *Acanthephippium bicolor* but this mechanism was absent in *Paphiopedilum villosum* as reported by Manoharachary (2019). It has been observed that lipids and polyphosphates are degraded in the degenerating hyphae of *Platanthera* and in *Spathoglottis plicata* (Senthilkumar and Krishnamurthy 1998a, b). Although epiphytic orchid growing in the tropical area depends less on mycotrophy as they are exposed to sunlight throughout the year but the association of mycorrhiza in some species developed an alternative source of nutritional requirement to enhance its survival. It is assumed that the presence of pelotons in older roots of *Paphiopedilum villosum* and *Acanthephippium bicolor* is to achieve a balance between digestion and reinfection (Manoharachary 2019).

Salome et al. (1983) through their cytochemical studies got some evidence that endomycorrhizal fungi invade host cells which produce active poly-pehnoloxidases and accumulate in the host fungal interface which promotes degradation and inactivation of fungistatic phenolics (phytoalexins) that are frequently found in orchids and accumulate after infection by mycorrhizae. The balance between the production of phytoalexin and its inactivation has been controlled by the fungus also happens to be another stabilizing factor to maintain orchid symbiosis. Production of various enzymes helps in degrading various substrates that produce essential minerals and a variety of carbon sources utilized by endophytic orchids (Hadley and Williamson 1972; Smith 1974).

According to Hadley and Williamson (1972), in the post-infection phase if no external nutrient is present, hyphae get infected and lyzed frequently and little or no stimulation of growth occurs. Thus, digestion of the endophyte is a defense mechanism not a prerequisite for growth stimulus and nutrients are transferred across the living fungal host interface that stimulates growth.

16.7 Role of Mycorrhiza

16.7.1 Nutrient Transfer by Orchid Mycorrhizal Fungi (OMF)

Mycorrhiza and orchids have a symbiotic relationship where mycorrhiza provides minerals like nitrogen, phosphorous, and potassium for plants and carbon (photosynthetic products) provided by plants to mycorrhiza (Garcia and Zimmermann 2014; Hijri and Ba 2018). It helps in increasing growth of the plants. Yeh et al. (2019) described the role of fungi which helps in increasing the tolerance level to abiotic stresses like inundation, salinity, temperature, and biotic stresses like pathogens and insects.

The old coils of hyphae collapse within autotrophic orchids and it is similar to the collapse of older arbuscules in Vesicular Arbuscular Mycorrhiza (VAM) but the significance of this process to plant nutrition is unknown. The digestion of hyphae allows successive waves of formation of pelotons along with digestion, reinfection within the same root of the cells and thus primarily increases the durations of active associations (Burgeff 1959; Smith and Read 1997; Brundrett 2002). The flux of substances across the host fungus interface would have to be much higher than occurs in other types of associations as the volume of plant tissues of some orchids are higher.

Orchids are highly dependent on mycorrhizal fungi as the orchid seeds are minute and they have limited food resources and they are inefficient to acquire nutrients from the substrates. Several studies have authenticated that in mixotrophic orchids OMF plays a role in transferring carbon, nitrogen, phosphorous, and other minerals to the adult photosynthetic green plants (Zimmer et al. 2007) and in epiphytic orchids, the OMF mycelium plays a vital role in decomposing organic matter like bark and nutrient availability to the orchids (Zhang et al. 2018). This nutrient transfer can be unidirectional (Hadley 1989), bidirectional (Cameron et al. 2006), or even mycorrhizal independent (Purves and Hadley 1975).

The fungi mycorrhiza in orchid plants at first penetrates through root hairs and then enters into parenchymal cortex cells and hyphae coiled forms pelotons. The pelotons help in transfer of nutrients between the orchid and mycorrhiza. The pelotons are connected to each other and also with other peloton in another cell. Peloton then lysis and released nutrient (Kaur 2020) (Fig. 16.5).

16.7.2 Carbon Transfer

Although the adult autotrophic orchid fulfills their C requirement by photosynthesis, but they might also supplement the requirement through the transfer of organic carbon through the OMF (Gebauer and Meyer 2003). The transfer of carbon can be both unidirectional and bidirectional, the translocation of carbon is stopped in

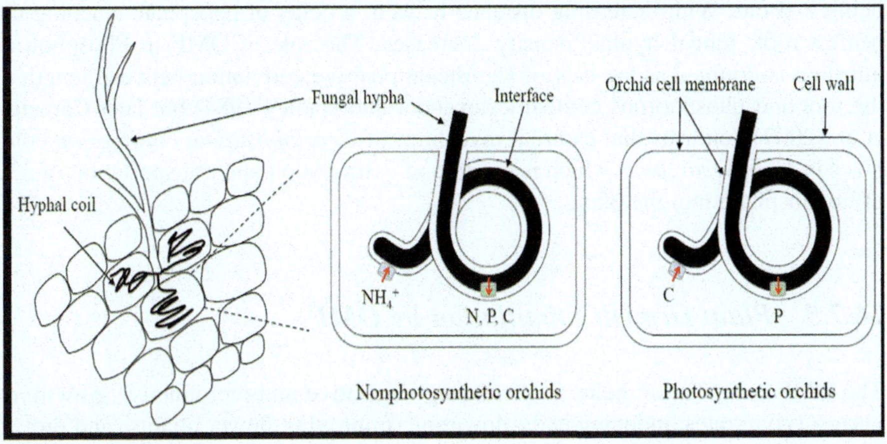

Fig. 16.5 The possible model of nutrient exchange in orchid mycorrhiza (Source: Chuan-Ming Yeh et al. 2019)

unidirectional carbon transfer when the orchids reach certain stage of development (Alexandar and Hadley 1984). The study of Cameron et al. (2006) confirms the bidirectional flow of carbon by studying the mycorrhizal mycelium of *Ceratobasidium cornigerum* transferring carbon to roots, rhizome, and green shoots of *Goodyera repens*. The presence of carbon in the shoots was incorporated and transferred to the rhizome and further external mycelium extending to the OMF, thus affirming the true mutualistic nature of OMF.

16.7.3 Nitrogen Transfer

The terrestrial and epiphytic orchid life forms absorb NO_3^- and NH_4^+ from the substrates, but terrestrial orchids tend to absorb more NO_3^- than NH_4^+. The rate of nitrogen transfer from the fungus to the plant is dependent on the activity and nature of the plant–fungal interface in the roots (Cameron et al. 2006). The mechanism of formation of flower and flower bud in *Cymbidium sinense* regulated by NO_3^- which is possibly carried out by mycorrhizal symbiosis (Pan and Chen 1994). This transfer of nitrogen from the fungus to the orchid was first demonstrated by Cameron et al. (2006).

16.7.4 Phosphorous Transfer

Alexandar and Hadley (1984) reported that translocation of phosphorous through fungal hyphae occurs up to a distance of 9 cm from the root with optimum distance being 2–3 cm. With increasing distance leads to a delay of phosphate reaching the nearest root, fungal hyphae density decreases. The role of OMF in Phosphorous nutrition is doubted as the lack of significant positive correlation between length of the root and phosphorous content (Alexandar and Hadley 1984) but later Cameron et al. (2007) confirms that external mycelium of *Ceratobasidium cornigerum* colonizes in *Goodyerarepens* which is capable of assimilation and transportation of 33P orthophosphate into the plant.

16.7.5 Plant Growth Stimulation by OMF

The OMF colonization helps in increasing vegetative and reproductive growth, ex vitro survival rates, inducing early flowering, improving flower quality, and protect seeds from diseases (Chang 2008). The colonization of *Rhizoctonia* sp. in the roots of *Cymbidium kanran* (Lee et al. 2003); *Cymbidium goeringii* (Wu et al. 2013), and *Cymbidium* sp. (Wu et al. 2010) enhances height of plant, biomass, and formation of root and length of the root. The plant biomass in orchids increases after colonization by *Tulasnella repens* was reported for *Dendrobium nobile*, *D. loddigessi* (Zaiqi and Yin 2008), *D. officinale* (Yang et al. 2008), and *Anoectochillus formosanus*. The *Epulorhiza* sp., *Mycena dendrobii*, *Moniliopsis* sp., *Gliocladium* sp., and *Mycena anoectochila* colonized in orchids like *Anoectochilus roxburghii* (Dan et al. 2012a), *Dendrobium candidum*, and *Dendrobium nobile* were taller and with higher biomass (Dan et al. 2012a). The *Mycena* sp. colonized in the roots of *Dendrobium officinale* which helps in increasing plant height, biomass, and the number of new buds (Zhang et al. 2012).

16.7.6 Phytohormone Production by OMF

Phytohormones responsible for orchid plant growth are produced by beneficial plant growth promoting organisms which are utilized by the host plants. Dan et al. (2012b) report the stimulatory activity of bioactive compounds like IAA (Indole-acetic acid), Gibberellic acid (GA), and Naphthalene acetic acid (NAA) produced by OMF that induces development in *Dendrobium candidum* and *D. nobile* plantlets. GA and NAA produced by OMF also elongation of stems and roots in *D. huoshanense* (Zhang et al. 1999). It has been observed that significant amount of IAA detected in the fungal mycelium and in also in the culture medium of experimental set ups highly influence in the colonization process (Barroso et al. 1986). OMF is also capable of producing heteroauxin, dormin, zeatin, and zeatin riboside that directly play important role in growth of orchids. Guo and Xu (1990) have reported the role of fungal metabolic products in seed germination and cell differentiation process in *Gastrodiaelata*. The compounds which are extracted and fungal mixture from mycelia extracts, hydrolytic products of cell walls, peptide and proteins acts as fungal elicitors (Smith 1996; Hahn 1996).

16.7.7 Role of OMF in Disease Resistance

Orchids are one of the most tradable commodities that contribute largely to the economy of the country. But the commercial orchid cultivation hugely suffers due to disease outbreaks caused by bacteria, fungi, and viruses leading to unhealthy plants and flowers. One of the commonest diseases, soft rot disease caused by *Erwinia* sp., creates devastation in the commercial orchid production (Liau et al. 2003). To avoid this loss, a hugely traded orchid, *Phalaenopsis* was inoculated with *Rhizoctonia solani* and *Ceratobasidium* and it showed reduced severity of soft rot, reducing disease symptoms and promoting plant growth (Wu et al. 2011). *Vanilla*, well-known for its production of vanillin is often affected by *Fusarium* causing root rot and can be controlled by colonization of *Ceratobasidium* as reported by Bayman et al. (2011). OMF provides growth promoters or systematic induced resistance that strengthens plant defense mechanism, making the plant less susceptible to pathogens and also challenges the pathogens by putting competition for space or nutrients for them.

16.8 Micro Seeds and Strategies Adopted for Germination

The loss of endosperm with reference to the evolution of small seeds in orchids demands inhibition in hormone of embryo differentiation (Eriksson and Kainulainen 2011). The orchid seeds with simple embryo are cleared a precondition for omitting the apical meristem of root and the radicle pole turns into a mycotrophic tissue. The seedling structure in microspermic plant species is documented in a scattered way so that the primary root preserved clearly in some microparasitic plant seedlings. The loss of radicle characterizes extant orchid seedlings, the seedling root of orchids was infected through root hairs. The radicle become fallen into disuse as the external mycelium helps in the uptake of water and the water requirement is reduced in orchid seedlings covered by soil or plant debris. After losing radicle there is another option for a seed suspensor as an entrance for hyphae. It is clear that some other seeds of orchid take water, germinate, and produce rhizoids before any incursion occurs and they can germinate without fungi. The more common state appears when small seedlings which are non-infected occur in seed packets together with larger infected ones, requirement of germination before invasion in fungi assumed. Asymbiotic seedlings survive only when the nutrient stores last unless infected in time (Rasmussen and Whigham 1993). Few data are available on distribution of rhizoid versus suspensor infection. It is cleared that suspensor infection is the evolutionarily derived condition because in the similar seedling, fungi will be subsequently entering through rhizoids. It is safe to assume that "'seedling mycorrhiza anticipated 'seed mycorrhiza'."

16.9 Role of Mycorrhiza Against Plant Stress

Plants face different stresses like heavy metal toxicity, salt, drought, biotic stresses, and also cause oxidative stress as all the stresses produce reactive oxygen species (ROS). The mutualistic association with mycorrhiza helps in amelioration of these stresses. ROS causes damage in cells by affecting the cellular activities like oxidation of proteins, peroxidation of lipids and the enzyme activity (Sharma et al. 2012). AM fungi helps in increasing the antioxidant defense level in plants and produces isoprenoid to defend plant from different stresses. Mycorrhizal roots are more permeable at lower water potential and increases the stomatal transmission which enhances the transpiration process during salt stress condition. Mycorrhizal symbiosis helps to increase resistance against salt stress in plants by assembling the solutes and enhance the osmotic adjustment and more P uptake helps to increase the production of antioxidants and also for the morphological and physiological advancement in host plants (Pandey et al. 2019). Heavy metal (HM) negatively affects in the growth, development of the plant, and causes some diseases like chlorosis, necrosis, senescence, turgor loss, and finally death of the plants occurs. Endomycorrhizal fungi detect heavy metals in the binding area and deactivate and cumulate in their mycelia and AM fungi produce glomalin protein which helps in bind the area where HM is affected (Pandey et al. 2019).

16.10 Possibility of Mycorrhizal Fungal Diversity in Orchids and Role in Seed Germination

By previous scientific evidences, mycorrhizal fungi help in improving crop yield and also enhance antioxidants, vitamins, and essential trace elements in plants. There is more indirect evidence that mycorrhizal fungi help to sterilize the soils polluted by heavy metals, they are also capable to remove the heavy metals from root to ground part and improving the food quality (Wu et al. 2014; Pandey et al. 2019). Mycorrhizal fungi give water and nutrients mainly phosphorous and also protect from abiotic and biotic stresses to the plants and receive carbon from the plants formed during photosynthesis (Tedersoo and Bahram 2019). The host plant dependent on fungi until adult stage comes. Since orchids are "generalized" so they interact with a particular set of mycorrhizal fungi as compared to the other mycorrhizal plants, specifically to the OMF than Ectomycorrhizae (ECM) and AM fungi. Some environmental factors like climatic condition, selection of hosts, evolutionary history, and geographical distribution controls the specificity of OMF and the density of

OMF in soil is also variable for epiphytes (Shefferson et al. 2019; Xing et al. 2019). Reiter et al. (2020) found that *Neottia* and *Caladenia* are associated with sebacinales fungi but the terrestrial species *Caladenia huegelli* is associated particularly with *Sebacina vermifera*; *Cypripedium*, *Ophrys*, and *Chiloglottis* favor Tulasnellaceae (Shefferson et al. 2019); *Dendrobium nobile* and *Liparis japonica* high chance of specificity with Tulasnellaceae (Xing et al. 2017); *Platanthera leucophaea* only associated with *Ceratobasidium* fungi for more than 10 years (Thixton et al. 2020); *Corallorhiza trifida* highly specified with *Thelephoraceae* in various countries; and with different habitats (Zimmer et al. 2008). Recently, Rammitsu et al. (2019) reported *Taeniophyllum glandulosum*, epiphytic orchids species associated with Ceratobasidiaceae and two rare orchid species of *Bipinnula* associated with *Ceratobasidium* spp. helps in germination of seed (Claro et al. 2020). OMF occurs in various habitats, orchids associate not only with *Rhizoctonia* taxa but also with some non-*Rhizoctonia* taxa like *Mycena*, *Thelephoraceae*, and *Russulaceae* are distributed (Li et al. 2021a, b). *Aerangisellisii* an epiphytic endemic orchid species found in Madagascar which will be on the verge of extinction if the fungi they are depending on become scarce. The conservation of this species has been undertaken by in vitro germination and seedling development by isolating the fungi which from *A. ellisii* and other species of *Aerangis*. After several experiments they concluded that *Ceratobasidium* is a specific mycorrhiza which associate with *A. ellisii* (Kendon et al. 2020). The *Dendrobium officinale* has high medicinal importance and due to over exploitation, it will become extinct. The mycorrhizal fungi help in the growth of *D. officinale* and also enhance the tolerance level to the environment (Zhang et al. 2020). As the specificity of mycorrhiza associating with orchids are not much known, but Li et al. (2021a, b) concluded that mycorrhizal fungi of *Cymbidium* is compatible with *D. officinale* as they could promote in the seedling development, also enhanced the resistance to defend against biotic and abiotic stresses. OMF plays a key role in the orchid population dynamic and also density of orchids during vegetative dormancy.

16.11 Conclusion

Symbiotics associations are generally considered to be essential throughout the Orchidaceae. However, we have a limited understanding of the ecology of the orchid, their associated symbionts or the interactions between them. The majority of current knowledge on orchid fungus and host plant interactions is based on in vitro studies using fungal isolates from mature orchid species. In recent years, researchers demonstrated the ability to unlock some of the mysteries of orchid mycorrhizae under field (in situ) situations. Most of our knowledge of the role of mycorrhizal associations from the studies of terrestrial orchids and very little known about their role in the establishment and growth of epiphytes.

Orchids are a highly commercial group that contribute largely in the economy of the country. Although historically the greater focus has been mainly on orchid pollination biology, likely for establishing seedlings rather than seed set is the limiting step in orchid life cycles. One pollination event will yield millions of seeds which are generally wind dispersed. Seed germination in contrast with seedling establishment may be more challenging for the orchid plant. The germination and seed development are very slow as it is devoid of endosperm thus orchid requires association with specific mycorrhizal fungi during seed germination. Orchid mycorrhizal fungi play an important role in the rehabilitation of threatened orchid species in their natural habitats. The endophytic orchid mycorrhizal fungi (OMF) are a true mycorrhizal fungus stimulating early germination of seed, enhance development of protocorm and improves the growth and reproduction of the adult plants. The diverse assemblage of fungi belonging to other taxonomic groups have also been isolated from orchid roots but all of them are may not beneficial. For effective germination we have to identify mycorrhiza or through cultural media and mechanism of its action will not only open the way for mass multiplication and conservation but also contribute largely for the economy and livelihood of mankind.

References

Aggarwal S, Zettler LW (2010) Reintroduction of an endangered terrestrial orchid *Dactylorhiza hatagirea* (D. Don) Soo, assisted by symbiotic seed germination: first report from the Indian sub-continent. Nat Sci 8:139–145

Alexandar C, Hadley G (1984) The effect of mycorrhizal infection of *Goodyera repens* and its control of fungicide. New Phytol 97:391–400

Alexandar C, Hadley G (1985) Carbon movement between host and mycorrhizal endophyte during the development of the orchid *Goodyera repens* Br. New Phytol 101:657–665

Bagyalakshmi G, Muthukumar T, Sathiyadash K, Muniappan V (2010) Mycorrhizal and dark septate fungal associations in shola species of Western Ghats, southern India. Mycoscience 51:44–52

Barroso J, Chaves NH, Pair MS (1986) Production of indole-3-ethanol and indole-3-acetic acid by the mycorrhizal fungus of *Ophrys lutea* (Orchidaceae). New Phytol 103:745–749

Batty AL, Dixon KW, Sivasithamparam K (2000) Soil seed bank dynamics of terrestrial orchids. Lindleyana 15:227–236

Bayman P, Mosquera-Espinosa AT, Porras-Alfaro A (2011) Mycorrhizal relationships of *Vanilla* and prospects for biocontrol of root rots. In: Havkin-Frenkel D, Belanger FC (eds) Handbook of *Vanilla* science and technology. Blackwell, West Sussex, pp 266–280

Braunberger PG, Abbott LK, Robson AD (1997) Early vesicular-arbuscular mycorrhizal colonisation in soil collected from an annual clover-based pasture in a Mediterranean environment: soil temperature and the time of autumn rains. Aust J Agric Res 48:103–110

Brundrett MC (2002) Coevolution of roots and mycorrhizas of land plants. New Phytol 154:275–304

Brundrett MC, Abbott LK (1994) Mycorrhizal fungus propagules in the Jarrah Forest: I. Seasonal study of inoculum levels. New Phytol 127:539–546

Bukhari MJ (2020) Mycorrhizal association in *Porpax reticulata* an epiphytic orchid from Western Ghats, India. Plant Archives 20(1):2614–2616

Bukhari MJ, Herlekar R (2015) Endosymbiotic association in the epiphytic orchids from Western Ghats. In: Krishnan S, Rodrigues BF (eds) Advances in plant sciences and biotechnology, pp 64–67

Burgeff H (1959) In: Withner CL (ed) Mycorrhiza of orchids. In: the orchids: a scientific survey. Ronald Press, New York, pp 361–395
Cameron DD, Leake JR, Read DJ (2006) Mutualistic mycorrhiza in orchids: evidence from plant-fungus carbon and nitrogen transfer in the green leaved terrestrial orchid *Goodyera repens*. New Phytol 171:405–416
Cameron DD, Johnson I, Leake JR, Read DJ (2007) Mycorrhizal acquisition of inorganic phosphorous by the green-leaved terrestrial orchid *Goodyera repens*. Ann Bot 99:831–834
Chang DCN (2008) Research and application of orchid mycorrhiza in Taiwan. Acta Hortic 766: 299–305
Chowdhery HJ, Agrawala DK (2013) A century of west Himalayan orchids. Bishen Singh Mahendra Pal Singh, Dehradun
Claro A, Mujica MI, Cisternas M, Armesto JJ, Pérez F (2020) Low mycorrhizal diversity in the endangered and rare orchids *Bipinnula volckmannii* and *B. apinnula* of Central Chile. Symbiosis 80:145–154. https://doi.org/10.1007/s13199-019-00648-w
Currah RS, Zelmer CA (1992) A key and notes for the genera of fungi mycorrhizal with orchids and a new species in the genus *Epulorhiza*. Rep Tottori Mycol Inst 30:43–59
Currah RS, Zelmer CD, Hanbleton S, Richardson KA (1997) Fungi from orchid mycorrhizas. In: Arditti, Pridgeon AM (eds) Orchid biology: reviews and perspectives, vol VII. Kluwer academic Publishers, Dordrecht, pp 117–170
Dan Y, Meng ZX, Guo SX (2012a) Effects of forty strains of Orchidaceae mycorrhizal fungus growth of protocorms and plantlets of *Dendrobium candidum* and *D. nobile*. Afr J Microbiol Res 6:34–39
Dan Y, Yu XM, Guo SX, Meng ZX (2012b) Effects of forty-two strains of orchid mycorrhizal fungi on growth of plantlets of *Anoectochilus roxburghii*. Afr J Microbiol Res 6:1411–1416
Dearnaley JDW (2007) Further advances in orchid mycorrhizal research. Mycorrhiza 17:475–486
Downie DG (1957) *Corticiumsolani* an orchid endophyte. Nature 179:160
Downie DG (1959) *Rhizoctonia solani* and orchid seed. Trans Bot Soc Edinb 37:279–285
Eriksson O, Kainulainen K (2011) The evolutionary ecology of dust seeds. Perspect Plant Ecol Evol Syst 13:73–87
Fan L, Guo SX, Cao WQ, Xiao PG, Xu JT (1996) Isolation, culture, identification and biological activity of *Mycenaorchidicola* sp. nov. in *Cymbidium sinense* (Orchidaceae). Acta Mycol Sin 15:251–255
Garcia K, Zimmermann SD (2014) The role of mycorrhizal associations in plant potassium nutrition. Front Plant Sci 5:1–9
Gebauer G, Meyer M (2003) ^{15}N and ^{13}C natural abundance of autotrophic and mycoheterotrophic orchids provides insight into nitrogen and carbon gain from fungal association. New Phytol 160: 209–223
Guo SX, Xu JT (1990) Isolation and culture of fungi promoting seed germination of Shihu etc. medicinal plants of orchid family (Orchidaceae). Chin Tradit Herb Drug 21:30–31
Guo SX, Fan L, Cao WQ, Xu JT, Xiao PG (1997) *Mycenaanoectochila* sp. nov. isolated from mycorrhizal roots of *Anoectochilus roxburghii* from Xishuangbanna, China. Mycologia 89: 952–954
Hadley G (1982) Orchid mycorrhiza. In: Arditti J (ed) Orchid biology: reviews and perspectives, II. Cornell University Press, Ithaca, pp 83–181
Hadley G (1984) Uptake of (^{14}C) glucose by asymbiotic and mycorrhizal orchid protocorms. New Phytol 96:263–273
Hadley G (1989) Host-fungus relationships in orchid mycorrhizal systems. In: Pritchard HW (ed) Modern methods in orchid conservation: the role of physiology, ecology and management. Cambridge University Press, Cambridge, pp 5–71
Hadley G, Williamson B (1972) Features of mycorrhizal infection in some Malayan orchids. New Phytol 71:1111–1118
Hahn MG (1996) Microbial elicitors and their receptors in plants. Annu Rev Phytopathol 34:387–412

Hijri M, Ba A (2018) Editorial: mycorrhiza in tropical and neotropical ecosystems. Front Plant Sci 9(308):1–3

Hooker JD (1888–1890) Orchidaceae. In: The flora of British India 5: 667–858 and 6: 1–198. Ashford, Kent

Hooker JD (1895) A century of Indian orchids. Ann Roy Bot Gard Calcutta 5(1):1–68

Jacquemyn H, Duffy K, Sellose M (2017) Biogeography of orchid mycorrhizas. In: Tedersoo L (ed) Biogeography of mycorrhizal symbiosis, ecological studies, vol 230. Springer, pp 159–177

Julou T, Burghardt B, Gebauer G et al (2005) Mixotrophy in orchids: insights from a comparative study of green individuals and non-photosynthetic individuals of *Cephalanthera damasonium*. New Phytol 166:639–653

Kaliamoorthy S (2007) Pattern of mycorrhizal infection in the roots of *Aerides maculosum* Lindl. and *Calanthe triplicata* (Willem.) Ames. Mycorrhiza News 19:14–18

Katiyar RS, Sharma GD, Mishra RR (1986) Studies on mycorrhizal associations in terrestrial orchids. In: Vij SP (ed) Biology conservation and culture of orchids. Affiliated East-West Press, New Delhi, pp 63–70

Kaur S (2020) Mycorrhiza in orchids. In: Mérillon JM, Kodja H (eds) Orchids phytochemistry, biology and horticulture. Springer, New York, pp 1–14

Kaushik P (1983) Ecological and anatomical marvels of the Himalayan orchids. Today and Tomorrow's Printers et Publishers, New Delhi

Kaushik P, Pal P (2011) *Rhizoctonia*: a genus of orchid symbionts. J Orchid Soc India 25(1–2):19–28

Kendon JP, Yokoya K, Zettler LW, Jacob AS, McDiarmid F, Bidartondo MI, Sarasan V (2020) Recovery of mycorrhizal fungi from wild collected protocorms of Madagascan endemic orchid *Aerangisellisii* (B.S. Williams) Schltr. and their use in seed germination in vitro. Mycorrhiza 30: 567–576

King G, Pantling R (1898) The orchids of Sikkim-Himalaya. Ann Roy Bot Gard Calcutta 8: 1–342, tt. 1–448

Kottke I, Suarez JP, Herrera P, Cruz D, Bauer R, Haug I, Garnica S (2009) Atractiellomycetes belonging to the 'rust' lineage (Pucciniomycotina) form mycorrhizae with terrestrial and epiphytic neotropical orchids. Proc R Soc B (Biol Sci) 277:1289–1298

Kumar SS, Krishnamurthy KV (1998) A cytochemical study on the mycorrhiza of *Spathoglottis plicata*. Biol Plantarum 41(1):111–119

Lee JK, Lee SS, Eom AH, Paek KY (2003) Interactions of newly isolated orchid mycorrhizal fungi with Korean *Cymbidium kanran* hybrid 'Chungsu'. Mycobiology 31:151–156

Li Y, Kang Z, Zhang X, Sun P, Jiang X, Han Z (2021a) The mycorrhizal fungi of *Cymbidium* promote the growth of *Dendrobium officinale* by increasing environmental stress tolerance. Peer J 9:e12555. https://doi.org/10.7717/peerj.12555

Li T, Yang W, Wu S, Selosse MA, Gao J (2021b) Progress and prospects of mycorrhizal fungal diversity in orchids. Front Plant Sci 12:646325

Liau CH, Lu JC, Prasad JC, Hsiao HH, You SJ, Lee JT, Yang NS, Huang HE, Feng TY, Chen WH, Chan MT (2003) The sweet pepper ferredoxin-like protein (pflp) conferred resistance against soft rot disease in *Oncidium* orchid. Transgenic Res 12:329–336

Lindley J (1830–1840) The genera and species of orchidaceous plants. J. Ridgways and Sons, London

Lindley J (1852–1859) Folia Orchidacea. An enumeration of the known species of orchids. London

Lucksom SZ (2007) The orchids of Sikkim and north east Himalaya. CONCEPT, Siliguri

Madhaiyan M, Santhanakrishnan P, Pragatheswari D (2003) Rapid detection and assessment of orchid mycorrhizal colonization in *Vanilla planifolia* and r. roots. Mycorrhiza News 14:10–13

Manoharachary C (2019) Orchids-Mycorrhizae. J Orch Soc India 33:23–29

Masuhara G, Katsuya K (1992) Mycorrhizal differences between genuine roots and tuberous roots of adult plants of *Spiranthessinensis* var. *amoena*(Orchidaceae). Bot MagazineTokyo 105:453–460

Masuhara G, Katsuya K, Yamaguchi K (1993) Potential for symbiosis of *Rhizoctoniasolani* and binucleate *Rhizoctonia* with seeds of *Spiranthessinensis* var. *amoena* in vitro. Mycol Res 97: 746–752

Misra S (2007) Orchids of India- a glimpse. Bishen Singh Mahendra Pal Singh, Dehradun

Pan RC, Chen JX (1994) Effects of nitrate-nitrogen and ammonium-nitrogen on growth and development in *Cymbidium sinense*. Acta Bot Yunnanica 16:285–290

Pandey D, Kehri HK, Zoomi I, Akhtar O, Singh AK (2019) Mycorrhizal fungi: biodiversity, ecological significance, and industrial applications. In: Yadav AN et al (eds) Recent advancement in white biotechnology through fungi, fungal biology, pp 181–199

Peterson RL, Uetake Y, Zelmer C (1998) Fungal symbiosis with orchid protocorms. Symbiosis 25: 29–55

Pradhan UC (1976) Indian orchids. Guide to identification and culture, vol I. Bharat Lithographing Co., Calcutta

Pradhan UC (1979) Indian orchids: guide to identification and culture, vol II. Thomson Press (India) Ltd, Faridabad

Purves S, Hadley G (1975) Movement of carbon compounds between the parents in orchid mycorrhizal. In: Sanders FE, Mosse B, Tinker PB (eds) Endomycorrhizas. Academic, London, pp 175–194

Radhika KP, Rodrigues BF (2007) Orchid mycorrhizal colonization in *Rhynchostylisretusa* (L.) Blume. Mycorrhiza News 19:22–23

Rajkumar, Kaushik P (2007) Isolation of *Rhizoctonia solani*- a cellulose producing mycorrhizal fungus from *Zeuxinestrateumatica* (L.) Sehltr. J Orchid Soc India 21(1–2):3–6

Rammitsu K, Yagame T, Yamashita Y, Yukawa T, Isshiki S, Ogura-Tsujita Y (2019) A leafless epiphytic orchid, *Taeniophyllum glandulosum* Blume (Orchidaceae), is specifically associated with the Ceratobasidiaceae family of basidiomycetous fungi. Mycorrhiza 29:159–166. https://doi.org/10.1007/s00572-019-00881-7

Rasmussen HN (1995) Terrestrial orchids: from seed to mycotrophic plant. Cambridge University Press, New York

Rasmussen HN, Whigham DF (1993) Seed ecology of dust seeds in situ: a new study technique and its application in terrestrial orchids. Am J Bot 80:1374–1378

Reiter N, Phillips RD, Swarts ND, Wright M, Holmes G, Sussmilch FC et al (2020) Specific mycorrhizal associations involving the same fungal taxa in common and threatened Caladenia (Orchidaceae): implications for conservation. Ann Bot 126:943–955. https://doi.org/10.1093/aob/mcaa116

Roxburgh W (1832) Flora Indica ed. 2. Parbury, Allen and Co., London

Saha D, Rao AN (2006) Studies on endophytic mycorrhiza of some selected orchids of Arunachal Pradesh – 1. Isolation and identification. Bull Arunachal For Res 22:9–16

Salome M, Pais S, Barroso J (1983) Localization of polyphenoloxidases during the establishment of *Ophrys luenta* endomycorrhizas. New Phytol 95:219–222

Sathiyadash K, Muthukumar T, Uma E, Pandey RR (2012) Mycorrhizal association and morphology in orchids. J Plant Interact 7(3):238–247

Sathiyadash K, Muthukumar T, Karthikeyan V, Rajendran K (2020) Orchid mycorrhizal fungi: structure, function, and diversity. In: Khasim SM et al (eds) Orchid biology: recent trends & challenges. Springer Nature Singapore Pte Ltd, Singapore, pp 239–280

Selosse MA, Roy M (2009) Green plants that feed on fungi: facts and questions about mixotrophy. Trends Plant Sci 14:64–70

Senthilkumar S, Krishnamurthy KV (1996) Certain peculiar features of mycorrhizal association in the ground orchid *Spathoglottis plicata* Blume. Mycorrhiza News 8:9–11

Senthilkumar S, Krishnamurthy KV (1998a) A cytochemical study on the mycorrhizae of *Spathoglottis plicata*. Biol Planta 41:111–119

Senthilkumar S, Krishnamurthy KV (1998b) The role of root hair in the mycorrhizal association of the ground orchid *Spathoglottis plicata* Blume. Mycorrhiza News 10:15–17

Senthilkumar S, Britto SJ, Krishnamurthy KV, Hariharan C (2000) Biochemical analysis of mycorrhizal roots of *Aerides maculosum*. Phytomorphology 50:273–279
Sharma M, Vij SP, Deep N (1988) Studies on mycorrhizal endophytes in some orchids. In: Vij SP, Khullar SP (eds) Abstracts of National Seminar on current research trends in Indian orchids with special reference to tissue culture technology. The Orchid Society of India, Chandigarh, pp 7–8
Sharma P, Jha AB, Dubey RS, Pessarakli M (2012) Reactive oxygen species, oxidative damage, antioxidative defense mechanism in plants under stressful conditions. J Bot 2012:26
Shefferson RP, Bunch W, Cowden CC, Lee YI, Kartzinel TR, Yukawa T et al (2019) Does evolutionary history determine specificity in broad ecological interactions? J Ecol 107:1582–1593. https://doi.org/10.1111/1365-2745.13170
Singh A, Varma A (2000) Orchidaceous mycorrhiza fungi. In: Mukerji KG, Chamola BP, Singh J (eds) Mycorrhizal biology. Kluwer Academic, Netherlands
Singh SK, Agrawala DK, Jalal JS, Dash SS, Mao AA, Singh P (2019) Orchids of India-a pictorial guide. Botanical Survey of India, Kolkata
Smith SE (1974) Mycorrhizal fungi. Crit Rev Microbiol 3(3):275–313
Smith CJ (1996) Accumulation of phytoalexins: defence mechanism and stimulus response system. New Phytol 132:1–45
Smith SE, Read DJ (1997) Mycorrhizal symbioses, 2nd edn. Academic, San Diego
Smith SE, Read DJ (2008) Mycorrhizal Symbiosis. Academic Press, Cambridge
Tedersoo L, Bahram M (2019) Mycorrhizal types differ in ecophysiology and alter plant nutrition and soil processes. Biol Rev 94:1857–1880. https://doi.org/10.1111/brv.12538
Thixton HL, Esselman EJ, Corey LL, Zettler LW (2020) Further evidence of *Ceratobasidium* D.P. Rogers (Basidiomycota) serving as the ubiquitous fungal associate of *Platantheraleucophaea* (Orchidaceae) in the north American tallgrass prairie. Bot Stud 61:12. https://doi.org/10.1186/s40529-020-00289-z
Varma A (1994) Beneficial role of mycorrhizal fungi in the early establishment of orchid seedling raised from tissue culture technique. In: Proceedings of national seminar on development, biology and commercialization of orchids and orchid show The orchid society of India, Chandigarh, pp 40–42
Vij SP, Sharma M (1983) Mycorrhizal association in North Indian Orchidaceae: a morphological study. Bibl Mycol 91:467–503
Vij SP, Sharma M (1988) Mycorrhizal endophytes of *Nepelaphyllum* Bl. (Orchidaceae). In: International conference on research in plant science and its relevance to future. New Delhi
Vij SP, Sharma M, Datta SS (1985) Mycorrhizal endophytes of *Sprianthes lancea* (Sw) Baker (Orchidacea). J Indian Bot Soc 64:175–179
Vij SP, Lakhanpal TN, Gupta A (2002) Orchidoid mycorrhiza and techniques to investigate. In: Mukerji KG, Manoharachary C, Chamola BP (eds) Techniques in Mycorrhizal studies. Kluwer Academic, Netherlands, pp 385–434
Wang B, Qiu YL (2006) Phylogenetic distribution and evolution of Mycorrhiza in land plants. Mycorrhiza News 16:299–363
Warcup JH, Talbot PHB (1967) Perfect states of *Rhizoctonias* associated with orchids. New Phytol 66:631–641
Warcup JH, Talbot PHB (1980) Perfect states of *Rhizoctonias* associated with orchids III. New Phytol 86:267–272
Weber RWS, Webster J (2001) Teaching techniques for mycology: 14. Mycorrhizal infection of orchid seedling in the laboratory. Mycologist 15(2):55–59
Wu J, Ma H, Lu M, Han S, Zhu Y, Jin H, Liang J, Liu L, Xu J (2010) *Rhizoctonia* fungi enhance the growth of the endangered orchid *Cymbidium goeringii*. Botany 88:20–29
Wu PH, Huang DD, Chang DCN (2011) Mycorrhizal symbiosis enhances *Phalaenopsis* orchid's growth and resistance to *Erwinia chrysanthemi*. Afr J Biotechnol 10:10095–10100
Wu J, Ma H, Deng Z, Chen J, Wu F, Guo S (2013) Specificity and preference of mycorrhizal associations in two species of the genus *Dendrobium* (Orchidaceae). Mycorrhiza 23:317–324

Wu QS, Cao MQ, Zou YN, He XH (2014) Direct and indirect effects of glomalin, mycorrhizal hyphae, and roots on aggregate stability in rhizosphere of trifoliate orange. Sci Rep 4:5823

Xing XK, Ma XT, Men JX, Chen YH, Guo SX (2017) Phylogenetic constrains on mycorrhizal specificity in eight *Dendrobium* (Orchidaceae) species. Sci China Life Sci 60:536–544. https://doi.org/10.1007/s11427-017-9020-1

Xing XK, Jacquemyn H, Gai XG, Gao Y, Liu Q, Zhao ZY et al (2019) The impact of life form on the architecture of orchid mycorrhizal networks in tropical forest. Oikos 128:1254–1264. https://doi.org/10.1111/oik.06363

Yang YL, Lai PF, Jiang SP (2008) Research development in *Dendrobium officinale*. J Shandong Univ TCM 32:82–85

Yeh CM, Chung KM, Liang CK, Tsai WC (2019) New insights into the symbiotic relationship between orchids and fungi. Appl Sci (Switzerland) 9(3):1–14

Zaiqi L, Yin Y (2008) Endophytic fungi and their growth effects on *Dendrobium nobile* Lindl. and *Dendrobium loddigessii* Rolfe. Guizhou For. Sci Technol 1:28–32

Zettler LW (1997) Orchid-fungal symbiosis and its value in conservation. Mcllvainea 13:40–45

Zhang JH, Wang CL, Guo SX, Chen JM, Xiao PG (1999) Studies on the plant hormones produced by 5 species of endophytic fungi isolated from medicinal plants (Orchidaceae). Acta Acad Medicinae Sinicae 21:460–465

Zhang L, Chen J, Lv Y, Gao C, Guo S (2012) *Mycena* sp., a mycorrhizal fungus of the orchid *Dendrobium officinale*. Mycol Prog 11(2):395–401

Zhang SB, Yang Y, Li J, Qin J, Zhang W, Huang W, Hu H (2018) Physiological diversity of orchids. Plant Divers 40:196–208

Zhang Y, Li YY, Chen XM, Guo SX, Lee Y (2020) Effect of different mycobionts on symbiotic germination and seedling growth of *Dendrobium officinale*, an important medicinal orchid. Bot Stud 61(2):1–10

Zimmer K, Hynson NA, Gebauer G, Allen MF, Read DJ (2007) Wide geographical and ecological distribution of nitrogen and carbon gains from fungi in pyroloids and monotropoids (Ericaceae) and in orchids. New Phytol 175:166–175

Zimmer K, Meyer C, Gebauer G (2008) The ectomycorrhizal specialist orchid *Corallorhizatrifida* is a partial myco-heterotroph. New Phytol 178:395–400. https://doi.org/10.1111/j.1469-8137.2007.02362.x

Chapter 17
Current Status of Mycorrhizal Biofertilizer in Crop Improvement and Its Future Prospects

Prashanta Kumar Mitra, Rajsekhar Adhikary, and Vivekananda Mandal

Abstract Improving crop yield and quality has always been a concern to meet the demands at the consumer level. But our current chemical fertilizer-based agricultural practices are breaking the bridge between the ecosystem and economy and raising the issue of sustainability in agriculture. Considering agriculture as biospheric inter-kingdom interaction, every organism is in some sort of interaction with others for its existence. When addressing the sustainability issue in agriculture, these fundamental aspects of the ecosystem should not be ignored. In this context, biofertilizers are not a new concept. At the application level, many workers around the world discover the deeper dynamics of the effects of biofertilizers on agroecosystems. So, it is very much essential and pertinent to rediscover the current trends and practices of biofertilizers in agroecosystems to meet the present demand for crop productivity. Presently, most of the biofertilizers being used in agroecosystems are microorganisms like phosphate-solubilizing bacteria, nitrogen-fixing bacteria, and mycorrhiza, out of which the major focus has gained by plant growth-promoting microbes (PGPR) and diazotrophic microbes, whereas mycorrhizal biofertilizers, specifically arbuscular mycorrhizal fungi (AMF), are relatively lagging at the application level. The major reasons could be the lack of knowledge on the basic aspects related to the persistence of these mycorrhizal fungi in field conditions, amplitude, and nature of interactions with other microorganisms, particularly the host-specific fungi for field use. Different studies conducted on these mycorrhizal biofertilizers show an approximate 12–15% increment in average yield in crops. The studies also reflect that the application has also resulted in rejuvenating fertility and soil health for a prolonged

P. K. Mitra
Plant and Microbial Physiology and Biochemistry Laboratory, Department of Botany, University of Gour Banga, Malda, West Bengal, India

Department of Botany, Kalyani University, Nadia, West Bengal, India

R. Adhikary · V. Mandal (✉)
Plant and Microbial Physiology and Biochemistry Laboratory, Department of Botany, University of Gour Banga, Malda, West Bengal, India
e-mail: vivek.bot@ugb.ac.com

P. Mathur et al. (eds.), *Microbial Symbionts and Plant Health: Trends and Applications for Changing Climate*, Rhizosphere Biology,
https://doi.org/10.1007/978-981-99-0030-5_17

time. Studies revealed that mycorrhiza formation induces an altered plant metabolome increasing crop yield, reducing weed community, increasing stress tolerance, and also playing a role in mycoremediation. The present literature reflects, in a nutshell, the current trends of research on how arbuscular mycorrhizae can contribute to establishing sustainable agriculture for future generations.

Keywords Agriculture · Arbuscular mycorrhiza · Biofertilizer · Mycoremediation · Sustainability

Abbreviations

AMF	Arbuscular mycorrhizal fungi
EPA	US Environmental Protection Agency
IARC	International Agency for Research on Cancer
NADPH	Nicotinamide adenine dinucleotide phosphate reduced
PGPR	Plant growth-promoting rhizobacteria
REEs	Rare earth elements
ROS	Reactive oxygen species
WHO	World Health Organization

17.1 Introduction

Agriculture is the practice of cultivating food crops to survive and evolve in nature through the proper understanding and interactions with the soil closely. With the development of civilization, agricultural practices became intensified by higher yield in quenching land resources with the reflection of global changes and leading to predominant environmental problems (Srivastava et al. 2016). For the last few decades, the use of commercial chemical fertilizers has become one of the major contributors to targeted crop yield. Crop productivity is dependent on the available nutrient uptake of chemical fertilizers. In a PubMed data search for the last few decades on mycorrhizae, pesticides and fertilizer represent the following number of publications indicating their degree of central focus on global research, as shown in Fig. 17.1. Commonly plants require a wide range of elements (18 nos.) which includes micro- and macronutrients, such as carbon, hydrogen, nitrogen, phosphorus, potassium, magnesium, manganese, boron, zinc, etc. to survive in nature and give the highest productivity. In 2017, the world scenario shows that about 10.9% and 10.2% of the population remained undernourished and had severe food insecurity, respectively. The dietary energy supply is an indicator of public health. In this respect, the overall average dietary energy supply had increased from 2716 kcal/cap/day in 1999–2001 to 2904 kcal/cap/day in 2015–2017. However, the average annual increase in the yield of cereals was 1.75% with an increase of less than 0.5% in land

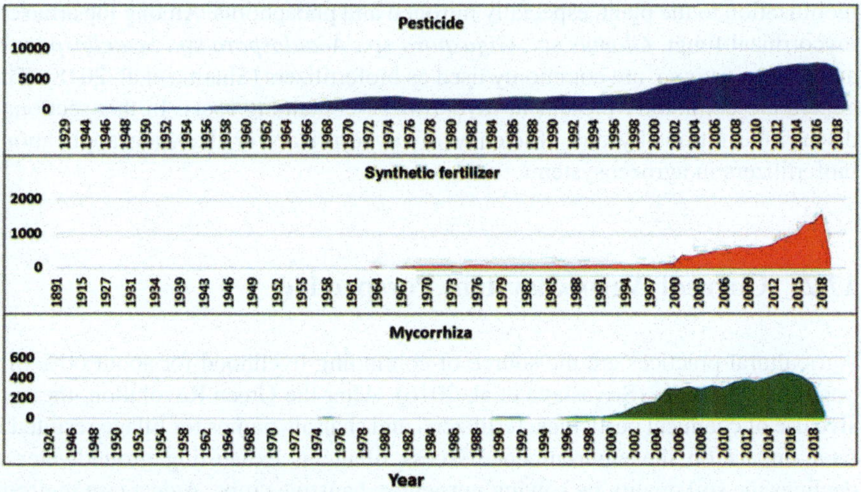

Fig. 17.1 Year-wise published papers were searched with the keywords “Pesticide,” “Synthetic fertilizer,” and “Mycorrhiza” on PubMed (https://www.ncbi.nlm.nih.gov/pubmed/)

uses in the last 10 years (FAO 2018). The use of lands for agricultural purposes became stringent with the gradual increase of urbanization and anthropogenic activities. In this food crisis condition, the extensive injudicious use of chemical fertilizers drastically changes the soil chemical properties that affect ecological integrity by destabilization of soil microbes and fauna and flora communities. These destabilizations of soil biodiversity face a great challenge to human civilization as well as a change in climatic conditions leading to polluting air, water, and soils. The development of the fertilizer industry releases excess metal elements which contains a large number of heavy metals, metal cations, phosphate, and ammonia which lead to developing toxicity to the environment. Additionally, the use of excess chemical fertilizers in the agricultural fields by the farmers desiring more production causes a hazardous impact on the geometric progressions of soil environments. Commonly, chemical fertilizers contain large numbers of manganese, magnesium, iron, boron, and calcium which cause high deposition in soils with their excess use than recommended doses. Excess application of nitrogen fertilizer causes carcinogenicity through nitrosamines in lettuce and spinach leaves (Savci 2012).

In this scenario, the development and application of a sustainable strategy to improve crop yield without affecting forests are important agendas worldwide, and biofertilizers are an alternative option for sustainable agriculture. Biofertilizers contain microorganisms’ living or latent cells that increase the uptake of nutrients by host plants when applied to the targeted crops. Biofertilizers are more cost-effective and environment friendly as compared to chemical fertilizers. Depending on the mode of action, different types of biofertilizers are available such as nitrogen fixers, phosphate solubilizers, compost, or nutrient-mobilizing biofertilizers. Endomycorrhizae are one of the important biofertilizers which help in nutrient mobilization to the plant, especially nitrogen and phosphorus. Among the arbuscular mycorrhizal fungi, *Glomus* sp., *Gigaspora* sp., *Acaulospora* sp., *Scutellospora* sp., and *Sclerocystis* sp. are commonly used as biofertilizers (Sharma et al. 2018). These genera are commonly present in mycorrhizae containing roots. In this section, we discuss the rationality and efficiency of mycorrhizae for the wide application of biofertilizers in agroecosystems.

17.2 Current Agroecosystem Perspective

Agricultural practices are the source of continuing livelihood for about 60% of the population of India (Srivastava et al. 2016). After the Green Revolution, the extensive use of chemical pesticides, fertilizers, and shallow pumps for tillage resulted in a fatal impact on the environment. Excess tillage is exhausting groundwater and declines the soil quality by sowing nutrient-exhaustive crops. A decrease in groundwater increases the contamination of arsenic, lead, or mercury in water, and when this contaminated water was applied for agricultural purposes, these heavy metals enter the food chains. Apart from that, the overapplication of chemical pesticides and fertilizers also decreases the soil quality and hampers the equilibrium of soil biodiversity. The impact of different components in the agricultural scenario is extensively elaborated in this section.

17.2.1 Heavy Metal Contamination

Heavy metal contamination in soil has become a critical environmental concern for its adverse effects on ecosystems. Rare earth elements (REEs) or heavy metals are being widely used in electronic gadgets and agriculture for over a few decades, and the subsequent accumulation of REEs in soils affects plant growth and became a threat to the agroecosystem as well as human health through food chains. Mercury (Hg) is a widely dispersed pollutant persisting in the environment in different forms such as HgO, Hg^{+}, Hg^{2+}, and methyl-Hg which drastically affect plants, microbes, and animals living in soil and water biome. Subsequently, Hg gets accumulated in edible plant parts at low concentrations and then gets transferred to animals and humans. Hg alters the cell membrane, reacts with thiol (–SH) groups of enzymes, reacts with phosphate groups and active sites of ADP or ATP, replaces essential cations, and eventually also binds to DNA, RNA, and ribosomes. It binds to sulfur ligands in cysteine, glutathione (GSH), and metallothioneins to form physiologically inactive complexes. In plants, Hg is mainly accumulated in roots through binding to the cell wall, sequestered in root cell vacuoles, and then translocated to the shoots (Debeljak et al. 2018; Fan et al. 2018). The transport of Hg from the roots to the shoots is plant species-specific. Cadmium (Cd) contamination in soil is rapidly increased by industrial waste and effluents. Cadmium contamination in the environment occurs through the wastes of industries and fertilizers. This gives a toxic effect after accumulation in plants and human metabolism in minimum quantity. In recent years, Cd has become a challenging pollutant in staple crops such as rice and maize to maintain biosafety and grain quality. Cadmium interferes with numerous biochemical and physiological processes of plant cells, such as plant growth, photosynthesis, nutrient uptake, and enzyme activity (Chang et al. 2018). Arsenic (As), another threatened heavy metal, is classified as a group 1 carcinogenic element by the US Environmental Protection Agency (EPA) and International Agency for Research on Cancer (IARC) of the World Health Organization (WHO). Arsenic contamination in the environment is caused by anthropogenic and many natural processes. Naturally, arsenic contamination occurs due to volcanic eruption and weathering of igneous and metamorphic rocks. It causes respiratory diseases, liver and cardiovascular complications, diabetes, high blood pressure, gastrointestinal diseases, and cancer in the skin, lungs, kidneys, and urinary bladder in humans. In plants, it causes physiological, morphological, and biochemical disorders by causing toxicity at low concentration. It also directly affects stomatal behavior, transpiration, shoot, and root development. This contamination also causes oxidative stress and lipid peroxidation due to the production of reactive oxygen species (ROS) such as O_2^{*-}, H_2O_2, and $^1O_2$ (Rafiq et al. 2017).

17.2.2 Pollution by Fertilizer

Nitrogenous fertilizers cause nitrate leaching from agricultural lands and accumulated in soils. Normally, plants use 50% of nitrogenous fertilizers applied in normal doses and the rest 50% is deposited in the environment. These large remnants cause nitrate contamination in water. In Europe, 22% of cultivated areas contain ≥11.3 mg/L nitrates in drinking water. Nitrates are absorbed in the intestinal tract from drinking water by the body within 4–12 h and excretion occurs from the kidneys. This nitrate is reduced to nitrite in the anaerobic environment which causes toxicological effects. Nitrates generate toxicity in three stages. The primary toxicity resulted in inflammation in the digestive and urinary systems. Secondary toxicity is caused by high nitrate concentration which caused methemoglobinemia in infants below the age of 6 months. In such cases, stomach acid does not occur, and in this condition, hemoglobin is reacted with nitrite in the blood and the patient is going to die due to impairment of blood oxygen transport function. Secondary and tertiary amines, tertiary nitrites, and alkyl ammonium bases react with amides in an acidic medium to form nitrosamines which have strong carcinogenic effects. Another negative impact of nitrate contamination in water is to cause eutrophication which induces the formation of a high amount of algae and aquatic plants due to high nitrogen and phosphorus contents resulting in the deterioration of water quality through dystrophication. In this environment, scarcity of dissolved oxygen occurred and the numbers of aerobic fauna have reduced drastically (Savci 2012).

The excess use of chemical fertilizers causes air pollution too by the emission of different oxides of nitrogen such as nitrous oxide, nitrogen dioxides, nitric oxide, etc. These oxides increase greenhouse effects through the association with water vapor, carbon dioxide, methane, hydrogen sulfide, chlorofluorocarbons, etc. Excessive use of nitrogenous fertilizers resulted in the emission of ammonia into the environment. In atmospheric condition, this ammonia is reduced to nitric acid and causes acid rain that affects vegetation, water resources, civilizations, aesthetics, and sculptures (Savci 2012). The effect of chemical fertilizers in the soil system is not so rapid because of the strong buffering nature of the soil. But after long-term exposure, the soil loses its fertility, and changes occur in the soil's physical and chemical properties. Apart from that, the deposited toxic compounds accumulate in the vegetables and transfer to animals and humans through the food chain. High-quantity application of alkaline fertilizers such as $NaNO_3$, NH_4NO_3, KCI, K_2SO_4, and NH_4Cl dismantles the soil equilibrium by altering soil pH, conductivity, and other physical and chemical characteristics. Overapplication of nitrogenous fertilizers and liming decrease soil pH and inhibit the growth of symbiotic (*Rhizobium* sp.) and free-living (*Azotobacter* sp., *Azospirillum* sp., etc.) bacteria, and in such case, the plants face scarcity of soluble nitrogen from the soil. Apart from that, excess use of potassium, calcium, iron, and zinc disrupts the nutrient balance in the soil which causes an unfavorable lethal environment for soil worms and mites (Savci 2012).

17.2.3 Nutrient Leaching and Availability

Nutrients can be lost from the agricultural ecosystem in different ways depending upon the mobility of nutrients. Mobile elements such as nitrate (NO_3^-) and sulfate (SO_4^{-2}) are easily leached out and become unavailable to the plant roots, while immobile nutrients such as P, K, and Zn form chelates with organic materials of clay soils and are precipitated as organic-metallic complexes. However, these chelated metals are available to plants slowly which is insufficient to replenish plants' needs. Nutrient loss determines land-use practices, climate variability, plant species composition, soil type, soil structure, and soil organisms. These components and nutrient loss influence nutrient cycling and nutrient mobility.

Phosphorus is an important plant nutrient, with little known about its uptake pattern in the plant system. Phosphorus is available to the plant through the sorption methods where the nutrient is rapidly sorbed to the soil matrix. Potassium, another essential macronutrient of the plant system, helps in the developmental processes of plants. Potassium helps to decrease the oxidative damage (ROS) and nicotinamide adenine dinucleotide phosphate (NADPH) oxidase activity of the plants and increase the photosynthetic rate and stress tolerance ability. The plant can easily uptake potassium as the cation (K^+) and many soluble salt forms such as potassium chloride (KCl), potassium nitrate (KNO_3), potassium sulfate (K_2SO_4), potassium carbonate (K_2CO_3), etc. Besides these potassium sources, potassium silicate is also present in the soil but the bioavailability in the plant is very low. Some bacteria and fungi, especially AM fungi, can solubilize these salts to the plants through acidification, chelation, and exchange reactions (Hasanuzzaman et al. 2018).

Zinc (Zn) malnutrition is a major global health problem for people relying on cereals for energy and mineral sources. It was estimated that approx. 17.3% of the world's population is at risk of inadequate dietary Zn intake (Coccina et al. 2019). In this condition, the improvement of grain quality insoluble zinc-deficient soil is a challenge to mitigate the food crisis for the growing populations throughout the globe. Insufficient Zn can affect the synthesis and function of a wide range of macromolecules in plants and consequently decrease the yield and quality of crops. Zinc deficiency reduces the body's immune system in adults, causes stunted growth in infants, and increases the risk of adverse pregnancy outcomes in women.

17.2.4 Drought Stress

Water scarcity in groundwater is becoming a major problem that will lead to a severe drought situation. Drought stress is one of the most challenging abiotic factors retarding the vegetative and reproductive growth of plants leading to declining productivity of crops. Plants often face several stresses simultaneously in the field rather than single stress conditions and combinations of stresses such as drought with high temperatures that are prevalent in many agricultural areas on the globe, particularly in the arid and semiarid areas of the tropics. The co-occurrence of drought and high temperature leads to maximum agricultural losses than when each stress causes individually. During drought stress, plants do not meet up the required nutrient demand by solute transfer from the soil and neighboring tissues. Drought stress leads to impairment of the metabolism of plants resulting in stunted growth, discontinuous CO_2 diffusion to chloroplast, stomata closure, reduced photosynthetic rate, and accelerated leaf and bud senescence. Long-term drought stress increases the motility rate of plants. It also leads to amplifying the frequency and amplitude of heat waves and thus expands the areas of desert in different parts of the globe. Concurrent occurrence of different stresses in changing climates, plants often deploy stress-adaptation strategies which are distinct and sometimes opposite to those observed under individual stresses (Mathur et al. 2018).

17.2.5 Soil Salinity Stress

Salinity in the soil is nowadays an emerging issue that causes stresses to plants in growth and metabolism. Soil salinity is rapidly increasing with an estimated yearly addition of 0.3–1.5 million ha of agricultural land resulting in a decrease in crop production by more than 20%. It also exposes another 20–46 million ha with decreased capacity for production. Soils are going to be rendered saline due to the deposition of salt either by natural (primary) or anthropogenic (secondary) processes. Natural processes include weathering of parent rocks, deposition from seawater, and atmospheric sedimentation. Secondary or anthropogenic processes include poor drainage facilities, irrigation with brackish groundwater, and continuous irrigation and tillage for long durations and water improper management. On the other hand, food security for the growing population throughout the world amidst the declining agricultural land is one of the most important challenges for modern agriculture. Sodium chloride (NaCl) is the most prevalent salt in this regard. High salt deposition (NaCl, $CaCO_3$, $CaCl_2$) in the soil results in osmotic stresses and specific ion effects, which lead to secondary oxidative stress in plants. Thus, salinity shows adverse effects on the germination, growth, and reproduction of plants that subsequently diminish crop yield. Plants have evolved to be highly flexible systems that can adjust to morphological, physiological, biochemical, and molecular mechanisms to survive and sustain the changing environment. To counter the problems of salinity, plants exhibit growth plasticity (morphological and developmental pattern change), accumulation of compatible osmolytes to maintain turgor as well as prevent ultrastructural damage, ion homeostasis, regulation of water uptake and enhanced water use efficiency, enhanced photosynthesis, detoxification of reactive oxygen species (ROS) through antioxidant enzymes and molecules, and induction of phytohormones. However, these adaptive strategies become inefficient to cope with the rapidly increasing salinity in soil and water (Kapoor et al. 2019).

17.2.6 Oxidative Stress

The oxidative stress is caused by the undesired superoxide radicals that become more abundant in the living system than the antioxidant molecules. The superoxide radicals contain a singlet oxygen ion with two impaired electrons with the same spin electron number. This spin rotation once allows oxygen to accept one electron to make it reactive species (ROS) that damage the cell. This ROS is produced as a by-product of the metabolic pathways operated by chloroplast, mitochondria, and peroxisome. These ROS molecules (O_2^{*-}, H_2O_2, $^1O_2$, HO_2^{*-}, OH^*, ROOH, ROO^*, and RO^*) are scavenged by different antioxidant molecules as the defense mechanism of the cell. The ROS molecules are highly toxic and reactive and cause damage to micro- and macronutrients such as proteins, lipids, carbohydrates, and nucleic acids that eventually causes cell death (Gill and Tuteja 2010).

17.2.7 Air Pollution

Air pollution is a universal issue correlated with the development of urbanization. In most industrial cities in the world, a large number of inhabitants suffer from air pollution due to the massive consumption of fossil fuels. It deteriorates the air quality which declines the equilibrium of gaseous and particle components of air resulting in dismantling of environmental structure. Air pollution directly affects the growth and metabolism of plants. The rise of CO_2, SO_2, and NO_X in the atmosphere due to the combustion of fossil fuels and the overapplication of chemical fertilizers and pesticides are rapidly increasing greenhouse effects, and after reacting with water vapor, these oxides are reduced to form various acids like nitric acid, sulfuric acid, etc. which after precipitation from acid rain damages the cell wall and membrane of plants and animals. This acid rain drastically reduces soil characteristics by altering pH, conductivity, and other physicochemical properties (Savci 2012).

17.2.8 Agricultural Practices

Agronomic management, including cropping systems, application of fertilizers, plant species, soil management, and tillage practices, can have strong effects on the physical, chemical, and biological properties of soil, including microbial community composition. These differentially affect the growth and yield of crops. Planning and application of proper cropping system, i.e., application of minimum fertilizers, depending on solar energy, water resources, and soil health increase yield percentage. The soil microbiome and soil properties are highly variable in different plant species at inter- and intrageneric levels. Soil microbes also have a high impact on productivity and yield enhancement and retention of soil fertility (Lehman et al. 2019).

17.2.9 Nanoparticle Pollution

Nanoparticles are commonly used as target-specific synthetic metalloid compounds applied over a wide arena, coming from agriculture to pharmaceuticals. These nanoparticles became toxic to cells after coagulation and dispose of in the environment. The commonly used nanoparticles are synthesized from zinc, and silver ions, which have high toxicity on the microbes and mites. In agriculture, nanoparticles are directly applied as fertilizer, and fungicides for their antifungal properties. Zinc and silver nanoparticles when applied in soil become stable in Zn^{+2} and Ag^{+} form. These ions when inducing toxicity in plants alter soil community composition, thus changing the entire function of the ecosystem resulting in a reduction of soil fertility and quality and plant growth and production. When exposed to ZnNP and AGNP, some crops can accumulate excess metals in their edible parts, which also enter into the food chain and lead to the bioaccumulation of toxic molecules (Wang et al. 2018).

17.2.10 Pollution by Radioactive Material

Uranium, one of the commonly used radioactive metals, is chemically toxic and highly carcinogenic. Soils are contaminated by radioactive molecule uranium through nuclear reactors, industries, refineries, and mining sites. It causes great harm to human and plant health. Engineering-based uranium remediation methods like excavation are not very efficient due to the high economic costs and requirement of scarce landfill space and additional site restoration. Among the various strategies for the remediation of uranium-polluted soil, bioremediation represents an eco-friendly and cost-efficient method. Some plants can accumulate a higher amount of toxic trace elements from contaminated soil, sediments, or water. This phytoremediation sometimes fails due to high concentrations of heavy metal toxicity and nutrient deficiency of these sites which resulted in reduced plant growth and impairment of metabolism due to toxic effects (Ren et al. 2019; Chang et al. 2018).

17.3 Current Perspective of Mycorrhizal Research

Mycorrhiza is a symbiosis between the majority of terrestrial plants and AM (arbuscular mycorrhizal) fungi belonging to the monophyletic *Glomeromycotina* (Spatafora et al. 2016). AM is an essential component of ecosystems as it contributes to biodiversity and ecosystem productivity. It also participates in the global carbon cycle by sequestering atmospheric carbon dioxide fixed by the host plant in the soil, thereby increasing soil organic carbon pools. Improving crop yield along with nutrient efficiency is an essential consideration for environment-friendly sustainable agriculture. Mycorrhizal research has gained some momentum in the past few decades, but is still low compared to other agriculture-related research. Most of the studies have focused on understanding the mechanisms of mycorrhizae infestation in suitable host plant and their plant health attributes; however, their diversity assessment on the cultivated field on a temporal and spatial scale has been ignored. In India, the studies on mycorrhizae have not spread well. Most of the studies are restricted to some regions and focused on a particular type of mycorrhizal fungi and even not related to crop plants directly. Deepika and Kothamasi (2015) studied the regulatory effect of soil moisture on arbuscular mycorrhizal fungal community assembly and symbiotic phosphorus uptake. Kumar et al. (2016) studied the species diversity of arbuscular mycorrhizal fungi in Litchi trees in India. Suchitra et al. (2012) studied morpho-typing and molecular diversity of arbuscular mycorrhizal fungi in Tamil Nadu. There are many kinds of literature available that cover the mycorrhizal association in the crop plants such as rice, groundnut, and mulberry (Bhattacharjee and Sharma 2011). Azaizeh et al. (1995) studied the effects of a vesicular-arbuscular mycorrhizal fungus and other soil microorganisms on the growth, mineral nutrient acquisition, and root exudation of soil-grown maize plants. Douds and Millner (1999) studied the biodiversity of arbuscular mycorrhizal fungi in agroecosystems. Augé (2001) studied the effect of vesicular-arbuscular mycorrhizal symbiosis in water relations and drought tolerance. Sundar et al. (2011) studied the AM fungal diversity in selected medicinal plants of Kanyakumari District, Tamil Nadu, India, while Sunil Kumar and Garampalli (2013) studied the status of AM fungi associated with some medicinal plants of Mysore. But none of the works focused on the large-scale characterization of mycorrhizae in selected crops (i.e., rice, jute, wheat, mustard, maize, pulses, potato, and sugarcane), and also no assessments have been done on the effect of pesticides/fungicides on the mycorrhizal population. Therefore, it is very much of the utmost need of the present agroeconomic system to validate the mycorrhizal population diversity and their fate in high fertilizer and pesticide residues. However, the research on mycorrhizae in different science laboratories in India has been reporting on the different perspectives of the mycorrhizal association on selected plant hosts. So, from the above literature study, it is evident that the mycorrhizal species diversity in crop fields is new in its concept and the study site, districts of North Bengal region of West Bengal is an unexplored region.

International status regarding the studies on mycorrhizae is mostly centered on the chemical component or molecular analysis. For eco-friendly and sustainable agricultural practices, the plant-mycorrhizae association has also attracted different laboratories abroad to exploit the detail of nutrient availability by the plant system (Ramaekers et al. 2010; Mack and Rudgers 2018). Enormous studies have been done to characterize the different effects of biomolecules in the plant-mycorrhizae association. Sawers et al. (2018) studied the variation in mycorrhiza among diverse plant varieties. A study by Wahbi et al. (2016) reported the plant-microbial association in crop fields. Duponnois Plenchette (2003) studied the mineral solubilization process of the rock phosphate to soluble and easily available forms to plants. Sun et al. (2018) studied the identification and functional characterization of a maize phosphate transporter induced by mycorrhiza formation. Corrêa et al. (2012) studied the nitrogen and carbon/nitrogen dynamics in arbuscular mycorrhiza. Lefebvre (2017) focused on the role of N-acetylglucosamine in the arbuscular mycorrhizal association. Garcia et al. (2016) studied the transportome of mycorrhiza during plant and fungal interactions. The study of Parniske (2008) hypothesized arbuscular mycorrhizae as the mother of plant root endosymbiosis. Fritz et al. (2006) reported that arbuscular mycorrhiza reduces the susceptibility of tomatoes to *Alternaria solani*. But none of these research works has focused on the mycorrhizal community dynamics in crop fields and their consequence due to the use of chemical fertilizers and pesticides. These literature surveys indicate that at both national and international levels, the studies on mycorrhizae are mostly associated with the assessment of impact analysis of mycorrhizae on selective crop productivity or molecular interactions of plant-fungal relations. Though there are some studies on changes in the VAM community concerning agricultural practices, none of these studies focused on changes in crop production due to the AMF community changes with agricultural practices and also the available data cannot be generalized because the studies are so discrete.

17.4 Mitigation of Challenged Agroecosystems with AM Fungi

Arbuscular mycorrhizal fungi belonging to the division *Glomeromycota* are a monophyletic, widespread group of filamentous fungi that form a mutualistic relationship with most land plants, including the majority of crops. They are predominantly renowned for their ability to increase plant nutrient uptake and mobilization and increase crop productivity and biomass enhancement which results in the establishment of a stable ecosystem.

17.4.1 Nutrient Uptake

Nutrient and mineral uptake is an essential phenomenon of plant growth and development. The availability of nutrients and their efficient mobilization throughout the plants increase crop productivity and yield. Nutrient uptake and mobilization are effectively done by the plant through diffusion and capillary interactions. This process is accelerated by the symbiotic association of several microbes and fungi, especially mycorrhizae. In this respect, the most efficient symbiont is arbuscular mycorrhizal (AM) fungi which help to increase plant performance through their ability to increase plant nutrient uptake, particularly phosphorus and nitrogen indirectly. AM fungi, soil bacteria, and fungal communities could solubilize a small portion of soil phosphorus and make it available to the plants. Apart from phosphate mobilization, AM fungi are also very efficient organisms for phosphorus uptake through their hyphae, which can easily reach the micropores with a high surface area to volume ratio (Spohn et al. 2018). Therefore, the application of chemical fertilizers can be reduced by the utilization of AM fungi for their efficiency to harness the nutrient supply from natural sources through asymmetric root size and high absorptive surface area.

Plants colonized by AM fungi have two soil nutrient uptake pathways simultaneously, viz., directly through the root epidermis and indirectly through fungal structures that form the mycorrhizal pathway of uptake. The symbiotic association between plant and arbuscular mycorrhizae stabilizes the community's nutritional skeleton with efficient nutrient cycling. In agriculture, this nutritional stabilization of crop plants increases yield and productivity. Apart from that, this increasing nutritional pool increases resistance to diseases and enhanced plant growth and development.

Mycorrhizal symbiosis depends upon nutritional understanding between the host and symbiont (AM fungi). This exchange of nutrition not only causes the nourishment of both symbionts but also acts as a signal for mycorrhizal development (Lanfranco et al. 2018). In this symbiotic relationship, mycorrhizae receive lipids and carbohydrates, whereas AM fungi transfer phosphate to host plants. From plants' perspective, phosphate, a low mobilizing macronutrient, is commonly present in the soil as dihydrogen phosphate forms. To overcome this phosphate starvation, the plant adopts several strategies to increase accessibility to phosphate. One of the strategies is to make symbiosis with AM fungi for optimum phosphate acquisition through the 187 kDa enzyme acid phosphatase. The arbuscular mycorrhizae also induce the expression and secretion of the plants to increase the ability of plants to solubilize organic phosphate (Lanfranco et al. 2018).

Mycorrhizae use P_i: H^+ symporter for inorganic phosphate uptake from soil using their extraradical mycelium. The extraradical mycelium, after absorption, quickly converts inorganic phosphate into the polyphosphate chain. This polyphosphate chain is broken down to form simple phosphate by acid and alkaline phosphatases. Simple phosphate (P_i) then enters into the cortical cells of host plants through periarbuscular spaces (Lanfranco et al. 2018). AM fungi take up nitrogen in the form of ammonium (NH_4^+) rather than nitrates which reduces nitrogen mobility. In subsoil conditions, AM fungi inhibit nitrogen to migrate down or immobilize it and deliver it to the host plant, thus nitrogen loss can be avoided by nitrogen mobilization to plants that reduce the need for artificial nitrogen fertilization.

Iron is also an essential micronutrient of the plant system. The plant cannot directly absorb iron from the environment. The chelation of iron insoluble form can be done by different microorganisms and fungi. AM fungi also chelate iron by secreting phytosiderophores and different organic acids such as citric acid, oxalic acid, etc. and phenolic compounds to chelate insoluble iron to the soluble form.

17.4.2 Water Uptake

Arbuscular mycorrhizal fungi efficiently help the host plants to uptake and mobilize water through the whole plant system. Plants adopt several mechanisms to tolerate water deficiency in various situations. Water scarcity changes the physiology of plants. AM fungi moderately change the physiology of the host plant by optimized stomata opening, optimum efficiency to use water, and efficient maintenance of enzymatic and nonenzymatic reactions of antioxidative reactions to prevent oxidative damage. The water retention ability of the soil was high in AM fungi-infected roots rather than in non-AM fungi roots. AM fungi improve water availability and water transport in the colonized roots which physiologically experience lesser stress conditions when soil water content declines (Bitterlich et al. 2018). Apart from that, AM fungi regulate transpiration rate with a slow translocation rate of polyphosphate molecules that moderate water flows through the expression of fungal aquaporin gene *RcAQP3* and mycorrhizae-inducible plant aquaporins (Lanfranco et al. 2018).

17.4.3 Abiotic Stress Tolerance

Abiotic stresses such as drought, salinity, and temperature stresses directly affect plant growth and development. Soil microbiome is an effective indicator and controller of these stresses to the plants. These stresses can be tolerated by altering photosynthetic carbon metabolism, change in the signaling cascade, and hormonal and antioxidant machinery. Mycorrhizae help to accumulate essential isoprenoids such as abscisic acid, chlorophyll, and carotenoids rather than nonessential monoterpenes and sesquiterpenes. Jasmonic acid is also nonessential for drought tolerance. Apart from that, AM fungi help the plant to accumulate nutrients and carbon sources, and plants eventually use this carbon and nutrient reservoir for their growth and development. AM fungi help to tolerate salinity through the upregulation of antioxidant enzymes and moderate the quantities of osmoprotectants such as glycine, betaine, proline, phenols, etc. *Glomus intraradices* have *GintAQPF1* and *GintAQPF2* aquaporin genes to combat drought stresses. They modulate water transport and nitrogen metabolism and produce signaling molecules. Glomalin, a glycoprotein, can also combat drought stress by increasing water use efficiency and conductance of stomata. Apart from that, AM fungi combat drought by the accumulation of water through the enhanced hyphal net and reduce the ROS accumulation in the hyphal and host system (Basu et al. 2018). AM fungi induce higher superoxide dismutase (SOD), ascorbate oxidase, and peroxidase and catalase activities in leaves, tubers, and roots of several dicots and monocot plants to ameliorate oxidative stresses (Basu et al. 2018).

AM fungi also ameliorate heavy metal stress through the prevention of entrance into the cytoplasm. They reduce the bioavailability of heavy metals by excreting glycoproteins like glomalin, which can help in the sequestration of large amounts of heavy metals in soil. AM fungi are also able to bind heavy metals to their cell walls to limit heavy metal transport to plant tissues. Sometimes, heavy metals are passed through external barriers that can be detoxified inside fungal cells by chelating agents, like glutathione or metallothionein. These chelated heavy metals are disposed of through the vacuole by pumping them out of the cells (Gupta et al. 2013).

17.4.4 Modulation of Plant Physiology

Mycorrhizae increase plant photosynthetic rate and reduce plant growth at elevated atmospheric CO_2 content to compensate for the carbon source through the plant source-sink relationships. This symbiotic association influences the photosynthesis and carbon cycle both above and below the ground and through this balanced pattern optimizes the atmospheric CO_2. During photosynthesis, the AM fungi upregulate and protect the *RubisCO* gene and D1/D2 proteins of the PSII system. AM fungi also regulate several plant hormones as well as signaling molecules. Strigolactone, a sesquiterpene, is one of the important hormones, synthesized through the carotenoid pathway. This molecule is attached to several physiological and cellular processes, which include respiration and hyphal branching of fungi. This regulation is caused by the mitochondria and lipid catabolism by inducing the expression of *the Myc* gene. Apart from strigolactone, AM fungi regulate the biosynthesis of abscisic acid, ethylene, cytokinin, brassinosteroids, jasmonic acid, and salicylic acid. These hormones help the plant to combat diseases and senescence as well as help to keep healthy by continuous cell division and dedifferentiation. However, the mycorrhizal infection does not directly influence cytokinin biosynthesis (Basu et al. 2018).

17.4.5 Nutrient Recycling and Leaching

In agriculture, the majority of the time without proper AM fungal infection, most of the nutrient is drained out without proper nutrient management. Nutrient drainage can be reduced by arbuscular mycorrhizal fungi not only expanding the nutritionally rich area by creating myco-rhizosphere but also improving nutrient uptake, organizing soil structure according to the plant's requirements, and encouraging the microbial community to trap the nutrients and immobilize it out of the plant system.

AM fungi create an anaerobic condition by developing macroaggregates, where they can manage nitrogen by denitrification. These macroaggregates change the carbon availability that shifts the denitrifying bacteria with the AM fungi, which accelerates the denitrification process. Therefore, in this mechanism, nitrogen loss can be restricted through several factors, and in this way, AM fungi efficiently manage the nitrogen cycle in nature (Parihar et al. 2019).

17.4.6 Soil Health and Plant-Soil Feedback (PSF)

Plant-soil feedback (PSF) has a large impact on plant growth and development in agriculture and natural system. This new concept has been developed for sustainable development and management of agriculture and natural system for its population load and ecosystem services that subsequently decline ecosystem health. PSF includes both the wild and cultivated plants that directly or indirectly influence the root and soil microbes which include soilborne pathogens, symbionts, and saprophytes. These microbes affect the plant's growth and development either positively or negatively. These effects help to determine the strength of PSF. In the plant system, plants generally have an association with fungal and bacterial endophytes, AM fungi, and plant growth-promoting bacteria and fungi. AM fungi, among them, have a direct relationship with soils and root systems of plants. In the natural system, the AM fungi increase plant diversity in the initial phase of ecological development, but after the climax condition of succession, a single plant species dominates the whole ecology. In this condition, diversity of both plant and mycorrhizae has reduced. In the agricultural system, unscientific tillage and application of fertilizers and pesticides reduce the fungal biomass and disrupt AM fungal network, which drastically affects nutrient leaching from soil and reduces plant productivity. Therefore, a proper soil management practice can effectively manage PSF through the utilization of soil organisms in both natural and agricultural systems. In agricultural systems, the PSF includes an optimized cropping system, well-managed disease resistance, and pest control system, and efficient resource use efficiency to improve the plant-soil feedback (Wang et al. 2019; Mariotte et al. 2018).

17.4.7 Agricultural Costs and Pollution

In agriculture, the cost of crop production is increasing rapidly per year. The cost of fertilizers has increased by 5–26% in last 5 years. In this scenario, the cost of agriculture is going to be more expensive. Additionally, at the same time, the overapplication of fertilizers decreases soil fertility which reduces crop yield. In this condition, food scarcity is an emerging problem throughout the world. In search for an alternative way to combat this issue, the application of biofertilizers and biopesticides became a suitable alternative. Biofertilizers can induce crop yield as well as maintain soil fertility for a long time. One of the effective biofertilizers is arbuscular mycorrhizal fungi, which effectively increase soil fertility by nutrient solubilization and water uptake as well as effectively mobilizing soluble nutrients to the plants. This soluble nutrients increase plant growth and development through the induction of metabolisms and photosynthetic rates. This plant's growth-promoting properties increase crop yield and productivity in a cost-effective way which can mitigate food scarcity at minimum expenses. Apart from these, AMF chelate and degrade different metals like iron, aluminum, heavy metals, and organic compounds to make a pollution-free soil.

17.4.8 Mycoremediation

Mycoremediation is a process of bioremediation where fungi take the leading role of decontaminating the environment. AM fungi colonize in the root of the plant using a symbiosis understanding. AM fungi generally bioremediate metal ions by improving the uptake of zinc, copper, iron, manganese, etc. and reducing the cadmium uptake of plants. AM fungi alleviate the barrier between root and shoot and also induce metal resistance to the host plants that reduce the metal toxicity level in the food chain. AM fungi also effectively remediate radioactive metals. Uranium- and lanthanum-mediated toxicity can be effectively healed by AM fungi associated with host roots. Apart from the metal toxicity, organic pollutants such as polycyclic aromatic hydrocarbon (PAH) can be degraded or converted into a simpler form by the arbuscular mycorrhizae in the rhizospheric region (Arya et al. 2018). Summarily, AM fungi systematically act as a plant growth regulator, environment monitor, and bioremediation in a cost-effective, sustainable manner (Figs. 17.2 and 17.3).

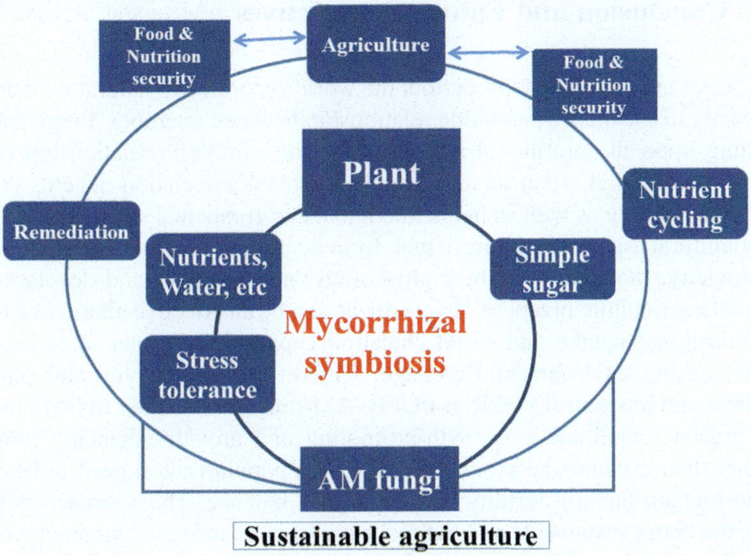

Fig. 17.2 The impact of AM fungi in maintaining sustainable agriculture

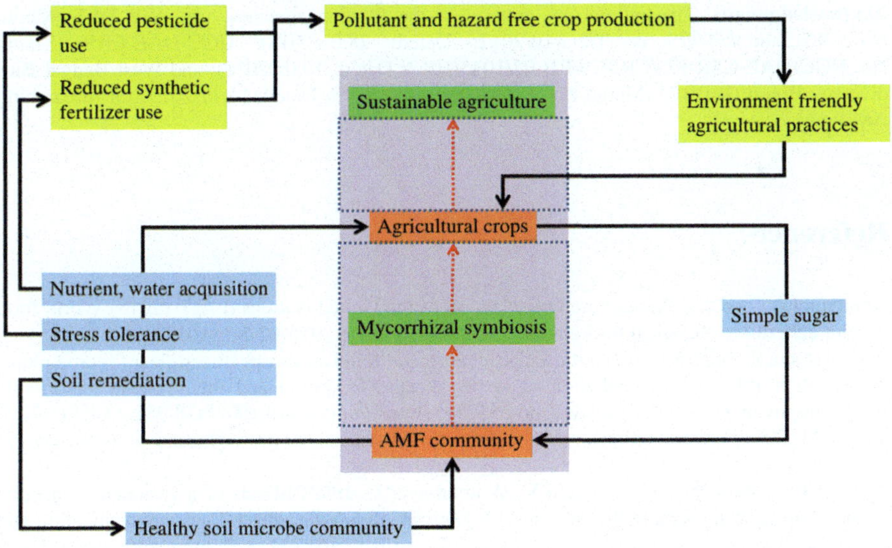

Fig. 17.3 The strategies of sustainable crop production through AM fungi and the factors to reconsider their viability and diversity

17.5 Conclusion and Future Perspectives

Plant species are dominated throughout the world according to natural selection with the capacity to establish a symbiotic relationship between microbes, fungi, and mites depending upon the nutritional requirement. The symbiotic relationship between plants and AM fungi is conserved. The mycorrhizal association benefits the plant system nutritionally as well as helps to combat environmental stresses, both natural and agricultural aspects. Apart from that, for reduction of food scarcity, mitigation of metal toxicity, modulation of host physiology, and behavior and development of sustainable agriculture practices, mycorrhizae can be an effective alternative because of its nutritional uptake and metal chelation capacity and water retention ability without causing environmental damage. Unlike the diazotrophic and phosphate solubilizer and biocontrol PGPR or PGPB, AM fungi have higher toxicity tolerance and a greater magnitude of growth-promoting and growth-enhancing properties; therefore, their commercial production, use, and popularization need to be emphasized to reclaim the soil fertility and ecological balance. The approach will bring hope in the better sustainable and judicious use of croplands with targeted crop yield and nutritional demand and concur with an everlasting sustainable agroecosystem. Further deeper studies need to develop a consortium of biofertilizers using AM fungi as a major component to revisit the Green Revolution sustainably.

Acknowledgments The authors are grateful to the funding agencies, viz., WB DST (Memo No. 285(Sanc.)/ST/P/S & T/2G-10/2017; Dated: 28.03.2018), UGC-DAE-CRS (Memo No. UGC-DAE-CSR-KC/CRS/19/RB-02/1045/1063; Dated: 10.05.2019), and West Bengal Biodiversity Board (WBBB) (Memo No. 1218/3 K(Bio)-6/2019; Dated: 11.11.2019) for researching AM fungi.

References

Anonymous, Food and Agriculture Organization of the United Nations (FAO) (2018) World food and agriculture statistical pocketbook 2018. Rome. ISBN: 978-92-5-131012-0

Arya A, Ojha S, Singh S (2018) Arbuscular mycorrhizal fungi as phosphate fertilizer for crop plants and their role in bioremediation of heavy metals. In: Fungi and their role in sustainable development: current perspectives, pp 255–265. https://doi.org/10.1007/978-981-13-0393-7_14

Augé RM (2001) Water relations, drought and vesicular-arbuscular mycorrhizal symbiosis. Mycorrhiza 11:3–42

Azaizeh HA, Marschner H, Römheld V, Wittenmayer L (1995) Effects of a vesicular-arbuscular mycorrhizal fungus and other soil microorganisms on growth, mineral nutrient acquisition and root exudation of soil-grown maize plants. Mycorrhiza 5(5):321–327. https://doi.org/10.1007/BF00207404

Basu S, Rabara RC, Negi S (2018) AMF: the future prospect for sustainable agriculture. Physiol Mol Plant Pathol 102:36–45. https://doi.org/10.1016/j.pmpp.2017.11.007

Bhattacharjee S, Sharma GD (2011) The vesicular-arbuscular mycorrhiza associated with three cultivars of rice (*Oryza sativa* L.). Indian J Microbiol 51:377–383. https://doi.org/10.1007/s12088-011-0090-9

Bitterlich M, Sandmann M, Graefe J (2018) Arbuscular mycorrhiza alleviates restrictions to substrate water flow and delays transpiration limitation to stronger drought in tomato. Front Plant Sci 9:154. https://doi.org/10.3389/fpls.2018.00154

Chang Q, Diao FW, Wang QF, Pan L, Dang ZH, Guo W (2018) Effects of arbuscular mycorrhizal symbiosis on growth, nutrient and metal uptake by maize seedlings (*Zea mays* L.) grown in soils spiked with Lanthanum and Cadmium. Environ Pollut 241:607–615. https://doi.org/10.1016/j.envpol.2018.06.003

Coccina A, Cavagnaro TR, Pellegrino E, Ercoli L, McLaughlin MJ, Watts-Williams SJ (2019) The mycorrhizal pathway of zinc uptake contributes to zinc accumulation in barley and wheat grain. BMC Plant Biol 19:133. https://doi.org/10.1186/s12870-019-1741-y

Corrêa AJ, Gurevitch MA, Martins-Loucao CC (2012) C allocation to the fungus is not a cost to the plant in ectomycorrhizae. Oikos 121:449–463. https://doi.org/10.1111/j.1600-0706.2011.19406.x

Debeljak M, van Elteren JT, Špruk A, Izmer A, Vanhaecke F, Vogel-Mikuš K (2018) The role of arbuscular mycorrhiza in mercury and mineral nutrient uptake in maize. Chemosphere 212:1076–1084. https://doi.org/10.1016/j.chemosphere.2018.08.147

Deepika S, Kothamasi D (2015) Soil moisture—a regulator of arbuscular mycorrhizal fungal community assembly and symbiotic phosphorus uptake. Mycorrhiza 25:67. https://doi.org/10.1007/s00572-014-0596-1

Douds DD, Millner P (1999) Biodiversity of arbuscular mycorrhizal fungi in agroecosystems. Agric Ecosyst Environ 74:77–93

Duponnois R, Plenchette C (2003) A mycorrhiza helper bacterium enhances ectomycorrhizal and endomycorrhizal symbiosis of Australian Acacia species. Mycorrhiza 13:85–91

Fan X, Chang W, Fenga F, Song F (2018) Responses of photosynthesis-related parameters and chloroplast ultrastructure to atrazine in alfalfa (*Medicago sativa* L.) inoculated with arbuscular mycorrhizal fungi. Ecotoxicol Environ Saf 166:102–108. https://doi.org/10.1016/j.ecoenv.2018.09.030

Fritz M, Jakobsen I, Lyngkjær MF, Thordal-Christensen H, Pons-Kühnemann J (2006) Arbuscular mycorrhiza reduces the susceptibility of tomato to *Alternaria solani*. Mycorrhiza 16:413. https://doi.org/10.1007/s00572-006-0051-z

Garcia K, Doidy J, Zimmermann SD, Wipf D, Courty PE (2016) Take a trip through the plant and fungal transportome of mycorrhiza. Trends Plant Sci 21:937–950

Gill SS, Tuteja N (2010) Reactive oxygen species and antioxidant machinery in abiotic stress tolerance in crop plants. Plant Physiol Biochem 48(12):909–930. https://doi.org/10.1016/j.plaphy.2010.08.016

Gupta DK, Vandenhove H, Inouhe M (2013) Role of phytochelatins in heavy metal stress and detoxification mechanisms in plants. Heavy Metal Stress Plants 73–94. https://doi.org/10.1007/978-3-642-38469-1_4

Hasanuzzaman M, Bhuyan MHMB, Nahar K, Hossain MS, Mahmud JA, Hossen MS, Masud AAC, Moumita Fujita M (2018) Potassium: a vital regulator of plant responses and tolerance to abiotic stresses. Agronomy 8:31. https://doi.org/10.3390/agronomy8030031

Kapoor R, Evelin H, Devi TS, Gupta S (2019) Mitigation of salinity stress in plants by arbuscular mycorrhizal symbiosis: current understanding and new challenges. Front Plant Sci 10:470. https://doi.org/10.3389/fpls.2019.00470

Kumar A, Choudhary AK, Suri VK (2016) Influence of AM fungi, inorganic phosphorus and irrigation regimes on plant water relations and soil physical properties in okra (*Abelmoschus esculentus* L.)–pea (*Pisum sativum* L.) cropping system in Himalayan acid Alfisol. J Plant Nutr 39:666–682

Lanfranco L, Fiorilli V, Gutjahr C (2018) Partner communication and role of nutrients in the arbuscular mycorrhizal symbiosis. New Phytol 220:1031. https://doi.org/10.1111/nph.15230

Lefebvre B (2017) Arbuscular mycorrhiza: a new role for N-acetylglucosamine. Nat Plants 3:17085. https://doi.org/10.1038/nplants.2017.85

Lehman RM, Osborne SL, Taheri WI, Buyer JS, Chim BK (2019) Comparative measurements of arbuscular mycorrhizal fungal responses to agricultural management practices. Mycorrhiza 29: 227. https://doi.org/10.1007/s00572-019-00884-4

Mack KML, Rudgers JA (2018) Balancing multiple mutualists: asymmetric interactions among plants, arbuscular mycorrhizal fungi, and fungal endophytes. Oikos 117:310–320. https://doi.org/10.1111/j.2007.0030-1299.15973.x

Mariotte P, Mehrabi Z, Bezemer TM, De Deyn GB, Kulmatiski A, Drigo B, Veen GFC, van der Heijden MGA, Kardol P (2018) Plant–soil feedback: bridging natural and agricultural sciences. Trends Ecol Evol 33(2):129–142. https://doi.org/10.1016/j.tree.2017.11.005

Mathur S, Tomar RS, Jajoo A (2018) Arbuscular mycorrhizal fungi (AMF) protects photosynthetic apparatus of wheat under drought stress. Photosynth Res 139:227. https://doi.org/10.1007/s11120-018-0538-4

Parihar M, Meena VS, Mishra PK, Rakshit A, Choudhary M, Yadav RP, Rana K, Bisht JK (2019) Arbuscular mycorrhiza: a viable strategy for soil nutrient loss reduction. Arch Microbiol 201: 723. https://doi.org/10.1007/s00203-019-01653-9

Parniske M (2008) Arbuscular mycorrhiza: the mother of plant root endosymbioses. Nat Rev Microbiol 6(10):763–775. https://doi.org/10.1038/nrmicro1987

Rafiq M, Shahid M, Shamshad S, Khalid S, Niazi NK, Abbas G, Saeed MF, Ali M, Murtaza B (2017) A comparative study to evaluate efficiency of EDTA and calcium in alleviating arsenic toxicity to germinating and young *Vicia faba* L. seedlings. J Soils Sediments 18(6):2271–2281. https://doi.org/10.1007/s11368-017-1693-5

Ramaekers L, Remans R, Rao IM, Blair MW, Vanderleydena J (2010) Strategies for improving phosphorus acquisition efficiency of crop plants. Field Crop Res 117:169–176

Ren CG, Kong CC, Wang SX, Xie ZH (2019) Enhanced phytoremediation of uranium-contaminated soils by arbuscular mycorrhiza and rhizobium. Chemosphere 217:773e779. https://doi.org/10.1016/j.chemosphere.2018.11.085

Savci S (2012) An agricultural pollutant: chemical fertilizer. Int J Environ Sci Dev 3(1):77–80

Sawers RJH, Ramírez-Flores MR, Olalde-Portugal V, Paszkowski U (2018) The impact of domestication and crop improvement on arbuscular mycorrhizal symbiosis in cereals: insights from genetics and genomics. New Phytol 220(4):1135–1140. https://doi.org/10.1111/nph.15152

Sharma S, Rana VS, Kumari M, Mishra P (2018) Biofertilizers: boon for fruit production. J Pharmacognosy Phytochem 7(5):3244–3247

Spatafora JW, Chang Y, Benny GL, Lazarus K, Smith ME, Berbee ML, Bonito G, Corradi N, Grigoriev I, Gryganskyi A, James TY, O'Donnell K, Roberson RW, Taylor TN, Uehling J, Vilgalys R, White MM, Stajich JE (2016) A phylum-level phylogenetic classification of zygomycete fungi based on genome-scale data. Mycologia 108(5):1028–1046. https://doi.org/10.3852/16-042

Spohn M, Zavisic A, Nassal P, Bergkemper F, Schulz S, Marhan S, Schloter M, Kandeler E, Polle A (2018) Temporal variations of phosphorus uptake by soil microbial biomass and young beech trees in two forest soils with contrasting phosphorus stocks. Soil Biol Biochem 117:191–202. https://doi.org/10.1016/j.soilbio.2017.10.019

Srivastava P, Singh R, Tripathi S, Raghubanshi AS (2016) An urgent need for sustainable thinking in agriculture – an Indian scenario. Ecol Indic 67:611–622. https://doi.org/10.1016/j.ecolind.2016.03.015

Suchitra R, Kumutha K, Balachandar D (2012) Morpho-typing and molecular diversity of arbuscular mycorrhizal fungi in sub-tropical soils of Coimbatore region, Tamil Nadu, India. Indian J Microbiol 52:145–152. https://doi.org/10.1007/s12088-011-0206-2

Sun Z, Song J, Xin X, Xie X, Zhao B (2018) Arbuscular mycorrhizal fungal 14-3-3 proteins are involved in arbuscule formation and responses to abiotic stresses during AM symbiosis. Front Microbiol 9:91. https://doi.org/10.3389/fmicb.2018.00091

Sundar SK, Palavesam A, Parthipan B (2011) AM fungal diversity in selected medicinal plants of Kanyakumari district, Tamil Nadu. India Indian J Microbiol 51(3):259–265. https://doi.org/10.1007/s12088-011-0112-7

Sunil Kumar PC, Garampalli R (2013) Diversity of arbuscular mycorrhizal fungi in irrigated and non-irrigated fields of southern Karnataka. India J Environ Biol 34(1):159–164

Wahbi S, Sanguin H, Baudoin E, Tournier E, Maghraoui T, Prin Y, Hafidi M, Duponnois R (2016) Managing the soil mycorrhizal infectivity to improve the agronomic efficiency of key processes from natural ecosystems integrated into agricultural management systems. In: Plant, soil and microbes, vol 1: implications in crop science. Springer, Cham, pp 17–27. ISBN: 978-3-319-27453-9

Wang F, Jing X, Adams CA, Shi Z, Sun Y (2018) Decreased ZnO nanoparticle phytotoxicity to maize by arbuscular mycorrhizal fungus and organic phosphorus. Environ Sci Pollut Res 25(24):23736–23747. https://doi.org/10.1007/s11356-018-2452-x

Wang XX, Hoffland E, Mommer L, Feng G, Kuyper TW (2019) Maize varieties can strengthen positive plant-soil feedback through beneficial arbuscular mycorrhizal fungal mutualists. Mycorrhiza 29(3):251–261. https://doi.org/10.1007/s00572-019-00885-3

Chapter 18
New Developments in Techniques Like Metagenomics and Metaproteomics for Isolation, Identification, and Characterization of Microbes from Varied Environment

Kruti Shah, Vijay Jagdish Upadhye, and Anupama Shrivastav

Abstract Most of the world's microbial diversity is comprised of culture-independent microorganisms. In the natural environment, 99% of microbes are unculturable, which means that only 1% of microbes are capable of being cultured in a laboratory setting. Consequently, it is necessary to develop culture-independent approaches for the identification and characterization of such microbes and the evaluation of the role they play in the environment. Metagenomics, metaproteomics, etc. technologies are being developed to tackle new challenges and provide alternatives to traditional microbiology. Metagenomics focuses on the genetic and evolutionary relationship, microbial diversity, functional activities, and interaction with varied environments. Similarly, metaproteomics focuses on the protein composition of a complex sample as well as helps in the identification and quantification of protein. Also, metaproteomics detects proteins at a posttranslational level that gives a broader idea of the protein's function. In this chapter, new developments in techniques like Oxford Nanopore sequencing with advanced bioinformatics tools, metagenome-assembled genomes (MAGs), and new techniques for the isolation, identification, and characterization of microbes from different environments will be discussed.

K. Shah
P.D. Patel Institute of Applied Sciences, CHARUSAT, Anand, Gujarat, India

V. J. Upadhye
Center of Research for Development (CR4D), Parul Institute of Applied Sciences (PIAS), Parul University (DSIR-SIRO Recognized), Vadodara, Gujarat, India

A. Shrivastav (✉)
Center of Research for Development (CR4D), Parul Institute of Applied Sciences (PIAS), Parul University (DSIR-SIRO Recognized), Vadodara, Gujarat, India

Department of Microbiology, Faculty of Life Health and Allied Science, ITM Vocational University, Vadodara, Gujarat, India

P. Mathur et al. (eds.), *Microbial Symbionts and Plant Health: Trends and Applications for Changing Climate*, Rhizosphere Biology,
https://doi.org/10.1007/978-981-99-0030-5_18

Keywords MAGs · Metagenomics · Metaproteomics · Oxford Nanopore

Abbreviations

BLAST	Basic local alignment search tool
DNA	Deoxyribonucleic acid
EggNOG	Evolutionary genealogy of genes non-supervised orthologous groups
ESI	Electrospray ionization
HPLC	High-performance liquid chromatography
KEGG	Kyoto Encyclopedia of Genes and Genomes
LAST	Local alignment search tool
LC-MS	Liquid chromatography-mass spectrometry
MAGs	Metagenome-assembled genomes
MALDI-TOF	Matrix-assisted laser desorption ionization-time-of-flight mass spectrometry
MAP	MinION Access Program
MS	Mass spectrometry
MudPIT	Multidimensional protein identification technology
NCBI	National Center for Biotechnology Information
NGS	Next-generation sequencing
PCR	Polymerase chain reaction
qRT-PCR	Quantitative reverse transcription-polymerase chain reaction
SAGs	Single amplified genomes
SCX	Strong cation exchange column
SDS-PAGE	Sodium dodecyl sulfate-polyacrylamide gel electrophoresis
SMRT	Single-molecule real time
ZMW	Zero-mode waveguide

18.1 Introduction

Microorganisms play a pivotal role in ecosystems such as in the biogeochemical cycle to perform various processes. However, numerous microbiological techniques are available that help in the identification of microbes from different environments. Still, only 1% of microorganisms are cultivable under standard laboratory conditions. Therefore, cultivating the vast majority of microbes is challenging nowadays. In order to examine diverse microbial communities, metagenomic analysis is often used without cultivating or isolating a single organism. In metagenomic analysis, the collection and analysis of DNA from a microbial community are explored in order to determine its complete genetic composition. In addition, it allows us to understand the biochemistry of microbes in atypical environments and their interactions with the environment (Zhang et al. 2021). Through sequencing and subsequent analysis, the metagenomic study examines the overall genetic makeup of the microbial populations. It contributes to our understanding of the biochemical makeup of microorganisms in unusual habitats as well as how they interact with other environmental elements (Liu et al. 2022). Metagenomics analyses typically begin with the isolation of metagenomics DNA from the samples, followed by the amplification of the 16S rRNA gene sequencing, a shotgun metagenome, analyzing amplicon reads, analyzing metagenomic reads, and finally determining the diversity of the microbial community as described in Fig. 18.1 (Liu et al. 2022).

Metagenomics application comprises an examination of microbes in soils, glaciers, oceans, deep oceans, and various other environments. It also provides greater insight into the field of medicine where the identification of microbes in the bloodstream, lungs, intestinal tract, etc. can be optimized. In the future, metagenomics may play a significant role in antibiotic-resistant genes or antibiotic-resistant bacteria (Ko et al. 2022).

The study of metaproteomics focuses on the entire proteome expressed by microbial communities. As represented in Fig. 18.2, metabolic pathways can be studied, and enzymes found in unculturable microbes can be identified. The basic steps of metaproteomics include sample collection, extraction of proteins,

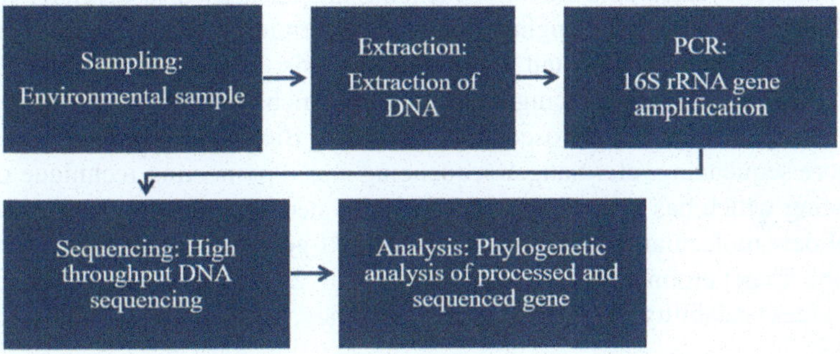

Fig. 18.1 Flowchart of metagenomics

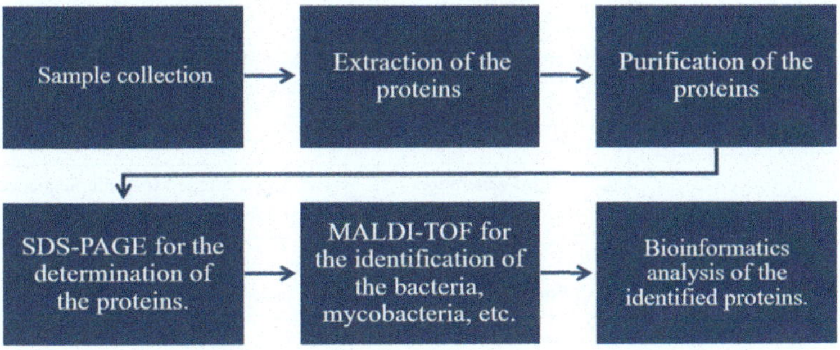

Fig. 18.2 Flowchart of metaproteomics

purification, etc. using techniques such as SDS-PAGE, MALDI-TOF, and LC-MS. Bioinformatics tools such as Mascot and SEQUEST are used for the analysis of proteins (Raza and Ameen 2017).

In last few decades, microorganisms can be isolated traditionally through first sample collection and then serial dilution, streaking; plating, enriched procedure, single-cell techniques, etc. were applied for isolation. Identification of microbes can be done through staining of cells, culturing the cells, and DNA sequencing, whereas characterization of microbes can be done through observing the morphology of the cell, analytical techniques, genotypic identification, as well as sometimes biotyping. The major disadvantages of these techniques are time-consuming, laborious, and requiring specialized expertise. Therefore, developing new techniques is challenging.

This chapter gives an overview of techniques of metagenomics and metaproteomics for the isolation, identification, characterization, and application of microbes from different environments such as water, soil, gut microbes, etc.

18.2 DNA Sequencing Methods

In the past few years, 16 s rRNA sequencing has been used to diversify microorganisms, although it is challenging to identify different bacterial species because of constraints in phylogenetic and taxonomy analysis. Figure 18.3 shows different generations of DNA sequencing technologies which include Sanger sequencing, Illumina sequencing, 454 pyrosequencing, Pacific Biosciences sequencing, Oxford Nanopore sequencing, etc. Sanger sequencing was a pioneering technique of DNA sequencing which has several drawbacks, which need to be resolved. The development of new molecular technologies, such as next-generation sequencing, has made it possible to get around these restrictions (Li et al. 2021). Next-generation sequencing provides scalability, high throughput, efficiency, and different platform to utilize

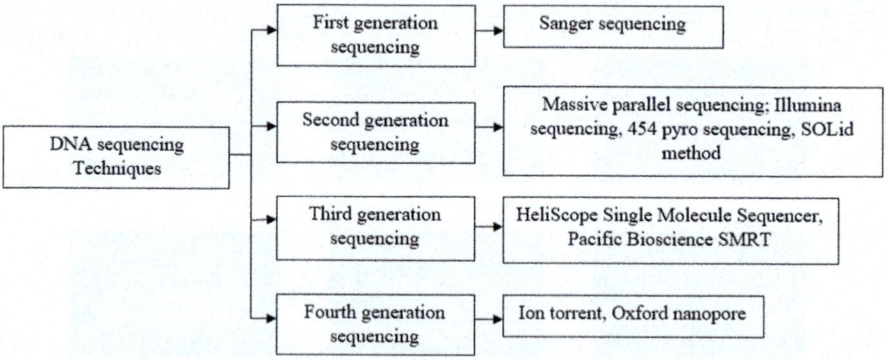

Fig. 18.3 Types of DNA sequencing techniques

for advancement. NGS technology is a second-generation sequencing technique. Researchers can work effectively and quickly to arrange microbial communities in various environmental samples by utilizing NGS technology (Moss et al. 2020). In 2014, Illumina sequencing technology was developed. It has greater efficiency to sequence the human genome in a single day. Nowadays, Illumina sequencing is one of the most effective high-throughput sequencing methods which is also known as the Illumina reversible terminator sequencing method (Li et al. 2021).

18.2.1 Illumina Sequencing

In the Illumina sequencing technique, a single molecule of DNA is attached to the flow cell while continuously amplified to form a clone construct. The reaction is continuous where a complementary strand of DNA is synthesized via the addition of nucleotides in every cycle of the technique, and because of this, fluorescently labeled nucleotide reads as well as the length of the DNA can be determined (Li et al. 2021). The major advantage is the long read length; the disadvantage is that it takes quite a long time like 2 to 3 days for the sequencing of DNA.

18.2.2 Pacific Biosciences SMRT

Technology for third-generation sequencing includes single-molecule sequencing. Eliminating the PCR amplification procedure improves the detection of nucleotides (ten nucleotides per second). Additionally, it can directly detect methylation and collect data about it. Third-generation sequencing by synthesis technology is a part of Pacific Biosciences SMRT. In the SMRT cell, a nanoscale zero-mode waveguide (ZMW) is present in the cell channel (Goodwin et al. 2016; Yu et al. 2019). Real-time detection of the sequencing process is made possible when the signal comes from the fluorescently labeled nucleotide attached to the growing chain. In order to continue DNA synthesis, tagging groups are detached as soon as a nucleotide connects to the next chain. SARS-CoV-2's genome was recently sequenced using the PacBio sequencing technique, which was crucial during the recent pandemic. To overcome the errors that occur during the sequencing, fourth-generation sequencing is necessary.

18.2.3 Nanopore Sequencing

Nanopore sequencing is a component of fourth-generation sequencing that outperforms second-generation sequencing methods. A long read can be covered at once by nanopore sequencing, a real-time sequencing analysis (Quick et al. 2015). The nanopore long-read sequencing approach enables the structuring of the genomes of isolated bacterial and viral cultures. Following a review of the literature, it is estimated that one flow cell can sequence 100,000 reads with an average length of 5 kilobytes in 24 h using metagenomic analysis. A Meta-PORE, a web-based application for real-time sequence analysis and visualization, was available for the identification of bacteria using nanopore data (Ashton et al. 2015).

18.2.3.1 Nanopore Sequencing Methodology

This review paper (Greninger et al. 2015) shows that the Oxford Nanopore MinION is accessible through the MinION Access Program (MAP). Following that, nucleic acid extraction was carried out utilizing a kit-based technique, such as the Direct-zol RNA MiniPrep Kit for viral samples or TRIzol LS. Similar to the viral sample, multiple Kit-based procedures were available for the nucleic acid extraction from the bacterial sample. Then, molecular confirmation was carried out by applying a quantitative reverse transcription-polymerase chain reaction (qRT-PCR). Using this technique, identification and quantity of the unknown sample can be estimated by calculating the standard curve. Then, metagenomics amplified library construction was made, where a specific amount of cDNA library was used for the nanopore sequencing as per the standard protocol.

Purification of amplified cDNA was accomplished using AMpure XP beads, as described in Greninger et al. (2015), and approximately 1 microgram of DNA was purified into the Oxford Nanopore genome Kit, along with other procedures for the construction of the libraries. Then, using the instructions provided on the kit, libraries were run on an Oxford Nanopore MinION flow cell for the sequencing procedure.

MegaBLAST or Bioinformatics Meta-PORE can be utilized for quicker sequencing that is aligned to NCBI (National Center For Biotechnology Information) and has greater sensitivity, which can also restrict the reference database for nucleotide BLAST (Frith et al. 2010). The LAST alignment approach may be more accurate at identifying nanopore reads, according to some research.

18.3 MAGs: Metagenome-Assembled Genomes and Reference Databases

Metagenome-assembled genomes, or MAGs, are constructed by binning assembled contigs and can be used to characterize and filter identical microorganism strains. Although it has several drawbacks, MAGs have been a key analysis tool for metagenome datasets (Venter et al. 2004). The quality of MAGs can be enhanced and improved by modern long-read sequencing methods, and new binning algorithms also offer entire circularized MAGs (Tyson et al. 2004). Additionally, it is possible to get single-cell amplified genomes, which aid in the recognition of uncultured organisms from a variety of environmental samples.

Various bioinformatics tools are available for functional annotation of MAGs and SAGs such as EggNOG (Huerta-Cepas et al. 2019), KEGG pathways (White et al. 2016), BioCyc, Rhea, EcoCyc, Gene Ontology, etc. (Harris et al. 2004). These can disclose the metabolic potential, various protein domains, functional categories, as well as traits of uncultured organisms.

There are many reference databases available, including the Genome database, GenBank database, Nonredundant Nucleotide Sequence Database (NT/NR), NCBI Reference Sequence (RefSeq), and National Center for Biotechnology Information (NCBI), all of which can be used to identify microbes from various environments more accurately (Haft et al. 2018). For the creation of a standard genome group, comparative genome analysis, genome quality evaluation, and phylogenetic techniques like CheckM were integrated (Parks et al. 2015).

18.4 Metaproteomics

The advancement of metagenomics sequencing and metagenomics research is necessary for the beginning of the metaproteomics study. Metaproteomics refers to the study of all expressed proteins from the ecosystem. This is crucial in determining how well-functioning microorganisms are. As previously mentioned in the introduction, the following phases are involved in metaproteomics techniques. Protein identification requires the use of de novo peptide sequencing in the absence of available metagenomics sequencing, which is necessary for protein identification. When metabolomics and metaproteomics work together, excellent information on active metabolic pathways is produced (Kleiner 2019). Even it helps in the characterization of the microbial community, the interaction between communities can be determined (Kleiner et al. 2017). Recent advancements in mass spectrometry play important role in metaproteomics.

The first step is to prepare the sample for pre-fractionation, which improves the sensitivity of the less abundant peptides by deepening the proteome. Proteins can be separated into one or two dimensions using gel electrophoresis. The peptides were separated using HPLC (Issa Isaac et al. 2019). The major goal of the HPLC is to prevent peptides from entering the mass spectrophotometer's ionization stage. Reverse-phase liquid chromatography is a newly developed method for peptide separation in the field of metaproteomics, which is advanced HPLC (Josic and Kovac 2010). Due to numerous steps, pre-fractionation may expand the cost of a single sample. Moreover, advancement in fraction continued and 2/3 phase MudPIT, strong cation exchange (SCX) column, etc. were developed (Tanca et al. 2015).

18.4.1 Mass Spectrometry

Proteomics based on MS aims to identify recommended proteins. Mass spectrometry is becoming a more acceptable method for protein analysis as procedures are improved and modified. The three components of mass spectrometry are an ion source, a mass analyzer, and an ion current detector. Tandem spectrometry (MS/MS) is the process where two analyzers work together to perform like one analyzer. It is feasible to learn the structural details of proteins attributed to these fragmenting ions and the mass of the fragments. The mass spectrum is assembled once the detector has detected the fragmented ions (Hillenkamp and Karas 1990; Fenn et al. 1990).

For the study of proteins or peptides in a sample, a variety of ionization methods are available, including electrospray ionization (ESI) and matrix-assisted laser desorption ionization (MALDI ionization). The mass analyzer receives an intact molecule via these ionizers from the gas phase. While MALDI needs co-crystallization of molecules with a matrix, ESI provides air pressure that makes it easier for molecules to link with liquid chromatography (Issa Isaac et al. 2019). The coated matrix is then exposed to a pulse laser beam while operating in a high vacuum. Multiple peptides from a complicated sample may be sufficiently differentiated using mass spectrometry (Graham et al. 2011).

18.4.2 Metaproteome Bioinformatics

Software for proteomics, such as OpenMS, Trans-Proteomic Pipelines, or Compomics, is currently trending toward the development of proteomics tool libraries. This library of proteomics software contains tools for every stage of processing, from data administration through data interpretation (Heyer et al. 2017).

A tiny collection of bioinformatics tools called Expasy is a type of web service used to analyze biochemical proteins. A long-term storage database called PRIDE makes data accessible to researchers so they can use it for their studies (Vizca et al. 2016).

Graph 2Pep and compile are the best tools for protein identification in protein databases. Protein grouping, taxonomic analysis, or functional data evaluation is available in Megan CE, Unipept, Prophane, etc.

For peptide spectral matching, many algorithms are utilized, such as OMSSA, MASCOT, SEQUEST, etc. Data from MASCOT searches are transferred to the MPA program (Geer et al. 2004). MetaProSIP is a quantitation tool that helps in the quantification and detection of isotope ratios for protein-SIP (Sachsenberg et al. 2015).

18.5 Conclusion

This entire study suggests that metagenomics and metaproteomics play an important role in isolating, identifying, and characterizing microorganisms from varied environments. Researchers should take note of recent advancements in metagenome databases, mass spectrometers, nanopore sequencing, and metaproteome databases. Recent studies have concentrated on the use of these technologies in agriculture, winemaking, and the verification of medical claims. Plant production can progressively rise due to the relationship between the soil microbial community and plant roots in agriculture. The metagenome and metaproteome may be sequenced using fourth-generation and next-generation methods to determine the wine's quality and flavor. Pharmaceutical value has increased due to an increase in the microbial composition in a specific environment. Nowadays, forensic investigation is one area where metagenomics has a substantial use. Despite the fact that the new methods have a number of limitations, recent developments in the field of sequence software analysis tools have the potential to encourage additional study in the metagenomics sector.

References

Ashton PM, Nair S, Dallman T et al (2015) MinION nanopore sequencing identifies the position and structure of a bacterial antibiotic resistance island. Nat Biotechnol 33:296–302

Fenn JB, Mann M, Meng CK et al (1990) ChemInform abstract: electrospray ionization for mass spectrometry of large biomolecules. ChemInform 246(4926):64–71

Frith MC, Hamada M, Horton P (2010) Parameters for accurate genome alignment. BMC Bioinformatics 11(1):1–14

Geer LY, Markey SP, Kowalak JA et al (2004) Open mass spectrometry search algorithm. J Proteome Res 3:958–964

Goodwin S, McPherson JD, McCombie WR (2016) Coming of age: ten years of next-generation sequencing technologies. Nat Rev Genet 17:333–351

Graham C, McMullan G, Graham RLJ (2011) Proteomics in the microbial sciences. Bioeng Bugs 2: 17–30

Greninger AL, Naccache SN, Federman S et al (2015) Rapid metagenomic identification of viral pathogens in clinical samples by real-time nanopore sequencing analysis. Genome Med 7:1–13

Haft DH, DiCuccio M, Badretdin A et al (2018) RefSeq: an update on prokaryotic genome annotation and curation. Nucleic Acids Res 46:D851–D860

Harris MA, Clark J, Ireland A et al (2004) The gene oncology (GO) database and informatics resource. Nucleic Acids Res 32:258–261

Heyer R, Schallert K, Zoun R et al (2017) Challenges and perspectives of metaproteomic data analysis. J Biotechnol 261:24–36

Hillenkamp F, Karas M (1990) Mass spectrometry of peptides and proteins by matrix-assisted ultraviolet laser desorption/ionization. Methods Enzymol 193:280–295

Huerta-Cepas J, Szklarczyk D, Heller D et al (2019) EggNOG 5.0: a hierarchical, functionally and phylogenetically annotated orthology resource based on 5090 organisms and 2502 viruses. Nucleic Acids Res 47:D309–D314

Issa Isaac N, Philippe D, Nicholas A et al (2019) Metaproteomics of the human gut microbiota: challenges and contributions to other OMICS. Clin Mass Spectrom 14:18–30

Josic D, Kovac S (2010) Reversed-phase high performance liquid chromatography of proteins. Curr Protoc Protein Sci 2010:1–22

Kleiner M (2019) Metaproteomics: much more than measuring gene expression in microbial communities. mSystems 4(3):e00115–e00119

Kleiner M, Thorson E, Sharp CE et al (2017) Assessing species biomass contributions in microbial communities via metaproteomics. Nat Commun 8(1):1–14

Ko KKK, Chng KR, Nagarajan N (2022) Metagenomics-enabled microbial surveillance. Nat Microbiol 7:486–496

Li N, Cai Q, Miao Q et al (2021) High-throughput metagenomics for identification of pathogens in the clinical settings. Small Methods 5:1–27

Liu S, Moon CD, Zheng N et al (2022) Opportunities and challenges of using metagenomic data to bring uncultured microbes into cultivation. Microbiome 10:1–14

Moss EL, Maghini DG, Bhatt AS (2020) Complete, closed bacterial genomes from microbiomes using nanopore sequencing. Nat Biotechnol 38:701–707

Parks DH, Imelfort M, Skennerton CT et al (2015) CheckM: assessing the quality of microbial genomes recovered from isolates, single cells, and metagenomes. Genome Res 25:1043–1055

Quick J, Ashton P, Calus S et al (2015) Rapid draft sequencing and real-time nanopore sequencing in a hospital outbreak of Salmonella. Genome Biol 16(1):1–14

Raza S, Ameen A (2017) International journal of advances in scientific research metaproteomics approaches and techniques: a review. Int J Adv Sci Res 3:49–51

Sachsenberg T, Herbst FA, Taubert M et al (2015) MetaProSIP: automated inference of stable isotope incorporation rates in proteins for functional metaproteomics. J Proteome Res 14:619–627

Tanca A, Palomba A, Pisanu S et al (2015) Enrichment or depletion? The impact of stool pretreatment on metaproteomic characterization of the human gut microbiota. Proteomics 15: 3474–3485

Tyson GW, Chapman J, Hugenholtz P et al (2004) Community structure and metabolism through reconstruction of microbial genomes from the environment. Nature 428:37–43

Venter JC, Remington K, Heidelberg JF et al (2004) Environmental genome shotgun sequencing of the Sargasso Sea. Science 304(5667):66–74

Vizca JA, Csordas A, Griss J et al (2016) Erratum: 2016 update of the PRIDE database and its related tools (Nucleic Acids Research (2016) 44:D1 D447-D456)

White RA, Bottos EM, Roy Chowdhury T et al (2016) Moleculo long-read sequencing facilitates assembly and genomic binning from complex soil metagenomes. mSystems 1:1–15

Yu X, Jiang W, Shi Y et al (2019) Applications of sequencing technology in clinical microbial infection. J Cell Mol Med 23:7143–7150

Zhang L, Chen FX, Zeng Z et al (2021) Advances in metagenomics and its application in environmental microorganisms. Front Microbiol 12:1–15

Chapter 19
Mushroom Metagenome: Tool to Unravel Interaction Network of Plant, Mycorrhiza, and Bacteria

Vineet Vishal, Sweta Sushmita Tigga, Sukanya Hembrom, Binit Baraik, Geetanjali Singh, and Shalini Lal

Abstract In a world of scientific breakthroughs, the solution to our most pressing problems could be right under our noses – microbes. The rhizospheric bacterial consortium works as an ecosystem engineer to grow and sustain ecosystems utilizing many eco-friendly ways. Mushroom rhizosphere is a core where plant roots and soil microorganisms actively interact (mycorrhizal fungi and bacteria). However, the mechanisms that regulate interactions between mycorrhizal fungi, soil bacteria, and plant roots are poorly understood. The goal of this metagenomics study is to open the Pandora box of rhizospheric bacterial consortiums in the mushroom's active zone. Community profiling using diversity indices provides information about diverse groups of phyla (*Proteobacteria*, *Acidobacteria*, *Actinobacteria*, *Firmicutes*, and *Bacteroidetes*) living in a specific niche. Moreover, functional roles of the microbes and potential interactions between fungi and soil bacteria, primarily nitrogen-fixing bacteria, phosphate-solubilizing bacteria, and other PGPR, aid plants in a variety of

V. Vishal
Department of Botany, Dr. Shyama Prasad Mukherjee University, Ranchi, Jharkhand, India

Department of Botany, Bangabasi Evening College, Kolkata, West Bengal, India

S. S. Tigga · S. Hembrom
Department of Microbiology, Dr. Shyama Prasad Mukherjee University, Ranchi, Jharkhand, India

B. Baraik
Department of Botany, Dr. Shyama Prasad Mukherjee University, Ranchi, Jharkhand, India

Department of Botany, J. N College, Ranchi, Jharkhand, India

G. Singh
Department of Botany, Dr. Shyama Prasad Mukherjee University, Ranchi, Jharkhand, India

S. Lal (✉)
Department of Botany, Dr. Shyama Prasad Mukherjee University, Ranchi, Jharkhand, India

Department of Microbiology, Dr. Shyama Prasad Mukherjee University, Ranchi, Jharkhand, India

P. Mathur et al. (eds.), *Microbial Symbionts and Plant Health: Trends and Applications for Changing Climate*, Rhizosphere Biology,
https://doi.org/10.1007/978-981-99-0030-5_19

areas such as bioremediation, biocontrol, and biofertilization and thus are notably important in agroecosystems for soil health.

Keywords Bioremediation · Metagenome · Mycorrhizae · Rhizosphere

Abbreviations

ARDRA	Amplified ribosomal DNA restriction analysis
ARISA	Automated method of ribosomal intergenic spacer analysis
CFU	Colony-forming unit
CLPP	Community-level physiological profiling
COGs	Clusters of orthologous genes
DGGE	Denaturing gradient gel electrophoresis
ePGPR	Plant growth-promoting rhizobacteria
FAME	Fatty acid methyl ester
FISH	Fluorescence in situ hybridization
HMM	Hidden Markov models
HRM	High-resolution melting assay
HTNGS	High-throughput next-generation sequencing
iPGPR	Intracellular plant growth-promoting rhizobacteria
KEGG	Kyoto Encyclopedia of Genes and Genomes
KO	KEGG Orthology
MHB	Mycorrhizae helper bacteria
MIAB	Mushroom-interacting associated bacteria
NGS	Next-generation sequencing
PCoA	Principal coordinate analysis
PICRUSt	Phylogenetic investigation of communities by reconstruction of unobserved states
PLFA	Phospholipid fatty acid
QIIME	Quantitative insights into microbial ecology
qPCR	Quantitative polymerase chain reaction
RAPD	Random amplified polymorphic DNA
RFLP	Restriction fragment length polymorphism
RISA	rRNA intergenic spacer analysis
RSGP	Reverse sample genome probing
SSCP	Single-strand conformation polymorphism states
T-RFLP	Terminal restriction fragment length polymorphism
TTGE	Temporal temperature gradient gel electrophoresis

19.1 Introduction

Microbes like bacteria and fungi are the cornerstone of all the ecosystems. These microorganisms evolved much before alga, land plants, or animals 3500 million years ago. They were discovered on rocks or in rocky belts, and they discharged macro- and micronutrients into the soil (Mojzsis et al. 1996). Around 500 million years ago, these surprisingly small microbes greatly helped plants in colonizing the land. Most of the terrestrial ecosystem plants now obtain water and essential nutrients by collaborating with microbes, particularly bacteria and fungi, which frequently grow in their rhizosphere. Astonishingly, more than 99.9% of the microbes in that rhizosphere are not culturable, and hundreds of millions of species remain undescribed to date (Bodor et al. 2020). Microbes may have a wide range of environmental requirements, including high to low nutrient media concentrations, aerobic or anaerobic conditions, and incubation period, and exhibit slower growth rate due to other factors such as moisture, pH, temperature, salinity, trace metals, etc. (Overmann 2013; Wu et al. 2020). Since only 0.9–1% of microbes can be cultured using culture-dependent traditional microbiological techniques, their physiologies and ecological roles remain largely unknown (Wu et al. 2020). There is a growing need to study and explore more about the missing taxon due to limited knowledge of microbes such as bacteria and fungus based on culture-dependent approaches. Advances in high-throughput sequencing have provided new tools to generate volume of data that supplement culture-based approaches. These microbes have the greatest potential for mycoremediation and are an excellent tool for green agricultural practices since they serve as ecosystem engineers (Epstein 2013). With the growing popularity of omics-based culture-independent techniques, metagenomics study of diverse environmental samples is useful in obtaining fundamental data about the genetic diversity and community structures of microbes, including unculturable bacteria. Metagenomics has the potential to greatly assist the researcher of agroecology (Bodor et al. 2020). It can aid in the knowledge of the vast soil microorganisms present in the soil environment, allowing for the development of new compounds with medicinal potential and biotechnological breakthroughs, for organic agriculture (Garrido-Oter et al. 2018).

19.2 Mushroom Taxonomy and Ecology

Mushroom taxonomy is a fascinating yet a difficult task. The classification has historically been based on morphological characteristics like size and color of basidiocarp, exoperidium layers, endoperidium surface, rhizomorph, structure of the hyphae, type of peristome (fibrillose or plicate), presence or absence of apophysis, and stipe between exoperidium and endoperidium. Agaricoid, corticioid, boletoid, polyporoid, and gasteroid sporocarps make up the majority of the 41,000 identified mushroom species worldwide (Manoharachary et al. 2005; Sjokvist et al.

2014). Only a small number of grown mushrooms are grown commercially around the world, and these include button mushrooms (*Agaricus* spp.), milky mushrooms (*Calocybe* spp.), shiitake mushrooms (*Lentinula* spp.), Jew's ears (*Auricularia* spp.), paddy straw mushrooms (*Volvariella* spp.), oyster mushroom (*Pleurotus* spp.), morel (*Morchella* spp.), and reishi (*Ganoderma* spp.) (Carrasco et al. 2018; Kertesz and Thai 2018; Kumla et al. 2020). Selective microbiota is present on the mycelial surface and within proximity to mycelia of the fungi in natural surrounding that plays important role in growth and development of mushroom (Halsey et al. 2016).

The influence of bacterial community on the mycosphere (environment within and surrounding fungal hyphae) is manifested in inability of many macrofungi to produce fruiting bodies in a sterile environment (Noble et al. 2003, 2009). Formation of ectomycorrhizal associations with host plants is observed in some fungi such as truffles and morels. For these fungi, the establishment of plant-fungal symbiosis is supported by helper bacteria. The fungus along with associated microorganisms, composed of the microbiota inhabiting the fruit bodies or interacting with fungal hyphae, makes a mushroom holobiont. It is determined by the selective processes acting on the microbiota present in substrates and soils (Carasco and Preston 2020).

19.3 Mushroom Biology and Their Potential Roles for Sustainable Agriculture

A mushroom is a macrofungi with a distinct fruiting body which may be above (epigenous) or below the ground (hypogenous), large enough to be seen by naked eyes and picked by the hand (Miles and Chang 1997). Mushrooms belonging to the group *Basidiomycota* and certain *Ascomycota* are the most evident fungi present today (Fig. 19.1). Although they have one of the largest fruit bodies among fungi, potentially harboring highly diverse bacteria, evidence of their microbiome is limited and mostly to the species that are grown in pure culture (Rangel-Castro et al. 2002; Tsukamoto et al. 2002; Kumari et al. 2013). Enormous increase in the agricultural activities across the globe has raised the extent of agro-product processing resulting in a huge agriculture residue generation, thus adding to the ever-increasing environmental pollution. A highly feasible approach with multiple benefits is the utilization of agro-residues from agricultural sectors as primary or secondary sources of energy (Gowda et al. 2021). Mushrooms can grow on a wide variety of lignocellulosic materials by virtue of enzyme complexes enabling them to attack the cellulose, hemicellulose, and lignin components of plant cell walls. Thus, cultivation of edible mushrooms presents an effective solution for utilization of the huge quantities of waste residues generated annually through the activities of the agricultural, forest, and food industries (Buswell and Chang 2018). Edible mushroom production technology offers partial but meaningful solutions by generating relatively cheap source of high-quality food protein and mushroom nutraceuticals and the maintenance of balanced ecosystems by bioremediation.

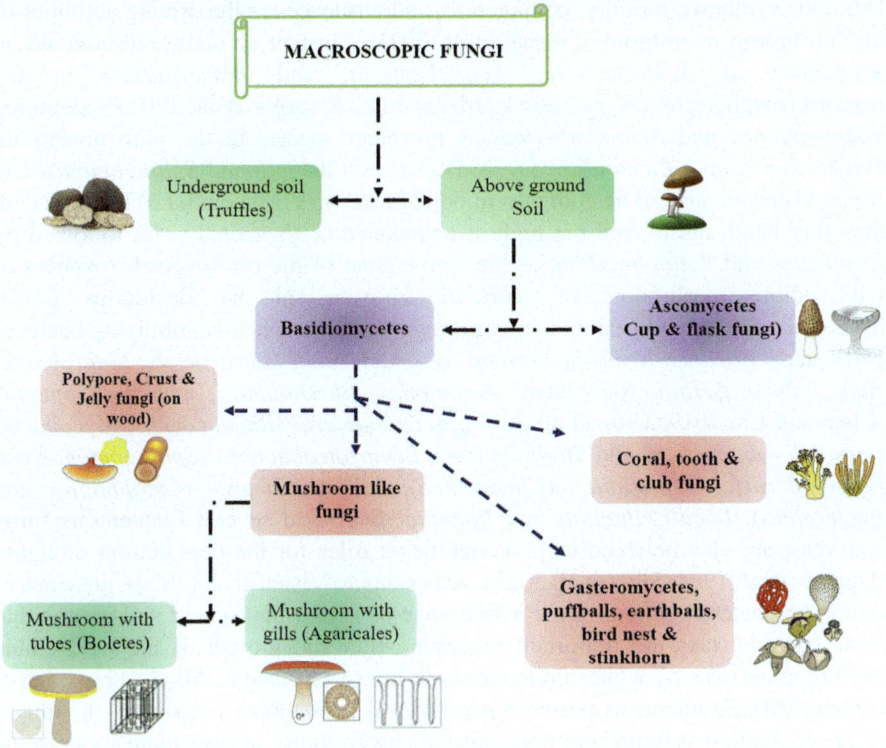

Fig. 19.1 Basidiomycetes and ascomycetes are the two groups where different types of macrofungi are most commonly seen

19.4 Cataloguing Rhizospheric Bacterial Consortium in the Active Zone of the Mushroom

Understanding mushroom rhizosphere, where various meta-organisms coexist, can greatly assist mushroom research and help mycologist find techniques to boost yield and mushroom quality. The mushroom rhizosphere is a point at which plant roots and soil microflora (endo- and ectomycorrhizal fungi and bacteria) work together in a variety of ways, ranging from mutualistic to saprophytic, antagonism, or competition. Many mycologists have reported that the mushroom microbiota plays a wide range of roles, including fruiting body maturation (Barbieri et al. 2007; Chen et al. 2022; Sillo et al. 2022), aroma production (Niimi et al. 2021), and probable nitrogen (Uroz et al. 2007) and phosphorus fixation (Chen and Liu 2019; Vishal et al. 2021). It has been proposed that colonization of the fertile glebal body (reproductive growth phase) by soil microflora may occur prior to exo- and endoperidium (vegetative growth phase) differentiation when the primordium (a yellowish mycelial pellet) would be in close contact with the soil (Sillo et al. 2022). Additionally, quorum sensing, microbial gene expression regulation, symbiosis, biofilm formation, antibiotic synthesis, motility, conjugation, and virulence are the driving activities of the mushroom microbiome (Vishal et al. 2021). Uroz et al. (2012) discovered an abundance of *Acidobacteria*, *Actinobacteria*, and *Bacteroidetes* in the mycorrhizosphere of an oak forest. Meanwhile, Kataoka et al. (2012) identified *Sphingomonas* and *Acidobacterium* as prevalent species in the pine mushroom *Tricholoma* from a *Picea abies* forest. Pent et al. (2017) found *Sphingomonas* and *Alphaproteobacteria* to be common in boreal forests. Vishal et al. (2021, 2022), on the other hand, discovered the highest abundance of *Proteobacteria*, followed by *Firmicutes* and *Actinobacteria*, in the active zone of the ectomycorrhizosphere of wild edible mushrooms of *Astraeus* from a sal dry deciduous forest. Mycorrhizosphere soil is abundant in nitrogen- and phosphate-solubilizing bacteria, particularly *Bacillus*, *Arthrobacterium*, *Agrobacterium*, *Micrococcus*, *Enterobacterium*, *Vibrio*, *Serbia*, *Rhizobium*, *Aeromonas*, *Burkholderia*, and *Pseudomonas* (Chen and Liu 2019; Uroz et al. 2007). *Actinobacteria* (*Rhodococcus*, *Mycobacterium*, *Microbacterium*, and *Streptomyces*), *Betaproteobacteria* (*Sphingomonas* and *Burkholderia*), *Clostridia* (*Clostridium*), *Bacteroidetes* (*Cytophaga* and *Bacteroides*), *Bacilli* (*Bacillus* and *Paenibacillus*), and several filamentous fungi and yeast are also believed to play significant roles for the degradation of lignin (Durham et al. 1984; Silva et al. 2021). According to Vishal et al. (2022), presence of lignolytic bacterial genera in the active zone of mycorrhizosphere soil asserts that these bacterial taxa are important in germination and growth of the wild edible mushroom, *Astraeus*, which flourishes in a litter-rich habitat. Mycorrhizae helper bacteria (MHB) intend to promote mycorrhizal formation by assisting in fungal-plant recognition systems and boosting the susceptibility of host plant roots to the mycorrhizal fungi (Garbaye 1991; Bonfante and Anca 2009).

19.4.1 Bacterial Population in the Gleba, Peridium, and Fruit Bodies of the Mushroom

It would undoubtedly be of enormous agronomic and economic significance to identify and characterize the microbial communities that promote the development of basidiomycetes or ascomycetes primordia into mature fruit bodies. Many mycologists have previously noted the existence of bacterial communities on the surface, internal tissue, peridium, gleba, and mycelial mat under the fruiting bodies (hyphosphere) of the mushrooms (Boer et al. 2005; Zagriadskaia et al. 2013; Antony-Babu et al. 2014; Liu et al. 2021; Sillo et al. 2022). In the forest ecosystem, mushroom-forming fungi have bacterial population ranging from 0.6 to 1.645 million CFU/g recorded in *Laccaria laccata*, 0.3 to 0.714 million CFU/g in *Coltricia perennis*, 0.9 to 0.378 million CFU/g in *Lactarius aurantiacus*, 4.1 to 0.228 million CFU/g in *Gymnopus confluens*, 7.09 to 0.021 million CFU/g in *Lycoperdon perlatum*, 17.0 to 0.006 million CFU/g in *Clavariadelphus ligula* (Zagriadskaia et al. 2013), and 107 to 108 CFU/g in *Tuber aestivum* (Gryndler et al. 2013),

Tuber borchii, (Barbieri et al. 2005), and *Tuber magnatum* (Barbieri et al. 2007). Additionally, diverse microbial habitats are developed by mushroom-forming fungus, which has an impact on the size and composition of the rhizospheric bacterial community. Zagriadskaia et al. (2013) recorded saprophytic bacteria, viz., *Myxococcus*, *Xanthomonas*, *Pseudomonas*, *Aeromonas*, and *Erwinia* from the surface of the fruiting bodies of *Gymnopus confluens*, *Lactarius aurantiacus*, *Coltricia perennis*, and *Clavariadelphus ligula*; *Streptomyces*, *Micrococcus*, *Xanthomonas*, *Myxococcus*, *Pseudomonas*, *Acinetobacter*, *Bacillus*, *Arthrobacter*, and *Erwinia* from the internal tissue of the fruiting bodies; and *Bacillus*, *Arthrobacter*, *Pseudomonas*, *Myxococcus*, *Rhodococcus*, *Cytophaga-Flavobacterium*, and *Streptomyces* genera from the mycelial mat under the fruiting bodies (hyphosphere) of ectomycorrhizal fungi *Gymnopus confluens*, *Laccaria laccata*, *Lactarius aurantiacus*, *Coltricia perennis*, *Clavariadelphus ligula*, and *Lycoperdon perlatum* using the fluorescence in situ hybridization (FISH) method. Zagriadskaia et al. (2013) further added that *Bacilli* of *Firmicutes* and *Streptomyces* of *Actinobacteria* are able to invade the internal tissues of the ectomycorrhizal fungi because of their chitinolytical behavior, whereas gram-negative *Myxococcus* of *Myxococcota*, which is prevalent in plant phyllosphere (Lysak et al. 2003), is found only on the surface and in the internal tissues of the ectomycorrhizal fungi. As bacteria in this genus show effective hydrolytic activity, their identification could indicate the start of ectomycorrhizal fungi fruiting body decay. In the most popular mushrooms *Agaricus* (white button mushroom), *Pseudomonas* sp. (Rainey 1991), *Bacillus*, *Alphaproteobacteria* and *Betaproteobacteria*, and *Firmicutes* have accelerated vegetative growth using a co-culture plate assay (Young et al. 2013). Similarly, it has been noted that *Pseudomonas* sp. in *Pleurotus ostreatus* (Cho et al. 2003) and *P. eryngii* (Kang and Cho 2014), *Glutamicibacter arilaitensis MRC119* in *P. ostreatus* (Kumari and Naraian 2021), and *Micromonospora lupini* in *Pleurotus florid* (Suarez et al. 2020) boost vegetative development and yield in oyster mushrooms. Recent research has focused on the microbiomes of different types of truffles, including black truffle *Tuber melanosporum* (Antony-Babu et al. 2014), Chinese black truffle *Tuber indicum* (Liu et al. 2021), and white truffle *Tuber magnatum* (Sillo et al. 2022) (Tables 19.1 and 19.2). Formation of gleba, peridium, and aroma productions of the truffle and their ecological functions such as nitrogen or potassium fixation are all dependent on the truffle microbiota. *Proteobacteria* are the most abundant phylum, followed by *Actinobacteria* and *Firmicutes*, and *Bradyrhizobium*, *Pseudomonas*, and *Flavobacterium* are the most prevalent genera in truffle fruiting bodies, according to metagenomic data (Antony-Babu et al. 2014; Liu et al. 2021; Sillo et al. 2022). *Actinobacteria*, *Acidobacteria*, and *Planctomycetes* are more prevalent phyla in peridium, whereas they were absent from gleba. In terms of the presence of bacterial populations throughout ascocarp maturity, gleba and peridium displayed a distinct trend – *Bacteroidetes* population surged in peridium and declined in gleba and *Firmicutes* stayed static in peridium and increased in gleba. Fungal diversity was also determined during mushroom cultivation and various compositing processes. The *Ascomycota* thermophilic fungi *Thermomyces ibadanensis*, *Chaetomium thermophilum*, *Mycothermus thermophilus*, *Talaromyces*

Table 19.1 Bacterial genus found in the fruiting body, peridium, and gleba of mushrooms

Mushroom body	Antony-Babu et al. (2014) *Tuber melanosporum* (Black truffle)	Liu et al. (2021) *Tuber indicum* (Chinese black truffle)	Sillo et al. (2022) *Tuber magnatum* (White truffle)
Fruit body	*Bradyrhizobium, Pseudomonas*	–	*Flavobacterium*
Peridium	*Pedobacter, Bradyrhizobium, Terrimonas, Acidovorax, Variovorax, Dyadobacter, Flavobacterium, Rhizobium, Sphingobacterium*	*Cupriavidus, Bradyrhizobium, Aminobacter, Mesorhizobium*	*Phenylobacterium Flavobacterium, Brochothrix, Phenylobacterium, Pseudomonas*
Gleba	*Variovorax, Bradyrhizobium, Mitsuaria, Sphingobacterium, Pedobacter, Myroides, Pseudomonas, Pseudomonas*	*Sphingomonas, Afipia, Amycolatopsis, Acinetobacter, Bradyrhizobium, Cupriavidus, and Pseudomonas*	*Flavobacterium, Paraburkholderia, Pseudomonas, Pedobacter, Sphingobacterium, Sphingobacterium*

thermophilum, *Scytalidium thermophilum*, *Malbranchea sulfurea*, *Thermomyces lanuginosus*, and *Torula thermophila* produce cellulase and hemicellulase enzymes which promote the growth of *A. bisporus* (Vieira and Pecchia 2021; Basotra et al. 2016; Romdhane et al. 2010) and *Cordyceps militaris* (Zhang et al. 2021).

19.5 Mushroom Metagenomics Cloud-Based Pipeline

For the identification and characterization of hidden microorganisms in soil, water, air, or human gut, high-throughput next-generation sequencing (HTNGS) provided excellent solution. However, there are significant challenges in analyzing and visualizing the results due to the complexity of metagenomic data analysis and lack of expertise in command-line-based processes. To overcome these issues, cloud-based metagenomics computing constructed one powerful remote server offer where researchers with no to little bioinformatics skills can use a click-based open-source microbiome platform for data analysis as well as visualization in a speedy and productive way, due to the advancement of cutting-edge technology (Fig. 19.2). Nephele (Weber et al. 2018), EDGE bioinformatics (Philipson et al. 2017), EasyMAP (Hung et al. 2021), BusyBee (Schmartz et al. (2022), and MiCloud (Gu et al. 2022) are the cloud-based microbial bioinformatics platform with inbuilt QIIME (Quantitative Insights into Microbial Ecology) and Mothur pipeline to analyzing uncultured hidden bacterial community in an 16S amplicon or shotgun metagenome dataset. DAnIEL (Loos et al. 2021) is an exclusive cloud platform for analyzing hidden fungal communities (Table 19.3).

Table 19.2 Mushroom-interacting associated bacteria (MIAB) with their microbiome characterization

Mushroom growth-promoting bacteria (MGPB)	Mushroom	Method of analysis	Source	Microbial function/roles	References
Pseudomonas putida	*A. bisporus*	16 s rRNA genes	Fruit bodies	Increase fruiting body yield	Zarenejad et al. (2012)
Pseudomonas, Bdellovibrio, Flavobacterium, Bacillus, Hyphomicrobium	*A. bisporus*	NGS (next generation sequencing-Illumina MiSeq)	Fruit bodies	Cellulose-degrading consortium	Carrasco et al. (2019)
Pseudomonas, Chryseobacterium, Herbaspirillum, Burkholderia	*I. terrigena*	16 s rRNA genes	Fruit bodies	Improving the fruit body yields and biological efficiency	Bahram et al. (2018)
Pseudomonas fluorescens	*P. eryngii*	16 s rRNA genes	Fruit bodies	Promote the formation of the primordium, the mycelial growth, and the fruiting body productivity	Kim et al. (2008)
Pseudomonas fluorescens	*P. ostreatus*	Fatty acid methyl ester (FAME)	Fruit bodies	Promote the formation of the primordium, the mycelial growth, and the fruiting body productivity	Cho et al. (2003)
Pedobacter, Sphingobacterium, Bradyrhizobium, Variovorax, and Pseudomonas	*T. melanosporum*	454-pyrosequencing of 16S rRNA gene + TTGE fingerprinting + FISH + functional GeoChip 3.0	Fruit bodies, peridium, and gleba	Nif genes assigned to uncultured bacteria, as well as genes involved in sulfur metabolism were detected in the ascocarp	Antony-Babu et al. (2014)
Pseudomonas, Alcaligenes faecalis, Microbacterium, Arthrobacter	*T. indicum*	NGS (next-generation sequencing-Illumina HiSeq)	Fruit bodies	Exhibited the potential abilities of nitrogen fixation and phosphorus solubilization	Chen et al. (2019)
Pedobacter, Burkholderia, Pseudomonas, Flavobacterium	*T. magnatum*	NGS (next-generation sequencing-Illumina MiSeq)	Fruit bodies, peridium, and gleba	Abundance of different microbial species seems to play an essential role in truffle production	Sillo et al. (2022)

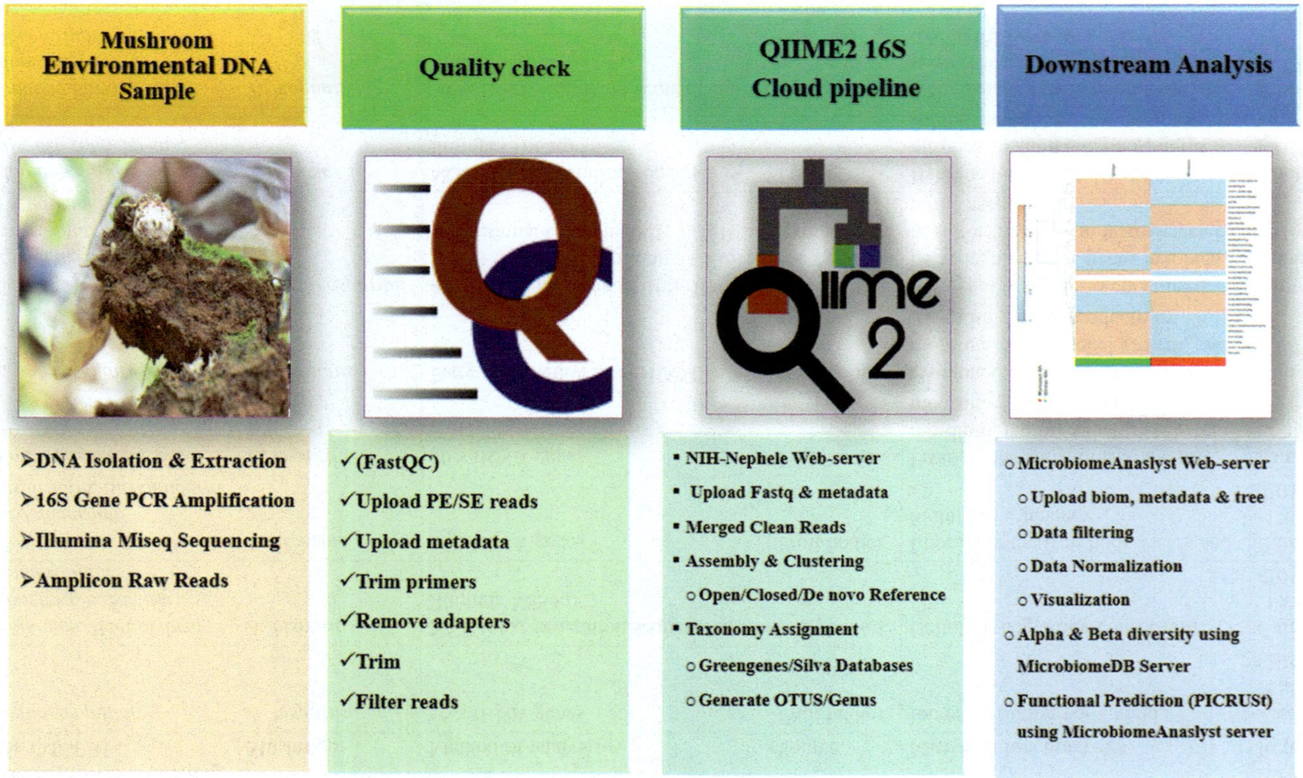

Fig. 19.2 Cloud-based 16S metagenomic workflow for identifying hidden bacterial communities in the mycorrhizosphere

Table 19.3 Cloud-based web server/online tool for amplicon and metagenomic analysis

Name	Link	Description	References
Nephele	https://nephele.niaid.nih.gov/	A web-based platform for microbiome data analysis	Weber et al. (2018)
DAnIEL	https://sbi.hki-jena.de/daniel/latest/	DAnIEL is a web server to analyze fungal ITS amplicon sequencing	Loos et al. (2021)
EDGE bioinformatics	https://edgebioinformatics.org/	EDGE provides an intuitive web-based interface for user input, allows users to visualize and interact with selected results	Philipson et al. (2017)
EasyMAP	http://easymap.cgm.ntu.edu.tw/	EasyMAP is a convenient 16S ribosomal DNA analysis platform	Hung et al. (2021)
CosmosID	https://app.cosmosid.com/	A range of metagenomics services, ranging from targeted tests (16S for prokaryotes and ITS2 for fungi) to whole genome analysis	Yan et al. (2019)
EzBioCloud	https://www.ezbiocloud.net/contents/16smtp	EzBioCloud MTP provides various publicly available datasets for you to compare with your own 16S data	Yoon et al. (2017)
OneCodex	https://www.onecodex.com/	A fast, easy-to-use platform for microbiome sequencing and analysis	Minot et al. (2015)
MiCloud	http://micloud.kr:3838/app/micloud	Unified web platform for comprehensive microbiome data analysis	Gu et al. (2022)
MG-RAST	https://www.mg-rast.org/	Has served as a public resource for annotation and analysis of metagenomic sequence data	Keegan et al. (2016)
KAUST	https://www.cbrc.kaust.edu.sa/aamg/kmap.start/	A new integrated open web-based tool for the comprehensive exploration of shotgun metagenomic data	Alam et al. (2021)
MGnify	https://www.ebi.ac.uk/metagenomics/	Provides a free-to-use platform for the assembly, analysis, and archiving of microbiome data	Mitchell et al. (2020)
NanoPipe	https://bioinformatics.uni-muenster.de/tools/nanopipe2/generate/index.pl?lang=en	NanoPipe pipeline analyzes reads generated by the Oxford Nanopore sequencing devices	Shabardina et al. (2019)
BusyBee	https://ccb-microbe.cs.uni-saarland.de/busybee-update	Enable the convenient analysis of metagenomic data in the form of assembled contigs or long reads (PacBio or ONT)	Schmartz et al. (2022)

(continued)

Table 19.3 (continued)

Name	Link	Description	References
iMicrobe	https://www.imicrobe.us/	iMicrobe is a platform that connects researchers' own data to published, curated, microbial metagenomic datasets and high-performance computing (HPC) methods for their analysis	Youens-Clark et al. (2019)
IDseq	https://czid.org/	An open-source cloud-based metagenomics pipeline and service for global pathogen detection and monitoring	Kalantar et al. (2020)

19.5.1 Community Profiling by Alpha Diversity: A Measure of Within-Sample

Community profiling of microbes gives us the deep insight of microorganisms in an ecosystem (Hamady et al. 2008). Community profiling basically uses screening and analyzing microbes to represent abundance of microbial species in a habitat (Nocker et al. 2007; Wagner et al. 2018). With reference to habitation and diverse quantification, community profiling is usually studied at three different levels: alpha, beta, and gamma diversity (Whittaker 1972). Alpha diversity measures the microbial community according to microbes' richness (number of taxonomic group) and evenness (distribution of abundance) (Willis 2019). Ecologists often use species richness indices to disclose species richness such as Margalef index, Shannon index, Brillouin index, Fisher's index, Gini index, and McIntosh index (Fig. 19.3). The very frequent used species richness indices in literature are Chao and Shannon richness, whereas Margalef index and McIntosh index are rarely used in literature (Fedor and Zvaríkova 2019; Magurran 1988). In recent researches, Shannon diversity has been used to analyze the relationships between plant diversity and microbial diversity in which microbial richness illustrated modest but positive correlation with plant diversity (Liu et al. 2020). Vishal et al. (2022) reported that bacterial diversity in the mycorrhizosphere was significantly high in winter season compared to monsoon season due to soil low moisture content, soil respiration, and high cellulase and lignin degradation activity. In ecosystem, microbes play an important role in biotransformation of inorganic carbon into biomass by primary producers, nutrient regeneration, and recycling. Therefore, species evenness is very important to understand interaction of microbes with plant and soil and eventually helps to device ways to maintain a healthy environment.

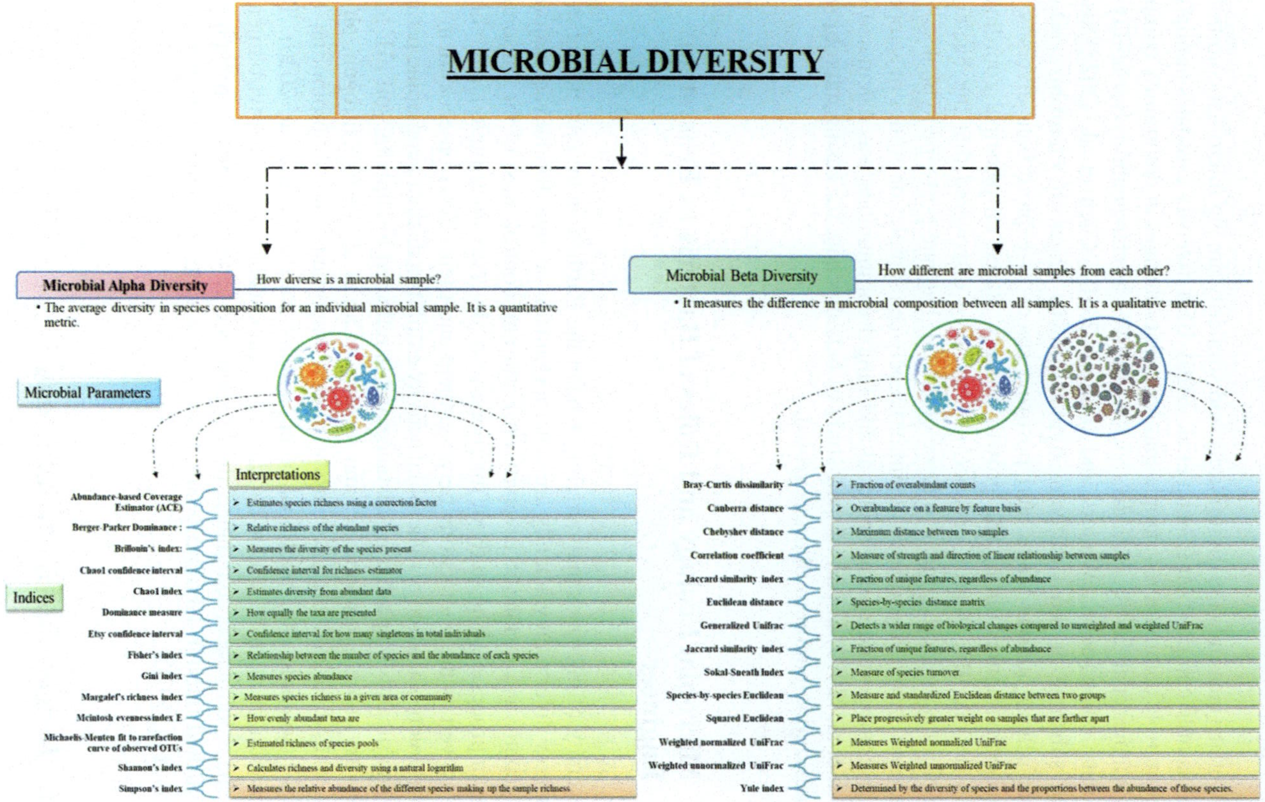

Fig. 19.3 Flowchart showing different parameters/indices of microbial diversity

19.5.2 Community Profiling by Beta Diversity: A Measure of Similarity Between Samples

Beta diversity measures the dissimilarity and similarity among microbial communities and sample composition (Cheng et al. 2020). Beta diversity estimates similarity among multiple communities/sample that represents changes in community composition (Wagner et al. 2018). Beta diversity is observed as changeability/variability in community composition (Walters and Martiny 2020). Chen et al. (2022) used cultured substrate at three different hyphal growth stages of *Pleurotus eryngii* (half-bag stage, full-bag stage, and post-ripe stage) to explore beta diversity in bacterial communities during the growth process. Based on principal coordinate analysis (PCoA) Bray-Curtis distance metrics, they discovered that the bacterial communities were more similar to those of the full-bag and post-ripe hyphal stages of *Pleurotus eryngii*. The degree of similarity between samples or groups of samples from the compost and casing bacterial and fungal communities was calculated by Carrasco et al. (2019) as part of their study of beta diversity for casing microbiome dynamics during button mushroom growth.

19.6 Microbial Ecology and Roles of Bacteria as Ecosystem Engineer

The diversity of microbes that currently inhabit the Earth is enormous, though it is unclear how diverse they actually are. Microorganisms account for the vast majority of the genetic and metabolic diversity on Earth and control most of the important ecosystem processes. The study of microorganism diversity, distribution, and abundance, as well as the interactions between them and their effects on ecosystems, is known as microbial ecology. Microbial activity exhibits a distinct geographic pattern that is significantly impacted by distance and is reflected by climate, soil pH, total phosphorus, rainfall, temperature, and location (Cao et al. 2016). Nitrogen, phosphorus, and dissolved organic carbon played important roles in promoting the bacterial taxonomic and functional compositional patterns (Liu et al. 2020). The identification of microbes, their evolutionary relationships, and the environmental factors that control their abundance, distribution, and particular activity are typically accomplished through laboratory culture as well as culture-independent molecular methods (Fig. 19.4).

19.6.1 Insights into Bacterial Ecology

Plant growth-promoting rhizobacteria (PGPR), a class of free-living soil bacteria, have the capacity to promote plant development by colonizing the plant root.

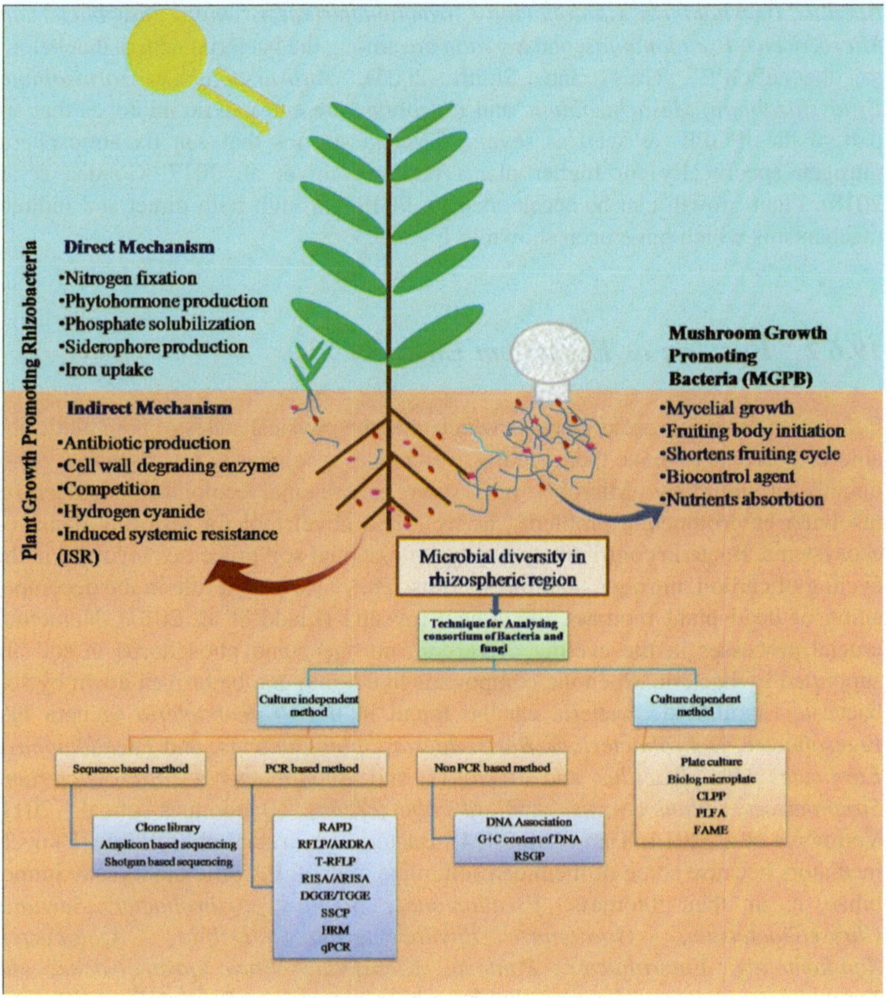

Fig. 19.4 Microbial diversity in the mycorrhizosphere and their probable ecological function with techniques for identifying and studying bacterial and fungal diversity consortia in the rhizosphere. The culture-dependent identification is based on the morphology, physiology, and phylogenetic analysis after isolation and culture, whereas culture-independent method identification is based on genetic biomarkers

Rhizobacteria that promote plant growth can be categorized into two main types: extracellular and intracellular. Extracellular plant growth-promoting rhizobacteria (ePGPR) are found in the rhizosphere (on the rhizoplane) or the spaces between the cells of the root cortex, whereas intracellular plant growth-promoting rhizobacteria (iPGPR) are primarily found inside the specialized nodular structures of root cells (Gray and Smith 2005). *Agrobacterium*, *Arthrobacter*, *Azotobacter*, *Azospirillum*,

Bacillus, *Burkholderia*, *Caulobacter*, *Chromobacterium*, *Erwinia*, *Flavobacterium*, *Micrococcus*, *Pseudomonas*, and *Serratia* are among the bacterial genera that belong to the ePGPR (Gray and Smith 2005). *Allorhizobium*, *Azorhizobium*, *Bradyrhizobium*, *Mesorhizobium*, and *Rhizobium* are endophytic microbes that are part of the iPGPR, as well as several *Frankia* species that can fix atmospheric nitrogen specifically for higher plants (Olanrewaju et al. 2017; Goudaa et al. 2018). Plant growth can be accelerated by PGPR through both direct and indirect mechanisms which have been shown in Fig. 19.3.

19.6.2 Bacteria as Ecosystem Engineer

Ecosystem engineers are organisms which alter the physical states of both biotic and abiotic materials to make them readily available as resources to other species, either directly or indirectly. Microbes may serve as principal ecological engineers in resolving environmental problems, as well as a novel tool for restoring damaged ecosystems. Bacteria contribute to a range of essential soil processes involved in the cycling of carbon, nitrogen, and phosphorus. They also play a role in the decomposition of dead plant biomass and fungal mycelia (Lladó et al. 2017). Numerous crucial processes in the cycling of carbon, nitrogen, and phosphorus in soil are supported by bacteria. Phenolic compounds like lignin can be broken down by soil bacteria. Ligninolytic bacteria can be found in the *Proteobacteria* (genera like *Enterobacter*, *Ochrobacterium*, *Burkholderia*, *Sphingomonas*, and *Pseudomonas*), *Firmicutes* (*Paenibacillus* and *Bacillus*), and *Actinobacteria* (*Microbacterium*, *Streptomyces*, *Rhodococcus*, and *Mycobacterium*) (Bandounas et al. 2011; Vetrovsky et al. 2014; Tian et al. 2014). Bacteria are crucial phosphorus (P) cycle mediators because some of them can solubilize mineral P, while others may immobilize it in their biomass. *Pseudomonas*, *Bacillus*, *Arthrobacter*, *Serratia*, *Chryseobacterium*, *Gordonia*, *Phyllobacterium*, *Delftia*, *Azotobacter*, *Xanthomonas*, *Enterobacter*, *Pantoea*, *Klebsiella*, *Vibrio proteolyticus*, and *Xanthobacter agilis* are bacteria that have been reported as P-solubilizers (Sharma et al. 2013; Alori et al. 2017). Bacteria are important for the health of the soil and the plants that grow within it. There are many different bacterial species in soil, and many of them not only perform crucial roles in the cycling of nutrients but can also shield crops from disease as well as protect them from pests and insects.

19.7 Functional Annotation of Mushroom Rhizospheric Bacteria Consortium

The endeavor to map genes to functions and, eventually, to higher-level biological, cellular, and molecular functions that reflect the mode of action of host-microbiome interaction is a crucial step in metagenomic analysis (Fig. 19.4). This is due to the fact that a sizable number of bacterial genomes that have not been cultured or characterized correspond to an equal portion of newly discovered bacteria whose functions are still unclear (Yen and Johnson 2021). A variety of biological functioning pathways, including quorum sensing, regulating microbial gene expression, symbiosis, biofilm formation, and acting as a first line of defense against various biotic and abiotic stresses, are thought to be induced by mushroom-interacting associated bacteria (MIAB) and other microbes. PICRUSt (phylogenetic investigation of communities by reconstruction of unobserved states), a bioinformatics program, has been primarily used to study the prediction of microbial functional profiling using the 1.5 kbp 16S rRNA gene (Morgan and Huttenhower 2012). Matching query sequences to groups and alignment to individual reference sequences using hidden Markov model (HMM)-based methods can be used to identify homologous groups of genes, which are presumed to have a shared function, in functional reference databases like KEGG, KOs (Kanehisa et al. 2010), and COGs (Tatusov et al. 2003).

19.7.1 KEGG Metabolism, Pathway, and Module

Large-scale molecular datasets, such as the genomic and metagenomic sequences, can be biologically interpreted using the reference-based database such as Kyoto Encyclopedia of Genes and Genomes (KEGG). It piles up new information of the high-level biological and molecular functions of the cell and the organism as they are portrayed by KEGG molecular networks, comprising KEGG pathway maps, BRITE hierarchies, and KEGG modules. Bahram et al. (2018) annotated 4467 microbial genera from dikaryotic fruiting body tissues from *Inocybe terrigena*. Vishal et al. (2022) further annotated 10 KEGG metabolisms, 138 KEGG pathways, and 191 KEGG modules of which nucleotide metabolism, cofactor and vitamin metabolism, glycan biosynthesis and metabolism, and xenobiotics biodegradation and metabolism, as well as lipid metabolism, were the most enriched metabolisms from the active zone of ectomycorrhizosphere of wild edible mushrooms *Astraeus*.

19.7.2 Cluster of Orthologous Groups of Protein

The clusters of orthologous gene (COG) database offers an evolutionary classification of protein families. COG is a small, manually curated, and therefore highly effective orthology resource. Orthologous interactions provide a wealth of information about how proteins work. In ectomycorrhizal fungi *Astraeus* carbohydrate transport and metabolism, cell wall membrane envelope biogenesis was highly enriched ortholog groups of genes found in the ectomycorrhizosphere (Vishal et al. 2022).

19.8 Cross Talk and Fungi-Bacteria Interaction

Plant-fungus interaction has played an important role in establishment of land plants. It has been hypothesized that the development of plant-fungal mutualism preceded the development of roots and therefore was crucial to the ability of plants to colonize the land (Brundrett 2002). During the evolution of terrestrial microbial life, bacteria have been confined to the degradation of simple substrates, whereas fungi evolved as decomposer of recalcitrant organic matter like lignin (Boer et al. 2005). Bacteria and fungi are the cornerstone of all the ecosystems. Both often share microhabitats where they coevolve and their interactions play a key role in functioning of the ecosystem.

Bacteria are present in different tissues of the fungi such as mushrooms. The association of bacteria and fungi is very specific and it has coevolved over millions of years (Uehling et al. 2017). Both interact with each other and modulate the behaviors of either one or both interacting partners. Depending upon the environment, the interactions can be mutualistic or antagonistic and diverse varying from simple to very complex. In *Ascomycota* and *Basidiomycota*, the endofungal bacteria are reported to be more transient in nature, and they influence the fitness of the host mushroom. Such transient bacterial fungal association may be ecologically important in local habitat-associated adaptation in mushrooms (Spraker et al. 2016). With the advent of molecular tools like next-generation sequencing methods, a culture-independent approach is being used for describing the microbial diversity in varied environment. The plethora of information generated has brought into light the huge diversity of microorganism and the potential interaction between them (Thompson et al. 2017). These data do not represent the true interaction, but they deal with potential interactions. However, focus is now shifting on multiple organismic interaction rather than bacteria and fungi interaction. The environment within and surrounding fungal hyphae (mycosphere) influences and itself gets influenced by the bacterial communities to the extent that many macrofungi are unable to produce fruiting bodies in a sterile environment (Noble et al. 2003, 2009). Commercially, it has been exploited to increase the productivity of mushrooms. It has been observed that bacteria associated with *Pinus densiflora* influence the mycelial growth of *Tricholoma matsutake*, an ectomycorrhizal fungus (Oh and Lim 2018).

In case of mushrooms, the composition and dynamics of microbial communities at different stages of edible mushroom such as *Agaricus* have shown that these microbes impact the quality of product and confer tolerance against viral, bacterial, as well as fungal diseases (Braat et al. 2022). Understanding the mode of action of the compound released by microbes and their specificity can contribute to future use of bacterial biocontrol against fungal pathogen. The bacteria may act directly by stimulating vegetative growth and mushroom formation or by inhibiting the pathogen. In *Agaricus* self-inhibitory substances 1-octen-3-ol and ethylene produced by *A. bisporus* mycelium are removed by *Pseudomonas* sp., thus promoting vegetative growth and primordia formation (Zhang et al. 2019). While conferring tolerance against disease such as brown blotch disease caused by *P. tolaasii* in *Agaricus* and *Pleurotus*, bacterium *Mycetocola* has the potential to control the spread of disease

(Ghasemi et al. 2020). They act by detoxifying tolaasin and by inhibiting growth and motility of *P. tolaasii* (Hermenau et al. 2020; Ghasemi et al. 2021). Use of *Bacillus velezensis* as a biocontrol agent against *Trichoderma aggressivum* causing the green mold disease in *Agaricus* has been reported. This disease suppression may involve the production of antifungal lipopeptides by *Bacillus velezensis* that are active against the pathogenic fungi without affecting *Agaricus* (Pandin et al. 2019). The study of the bacteria-fungi interaction will help in understanding the role of bacteria in growth of mushroom, and the information can be exploited to facilitate the commercial harvesting of mushrooms. Although many interactions are being studied among bacteria and fungi, a lot more is to be explored. With the advent of new molecular tools and advance knowledge on plant-microbe interaction, microbiome engineering can boost the plant productivity, thus beginning a new Green Revolution.

19.9 Conclusion

High-throughput metagenome investigations offer a more comprehensive methodology with a more potent workflow than conventional methods like DGGE and T-RFLP. It is possible to work with unrefined environmental samples and characterize the non-culturable microbes using NGS-based metagenomics because it does not rely on enrichment or isolation. It directly identifies entire gene collections in environmental samples, such as soil, water, atmosphere, plant, and animal systems, and examines their biochemical activities and complexes. It also aids in the discovery of novel natural products, antibiotics, and enzymes with novel functions in a variety of fields, including medicine, agriculture, energy, food and nutrition, etc.

References

Alam I, Kamau AA, Ngugi DK, Gojobori T, Duarte CM, Bajic VB (2021) KAUST metagenomic analysis platform (KMAP), enabling access to massive analytics of re-annotated metagenomic data. Sci Rep 11(1):1–14

Alori ET, Glick BR, Babalola OO (2017) Microbial phosphorus solubilization and its potential for use in sustainable agriculture. Front Microbiol 8:971

Antony-Babu S, Deveau A, Van Nostrand JD, Zhou J, Le Tacon F, Robin C, Uroz S (2014) Black truffle-associated bacterial communities during the development and maturation of *Tuber melanosporum* ascocarps and putative functional roles. Environ Microbiol 16(9):2831–2847

Bahram M, Vanderpool D, Pent M, Hiltunen M, Ryberg M (2018) The genome and microbiome of a dikaryotic fungus (*Inocybe terrigena*, Inocybaceae) revealed by metagenomics. Environ Microbiol Rep 10(2):155–166

Bandounas L, Wierckx NJ, de Winde JH, Ruijssenaars HJ (2011) Isolation and characterization of novel bacterial strains exhibiting ligninolytic potential. BMC Biotechnol 11:94

Barbieri E, Bertini L, Rossi I, Ceccaroli P, Saltarelli R, Guidi C, Stocchi V (2005) New evidence for bacterial diversity in the ascoma of the ectomycorrhizal fungus *Tuber borchii* Vittad. FEMS Microbiol Lett 247(1):23–35

Barbieri E, Guidi C, Bertaux J, Frey-Klett P, Garbaye J, Ceccaroli P, Stocchi V (2007) Occurrence and diversity of bacterial communities in *Tuber magnatum* during truffle maturation. Environ Microbiol 9(9):2234–2246

Basotra N, Kaur B, Di Falco M, Tsang A, Chadha BS (2016) *Mycothermus thermophilus* (Syn. *Scytalidium thermophilum*): repertoire of a diverse array of efficient cellulases and hemicellulases in the secretome revealed. Bioresour Technol 222:413–421

Bodor A, Bounedjoum N, Vincze GE, Erdeine Kis A, Laczi K, Bende G, Rakhely G (2020) Challenges of unculturable bacteria: environmental perspectives. Rev Environ Sci Biotechnol 19(1):1–22

Boer W, Folman LB, Summerbell RC, Boddy L (2005) Living in a fungal world: impact of fungi on soil bacterial niche development. FEMS Microbiol Rev 29(4):795–811

Bonfante P, Anca IA (2009) Plants, mycorrhizal fungi, and bacteria: a network of interactions. Annu Rev Microbiol 63:363–383

Braat N, Koster MC, Wosten HAB (2022) Beneficial interactions between bacteria and edible mushrooms. Fungal Biol Rev 39:60–72

Brundrett MC (2002) Coevolution of roots and mycorrhizas of land plants. New Phytol 154:275–304

Buswell JA, Chang ST (2018) Edible mushrooms: attributes and applications. In: Genetics and breeding of edible mushrooms. Routledge, pp 297–324

Cao H, Chen R, Wang L, Jiang L, Yang F, Zheng S, Wang G, Lin X (2016) Soil pH, total phosphorus, climate and distance are the major factors influencing microbial activity at a regional spatial scale. Sci Rep 6:25815

Carasco J, Preston GM (2020) Growing edible mushrooms: a conversation between bacteria and fungi. Environ Microbiol 22(3):858–872

Carrasco J, Zied DC, Pardo JE, Preston GM, Pardo-Giménez A (2018) Supplementation in mushroom crops and its impact on yield and quality. AMB Express 8:e146

Carrasco J, Tello ML, de Toro M, Tkacz A, Poole P, Pérez-Clavijo M, Preston G (2019) Casing microbiome dynamics during button mushroom cultivation: implications for dry and wet bubble diseases. Microbiology 165(6):611–624

Chen Q, Liu S (2019) Identification and characterization of the phosphate-solubilizing bacterium *Pantoea* sp. S32 in reclamation soil in Shanxi, China. Front Microbiol 10:2171

Chen J, Li JM, Tang YJ, Xing YM, Qiao P, Li Y, Guo SX (2019) Chinese black truffle-associated bacterial communities of *Tuber indicum* from different geographical regions with nitrogen fixing bioactivity. Front Microbiol 10:2515

Chen L, Yan M, Qian X, Yang Z, Xu Y, Wang T, Sun S (2022) Bacterial community composition in the growth process of *Pleurotus eryngii* and growth-promoting abilities of isolated bacteria. Front Microbiol 13:787628

Cheng Z, Lei S, Li Y, Huang W, Ma R, Xiong J, Zhang T, Jin L, Haq HU, Xu X, Tian B (2020) Revealing the variation and stability of bacterial communities in tomato rhizosphere microbiota. Microorganisms 8(2):170

Cho YS, Kim JS, Crowley DE, Cho BG (2003) Growth promotion of the edible fungus *Pleurotus ostreatus* by fluorescent pseudomonads. FEMS Microbiol Lett 218(2):271–276

Durham DR, McNameen CG, Stewart DB (1984) Dissimilation of aromatic compounds in *Rhodotorula graminis*: biochemical characterization of pleiotropically negative mutants. J Bacteriol Res 160(2):771–777

Epstein SS (2013) The phenomenon of microbial uncultivability. Curr Opin Microbiol 16(5):636–642

Fedor P, Zvaríkova M (2019) Biodiversity indices. Encycl Ecol 2:337–346

Garbaye J (1991) Mycorrhization helper bacteria: a new dimension to the mycorrhizal symbiosis. Acta Bot Gall 141(4):517–521

Garrido-Oter R, Nakano RT, Dombrowski N, Ma KW, Team TA, McHardy AC, Schulze-Lefert P (2018) Modular traits of the Rhizobiales root microbiota and their evolutionary relationship with symbiotic rhizobia. Cell Host Microbe 24(1):155–167

Ghasemi S, Harighi B, Azizi A, Mojarrab M (2020) Reduction of brown blotch disease and tyrosinase activity in *Agaricus bisporus* infected by *Pseudomonas tolaasii* upon treatment with endofungal bacteria. Physiol Mol Plant Pathol 110:101474

Ghasemi S, Harighi B, Azizi A, Mojarrab M (2021) Response of *Pseudomonas tolaasii*, the causal agent of mushroom brown blotch disease to the volatile compounds produced by endofungal bacteria. BioControl 66(3):421–432

Goudaa S, Kerryb RG, Dasc G, Paramithiotisd S, Shine HS, Patrac JK (2018) Revitalization of plant growth promoting rhizobacteria for sustainable development in agriculture. Microbiol Res 206:131–140

Gowda NA, Gurikar C, Lokesh AC (2021) Recent advances in mushroom cultivation technology and its application. In: Khulshreshtha et al (eds) Bright sky publications, New Delhi

Gray EJ, Smith DL (2005) Intracellular and extracellular PGPR: commonalities and distinctions in the plant–bacterium signaling processes. Soil Biol Biochem 37(3):395–412

Gryndler M, Soukupová L, Hrselova H, Gryndlerová H, Borovička J, Streiblová E, Jansa J (2013) A quest for indigenous truffle helper prokaryotes. Environ Microbiol Rep 5(3):346–352

Gu W, Moon J, Chisina C, Kang B, Park T, Koh H (2022) MiCloud: a unified web platform for comprehensive microbiome data analysis. PLoS One 17(8):e0272354

Halsey JA, Silva MDCP, Andreote FD (2016) Bacterial selection by mycospheres of Atlantic rainforest mushrooms. Anton Leeuw Int J G 109:1353–1365

Hamady M, Walker JJ, Harris JK, Gold NJ, Knight R (2008) Error-correcting barcoded primers for pyrosequencing hundreds of samples in multiplex. Nat Methods 5(3):235–237

Hermenau R, Kugel S, Komor AJ, Hertweck C (2020) Helper bacteria halt and disarm mushroom pathogens by linearizing structurally diverse cyclolipopeptides. Proc Natl Acad Sci 117(38): 23802–23806

Hung YM, Lu TP, Tsai MH, Lai LC, Chuang EY (2021) EasyMAP: a user-friendly online platform for analyzing 16S ribosomal DNA sequencing data. New Biotechnol 63:37–44

Kalantar KL, Carvalho T, de Bourcy CF, Dimitrov B, Dingle G, Egger R, DeRisi JL (2020) IDseq—an open-source cloud-based pipeline and analysis service for metagenomic pathogen detection and monitoring. Gigascience 9(10):giaa111

Kanehisa M, Goto S, Furumichi M, Tanabe M, Hirokawa M (2010) KEGG for representation and analysis of molecular networks involving diseases and drugs. Nucleic Acids Res 38:D355–D360

Kang YM, Cho KM (2014) Identification of auxin from *Pseudomonas* sp. P7014 for the rapid growth of *Pleurotus eryngii* mycelium. Korean. J Microbiol 50(1):15–21

Kataoka R, Siddiqui ZA, Kikuchi J, Ando M, Sriwati R, Nozaki A, Futai K (2012) Detecting nonculturable bacteria in the active mycorrhizal zone of the pine mushroom *Tricholoma matsutake*. J Microbiol 50(2):199–206

Keegan KP, Glass EM, Meyer F (2016) MG-RAST, a metagenomics service for analysis of microbial community structure and function. In: Microbial environmental genomics (MEG). Humana Press, New York, NY, pp 207–233

Kertesz MA, Thai M (2018) Compost bacteria and fungi that influence growth and development of *Agaricus bisporus* and other commercial mushrooms. Appl Microbiol Biotechnol 102:1639–1650

Kim MK, Math RK, Cho KM, Shin KJ, Kim JO, San Ryu J, Yun HD (2008) Effect of Pseudomonas sp. P7014 on the growth of edible mushroom *Pleurotus eryngii* in bottle culture for commercial production. Bioresour Technol 99(8):3306–3308

Kumari S, Naraian R (2021) Enhanced growth and yield of oyster mushroom by growth-promoting bacteria *Glutamicibacter arilaitensis* MRC119. J Basic Microbiol 61(1):45–54

Kumari D, Reddy MS, Upadhyay R (2013) Diversity of cultivable bacteria associated with fruiting bodies of wild Himalayan *Cantharellus* spp. Ann Microbiol 63:845–853

Kumla J, Suwannarach N, Sujarit K, Penkhrue W, Kakumyan P, Jatuwong K, Vadthanarat S, Lumyong S (2020) Cultivation of mushrooms and their lignocellulolytic enzyme production through the utilization of agro-industrial waste. Molecules 25:2811

Liu L, Zhu K, Wurzburger N, Zhang J (2020) Relationships between plant diversity and soil microbial diversity vary across taxonomic groups and spatial scales. Ecosphere 11(1):e02999

Liu D, Pérez-Moreno J, He X, Garibay-Orijel R, Yu F (2021) Truffle microbiome is driven by fruit body compartmentalization rather than soils conditioned by different host trees. Msphere 6(4): e00039–e00021

Lladó S, López-Mondéjar R, Baldrian P (2017) Forest soil bacteria: diversity, involvement in ecosystem processes, and response to global change. Microbiol Mol Biol Rev 81(2):e00063–e00016

Loos D, Zhang L, Beemelmanns C, Kurzai O, Panagiotou G (2021) DAnIEL: a user-friendly web server for fungal ITS amplicon sequencing data. Front Microbiol 2337

Lysak LV, Dobrovol'skaya TG, Skvortsova IN (2003) Metody otsenki bakterial'nogo raznoobraziya pochv i identifikatsii pochvennykh bakterii (Methods for assessing soil microbial diversity and identification of soil bacteria). MAKS-Press, Moscow

Magurran AE (1988) Ecological diversity and its measurement. Princeton University Press

Manoharachary C, Sridhar K, Singh R, Adholeya A, Suryanarayanan TS, Rawat S, Johri BN (2005) Fungal biodiversity: distribution, conservation and prospecting of fungi from India. Curr Sci 58–71

Miles PG, Chang ST (1997) Mushroom biology: concise basics and current developments. World Scientific Press, pp 161–175

Minot SS, Krumm N, Greenfield NB (2015) One codex: a sensitive and accurate data platform for genomic microbial identification. BioRxiv 027607

Mitchell AL, Almeida A, Beracochea M, Boland M, Burgin J, Cochrane G, Finn RD (2020) MGnify: the microbiome analysis resource in 2020. Nucleic Acids Res 48(D1):D570–D578

Mojzsis SJ, Arrhenius G, McKeegan KD, Harrison TM, Nutman AP, Friend CRL (1996) Evidence for life on Earth before 3,800 million years ago. Nature 384(6604):55–59

Morgan XC, Huttenhower C (2012) Chapter 12: human microbiome analysis. PLoS Comput Biol 8(12):e1002808

Niimi J, Deveau A, Splivallo R (2021) Aroma and bacterial communities dramatically change with storage of fresh white truffle *Tuber magnatum*. LWT 151:112125

Noble R, Fermor TR, Lincoln S, Dobrovin-Pennington A, Evered C, Mead A, Li R (2003) Primordia initiation of mushroom (*Agaricus bisporus*) strains on axenic casing materials. Mycologia 95:620–629

Noble R, Dobrovin-Pennington A, Hobbs PJ, Pederby J, Rodger A (2009) Volatile C8 compounds and pseudomonads influence primordium formation of *Agaricus bisporus*. Mycologia 101:583–591

Nocker A, Burr M, Camper AK (2007) Genotypic microbial community profiling: a critical technical review. Microb Ecol 54(2):276–289

Oh SY, Lim YW (2018) Root-associated bacteria influencing mycelial growth of *Tricholoma matsutake* (pine mushroom). J Microbiol 56:399–407

Olanrewaju OS, Glick BR, Babalola OO (2017) Mechanisms of action of plant growth promoting bacteria. World J Microbiol Biotechnol 33:197

Overmann J (2013) Principles of enrichment, isolation, cultivation, and preservation of prokaryotes. In: Dworkin M, Falkow S, Rosenberg E, Schleifer KH, Stackebrandt E (eds) The prokaryotes. Springer, New York, NY

Pandin C, Darsonval M, Mayeur C, Le Coq D, Aymerich S, Briandet R (2019) Biofilm formation and synthesis of antimicrobial compounds by the biocontrol agent *Bacillus velezensis* QST713 in an *Agaricus bisporus* compost micromodel. Appl Environ Microbiol 85(12):e00327–e00319

Pent M, Põldmaa K, Bahram M (2017) Bacterial communities in boreal forest mushrooms are shaped both by soil parameters and host identity. Front Microbiol 8:836

Philipson C, Davenport K, Voegtly L, Lo CC, Li PE, Xu Y, Chain PS (2017) Brief protocol for EDGE bioinformatics: analyzing microbial and metagenomic NGS data. Bio-protocol 7(23): e2622–e2622

Rainey PB (1991) Effect of Pseudomonas putida on hyphal growth of *Agaricus bisporus*. Mycol Res 95(6):699–704

Rangel-Castro JI, Levenfors JJ, Danell E (2002) Physiological and genetic characterization of fluorescent Pseudomonas associated with *Cantharellus cibarius*. Can J Microbiol 48:739–748

Romdhane IBB, Achouri IM, Belghith H (2010) Improvement of highly thermostable xylanases production by *Talaromyces thermophilus*

Schmartz PG, Hirsch P, Amand J, Dastbaz J, Fehlmann TKF, Müller R, Keller A (2022) BusyBee web: towards comprehensive and differential composition-based metagenomic binning. Nucleic Acids Res.gkac 298

Shabardina V, Kischka T, Manske F, Grundmann N, Frith MC, Suzuki Y, Makałowski W (2019) NanoPipe—a web server for nanopore MinION sequencing data analysis. GigaScience 8(2): giy169

Sharma SB, Sayyed RZ, Trivedi MH, Gobi TA (2013) Phosphate solubilizing microbes: sustainable approach for managing phosphorus deficiency in agricultural soils. Springer plus 2:587

Sillo F, Vergine M, Luvisi A, Calvo A, Petruzzelli G, Balestrini R, Vita F (2022) Bacterial communities in the fruiting bodies and background soils of the white truffle *Tuber magnatum*. Front Microbiol 13

Silva JP, Ticona AR, Hamann PR, Quirino BF, Noronha EF (2021) Deconstruction of lignin: from enzymes to microorganisms. Molecules 26(8):2299

Sjokvist E, Pfeil BE, Larsson E, Larsson KH (2014) Stereopsidales-a new order of mushroom-forming fungi. PLoS One 9(4):e95227

Spraker JE, Sanchez LM, Lowe TM, Dorrestein PC, Keller NP (2016) *Ralstonia solanacearum* lipopeptide induces chlamydospore development in fungi and facilitates bacterial entry into fungal tissues. ISME J 10(9):2317–2330

Suarez C, Ratering S, Weigel V, Sacharow J, Bienhaus J, Ebert J, Schnell S (2020) Isolation of bacteria at different points of *Pleurotus ostreatus* cultivation and their influence in mycelial growth. Microbiol Res 234:126393

Tatusov RL, Fedorova ND, Jackson JD, Jacobs AR, Kiryutin B, Koonin EV, Natale DA (2003) The COG database: an updated version includes eukaryotes. BMC Bioinformatics 4(1):1–14

Thompson LR, Sanders JG, McDonald D, Amir A, Ladau J, Locey KJ, Knight R (2017) A communal catalogue reveals Earth's multiscale microbial diversity. Nature 551(7681):457–463

Tian JH, Pourcher AM, Bouchez T, Gelhaye E, Peu P (2014) Occurrence of lignin degradation genotypes and phenotypes among prokaryotes. Appl Microbiol Biotechnol 98:9527–9544

Tsukamoto T, Murata H, Shirata A (2002) Identification of nonpseudomonad bacteria from fruit bodies of wild agaricales fungi that detoxify tolaasin produced by *Pseudomonas tolaasii*. Biosci Biotechnol Biochem 66:2201–2208

Uehling J, Gryganskyi A, Hameed K, Tschaplinski T, Misztal PK, Wu S, Desirò A, Vande Pol N, Du Z, Zienkiewicz A, Zienkiewicz K (2017) Comparative genomics of *Mortierella elongata* and its bacterial endosymbiont *Mycoavidus cysteinexigens*. Environ Microbiol 19(8): 2964–2983

Uroz S, Calvaruso C, Turpault MP, Pierrat JC, Mustin C, Frey-Klett P (2007) Effect of the mycorrhizosphere on the genotypic and metabolic diversity of the bacterial communities involved in mineral weathering in a forest soil. Appl Environ Microbiol 73(9):3019–3027

Uroz S, Oger P, Morin E, Frey-Klett P (2012) Distinct ectomycorrhizospheres share similar bacterial communities as revealed by pyrosequencing-based analysis of 16S rRNA genes. Appl Environ Microbiol 78(8):3020–3024

Vetrovsky T, Steffen KT, Baldrian P (2014) Potential of cometabolic transformation of polysaccharides and lignin in lignocellulose by soil Actinobacteria. PLoS One 9:e89108

Vieira FR, Pecchia JA (2021) Fungal community assembly during a high-temperature composting under different pasteurization regimes used to elaborate the *Agaricus bisporus* substrate. Fungal Biol 125(10):826–833

Vishal V, Munda SS, Singh G, Lal S (2021) Cataloguing the bacterial diversity in the active ectomycorrhizal zone of *Astraeus* from a dry deciduous forest of Shorea. Biodivers Data J 9: e63086

Vishal V, Singh G, Lal S (2022) Metagenomic analysis of active ectomycorrhizal zone of Astraeus using cloud computing approach from two seasons. Kavaka 58(2):73–81

Wagner BD, Grunwald GK, Zerbe GO, Mikulich-Gilbertson SK, Robertson CE, Zemanick ET, Harris JK (2018) On the use of diversity measures in longitudinal sequencing studies of microbial communities. Front Microbiol 9:1037

Walters KE, Martiny JB (2020) Alpha-, beta-, and gamma-diversity of bacteria varies across habitats. PLoS One 15(9):e0233872

Weber N, Liou D, Dommer J, MacMenamin P, Quiñones M, Misner I, Hurt DE (2018) Nephele: a cloud platform for simplified, standardized and reproducible microbiome data analysis. Bioinformatics 34(8):1411–1413

Whittaker RH (1972) Evolution and measurement of species diversity. Taxon 21(2/3):213–251

Willis AD (2019) Rarefaction, alpha diversity, and statistics. Front Microbiol 10:2407

Wu X, Spencer S, Gushgari-Doyle S, Yee MO, Voriskova J, Li Y, Chakraborty R (2020) Culturing of "Unculturable" subsurface microbes: natural organic carbon source fuels the growth of diverse and distinct bacteria from groundwater. Front Microbiol 3171

Yan Q, Wi YM, Thoendel MJ, Raval YS, Greenwood-Quaintance KE, Abdel MP, Patel R (2019) Evaluation of the CosmosID bioinformatics platform for prosthetic joint-associated sonicate fluid shotgun metagenomic data analysis. J Clin Microbiol 57(2):e01182–e01118

Yen S, Johnson JS (2021) Metagenomics: a path to understanding the gut microbiome. Mamm Genome 32(4):282–296

Yoon SH, Ha SM, Kwon S, Lim J, Kim Y, Seo H, Chun J (2017) Introducing EzBioCloud: a taxonomically united database of 16S rRNA gene sequences and whole-genome assemblies. Int J Syst Evol Microbiol 67(5):1613

Youens-Clark K, Bomhoff M, Ponsero AJ, Wood-Charlson EM, Lynch J, Choi I, Hurwitz BL (2019) iMicrobe: tools and data-driven discovery platform for the microbiome sciences. GigaScience 8(7):giz083

Young LS, Chu JN, Hameed A, Young CC (2013) Cultivable mushroom growth-promoting bacteria and their impact on *Agaricus blazei* productivity. Pesqui Agropecu Bras 48:636–644

Zagriadskaia YA, Lysak LV, Sidorova II, Aleksandrova AV, Voronina EY (2013) Bacterial complexes of the fruiting bodies and hyphosphere of certain basidiomycetes. Biol Bull 40(4): 358–364

Zarenejad F, Yakhchali B, Rasooli I (2012) Evaluation of indigenous potent mushroom growth promoting bacteria (MGPB) on *Agaricus bisporus* production. World J Microbiol Biotechnol 28(1):99–104

Zhang C, Zhang G, Wen Y, Li T, Gao Y, Qiu MF, Ai L (2019) *Pseudomonas* sp. UW4 acdS gene promotes primordium initiation and fruiting body development of *Agaricus bisporus*. World J Microbiol Biotechnol 35(11):1–10

Zhang XM, Tang DX, Li QQ, Wang YB, Xu ZH, Li WJ, Yu H (2021) Complex microbial communities inhabiting natural *Cordyceps militaris* and the habitat soil and their predicted functions. Anton Leeuw Int J G 114(4):465–477

Chapter 20
Extremophile Bacterial and Archaebacterial Population: Metagenomics and Novel Enzyme Reserve

Jayjit Majumdar, Debojyoti Moulik, S. C. Santra, and Akbar Hossain

Abstract Extreme environments, marine hydrothermal vents, hot springs, hypersaline conditions, deep marine environments, extreme cold, extreme hot, extreme low or high pH conditions, and radionuclide-enriched sites are the residence of several genera of a distinctive bacterial population. Polyextremophiles as well as a polyextremotolerant bacterial population have their unique niche. During the course of time and evolutionary advancements, they develop special physiology, morphology, metabolism, and several other biochemical pathways. Due to their unique physiology and biochemical machinery, these extremophiles have tremendous potential for future research and industrial application in severe environmental conditions (such as *Thermus aquaticus*, *Acidithiobacillus*, *Arthrobacter*, *Geobacillus*, and several other genera). Traditional bacteriological methods of isolation, culture, and the study of colony morphology, genomics, and proteomics analysis in a lab are abstruse or unachievable for extremophiles. Metagenomics and metaproteomics are proven and convenient tools to understand their genetic, physiological, and biochemical makeup and optimization for their potential applications. These sets of culture-independent methods not only avoid hectic time-consuming traditional techniques but also help direct the extraction of knowledge of the

J. Majumdar (✉)
Department of Ecological Studies, University of Kalyani, Nadia, West Bengal, India
e-mail: jayjitecostud21@klyuniv.ac.in

D. Moulik (✉)
Plant Stress Biology and Metabolomics Laboratory, Department of Life Science and Bioinformatics H.G. Khorana School of Life Sciences, Assam University, Silchar, India
e-mail: drubha31@gmail.com

S. C. Santra
Department of Environmental Science, University of Kalyani, Nadia, West Bengal, India
e-mail: scsantra@yahoo.com

A. Hossain
Bangladesh Wheat and Maize Research Institute, Dinajpur, Bangladesh
e-mail: akbar.hossain@bwmri.gov.bd

P. Mathur et al. (eds.), *Microbial Symbionts and Plant Health: Trends and Applications for Changing Climate*, Rhizosphere Biology,
https://doi.org/10.1007/978-981-99-0030-5_20

presence and functionality of microbes in their functional and realized niche. The application of screening, cloning, and expression of genomic information can lead to the discovery of novel enzymes and other proteins for future research and industrial applications.

Keywords Archaebacteria Bacteria · Extreme environment · Extremophile · Extremophile metagenomics · Novel enzymes

Abbreviations

PCR	Polymerase chain reaction
Taq DNA polymerase	A DNA polymerase enzyme derived from the species *Thermus aquaticus*
T/2	Thermostability/half-life

20.1 Introduction

Microorganisms thrive in a wide range of environmental conditions. They not only thrive in a broad range of earth environmental conditions but to some extent can also survive in extreme environments and physicochemical conditions such as space, high pressure, very low and very high pH, extreme salinity, high gravitational force, and numerous other extreme conditions. They have their unique sets of enzymes and metabolism modified according to their surrounding conditions. Microbes that can survive in extreme environmental conditions can be subdivided into two major groups; they are extremotolerant (organisms that can withstand harsh to severe physicochemical and environmental conditions) and extremophiles (those organisms are highly adapted to extreme conditions under which they can physiologically and biochemically operate). There are multitudinous extreme environments found on planet earth (especially subsurface earth), and extremophiles abode these loci. The majority of these extremophiles can thrive and perform in conglomerate hostile conditions to common mesophiles, thus commonly known as polyextremophiles. Extremophiles can be categorized according to the hostile environment they reside, grow, and prosper: thermophiles (growing at 45–80 °C) and hyperthermophiles (growing at >80 °C), psychrophiles (grows ≤10 °C), acidophiles (adapted to pH ≤ 5), alkanophiles (adapted to pH ≥ 9), halophiles (2–5 mM NaCl), radioresistant (grows in 40–400 nM wavelength), dehydration tolerant (water activity aW = 0.7), pressure (peizotolerant/barotolerant to hyperpiezophile or hyperbarophile (≥50 MPa), and polyextremophiles (Kristjánsson and Hreggvidsson 1995; Di Donato et al. 2019; Merino et al. 2019; Orellana et al. 2013, 2018). Some examples of extreme conditions and extremophiles are briefly summarized in Table 20.1. Most extremophiles are microorganisms and archaea contribute to this

Table 20.1 Extremophiles (Bacteria and Archea): Nomenclature, conditions, and examples

Parameter	Extremophiles	Ranges	Examples	References
pH	Acidophile	≤3	*Picrophilus oshimae and P. torridus, Acidianus infernus*	Feyhl-Buska et al. (2016), Schleper et al. (1995)
	Alkanophile	≥9	*Halomonas campisalis*	Bowman (2017)
Salinity	Extreme halophile	10–26%	*Halorhodospira halophila, Haloanaerobium lacusroseus, Halarsenatibacter silvermanii, Halomonas campisalis, Salinibacter ruber*	Orellana et al. (2018)
Temperature	Psychrophile/ cryophile	−15 °C–20 °C	*Devosia psychrophila, Robiginitomaculum antarcticum, Phaeobacter arcticus, Polaromonas glacialis, P. naphthalenivorans, Simplicispira psychrophila Pseudoalteromonas arctica, Oleispira ant-arctica, Colwellia psychrerythraea, Psychrobacter spp., Psychromonas ingrahamii, Shewanella benthica, Thiomicrospira arctica Cryobacterium arcticum, C. flavum, C. levicorallinum, C. levicorallinum, Demequina lutea, Glaciibacter superstes, Psychrobacter arcticus, Acidianus infernus*	Bergholz et al. (2009), Mykytczuk et al. (2013), Ingraham (1958)
	Thermophile	50–100 °C	*Sulfolobus acidocaldarius, Thermocrinis ruber Thermotoga, Thermosipho, Aquifex*	Brock et al. (1972), Huber et al. (1998), Stetter (2006)
	Hyperthermophile	>100 °C	*Aquifex pyrophilus Thermus aquaticus, Thermoanaerobacter, Geothermobacterium*	Kristjánsson and Hreggvidsson 1995, Merino et al. (2019),

(continued)

Table 20.1 (continued)

Parameter	Extremophiles	Ranges	Examples	References
			ferrireducens, Thermotoga maritima, Methanopyrus, Pyrodictium occultum, Pyrobaculum islandicum, Pyrococcus furiosus, Methanopyrus kandleri 116	Orellana et al. (2018), Stetter (2006), Zeikus (1979)
Water scarcity	Xerophile	0.6 water activity (a_w)	*Deinococcus geothermalis DSM 11300, Deinococcus hohokamensis, Halobacterium salinarum NRC-1*	Merino et al. (2019)
Pressure	Piezophile/ barophile	0.1–135 MPa	*Acidianus infernus, Thermococcus piezophilus, Halarsenatibacter silvermanii*	Merino et al. (2019), Orellana et al. (2018)

as a major group along with cyanobacteria. Microorganisms have declared their presence in a vivid range of extreme environments where water is available in a liquid state. The liquid form of water was found to be the main governing factor controlling the presence and abundance of the life forms on earth. Water is a prime parameter to change or control the surrounding physicochemical properties of the environment shaping the realized niche of the life forms and controlling species composition, diversity of microorganism's population, and biogeochemical cycling of the materials and energy (Kristjánsson and Hreggvidsson 1995; Merino et al. 2019; Orellana et al. 2018).

In the last few decades, understanding the life and physiology of extremophiles has been in focus as they provide linkages between nonliving environments toward the formation of life on this planet. The study of extremophiles has become a key area of astrobiology, and findings of their physiology, metabolism, biochemistry, and ecosystem grant the development of a hypothesis on the origin of life elsewhere in this universe and made researchers rethink the boundaries and fundamental features of life. They provide the "missing links" of the probability of life on other planets and celestial bodies across the universe. The past and present research work on the culture of extremophiles and several other culture-independent methods to identify them, monitor their physiology and understand their biochemistry and their interaction with the surrounding environment not only provides insight into the perimeters of life subsequently rewriting the theory of the origin of life and evolution of life forms (Rampelotto 2013; Merino et al. 2019).

Creating extreme environmental conditions in the laboratory is a hellacious, herculean task and indispensable to a good sum of finance. Where as metagenomics provides direct isolation of genetic material from environmental samples, subsequent PCR, 16 s rRNA phylogenetic analysis, and culture-independent identification of microbial diversity within a short time frame. This is a comparatively cost-effective assessment of microbial diversity. Identification of novel enzyme reserves and other functional biomolecules found in these extremophiles and polyextremophiles and subsequent molecular cloning of responsible genes to industrially culturable microbes, or undeviating laboratory production of these enzymes as well as biomolecules, have the potential to revolutionize biotechnological applications. Research in this field may have the potential for the discovery of enzymes such as *Taq* DNA polymerase and other extreme biomolecules which can restructure future basic and industrial application-oriented research and production.

20.2 Bacteria and Archaea in Extreme Environments

A giant leap forward took place in the research of extremophiles with the advent of molecular biology techniques. A major portion of bacterial and archaeal biodiversity is still unknown to mankind, and a huge number of these unknown species are living in extremes, sometimes surviving in a multitude of harsh ecological, physical, and chemical conditions. The phylogenetic lineage of bacterial and archaebacterial species with extremophilic properties belongs to a diverse taxonomic background (Fig. 20.1).

20.2.1 Bacteria and Archaea of Saline to Hypersaline Environment

Hypersaline conditions are found in landlocked hypersaline alkaline soda lakes, acidic hypersaline lakes, salterns, salt plans, and deep ocean brine pools. They are globally distributed and stretch according to their salt concentration, hydrogeochemical composition, physical factors, topography, and geological influence. Some examples of hypersaline lakes are the Dead Sea (Israel and Jordan), Great Salt Lake (USA), Lake Assal (Djibouti), Gaet'ale Pond (Danakil Depression, Ethiopia), Lake Vida (Victoria Valley, McMurdo dry valley of Antarctica), Don Juan Pond (Wright Valley, Victoria Land, Antarctica), hypersaline solar ponds, etc. Hypersaline environments in general have low oxygen concentration, low organic matter, high or low to very low temperatures according to geographical position, low nutrient availability, high solar radiation, and high barometric pressure and are sometimes very alkaline (Bachofen 1986; Ventosa et al. 2014, 2015; Naghoni et al. 2017; Edwardson and Hollibaugh 2018; Shurigin et al. 2019). High salinity

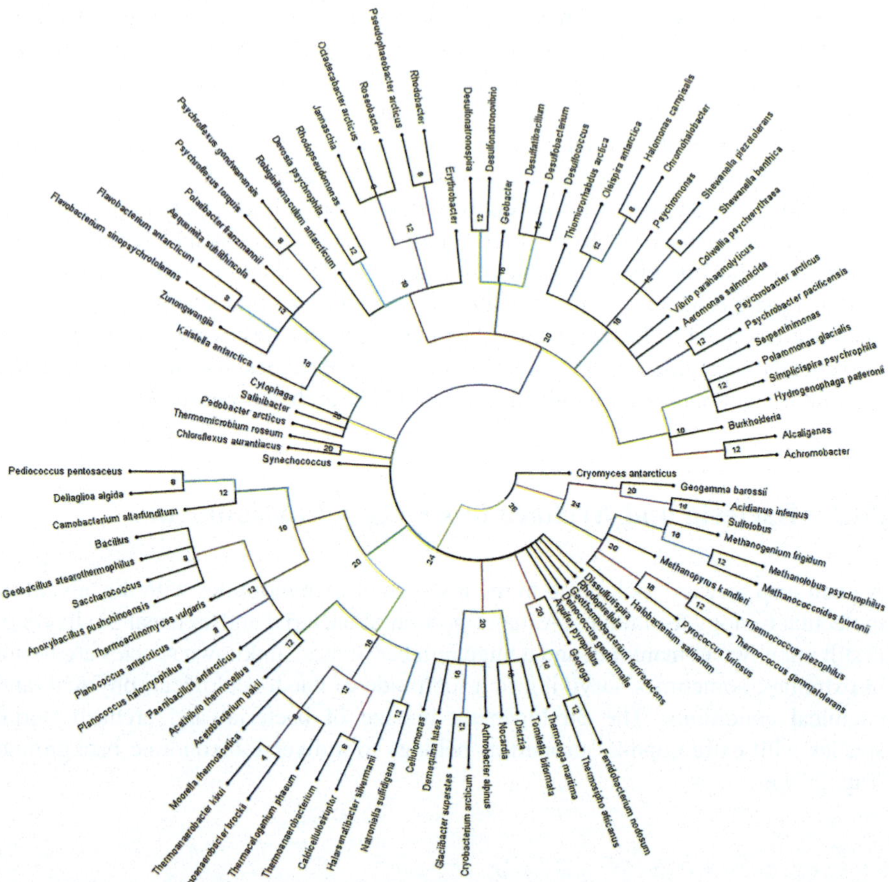

Fig. 20.1 Phylogenetic relationship among different species of extremophiles and polyextremophiles

is toxic for most organisms as it restricts the availability of water. Hence, hypersaline (salinity >10%) environments have low biological diversity. Halophily gradually developed and evolved over the ages. Although three domains of life forms have their halophilic members, Archaea dominates the population (especially members of the class Haloarchaea aka halobacteria and phylum Euryarchaeota). Bacteria are not far behind in terms of their halophily and abundance. Halophilic bacteria may have their territory in hypersaline environments from polar to tropical, the surface to the subsurface, submarine to terrestrial, and anaerobic to aerobic conditions (Fernandez et al. 2013; Javor 2011; Ventosa 2006; Ventosa et al. 2014, 2015). *Salinibacter ruber* is a rod-shaped, red-pigmented, motile, and extremely halophilic bacteria. It was isolated from the crystalized pond in Alicante and Mallorca (Spain). This species requires at least 15% salt and optimally grows at 20%–30% salt as halophilic as archaea belonging to the family Halobacteriaceae (Antón et al. 2002; Ventosa 2006). Most of the halophilic bacterial species belong to the phylum *Pseudomonadota* aka *Proteobacteria* although several other phyla like *Bacillota* aka *Firmicutes*, *Spirochaetota* aka Spirochaetes, *Actinobacteria* aka *Actinomycetes*, etc. have their representatives in this genre (Pandit et al. 2015). Species like *Halorhodospira halophila* (halophilic phototrophic) and *Haloanaerobium lacusroseus* (anaerobic) can grow optimally in saturated salt conditions (Cayol et al. 1995; Antón et al. 2002). Several other genera like *Rhodothermus* (Class: *Rhodothermia*), *Desulfohalobium* (Class: *Desulfovibrionia*), *Gemmatimonas* (Class: *Gemmatimonadetes*), *Marinobacter*, *Alkalilimnicola*, *Alkalispirillum*, *Thioalkalivibrio*, *Alkalimonas*, *Halomonas*, *Vibrio parahaemolyticus*, and *Chromohalobacter* (Class: *Gammaproteobacteria*); *Roseobacter*, *Rhodobacter*, *Rhodopseudomonas*, *Jannaschia* (Class: *Alphaproteobacteria*), *Burkholderia* (Class: *Betaproteobacteria*), *Geobacter*, and *Desulfobacterium* (Class: *Deltaproteobacteria*); *Bacillus* (Class: *Bacilli*), *Rhodopirellula* (Class: *Planctomycetia*), *Zunongwangia* (Class: *Flavobacteriia*), *Desulfatibacillum*, *Desulfobacterium*, and *Desulfococcus* (Class: *Desulfobacteria*); and *Desulfonatronospira*, *Desulfonatronovibrio* (Class: *Desulfovibrionia*), *Natroniellasulfidigena* (Class: *Halanaerobiia*), *Dietzia*, and *Cellulomonas* (Class: *Actinomycetia*), and several other genera have been reported to grow and prosper in saline to hypersaline environment and from marine to salt pan to saltern ecosystem (Pandit et al. 2015; Narayan et al. 2018). It is very difficult to cultivate extreme halophilic bacteria in laboratory conditions or on large scale. Maintenance of hypersaline conditions is very expensive as it needs frequent maintenance of culture system, low energy culture ecosystem, and types of equipment and nutrient/ion composition supplement than that of mesophiles. Many extreme halophiles are resistant to the conventional culturing approach. Maintenance of pure culture of the target organism is often difficult as some of the associated microbes from the same sample are polyextremophiles and they can resist conventional heat and pressure sterilization methods. It is also very difficult to isolate particular DNA/RNA fragments and manipulation of plasmid or nuclear DNA such as in the case of *Halomonas* and several other species (Ventosa et al. 1998; Madigan 2000; Hedlund et al. 2015; Zhu et al. 2020).

20.2.2 Thermophilic Bacterial Diversity and Enzymatic Potential

Thermophiles or thermophilic bacteria are those bacterial species that chose to grow in an environment that is high in temperature. The thermophilic bacterial life goes back to the period of origin of life on planet earth, which is approximately 4 billion years ago. They can be further classified according to their growth in optimum temperature ranges, like thermophiles (50–64 °C), extreme thermophiles (65–100 °

C), and hyperthermophiles (>100 °C). The history of thermophiles goes back to the 1800s when Louis Pasteur discovered the method of disinfection by heat. He suggested that most bacterial pathogens can be killed by a temperature over 80 °C; this process is later termed pasteurization. The early notable work on thermophiles goes back to the first quarter of the last century. In his article on thermophiles, David Hendricks Bergey (Bergey 1919) suggested that normal pathogenic bacteria can thrive up to the temperature of 45 °C, whereas true thermophiles merely grow in the temperature range of 35–45 °C. They develop at a temperature above 50 °C, some of them develop at a temperature beyond 80 °C, and most of them are abundant in the temperature ranges of 65–75 °C. In his experiments, Bergey isolated two strains which have similarities with *Nocardia*. Their growth was stopped at 70 °C and both were killed at 100 °C. After this pioneering work on thermophiles, a new field in understanding biology, survival, and physiology and the application of these thermophiles shoot off. Thermophiles have been studied and isolated from different environments such as hot springs, acidic and alkaline hot ponds/lakes, salterns, solar disinfection ponds, deep ocean hydrothermal vents, rocky surfaces, volcanoes, volcanic islands, volcanic lakes, and even from surfaces of solar panels, and other varied environments. Thermophiles range from anaerobic to aerobic with most of them belonging to chemolithoautotrophs using CO_2, $Fe(OH)_2$, SO^-, SO_4^{-2}, NO_3^-, and O_2 as electron acceptors and pyrite and H_2 as electron donors. Their metabolic products range from methane, magnetite, acetate, H_2S, NH_3, and H_2O with H_2 as the electron donor to H_2SO_4 and $FeSO_4$ with pyrite as the electron donor (Brock et al. 1972; Huber et al. 1998; Stetter 2006; Basen and Müller 2017). Few bacterial photoautotrophs have been reported to thrive and prosper in hyperthermophilic conditions, such as *Chloroflexus aurantiacus* and *Synechococcus lividus* (*Cyanobacteria*) along with some chemoheterotrophic species, such as *Bacillus stearothermophilus*, *B. acidocaldarius*, *B. caldotenax*, *Clostridium thermocellum*, *C. thermohydrosulphuricum*, *Thermoanaerobacter brockii*, *Desulfovibrio thermophiles*, *Thermomicrobium roseum*, *Thermus aquaticus*, and *Thermoactinomyces vulgaris* (Zeikus 1979; Basen and Müller 2017). Thermophilic and hyperthermophilic bacteria belong to many different taxonomic groups such as *Bacillus caldotenax*, *B. pallidus*, *B. thermophilus*, *Geobacillus thermodenitrificans*, *G. kaustophilus*, *G. stearothermophilus*, *G. thermoleovorans*, *G. thermocatenulatus*, *G. thermoglucosidasius*, *Saccharococcus thermophiles*, *Thermoactinomyces vulgaris* (Class: *Bacilli*), *Moorella thermoacetica*, *Thermoanaerobacter kivui*, *Thermacetogenium phaeum*, *Clostridium thermocellum*, *C. thermohydrosulphuricum*, *Thermoanaerobacter brockii* (Class: *Clostridia*), *Chloroflexus aurantiacus* (Class: *Chloroflexia*), *Desulfovibrio thermophiles* (Class: *Desulfovibrionia*), *Thermomicrobium roseum* (Class: *Thermomicrobia*), *Thermus aquaticus* (Class: *Deinococci*), *Thermotoga maritima*, *T. neapolitana*, *T. thermarum*, *T. elfii*, *Thermosipho africanus*, *Fervidobacterium nodosum*, *F. islandicum*, *F. pennovorans*, *Geotoga petrea*, *G. subterranean*, *Petrotoga miotherma* (Class: *Thermotogae*), *Aquifex pyrophilus*, *A. profundus*, (Class: *Aquificia*), *Flavobacterium autothermophilum* (Class: *Flavobacteriia*), *Pseudomonas palleronii*, *P. hydrogenothermophila*, *P. ruhlandii* (Class:

Gammaproteobacteria), *Achromobacter*, *Alcaligenes* (Class: *Betaproteobacteria*), *Cytophaga* (Class: *Cytophagia*), *Thermus aquaticus*, *T. thermophilus* (Class: *Deinococci*), *Nocardia* sp. (Class: *Actinomycetia*), *Dissulfurispira thermophile* (Class: *Thermodesulfovibrionia*), and several other groups (Goto et al. 1978; Maugeri et al. 2001; Stetter 2006; Basen and Müller 2017; Bala and Singh 2019; Umezawa et al. 2021). The key to survival at high temperatures for thermophiles relies on their stabilization of biomolecules at high temperatures. This is achieved through thermostabilizing cofactors in enzymes and by the presence of polyamines like thermine and thermospermine. These thermophiles also have straight and more saturated fatty acids in their membrane lipids which allow a correct degree of fluidity in the membranes, and they can also alter the fatty acids in their membranes (Oshima 1975; Zeikus 1979). *Thermus aquaticus* was isolated from the Mushroom Spring (Yellowstone National Park, USA) by Thomas D. Brock and Hudson Freeze, revolutionizing the field of molecular biology as it is the source of thermostable DNA polymerase enzyme named Taq DNA polymerase, one of the most important enzymes in molecular biology to perform polymerase chain reaction to amplify DNA sequences (Brock and Freeze 1969). Thermophiles can also act as polyextremophiles which can thrive and prosper in highly alkaline and acidic conditions. Several other thermostable enzymes have been isolated and commercially utilized as they are stable in engineered, industrial, and mass-scale applications. Enzymes such as cellulase, ligninase, and other cellulolytic and xylanolytic enzymes have the potential for the production of biofuel from long- and branched-chain carbohydrates. Isolates such as *Clostridium thermocellum* strain LQRI and strain 39E, *Geobacillus thermoglucosidasius* TM242, *Thermoanaerobacter saccharolyticum* (strains HK07, AK17, TD1), *Thermoanaerobacterium saccharolyticum* (strain M0355, M1051), *Thermoanaerobacter mathranii* strains BG1L1 and DIB 097X, *Caldicellulosiruptor* DIB 087C, *Thermoanaerobacter* DIB 097X, and *Pediococcus pentosaceus* can produce ethanol from varieties of substrates such as glucose, cellobiose, cellulose, xylose, glycerol, paddy straw, wheat straw, grass, rapeseed straw, Avicel®, sorghum stover, and different combination actions of these substrates even from Whatman paper with operational temperature ranging from 60 to 80 °C (Ng et al. 1981; Sudha Rani et al. 1997; Cripps et al. 2009; Yao and Mikkelsen 2010; Argyros et al. 2011; Almarsdottir et al. 2012; Svetlitchnyi et al. 2013; Tomás et al. 2013; Biswas et al. 2014; Bala and Singh 2019). Bio-butanol can be produced from substrates cellulose, Avicel cellulose, and xylose at a temperature of 60–70 °C using strains like *Thermoanaerobacterium thermosaccharolyticum* W/SL-YS485, *Clostridium saccharoperbutylacetonicum*, and *Clostridium thermocellum* (Nakayama et al. 2011; Bhandiwad et al. 2014). A comparative newer fuel biohydrogen can be produced by bacterial/archaeal species such as *Pyrococcus furiosus*, *Caldicellulosiruptor saccharolyticus*, *Thermotoga neapolitana*, *Thermotoga maritima*, and *Thermoanaerobacterium thermosaccharolyticum* from substrates such as glucose, maltose, xylose, starch, cellulose, carrot pulp hydrolyzate, and different combinations of those substrates at a temperature as high as 70–98 °C (Schicho et al. 1993; Kádár et al. 2004; Nguyen et al. 2008; O-Thong et al. 2008; De Vrije et al. 2010).

20.2.3 Psychrophilic Bacterial Diversity and Enzymatic Potential

Bacteria and archaeal populations in a cold environment open up a portal for enzymatic activity in low temperature and their application in agriculture in cold environments and industries such as cold fermentation. Microbes growing in low temperatures are also sometimes called cryophiles or rhigophiles. Psychrophilic bacterial and archaeal populations dwell in the Antarctic, arctic soil, glacier, polar ice caps, cold deserts, cold marine, frozen lakes, lake and marine sediments, frozen food and beverages, and several other environments. Some of the extreme psychrophiles are unable to grow at a temperature above 20 °C. Some of them have a comparatively broad range of temperatures to grow such as *Planococcus halocryophilus* (−15 °C > °C) and *Psychrobacter arcticus* (<−5 °C to ~30 °C) (Bergholz et al. 2009; Mykytczuk et al. 2013). Psychrophiles experience a variety of stumbling blocks such as freezing, extreme pH, oxidative stress (O_2 supersaturation), reduced growth rates, salinity, water availability, nutrient availability, osmolytes, and low molecular weight compounds (Bowman 2001; Roberts 2005). Psychrophiles represent different taxonomic groups. Examples of such psychrophilic bacteria are as follows: *Devosia psychrophila*, *Erythrobacter* sp., *Robiginitomaculum antarcticum*, *Phaeobacter arcticus* (Class: *Alphaproteobacteria*), *Polaromonas glacialis*, *P. naphthalenivorans*, *Simplicispira psychrophila* (Class: *Betaproteobacteria*), *Pseudoalteromonas arctica*, *Oleispira antarctica*, *Colwellia psychrerythraea*, *Psychrobacter* spp., *Psychromonas ingrahamii*, *Shewanella benthica*, *Thiomicrospira arctica* (class: *Gammaproteobacteria*), *Arthrobacter alpinus*, *Cryobacterium arcticum*, *C. flavum*, *C. levicorallinum*, *C. levicorallinum*, *Demequina lutea*, *Glaciibacter superstes*, *Tomitella biformata* (Class: *Actinomycetia*), *Bacillus cecembensis*, *Bacillus psychrosaccharolyticus*, *Lactobacillus algidus*, *Paenibacillus antarcticus*, *Planococcus antarcticus*, *Planococcus halocryophilus*, *Carnobacterium alterfunditum*, *Acetobacterium bakii* (Class: *Bacilli*), *Aequorivita sublithincola*, *Chryseobacterium antarcticum*, *Flavobacterium antarcticum*, *Flavobacterium sinopsychrotolerans*, *Polaribacter franzmannii*, *Psychroflexus gondwanensis*, *Pedobacter arcticus* (Class: *Flavobacteriia*), and representatives of several other different taxonomic groups. Archaea such as *Methanococcoides burtonii*, *Methanogenium frigidum*, and *Methanolobus psychrophilus* (Class: Methanomicrobia) have been found in sediments and ice in the polar region (Bakermans et al. 2003; Junge et al. 2003; Breezee et al. 2004; Bergholz et al. 2009; Mykytczuk et al. 2013; Bowman 2017). Psychrophiles have several different strategies to cope with their environment. Changes in membrane fluidity by altering fatty acid composition in membrane lipids is one of them. This is achieved through desaturation of membrane fatty acids or by changing branched-chain fatty acids. Another mechanism involves the production and release of antifreeze proteins (AFP) of DUF3494 domain in the cytoplasm. This does not allow water to freeze inside the cells and helps to continue the metabolic activity (Bowman 2017). From the studies on species like *Psychroflexus torquis*, *Octadecabacter arcticus*, *O. antarcticus*, and *Aeromonas salmonicida* (fish pathogen), the defense mechanism against cold was found to be the production of antifreeze proteins and cold shock proteins, production of polyunsaturated fatty acids to maintain membrane fluidity, and production of cryoprotectant extracellular polysaccharides (EPS). The presence of xanthorhodopsin was also found to be present in these cryophiles to vitalize the membrane transport chain. Production of storage polymers, like cyanophycin (commonly present in algae) to obtain nitrogen from poly-L-aspartic acid to replenish their requirement of nitrogen found in psychrophiles. In most cases, the major osmolyte present is glycine betaine to maintain a liquid environment within the cell to perform necessary metabolical and physiological functions within the cells (Vollmers et al. 2013; Feng et al. 2014; Vincent et al. 2016; Bowman 2017).

20.2.4 Polyextremophiles and Other Realms of Extreme Conditions

Most extremophiles act as polyextremophiles as their acquired defense mechanism acts as a protectant from other environmental harsh conditions. Microbes residing in saline to hypersaline, cold, and hot environments develop a mechanism to thrive in physiological water stress conditions by the production of EPS and regulation of membrane transport along with solutes in the cytoplasm. Cold, hot, and hypersaline environments are often associated with extreme pH conditions (<3 and >9) and low availability of oxygen. Such archaea as *Picrophilus oshimae* and *P. torridus* that optimally grow at pH 0.7 and are known to thrive at pH 0.6 were isolated from hot springs (Noboribetsu, Hokkaido, Japan). These heterotrophic polyextremophiles optimally grow in temperatures up to 65 °C with optimal growing temperatures of 45–60 °C (Schleper et al. 1995). Their response to pH, temperature, and nutrient stress was suggested to be achieved by the cyclization of tetraether membrane lipids (Feyhl-Buska et al. 2016). Several other polyextremophiles have been discovered, such as *Halomonas campisalis* (temperature 4–50 °C opt temp. 30 °C, salinity 1–26.3%, and optimum pH 6–12, opt 9.5), *Acidianus infernus* (temperature 65–96 °C opt temp. 90 °C, pressure 40–80 Mpa with opt. pressure 60 Mpa, and pH 1–5.5), *Serpentinomonas* sp. B1 (pH optimum 11 with pH max of 12.5), haloalkaliphilic *Halomonas campisalis* (pH range 6–12 and salinity up to 26.3%), radiotolerant *Thermococcus gammatolerans* (temperature 55–95 °C opt temp. 88 °C, salinity 20%, and optimum pH 5.5–6.6), *Pedobacter arcticus* (temperature 4–25 °C opt temp. 18 °C and optimum pH 6–9), *Methanopyrus kandleri 116* (temperature 90–122 °C opt temp. 106 °C, salinity 0.5–4.5% optimum salinity 3.0, and pH 6.3–6.6), *Halarsenatibacter silvermanii* SLAS-1 (temperature 28–55 °C, opt, temp. 44 °C, salinity 20–35% optimum salinity 30%, pressure 0.1–135 MPa, optimum pressure 50 Mpa, and pH 6.3–6.6), and several other genera and species, viz., *Halarsenatibacter silvermanii*, *Sulfolobus*, *Salinibacter*, *Anoxybacillus pushchinensis*, *Geothermobacterium ferrireducens*, *Shewanella piezotolerans*, *Deinococcus geothermalis*, *Halobacterium salinarum*, *Cryomyces antarcticus*, *Colwellia* sp., *Geogemma barossii*, *Methanopyrus kandleri*, *Thermococcus piezophilus*, *Psychrobacter pacificensis*, etc. can tolerate different extreme environmental conditions concomitantly (Segerer et al. 1986; Schleper et al. 1995; Mormile et al. 1999; Oremland et al. 2005; Aston and Peyton 2007; Mesbah and Wiegel 2008; Takai et al. 2008; Orellana et al. 2013; Dalmasso et al. 2016; Bowman 2017; Singh et al. 2018; Merino et al. 2019; Orellana et al. 2018). It is challenging to create laboratory conditions to simulate polyextremes, and maintaining cultures and conditions is very burdensome on time, money, and physical labor, so researchers nowadays utilize the full potential of molecular biology and bioinformatics to understand overall biochemistry, genetics, and physiology of these polyextremophiles.

20.3 Enzymatic Potential of Extremophiles

The industrial application potential of enzymes derived from extremophiles has increased many folds in the last few decades as these enzymes can thrive in harsh, far-reaching operational conditions such as pH, temperature, salinity, pressure, etc., which are not suitable for mesophilic enzymes. Several groups of enzymes such as catalase, esterase, lipase, glycosidase, phosphatase, aldolase, nitrilase, amidase, etc. are being used in a plethora of industries such as bioremediation, bleaching, dairy, detergent, food, ore processing, polymer, etc. (Nikolaivits et al. 2017; Cabrera and Blamey 2018; Atalah et al. 2019; Di Donato et al. 2019). Novel enzymes especially extracellular enzymes produced by extremophiles differ in their amino acid composition, orientation, and 3D structure which provide their stability on challenging operational conditions as they can operate at very low to very high pH (enzymes from acidophiles and alkanophiles) and very low to very high temperatures (psychrophiles and thermophiles). The discovery of thermostable DNA polymerase (Taq polymerase) and the invention of PCR (Mullis 1990) revolutionized the path of biotechnology and molecular cloning. This eventually raised the research and findings of extreme enzymes from different environments and different organisms in the last few decades. A majority of these organisms are polyextremophilic, and they can produce different groups of enzymes simultaneously to utilize the deficit amount of substrates in extreme conditions. For example, Choure et al. (2021) have isolated different species of bacilli including *Bacillus licheniformis CAP1*, *Bacillus licheniformis CAP3*, *Alkalihalobacillus clausii CAP7*, *Bacillus subtilis BAC18*, *Alkalihalobacillus clausii BAC23*, *Bacillus haynesii BAC26*, and *Bacillus subtilis BAC28* from two different hot springs situated in Madhya Pradesh (India). These species can produce a decent amount of thermostable enzymes such as amylase (51.85–61.26 IU ml^{-1}), lipase (33.09–47.36 IU ml^{-1}), cellulase (38.20–60.11 IU ml^{-1}), protease (31.15–62.14 IU ml^{-1}), lecithinase (27.15–37.39 IU ml^{-1}), and gelatinase (24.18–51.23 IU ml^{-1}) in growth condition favorable for large-scale industrial production such pH 8.5–9.5 and temperature 55 °C–65 °C. Countless extreme enzymes are discovered in the last few decades and isolated from different genres of extremophiles or their genetic code cloned in favorable organisms for lab scale or industrial production (Cabrera and Blamey 2018; Demirjian et al. 2001; Di Donato et al. 2019; Kim et al. 2000; Kour et al. 2019; Orellana et al. 2018; Singh and Ray 2021). They have tremendous research and industrial potential, some of which are summarized in Table 20.2.

20.4 Metagenomics of Extremophiles

Cultivable bacteria represent only a small fraction of the population in environmental samples. This creates an enormous potential for enhanced methods for the cultivation of the so-called "uncultivable" portion of diversity. Most of the time, the traditional culture by Petri dish method only represents a minor fraction of the maximum probable number leading to "the great plate count anomaly." A major problem of an extreme environmental sample is low population density along with their low biomass. Apart from these, cloning of functional genes of novel enzymes from these extremophiles is very difficult to achieve, and the scarcity of vector and host organisms and expression of these novel genes are also limited (Amann et al. 1995; Mirete et al. 2016; Staley 1985; Vester et al. 2015). In natural conditions, bacteria live as a consortium rather than isolated. Individual isolate merely gives genetic as well as metabolic representation of their potential as a naturally associated population (Allen and Banfield 2005). Metagenomics is a culture-independent method for identification of the biodiversity from environmental samples either by genome sequencing or expression. Genome sequencing may be based on the target genome sequencing (such as 16srRNA) or the shotgun method of sequencing DNA/RNA fragments and subsequent reconstruction by identifying overlapping sequences (Schloss and Handelsman 2005; Sahm et al. 2013; Mirete et al. 2016). The advent of advanced molecular biological techniques, such as isolation of environmental DNA samples, amplification of 16 s rRNA, and screening of functional genes bring into being culture-independent identification (metagenomics) of those bacteria that were undiscovered before in less time and cost. The overall process of metagenomics analysis of environmental samples is summarized in Fig. 20.2. Because of the comparatively less quantity of genomes, and highly branched taxonomy, extremophiles were the first organism whose genome was sequenced. This initiated a new dawn of molecular microbiology, environmental metagenomics, and even single-cell genomics, and extremophiles became the dark matter of microbiology (Fraser et al. 2000; Hedlund et al. 2014). In addition, these culture-independent methods reveal the biochemistry and physiology of those organisms and identification of functional enzymes responsible for their survival in extreme environmental conditions, and application of recombinant DNA technology to utilize their potential on large-scale culturable bacterial species or for industrial applications. Analysis of genome sequence obtained from environmental

Table 20.2 Enzymes from extremophiles and their potential application

Extremophiles	Organism	Enzyme	Condition	Application	References
Thermophile	*Bacillus thermocatenulatus*	Lipase	pH 8–9 (max pH 11); temp. 60°–70°	Detergent manufacturing	Demirjian et al. (2001)
	Bacillus stearothermophilus	Lipase	Temp. 55–68 °C; pH 6–8.8; T/2 30 min at 68 °C	Olive oil hydrolysis	Kim et al. (2000)
	Thermus aquaticus	DNA polymerase	Temp >85°	Polymerase chain reaction, genetic engineering, biotechnology	Miyazaki (2022)
	Thermococcus profundus	α-Amylase	Temp. 80–90 °C; pH 5.5–6,T/2 4 h at 90 °C and T/2 6.7 h at 80 °C	Food, medicine, biotechnology	Kwak et al. (1998)
	Thermotoga neapolitana	Xylanase (EC 3.2.1.8)	Temp. 102 °C; pH 5.5–6; T/2 2 h at 100 °C	Paper pulp industries, biofuel	Zverlov et al. (1996)
	Pyrodictium abyssi	Xylanases (EC 3.2.1.8)	Temp. 110 °C; pH 5.5; T/2 1.2 at 105 °C	Paper pulp industries, biofuel	Andrade et al. (2001)
	Pyrococcus sp. M24D13	Nitrilase (EC 3.5.5.1)	Temp. 85 °C; pH 7.5; T/2 8 h at 85 °C	Drug, pesticide	Dennett and Blamey (2016)
	Bacillus sp. PC-3	*Laccase (EC 1.3.10.2)*	Temp. 60 °C; pH 7; T/2 3.5 h at 60 °C	Textile, polymer, beverage, bioremediation	Sharma et al. (2019)
Psychrophile	*Pseudoalteromonas sp.*	α-Amylase	Temp. 20 °C; pH 8.0; T/2 10 at 40 °C	Food industry	Sanchez et al. (2019)
	Catenovulum sp. X3	Alkaline α-amylase	Temp. 25 °C; pH 9; T/2 1 h at 20–50 °C	Biohydrogen	Wu et al. (2017)
Acidophile	*Sulfolobus solfataricus P2*	Acidic protease	Temp. 70 °C pH 2.0	Medicine, biotechnology	Gogliettino et al. (2014)
Alkaliphiles	*Thermococcus kodakaraensis*	Serine proteases	Temp. 100 °C; pH 7–11.5; T/2 3.5 h at 60 °C	Medicine, biotechnology	Foophow et al. (2010)
Saline/hypersaline	*Bacillus subtilis subsp. subtilis A-53*	Carboxymethylcellulase	Temp. 50 °C; pH 6.5; T/2 24 h at 50 °C	Bioethanol production	Kim et al. (2009)

	Bacillus safensis	Catalase/oxidoreductase	Temp. 4–28 °C	Petroleum biodegradation	Da Fonseca et al. (2015)
Barophiles	*Nocardioides sp. strain KP7*	*Dioxygenase*	Temp. 45 °C; pressure 45 kDa	Biodegradation	Iwabuchi and Harayama (1998)
	Pseudomonas stutzeri DEH130	Dehalogenase	Temp. 40 °C; pH 7–10; pressure 109.9 kDa	Bioremediation; aromatic biodegradation	Zhang et al. (2013)

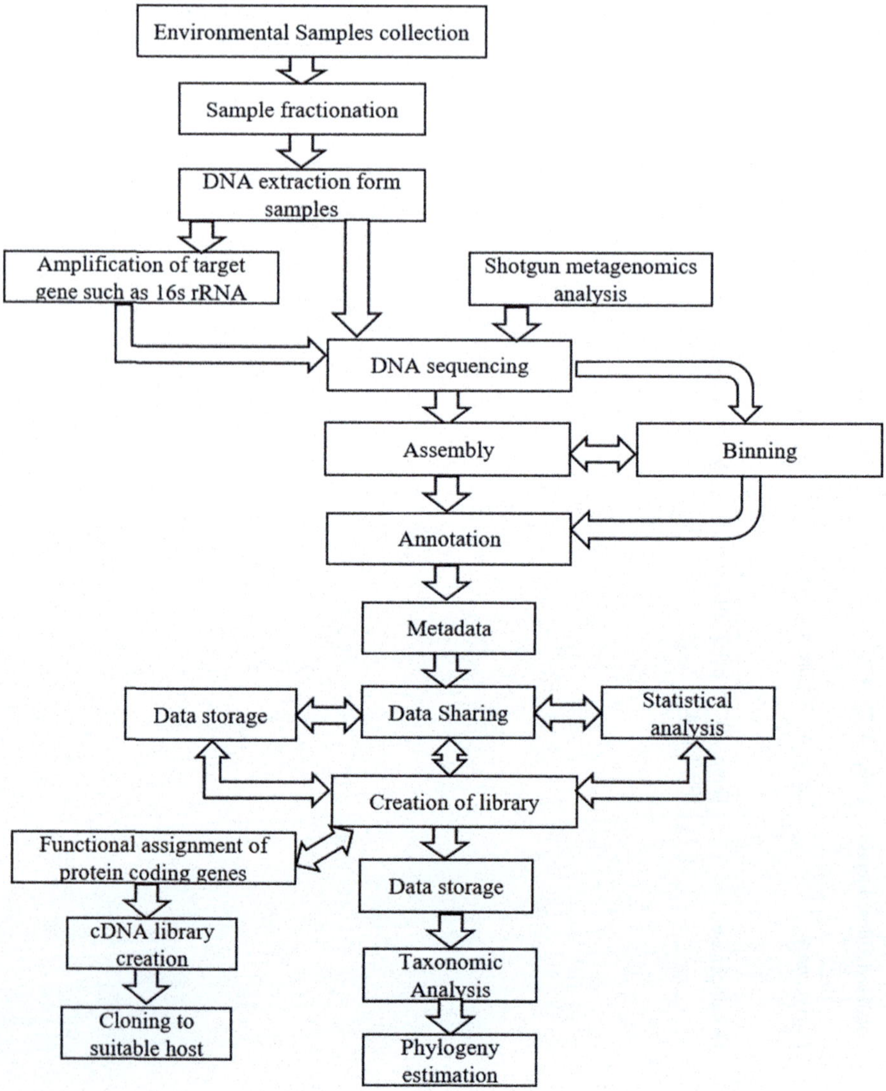

Fig. 20.2 Summarized method of metagenomics analysis of environmental samples

samples is impelled by the creation of a gene sequence library/inventory and search for new enzymes, other proteins, and other distinctive metabolites. Metagenomics provides a plan of action to understand the community ecology, evolutionary process, symbiosis, and biogeochemistry of the target environment. Extraction of high-quality DNA from environmental samples is the key to any sequencing analysis. Large inserts (~40 kb) provide better proximate analysis of genetic information of community members, but it needs Augean effort to obtain high molecular weight

DNA from environmental samples. Most of the cases, researchers depend upon shotgun sequences (3–4 Kb) for further statistical analysis as the DNA extraction process from environmental samples may nick longer fragments (Allen and Banfield 2005).

20.5 Limitations of Metagenomics

In community genomic study, the aim is to assign particular genome fragments to certain individuals. In environmental samples, each fragment or clone represents one particular individual which may be related to a gene pool of different related species. Metagenomics study generates huge databases of sequences which are burdensome to keep track of. Sequence database represents a mixed bag of gene fragments of different quantities and abundance so the data becomes very much troublesome. However, because the resulting sequence data represent a mixture of genomic fragments from different organisms at varying abundances, analysis of data at the organism level possesses lacunae; thus, phylogenetic context becomes quite inconsequential. This type of problem especially arises for organisms where data on gene fragments below ribosomal DNA is scarce. So researchers need to depend on codon usage, G+C content, repetitive segments, frequency, and homology using different statistical and bioinformatics techniques (Dick et al. 2009; Hedlund et al. 2014). Single-cell genomics involves whole genome mapping to help in better understanding the phylogeny, genomics, and proteomics of the particular organism. Single-cell genomics also has its restraints as among of DNA is low and it needs enormity in amplification, clean sorting of cells, and handling of samples, equipment, and operations need much more precision (Woyke et al. 2011; Stepanauskas 2012). Metagenomics and single-cell genomics rely on sequence similarities with an existing genomic database, but a majority of bacterial and archaeal species do not have sequenced representatives, in spite of a huge number of sequences of these microbes getting deposited in the databases on a single day (Baker and Dick 2013; Rinke et al. 2013).

20.6 Conclusion

Extremophiles especially polyextremophiles possess missing links to the origin of life on this planet. They have their unique physiology, metabolism, enzyme reserve, and biochemical properties to cope with their surrounding environmental conditions. By struggling to thrive and prosper in such environmental conditions, they acquire a unique set of the genome and produces outlandish sets of proteins and other biomolecules. These features not only can be used in developing new techniques in the laboratory but also have huge potential in industrial and commercial applications. But maintenance of the extremophiles in laboratory cultures is tedious. With the advent of advanced molecular biological technique culture, independent methods have been developed, and this initiated the boom for research on extremophiles. Although metagenomics and single-cell genomics have been in full progress in the last few decades, still it is a tiresome job as it requires an enormous amount of time, automated or manual assembly of DNA or protein sequences, and the creation and maintenance of huge databases. Worldwide databases on genome and protein sequences are getting enriched every day, and their repositories are getting huge amounts of new datasets, new bioinformatics tools, and applications of artificial intelligence luring researchers from all over the world to proceed in these fields. Metagenomics, single-cell genomics, and metaproteomics are not only bound to research laboratories now, but their impacts are also being seen in biotechnological, industrial, and commercial applications. This is still a growing field and in the future, it will enrich other fields of biological, clinical, and environmental research.

References

Allen EE, Banfield JF (2005) Community genomics in microbial ecology and evolution. Nat Rev Microbiol 3(6):489–498. https://doi.org/10.1038/nrmicro1157

Almarsdottir AR, Sigurbjornsdottir MA, Orlygsson J (2012) Effect of various factors on ethanol yields from lignocellulosic biomass by *Thermoanaerobacterium* AK 17. Biotechnol Bioeng 109 (3):686–694. https://doi.org/10.1002/bit.24346

Amann RI, Ludwig W, Schleifer KH (1995) Phylogenetic identification and in situ detection of individual microbial cells without cultivation. Microbiol Rev 59(1):143–169. https://doi.org/10.1128/mr.59.1.143-169.1995

Andrade CMMC, Aguiar WB, Antranikian G (2001) Physiological aspects involved in production of xylanolytic enzymes by deep-sea hyperthermophilic archaeon *Pyrodictium abyssi*. Appl Biochem Biotechnol – Part A Enzym Eng Biotechnol 91–93:655–669. https://doi.org/10.1385/ABAB:91-93:1-9:655

Antón J, Oren A, Benlloch S, Rodríguez-Valera F, Amann R, Rosselló-Mora R (2002) *Salinibacter ruber* gen. nov., sp. nov., a novel, extremely halophilic member of the bacteria from saltern crystallizer ponds. Int J Syst Evol Microbiol 52(2):485–491. https://doi.org/10.1099/00207713-52-2-485

Argyros DA, Tripathi SA, Barrett TF, Rogers SR, Feinberg LF, Olson DG, Foden JM, Miller BB, Lynd LR, Hogsett DA, Caiazza NC (2011) High ethanol titers from cellulose by using metabolically engineered thermophilic, anaerobic microbes. Appl Environ Microbiol 77(23): 8288–8294. https://doi.org/10.1128/AEM.00646-11

Aston JE, Peyton BM (2007) Response of Halomonas campisalis to saline stress: changes in growth kinetics, compatible solute production and membrane phospholipid fatty acid composition. FEMS Microbiol Lett 274(2):196–203. https://doi.org/10.1111/j.1574-6968.2007.00851.x

Atalah J, Cáceres-Moreno P, Espina G, Blamey JM (2019) Thermophiles and the applications of their enzymes as new biocatalysts. Bioresour Technol 280:478–488. https://doi.org/10.1016/j.biortech.2019.02.008

Bachofen R (1986) Microorganisms in extreme environments introduction. Experientia 42(11): 1179–1182. https://doi.org/10.1007/BF01946387

Baker BJ, Dick GJ (2013) Omic approaches in microbial ecology: charting the unknown. Microbe 8(9):353–360. https://doi.org/10.1128/microbe.8.353.1

Bakermans C, Tsapin AI, Souza-Egipsy V, Gilichinsky DA, Nealson KH (2003) Reproduction and metabolism at −10°C of bacteria isolated from Siberian permafrost. Environ Microbiol 5(4): 321–326. https://doi.org/10.1046/j.1462-2920.2003.00419.x

Bala A, Singh B (2019) Cellulolytic and xylanolytic enzymes of thermophiles for the production of renewable biofuels. Renew Energy 136:1231–1244. https://doi.org/10.1016/j.renene.2018.09.100

Basen M, Müller V (2017) "Hot" acetogenesis. Extremophiles 21:15–26. https://doi.org/10.1007/s00792-016-0873-3

Bergey DH (1919) Thermophilic bacteria. J Bacteriol IV(4):301–306. https://doi.org/10.1128/jb.4.4.301-306.1919

Bergholz PW, Bakermans C, Tiedje JM (2009) *Psychrobacter arcticus* 273-4 uses resource efficiency and molecular motion adaptations for subzero temperature growth. J Bacteriol 191 (7):2340–2352. https://doi.org/10.1128/JB.01377-08

Bhandiwad A, Shaw AJ, Guss A, Guseva A, Bahl H, Lynd LR (2014) Metabolic engineering of *Thermoanaerobacterium saccharolyticum* for n-butanol production. Metab Eng 21:17–25. https://doi.org/10.1016/j.ymben.2013.10.012

Biswas R, Prabhu S, Lynd LR, Guss AM (2014) Increase in ethanol yield via elimination of lactate production in an ethanol-tolerant mutant of Clostridium thermocellum. PLoS One 9(2):1–7. https://doi.org/10.1371/journal.pone.0086389

Bowman JP (2001) Methods for psychrophilic bacteria. Methods Microbiol 30:591–614. https://doi.org/10.1016/s0580-9517(01)30064-8

Bowman JP (2017) Genomics of psychrophilic bacteria and archaea. In: Margesin R (ed) Psychrophiles: from biodiversity to biotechnology. Springer, Cham, pp 345–387. https://doi.org/10.1007/978-3-319-57057-0_15

Breezee J, Cady N, Staley JT (2004) Subfreezing growth of the sea ice bacterium "*Psychromonas ingrahamii*". Microb Ecol 47(3):300–304. https://doi.org/10.1007/s00248-003-1040-9

Brock TD, Freeze H (1969) *Thermus aquaticus* gen. n. and sp. n., a nonsporulating extreme thermophile. J Bacteriol 98(I):289–297. https://doi.org/10.1128/jb.98.1.289-297.1969

Brock TD, Brock KM, Belly RT, Weiss RL (1972) *Sulfolobus*: a new genus of sulfur-oxidizing bacteria living at low pH and high temperature. Arch Mikrobiol 84(1):54–68. https://doi.org/10.1007/BF00408082

Cabrera MÁ, Blamey JM (2018) Biotechnological applications of archaeal enzymes from extreme environments. Biol Res 51(1):1–15. https://doi.org/10.1186/s40659-018-0186-3

Cayol J, Ollivier B, Patel BK, Ageron E, Grimont PA, Prensier G, Garcia JL (1995) *Haloanaerobium lacusroseus* sp. nov., an extremely halophilic fermentative bacterium from the sediments of a hypersaline lake. Int J Syst Bacteriol 45(4):790–797. https://doi.org/10.1099/00207713-45-4-790

Choure K, Parsai S, Kotoky R, Srivastava A, Tilwari A, Rai PK, Sharma A, Pandey P (2021) Comparative metagenomic analysis of two alkaline hot springs of Madhya Pradesh, India and deciphering the extremophiles for industrial enzymes. Front Genet 12:643423. https://doi.org/10.3389/fgene.2021.643423

Cripps RE, Eley K, Leak DJ, Rudd B, Taylor M, Todd M, Boakes S, Martin S, Atkinson T (2009) Metabolic engineering of *Geobacillus thermoglucosidasius* for high yield ethanol production. Metab Eng 11(6):398–408. https://doi.org/10.1016/j.ymben.2009.08.005

Da Fonseca FSA, Angolini CFF, Arruda MAZ, Junior CAL, Santos CA, Saraiva AM, Pilau E, Souza AP, Laborda PR, De Oliveira PFL, De Oliveira VM, Reis FDAM, Marsaioli AJ (2015) Identification of oxidoreductases from the petroleum *Bacillus safensis* strain. Biotechnol Rep 8:152–159. https://doi.org/10.1016/j.btre.2015.09.001

Dalmasso C, Oger P, Selva G, Courtine D, L'Haridon S, Garlaschelli A, Roussel E, Miyazaki J, Reveillaud J, Jebbar M, Takai K, Maignien L, Alain K (2016) *Thermococcus piezophilus* sp. nov., a novel hyperthermophilic and piezophilic archaeon with a broad pressure range for growth, isolated from a deepest hydrothermal vent at the Mid-Cayman rise. Syst Appl Microbiol 39(7):440–444. https://doi.org/10.1016/j.syapm.2016.08.003

De Vrije T, Budde MAW, Lips SJ, Bakker RR, Mars AE, Claassen PAM (2010) Hydrogen production from carrot pulp by the extreme thermophiles *Caldicellulosiruptor saccharolyticus* and *Thermotoga neapolitana*. Int J Hydrog Energy 35(24):13206–13213. https://doi.org/10.1016/j.ijhydene.2010.09.014

Demirjian DC, Morís-Varas F, Cassidy CS (2001) Enzymes from extremophiles. Curr Opin Chem Biol 5(2):144–151. https://doi.org/10.1016/S1367-5931(00)00183-6

Dennett GV, Blamey JM (2016) A new thermophilic nitrilase from an antarctic hyperthermophilic microorganism. Front Bioeng Biotechnol 4:1–9. https://doi.org/10.3389/fbioe.2016.00005

Di Donato P, Buono A, Poli A, Finore I, Abbamondi GR, Nicolaus B, Lama L (2019) Exploring marine environments for the identification of extremophiles and their enzymes for sustainable and green bioprocesses. Sustainability 11(1):149. https://doi.org/10.3390/su11010149

Dick GJ, Andersson AF, Baker BJ, Simmons SL, Thomas BC, Yelton AP, Banfield JF (2009) Community-wide analysis of microbial genome sequence signatures. Genome Biol 10(8):R85. https://doi.org/10.1186/gb-2009-10-8-r85

Edwardson CF, Hollibaugh JT (2018) Composition and activity of microbial communities along the redox gradient of an alkaline, hypersaline, Lake. Front Microbiol 9:1–18. https://doi.org/10.3389/fmicb.2018.00014

Feng S, Powell SM, Wilson R, Bowman JP (2014) Extensive gene acquisition in the extremely psychrophilic bacterial species *Psychroflexus torquis* and the link to sea-ice ecosystem specialism. Genome Biol Evol 6(1):133–148. https://doi.org/10.1093/gbe/evt209

Fernandez AB, Ghai R, Martin-Cuadrado AB, Sanchez-Porro C, Rodriguez-Valera F, Ventosa A (2013) Metagenome sequencing of prokaryotic microbiota from two hypersaline ponds of a marine saltern in Santa Pola, Spain. Genome Announc 1(6). https://doi.org/10.1128/genomeA.00933-13

Feyhl-Buska J, Chen Y, Jia C, Wang JX, Zhang CL, Boyd ES (2016) Influence of growth phase, pH, and temperature on the abundance and composition of tetraether lipids in the thermoacidophile *Picrophilus torridus*. Front Microbiol 7(AUG). https://doi.org/10.3389/fmicb.2016.01323

Foophow T, Tanaka S, Koga Y, Takano K, Kanaya S (2010) Subtilisin-like serine protease from hyperthermophilic archaeon *Thermococcus kodakaraensis* with N- and C-terminal propeptides. Protein Eng Des Sel 23(5):347–355. https://doi.org/10.1093/protein/gzp092

Fraser CM, Eisen JA, Salzberg SL (2000) Methods for whole-genome analysis. Nature 406:799–803. https://doi.org/10.1038/35021244

Gogliettino M, Riccio A, Cocca E, Rossi M, Palmieri G, Balestrieri M (2014) A new pepstatin-insensitive thermopsin-like protease overproduced in peptide-rich cultures of *Sulfolobus solfataricus*. Int J Mol Sci 15(2):3204–3219. https://doi.org/10.3390/ijms15023204

Goto E, Kodama T, Minoda Y (1978) Growth and taxonomy of thermophilic hydrogen bacteria. Agric Biol Chem 42(7):1305–1308. https://doi.org/10.1080/00021369.1978.10863157

Hedlund BP, Dodsworth JA, Murugapiran SK, Rinke C, Woyke T (2014) Impact of single-cell genomics and metagenomics on the emerging view of extremophile "microbial dark matter". Extremophiles 18:865–875. https://doi.org/10.1007/s00792-014-0664-7

Hedlund BP, Murugapiran SK, Alba TW, Levy A, Dodsworth JA, Goertz GB, Ivanova N, Woyke T (2015) Uncultivated thermophiles: current status and spotlight on "Aigarchaeota". Curr Opin Microbiol 25:136–145. https://doi.org/10.1016/j.mib.2015.06.008

Huber R, Eder W, Heldwein S, Wanner G, Huber H, Rachel R, Stetter KO (1998) Hyperthermophilic bacterium isolated from Yellowstone National Park. Appl Environ Microbiol 64(10):3576–3583. https://doi.org/10.1128/aem.64.10.3576-3583.1998

Ingraham JL (1958) Growth psychrophilic bacteria. J Bacteriol 76(1):75–80. https://doi.org/10.1128/jb.76.1.75-80.1958

Iwabuchi T, Harayama S (1998) Biochemical and molecular characterization of 1-hydroxy-2-naphthoate dioxygenase from *Nocardioides sp.* KP7. J Biol Chem 273(14):8332–8336. https://doi.org/10.1074/jbc.273.14.8332

Javor B (2011) Hypersaline environments. Springer, Berlin. https://doi.org/10.1007/978-3-642-74370-2

Junge K, Eicken H, Deming JW (2003) Motility of *Colwellia psychrerythraea* strain 34H at subzero temperatures. Appl Environ Microbiol 69(7):4282–4284. https://doi.org/10.1128/AEM.69.7.4282-4284.2003

Kádár Z, De Vrije T, Van Noorden GE, Budde MAW, Szengyel Z, Réczey K, Claassen PAM (2004) Yields from glucose, xylose, and paper sludge hydrolysate during hydrogen production by the extreme thermophile Caldicellulosiruptor saccharolyticus. Appl Biochem Biotechnol - Part A Enzym Eng Biotechnol 114(1–3):497–508. https://doi.org/10.1385/ABAB:114:1-3:497

Kim MH, Kim HK, Lee JK, Park SY, Oh TK (2000) Thermostable Lipase of *Bacillus stearothermophilus*: high-level production, purification, and calcium-dependent thermostability. Biosci Biotechnol Biochem 64:280–286. https://doi.org/10.1271/bbb.64.280

Kim BK, Lee BH, Lee YJ, Jin IH, Chung CH, Lee JW (2009) Purification and characterization of carboxymethylcellulase isolated from a marine bacterium, *Bacillus subtilis* subsp. subtilis A-53. Enzym Microb Technol 44(6–7):411–416. https://doi.org/10.1016/j.enzmictec.2009.02.005

Kour D, Rana KL, Kaur T, Singh B, Chauhan VS, Kumar A, Rastegari AA, Yadav N, Yadav AN, Gupta VK (2019) Extremophiles for hydrolytic enzymes productions: biodiversity and potential biotechnological applications. In: Molina M, Gupta V, Singh B, Gathergood N (eds) Bioprocessing for biomolecules production, pp 321–372. https://doi.org/10.1002/9781119434436.ch16

Kristjánsson JK, Hreggvidsson GO (1995) Ecology and habitats of extremophiles. World J Microbiol Biotechnol 11(1):17–25. https://doi.org/10.1007/BF00339134

Kwak YS, Akiba T, Kudo T (1998) Purification and characterization of α-amylase from hyperthermophilic archaeon *Thermococcus profundus*, which hydrolyzes both α-1,4 and α-1,6 glucosidic linkages. J Ferment Bioeng 86(4):363–367. https://doi.org/10.1016/S0922-338X(99)89005-9

Madigan MT (2000) Extremophilic bacteria and microbial diversity. Ann Missouri Bot Gard 87 (1):3–12. https://doi.org/10.2307/2666205

Maugeri TL, Gugliandolo C, Caccamo D, Stackebrandt E (2001) A polyphasic taxonomic study of thermophilic bacilli from shallow, marine vents. Syst Appl Microbiol 24(4):572–587. https://doi.org/10.1078/0723-2020-00054

Merino N, Aronson HS, Bojanova DP, Feyhl-Buska J, Wong ML, Zhang S, Giovannelli D (2019) Living at the extremes: extremophiles and the limits of life in a planetary context. Front Microbiol 10(MAR):780. https://doi.org/10.3389/fmicb.2019.00780

Mesbah NM, Wiegel J (2008) Life at extreme limits: the anaerobic halophilic alkalithermophiles. Ann N Y Acad Sci 1125:44–57. https://doi.org/10.1196/annals.1419.028

Mirete S, Morgante V, González-Pastor JE (2016) Functional metagenomics of extreme environments. Curr Opin Biotechnol 38:143–149. https://doi.org/10.1016/j.copbio.2016.01.017

Miyazaki K (2022) Complete genome sequences of thermus strains isolated from. Microbiol Resour Announc 1(14):20–22. https://doi.org/10.1128/mra.00354-22

Mormile MR, Romine MF, Garcia MT, Ventosa A, Bailey TJ, Peyton BM (1999) *Halomonas campisalis* sp, nov., a denitrifying, moderately haloalkaliphilic bacterium. Syst Appl Microbiol 22(4):551–558. https://doi.org/10.1016/S0723-2020(99)80008-3

Mullis KB (1990) The unusual origin of the polymerase chain reaction. Sci Am 262(4):56–65. https://doi.org/10.1038/scientificamerican0490-56

Mykytczuk NCS, Foote SJ, Omelon CR, Southam G, Greer CW, Whyte LG (2013) Bacterial growth at −15 °C; molecular insights from the permafrost bacterium *Planococcus halocryophilus* Or1. ISME J 7(6):1211–1226. https://doi.org/10.1038/ismej.2013.8

Naghoni A, Emtiazi G, Amoozegar MA, Cretoiu MS, Stal LJ, Etemadifar Z, Shahzadeh Fazeli SA, Bolhuis H (2017) Microbial diversity in the hypersaline Lake Meyghan. Iran Sci Rep 7(1):1–13. https://doi.org/10.1038/s41598-017-11585-3

Nakayama S, Kiyoshi K, Kadokura T, Nakazato A (2011) Butanol production from crystalline cellulose by Cocultured *Clostridium thermocellum* and *Clostridium saccharoperbutylacetonicum* N1-4. Appl Environ Microbiol 77(18):6470–6475. https://doi.org/10.1128/AEM.00706-11

Narayan A, Patel V, Singh P, Patel A, Jain K, Karthikeyan K, Shah A, Madamwar D (2018) Response of microbial community structure to seasonal fluctuation on soils of Rann of Kachchh, Gujarat, India: representing microbial dynamics and functional potential. Ecol Genet Genomics 6:22–32. https://doi.org/10.1016/j.egg.2017.11.001

Ng TK, Ben-Bassat A, Zeikus JG (1981) Ethanol production by thermophilic bacteria: fermentation of cellulosic substrates by cocultures of *Clostridium thermocellum* and *Clostridium thermohydrosulfuricum*. Appl Environ Microbiol 41(6):1337–1343. https://doi.org/10.1128/aem.41.6.1337-1343.1981

Nguyen TAD, Kim JP, Kim MS, Oh YK, Sim SJ (2008) Optimization of hydrogen production by hyperthermophilic eubacteria, *Thermotoga maritima* and *Thermotoga neapolitana* in batch fermentation. Int J Hydrog Energy 33(5):1483–1488. https://doi.org/10.1016/j.ijhydene.2007.09.033

Nikolaivits E, Dimarogona M, Fokialakis N, Topakas E (2017) Marine-derived biocatalysts: importance, accessing, and application in aromatic pollutant bioremediation. Front Microbiol 8:1–17. https://doi.org/10.3389/fmicb.2017.00265

Orellana R, Leavitt JJ, Comolli LR, Csencsits R, Janot N, Flanagan KA, Gray AS, Leang C, Izallalen M, Mester T, Lovley DR (2013) U(VI) reduction by diverse outer surface c-Type cytochromes of Geobacter sulfurreducens. Appl Environ Microbiol 79(20):6369–6374. https://doi.org/10.1128%2FAEM.02551-13

Orellana R, Macaya C, Bravo G, Dorochesi F, Cumsille A, Valencia R, Rojas C, Seeger M (2018) Living at the frontiers of life: extremophiles in Chile and their potential for bioremediation. Front Microbiol 9:2309. https://doi.org/10.3389/fmicb.2018.02309

Oremland RS, Kulp TR, Blum JS, Hoeft SE, Baesman S, Miller LG, Stolz JF (2005) Microbiology: a microbial arsenic cycle in a salt-saturated, extreme environment. Science (80-) 308(5726):1305–1308. https://doi.org/10.1126/science.1110832

Oshima T (1975) Thermine: a new polyamine from an extreme thermophile. Biochem Biophys Res Commun 63(4):1093–1098. https://doi.org/10.1016/0006-291X(75)90681-6

O-Thong S, Prasertsan P, Karakashev D, Angelidaki I (2008) Thermophilic fermentative hydrogen production by the newly isolated *Thermoanaerobacterium thermosaccharolyticum* PSU-2. Int J Hydrog Energy 33(4):1204–1214. https://doi.org/10.1016/j.ijhydene.2007.12.015

Pandit AS, Joshi MN, Bhargava P, Shaikh I, Ayachit GN, Raj SR, Saxena AK, Bagatharia SB (2015) A snapshot of microbial communities from the Kutch: one of the largest salt deserts in the world. Extremophiles 19(5):973–987. https://doi.org/10.1007/s00792-015-0772-z

Rampelotto PH (2013) Extremophiles and extreme environments. Life 3(3):482–485. https://doi.org/10.3390/life3030482

Rinke C, Schwientek P, Sczyrba A, Ivanova NN, Anderson IJ, Cheng JF, Darling A, Malfatti S, Swan BK, Gies EA, Dodsworth JA, Hedlund BP, Tsiamis G, Sievert SM, Liu WT, Eisen JA, Hallam SJ, Kyrpides NC, Stepanauskas R, Rubin EM, Hugenholtz P, Woyke T (2013) Insights into the phylogeny and coding potential of microbial dark matter. Nature 499(7459):431–437. https://doi.org/10.1038/nature12352

Roberts MF (2005) Organic compatible solutes of halotolerant and halophilic microorganisms. Saline Syst 1(1):1–30. https://doi.org/10.1186/1746-1448-1-5

Sahm K, John P, Nacke H, Wemheuer B, Grote R, Daniel R, Antranikian G (2013) High abundance of heterotrophic prokaryotes in hydrothermal springs of the Azores as revealed by a network of 16S rRNA gene-based methods. Extremophiles 17(4):649–662. https://doi.org/10.1007/s00792-013-0548-2

Sanchez AC, Ravanal MC, Andrews BA, Asenjo JA (2019) Heterologous expression and biochemical characterization of a novel cold-active α-amylase from the Antarctic bacteria *Pseudoalteromonas sp.* 2-3. Protein Expr Purif 155:78–85. https://doi.org/10.1016/j.pep.2018.11.009

Schicho RN, Ma K, Adams MWW, Kelly RM (1993) Bioenergetics of sulfur reduction in the hyperthermophilic archaeon *Pyrococcus furiosus*. J Bacteriol 175(6):1823–1830. https://doi.org/10.1128/jb.175.6.1823-1830.1993

Schleper C, Puehler G, Holz I, Gambacorta A, Janekovic D, Santarius U, Klenk HP, Zillig W (1995) Picrophilus gen. nov., fam. nov.: a novel aerobic, heterotrophic, thermoacidophilic genus and family comprising archaea capable of growth around pH 0. J Bacteriol 177(24):7050–7059. https://doi.org/10.1128/jb.177.24.7050-7059.1995

Schloss PD, Handelsman J (2005) Metagenomics for studying unculturable microorganisms: cutting the Gordian knot. Genome Biol 6(8):6–9. https://doi.org/10.1186/gb-2005-6-8-229

Segerer A, Neuner A, Kristjansson IJK, Stetter K (1986) Acidianus infernus gen. nov.? sp. nov. and Acidianus brierleyi comb. nov.: facultatively aerobic, extremely acidophilic thermophilic sulfur-metabolizing archaebacteria. Int J Syst Bacteriol 36(4):559–564. https://doi.org/10.1099/00207713-36-4-559

Sharma V, Ayothiraman S, Dhakshinamoorthy V (2019) Production of highly thermo-tolerant laccase from novel thermophilic bacterium Bacillus sp. PC-3 and its application in functionalization of chitosan film. J Biosci Bioeng 127(6):672–678. https://doi.org/10.1016/j.jbiosc.2018.11.008

Shurigin V, Hakobyan A, Panosyan H, Egamberdieva D, Davranov K, Birkeland NK (2019) A glimpse of the prokaryotic diversity of the large Aral Sea reveals novel extremophilic bacterial and archaeal groups. Microbiology 8(9):1–13. https://doi.org/10.1002/mbo3.850

Singh KK, Ray L (2021) Extremozymes: biocatalysts from extremophilic microorganisms and their relevance in current biotechnology. Environ Agric Microbiol Adv Appl 293–311. https://doi.org/10.1002/9781119525899.ch14

Singh P, Jain K, Desai C, Tiwari O, Madamwar D (2018) Microbial community dynamics of extremophiles/extreme environment. In: Das S, Dash HR (eds) Microbial diversity in the genomic era. Academic Press, Elsevier Inc., pp 323–332. https://doi.org/10.1016/B978-0-12-814849-5.00018-6

Staley J, Konopka A (1985) Measurement of in situ activities of nonphotosynthetic microorganisms in aquatic and terrestrial habitats. Annu Rev Microbiol 39(1):321–346. https://doi.org/10.1146/annurev.micro.39.1.321

Stepanauskas R (2012) Single cell genomics: an individual look at microbes. Curr Opin Microbiol 15(5):613–620. https://doi.org/10.1016/j.mib.2012.09.001

Stetter KO (2006) History of discovery of the first hyperthermophiles. Extremophiles 10(5): 357–362. https://doi.org/10.1007/s00792-006-0012-7

Sudha Rani K, Swamy MV, Seenayya G (1997) Increased ethanol production by metabolic modulation of cellulose fermentation in *Clostridium thermocellum*. Biotechnol Lett 19 (8):819–823. https://doi.org/10.1023/A:1018312931542

Svetlitchnyi VA, Kensch O, Falkenhan DA, Korseska SG, Lippert N, Prinz M, Sassi J, Schickor A, Curvers S (2013) Single-step ethanol production from lignocellulose using novel extremely thermophilic bacteria. Biotechnol Biofuels 6(1):31. https://doi.org/10.1186/1754-6834-6-31

Takai K, Nakamura K, Toki T, Tsunogai U, Miyazaki M, Miyazaki J, Hirayama H, Nakagawa S, Nunoura T, Horikoshi K (2008) Cell proliferation at 122°C and isotopically heavy CH4 production by a hyperthermophilic methanogen under high-pressure cultivation. Proc Natl Acad Sci U S A 105(31):10949–10954. https://doi.org/10.1073/pnas.0712334105

Tomás AF, Karagöz P, Karakashev D, Angelidaki I (2013) Extreme thermophilic ethanol production from rapeseed straw: using the newly isolated *Thermoanaerobacter pentosaceus* and combining it with *Saccharomyces cerevisiae* in a two-step process. Biotechnol Bioeng 110 (6):1574–1582. https://doi.org/10.1002/bit.24813

Umezawa K, Kojima H, Kato Y, Fukui M (2021) *Dissulfurispira thermophila* gen. nov., sp. nov., a thermophilic chemolithoautotroph growing by sulfur disproportionation, and proposal of novel taxa in the phylum Nitrospirota to reclassify the genus *Thermodesulfovibrio*. Syst Appl Microbiol 44(2):126184. https://doi.org/10.1016/j.syapm.2021.126184

Ventosa A (2006) Unusual micro-organisms from unusual habitats: hypersaline environments. In: Logan N, Lappin-Scott H, Oyeston P (eds) Prokaryotic diversity: mechanisms and significance. Publ Soc Gen Microbiol (1988):223–254. Cambridge University press. https://doi.org/10.1017/CBO9780511754913.015

Ventosa A, Nieto JJ, Oren A (1998) Biology of moderately halophilic aerobic bacteria. Microbiol Mol Biol Rev 62(2):504–544. https://doi.org/10.1128/mmbr.62.2.504-544.1998

Ventosa A, Fernández AB, León MJ, Sánchez-Porro C, Rodriguez-Valera F (2014) The Santa Pola saltern as a model for studying the microbiota of hypersaline environments. Extremophiles 18:811–824. https://doi.org/10.1007/s00792-014-0681-6

Ventosa A, de la Haba RR, Sánchez-Porro C, Papke RT (2015) Microbial diversity of hypersaline environments: a metagenomic approach. Curr Opin Microbiol 25:80–87. https://doi.org/10.1016/j.mib.2015.05.002

Vester JK, Glaring MA, Stougaard P (2015) Improved cultivation and metagenomics as new tools for bioprospecting in cold environments. Extremophiles 19:17–29. https://doi.org/10.1007/s00792-014-0704-3

Vincent AT, Trudel MV, Freschi L, Nagar V, Gagné-Thivierge C, Levesque RC, Charette SJ (2016) Increasing genomic diversity and evidence of constrained lifestyle evolution due to insertion sequences in *Aeromonas salmonicida*. BMC Genomics 17(1):1–12. https://doi.org/10.1186/s12864-016-2381-3

Vollmers J, Voget S, Dietrich S, Gollnow K, Smits M, Meyer K, Brinkhoff T, Simon M, Daniel R (2013) Poles apart: Arctic and Antarctic Octadecabacter strains share high genome plasticity and a new type of Xanthorhodopsin. PLoS One 8(5):e63422. https://doi.org/10.1371/journal.pone.0063422

Woyke T, Sczyrba A, Lee J, Rinke C, Tighe D, Clingenpeel S, Malmstrom R, Stepanauskas R, Cheng JF (2011) Decontamination of MDA reagents for single cell whole genome amplification. PLoS One 6(10):2–6. https://doi.org/10.1371/journal.pone.0026161

Wu YR, Mao A, Sun C, Shanmugam S, Li J, Zhong M, Hu Z (2017) Catalytic hydrolysis of starch for biohydrogen production by using a newly identified amylase from a marine bacterium *Catenovulum sp.* X3. Int J Biol Macromol 104:716–723. https://doi.org/10.1016/j.ijbiomac.2017.06.084

Yao S, Mikkelsen MJ (2010) Identification and overexpression of a bifunctional aldehyde/alcohol dehydrogenase responsible for ethanol production in Thermoanaerobacter mathranii. J Mol Microbiol Biotechnol 19(3):123–133. https://doi.org/10.1159/000321498

Zeikus JG (1979) Thermophilic bacteria: ecology, physiology and technology. Enzym Microb Technol 4(1):243–252. https://doi.org/10.1016/0141-0229(79)90043-7

Zhang J, Cao X, Xin Y, Xue S, Zhang W (2013) Purification and characterization of a dehalogenase from *Pseudomonas stutzeri* DEH130 isolated from the marine sponge *Hymeniacidon perlevis*. World J Microbiol Biotechnol 29(10):1791–1799. https://doi.org/10.1007/s11274-013-1340-2

Zhu D, Adebisi WA, Ahmad F, Sethupathy S, Danso B, Sun J (2020) Recent development of extremophilic bacteria and their application in biorefinery. Front Bioeng Biotechnol 8(June):1–18. https://doi.org/10.3389/fbioe.2020.00483

Zverlov V, Piotukh K, Dakhova O, Velikodvorskaya G, Borriss R (1996) The multidomain xylanase A of the hyperthermophilic bacterium *Thermotoga neapolitana* is extremely thermoresistant. Appl Microbiol Biotechnol 45(1–2):245–247. https://doi.org/10.1007/s002530050678

Chapter 21
Microbial Nanotechnology: A Biocompatible Technology for Sustainable and Green Agriculture Practice

Md Asif Amin

Abstract Microbial nanotechnology is an emerging field because of its various applications in material sciences, agricultural sciences, and medical sciences. It is a more eco-friendly, biocompatible, invasive, and cost-effective technology. Different nanomaterials including metal and metal oxide nanoparticles for different uses can be synthesized using selective microorganism. Metabolites of bacteria, fungi, and algae may replace expensive and harmful chemicals to produce nanomaterials. The most important aspect of microbial nanotechnology is its application in plant growth. Here we discussed about different symbiotic processes between beneficial soil-dwelling microbes and nanoparticles to enhance sustainable plant growth. Toxic metal may lead to various plant diseases and inhibit plant growth. Microbes may remove that toxicity from various sources by forming nanostructure of those metals. Threshold concentration of nanoparticles with optimum properties (size, shape, dissolution, aggregation) should be used to avoid ecotoxicological effect as toxicity depends on the above said factors.

Keywords Microorganism · Microbial nanotechnology · Nanoparticles · Plant growth

Abbreviations

AgNPs	Silver nanoparticles
Al_2O_3 NPs	Alumina nanoparticles
CB	Cell biomass
CFS	Cell-free supernatant
CFU	Colony-forming unit

M. A. Amin (✉)
Department of Chemistry, Balurghat Mahila Mahavidyalaya, Balurghat, West Bengal, India

P. Mathur et al. (eds.), *Microbial Symbionts and Plant Health: Trends and Applications for Changing Climate*, Rhizosphere Biology,
https://doi.org/10.1007/978-981-99-0030-5_21

CoONPs	Cobalt oxide nanoparticles
CuNPs	Copper nanoparticles
DMRB	Dissimilatory metal-reducing bacteria
HA	Hydroxyapatite
IOMNPs	Iron oxide magnetic nanoparticles
PbNPs	Lead nanoparticles
PbSNPs	Lead sulfide nanoparticles
PdNPs	Pd(0) nanoparticles
PGPR	Plant growth-promoting factor
PtNPs	Platinum nanoparticles
SAED	Selected area electron diffraction
SEM	Scanning electron microscope
TEM	Transmission electron microscope
TiBALDH Titanium	bis-ammonium lactato dihydroxide
TNs	Titania nanoparticles
ZnO NPs	Zinc oxide nanoparticles

21.1 Introduction

Nanoparticles as well as nanomaterials are drawing attention of researchers for their unique physical, chemical, and biological properties compared to their large aggregated forms. Those properties arise due to increasing number of surface atoms resulting in high surface-to-volume ratio. Significant increase in number of surface atom and contraction of lattice parameter on forming nanometric particle from macroscopic particle causes broadening of phonon spectrum (Teo 1986; Cresnzi 1995; Kara and Rahman 1998; Meyer et al. 2002). The high surface area makes it unique from other macroscopic materials leading to various uses in every field of sciences (Grasso et al. 2020). There are many soil-dwelling beneficial microbes which help in plant growth. Previously, nanoparticles are known to be inhibitors of growth of different microorganisms (Rousk et al. 2012; Antisari et al. 2013). Se nanoparticles can be used as a medicine against *Staphylococcus aureus* as it inhibits the growth of *Staphylococcus aureus* (Husen and Siddiqi 2014). Available large surface area of nanomaterials may promote interaction of bacteria with nanomaterials, thus helping bacterial growth. Microbial biotechnology or industrial biotechnology is an important aspect of biotechnology because it uses living microorganisms to synthesize very useful and daily life products in quasi-natural method. Incorporation of nanotechnology into microbial biotechnology is called microbial nanotechnology. Microbial nanotechnology is important because it expanded the application of microbial biotechnology along with usage of very little amount of starting materials.

Here we discuss about assistance of nanomaterials in growth of beneficial microbes as well as their metabolism and physiology. Extension in population and increasing metabolism may boost up plant growth as a synergistic effect of nanomaterials and microorganism symbiosis effect. Finding out of such symbiotic phenomena will definitely speed up agricultural activities by proper application of those processes. Separation and identification of different intermediates, i.e., detailed mechanism, become a challenging job. The pathways and mechanisms that should be traversed of this nanoparticle-microorganism symbiotic plant growth become interesting to chemists, biologists, and physicists as it is an unexplored field.

21.2 Synthesis of Nanomaterials by Microorganism

Eco-friendly synthesis of various materials becomes important nowadays for their invasive application in biological systems. Bio-assisted synthesis becomes important for their enhanced activity in biological processes rather than nonbiological-assisted synthesis (Bhattacharya and Mukherjee 2008; Simkiss and Wilbur 1989). Mukherjee et al. synthesize gold nanoparticles by treating aqueous solution of $AuCl_4^-$ ions with fungus *Fusarium oxysporum*. This in situ reduction of gold ions is assisted by reductases released by the fungus in the solution. This method of extracellular synthesis can be applied commercially due to ease of synthesis, and the nanoparticles have good monodispersity (Mukherjee et al. 2002). Actinomycetes are important class of microorganism as they have characteristics of both fungi and prokaryotes such as bacteria (they have close affinity with mycobacteria and the coryneforms). Their most important aspect is they produce secondary metabolites such as antibiotics. Alkalothermophilic (extremophilic) actinomycete *Thermomonospora* sp. can be used for the extracellular synthesis of gold nanoparticles by reaction of its biomass with aqueous chloroaurate ions (Ahmad et al. 2003). Microorganism becomes potential nanofactory for production of silver nanoparticles.

Microbial resistance mechanisms take place at periplasmic space, the region of a variety of enzymes and functions. External toxic metals can be reduced and hence concentrated at periplasmic space by these resistance mechanisms. Bacterium *Pseudomonas stutzeri* AG259, isolated from a silver mine, when placed in a concentrated aqueous solution of silver nitrate, reduces $Ag+$ ions into silver nanoparticles (AgNPs) of well-defined size and distinct topography outside the cytoplasmic membrane (Klaus et al. 1999). Mikhailov and Mikhailova (2019) showed various methods of biosynthesis of elemental silver nanoparticles (AgNPs) and their applications.

Toxic selenate and selenite oxoanions may be converted to nontoxic and insoluble Se nanoparticles using bacterial culture. In anaerobic condition, Tam et al. synthesize selenium nanoparticles from aqueous selenite using *Shewanella* sp. HN-41, and from HR TEM and SAED image, it is clear that those nanoparticles are amorphous (Tam et al. 2010). However, fruitful efforts have also been given to synthesize Se nanoparticles with definite morphology (nanowires and nanorods) biologically from *Rhizobium selenireducens* sp., *Dechlorosoma* sp., *Pseudomonas* sp., *Paracoccus* sp., *Enterobacter* sp., *Thaurea* sp., *Sulfurospirillum* sp.,

Desulfovibrio sp., and *Shewanella* sp. (Rathgeber et al. 2002; Morita et al. 2007; Klonowska et al. 2005).

However, there are natural mechanisms, i.e., biomineralization, by virtue of which nanoparticles can be synthesized. Iron-reducing bacterium *Shewanella oneidensis* extracellulary synthesizes magnetite particles by combined active and passive mechanisms. In active mechanism, it utilizes ferrihydrite as a terminal electron acceptor to produce Fe^{2+}. In passive mechanism net negatively charged cell wall, cell structures induce positively charged Fe^{2+} and Fe^{3+}. So local rise of the supersaturation of the system with respect to magnetite produced such particles (Gonzalez et al. 2010).

TiO_2 nanoparticles can be synthesized by treating $TiO(OH)_2$ solution with culture solution of very-low-cost green and reproducible microbes (*Lactobacillus* sp. and *Saccharomyces cerevisiae*) (Jha et al. 2009).

ZnO nanoparticles (ZnO NPs) can be synthesized by incubating Zn^{2+} salt with cell biomass (CB) and supernatant (CFS) of zinc-tolerant *Lactobacillus plantarum* TA4. Biosynthesized ZnO NPs that showed antibacterial activity against Gram-positive and Gram-negative pathogens also revealed that the inhibitory and bactericidal efficacy of both biosynthesized ZnO NPs were concentration-dependent. This ZnO NPs are also biocompatible with Vero cell line at specific concentrations (Yusof et al. 2020).

21.3 Microorganism-Assisted Nanomaterials in Plant Growth

It is well known that nanoparticles have antimicrobial property. Various Zn, Se, Ti, and Cu nanoparticles inhibit microbial growth. Nanoparticles reduce a number of beneficial microbes or inhibit their growth also. However, there are some evidences where nanoparticles help in bacterial growth. Fe_3O_4 nanoparticles expand the population of soil microbes (bacteria and actinomycetes but decreased growth of fungi) by enhancing their enzyme (amylase, phosphatase, catalase, urease) activities (Fang et al. 2012). Iron oxide magnetic nanoparticles (IOMNPs, Fe_3O_4, and γ Fe_2O_3) also stimulated beneficial actinobacterial (*Duganella*, *Streptomycetaceae*, or *Nocardioides*) growth and change the soil bacterial community structure as the result of concomitant increase of soil urease and invertase activities (He et al. 2011). Fe_3O_4 and SnO_2 nanoparticles increase microbial C/N ratio of beneficial microorganism like ectomycorrhizae and increase their metabolism quotient (qCO_2). Ectomycorrhiza is a form of symbiotic relationship that occurs between a fungal symbiont, or mycobiont, and the roots of various plant species. Probable reason of this phenomenon is due to changes in the composition of microbial communities inhabiting soil and microbial stress (Antisari et al. 2013).

21.3.1 Amplification of Adhesion of Beneficial Bacteria by Nanoparticle

Attachment of beneficial bacteria on the surface of plant root is very important for plant growth in a symbiotic process. This is because most of the bacterial cell walls carry net negative charge though a few candidates carry positive charge (Dickson and Koohmaraie 1989). The roots' surface also carries negative charge (mean value $-30\ mCm^{-1}$) (Kinraide and Wang 2010). So an average repulsive force dominates over attractive factors. Naturally occurring metal oxide nanoparticles may be a promising agent for enhancing aggregation of single cell organisms and function as glue between bacteria and roots as they possess enhanced affinity to phosphate and phosphonate ligands (Cabrera et al. 1977; Pązik et al. 2011). Nanoparticles also act as glue between bacterial cells and enhance bacterial colony growth (Fig. 21.1). Palmqvist et al. (2015) synthesized titanium oxide nanoparticles using patented Captigel and TiBALDH (titanium bis-ammonium lactato dihydroxide) precursor. Titania (TiO_2) nanoparticle becomes an emerging material among various nanoparticles for its extensive and invasive application, easy synthesis, biocompatibility, and low cost. They used this titania nanoparticles (TNs) in the nanointerface interaction between a beneficial plant growth-promoting bacterium (*Bacillus amyloliquefaciens* UCMB5113) and oilseed rape plants (*Brassica napus*) for protection against the fungal pathogen *Alternaria brassicae* (Fig. 21.2). In the rhizosphere species *Bacillus amyloliquefaciens*, several strains act as plant growth-

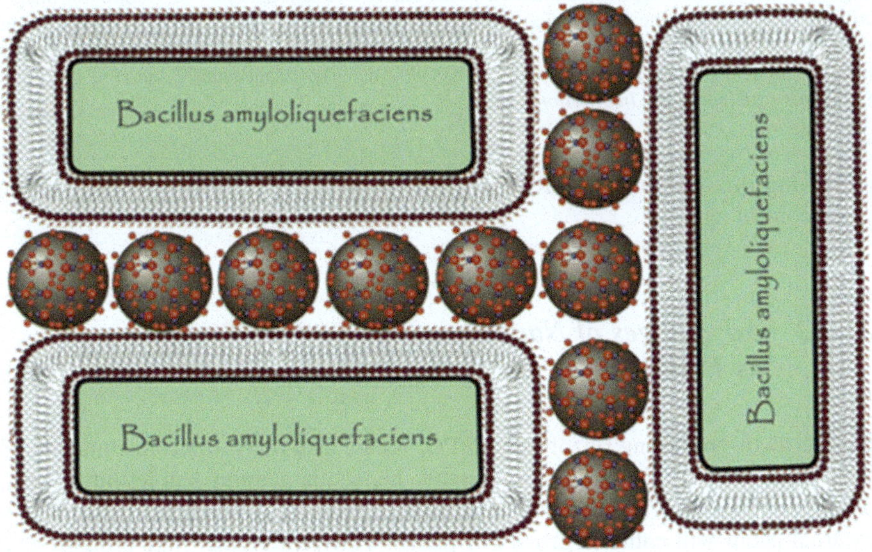

Fig. 21.1 Schematic representation of cluster formation for bacteria initiated by TNs (Palmqvist et al. 2015, *Reproduced under guideline of Creative Commons Attribution 4.0 International License*)

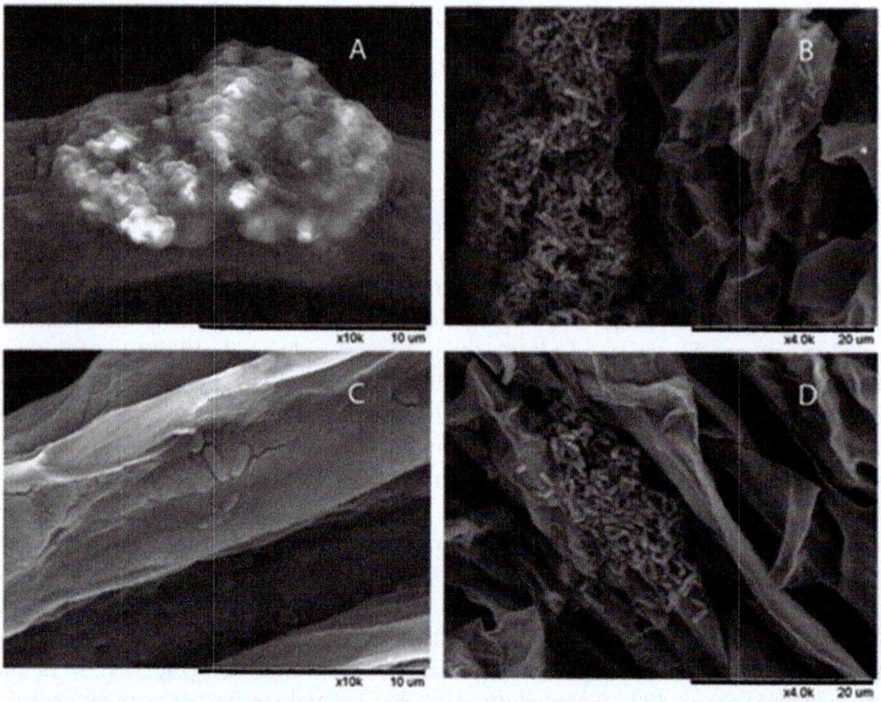

Fig. 21.2 SEM images of *B. amyloliquefaciens* on oilseed rape roots. (**a**) Bacteria together with Captigel nanoparticles. (**b**) Bacteria together with TiBALDH nanoparticles. (**c**) Control for image A with bacteria found in the same region of the root, namely, the base, of plant treated with bacteria only. (**d**) Control for image (**b**) with bacteria found in the same region of the root, namely, the base, of a plant treated with bacillus only. (Palmqvist et al. 2015, *Reproduced under guideline of Creative Commons Attribution 4.0 International License*)

promoting rhizobacteria (PGPR) and improve stress management of plants serving as biocontrol agents.

21.3.2 Advantages of Nanosilica over Sodium Silicate as Fertilizer

Plant growth-promoting rhizobacteria (PGPR) act as biofertilizer, biostimulant, and bioprotectant by recycling the soil nutrients, hence maintaining soil fertility. Thus, it promotes plant growth (Gholami et al. 2009). Increasing population of PGPR by external agent will enhance growth of plants. Thus, a fertilizer may promote plant growth without affecting soil biology. In the presence of nanosilica, colony-forming unit (CFU) of PGPR was doubled in the presence of from 4×10^5 CFU (control) to 8×10^5 CFU per gram of soil (Karunakaran et al. 2013). Probable mechanism of this

Scheme 21.1 Comparative reactions of sodium silicate and nanosilica in soil to form silicic acid

$$\underset{\text{Sodium Silicate}}{2Na_2SiO_3} + 4H_2O \xrightarrow{\text{Soil}} \underset{\text{Silicic acid}}{2H_4SiO_4} + \underset{\text{Sodium Oxide}}{2Na_2O}$$

$$\underset{\text{Nanosilica}}{2SiO_2} + 4H_2O \xrightarrow{\text{Soil}} \underset{\text{Silicic acid}}{2H_4SiO_4}$$

growth may be due to interaction between bacteria and nanosilica arising from hydration property of nanosilica surface, which could facilitate the attraction of silica on the microbial surface (Gordienko and Kurdish 2007). In contrast, sodium silicate depleted bacterial communities, thus indirectly affecting fertility (N, P, and K) of the soil. This can be explained by pH of soil. Optimal pH assists in microbial growth and ameliorates soil nutrient value. When Na_2SiO_3 is applied into soil, it absorbs soil moisture to form alkaline sodium oxide and two molecules of silicic acid and turned pH of soil to alkaline condition (Scheme 21.1). In this alkaline pH (pH 9.9), bacterial communities are diminished. Nanosilica does not influence soil pH to such extent, thus sustaining optimal pH for bacterial growth. Nanosilica thus increases soil NPK values but sodium silicate inclusion in soil decreases available NPK in soil.

There is also a correlation between silica uptake and protein content of bacterial cell which could be further evidence of nanosilica-induced bacterial growth.

21.3.3 Uses of Nano-hydroxyapatite to Increase Soil Quality Along with Microbial Growth

Nanoform of naturally occurring mineral hydroxyapatite ($Ca_{10}(PO_4)_6(OH)_2$) is a potential phosphorous-based nanofertilizer. Jia et al. (2022) showed that nano-hydroxyapatite (nano-HA) improves soil quality by elevating soil bioavailable phosphorous, electrical conductivity, and soil organic matter. Fresh biomass, percentage of bioavailable phosphorus, and chlorophyll content of maize plants have been magnified on application of nano-HA. Especially, beneficial P-solubilizing bacterial growth (e.g., *Burkholderiaceae*, *Massilia*, *Sinomonas*, and *Streptomyces*) was observed on incorporation of nano-HA into soil. Abundance of *Sinomonas* and *Streptomyces* (67.1%), and the family *Micrococcaceae* (77.0%), which belongs to *Actinobacteria* were also amplified. Overall nano-HA didn't alter soil biodiversity of the soil microbial communities (Shannon index remain unchanged). This may be due to increase of small biomolecules (e.g., D-arabinose, L-arabitol, palatinitol, and sorbitol) into soil which are important for microbial growth (C source of microbes). Extension of this scheme in agriculture may be an important move for sustainable plant growth.

21.4 Toxic Metal Removal by Microbial Nanotechnology

Metal plays an important role in plant growth in virtue of various biological processes. Several metal ions are crucial part of various biomolecules (e.g., hemoglobin, chlorophyll, vitamin B_{12}) and cofactor of many biologically important enzymes or proteins. However, heavy metals as well as essential metals may be toxic when threshold limits exceed. Metal toxicity causes low biomass accumulation, chlorosis, altered water balance, and nutrient assimilation and senescence resulting in plant death. Supplementation of different inorganic and organic agents such as salts, biochar, zeolites, manure, and lime becomes popular to remove heavy metals by precipitation, complex formation, and adsorption. Major drawbacks of such materials are accumulation of them in soil and water and degradation of plant health. Here microbial nanotechnology could be used as the fittest technology as it is biocompatible and recyclable. Microbes may be used as efficient biofactories for reduction and recovery of heavy metals from industrial waste (Saba et al. 2021). We only focus on removal of toxic metals by formation of nanostructure assisted by microbes.

21.4.1 Metal-Removing Microbes

Bacteria isolated from activated sludge such as *Enterobacter* sp., *Stenotrophomonas* sp., *Providencia* sp., *Chryseobacterium* sp., *Comamonas* sp., *Ochrobactrum* sp., and *Delftia* sp. are found to be efficient in heavy metal removal (Bestawy et al. 2013). *Ralstonia eutropha* CH34 (former name *Alcaligenes eutrophus*) is known to be heavy metal-resistant and bioprecipitator. Zn, Cu, Co, Fe, Al, Ag, Cr, As, and Se can be removed by metal biosorbing *Pseudomonas mendocina* AS302 and *Arthrobacter* sp. BP7/26 (Diels et al. 2003) (Table 21.1).

21.4.2 Conversion to Nanostructure of Toxic Metal by Microbes

Microbes can convert toxic metal ions into metal nanostructure by using their proteins, enzymes. Major advantages of this metal removal process are devoid of any external harmful chemicals. *Saccharomyces cerevisiae* (ATCC9763) can convert selenite (Na_2SeO_3) and Cd into CdSe quantum dots (QDs) (Wu et al. 2015). Chromium toxicity can be removed by bacteria via reduction of Cr (VI) to Cr (III) species through Cr (IV) and Cr (V) intermediates. Thus, Cr (VI)-affected environment may be rejuvenated by various species of *Pseudomonas* species, including *Pseudomonas aeruginosa*, *P. synxantha*, *P. putida*, *P. ambigua*, *P. fluorescens*, *P. dechromaticans*, and *P. chromatophila* (Cheung et al. 2006). Gram-negative

Table 21.1 Example of some microbes which convert toxic metal into their nanostructure

S. no	Starting materials	Final materials	Microbes
1	Selenite (Na_2SeO_3), Cd	CdSe quantum dots	*Saccharomyces cerevisiae*
2	Cr (VI)	Cr (III)	Various species of *Pseudomonas*
3	Co	CoONPs	*Micrococcus lylae, Bacillus subtilis, Escherichia coli, Paracoccus* sp., and *Haloarcula vallismortis*
4	Copper salt	CuNPs	*Thermoanaerobacter* sp., *Micrococcus* sp.
5	Pb compounds	PbSNPs	*Rhodosporidium diobovatum*
6	Pb compounds	PbNPs	*Aspergillus species*
7	Mn	MnO_2 nanoparticles	*Saccharophagus degradans* and *Saccharomyces cerevisiae*
8	Hg	Hg nanoparticles	*Bacillus cereus, Lysinibacillus* sp., *Bacillus* sp., *Kocuria rosea, Microbacterium oxydans, Serratia marcescens,* and *Ochrobactrum* sp.
9	Pd (II)	PdNPs	*Geobacter sulfurreducens*
10	Pt	PtNPs	*Plectonema boryanum, Escherichia coli, Shewanella* algae, *Desulfovibrio alaskensis*

bacteria such as *Micrococcus lylae*, *Bacillus subtilis*, *Escherichia coli*, *Paracoccus* sp., and *Haloarcula vallismortis* can convert various cobalt species' toxicity into cobalt oxide nanoparticles (CoONPs). Those nanoparticles possess well-defined sizes and shapes. The sizes of nanoparticles are also dependent on bacteria. *E. coli*-mediated synthesized rod-shaped nanostructures have size of 473 ± 54 nm and *M. lylae* media yield rod-shaped nanostructure of 356 ± 55 nm (Jang et al. 2015a). Toxic copper salt can be reduced to copper nanoparticles (CuNPs) by extracellular metal reduction mechanism of anaerobic *Thermoanaerobacter* sp. X513 (Jang et al. 2015b). *Micrococcus* sp. isolated from activated sludge can remove and recover copper from effluents in a cost-effective manner (Wong et al. 2001). Removal of toxic metals through nanoparticle formation can be done both inside and outside the cell. Seshadri et al. (2011) showed that toxic lead compound can be internalized through formation of lead sulfide nanoparticles (PbSNPs, average size 2–5 nm) inside marine yeast cell *Rhodosporidium diobovatum.* Intracellular sulfur-rich protein did the job of reduction of Pb cation followed by stabilization of PbSNPs. Pavani et al. (2012) showed lead nanoparticles (PbNPs, with average size 5–20 nm) can be formed inside and outside *Aspergillus* species. It is obvious that Pb nanoparticles formed outside the cell is larger than PbNPs formed inside the cell. Marine bacterium *Saccharophagus degradans* ATCC 43961 (strain 2–40) and yeast *Saccharomyces cerevisiae* remove and recover manganese toxicity and form MnO_2 nanoparticles (Salunke et al. 2015). Mercury (Hg) can be removed through the mechanism of exopolysaccharide-associated process by microbes such as *Bacillus cereus*, *Lysinibacillus* sp., *Bacillus* sp., *Kocuria rosea*, *Microbacterium oxydans*,

Serratia marcescens, and *Ochrobactrum* sp. Hg elimination can be done by precipitation or by sticking Hg on the surface of bacterial cell (François et al. 2012). Similarly, in this process, dissimilatory metal-reducing bacteria (DMRB) *Geobacter sulfurreducens* can reduce soluble toxic Pd(II) to Pd(0) nanoparticles (PdNPs) at outer cell surface. Extracellular reduction followed by nanoparticle formation or other association has the advantage of recovery of the metal in a cost-effective manner (Yates et al. 2013). Platinum toxicity can be removed by amorphous spherical nanoparticle (PtNP) formation within bacteria cells of *Cyanobacteria Plectonema boryanum* UTEX 485. *Escherichia coli* MC4100, *Shewanella algae*, and *Desulfovibrio alaskensis* G20 are also potential microbes that can synthesize PtNPs (Lengke et al. 2006; Attard et al. 2012; Konishi et al. 2007; Capeness et al. 2015).

21.5 Environmental Issues and Optimal Use of Nanoparticles in Microbial Nanotechnology

Nowadays, effect of modern engineered materials (polymers, nanoparticles, photosensitive materials, piezoelectric materials, etc.) on environment becomes a major concern among scientific communities. Nanoparticle accumulation in environment is an obvious consequence of application of microbial nanotechnology in plant growth, though there are insufficient studies on ecotoxicological effect of nanoparticles. Excess nanoparticles may accumulate into solid soil matrix and various colloidal suspensions into water inside soil. Those particles may penetrate into plant through root tissue. Then it may enter into human or any other animal body while consuming that plant derivatives. Thus, nanoparticles may enter into food chain and show various negative effects. ZnO nanoparticles damage skin cell DNA (Sharma et al. 2009). Increased reactivity of those nanoparticles may affect important biological processes such as photosynthesis and respiration with a ramification of reduced plant growth. This may also damage the beneficial soil microbes. Yang and Watts (2005) showed alumina nanoparticles (Al_2O_3 NPs) can reduce root growth of *Crocus sativus* as a concomitant effect of microbial growth. Lin and Xing (2007) observed that seed germination and root growth may be inhibited by zinc and zinc oxide nanoparticles (ZnO NPs). Effect on environment may depend on various properties of nanoparticles such as size, shape, concentration, dissolution, and aggregation as reactivity of nanoparticles depends on those properties. There are various aspects to control eco-toxicity of those nanoparticles. For sustainable plant growth with least effect on ecosystem, we should find out at which concentration we should apply nanoparticles, i.e., threshold value should be determined before application. Besides this, we have to also detect the chemical and physical properties for which the toxicity generates. Then we can easily synthesize the least toxic nanoparticles which show symbiotic effect along with microbes for sustainable plant growth.

21.6 Conclusion

Microbial nanotechnology is a revolutionary tool in agriculture as it is an invasive and eco-friendly and a cost-effective and biocompatible process. Microbe-assisted biosynthesis of nanomaterials is a modern, green, and cost-effective method because it's devoid of harmful chemicals (reducing agent) and there is no need of expensive stabilizing agent (protein, DNA, peptides, etc.). Intracellular proteins act as reducing as well as stabilizing agent in microbial synthesis. There are many beneficial microbes in the soil but both bacteria and root surface are negatively charged. Here TiO_2nanoparticles act as a nanoadhesive between bacterial cell and plant root tissue. Iron oxide and tin oxide nanoparticles increase the C/N ratio of various microbes which help in plant growth. Nanosilica increases the population of beneficial bacterial community. Growth of bacterial colony enhances the recycling process of soil nutrients (N, P, K). Thus, nanosilica acts as nanofertilizer. Microbes also remove toxic metals in the form of nanostructure enabling recovery and recycling of metals. We have discussed briefly how microbial nanotechnology enables plant growth and revolutionizes sustainable and green agriculture practice.

References

Ahmad A, Senapati S, Khan MI, Kumar R, Sastry M (2003) Extracellular biosynthesis of monodisperse gold nanoparticles by a novel extremophilicactinomycete, thermomonospora sp. Langmuir 19(8):3550–3553

Antisari LV, Carbone S, Gatti A, Vianello G, Nannipieri P (2013) Toxicity of metal oxide (CeO_2, Fe_3O_4, SnO_2) engineered nanoparticles on soil microbial biomass and their distribution in soil. Soil Biol Biochem 60:87

Attard G, Casadesus M, Macaskie LE, Deplanche K (2012) Biosynthesis of platinum nanoparticles by *Escherichia coli* MC4100: can such nanoparticles exhibit intrinsic surface enantioselectivity? Langmuir 28:5267–5274

Bestawy EE, Helmy S, Hussien H, Fahmy M, Amer R (2013) Bioremediation of heavy metal-contaminated effluent using optimized activated sludge bacteria. Appl Water Sci 3:181–192

Bhattacharya R, Mukherjee P (2008) Biological properties of "naked" metal nanoparticles. Adv Drug Deliv Rev 60(11):1289–1306

Cabrera F, Madrid L, Arambarri DP (1977) Adsorption of phosphate by various oxides: theoretical treatment of the adsorption envelope. J Soil Sci 28:306–313

Capeness MJ, Edmundson MC, Horsfall LE (2015) Nickel and platinum group metal nanoparticle production by *Desulfovibrio alaskensis* G20. New Biotechnol 32:727–731

Cheung KH, Lai HY, Gu JD (2006) Membrane-associated hexavalent chromium reductase of Bacillus megaterium TKW3 with induced expression. J Microbiol Biotechnol 16:855–862

Cresnzi MD (1995) Structural surface investigations with low-energy backscattered electrons. Surf Sci Rep 21:89–175

Dickson JS, Koohmaraie M (1989) Cell surface charge characteristics and their relationship to bacterial attachment to meat surfaces. Appl Environ Microb 55:832–836

Diels L, Spaans PH, Roy SV, Hooyberghs L, Ryngaert A, Wouters H, Walter E, Winters J, Macaskie L, Finlay J, Pernfuss B, Woebking H, Pumpel T, Tsezos M (2003) Heavy metals removal by sand filters inoculated with metal sorbing and precipitating bacteria. Hydrometallurgy 71:235–241

Fang GD, Si YB, Tian C, Zhang GY, Zhou DM (2012) Degradation of 2,4-D in soils by Fe_3O_4 nanoparticles combined with stimulating indigenous microbes. Environ Sci Pollut Res 19:784–793

François F, Lombard C, Guigner JM, Soreau P, Jaisson FB, Martino G, Vandervennet M, Garcia D, Molinier AL, Pignol D, Peduzzi J, Zirah S, Rebuffat S (2012) Isolation and characterization of environmental bacteria capable of extracellular biosorption of mercury. Appl Environ Microbiol 78:1097–1106

Gholami A, Shahsavani S, Nezarat S (2009) The effect of plant growth promoting rhizobacteria (PGPR) on germination, seedling growth and yield of maize. Int J Biol Life Sci 1:35–40

Gonzalez TP, Lopez CJ, Neal AL (2010) Magnetite biomineralization induced by *Shewanella oneidensis*. Geochimicaet Cosmochimica Acta 74(3):967–979

Gordienko AS, Kurdish IK (2007) Surface electrical properties of Bacillus subtilis cells and the effect of interaction with silicon dioxide particles. Biophysics 52:217–219

Grasso G, Zane D, Dragone R (2020) Microbial nanotechnology: challenges and prospects for green biocatalytic synthesis of nanoscale materials for sensoristic and biomedical applications. Nano 10:11

He SY, Feng YZ, Ren HX, Zhang Y (2011) The impact of iron oxide magnetic nanoparticles on the soil bacterial community. J Soils Sediments 11:1408–1417

Husen A, Siddiqi KS (2014) Plants and microbes assisted selenium nanoparticles: characterization and application. J Nanobiotechnol 12:28

Jang E, Shim HW, Ryu BH, An DR, Yoo WK, Kim KK, Kim DW, Kim TD (2015a) Preparation of cobalt nanoparticles from polymorphic bacterial templates: a novel platform for biocatalysis. Int J Biol Macromol 81:747–753

Jang GG, Jacobs CB, Gresback RG, Ivanov IN, Meyer HM, Kidder M, Joshi PC, Jellison GE, Phelps TJ, Graham DE, Moon JW (2015b) Size tunable elemental copper nanoparticles: extracellular synthesis by thermoanaerobic bacteria and capping molecules. J Mater Chem C 3:644–650

Jha AK, Prasad K, Kulkarni AR (2009) Synthesis of TiO_2 nanoparticles using microorganisms. Colloids Surf B 71(2):226–229

Jia X, Shi N, Tang W, Su Z, Chen H, Tang Y, Sun B, Zhao L (2022) Nano-hydroxyapatite increased soil quality and boosted beneficial soil microbes. Plant Nano Biol 1:100002

Kara A, Rahman TS (1998) Vibrational properties of metallic nanocrystals. Phys Rev Lett 81:1453–1456

Karunakaran G, Suriyaprabha R, Manivasakan P, Yuvakkumar R, Rajendran V, Prabu P, Kannan N (2013) Effect of nanosilica and silicon sources on plant growth promoting rhizobacteria, soil nutrients and maize seed germination. IET Nanobiotechnol 7:70–77

Kinraide TB, Wang P (2010) The surface charge density of plant cell membranes (σ): an attempt to resolve conflicting values for intrinsic σ. J Exp Bot 61:2507–2518

Klaus T, Joerger R, Olsson E, Granqvist CG (1999) Silver based crystalline nanoparticles, microbially fabricated. Proc Natl Acad Sci U S A 96(24):13611–13614

Klonowska A, Heulin T, Vermeglio A (2005) Selenite and tellurite reduction by Shewanella oneidensis. Appl Environ Microbiol 71:5607–5609

Konishi Y, Ohno K, Saitoh N, Nomura T, Nagamine S, Hishida H, Takahashi Y, Uruga T (2007) Bioreductive deposition of platinum nanoparticles on the bacterium Shewanella algae. J Biotechnol 128:648–653

Lengke MF, Fleet ME, Southam G (2006) Synthesis of platinum nanoparticles by reaction of filamentous cyanobacteria with platinum (IV) chloride complex. Langmuir 22:7318–7323

Lin D, Xing B (2007) Phytotoxicity of nanoparticles: inhibition of seed germination and root growth. Environ Pollut 150:243–250

Meyer R, Prakash S, Entel P (2002) Capillary pressure and phonons in Ag, Au, Cu and Ni nanoparticles. Phase Transit 75(1–2):51–58

Mikhailov OV, Mikhailova EO (2019) Elemental silver nanoparticles: biosynthesis and bio applications. Materials 12:3177

Morita M, Uemoto H, Watanabe A (2007) Reduction of selenium oxyanions in wastewater using two bacterial strains. Eng Life Sci 7:235–240

Mukherjee P, Senapati S, Mandal D (2002) Extracellular synthesis of gold nanoparticles by the fungus *Fusarium oxysporum*. Chembiochem 3(5):461–463

Palmqvist NGM, Bejai S, Meijer J, Seisenbaeva GA, Kessler VG (2015) Nano titania aided clustering and adhesion of beneficial bacteria to plant roots to enhance crop growth and stress management. Sci Rep 5:10146

Pavani KV, Kumar NS, Sangameswaran BB (2012) Synthesis of lead nanoparticles by Aspergillus species. P J Microbiol 61:61–63

Pązik R, Andersson R, Kępiński L, Nedelec JM, Kessler VG, Seisenbaeva GA (2011) Surface functionalization of the metal oxide nanoparticles with biologically active molecules containing phosphonate moieties. Case study of $BaTiO_3$. J Phys Chem C 115:9850–9860

Rathgeber C, Yurkova N, Stackebrandt E, Beatty JT, Yurkov V (2002) Isolation of tellurite- and selenite-resistant bacteria from hydrothermal vents of the juan de fuca ridge in the pacific ocean. Appl Environ Microbiol 68:4613–4622

Rousk J, Ackermann K, Curling SF, Jones DL (2012) Comparative toxicity of nanoparticulate CuO and ZnO to soil bacterial communities. PLoS One 7(3):e34197. https://doi.org/10.1371/journal.pone.0034197.94

Saba I, Wani K, Syed A, Rehman S (2021) Role of microbial nanotechnology in bioremediation of heavy metals. In: Ansari MA, Rehman S (eds) Microbial nanotechnology: green synthesis and applications. Springer, Singapore. https://doi.org/10.1007/978-981-16-1923-6_15

Salunke BK, Sawant SS, Lee SI, Kim BS (2015) Comparative study of MnO_2 nanoparticle synthesis by marine bacterium *Saccharophagus degradans* and yeast *Saccharomyces cerevisiae*. Appl Microbiol Biotechnol 99:5419–5427

Seshadri S, Saranya K, Kowshik M (2011) Green synthesis of lead sulfide nanoparticles by the lead resistant marine yeast, *Rhodosporidium diobovatum*. Biotechnol Prog 27:1464–1469

Sharma V, Shukla RK, Saxena N, Parmar D, Das M, Dhawan A (2009) DNA damaging potential of zinc oxide nanoparticles in human epidermal cells. Toxicol Lett 185(3):211–218

Simkiss K, Wilbur KM (1989) Biomineralization. Elsevier Academic, New York

Tam K, Ho CT, Lee JH, Lai M, Chang CH, Rheem Y, Chen W, Hur HG (2010) Growth mechanism of amorphous selenium nanoparticles synthesized by Shewanella sp. HN-41. Biosci Biotech Bioch 74:696–700

Teo BK (1986) EXAFS: basics principle, data analysis. Springer, Berlin

Wong MF, Chua H, Lo W, Leung CK, Yu PH (2001) Removal and recovery of copper(II) ions by bacterial biosorption. Appl Biochem Biotechnol 91:447–457

Wu SM, Su Y, Liang RR, Ai XX, Qian J, Wang C, Chen JQ, Yan ZY (2015) Crucial factors in biosynthesis of fluorescent CdSe quantum dots in Saccharomyces cerevisiae. RSC Adv 5: 79184–79191

Yang L, Watts DJ (2005) Particle surface characteristics may play an important role in phytotoxicity of alumina nanoparticles. Toxicol Lett 158:122–132

Yates MD, Cusick RD, Logan BE (2013) Extracellular palladium nanoparticle production using Geobacter sulfurreducens. ACS Sustain Chem Eng 1:1165–1171

Yusof HM, Rahman NAA, Mohamad R, Zaidan UH, Samsudin AA (2020) Biosynthesis of zinc oxide nanoparticles by cell-biomass and supernatant of *Lactobacillus plantarum* TA4 and its antibacterial and biocompatibility properties. Sci Rep 10:19996

Chapter 22
Bacteriophage-Assisted Diagnostics and Management of Plant Diseases

Sanghmitra Aditya, Bhagyashree Bhatt, Yaratha Nishith Reddy, Ajay Singh Sindhu, and Gurudatt M. Hegde

Abstract The threat of insects and diseases to plants is constant in the natural world. One of the major pathogen groups that infect different plant species and cause diseases having detrimental impact on plant development and agricultural productivity is pathogenic bacteria. In order to manage bacterial plant diseases, chemical bactericides and antibiotics have been utilised extensively. However, the emergence of bacteria that are resistant to traditional antibiotics and bactericides, as well as their potential adverse effects on the environment and human health, necessitate the development of alternative and ecologically safe control strategies by the scientific community. Bacteriophages, or bacteria eater, have been shown to be potential agents for management of bacterial diseases as they are not harmful to the environment or human health and can only infect and destroy the target pathogenic bacteria very specifically. The ability to target certain disease-causing microorganisms with customised phage mixtures makes phage biocontrol superior than chemical controls. Phage combinations, as opposed to conventional pesticides, are easily adaptable to bacterial resistance that may arise over time. Moreover, this ability of bacteriophages to target specific bacterial species has also been used to create diagnostic tools for the detection of bacteria that cause plant diseases. We will provide a summary of how bacteriophages are used to control and detect plant diseases brought on by plant-pathogenic bacteria in this chapter. We will also discuss the benefits and drawbacks of utilising bacteriophages as management and detection tools.

S. Aditya (✉) · Y. N. Reddy
Division of Plant Pathology, Indian Agricultural Research Institute, New Delhi, India

B. Bhatt
Shoolini University, Solan, Himachal Pradesh, India

A. S. Sindhu
Division of Nematology, Indian Agricultural Research Institute, New Delhi, India

G. M. Hegde
University of Agricultural Sciences, Dharwad, Karnataka, India
e-mail: hegdegm@uasd.in

P. Mathur et al. (eds.), *Microbial Symbionts and Plant Health: Trends and Applications for Changing Climate*, Rhizosphere Biology,
https://doi.org/10.1007/978-981-99-0030-5_22

Keywords Bacteriophage · Detection · Diagnosis · Eco-friendly · Management · Pathogens

Abbreviations

q-PCR	Quantitative polymerase chain reaction
CFU/ml	Colony-forming unit per millilitre
PCR	Polymerase chain reaction
PFU/ml	Plaque-forming unit per millilitre

22.1 Introduction

Worldwide, bacterial infections continue to be a danger to agricultural output (Sharma et al. 2022). Even though they are less common than fungal diseases, effective management of bacterial diseases is difficult (Oerke 2006) because there are few effective bactericides available, high mutation rates or high rates of horizontal gene transfer, high variability and rapid population growth rates (Balogh et al. 2010). As a result, it's critical to have precise diagnosis and efficient control measures in order to minimise losses and prevent the disease from spreading (Born et al. 2017).

Rapid, cost-effective and reliable pathogen detection methods are needed to combat bacterial infections. The standard technique for detecting and identifying pathogenic bacteria remains culture-based approaches (Bell et al. 2016), but this method lacks sensitivity, specificity and accuracy and requires a lot of labour (Hagens et al. 2011; Cho and Ku 2017). Despite the recent surge in popularity of non-cultural molecular approaches for detection (Lau and Botella 2017), they are quite costly, demand a high degree of skill and necessitate a robust laboratory infrastructure (Lazcka et al. 2007). Serological techniques using antibodies lack sensitivity and can react with organisms with a similar structure (Law et al. 2015). In this situation, phage-based diagnostics provide a reliable and effective substitute for managing bacterial infections (Meile et al. 2020). Due to their great sensitivity to host bacterial strains, capacity to be produced in large quantities and ability to discriminate between living and dead bacteria, bacteriophages are potential tools for bacterial detection (Farooq et al. 2018). Phage typing techniques allow for the rapid and accurate identification of live bacterial cells since phages only infect living hosts (Eriksson and Lindberg 1977; Sun and Webster 1987). Phages are immensely helpful detection techniques because of their capacity to modify, lyse, isolate and eliminate their bacterial hosts (Jones et al. 2020; Smartt and Ripp 2011).

Antibiotics (such as streptomycin and oxytetracycline) and copper have historically been used as chemical-based bactericides to treat bacterial infections (McManus et al. 2002). However, due to the abuse or overuse of antibiotics or copper, many plant-pathogenic bacteria have developed resistance to these substances, which has led to a decline in their usage (Svircev et al. 2018). Sustainable and eco-friendly management techniques were implemented to solve this problem. Many substances, such as antagonists, naturally occurring molecules derived from antagonists and agents that boost plant immunity, have been created and are frequently used to combat plant diseases (Dennis and Webster 1971; Compant et al. 2005; Wiesel et al. 2014). About 100 years have passed since d'Herelle first envisioned the use of phages, or phage therapy, to the management of harmful microorganisms (d'Herelle 1917). Unfortunately, phage therapy research became obsolete by discovery of antibiotics in 1929 (Fleming 1929). Phage treatment research has been revitalised in response to the global issue of antibiotic resistance (Hagens and Loessner 2007). Even a century after their discovery, phages are prepared to fulfil their early promise and significantly contribute to the control and detection of bacterial infections (Keen 2014).

Bacteriophages, or bacterial viruses, are the most common biological organisms in the biosphere. They are made up of a protein capsule that contains either a DNA or RNA genome (Whitman et al. 1998; Hendrix 2002). Phages having DNA as their genetic material are more prevalent in nature. These bacterial viruses were known as "bacteriophages" (Greek, "bacteria eaters") because of their capacity to lyse bacterial cells (Citorik et al. 2014). Bacteriophages are parasitic obligatory intracellular organisms without a functioning metabolism. While some phages only have ten genes and are almost entirely reliant on bacterial cellular processes, others contain hundreds of genes and rely on proteins encoded by their own genetic material (Birge 1994). With a few noteworthy exceptions, such as Listeria phage A511, which is able to infect and control bacteria throughout an entire genus, they have a limited host range, indicating that they selectively infect and eliminate target bacteria without impacting others (Zink and Loessner 1992; Loc-Carrillo and Abedon 2011).

Phage biotechnology has grown out of its infancy and may now profit from a true interdisciplinary approach to bacterial detection and management, employing the advantages and methods of modern molecular microbiology and nanotechnology (Rider et al. 2003; Radke and Alocilja 2005). In this chapter, we place particular emphasis on the use of bacteriophages for the detection and management of plant-pathogenic bacteria.

22.2 Historical Background

In an attempt to propagate the vaccinia virus, Frederick William Twort provided the first indisputable description of bacterial viruses (Twort 1915). However, it was Canadian-French scientist Félix d'Herelle who is credited with making the "final discovery" of bacterial viruses and coining the term "bacteriophage" (d'Herelle 1917). In 1921, he also released a book on bacteriophages titled *The Bacteriophage: Its Role in Immunity*, which was later translated into Russian in 1926 (d'Herelle 1926). In reality, Ernest Hanbury Hankin first reported the occurrence of bacteriophages in the literature in the nineteenth century in the Indian rivers Jumna and Ganges (Hankin 1896); however, the conclusions drawn from this research were not helpful for further research.

Early in the twentieth century, it was thought that phages, also known as "exogenous agents of immunity" (d'Herelle 1917), were caused by bacteria that prompted autolytic enzyme activation (Northrop 1937; Stent 1960; Billiau 2016). The ability to directly observe phages using electron microscopy, which was developed in 1940, led to the broad acceptance of d'Herelle's theory that bacteriophages are viruses that infect bacteria (Summers 2005). The arrival of German physicist Max Ludwig Henning Delbruck is widely recognised as the pivotal event in contemporary phage research (Stent et al. 1966; Fischer et al. 1989). Delbruck and Salvador Luria began working together in the 1940s, and Alfred Hershey subsequently joined them in 1943 (Summers 1993). These three researchers were the original founders of the phage group, an unofficial organisation of researchers. Later, more members of their respective faculties joined (Letarov 2020). Delbruck chose a group of "authorised phages," later known as T-phages (Demerec and Fano 1945), realising the importance of doing research on a small number of phages in order to compare the results from various experimental laboratories. He then invited researchers to join the so-called phage treaty and conduct research on these particular viruses (Letarov 2020).

Since the 1960s, research on bacteriophages has swiftly grown from a niche area within the domains of microbiology, virology, and genetics to one of the most significant and actively researched areas in biology (Letarov 2020). By the end of the 1990s, bacteriophage biology research has transitioned from using a few traditional model systems to prokaryote viral biodiversity. This rapid expansion of examined objects can be attributed to high-throughput genomes and metagenomics technologies, bioinformatics analysis, and more productive structural virology approaches like cryo-electron microscopy reconstruction and cryo-electron tomography (Grafe 1991). The majority of non-tailed bacteriophage families were identified in the 1960s and 1980s, and a few of them have since emerged as significant molecular biology model organisms. In fact, the X174 phage was the first organism whose whole genome was sequenced in 1977 (Sanger et al. 1977).

However, research into the biology of phage has been overshadowed in the latter half of the twentieth century by the therapeutic potential of phage. The first study to describe the use of phage as a treatment for cutaneous boils was published in 1921 by Bruynoghe and Maisin. In the context of plant-pathogenic bacteria, affiliation of bacteriophages in management was originally documented in *Xanthomonas campestris* pv. *campestris* (Mallmann and Hemstreet 1924), and later in *Pectobacterium atrosepticum*, *Pectobacterium carotovorum* subsp. *carotovorum* (Coons and Kotila 1925; Kotila 1920) and *Pantoea stewartii* (Thomas 1935).

Several studies have demonstrated the use of bacteriophages as biosensors for detecting plant pathogens during host infections (Vu and Oh 2020). Sutton and Katznelson (1953) were among the first to use bacteriophages as biosensors to detect pathogenic bacteria in seed. Bacteriophages may be altered to act as reporter phages, which transmit or insert a reporter gene into the bacteria they infect, using cutting-edge genetic engineering techniques (Farooq et al. 2018). These genes' expression distinguishes bacteria as diagnostic signal indicators (Burnham et al. 2014).

Since the history of these studies is still developing, according to Letarov (2020), it is yet too early to analyse them in detail. In conclusion, studies on the bacteriophage life cycle have shown certain traits that distinguish viruses from other living things.

22.3 Types of Bacteriophages

The discovery of bacteriophages is attributed to Felix d'Herelle (d'Herelle 1917) and Frederick Twort (Twort 1915). It is well known that bacteriophages are exceedingly prevalent throughout the biosphere; they may be found anywhere that host bacteria continue to exist. Their density in a particular environment might reach 10^8–10^9 phage particles per millilitre of water or per gram of soil (Havelaar 1987; Marsh and Wellington 1994; Seeley and Primrose 1982).

Bacteriophages are made up of linear or circular, single- or double-stranded DNA or RNA molecules that are wrapped in capsids, which are protein or lipoprotein coverings. Additionally, certain bacteriophages may have specific features that make it easier for them to engage with bacterial hosts or allow them to introduce genetic information into host cells (Bradley 1965).

According to the nature of infection, bacteriophages are divided into two groups. Lytic infection distinguishes the first group, whereas a lysogenic, or temperate, kind of infection characterises the second (Fig. 22.1). In the first stage of infection, the release of DNA causes the host bacterium's protein machinery to turn to their advantage, producing 50–200 new phages. Nearly all of the cell's resources are used to produce so many new phages, causing the cell to become fragile and explode. In other words, lysis occurs, which results in the host bacterial cell dying. New phages are consequently discharged into the extracellular environment. The other mode of infection, called lysogenic, is characterised by the integration of the phage DNA into the genome of the host cell; however, it can also exist as a plasmid. Phage DNA that has been incorporated will multiply alongside the host bacterium's genome, and any new bacteria will take on the viral DNA. The bacteria might undergo such a viral DNA shift over a number of generations without suffering any significant metabolic repercussions. Eventually, under specific conditions that hinder the bacterial state, the phage genes will switch back to the lytic cycle, resulting in the release of fully completed phages. Analysis of phages with lysogenic or lytic modes of infection revealed a vast array of bacteriophages with distinct features for each form of infection.

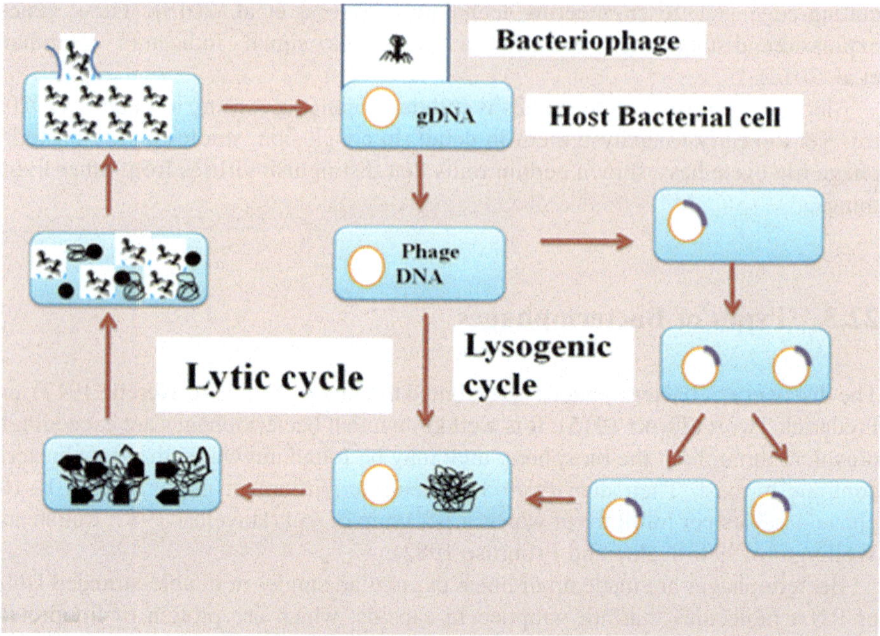

Fig. 22.1 General life cycles of bacteriophage

22.4 Role of Bacteriophages in Plant Disease Diagnostics

The traditional methods for detecting bacteria rely mostly on methods for isolating and cultivating them, which are then verified using various biochemical assays. For various purposes, these methods are mostly arduous and time-consuming. The polymerase chain reaction (PCR) (Berg et al. 2005; Dreo et al. 2014), enzyme-linked immunosorbent assay (ELISA) (Wu et al. 2015), lateral flow immunoassay (LFIA) (Soriano et al. 2017) and mass spectrometry (MS) (Bilt et al. 2018) are the main components of traditional bacterial detection methods. These traditional methods are typically expensive and rely on bacterial supplementation, specialised equipment and trained operators. These methods have a number of other drawbacks. For instance, nucleic acid (NA) and MS-based methods can yield false-positive results on encountering dead bacteria. Similar to this, ELISA is unable to distinguish between living and dead bacteria, and the scarcity of specific antibodies further limits its use (Farooq et al. 2018).

Bacteriophage are naturally occurring bacteria-eating organisms, and because of their high sensitivity, selectivity, cheap cost and better thermostability, they are emerging as a feasible option for pathogen detection (Fang and Ramasamy 2015; Hoang and Nhung 2018). Bacteriophages (phages) have developed the ability to quickly deliver their viral genomes to their hosts and reproduce with unmatched specificity. It should come as no surprise that phages and proteins encoded by phages have been employed to create a wide range of diagnostic tests, many of which perform better than traditional culture-based and molecular detection techniques (Meile et al. 2020).

Bacteriophages are the most prevalent kind of biological organisms on Earth and are ubiquitous bacterial viruses (Güemes et al. 2016). They typically exclusively infect specific strains of host bacteria and hence become effective and ecologically acceptable solution for bacterial pathogen identification and management. The likelihood of false-positive diagnostic results is decreased by their specificity, which prevents infection of commensal environmental microorganisms (Born et al. 2017). The bacteriophages have emerged as possible instruments for bacterial detection due to their great specificity to host bacterial strains, the capacity for huge production, tolerance to adverse environments and the ability to distinguish between living and dead bacteria (Farooq et al. 2018).

The models that are used for detection of plant-pathogenic bacteria are phage typing, reporter phages and phage progeny-based detection (Fig. 22.2).

22.4.1 Phage Typing

It is a phenotypic technique where bacteriophages (phages) are employed to precisely infect their bacterial host and induce cell lysis as a means to specifically identify target bacteria. The precise binding of phages to antigens and receptors on the surface of bacteria and the subsequent bacterial lysis (or lack thereof) are the basis for phage typing. Adsorption is a term used to describe the binding process. A phage may go through either the lytic cycle or the lysogenic cycle after adhering to a bacterial surface. Based on its lysis pattern, the bacterial strain is given a type. Throughout the 1900s, phage typing was employed to determine how infectious epidemics originated; however, genotypic techniques have since replaced this technique. A number of bacteriophages are plated together with the bacterial sample to be typed to generate a bacterial lawn, where clearings and plaque development indicate bacteriophage replication and host sensitivity to the particular bacteriophage (Pradeep and Nargund 2015). Most bacterial species have phage typing schemes, which are occasionally used to identify certain strains within a species. Phage typing is a rather easy operation. Despite being reliable, the process necessitates the maintenance of several phage stocks and bacterial propagation strains. Phage typing requires the use of pure bacterial cultures; hence, complicated clinical or environmental samples cannot be used with this procedure (Schofield et al. 2013).

22.4.2 Reporter Phages

Reporter phage methods may be thought of as a more sophisticated kind of phage typing techniques based on the use of genetically altered phages (Singh et al. 2012).

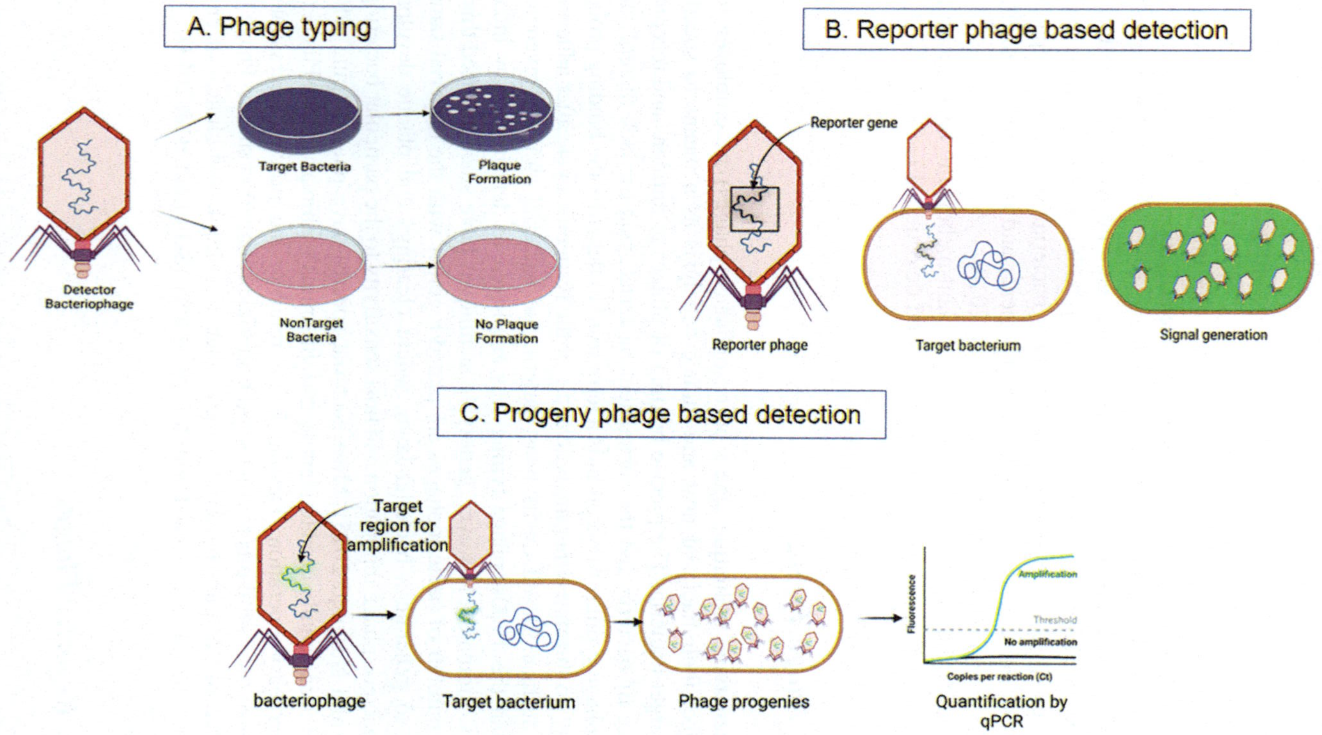

Fig. 22.2 The working models of bacteriophages to detect plant-pathogenic bacteria

In order to develop a genetically altered phage that can emit a detectable signal, the method depends on integrating reporter genes into the phage genome (usually colorimetric, fluorescent or bioluminescent). The reporter phage is unable to generate a signal in the absence of a target cell. The reporter phage attaches to certain receptors on the bacterial cell wall and infects the cell if the target cell is present. Following the expression of phage and reporter genes, a signal enables the bacteria to be positively identified (Schofield et al. 2013). In this method, phage characteristics may be deliberately modified to achieve desired outcomes, including improved killing or detection. Reporter phages promote the synthesis of a readily traceable enzyme or protein for detecting purposes. Common reporters include fluorescent proteins, glycosidases, luciferases from bacteria or insects and others (Born et al. 2017).

22.4.3 Phage Progeny-Based Detection

This method employs a particular bacteriophage to produce fast amplification of progenies prior to identification (Kutin et al. 2009). The bacterial host's strain can be identified later on using released progeny phages or the contents of the bacterial cell (such as ATP, DNA, RNA and bacterial proteins). A phage usually injects its genome and internal proteins into the bacterium after adhering to the host cell successfully. It then manipulates the host's metabolic processes to start producing progeny phages, which are then expelled upon host cell lysis. A rise in phage counts suggests that the sample has a vulnerable host. Traditional plaque tests with proper indicator strains can be used to evaluate the change in phage titre. ELISA-based assays can also be used to physically detect the phage particles and measure the rise in phage nucleic acid content. Real-time PCR techniques have also been developed for detecting the rise in phage DNA produced during infection (Meile et al. 2020).

22.5 Successful Detection and Diagnosis of Plant Diseases Using Bacteriophages

*22.5.1 Fire Blight (*Erwinia amylovora*)*

Phage engineering requires the availability of genome sequences, and fully sequenced *E. amylovora* viruses allow for the creation of recombinant phages for quick diagnosis. The purpose of this work was to generate a bioluminescent reporter phage by introducing several heterologous genes into the virulent *E. amylovora* phage Y2 (Born et al. 2017). Due to its wide host range and excellent specificity for *E. amylovora*, Y2 is an ideal option. In terms of the reporter phage, the preferred reporter was a *Vibrio harveyi* luciferase that was delivered as a luxAB fusion. The bioluminescent detection of viable *E. amylovora* cells was quick, precise and sensitive due to the recombinant reporter phage. A reporter phage was created by introducing the luxAB fusion gene into the Y2 genome. Y2::luxAB. Y2::luxAB infection allows for the quick and accurate identification of *E. amylovora* viable cells within 1 h. By monitoring the luciferase activity that Y2::luxAB transduces in the bacterium, *E. amylovora* cells may be quickly identified. Following infection with Y2::luxAB, it was discovered that the threshold for direct detection (without enrichment) of *E. amylovora* cells was as low as 3.8103 CFU/ml. The identification of *E. amylovora* by Y2::luxAB from genuine field samples was extremely effective and precise, and the phage activity was unaffected by other bacteria that were present on the tissue or plant extracts. Because of Y2's wide range of hosts, the trials showed that random wild-type *E. amylovora* field isolates are all susceptible to Y2::luxAB infection, which reduces the likelihood of false-negative findings. Y2 has an exceptionally minimal probability of producing false-positive results due to its high host specificity. Furthermore, it appears that the Y2::luxAB assay's sensitivity is enough for in vivo detection in most cases. Luciferase reporter phage Y2::luxAB appears to be superior to ELISA-based testing and offers a valuable mix of viable cell selectivity, high sensitivity and speed (Born et al. 2017).

22.5.2 *Bacterial Blight of Crucifers* (**Pseudomonas cannabina** *pv. alisalensis*)

A recombinant "luxAB-tagged" reporter phage was created by modifying the *P. cannabina* pv. *alisalensis* phage PBSPCA1 genome to include the bacterial luxAB reporter genes (encoding the luciferase protein). The reporter phage infects the cell after attaching to a certain receptor. With the addition of the substrate n-decanal and in the presence of oxygen and a flavin mononucleotide, the luciferase catalyses a complex reaction that results in light emission. As a result, signal generation is strictly dependent on host metabolic activity. The host's transcriptional and translational machinery is used to express the luxAB genes, which are regulated by phage promoters, and ultimately result in the production of the luciferase enzyme. It is the first experiment using a bioluminescent reporter phage for a significant plant pathogen. The test works well with cultured isolates as well as naturally infected specimens. The reporter phage is unable to generate a signal in the absence of a target cell. The reporter phage attaches to certain receptors on the bacterial cell wall and infects the cell if the target cell is present. Following the production of phage and reporter genes, a signal enables the bacteria to be positively identified. Upon reporter phage addition, PBSPCA1::luxAB identifies cultured *P. cannabina* pv. *alisalensis* in 20 min. The detection limits are around 10^3 CFU/mL, which is adequate to detect *P. cannabina* pv. *alisalensis* in sick plant specimens. Reporter phage technology, in contrast to other detection techniques, such as PCR, hybridisation and immunoanalysis, will only identify alive cells, whereas the others will detect both living and dead cells (Schofield et al. 2013).

22.5.3 *Bacterial Wilt (*Ralstonia solanacearum*)*

Based on the combination of the quick self-replication capacity of bacteriophages with quantitative PCR, a sensitive, specific and quick procedure for identifying *Ralstonia solanacearum* from soil and plant tissues was created (q-PCR). For their capacity to exclusively infect and lyse *R. solanacearum*, six bacteriophages were identified and chosen. The sensitivity of 63 *R. solanacearum* strains and 72 isolates of different bacterial species to the bacteriophages was examined. For the creation of the bacteriophage-indirect assay, the bacteriophage M DS1, which specifically infected 61 of 63 *R. solanacearum* isolates, was chosen. PCR primers were synthesised using distinct *R. solanacearum*-specific bacteriophage gene sequences, and the method was applied to identify *R. solanacearum* from a variety of substrates. After an hour of incubation with 5.3×10^2 PFU/ml M DS1, the procedure consistently detected about 3.3 CFU/ml in pure *R. solanacearum* cultures. The methodology was also utilised to find *R. solanacearum* in effluent water from pots containing infected ginger plants, in various tissues of infected potted ginger (leaves and roots) and in soil contaminated with *R. solanacearum* after a 100-fold dilution of the sample extracts. By employing *R. solanacearum*-specific bacteriophages to quickly achieve target amplification before molecular detection, the goal of this study is to increase sensitivity and speed of the detection process (Kutin et al. 2009).

22.6 Advantages of Bacteriophage-Mediated Diagnostics

1. Extremely specific to host bacteria, high selectivity and low cost of the phage are advantages of using it as a tool for detection. In addition, bacteriophage-based sensors are more thermostable than antibody-based ones, allowing for detection over a wider range of temperatures and a longer shelf life (Bárdy et al. 2016).
2. Due to their natural ability to divide exclusively in a living bacterium, bacteriophage-based biosensors can distinguish between live and dead bacterial pathogens (Altintas et al. 2015), which reduces false-positive signals during testing (Fang and Ramasamy 2015).
3. The capacity to create high titres of offspring phages: Infecting a single host bacterium with a phage particle results in hundreds of particles, aiding in pathogen identification.
4. The ability to withstand challenging circumstances, such as chemical solvents, extremely high temperatures, and a range of pH, in comparison to Abs, as well as storage at room temperature without endangering their activity (Bárdy et al. 2016).
5. Easy and affordable mass manufacturing (Farooq et al. 2018).

22.7 Disadvantages of Bacteriophage-Mediated Diagnostics

1. Because of the difficult sample preparation requirements, measuring the pathogen in an actual sample using phage-based biosensors could take longer (Tlili et al. 2013).
2. The time needed for the isolation and selection of a phage that exhibits the appropriate host range, i.e. a phage that exhibits species specificity and broad-strain infectivity, is a limiting factor in the development of a diagnostic test (Schofield et al. 2013).
3. In addition, the utility of bacteriophage-based sensors for the bulk of crops afflicted by fungus infections is highly constrained because they can only be used to detect bacteria rather than fungi or viruses (Fang and Ramasamy 2015; Vu and oh 2020).
4. The increased size of the phages also works against their ability to be integrated into some detection platforms (Singh et al. 2012).
5. The relatively small lambda phage capsid (53 kb) packaging capacity is a major drawback of this system, making larger genomes unsuitable for plasmid construction, especially for simultaneous insertion of reporter genes (Marinelli et al. 2012).
6. Adding new genetic material to a phage genome without deleting any existing sequences may potentially go beyond the capsid size restrictions and result in the uncontrolled loss of other genes, which might diminish the phage's infectivity or perhaps jeopardise its viability (Born et al. 2017).

22.8 Role of Bacteriophages in Plant Disease Management

The production of food from agricultural products has long been hampered by plant diseases. Bacterial plant diseases also significantly contribute to the decline in crop productivity in a number of crops. Although there are several strategies for controlling bacterial plant diseases, including the use of antibiotics and copper-based compounds (McManus et al. 2002), there are certain restrictions on their routine application. The abuse of antibiotics has sparked worries about the emergence of resistance genes in the dominant bacterial populations, and the overuse of copper sprays has a detrimental impact on the environment. As a result, the use of biological control agents is being emphasised increasingly, and bacteriophages are emerging as effective biocontrol agents with few side effects. Bacteriophages have been used in a variety of applications, including soil soaking, foliar spraying and seed treatment. Phages, in combination with biocontrol agents and plant inducers, were proven to be efficient in integrated disease management (Obradovic et al. 2005) and are now officially recommended against bacterial spot of tomato in Florida. There are also commercially available bacteriophage mixtures. Bacteriophage spraying on geranium leaves also lessens bacterial blight (Flaherty et al. 2000). Phages have been used to treat potato tubers infected with *Streptomyces scabies* (McKenna et al.

2001), reduce the incidence of *Xanthomonas* blight on onions (Lang et al. 2004) and treat citrus bacterial canker (Balogh et al. 2005). Bacteriophage use has reduced illnesses in pear blossom and orchids with effectiveness comparable to streptomycin (Svircev et al. 2006). Soil drenching in the tomato rhizosphere decreased the disease incidence of *Ralstonia solanacearum*-caused tomato wilt (Elhalag et al. 2018). Infection with XacF1 decreased the growth rate, motility and exopolysaccharide synthesis of *Xanthomonas campestris* pv. *citri*, reducing its potential for pathogenicity (Ahmad et al. 2014). Using foliar phage spray, disease incidence of *Xanthomonas campestris* pv. *campestris*, *Xanthomonas euvesicatoria* and *Pectobacterium carotovorum* subsp. *carotovorum* was also reduced (Lim et al. 2013). RSM phages increased the expression of pathogenesis-related genes in tomato plants, causing loss of variance in *Ralstonia solanacearum* (Addy et al. 2012). Under glasshouse and field conditions, a mixture of phages was shown to be efficient against the two main races of the pathogen *X. campestris* pv. *vesicatoria* (Flaherty et al. 2000). Table 22.1 presents list of different bacteriophages used for the management of diseases.

22.8.1 Xanthomonas

There are as many as 148 phages that infect *Xanthomonas*, and they are all described in the literature as having different host ranges, including restricted, wide and polyvalent host ranges. Not only does the host range of phages vary, but so does their lytic activity, which ranges from 13 to 100% for various phages. Numerous isolates of the genus *Xanthomonas* are susceptible to the polyvalent *Xanthomonas* phage Pg125 (Sutton et al. 1958). The additional phages are Xcu-Pl, Xcu-P3, Xve-P1 and Xca-P1 which are lytic to the pathovars of *Xanthomonas campestris*. An efficient alternate strategy for managing plant diseases is to use lytic *Xanthomonas* phages, which have various host ranges. The likelihood of a phage attack on beneficial bacteria is reduced by the high level of host specificity (Gašić et al. 2011). Additionally, *Xanthomonas* phages are available for the biocontrol of citrus canker and pepper spot. The first application of *Xanthomonas* phage occurred in the early nineteenth century by Mallmann and Hemstreet (Mallmann and Hemstreet 1924). They noticed that *Xanthomonas campestris* pv. *campestris* was inhibited by the filtrate from decaying cabbage. The phage XOF4 inhibited the development of *Xanthomonas oryzae* pv. *oryzae*, the cause of bacterial leaf blight of paddy. When compared to untreated seedlings, treatment with XOF4 at a titre of 1×10^8 pfu/ml prevented the development of symptoms. All strains of *Xanthomonas oryzae* pv. *oryzae* were lysed by the *Siphoviridae*-related phage XOF4, which has a burst size that results in a titre of 1.8×10^7 pfu/ml and a short latent period (Ranjani et al. 2018). Although certain studies (Ranjani et al. 2018; Nga et al. 2021) have demonstrated the successful management of disease with a single phage, a combination of phages must be employed for the management of bacterial plant diseases.

Table 22.1 List of bacteriophages used in the diagnosis and management of plant pathogens

Pathogen	Bacteriophage	References
Xanthomonas sp.	*Xanthomonas* phage Pg125	Sutton et al. (1958)
Pseudomonas aeruginosa	Phages B3 and D3112	Heo et al. (2007)
Pectobacterium carotovorum subsp. *carotovorum*	ZF40 bacteriophage	Panshchina et al. (2007)
Ralstonia solanacearum	M DS1	Kutin et al. (2009)
Ralstonia solanacearum	RSM phages	Addy et al. (2012)
P. cannabina pv *alisalensis*	PBSPCA1::luxAB	Schofield et al. (2013)
Xanthomonas campestris pv. *citri*	XacF1 decreased	Ahmad et al. (2014)
Dickeya solani	ΦD3 and ΦD5 bacteriophages	Czajkowski et al. (2014)
Pectobacterium carotovorum subsp. *carotovorum*	PM1 and PM2 bacteriophages	Lim et al. (2014, 2015)
Xylella fastidiosa	Xf phage	Das et al. (2015)
Pseudomonas syringae pv. *actinidiae*	KHUφ38 phage	Yu et al. (2016)
Pectobacterium carotovorum subsp. *carotovorum*	PP2 phage	Lim et al. (2017)
Xanthomonas oryzae pv. *oryzae*	XOF4 phage	Ranjani et al. (2018)
Pectobacterium carotovorum	Nepra, Nobby, Slant, Lelidair, Gaspode and Momine phage	Carstens et al. (2019)
Erwinia amylovora	Hena 1 phage	Besarab et al. (2020)
Ralstonia solanacearum	Phage RpY1	Lee et al. (2021)
Xanthomonas campestris pv. *campestris*	Seregon phage	Erdrich et al. (2022)

The selection of phages for the mixture is based on their effectiveness to reduce pathogens and avoid the formation of resistance in pathogens (Chan et al. 2013).

22.8.2 *Ralstonia solanacearum*

Ralstonia solanacearum is a gram-negative soil-borne bacterium that causes bacterial wilt in its solanaceous host crops. *R. solanacearum* has a few bacteriophages that have been identified and studied. Only a few strains of the pathogens are susceptible to infection by the virulent phage P4282 (Ozawa et al. 2001) and PK-101 (Toyoda et al. 1991), which also has a limited host range. By infecting the pathogen's strain

M4S, phage P4282 demonstrated how phage-mediated protection might suppress the bacterial wilt of tobacco (Tanaka et al. 1990). The strains of *R. solanacearum* are classified into five races according to host range and five biovars according to biochemical and physiological characteristics (Hayward 2000). Several phages that demonstrate selectivity in infecting certain races or biovars of *R. solanacearum* strains have been identified and classified, and two of the phages, RSS1 and RSM1, were filamentous (Yamada et al. 2007). For diagnostics, RSS1 and RSM1 can be employed (Kawasaki et al. 2007). All but one of the 18 examined strains of races 1, 3 or 4 and biovars 1, N2, 3 or 4 were sensitive to this phage, which is a P2-like head-tailed virus (*Myoviridae*). Another myovirus, RSL1, significantly lysed 17 out of 18 distinct strains and has a relatively large genomic DNA of about 240 kbp. RSB1 actively kills host cells and creates enormous, transparent plaques. As an alternative to utilising a phage cocktail made up of highly lytic phages, Fujiwara et al. (2011) suggested using a unique phage, such as RSL1.

22.8.3 Dickeya *and* Pectobacterium

Dickeya and *Pectobacterium* are the most important bacterial plant pathogens that cause severe losses (Mansfield et al. 2012). Bacteriophages can be readily isolated for their management. Lytic bacteriophages infecting soft rot-causing bacteria, *D. solani* (Adriaenssens et al. 2012) and *P. carotovorum* subsp. *carotovorum* (Lee et al. 2012), have been sequenced recently. All the phages have some common features in the genome organisation but no homology can be found in sequence databases. Lack of similarity to any known sequences is a major obstacle in bacteriophage research (Bossche et al. 2014). Czajkowski et al. (2014) described nine lytic bacteriophages infecting various *Dickeya* species, belonging to *Myoviridae* family with identical morphologies of capsid. ΦD3 and ΦD5 bacteriophages had wider host ranges infecting isolates of *D. dianthicola*, *D. solani*, *D. dadantii* and *D. zeae* but are unable to infect *Pectobacterium* spp. The bacteriophages reduced the population of *D. solani*, in vitro and on potato tuber surfaces, thus retaining the potential to be effective biocontrol agents. PM1 (Lim et al. 2014), PM2 (Lim et al. 2015), PP1 (Lee et al. 2012; Lim et al. 2013) and ZF40 (Korol and Tovkach 2012) bacteriophages are also effective against *Pectobacterium* spp. PM1 and PM2 bacteriophages are lytic to *P. carotovorum* subsp. *carotovorum* (Lim et al. 2014, 2015). The host range of PM1 is not known but PM2 is lytic to *P. carotovorum* subsp. *carotovorum* and *P. carotovorum* subsp. *brasiliense* strains, and it is unable to *infect P. wasabiae*, *P. betavasculorum*, *P. atrosepticum* and *P. carotovorum* subsp. *odoriferum*. ZF40 bacteriophage is the temperate phage infecting *P. carotovorum* subsp. *carotovorum.* ZF40 bacteriophages are P2-like phages belonging to the *Myoviridae* family in the order *Caudovirales* (Panshchina et al. 2007).

22.8.4 *Xylella fastidiosa*

The xylem-limited bacterium *Xylella fastidiosa* is the causal agent of several plant diseases, Siphophages Sano and Salvo and podophages Prado and Paz are *Xanthomonas* spp. phages. The Nazgul-like phage type contains a gene that codes for lipoprotein endolysin which indicate the lysis pathway for phages of Gram-negative hosts. The four phages are type IV pilus dependent for infection of both *X. fastidiosa* and *Xanthomonas* (Ahern et al. 2014). It has also been reported that a combination of four virulent (lytic) Xf phages is effective against *X. fastidiosa* (Das et al. 2015). Four virulent phages for *X. fastidiosa* isolated were Xfas103, Xfas106, Xfas303 and Xfas304. Siphophages Xfas103 and Xfas106 exhibit over 80% nucleotide identity to each other and are syntenic to phage BcepNazgul. The phages may be useful as bioagents for an effective and environmentally friendly strategy for the control of diseases caused by *X. fastidiosa* (Ahern 2013). The *Xylella fastidiosa* phage Xfas53, which produces plaques on the sequenced strain, was obtained from the supernatants of strain 53 of the bacterium. Xfas53 generates short-tailed virions that resemble podophage P22 in appearance. It contains two BroN domain-containing proteins possibly involved in lysogenic control (Summer et al. 2010).

22.8.5 *Erwinia amylovora*

Erwinia amylovora, a member of the *Enterobacteriaceae*, is the causal organism of fire blight, a serious disease of the pome fruit (Vanneste 2000). The pathogen is controlled by the antibiotic streptomycin and pruning. Alternative management is through bacteriophages which are able to infect *E. amylovora*, and an avirulent yellow bacterium commonly found in the orchard ecosystem was discovered (Hendry et al. 1967). The yellow bacterium was subsequently identified as *Pantoea agglomerans* (*Erwinia herbicola*). *E. amylovora* phages may play an important role in the epidemiology of fire blight and phages can be efficiently utilised as biological control agents. *E. amylovora* phages isolated fire blight-infected trees from *E. amylovora* 110, and phages PEa1 and PEa7 were categorised into two distinct groups on the basis of chemical and physiological properties. *E. amylovora* phages collected from orchards with fire blight symptoms were characterised by plaque morphology, PCR, restriction fragment polymorphisms, pulsed-field gel electrophoresis and host range studies (Schnabel and Jones 2001).

22.8.6 *Pseudomonas* Phages

Phage fKZ was isolated in Kazakhstan in 1975, which belongs to unusually large virulent phages that specifically infect *Pseudomonas* species. Twenty giant phages have been reported, which are able to infect *Pseudomonas stutzeri*, *P. fluorescens*, *P. aeruginosa* and *Pseudomonas chlororaphis* (Krylov et al. 2007). They are the largest sequenced. They are packed into a series of highly condensed layers, separated by 24 Å, which follow the contour of the inner wall of the capsid (Fokine et al. 2005). Phages have a wide host range and form a distant branch in *Myoviridae* family (Mesyanzhinov et al. 2002; Thomas et al. 2008). fCTX is a temperate phage of *Pseudomonas*, cytotoxin (CTX)-converting myovirus, which infects the PAS10 strain of *P. aeruginosa*. Upon integration at the 3′-end of the serine tRNA gene of the bacterial genome, the phage-encoded ctx gene enables *P. aeruginosa* to produce CTX, a pore-forming cytotoxin that increases host virulence. Phages B3 and D3112 are the best studied temperate, transposable phages of *P. aeruginosa*. They are easily isolated from natural populations of this bacterium. Although few isolates are sequenced, over 100 D3112-related phages have been reported in the literature (Heo et al. 2007).

22.9 Advantages of Using Bacteriophage Over Other Biocontrol Agents

Phages may be easily isolated from bacteria found in soil, water, plants and animals and are present in nature (Adams 1959). Bacteriophages exclusively harm the specific target bacteria they are designed to attack, not other bacteria (Loc-Carrillo and Abedon 2011). In the presence of the host bacteria, they self-replicate, and when the host bacteria are absent, they self-limit. The identification of plant pathogens is aided by this specificity (Farooq et al. 2018). As a result, they can be utilised in conjunction with bacteria that are antagonistic to the pathogen to inhibit it or to encourage beneficial microbes. Bacterial receptors required for pathogenesis can be targeted by phages, which results in decreased virulence in resistant mutants (Lang et al. 2004). The eukaryotes are not harmed by the phage (Greer 2005). Bacteriophages can be used in situations where chemical control cannot be used because of legal restrictions. Phage preparations can be made quite quickly and cheaply, and they can be kept at 4 °C for months without significantly losing titre. Application may be done with ordinary farm equipment. Due to their insensitivity to agrochemicals, phages can be tank-mixed with a variety of agrochemicals (Balogh et al. 2005). Bacteriophages have been shown to have no harmful effects on plants or animals (Greer 2005), making them potentially useful in the future. Phages and plants are thought to not interact directly. The presence of phage-like genes has also been reported in wheat, corn and *Arabidopsis* cress (Hedtke et al. 1997; Ikeda and Gray 1999), suggesting that phage DNA can be absorbed into the genomes of the crops to reduce disease losses.

22.10 Disadvantages of Using Bacteriophage Over Other Biocontrol Agents

A sufficient phage population must be available at key points in a pathogen's disease cycle in order to manage plant disease effectively. As a result, another crucial element is the bacteriophage's short-term inability to adhere to the surfaces of the host plant, which is caused by a number of environmental conditions. Use of bacteriophages for effective disease control is greatly hampered by the phages' deactivation by UV radiation.

The bacteriophage must touch the target in sufficient numbers for the biocontrol therapy to be successful. The initial phage concentration, rates of virion decay, the ability of the phage to replicate in the target environment, the concentration and location of the target bacteria and the availability of sufficient water as a medium for phage diffusion all affect the likelihood of phage-bacterium interaction (Gill and Abedon 2003). The timing of phage application, the relative fitness of phage-resistant bacterial mutations and the environment can all have an impact on how effectively a disease is controlled.

In the rhizosphere or phyllosphere, phages are used to suppress plant diseases. Low rates of diffusion occur via the heterogeneous soil matrix, and they vary depending on the amount of free water that is accessible. Phages may become stuck in biofilms (Storey and Ashbolt 2001), reversibly bind to soil particles like clay or become inactive due to low soil pH. Physical obstacles may stop bacteria from coming into touch with phages. Only a few live phages are available to lyse target bacteria due to low phage diffusion rates and high phage inactivation rates. The requirement for large populations of both phage and bacteria in order to initiate the cascade of bacterial lysis is another issue. The phyllosphere is an unforgiving environment, and phages used on aerial tissues quickly lose their effectiveness (Balogh 2002). One of the main restrictions on the use of bacteriophages is their temporary survival on plant leaf surfaces. High temperatures, extreme pH levels, sunshine exposure and rain draining render viruses inactive. Sunlight's UV-A and UV-B spectra (280–400 nm) are the most harmful environmental component. The effectiveness of phage therapy was hampered by the short residual activity of the control agents.

22.11 Conclusion

Bacteriophages are bacteria-infecting viruses. Since their discovery at the beginning of the twentieth century, they have undergone a thorough evaluation for the detection and management of all bacterially induced plant diseases. Because they are common in nature, self-replicating, targetable against bacterial receptors necessary for pathogenesis, nontoxic to eukaryotes and specific to particular bacterial species or strains, without harming other potentially beneficial members of the local flora, bacteriophages have a great deal of potential. Phages are quite simple to make, store and isolate. They could also be useful in places with less advanced technology infrastructure. The utilisation of phages comes with a number of difficulties. The bacteria can quickly change and develop a resistance to a certain phage. Consequently, it is dangerous to use a single phage for the intended purpose. Phage populations may be lowered by environmental conditions such as temperature changes, leaching, desiccation, exposure to UV radiation from the sun and exposure to sunlight in general. Particularly, UV will result in a significant decline. Phages can be used with protective coatings or carrier bacteria in which they can proliferate to extend their lifespan. Applying the phage in the evening or early morning as opposed to during times of intense solar irradiation will also help to reduce the decline. These strategies increase the chance of control by prolonging the duration of high phage populations in the target location. Phages have the potential to be used in the detection and diagnosis of bacterial plant diseases, as well as in integrated disease management methods, particularly when combined with other biocontrol agents.

References

Adams MH (1959) Bacteriophages. Interscience, New York

Addy HS, Askora A, Kawasaki T, Fujie M, Yamada T (2012) Utilization of filamentous phage φRSM3 to control bacterial wilt caused by *Ralstonia solanacearum*. Plant Dis 96(8):1204–1209

Adriaenssens E, Ackermann HW, Anany H (2012) A suggested new bacteriophage genus: "Viunalike virus". Arch Virol 157:2035–2046

Ahern, Stephen J (2013) Novel virulent phages for *Xylella fastidiosa* and other members of the Xanthomonadaceae. Doctoral dissertation, Texas A & M University

Ahern SJ, Das M, Bhowmick TS, Young R, Gonzalez CF (2014) Characterization of novel virulent broad-host-range phages of *Xylella fastidiosa* and Xanthomonas. J Bacteriol 196(2):459–471

Ahmad AA, Askora A, Kawasaki T, Fujie M, Yamada T (2014) The filamentous phage XacF1 causes loss of virulence in *Xanthomonas axonopodis* pv. *citri,* the causative agent of citrus canker disease. Front Microbiol 5:321

Altintas Z, Pocock J, Thompson KA, Tothill IE (2015) Comparative investigations for adenovirus recognition and quantification: plastic or natural antibodies? Biosens Bioelectron 74:996–1004

Balogh B (2002) Strategies of improving the efficacy of bacteriophages for controlling bacterial spot of tomato. MS thesis Univ FL, Gainesville

Balogh B, Jones JB, Momol MT, Olson SM (2005) Persistence of bacteriophages as biocontrol agents in the tomato canopy. In: Proceedings of international symposium on tomato diseases, 1st, Orlando, FL. ISHS Acta Hortic, vol 695, p 299

Balogh B, Jones JB, Iriarte FB, Momol MT (2010) Phage therapy for plant disease control. Curr Pharm Biotechnol 11:48–57

Bárdy P, Pantůček R, Benešík M, Doškař J (2016) Genetically modified bacteriophages in applied microbiology. J App Microbiol 121(3):618–633

Bell RL, Jarvis KG, Ottesen AR, McFarland MA, Brown EW (2016) Recent and emerging innovations in Salmonella detection: a food and environmental perspective. Microb Biotechnol 9:279–292

Berg T, Tesoriero L, Hailstones DL (2005) PCR-based detection of *Xanthomonas campestris* pathovars in Brassica seed. Plant Pathol 54(3):416–427

Besarab NV, Akhremchuk AE, Zlatohurska MA, Romaniuk LV, Valentovich LN, Tovkach FI, Lagonenko AL, Evtushenkov AN (2020) Isolation and characterization of Hena1 – a novel *Erwinia amylovora* bacteriophage. FEMS Microbiol Lett 367(9):070

Billiau A (2016) At the centennial of the bacteriophage: Reviving the overlooked contribution of a forgotten pioneer, Richard Bruynoghe (1881–1957). J Hist Biol 49:559–580

Bilt JLJ, Wolsink MHL, Smits GPPMA, Landman NM, Vlami BM (2018) Application of matrix-assisted laser desorption ionization time-of-flight mass spectrometry for rapid and accurate identification of *Ralstonia solanacearum* and *Ralstonia pseudosolanacearum*. Eur J Plant Pathol 152(4):921–931

Birge EA (1994) Bacterial and bacteriophage genetics, 3rd edn. Springer, New York, pp 264–266

Born Y, Fieseler L, Thöny V, Leimer N, Duffy B, Loessner MJ (2017) Engineering of bacteriophages Y2:: dpoL1-C and Y2:: luxAB for efficient control and rapid detection of the fire blight pathogen, *Erwinia amylovora*. Appl Environ Microbiol 83(12):e00341–e00317

Bossche VDA, Ceyssens PJ, De Smet J (2014) Systematic identification of hypothetical bacteriophage proteins targeting key protein complexes of *Pseudomonas aeruginosa*. J Proteome Res 13:4446–4456

Bradley DE (1965) The morphology and physiology of bacteriophages as revealed by the electron microscope. J R Microsc Soc 84:257–316

Burnham S, Hu J, Anany H, Brovko L, Deiss F, Derda R, Griffiths MW (2014) Towards rapid on-site phage-mediated detection of generic *Escherichia coli* in water using luminescent and visual readout. Anal Bioanal Chem 406(23):5685–5693

Carstens AB, Djurhuus AM, Kot W, Hansen LH (2019) A novel six-phage cocktail reduces *Pectobacterium atrosepticum* soft rot infection in potato tubers under simulated storage conditions. FEMS Microbiol Lett 366(9):101

Chan BK, Abedon ST, Loc-carrillo C (2013) Phage cocktails and the future of phage therapy. Future Microbiol:769–783

Cho IH, Ku S (2017) Current technical approaches for the early detection of foodborne pathogens: challenges and opportunities. Int J Mol Sci 18:2078

Citorik RJ, Mimee M, Lu TK (2014) Bacteriophage-based synthetic biology for the study of infectious diseases. Curr Opin Microbiol 19:59–69

Compant S, Duffy B, Nowak J, Clement C, Barka EA (2005) Use of plant growth-promoting bacteria for biocontrol of plant diseases: principles, mechanisms of action, and future prospects. App Environ Microbiol 71:4951–4959

Coons GH, Kotila JE (1925) The transmissible lytic principle (bacteriophage) in relation to plant pathogens. Phytopathol 15:357–370

Czajkowski R, Ozymko Z, Lojkowska E (2014) Isolation and characterization of novel soilborne lytic bacteriophages infecting Dickeya spp. biovar 3 ('*D. solani*'). Plant Pathol 64:758–772

d'Herelle F (1917) On an invisible microbe antagonistic to dysentery bacilli. CR Acad Sci Paris 165:373–375

d'Herelle F (1926) Bacteriophage and its significance for immunity Gosudarstvennoye Izdatelstvo. Moscow-Leningrad

Das M, Bhowmick TS, Ahern SJ, Young R, Gonzalez CF (2015) Control of Pierce's disease by phage. PLoS One 10(6):e0128902

Demerec M, Fano U (1945) Bacteriophage-resistant mutants in *Escherichia coli*. Genetics 30:119

Dennis C, Webster J (1971) Antagonistic properties of species-groups of Trichoderma: I. Production of non-volatile antibiotics. Trans Br Mycol Soc 57:25–39

Dreo T, Pirc M, Ramšak Ž, Pavšič J, Milavec M, Žel J, Gruden K (2014) Optimising droplet digital PCR analysis approaches for detection and quantification of bacteria: a case study of fire blight and potato brown rot. Anal Bioanal Chem 406(26):6513–6528

Elhalag K, Nasr-Eldin M, Hussien A, Ahmad A (2018) Potential use of soilborne lytic Podoviridae phage as a biocontrol agent against *Ralstonia solanacearum*. J Basic Microbiol 58(8):658–669

Erdrich SH, Sharma V, Schurr U, Arsova B, Frunzke J (2022) Isolation of novel *Xanthomonas* phages infecting the plant pathogens *X. translucens* and *X. campestris*. Viruses 14(7):1449

Eriksson U, Lindberg AA (1977) Adsorption of phage P22 to *Salmonella typhimurium*. J Gen Virol 34:207–221

Fang Y, Ramasamy RP (2015) Current and prospective methods for plant disease detection. Biosensors 5(3):537–561

Farooq U, Yang Q, Ullah MW, Wang S (2018) Bacterial biosensing: recent advances in phage-based bioassays and biosensors. Biosens Bioelectron 118:204–216

Fischer EP, Lipson C, Norton WW (1989) Thinking about science. Eng Sci 39

Flaherty JE, Jones JB, Harbaugh BK, Somodi GC, Jackson LE (2000) Control of bacterial spot on tomato in the greenhouse and field with H-mutant bacteriophages. HortScience 35:882–884

Fleming A (1929) On the antibacterial action of cultures of a Penicillium, with special reference to their use in the isolation of *B. influenzae*. Br J Exp Pathol 10:226

Fokine A, Kostyuchenko VA, Efimov AV (2005) A three-dimensional cryo-electron microscopy structure of the bacteriophage phiKZ head. J Mol Biol 352(1):117–124

Fujiwara A, Fujisawa M, Hamasaki R, Kawasaki T, Fujie M, Yamada T (2011) Biocontrol of *Ralstonia solanacearum* by treatment with lytic bacteriophages. Appl Environ Microbiol 77: 4155–4162

Gašić K, Ivanović MM, Ignjatov M, Calić A, Obradović A (2011) Isolation and characterization of *Xanthomonas euvesicatoria* bacteriophages. Plant Pathol 93(2):415–423

Gill JJ, Abedon ST (2003) Bacteriophage ecology and plants. APSnet

Grafe A (1991) A history of experimental virology. Springer Science & Business Media

Greer GG (2005) Bacteriophage control of foodborne bacteria. J Food Prot 68:1102–1111

Güemes CAG, Youle M, Cantú VA, Felts B, Nulton J, Rohwer F (2016) Viruses as winners in the game of life. Annu Rev Virol 3:197–214

Hagens S, Loessner MJ (2007) Application of bacteriophages for detection and control of foodborne pathogens. Appl Microbiol Biotechnol 76:513–519

Hagens S, de Wouters T, Vollenweider P, Loessner MJ (2011) Reporter bacteriophage A511: celB transduces a hyperthermostable glycosidase from *Pyrococcus furiosus* for rapid and simple detection of viable Listeria cells. Bacteriophage 1:143–151

Hankin ME (1896) The bactericidal action of the waters of the Jamuna and Ganges rivers on Cholera microbes. Ann Inst Pasteur 10:511–523

Havelaar AH (1987) Bacteriophages as model organisms in water treatment. Microbiol Sci 4:362–364

Hayward AC (2000) *Ralstonia solanacearum*. In: Lederberg J (ed) Encyclopedia of microbiology, vol 4. Academic Press, San Diego, pp 32–42

Hedtke B, Borner T, Weihe A (1997) Mitochondrial and chloroplast phage- type RNA polymerases in Arabidopsis. Science 277:809–811

Hendrix RW (2002) Bacteriophages: evolution of the majority. Theor Popul Biol 61:471–480

Hendry AT, Carpenter JA, Gerrard EH (1967) Bacteriophage studies of isolates from fire blight sources. Can J Microbiol 13:1357–1364

Heo YJ, Chung IY, Choi KB, Lau GW, Cho YH (2007) Genome sequence comparison and superinfection between two related *Pseudomonas aeruginosa* phages, D3112 and MP22. Microbiol 153(9):2885–2895

Hoang HA, Nhung NT (2018) Development of a bacteriophage-based method for detection of *Escherichia coli* O157: H7 in fresh vegetables. Food Saf 6(4):143–150

Ikeda TM, Gray MW (1999) Identification and characterization of T3/T7 bacteriophage-like RNA polymerase sequences in wheat. Plant Mol Biol 40:567–578

Jones HJ, Shield CG, Swift BM (2020) The application of bacteriophage diagnostics for bacterial pathogens in the agricultural supply chain: from farm-to-fork. PHAGE 1:176–188

Kawasaki T, Nagata S, Fujiwara A, Satsuma H, Fujie M, Usami S, Yamada T (2007) Genomic characterization of the filamentous integrative bacteriophages RSS1 and RSM1, which infect *Ralstonia solanacearum*. J Bacteriol 189:5792–5802

Keen EC (2014) A century of phage research: bacteriophages and the shaping of modern biology. Bioessays 37:6–9

Korol NA, Tovkach FI (2012) Identification of the major proteins of the virions of bacteriophage ZF40 *Pectobacterium carotovorum*. Microbiologia 74:64–70

Kotila JE (1920) Investigations of the blackleg disease of the potato. Michigan Agricultural College

Krylov VN, Dela Cruz DM, Hertveldt K, Ackermann HW (2007) "phiKZ-like viruses", a proposed new genus of myovirus bacteriophages. Arch Virol 152(10):1955–1959

Kutin RK, Alvarez A, Jenkins DM (2009) Detection of *Ralstonia solanacearum* in natural substrates using phage amplification integrated with real-time PCR assay. J Microbiol Methods 76(3):241–246

Lang JM, Schwartz HF, Gent DH (2004) Alternative strategies for onion Xanthomonas leaf blight management. Phytopathol 94:S6

Lau HY, Botella JR (2017) Advanced DNA-based point-of-care diagnostic methods for plant diseases detection. Front Plant Sci 8:2016

Law JWF, Ab Mutalib NS, Chan KG, Lee LH (2015) Rapid methods for the detection of foodborne bacterial pathogens: principles, applications, advantages and limitations. Front Microbiol 5:770

Lazcka O, Del Campo FJ, Munoz FX (2007) Pathogen detection: a perspective of traditional methods and biosensors. Biosens Bioelectron 22:1205–1217

Lee DH, Lee JH, Shin H (2012) Complete genome sequence of *Pectobacterium carotovorum* subsp. *carotovorum* bacteriophage My1. J Virol 86:11410–11411

Lee SY, Thapa MR, Kim HJ, Choi K, Lee SW (2021) Complete genome sequence of a novel bacteriophage RpY1 infecting *Ralstonia solanacearum* strains. Curr Microbiol 78:2044–2050

Letarov AV (2020) History of early bacteriophage research and emergence of key concepts in virology. Biochemistry (Moscow) 85:1093–1112

Lim JA, Lee DH, Jung KS (2013) Biocontrol of *Pectobacterium carotovorum* subsp. *carotovorum* using bacteriophage PP1. J Microbiol Biotechnol 23:1147–1153

Lim J-A, Shin H, Lee DH (2014) Complete genome sequence of the *Pectobacterium carotovorum* subsp. *carotovorum* virulent bacteriophage PM1. Arch Virol 159:2185–2187

Lim JA, Lee DH, Heu S (2015) Isolation and genomic characterization of the T4-Like bacteriophage PM2 infecting *Pectobacterium carotovorum* subsp. *carotovorum*. Plant Pathol J 31:83–89

Lim JA, Heu S, Park J et al (2017) Genomic characterization of bacteriophage vB_PcaP_PP2 infecting *Pectobacterium carotovorum subsp. carotovorum*, a new member of a proposed genus in the subfamily Autographivirinae. Arch Virol 162:2441–2444

Loc-Carrillo C, Abedon ST (2011) Pros and cons of phage therapy. Bacteriophage 1:111–114

Mallmann WL, Hemstreet CARL (1924) Isolation of an inhibitory substance from plants. Agric Res 28:599–602

Mansfield J, Genin S, Magori S (2012) Top 10 plant pathogenic bacteria in molecular plant pathology. Mol Plant Pathol 13:614–629

Marinelli LJ, Hatfull GF, Piuri M (2012) Recombineering: a powerful tool for modification of bacteriophage genomes. Bacteriophage 2(1):5–14

Marsh P, Wellington EMH (1994) Phage-host interactions in soil. FEMS Microbiol Ecol 15:99–107

McKenna F, El-Tarabily KA, Hardy GESTJ, Dell B (2001) Novel in vivo use of a polyvalent streptomyces phage to disinfest streptomyces scabies-infected seed potatoes. Plant Pathol 50: 666–675

McManus PS, Stockwell VO, Sundin GW, Jones AL (2002) Antibiotic use in plant agriculture. Annu Rev Phytopathol 40:443–465

Meile S, Kilcher S, Loessner MJ, Dunne M (2020) Reporter phage-based detection of bacterial pathogens: design guidelines and recent developments. Viruses 12(9):944

Mesyanzhinov VV, Robben J, Grymonprez B (2002) The genome of bacteriophage phiKZ of *Pseudomonas aeruginosa*. J Mol Biol 317(1):1–19

Nga NTT, Tran TN, Holtappels D, Kim Ngan NL, Hao NP, Vallino M (2021) Phage biocontrol of bacterial leaf blight disease on welsh onion caused by *Xanthomonas axonopodis* pv *allii*. Antibiotics 10:517

Northrop JH (1937) Chemical nature and mode of formation of pepsin, trypsin and bacteriophage. Science 86:479–483

Obradovic A, Jones JB, Momol MT, Olson SM, Jackson LE (2005) Integration of biological control agents and systemic acquired resistance inducers against bacterial spot on tomato. Plant Dis 89: 712–716

Oerke EC (2006) Crop losses to pests. J Agric Sci 144:31–43

Ozawa, H, Tanaka, H, Ichinose, Y, Shiraishi, T, Yamada, T (2001) Bacteriophage P4282, a parasite of *Ralstonia solanacearum*, encodes a bacteriolytic protein important for lytic infection of its host. Mol Genet Genom 265: 95–101

Panshchina AI, Tovkach FI, Romaniuk LV (2007) Physico-chemical properties of temperate bacteriophage ZF40 of *Erwinia carotovora*. Microbiol Z 69:15–22

Pradeep M, Nargund VB (2015) Characterization of *Ralstonia solanacearum* isolates, the causal agent of brinjal vascular wilt by Flic locus and phage typing. BioSci Trends 8(19):5244–5250

Radke SM, Alocilja EC (2005) A high density microelectrode array biosensor for detection of *E. coli* O157: H7. Biosens Bioelectron 20:1662–1667

Ranjani P, Gowthami Y, Gnanamanickam SS, Palani P (2018) Bacteriophages: a new weapon for the control of bacterial blight disease in rice caused by *Xanthomonas oryzae*. Microbiol Biotechnol Lett 46(4):346–359

Rider TH, Petrovick MS, Nargi FE, Harper JD, Schwoebel ED, Mathews RH, Blanchard DJ, Bortolin LT, Young AM, Chen J, Hollis MA (2003) AB cell-based sensor for rapid identification of pathogens. Science 301:213–215

Sanger F, Air GM, Barrell BG, Brown NL, Coulson AR, Fiddes CA, Hutchison CA, Slocombe PM, Smith M (1977) Nucleotide sequence of bacteriophage phi X 174 DNA. Nature 265:687–695

Schnabel EL, Jones AL (2001) Isolation and characterization of five *Erwinia amylovora* bacteriophages and assessment of phage resistance in strains of *Erwinia amylovora*. Appl Env Microbiol 67:59–64

Schofield DA, Bull CT, Rubio I, Wechter WP, Westwater C, Molineux IJ (2013) "Light-tagged" bacteriophage as a diagnostic tool for the detection of phytopathogens. Bioeng 4(1):50–54

Seeley ND, Primrose SB (1982) A review: the isolation of bacteriophages from the environment. J Appl Microbiol 53:1–17

Sharma A, Abrahamian P, Carvalho R, Choudhary M, Paret ML, Vallad GE, Jones JB (2022) Future of bacterial disease management in crop production. Ann Rev Phytopathol 60:259–282

Singh A, Arutyunov D, Szymanski CM, Evoy S (2012) Bacteriophage based probes for pathogen detection. Analyst 137(15):3405–3421

Smartt AE, Ripp S (2011) Bacteriophage reporter technology for sensing and detecting microbial targets. Anal Bioanal Chem 400:991–1007

Soriano LP, Noguera P, Gorris MT, Puchades R, Maquieira A, Noales ME, López MM (2017) Lateral flow immunoassay for on-site detection of *Xanthomonas arboricola* pv. *pruni* in symptomatic field samples. PLoS One 12(4):e0176201

Stent GS (1960). Papers on bacterial viruses. Papers on bacterial viruses

Stent G, Watson JD, Cairns J (1966) Phage and the origins of molecular biology. Cold Springs Harbor Laboratory of Quantitative Biology, New York

Storey MV, Ashbolt NJ (2001) Persistence of two model enteric viruses (B40–8 and MS-2 bacteriophages) in water distribution pipe biofilms. Water Sci Technol 43(12):133–138

Summer EJ, Enderle CJ, Ahern SJ, Gill JJ, Torres CP, Appel DN, Black MC, Young R, Gonzalez CF (2010) Genomic and biological analysis of phage Xfas53 and related prophages of *Xylella fastidiosa*. J Bacteriol 192(1):179–190

Summers WC (1993) How bacteriophage came to be used by the phage Group. J Hist Biol:255–267

Summers WC (2005) Bacteriophage research: early history. Bacteriophages: Biol Appl 5–27

Sun TP, Webster RE (1987) Nucleotide sequence of a gene cluster involved in entry of E colicins and single-stranded DNA of infecting filamentous bacteriophages into *Escherichia coli*. J Bacteriol 169:2667–2674

Sutton MD, Katznelson H (1953) Isolation of bacteriophages for the detection and identification of some seed-borne pathogenic bacteria. Can J Bot 31(2):201–205

Sutton MD, Katznelson H, Quadling C (1958) A bacteriophage that attacks numerous phytopathogenic Xanthomonas species. Can J Microbiol 4(5):493–497

Svircev AM, Lehman SM, Kim W, Barszcz E, Schneider KE, Castle AJ (2006) Control of the fire blight pathogen with bacteriophages. In: Proceedings of international symposium on biological control of bacterial plant diseases, 1st, Seeheim/Darmstadt, Germany, 259. Berlin: Dtsch. Bibl., CIP-Einh.aufn

Svircev A, Roach D, Castle A (2018) Framing the future with bacteriophages in agriculture. Viruses 10:218

Tanaka H, Negishi H, Maeda H (1990) Control of tobacco bacterial wilt by an avirulent strain of *Pseudomonas solanacearum* M4S and its bacteriophage. Ann Phytopathol Soc Japan 56:243–246

Thomas R (1935) A bacteriophage in relation to Stewart's disease of corn. Phytopathology 25:371–372

Thomas JA, Rolando MR, Carroll CA (2008) Characterization of *Pseudomonas chlororaphis* myovirus 201varphi2–1 via genomic sequencing, mass spectrometry, and electron microscopy. Virology 376(2):330–338

Tlili C, Sokullu E, Safavieh M, Tolba M, Ahmed MU, Zourob M (2013) Bacteria screening, viability, and confirmation assays using bacteriophage-impedimetric/loop-mediated isothermal amplification dual-response biosensors. Anal Chem 85(10):4893–4901

Toyoda H, Kakutani K, Ikeda S, Goto S, Tanaka H, Ouchi S (1991) Characterization of deoxyribonucleic acid of virulent bacteriophage and its infectivity to host bacterium, *Pseudomonas solanacearum*. J Phytopathol 131:11–21

Twort F (1915) An investigation on the nature of ultra-microscopic viruses. Lancet 186:4814

Vanneste JL (2000) Fire blight: the disease and its causative agent. *Erwinia amylovora*, CAB International, New York, NY

Vu NT, Oh CS (2020) Bacteriophage usage for bacterial disease management and diagnosis in plants. Plant Pathol J 36(3):204

Whitman WB, Coleman DC, Wiebe WJ (1998) Prokaryotes: the unseen majority. Proc Natl Acad Sci 95:6578–6583

Wiesel L, Newton AC, Elliott I, Booty D, Gilroy EM, Birch PR, Hein I (2014) Molecular effects of resistance elicitors from biological origin and their potential for crop protection. Front Plant Sci 5:655

Wu X, Kong D, Song S, Liu L, Kuang H, Xu C (2015) Development of sandwich ELISA and immunochromatographic strip methods for the detection of *Xanthomonas oryzae* pv. *oryzae*. Anal Methods 7(15):6190–6197

Yamada T, Kawasaki T, Nagata S, Fujiwara A, Usami S, Fujie M (2007) New bacteriophages that infect the phytopathogen *Ralstonia solanacearum*. Microbiol 153:2630–2639

Yu JG, Lim JA, Song YR, Heu S, Kim JH, Koh YJ, Oh CS (2016) Isolation and characterization of bacteriophages against *Pseudomonas syringae pv. actinidiae* causing bacterial canker disease in kiwifruit. J Microbiol Biotechnol 26(2):385–393

Zink RALF, Loessner MJ (1992) Classification of virulent and temperate bacteriophages of Listeria spp. on the basis of morphology and protein analysis. Appl Environ Microbiol 58:296–302

The manufacturer's authorised representative in the EU is Springer Nature Customer Service Centre GmbH, Europaplatz 3, 69115 Heidelberg, Germany. If you have any concerns regarding our products, please contact ProductSafety@springernature.com

Printed and bound by CPI Group (UK) Ltd, Croydon, CR0 4YY

01/10/2026

02225870-0019